Cardiovascular and Neurovascular Imaging

IMAGING IN MEDICAL DIAGNOSIS AND THERAPY

Series Editors: Andrew Karellas and Bruce R. Thomadsen

Published titles

IMAGING IN MEDICAL DIAGNOSIS AND THERAPY

Series Editors: Andrew Karellas and Bruce R. Thomadsen

Published titles

Ultrasound Imaging and Therapy
Aaron Fenster and James C. Lacefield, Editors
ISBN: 978-1-4398-6628-3

Cardiovascular and Neurovascular Imaging: Physics and Technology
Carlo Cavedon and Stephen Rudin, Editors

Forthcoming titles

Handbook of Small Animal Imaging: Preclinical Imaging, Therapy, and Applications
George Kagadis, Nancy L. Ford,
George K. Loudos, and Dimitrios Karnabatidis,
Editors

Physics of PET and SPECT Imaging
Magnus Dahlbom, Editor

Hybrid Imaging in Cardiovascular Medicine
Yi-Hwa Liu and Albert Sinusas, Editors

Scintillation Dosimetry
Sam Beddar and Luc Beaulieu, Editors

Cardiovascular and Neurovascular Imaging

Physics and Technology

Edited by

Carlo Cavedon
Azienda Ospedaliera University, Verona, Italy

Stephen Rudin
University at Buffalo, New York, USA

CRC Press
Taylor & Francis Group
Boca Raton London New York

CRC Press is an imprint of the
Taylor & Francis Group, an **informa** business

A TAYLOR & FRANCIS BOOK

First published 2016 by Taylor & Francis Group, LLC

Published 2019 by CRC Press
Taylor & Francis Group
6000 Broken Sound Parkway NW, Suite 300
Boca Raton, FL 33487-2742

© 2016 by Taylor & Francis Group, LLC
CRC Press is an imprint of Taylor & Francis Group, an Informa business

First issued in paperback 2019

No claim to original U.S. Government works

ISBN 13: 978-0-367-44581-2 (pbk)
ISBN 13: 978-1-4398-9056-1 (hbk)

Library of Congress Cataloging-in-Publication Data

Cardiovascular and neurovascular imaging : physics and technology / editors, Carlo Cavedon, Stephen Rudin.
   p. ; cm. -- (Imaging in medical diagnosis and therapy)
   Includes bibliographical references.
   ISBN 978-1-4398-9056-1 (hardcover : alk. paper)
   I. Cavedon, Carlo, editor. II. Rudin, Stephen, 1943- , editor. III. Series: Imaging in medical diagnosis and therapy.
   [DNLM: 1. Diagnostic Imaging. 2. Diagnostic Techniques, Cardiovascular. 3. Neuroimaging. WG 141]

RC683.5.I42
616.1'0754--dc23
                                          2015026758

Visit the Taylor & Francis Web site at
http://www.taylorandfrancis.com

and the CRC Press Web site at
http://www.crcpress.com

To Ann, Cynthia, Robert, Lance, Liam, and Sylvie with love.

Stephen Rudin

To the best part of my four-people family: Elena, Caterina, and Agnese.
To my father, Giuseppe, and in memory of my mother, Caterina.

Carlo Cavedon

Contents

SECTION I   Physical Basis and Clinical Introduction

SECTION II   Physics and Technology: Principal Applications

SECTION III Focused Applications and Dedicated Technology: Geometries, Sources, Detectors, Advanced Image Reconstruction, and Quantitative Analysis

SECTION IV Time-Resolved Imaging

SECTION V Image-Guided Therapeutic Procedures

SECTION VI Dosimetry and Radiation Protection

SECTION VII Trends

Series Preface

Advances in the science and technology of medical imaging and radiation therapy are more profound and rapid than ever before since their inception over a century ago. Further, the disciplines are increasingly cross-linked as imaging methods become more widely used to plan, guide, monitor, and assess treatments in radiation therapy. Today, the technologies of medical imaging and radiation therapy are so complex and so computer driven that it is difficult for the persons (physicians and technologists) responsible for their clinical use to know exactly what is happening at the point of care, when a patient is being examined or treated. The persons best equipped to understand the technologies and their applications are medical physicists, and these individuals assume greater responsibilities in the clinical arena to ensure that what is intended for the patient is actually delivered in a safe and effective manner.

The growing responsibilities of medical physicists in the clinical arenas of medical imaging and radiation therapy are not without their challenges, however. Most medical physicists are knowledgeable in either radiation therapy or medical imaging and expert in one or a small number of areas within their discipline. They sustain their expertise in these areas by reading scientific articles and attending scientific talks at meetings. In contrast, their responsibilities increasingly extend beyond their specific areas of expertise. To meet these responsibilities, medical physicists periodically must refresh their knowledge of advances in medical imaging or radiation therapy, and they must be prepared to function at the intersection of these two fields. How to accomplish these objectives is a challenge.

At the 2007 annual meeting of the American Association of Physicists in Medicine in Minneapolis, this challenge was the topic of conversation during a lunch hosted by Taylor & Francis Group involving a group of senior medical physicists (Arthur L. Boyer, Joseph O. Deasy, C.M. Charlie Ma, Todd A. Pawlicki, Ervin B. Podgorsak, Elke Reitzel, Anthony B. Wolbarst, and Ellen D. Yorke). The conclusion of this discussion was that a book series should be launched under the Taylor & Francis Group banner, with each volume in the series addressing a rapidly advancing area of medical imaging or radiation therapy of importance to medical physicists. The aim would be for each volume to provide medical physicists with the information needed to understand technologies driving a rapid advance and their applications to safe and effective delivery of patient care.

Each volume in the series is edited by one or more individuals with recognized expertise in the technological area encompassed by the book. The editors are responsible for selecting the authors of individual chapters and ensuring that the chapters are comprehensive and intelligible to someone without such expertise. The enthusiasm of volume editors and chapter authors has been gratifying and reinforces the conclusion of the Minneapolis luncheon that this series of books addresses a major need of medical physicists.

Imaging in Medical Diagnosis and Therapy would not have been possible without the encouragement and support of the series manager of Taylor & Francis Group, Luna Han. The editors and authors, and most of all I, are indebted to her steady guidance of the entire project.

William Hendee
Founding series editor
Rochester, Minnesota

Preface

The physics and technology underpinning vascular imaging is a specialized yet interdisciplinary branch of medical imaging. The physics of cardiovascular and neurovascular imaging includes principles of image formation and analysis, and aspects related to image quality, accuracy of the information achievable and hence of the related medical procedures, and the safety of patients and operators. It is uncommon for a student or professional in this field to find comprehensive information gathered together in a single source—book, review article, or other publication—rather than scattered throughout scientific papers each quite different in nature. This book is aimed at providing, from a physicist's standpoint, a single source for individuals who specialize in cardiovascular and neurovascular imaging with an attempt to find a convenient balance between up-to-date material and an overall summary. It is the editors' intent that this should become a useful single source for students, professionals, and researchers.

Although this book is intended primarily for physicists and biomedical engineers who may be relatively new to the field and who wish to benefit from a comprehensive resource on the subject, other professionals should find the book useful as well. These might include physicians who perform cardiovascular and neurovascular imaging procedures but may be interested in going more thoroughly into the underlying physics and technology of the various imaging modalities involved with an up-to-date picture of the safety aspects of their work.

The text is divided into seven major sections, starting with an introductory overview. The next five core sections are devoted to the main technologies, special applications, time-resolved techniques, image-guided therapy, and dosimetry plus radiation safety aspects. The final section provides some perspectives on current and future trends. Not only each section but also each chapter is self-contained in order to provide the reader with a reference-like textbook that does not require previous knowledge of other material. Authors were selected from among the major experts in the field in order to provide the reader with a comprehensive picture from the academic, clinical, and industrial points of view.

Editors

Dr. Carlo Cavedon is director of the Medical Physics Unit at Verona University Hospital in Italy, where he also serves as chief radiation safety officer. His scientific and professional interests cover image-guided interventions, image-guided radiation therapy and radiosurgery, quantitative techniques in MRI and metabolic imaging, 4D techniques in diagnostic and therapeutic procedures, Monte Carlo simulation, small-field radiation dosimetry, and radiation safety. He has been serving as professor of medical physics in several courses at the Universities of Verona, Padova, and Trieste in Italy since 1998. He is a full member of the AAPM, serves in the scientific committee of the Italian Association of Medical Physics (AIFM), and is an active member of several other scientific societies, including the European Society for Radiotherapy and Oncology (ESTRO). Dr. Cavedon has authored more than 150 publications and is frequently invited to speak at national and international meetings. He was a member of the *Medical Physics* editorial board from January 2005 to December 2013 and is currently a senior associate editor of the journal.

Stephen Rudin, PhD, is director of Radiation Physics at the University at Buffalo, The State University of New York (SUNY) and SUNY distinguished professor in the departments of radiology, neurosurgery, physiology and biophysics, biomedical engineering, electrical engineering, and mechanical and aerospace engineering. He is also the founding director of the Medical Physics Graduate Program at University at Buffalo, a founding codirector of the University at Buffalo-Toshiba Stroke and Vascular Research Center, and the radiation safety officer at the Erie County Medical Center. Dr. Rudin is a fellow of the AAPM, is certified by the American Board of Radiology and the American Board of Health Physics, serves on the board of editors of the journal *Medical Physics*, and is a member of 12 professional societies. He has authored more than 400 publications and won numerous awards and honors in the fields of medical imaging, image-guided endovascular interventions, and radiation safety. Dr. Rudin's research is supported by grants from the U.S. National Institutes of Health and the Toshiba Corporation.

Contributors

Amir A. Amini
Department of Electrical and Computer
  Engineering
University of Louisville
Louisville, Kentucky

Octavia Bane
Northwestern University Medical School
Evanston, Illinois

Daniel R. Bednarek
School of Medicine and Biomedical Sciences
University at Buffalo
Buffalo, New York

Alberto Beltramello
Department of Neuroradiology
Verona University Hospital
Verona, Italy

Filippo Cademartiri
Department of Radiology
Erasmus Medical Center University Hospital
Rotterdam, the Netherlands

Timothy J. Carroll
Northwestern University Medical School
Evanston, Illinois

Carlo Cavedon
Medical Physics Unit
Verona University Hospital
Verona, Italy

Juan R. Cebral
George Mason University
Fairfax, Virginia

Guang-Hong Chen
University of Wisconsin–Madison
Madison, Wisconsin

Periklis Davlouros
Department of Radiology
School of Medicine
University of Patras
Patras, Greece

Robert deKemp
University of Ottawa Heart Institute
Ottawa, Ontario, Canada

Robert C. Detrano
University of California, Irvine
Irvine, California

Rebecca Fahrig
School of Medicine
Stanford University
Stanford, California

Arundhuti Ganguly
Varian Medical Systems
Palo Alto, California

and

Department of Radiology
Stanford University
Stanford, California

Guido Germano
Cedars-Sinai Medical Center
and
School of Medicine
University of California, Los Angeles
Los Angeles, California

Michael Grass
Philips Technologie GmbH Innovative
  Technologies
Philips Research
Hamburg, Germany

Scott S. Hsieh
School of Medicine
Stanford University
Stanford, California

Ciprian N. Ionita
Departments of Biomedical Engineering and
    Neurosurgery
Toshiba Stroke And Vascular Research Center
University at Buffalo
Buffalo, New York

Amit Jain
School of Medicine and Biomedical Sciences
University at Buffalo
Buffalo, New York

Mo Kadbi
University of Louisville
Louisville, Kentucky

George C. Kagadis
Department of Medical Physics
School of Medicine
University of Patras
Patras, Greece

Dimitris Karnabatidis
Department of Radiology
School of Medicine
University of Patras
Patras, Greece

Konstantinos Katsanos
Department of Radiology
School of Medicine
University of Patras
Patras, Greece

Andrew Kuhls-Gilcrist
Toshiba America Medical Systems
Tustin, California

Erica Maffei
Cardiovascular Unit
Giovanni XXIII Clinic
Monastier di Treviso, Italy

Cesare Mantini
Department of Neuroscience and Imaging
Chieti University Hospital
Chieti, Italy

Gabriele Meliadò
Medical Physics Unit
Verona University Hospital
Verona, Italy

Sabee Molloi
Department of Radiological Sciences
University of California, Irvine
Irvine, California

M.J. Negahdar
University of Louisville
Louisville, Kentucky

Renato Padovani
Department of Medical Physics
Udine University Hospital
Udine, Italy

Norbert J. Pelc
School of Medicine
Stanford University
Standford, California

Giovanni Puppini
Department of Radiology
Verona University Hospital
Verona, Italy

Madan Rehani
International Atomic Energy Agency
Vienna, Austria

Giuseppe Kenneth Ricciardi
Department of Neuroradiology
Verona University Hospital
Verona, Italy

Stephen Rudin
School of Medicine and Biomedical Sciences
University at Buffalo
Buffalo, New York

Dirk Schäfer
Philips Technologie GmbH Innovative
   Technologies
Philips Research
Hamburg, Germany

Michael D. Silver
Toshiba Medical Research Institute USA, Inc.
Vernon Hills, Illinois

Piotr J. Slomka
Cedars-Sinai Medical Center
and
School of Medicine
University of California, Los Angeles
Los Angeles, California

Michael Speidel
University of Wisconsin–Madison
University of Wisconsin
Madison, Wisconsin

Stavros Spiliopoulos
Department of Radiology
School of Medicine
University of Patras
Patras, Greece

Joseph Stancanello
GE Healthcare
Vélizy-Villacoublay, France

Pascal Thériault-Lauzier
University of Wisconsin–Madison
Madison, Wisconsin

Kai Erik Thomenius
Institute of Medical Engineering and Science
Massachusetts Institute of Technology
Cambridge, Massachusetts

Parmede Vakil
Northwestern University Medical School
Evanston, Illinois

Michael Van Lysel
University of Wisconsin–Madison
Madison, Wisconsin

Eliseo Vano
Department of Radiology
School of Medicine
Complutense University
Madrid, Spain

Adam S. Wang
School of Medicine
Varian Medical Systems
Palo Alto, California

Lei Zhu
Nuclear and Radiological Engineering and Medical
   Physics Programs
The George W. Woodruff School of Mechanical
   Engineering
Georgia Institute of Technology
Atlanta, Georgia

Physical Basis and Clinical Introduction

1. Introduction to the Physics of Vascular Imaging

Stephen Rudin and Carlo Cavedon

The ability of medical professionals to meet the challenges of cardio- and neurovascular imaging requires an understanding of the physical principles of the many available modalities so as to be able to determine the range of applicability and the limitations of each. Modalities include radiography, fluoroscopy, and computed tomography (CT) where x-rays are generated, differentially attenuated, and detected; magnetic resonance imaging (MRI) where magnetic fields are used to modulate the use of radio waves for imaging; ultrasound where mechanical vibrations are transmitted and reflected; optical imaging where light is sensed; or nuclear medicine where radioisotopes emit gamma rays for imaging. The physical principles governing each of these modalities must be understood to optimally apply them to the appropriate clinical problem and, after a brief review in the following text, more details will be discussed in the chapters that follow.

1.1 Imaging Objectives

The vascular clinical challenges where these modalities are applied can be classified into different groups depending upon the main imaging objectives. Whether the imaging objective is for diagnosis where the underlying pathology is sought or for guiding interventions where a therapeutic treatment is to be carried out, may affect the choice of modality to be used as well as the way in which it is to be used. Whether the imaging objective is to visualize the morphology of a structure or to determine the blood flow or function may also affect the modality choices.

1.1.1 Diagnosis versus Intervention

Factors affected by whether the imaging procedure involves diagnosis or interventional guidance could include the field of view (FOV), spatial resolution and pixel size, timing, dose distribution, number of modalities needed, etc. For example, in diagnosis one might wish to have as large an FOV as possible because one does not know a priori where or what exactly one looks for. One might be satisfied with a moderate spatial resolution, depending upon the possible likely pathologies, and one might accept a static image that provides localization information sufficient to form an opinion as to what treatment might be appropriate. One might also want to use multiple modalities to confirm the diagnosis and one would want to be cautious about the amount of ionizing radiation dose to deliver to the patient, considering that other diagnostic procedure may be ordered for the patient in the future. On the other hand during image-guided interventions, a limited or small FOV may be acceptable while higher spatial resolution may be needed if the accuracy of the intervention may depend upon the quality of the images. Similarly, if the image guidance is happening during the course of the intervention, real-time imaging sequences may be critical and for serious clinical conditions, higher localized radiation doses may be acceptable.

1.1.2 Visualization of Morphology versus Determination of Flow or Function

Selection of an appropriate modality and parameters used during an imaging procedure are clearly affected by whether the imaging objective involves an evaluation of a structure in the patient or whether it involves a determination of blood flow or vascular function such as perfusion. In the first case, visualization of morphology may depend upon attenuation of externally imposed radiation such as x-rays, light, or ultrasound. Detecting how much of the radiation is transmitted, emitted, or reflected may provide information about the structure, size, and shape of objects of interest but not directly about function or motion. On the other hand, factors such as frame rates or frequency shifts may be essential for flow or function determination. Ultimately since details of blood flow are determined by physical principles such as solutions of the Navier–Stokes equation, modalities such as MRI and ultrasound may be unique selections to provide initial condition velocity distribution information, which not available from other modalities.

1.2 Clinical Modality Review

It may be helpful to briefly summarize and compare the physical features of each of the five generally accepted clinical imaging modalities: x-ray, nuclear medical, ultrasound, magnetic resonance, and optical imaging. All of them involve the detection of radiation emanating from a source where radiation is the transport of energy from the source outward to enable some of it to be detected by an imaging device.

1.2.1 Accepted Clinical Modalities

The five imaging modalities can be divided into two general classes according to whether the radiation that is used is either ionizing (x-rays in radiography or gamma rays in nuclear medicine) or nonionizing (ultrasound, MR, or optical imaging). Modalities involving ionizing radiation clearly have limitations in the quantity of radiation that can be efficaciously applied to a patient. The energy in quanta of x-rays and gamma rays is typically in the orders of magnitude larger than the binding energy of electrons in molecules of living tissue and are hence ionizing. The energy in packets of ultrasound, light, and radio waves is below the threshold energy to remove such an electron and hence these are nonionizing modalities. Because the energy associated with photons of x-rays and gamma rays is sufficiently large to cause damage to key molecules of cells in human tissue such as DNA, which nonionizing radiation cannot, the radiation dose imposed on a patient by ionizing radiation must be considered and controlled much more carefully. The result, for example, is that it would be more preferable and safer to deposit much more nonionizing ultrasound energy in an examination of a fetus than it would be to deposit a smaller amount of ionizing energy in the form of a pelvic x-ray.

1.2.2 Modality Image Formation Comparisons

For four of the modalities the radiation source is separated from the patient although it may be emitting internally such as during intravascular ultrasound or optical imaging. For MRI and x-ray including CT, the source of radiation is external to the patient, a radio antenna for MRI, and an x-ray tube for radiography. Since radiation from the x-ray source that passes unattenuated through the patient is used to form an imaging during radiography, this modality can be considered transmission imaging. MRI is a bit more complicated to categorize since radio waves are absorbed that cause resonances to occur in the patient's tissue resulting in turn in the emission of radio waves. In nuclear medical imaging, the radionuclide attached molecularly to a pharmaceutical agent is injected or ingested into the patient being taken up differentially in the organ or tissue to be imaged; hence this modality may be designated as emission imaging. Three-dimensional imaging variants such as single photon emission tomography (SPECT) or positron emission tomography (PET) are included in this description in that the objective is to use the gamma ray emissions to localize the radiopharmaceutical but now in a 3D volume. Since the physics of ultrasound enables pulses of radiation to be reflected back from interfaces where the properties of the two materials that meet to form the interface are different in their acoustic impedance, ultrasound imaging can be referred to as reflection imaging. Optical imaging also involves the reflection or scattering of light from a source back to a detector where a range of simple to complex processes can occur to form an image.

1.2.3   Radiation and Radiation Sources

The essence of radiation used for medical imaging is that it is some form of energy moving outwardly from a source. The radiation itself can generally be characterized as having a wave nature to it with features of wavelength, frequency, and speed although there could also be a particulate nature characterized by quantized packets of energy. This duality is most appreciated for the two modalities employing the ionizing radiation of x-rays and gamma rays where the particulate quantized energy is proportional in magnitude to the wave nature's frequency by Plank's constant. The four modalities of radiography, nuclear medical imaging, MRI, and optical imaging use electromagnetic waves traveling at the speed of light for radiation; however, the key difference is the wavelength and hence frequency or energy of each quantum. In order of increasing energy, this list would be radiofrequency waves used in MRI (for 3 T: 0.5×10^{-6} eV, 128 MHz), optical light (~2 eV, 5×10^{14} Hz), x-rays (30–100 keV, $7 - 24 \times 10^{18}$ Hz), and gamma rays (50–511 keV, $1 - 12 \times 10^{19}$ Hz) although once created x-rays and gamma rays are indistinguishable and may have an overlapping range of energies. It is these energy differences that ultimately control the difference features and capabilities of each modality. The one nonelectromagnetic radiation-based modality is of course ultrasound imaging; however, the frequency of the ultrasound mechanical vibrations also determines under what circumstances they may be used for medical imaging.

1.2.4   Radiation Interactions with Matter: Tissue for Various Radiations

Although the radiation used in four of the five imaging modalities may be electromagnetic in nature, the differing frequencies of the radiation used to determine in part how each of them may interact with matter. The other part is controlled by the characteristics of the matter itself such as physical density, electron density (electrons per gram of matter), Z-number (number of protons in the nuclei of the matter), nuclear magnetic moments, magnetic permeability, electrical conductivity, optical properties, ionization energy, and for living matter the radiation sensitivity. For ultrasound, the properties of matter that play a crucial part in determining the interaction with matter includes the acoustic impedance determined by the product of the density and speed and the attenuation properties as a function of frequency and sound intensity.

1.2.5   Use of Contrast Material to Enable Viewing Vasculature

Often the image contrast resulting from the use of a particular radiation may not be sufficient or optimal especially for visualizing the vasculature. All the modalities where the source is either discrete or external to the patient have the option of increasing the image contrast by the introduction of a contrast agent with suitable physical properties to enable a major change in the interaction of the radiation with the patient's tissue and in particular with the blood. Examples of such contrast agents include gadolinium for MRI, iodine for x-ray, air bubbles for US, and fluorescent dye for optical techniques.

1.2.6   Detection Techniques

Because of the physics of interaction for the different radiations associated with the different modalities, the means of detection and thus image formation will differ. Detectors are designed generally to optimize the interaction of the radiation with the material in the detector while enabling spatial determination of the location of the interaction. The result for all the modalities is somehow to turn the detected radiation into an electronic signal that can be used to form a digitized computer compatible image. In some cases, the detectors may use an intermediate radiation prior to the conversion to an electrical signal. For example, in radiography and nuclear medical imaging the detectors can be either direct where the photon energy is converted into charge or indirect where the photon energy in converted to light using a phosphor after which the light is converted to an electronic signal and then digitized to form the image. In some cases, the same device is used as the source and detector as is the case in ultrasound and sometimes in MRI. For the various modalities, the sensitivity for detection may vary greatly; however, ultimately, the quality of the image will be determined by the resulting signal compared to the noise and the system resolution. The noise can consist of system instrumentation noise determined by the design of the detector and the shot or quantum noise, which is an inherent characteristic of the radiation if the number of quanta is limited. The system spatial resolution is determined partly by the pixel size or sampling granularity of the image, and for dynamic image sequences the temporal resolution is determined by the lag or by temporal filtering. System metrics have now become available to quantitatively parameterize the imaging characteristics of many of the modality detector systems using standard linear systems analytic concepts such as modulation transfer function (MTF) for resolution assessment and detective quantum efficiency (DQE) for signal to noise ratio (SNR) as a function of spatial frequency where quantum noise can play a significant factor.

1.3 Conclusion

The specific radiations and the physics of how they interact with matter are the underpinning of each of the basic vascular imaging modalities, and determine the way in which these modalities are applied. While there may be some overlap in the physics behind each modality, each must be characterized in detail to fully understand their potentials and their limitations. In the next sections of this book, after a review of clinical applications in the heart and brain are presented, the physics and technology is discussed in detail as are applications involving blood flow finally concluding with discussions of radiation safety and dose.

2. Neurovascular Imaging
State of the Art and Clinical Challenges

Alberto Beltramello and Giuseppe Kenneth Ricciardi

2.1   Introduction

Over the years, cerebrovascular pathology has rapidly increased both in terms of frequency and importance.

Stroke is defined as a clinical syndrome consisting of rapidly developing clinical signs of focal (or global in the case of coma) disturbance of cerebral function lasting more than 24 h or leading to death with no apparent cause other than a vascular origin (Hatano 1976, Royal College of Physicians 2008). Stroke is the second most common cause of death worldwide and the third in the United States (Osborn 1994). Stroke can be subclassified into two major categories: ischemic, caused by impaired perfusion to the brain, (accounting for 80% of all acute stroke events) and hemorrhagic, characterized my intra- or extracerebral blood extravasation (20%) (Osborn 1994).

Nowadays, not only hemorrhagic but also ischemic stroke can be actively managed by neuroteams (diagnostic neuroradiologist, interventionalist, neurologist, and neurosurgeon) so that the exact delineation of the pathological picture is of crucial importance.

In *ischemic stroke*, the aim of early brain imaging is to exclude intracranial hemorrhage, identify ischemic change, and exclude stroke mimics. CT angiography (CTA) gives a rapid noninvasive assessment of the intracranial and extracranial circulation at adequate resolution with multislice technology. By using a contrast bolus tracking technique, a chosen volume (e.g., circle of Willis or from the aortic arch to vertex) can be obtained within seconds. CTA permits the assessment of arterial stenosis, occlusion, craniocervical arterial dissection (blood collection within the vascular walls), and identification of stroke mimics such as symptoms caused by arteriovenous malformations (AVMs) (i.e., angiomas of the brain).

Imaging also facilitates delineation of the status of cerebral perfusion, demonstrating the infarct core and also the penumbra (potentially salvageable parenchyma), the identification of which may aid effective management strategies. This information can nowadays be obtained by recent sophisticated diffusion weighted imaging–perfusion weighted imaging (DWI-PWI) algorithms (see Chapter 18) both with CT or magnetic resonance (MR) scanners and digital subtraction angiography (DSA) (see Chapter 17).

It is now possible, in fact, to perform perfusion imaging using cone beam CT (CBCT) in the angio suite. In case of local thrombolysis, new software can calculate the cerebral blood volume (CBV) from the contrast-enhanced 3D data set acquired with the state-of-the-art flat panel detector angiography systems. Using dedicated software, cerebral blood volume (CBV) DSA maps are generated, which can assist in clinical decision making (differentiating between potentially salvageable brain tissue and irreversibly damaged areas) that can help in deciding whether to apply systemic thrombolysis or undertake endovascular procedures. Blood volume imaging is already implemented in the clinical workflow, especially in the treatment of patients with acute stroke or vasospasms who are treated using endovascular means. Catheter angiography is until now unmatched in morphological and hemodynamic evaluation of AVMs, making in-depth analysis of arterial, nidal, and venous compartments possible; similarly, collateral flow in cases of cerebro-afferent vessels obstruction can be adequately evaluated only by the means of this technique. Already some advantages of the new application are evident: It gives the neuro-interventionalist the means to check brain viability just before, during, or after an intervention without the need for patient relocation to another imaging suite.

In hemorrhagic conditions CT, CT-angio (CTA), MRI, and MR-angio (MRA) are essential for an accurate morphological diagnosis, while catheter angiography is fundamental for a thorough insight into the lesion both for higher vessel resolution and precise flow dynamics.

As regards noninvasive imaging, 4D CTA and MRA are now useful clinical tools thanks to the availability of faster and more powerful scanners. They enable time resolved studies, where several 3D volumes (sometimes 2D in case of MRA) can be acquired at short temporal intervals one from another, in order to obtain a dynamic representation of arterial, parenchymal, and venous phases. By this means, much information previously available only by performing digital subtraction angiography (DSA) where arterial catheterization is necessary can now be obtained noninvasively with only a venous injection. The two main drawbacks of these techniques are, at the time of writing this chapter, their low spatial and temporal resolution, but considering the speed of recent technological improvements this may soon change. Nevertheless, further advances

in DSA with arterial catheterization are also being made (see Chapter 13).

Biplane or 3D high resolution road-map angiography using CBCT can be of great importance for the guidance of the microcatheter inside the brain vasculature. This technique improves results in the following conditions: depositing embolizing coils inside an aneurysm with better control of the progressive filling of its sac; releasing embolizing agents, such as Onyx or polymerizing glues, in brain AVMs obliteration; approaching the thrombus during local arterial thrombolysis within obstructed vessels in ischemic stroke; and deploying intracranial stents both when recanalyzing obstructed vessels or when aiding the safe deposition of coils in wide-neck aneurysms in order to prevent undesired coil prolapse into the vessels.

Thorough knowledge of complex vessel structure, in fact, enables effective device guidance that allows more rapid procedures.

Three-dimensional angiography is of paramount importance in deciding the most effective treatment strategy by enabling visualization of a high contrast 3D-volume; this technique enables positioning of C-arms for optimal projections to be obtained for 2D image guidance during the intervention permitting an accurate understanding of the aneurysm neck (Figure 2.1) and of the behavior of tortuous AVMs afferent vessels (Figure 2.2); moreover, it is of great assistance in searching for the optimal projection to release embolizing material (working position).

CBCT, drastically improved by flat-panel technology, is a valid tool to gain more anatomical information to support treatment decisions. Checking for bleeding or planning for surgery can be done immediately in the angio-room, enabling on the spot decisions regarding the treatment to be performed (Figure 2.3). Stent and coil deployment through visualization of 3D volumes can be performed more easily since with this technique one can better distinguish vessels from stent struts, coils, calcified plaques, or bone.

2.2 Ischemic Stroke

Ischemic stroke encompasses multiple etiologies including thrombosis, embolism, venous thrombosis, and systemic hypoperfusion. Thrombolytic therapy with recombinant tissue plasminogen activator (rt-PA) is the treatment of choice for ischemic stroke presenting within a variable time window between 3 and 6 h of clinical onset, provided that there is no contraindication for this treatment, including

exclusion of intracranial hemorrhage by CT. Using a 3 h time period cutoff for thrombolysis, the National Institute of Neurological Disorders and Stroke (NINDS) trial (Stroke Study Group 1995) showed efficacy, with the number of patients needed to treat to benefit a single subject being eight; however, there was a 6.4% treatment-related incidence of intracranial hemorrhage.

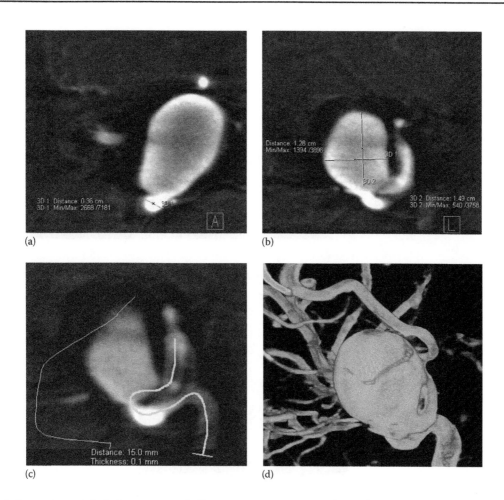

(a)   (b)

(c)   (d)

FIGURE 2.1 Left carotico-ophthalmic aneurysm. Different reconstructions are shown from a volumetric dataset obtained by the means of CBCT evaluation of the parent vessel (a) at the neck of the aneurysm and (b) of the dome diameters on 2D multiplanar reconstruction. (c) Simulation of the course of the parent vessel (thick intraluminal line) for precise calibration of the stent size. (d) 3D volume rendering technique for the evaluation of the relationship of the aneurysm with surrounding vessels.

Optimization of management of patients with acute stroke relies on prompt clinical assessment and stroke diagnosis followed by urgent brain imaging.

2.2.1 CT

CT is the imaging technique of choice in many institutions for the initial assessment of suspected stroke. It is readily available and permits rapid assessment of patients with acute stroke. Multidetector technology permits image acquisition within seconds and at submillimeter image resolution (Latchaw et al. 2009). Noncontrast CT consents the detection of intracranial hemorrhage with a high degree of sensitivity, identification of stroke mimics, and also the identification of early parenchymal ischemic change.

Major current guidelines on the use of thrombolysis include noncontrast CT as an acceptable investigation after which to make a decision about the benefits of thrombolysis can be made.

Approximately three-quarters of all ischemic cerebral infarcts occur within the territory supplied by the main arterial trunk perfusing each cerebral hemisphere (middle cerebral artery designated MCA). Detection of early ischemic change on noncontrast CT can be difficult, since many of the early radiological findings can be normal (Figures 2.4a and 2.5a) or subtle (Figure 2.5c). Early radiological signs indicate parenchymal infarction associated with pathological vessel occlusion (Wardlaw and Meilke 2005).

The absence of adequate perfusion to the brain tissue leads to impairment of transmembrane metabolic exchanges, with the accumulation of sodium inside the cell, actually provoking cellular swelling and death due to the increased water intake.

Chapter 2

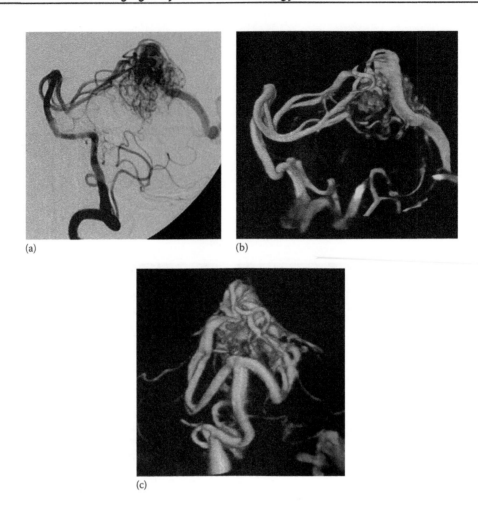

FIGURE 2.2 Superior cerebellar AVM. Conventional DSA angiography in the lateral projection (a) delineating the afferent vessels, the nidus, and the draining vein. 3D angiography, in the lateral (b) and antero-posterior (c) projections, better depicts the distal portion of afferent vessels.

The CT manifestation of this compartmental water shift is a focal mass effect with local cortical/gyral swelling and sulcal effacement; furthermore, loss of the gray/white interface evolves. The most common areas of cerebral hypoattenuation identified are in the MCA territory, at the level of the basal ganglia and cerebral cortex. Often the noncontrast scan in the hyperacute phase is normal, and, as such, the earlier these ischemic parenchymal changes are evident, the more severe the degree of ischemia is (Latchaw et al. 2009).

An international multicenter study (ECASS I: European Cooperative Acute Stroke Study) highlighted that early evidence of cortical hypoattenuation affecting over one-third of the MCA territory was a predictor of poor outcome after intravenous thrombolysis within 6 h of stroke onset, with increased risk of symptomatic intracranial hemorrhage (Hacke et al. 1995).

A further important sign that can be identified on the noncontrast CT study is vessel occlusion, seen as hyperdensity within the lumen of the vessel due to blood stagnation (hyperdense MCA). This is most often identified as linear hyperdensity within the proximal M1 segment of the MCA (Hill et al. 2003, Latchaw et al. 2009). Hyperdensity can be detected within any occluded vessel because of clot. A further example encountered quite commonly is thrombosis of the basilar artery, although identification of this sign can be difficult owing to vessel wall calcification and lack of a paired vessel for comparison.

2.2.2 MRI

MRI can be used as an alternative or an adjunct to noncontrast CT, as much of the information is complementary to the noncontrast CT study. CT has the benefit of being readily available, inexpensive, and

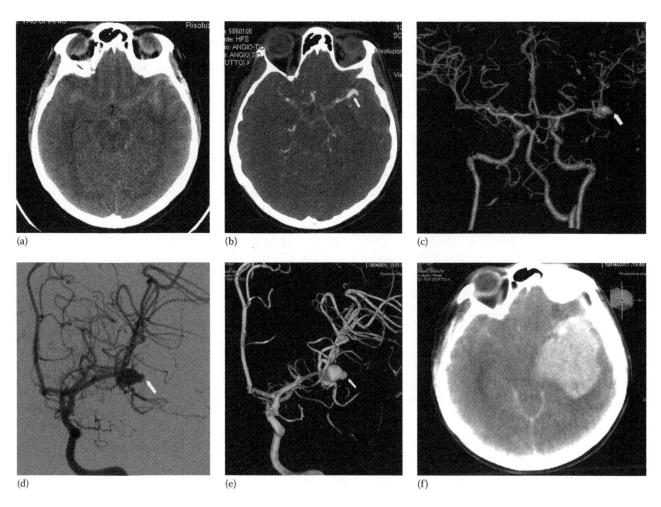

(a)   (b)   (c)

(d)   (e)   (f)

FIGURE 2.3 Acutely ruptured left middle cerebral artery aneurysm with rebleeding in the angio-suite. Plain CT scan (a) 1 h after an episode of acute headache, vomiting, and loss of consciousness in a 46-year female: evidence of diffuse subarachnoid hemorrhage. CT-angio source image (b) and CT-angio volume-rendering technique (VRT) (c) of the intracranial arteries in the antero-posterior projection: evidence of MCA bifurcation aneurysm (arrow). Oblique DSA angiography (d) showing the aneurysmal sac with a bleb on its dome (arrow) probable source of the hemorrhage in (a). This finding can be better recognized on the VRT 3D-angiography (arrow in e). At the end of the DSA study (f), before the beginning of treatment, the patient acutely deteriorated with left pupil mydriasis. Immediate cone-beam CT performed in the angio-suite reveals rebleeding with hyperacute left-temporal hematoma.

fast; however, increasingly out-of-hours MRI can be provided, and continued pulse sequence refinement is leading to reduced MRI acquisition time, and improved sensitivity in detection of acute ischemic change over noncontrast CT, especially in the detection of early ischemic change and identification of small infarcts and those within the posterior fossa. MRI provides improved anatomical detail and does not use ionizing radiation.

At present 1.5 T systems represent the standard for evaluation of patients with suspected acute ischemic stroke. 3 T scanners have been increasingly used, but in the acute setting of unstable patients they still present important disadvantages linked to increased movement-related artefacts and to safety concerns for the possible presence of ferromagnetic materials

(resuscitation trolleys, oro-tracheal cannulas, and metallic prosthesis).

The single most significant MRI pulse sequence in the imaging of acute stroke is DWI (Chung et al. 2002, Schellinger et al. 2003) (Figures 2.4c, d and 2.5d, e).

DWI uses the measurement of Brownian motion of molecules. As mentioned earlier for early CT detection of ischemic injury, the cellular process during a cerebral infarct is rapid disruption of the sodium/potassium pump and influx of water molecules into the intracellular compartment–cytotoxic edema. There is a resultant reduction in hydrogen ion diffusion ("restricted diffusion") and hence a decreased apparent diffusion coefficient (ADC), which manifests as a hypointensity area

Chapter 2

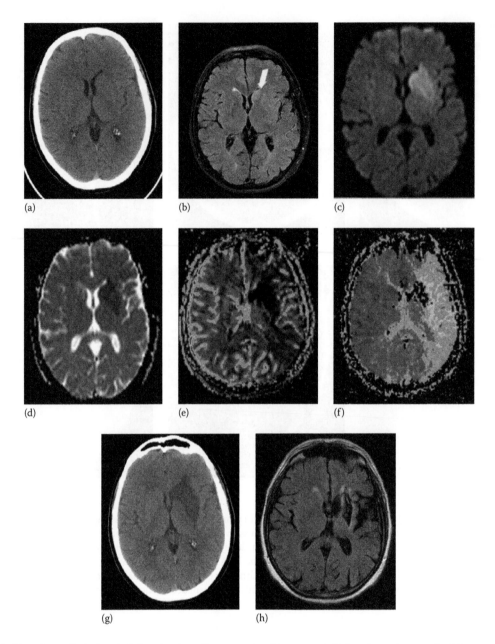

FIGURE 2.4   Left basal ganglia ischemic infarction in a 60-year male with severe right hemiparesis treated with systemic fibrinolysis. At plain CT (a) 2 h after stroke no clear abnormality is evident. FLAIR MRI image (b) at the same level reveals slight hyperintensity in the region of left basal ganglia (arrow). B1000 DWI sequence image (c) and corresponding ADC map (d) show restricted diffusion in the left basal ganglia appearing hyperintense in DWI and dark in ADC. MR perfusion cerebral blood volume (CBV) (e) and mean transit time (MTT) (f) maps show clear evidence of the presence of extensive penumbra in the left cerebral hemisphere. Systemic fibrinolysis was therefore performed. The definite infarct, as shown in the control plain CT (g) 14 days after, nearly matches the core of restricted diffusion as shown in (c) and (d). Follow-up control at 2 years with MR FLAIR image (h) shows the stabilized lesion with a cavity in the left basal ganglia.

on the ADC map (Figures 2.4d and 2.5e) and as hyperintensity on DWI (Figures 2.4c and 2.5d) (Minematsu et al. 1992, Rowley et al. 1999). Acquisition time for this pulse sequence is under 1 min and detection of acute/hyperacute infarction is possible within minutes of the event (Schellinger et al. 2003). Other MR pulse sequences offer complementary information in the evaluation of patients with acute stroke. Multiplanar spin-echo scans (T2-weighted and proton density imaging) identify the same early signs as those depicted on the noncontrast CT study but with increased sensitivity and specificity (Provenzale et al. 2003). Increased T2-weighted signal secondary to cytotoxic edema results in loss of gray/white differentiation.

Gradient-recalled echo (GRE) or, more recently, susceptibility-weighted imaging (SWI) (Ong and Stuckey 2010) is highly sensitive to the paramagnetic effect of the blood product deoxyhemoglobin, and hence this sequence is highly specific for identifying areas of microscopic and macroscopic hemorrhage.

Fluid-attenuated inversion recovery (FLAIR) MR, a heavily T2-weighted sequence, suppresses the cerebral spinal fluid signal and shows other fluid with high conspicuity; hence subarachnoid hemorrhage (SAH) is readily depicted (Kidwell et al. 2004, Makkat et al. 2002). The development of signal change within infarcted cortical or central gray matter is consistently seen earlier on FLAIR imaging (within 3 h) than it is with conventional spin-echo imaging (Gasparotti et al. 2009, Kamran et al. 2000) (Figure 2.4b).

MRI does offer better anatomical detail and increased sensitivity for acute infarct detection than CT. However, it is less readily available than CT and its usage in the acute stroke setting can be hampered by contraindications (such as cardiac prostheses) or confused patient status.

2.2.3 Delineation of the Ischemic Penumbra with Advanced CT and MRI Techniques

CTA and magnetic resonance angiography (MRA) allow noninvasive assessment of both the intracranial and extracranial vessels. Guidance from the American Heart Association recommends that, in the acute setting, MRA should only be performed if it does not delay intravenous thrombolysis within 0–3 h of stroke or if intra-arterial thrombolysis or mechanical thrombectomy is being considered and this facility is available (Latchaw et al. 2009, Ohue et al. 1998). Time of flight-MR angiography (TOF-MRA) is limited by its spatial resolution, and, as such, its depiction of more distal vessel occlusion is inferior to conventional catheter digital subtraction angiography.

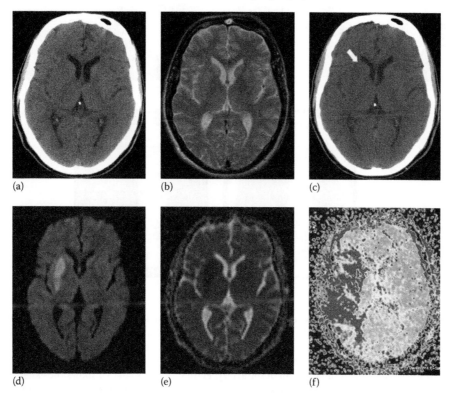

(a) (b) (c)

(d) (e) (f)

FIGURE 2.5 Right basal ganglia ischemic infarction in a 64-year male with left hemiparesis. Systemic fibrinolysis followed by late endovascular treatment according to "instrumental" rather than "time" window. Plain CT (a) and T2-W MR image (b) 150 min after stroke: no significant parenchymal abnormality is evident. Black spots due to flow void in normal patent vessels (rapidly moving spins in normal arteries do not give a detectable MR signal) are well recognizable in the left Sylvian fissure (arrows) but not in the right one. Systemic fibrinolysis was performed with clinical improvement. Eleven hours later the patient showed clinical worsening. A new plain CT (c) reveals only subtle indirect findings with loss of gray-white matter differentiation in the basal ganglia region (arrow), very difficult to recognize. B 1000 DWI (d) and corresponding ADC map (e) at the same level show a core of restricted diffusion that appears hyperintense in (d) and dark in (e) MR perfusion MTT map (f) shows evidence of extensive penumbra (see Chapter 18). *(Continued)*

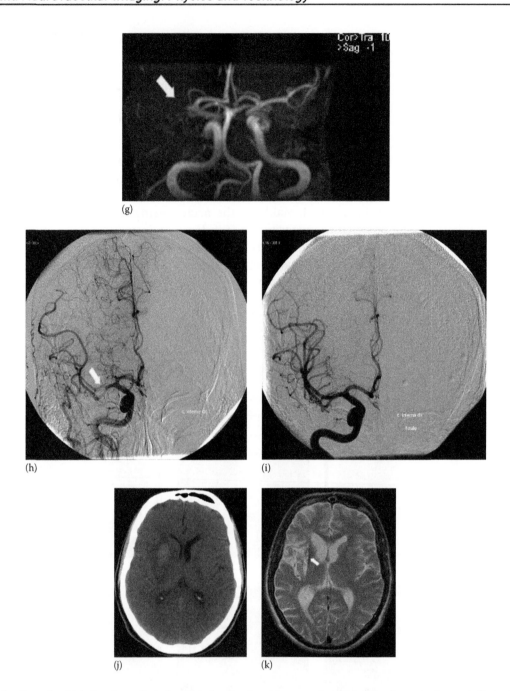

FIGURE 2.5 (*Continued*)   Right basal ganglia ischemic infarction in a 64-year male with left hemiparesis. Systemic fibrinolysis followed by late endovascular treatment according to "instrumental" rather than "time" window. Intracranial MR angiography (g) demonstrates right MCA occlusion (arrow). Despite that many hours had elapsed from the onset of symptoms, the clear presence of a mismatch core/penumbra was in favor of endovascular treatment. Pretreatment DSA (h) better shows the nearly occluded right MCA (arrow). Posttreatment DSA (i) shows complete recanalization of the artery. Plain CT 7 days later (j) shows the extension of the definite lesion, slightly greater than the ischemic core as shown in (d and e), but significantly smaller than the extensive penumbra as shown in (f). Slight hyperdensity in the core testifies to a small hemorrhagic component inside the infarct. Two years later a control T2 MR (k) shows the stabilized lesion in the basal ganglia with a small amount of hypointense hemosiderin deposit (arrow).

Multimodal imaging is being used more and more to tailor optimal management strategies in the acute stroke setting (Ogawa et al. 2007, Thomalla et al. 2006). Advanced imaging allows the assessment of the intracranial and extracranial vasculature and facilitates delineation of the status of cerebral perfusion, revealing the infarct core and the penumbra.

The ischemic penumbra can be defined as "ischemic tissue potentially destined for infarction (reversible sodium/potassium cell pumps impairment) but not yet irreversibly injured (cell death)"; this is the target of therapies in the acute setting (Fisher and Ginsberg 2004). The ischemic penumbra is therefore hypoperfused parenchyma in which neuronal electrical activity is interrupted but remains salvageable with the restoration of blood flow. Imaging evaluation of cerebral perfusion (see Chapter 18) is hence critical in the determination and calculation of the penumbra (Figure 2.4): for this purpose the most used parameter is the increase of blood mean transit time (MTT). The infarct core, which is probably irreversible, displays low cerebral blood flow (CBF) and volume (CBV) (Figure 2.4e). Several imaging techniques have been used to define the ischemic penumbra including positron emission tomography (PET). Access and availability of PET are limited, and hence most concerning mapping of the penumbra is focused on CT and MRI acquisition (Fisher and Ginsberg 2004).

Both MR and CT perfusion imaging use a contrast bolus tracking technique: CBV and MTT can be extrapolated from both techniques. In contrast with MR perfusion, CT offers both qualitative and quantitative measurements (Cianfoni et al. 2007). A series of images is acquired encompassing anterior, middle, and posterior cerebral arteries territories. MR perfusion imaging (PWI) is a qualitative study, and selected regions of interest are compared with the contralateral hemisphere. DWI depicts what is deemed to be the infarct core, and hence the "diffusion–perfusion" mismatch potentially represents salvageable parenchyma. For treatment purposes, CT scan CBV/MTT mismatch can be assimilated to MR DWI/PWI mismatch (i.e., core/penumbra) (Figure 2.4e and f).

Because of the narrow time window for patient presentation, assessment and start of therapy results in only a small proportion of eligible patients receiving thrombolysis. Consequently, fewer than 10% of patients with acute stroke actually receive rt-PA. It has been argued that the therapeutic window should be widened, with decisions to treat being guided by improved and integrated imaging strategies (Latchaw et al. 2009). This is especially true in cases of unsuccessful fibrinolysis, with no improvement or worsening of the neurological conditions in which the patient may be submitted to local fibrinolysis by endovascular route. In this way the concept of "instrumental

window" can be considered superior to "time window" in individual patients (Figure 2.5).

2.2.4 Neck Vessel Assessment

Assessment of the extracranial cerebro-afferent vasculature is important both when craniocervical dissection is considered to be the underlying etiology of stroke and when vascular stenosis (predominantly carotid) is suspected.

Where spontaneous craniocervical arterial dissection is considered to be the underlying etiology, MRA assessment of the carotid and vertebral arteries is advised from the level of the aortic arch to the circle of Willis, and this can be easily added to the initial brain MRI assessment (Thanvi et al. 2005). Although these lesions account for only 2% of all ischemic strokes (Schievink 2001), they are responsible for up to 20% of such strokes in patients under 45 years of age (Thanvi et al. 2005). Routinely adopted sequences include axial T1 and T2 and MRA, either with or without contrast enhancement. Findings include detection of an intimal flap; loss of normal flow void; intraluminal high signal on T1 consistent with stasis; classical tapering of the vessel lumen on MRA. Conventional catheter digital subtraction angiography was previously the reference standard technique for the evaluation of craniocervical dissection. However, the evolution of sensitive noninvasive techniques has superseded this technique in the routine setting; both MRA and CTA have been shown to be highly sensitive in the evaluation of this condition. For detection of lesions of epi-aortic vessels at the level of the neck, Doppler ultrasound is the most widely available, cheapest, and minimally invasive technique, but CTA and MRA, as mentioned earlier, can also be used. In the acute stroke setting, the principal role of neck vessel imaging is to identify an underlying mechanism, such as arterial occlusion (secondary to either thrombus or dissection), or vessel stenosis, the demonstration of which may influence therapy (Latchaw et al. 2009). Noninvasive imaging techniques for neck vessel assessment, including Doppler ultrasound, CTA, and MRA, offer similar levels of accuracy, therefore the choice of technique often is made according to local availability and preference.

2.3 Hemorrhagic Stroke

Intracranial hemorrhages account for 20% of strokes, three-fourth of which are intraparenchymal (within brain tissue) rather than subarachnoid (within the

meningeal spaces surrounding intracranial arteries) in location. The most frequent cause of intracranial hemorrhage is, in 50% of the cases, systemic hypertension,

Chapter 2

which is responsible for chronic lesions of vessel wall leading to rupture and consequently hemorrhage. In order of frequency, other possible causes of hemorrhage can be represented from rupture of vascular malformations, including aneurysms and AVMs, coagulation diseases, intracranial neoplasm, and trauma.

In acute settings, CT scan is the procedure of choice if hemorrhage is suspected. MRI plays a secondary role, while in sub acute and chronic stages it can be useful for a deeper insight into the underlying etiology.

2.3.1 Intracranial Aneurysms

There is no single cause for intracranial aneurysms. They are not really congenital; however, there are congenital disorders that affect vessel wall architecture and predispose to aneurysm formation. Disorders associated with an increased incidence of aneurysms include hypertension, aortic coarctation, adult polycystic kidney disease, fibromuscular dysplasia, connective tissue disorders (e.g., Marfan syndrome, Ehlers–Danlos syndrome), and moyamoya disease.

The relative incidence of aneurysms in different locations is given in Table 2.1 (Osborn 1994).

The frequency of intracranial aneurysms in the general population has been estimated between 1.5% and 8% (Jellinger 1979) (Bannerman et al. 1970). The peak age for rupture is between 40 and 70. The mean age of occurrence of fatal hemorrhage is 50.2 (Stehbens et al. 1963).

Aneurysms less than 7 mm in supratentorial locations (at the level of cerebral hemispheres) rarely bleed, whereas larger ones and those located infratentorially (at the level of cerebellum and brain stem) often occur with bleeding (Graf et al. 1971, Wiebers et al. 2003) (Table 2.2).

Multiple aneurysms are present in 20% of hemorrhagic stroke patients. Females are affected more often

Table 2.1 Relative Incidence of Intracranial Aneurysms by Location

Location	Incidence (%)
Anterior communicating artery	30–35
Internal carotid artery or posterior communicating artery	30–35
Middle cerebral artery bifurcation	20
Posterior fossa	2–10
Distal basilar	5
Miscellaneous sites distal to circle of Willis	1–3

Table 2.2 Aneurysm Bleeding Rates according to Location and Size

Location	Size (mm)	Bleeding Rate (%)
Supratentorial (ICA, AcoA, ACA, MCA)	<7	0
	7–12	2.6
	13–24	14.5
	25 or >	40
Infratentorial + PCoA	<7	2.5
	7–12	14.5
	13–24	18.4
	25 or >	50

than males (56% compared with 44%) (Fox 1983) and aneurysms are rarely found in children. Giant aneurysms frequently simply grow; they do not bleed as often as smaller aneurysms, but their long-term outcome is poor. In the posterior fossa (typically basilar tip), the mortality rates for giant aneurysms approach 100% at 5 years, whereas in the anterior circulation, the morbidity and mortality rates for giant aneurysms can reach nearly 80% of cases.

About one-third of patients experience symptoms of postoperative vasospasm, which can complicate their recovery. Vasospasm is defined as arterial narrowing of at least 50% and is demonstrated angiographically in up to 70% of posthemorrhage patients; only one-third of patients have symptomatic vasospasm. Ventricular dilation leading to hypertensive hydrocephalus can arise; in these cases CSF diversion is needed.

Computed tomography (CT) scan is positive in up to 90% of patients with intracranial aneurysm hemorrhages when performed within 1 day; it also yields diagnostic clues in about half the cases.

CT can also reveal important prognostic information. The grading system developed by Fisher et al. is based on retrospective and prospective studies demonstrating that the amount and location of intracranial blood in patients with aneurysmal SAH correlates highly with the incidence, severity, and location of subsequent vasospasm (Fisher et al. 1980, Kistler et al. 1983). This Fisher grading system is summarized in Table 2.3.

CTA, nowadays the imaging technique of choice in the assessment of patients with SAH, can be performed immediately after the plain examination. Spiral CTA is not dependent on flow rate, but rather on the presence or absence of nonclotted blood (and thus contrast); it is also not susceptible to artefact due to pulsation.

Table 2.3 Fisher Grading System for Computed Tomographic Scan Findings in Aneurysmal Bleeding

Grade/Group	Findings on CT	Predicted Severity of Vasospasm
1	No detectable blood	No severe spasm
2	Diffuse blood not dense enough to represent a large, thick homogeneous clot	No severe spasm
3	Dense collection of blood appearing to represent a clot >1 mm thick in the vertical plane (interhemispheric fissure, insular cistern, ambient cistern) or >5 × 3 mm in longitudinal and transverse dimension in the horizontal plane (stem of the Sylvian fissure, Sylvian cistern, interpeduncular cistern)	Severe spasm
4	Intracerebral or intraventricular clots but with only diffuse blood or no blood in the basal cisterns	No severe spasm

High-resolution MRA, on the other hand, can detect aneurysms as small as 3–4 mm but can be suboptimal because of poor flow characteristics within aneurysms, rendering visualization dependent on flow characteristics rather than the size of the aneurysm.

Three-dimensional reconstructions can be performed using either modality.

Whatever the technique used, these important questions must be answered: Is there an aneurysm? Are there multiple aneurysms? Which one was the source of bleeding? Is there any vasospasm? Is the neck wide or narrow? Are there any vessels coming from the neck or dome? Are there any efficient collateral anastomoses in case of vessel occlusion during intervention (Figures 2.1 and 2.6)? Recently sophisticated computerized flow simulation studies for improved understanding of intravascular fluid dynamics have been used to identify the exact weakness point in individual aneurysms so that a tailored treatment can be planned for each patient (Szikora et al. 2008).

Screening for unruptured aneurysms is advisable in a family with a history of aneurysms, but the choice of MRA, CTA, or catheter DSA should be made by the patient and his/her attending physician.

The ideal solution for the treatment of an intracranial aneurysm is the exclusion of this abnormality from the circulation. The gold standard for this therapy is open neurosurgical clipping of the neck of the aneurysm or endovascular embolization.

2.3.2 Brain Arteriovenous Malformations

Morphologically, brain AVMs are direct connections of one or more arteries to one or more draining veins, without an intervening capillary bed subdivided into three major components: arterial feeders, nidus, and venous drainage, resembling a short circuit. They usually occur in a sporadic fashion in approximately 0.5%–1% of the population (about one-tenth the frequency of intracranial arterial aneurysms) (Johnson et al. 1993, Rhoten et al. 1997). The association with aneurysms is well documented, occurring in between 6% and 20% of patients (Fisher 1996, Garretson 1996, Pollock et al. 1996, Solomon 1994). As knowledge regarding AVMs has improved, classification systems have been proposed and refined (Houdart et al. 1993, Valavanis et al. 1986).

The most frequent clinical presentation of a patient with an AVM is intracranial bleeding, followed by headache, seizures, and neurological deficits.

The first assessment is usually made by CT or MR, while CTA and MRA confirm the diagnosis and provide initial clues like lesion volume, principal afferents, and draining veins. They are frequently followed by catheter DSA (Figure 2.2), which even now is the method of choice for the precise delineation of the following: all the feeding vessels, including small branches directly penetrating brain tissue; AVM nidus (tangle of vessels site of artero-venous short-circuit), especially for the disclosure of intranidal aneurysms and/or varices that represent the components more likely to bleed; surrounding draining veins, distinguishing between those specific to the AVM and the normal ones that are recruited secondarily; and flow dynamics in the AVM and possible afferents from the external circulation.

MR advanced techniques, such as functional-MRI (f-MRI) (Figure 2.7) and diffusion tensor tractography (DTI) (Figure 2.8) can help the therapist significantly (neurosurgeon or neuroradiologist) to understand the relation of the AVM with the surrounding eloquent areas of the brain.

Chapter 2

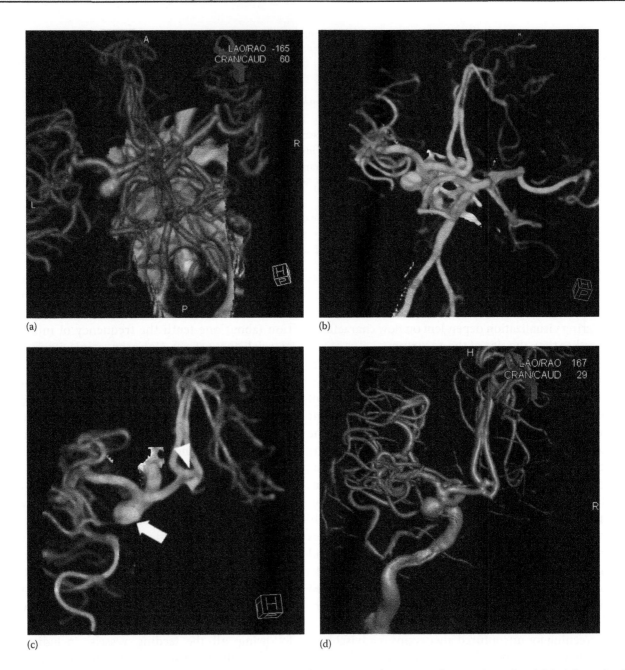

FIGURE 2.6   Multiple intracranial aneurysms as evidenced by different angio techniques. Right MCA (arrow) and ACoA (arrowhead) aneurysms denoted in C. Angio-CT volume rendering technique (VRT) reconstruction (a) of the entire intracranial vasculature; left postero-superior collapsed view with partial bone removal. VRT reconstruction (b) with different filter and complete bone removal in a similar projection. Close-up (c) of the same reconstruction as (b), with removal of uninvolved vessels. VRT reconstruction (d) of 3D DSA obtained by right internal artery selective injection. CT-angio, which is significantly less invasive than 3D-DSA, requires more elaborated and time-consuming postprocessing in order to produce comparable results.

The preceding pathologies are complex. Neither the real pathophysiology nor the risk associated with these intriguing lesions have been fully clarified as yet (Brown et al. 1988, Mast et al. 1997, Ondra et al. 1990) so that not infrequently the correct therapeutic management for an individual patient may be cumbersome (Choi et al. 2006). A number of factors depending on both the patient and the specific type of AVM must be taken into consideration when deciding firstly whether to treat a patient or not and secondly which is the best option of treatment. Thanks to the development of modern catheter angiography and MR techniques, over the years a number of classification systems have been proposed and refined (Houdart et al. 1993, Valavanis et al. 1986).

The different classifications proposed have considered various angio-architectural features such as size,

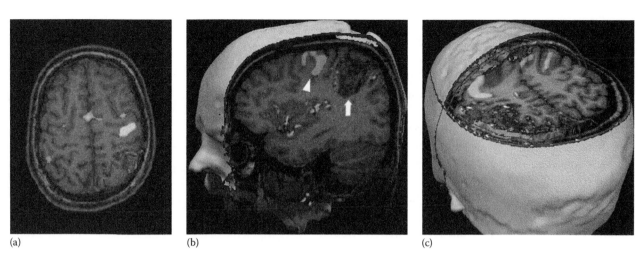

(a)   (b)   (c)

FIGURE 2.7   Left rolandic area AVM. Hand motor task f-MRI in the axial (a), sagittal (b), and composite coronal (c) views clearly shows that the AVM (arrow in b) is located at a safe distance and posteriorly to the cortical primary motor area (arrowhead in b). (By permission from *Interventional Neuroradiology*.)

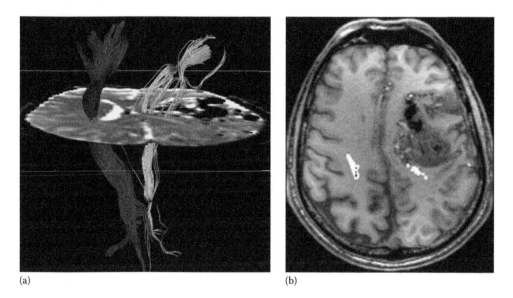

(a)   (b)

FIGURE 2.8   Left temporo-insular AVM. DTI (a) and its projection on axial T1-w image (b) clearly show the displacement and dispersion of the left (right in figure) cortico-spinal tract CST (light gray), caused by the AVM. Right (left in figure) CST (dark grey) is normal.

site, number and distribution of afferent vessels, patterns of venous drainage, flow velocity, and the amount of blood steal to the adjacent normal brain tissue in order to obtain a possible association with the bleeding risk (Berenstein et al. 2004, Brown et al. 1996, Graf et al. 1983, Kondziolka et al. 1995, Marks et al. 1990, Miyasaka et al. 1992, Stapf et al. 2006, Turjman et al. 1994). Nevertheless, the available classifications are so different that comparison between various series is extremely difficult.

Most relevant papers dealing with predictors of unfavorable outcome identified separately several factors for surgery (Fisher 1996, Hamilton and Sptelzer 1994, Spetzler and Martin 1986), radiosurgery (Andrade-Souza et al. 2005, Liscak et al. 2007, Nicolato et al. 2006a,b), or embolization (Hartmann et al. 2002, 2005, Haw et al. 2006, Jayaraman et al. 2008, Ledezma et al. 2006). In the case of endovascular treatment adverse factors linked to neurological complications were identified by means of clinical characteristics such as age or the absence of pretreatment neurological deficits (Hartmann et al. 2002) or morphological AVM characteristics such as deep venous drainage and the size of the AVM, basal ganglia location (Jayaraman et al. 2008), eloquent areas, or fistulas (Haw et al. 2006).

Other adverse factors identified such as periprocedural hemorrhage (Ledezma et al. 2006), venous deposition of glue (Haw et al. 2006), or number of embolizations (Hartmann et al. 2002) do not appear useful for pretreatment identification of possible risks. Moreover, in literature morbi/mortality rates deal exclusively with a single type of treatment, hardly taking into consideration the overall rate on completion (i.e., obliteration of the AVM) of combined treatments.

Relying on the data available in literature and on our personal experience (Beltramello et al. 2005), an operative classification of brain AVMs was developed in our department (Beltramello et al. 2008, 2009) with the main aim of helping the physician and the patient to reach informed consent on the treatability of his/her AVM (Tables 2.4 and 2.5).

This classification enables the clinician to reach a cumulative score (CS), made up of the sum of intention to treat score (ITS), depending on patient features and the individual AVM characteristics, and treatment risk score (TRS), depending mainly on the AVM size and other important parameters, that is, eloquence, according to the various types of treatments.

Table 2.4 Brain AVMs Intention to Treat Score (ITS: 0–12)

(A) Patient Features (0–6)		Score
Age (years)	≥60	2
	≥40 < 60	1
	<40	0
Previous hemorrhages	No	2
	Yes	0
Neurological deficits	No	1
	Yes	0
Patient's firm intention to be treated	No	1
	Yes	0
(B) AVM Characteristics (0–6)		Score
Small size	No	1
	Yes	0
Deep brain location	No	1
	Yes	0
Exclusive deep venous drainage	No	2
	Yes	0
Associated aneurysm/varix	No	2
	Yes	0

By permission from *Interventional Neuroradiology*.

Table 2.5 Brain AVMs Treatment Risk Score (TRS: 1–5)

(A) Surgery (1–5)	Grade	Score
Spetzler–Martin modified (Spetzler and Martin 1986)	I	1
	II	2
	III	3
	IV	4
	V	5

(B) Radiosurgery (1–5)		
Volume cm^3		Score
<5		1
>5 < 10		2
>10 < 20		3
>20 < 30		4
>30		5
One point minus if low flow		

(C) Embolization (1–5)		Score
Volume cm^3	<10	1
	> 10 < 20	2
	>20	3
Eloquence	No	0
	Yes	1
Perforators	No	0
	Yes	1

By permission from *Interventional Neuroradiology*.

Treatment is highly recommended for patients with CS ranging from 1 to 10; for those with CS 11 or 12, it is offered at a significant risk; while no treatment at all is advisable for those patients whose CS ranges between 13 and 17 (Table 2.6).

Together with surgery and endovascular techniques, radiosurgery is an effective treatment option for brain

Table 2.6 Brain AVMs CS with the Recommended Strategy

1–10
Treatment recommended
11, 12
Treatment offered with significant risk
13–17
Treatment not recommended

By permission from *Interventional Neuroradiology*.

lesions in eloquent or deep-seated locations and it has been increasingly used for cerebral AVMs.

Even if it is a noninvasive procedure, the rate of permanent radiation-related complications is low but not negligible, especially for AVMs located at critical sites. Some critical structures, such as the cranial nerves and brainstem are directly visible on conventional imaging, others such as cortical activation areas and white matter tracts can only be guessed on the basis of our knowledge of normal anatomy.

These assumptions however cannot be held in the case of pathologically induced modifications (Maruyama et al. 2005).

Cortico-spinal tract (CST) representation obtained by 3-T diffusion tensor imaging (DTI) tractography (DTI-t) in gamma knife treatment planning was implemented in our department, in order to evaluate its exposure to radiation in patients with cerebral AVMs located near this important motor function pathway (Foroni et al. 2010) (Figure 2.9).

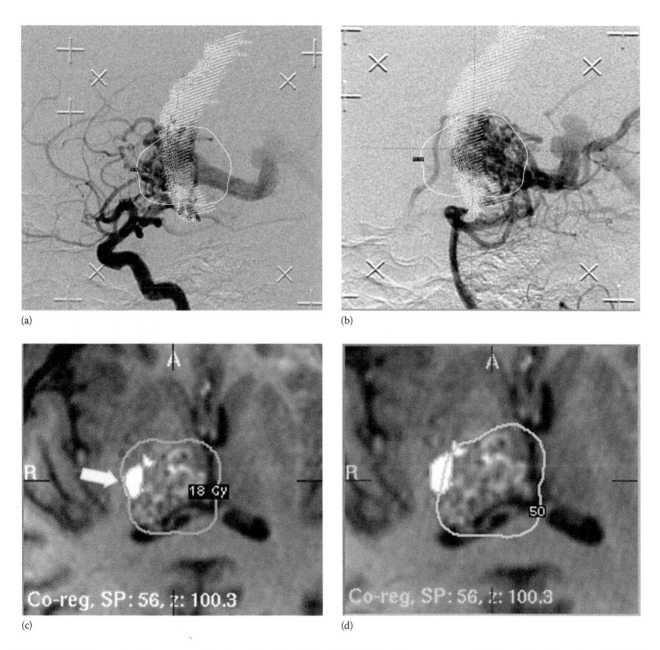

(a)

(b)

(c)

(d)

FIGURE 2.9 Gamma-knife treatment planning of a right thalamic AVM. The dose contours are indicated in yellow. Lateral DSA in stereotactic conditions showing the anterior (a) and posterior (b) compartments of the AVM. The CST is represented in light blue. Treatment planning as usual (c) includes the CST (arrow). In the optimized treatment planning (d) the CST is almost completely excluded by the dose contour.

Chapter 2

References

Andrade-Souza, Y. M., G. Zadeh, D. Scora, M. N. Tsao, and M. L. Schwartz. 2005. Radiosurgery for basal ganglia, internal capsule, and thalamus arteriovenous malformation: Clinical outcome. *Neurosurgery* 56:56–63; discussion 63–54.

Bannerman, R. M., G. B. Ingall, and C. J. Graf. 1970. The familial occurrence of intracranial aneurysms. *Neurology.* 20:283–292.

Beltramello, A., G. K. Ricciardi, E. Piovan et al. 2009. Operative classification of brain arteriovenous malformations. Part two: Validation. *Interventional Neuroradiology* 15:259–265.

Beltramello, A., P. Zampieri, G. K. Ricciardi et al. 2005. Combined treatment of brain AVMs: Analysis of five years (2000–2004) in the verona experience. *Interventional Neuroradiology* 11:63–72.

Beltramello, A., P. Zampieri, G. K. Ricciardi et al. 2008. Operative classification of brain arteriovenous malformations. *Interventional Neuroradiology* 14:9–19.

Berenstein, A., P. L. Lasjaunias, and K. Brugge. 2004. *Surgical Neuroangiography: Clinical and Endovascular Treatment Aspects in Adults.* Heidelberg, Germany: Springer-Verlag.

Brown, R. D. Jr, D. O. Wiebers, G. Forbes et al. 1988. The natural history of unruptured intracranial arteriovenous malformations. *Journal of Neurosurgery* 68:352–357.

Brown, R. D. Jr, D. O. Wiebers, J. C. Torner, and W. M. O'Fallon. 1996. Frequency of intracranial hemorrhage as a presenting symptom and subtype analysis: A population-based study of intracranial vascular malformations in Olmsted Country, Minnesota. *Journal of Neurosurgery* 85:29–32.

Choi, J. H., H. Mast, R. R. Sciacca et al. 2006. Clinical outcome after first and recurrent hemorrhage in patients with untreated brain arteriovenous malformation. *Stroke* 37:1243–1247.

Chung, S. P., Y. R. Ha, S. W. Kim, and I. S. Yoo. 2002. Diffusion-weighted MRI as a screening tool of stroke in the ED. *American Journal of Emergency Medicine* 20:327–331.

Cianfoni, A., C. Colosimo, M. Basile, M. Wintermark, and L. Bonomo. 2007. Brain perfusion CT: Principles, technique and clinical applications. *Radiologia Medica* 112:1225–1243.

Fisher, C. M., J. P. Kistler, J. M. Davis et al. 1980. Relation of cerebral vasospasm to subarachnoid hemorrhage visualized by computerized tomographic scanning. The relation of cerebral vasospasm to the extent and location of subarachnoid blood visualized by CT scan: A prospective study. *Neurosurgery* 6:1–9.

Fisher, M. and M. Ginsberg. 2004. Current concepts of the ischemic penumbra. *Stroke* 35:2657–2658.

Fisher, W. S. 1996. Concomitant intracranial aneurysms and arteriovenous malformations. In *Neurosurgery*, eds. Wilkins, R. H., and Rengachary, S. S., pp. 2429–2432: New York: McGraw-Hill.

Foroni, R. I., G. K. Ricciardi, F. Lupidi et al. 2010. Diffusion-tensor imaging tractography of the corticospinal tract for evaluation of motor fiber tract radiation exposure in gamma knife radiosurgery treatment planning. In *Radiosurgery*, ed. MW, M., pp. 128–138: Basel, Switzerland: Karger.

Fox, J. L. 1983. *Intracranial Aneurysms.* New York: Springer-Verlag.

Garretson, H. D. 1996. Intracranial arteriovenous malformations. In *Neurosurgery*, eds. Wilkins, R. H., and Rengachary, S. S., pp. 2433–2442: New York: McGraw-Hill.

Gasparotti, R., M. Grassi, D. Mardighian et al. 2009. Perfusion CT in patients with acute ischemic stroke treated with intra-arterial thrombolysis: Predictive value of infarct core size on clinical outcome. *American Journal of Neuroradiology* 30:722–727.

Graf, C. J., G. E. Perret, and J. C. Torner. 1983. Bleeding from cerebral arteriovenous malformations as part of their natural history. *Journal of Neurosurgery* 58:331–337.

Graf, C. J., D. O. Wiebers, J. P. Whisnant et al. 1971. Prognosis for patients with nonsurgically-treated aneurysms. Analysis of the Cooperative Study of Intracranial Aneurysms and Subarachnoid hemorrhage. Unruptured intracranial aneurysms: Natural history, clinical outcome, and risks of surgical and endovascular treatment. Early management of aneurysmal subarachnoid hemorrhage. A report of the Cooperative Aneurysm Study Surgical risk as related to time of intervention in the repair of intracranial aneurysms. Relation of cerebral vasospasm to subarachnoid hemorrhage visualized by computerized tomographic scanning. The relation of cerebral vasospasm to the extent and location of subarachnoid blood visualized by CT scan: A prospective study. *Journal of Neurosurgery* 35:438–443.

Hacke, W., M. Kaste, C. Fieschi et al. 1995. Intravenous thrombolysis with recombinant tissue plasminogen activator for acute hemispheric stroke. The European Cooperative Acute Stroke Study (ECASS). *The Journal of the American Medical Association* 274:1017–1025.

Hamilton, M. G. and R. F. Spetzler. 1994. The prospective application of a grading system for arteriovenous malformations. *Neurosurgery* 34:2–6; discussion 6–7.

Hartmann, A., H. Mast, J. P. Mohr et al. 2005. Determinants of staged endovascular and surgical treatment outcome of brain arteriovenous malformations. *Stroke* 36:2431–2435.

Hartmann, A., J. Pile-Spellman, C. Stapf et al. 2002. Risk of endovascular treatment of brain arteriovenous malformations. *Stroke* 33:1816–1820.

Hatano, S. 1976. Experience from a multicentre stroke register: A preliminary report. *Bull World Health Organ* 54:541–553.

Haw, C. S., K. terBrugge, R. Willinsky, and G. Tomlinson. 2006. Complications of embolization of arteriovenous malformations of the brain. *Journal of Neurosurgery* 104:226–232.

Hill, M. D., H. A. Rowley, F. Adler et al. 2003. Selection of acute ischemic stroke patients for intra-arterial thrombolysis with pro-urokinase by using ASPECTS. *Stroke* 34:1925–1931.

Houdart, E., Y. P. Gobin, A. Casasco et al. 1993. A proposed angiographic classification of intracranial arteriovenous fistulae and malformations. *Neuroradiology* 35:381–385.

Jayaraman, M. V., M. L. Marcellus, S. Hamilton et al. 2008. Neurologic complications of arteriovenous malformation embolization using liquid embolic agents. *American Journal of Neuroradiology* 29:242–246.

Jellinger, K. 1979. Pathology and aetiology of intracranial aneurysms. In *Cerebral Aneurysms: Advances in Diagnosis and Therapy*, eds. Pia, H., Langmaid, C., and Zierski, J., pp. 5–19: New York: Springer.

Johnson, P. C., T. M. Wascher, J. Golfinos, and R. F. Spetzler. 1993. Definition and pathologic features. In *Cavernous Malformation*, eds. Awad, I. A., Barrow, D. L., and Park Ridge, I. L., pp. 1–11: Park Ridge, IL: American Association of Neurological Surgeons.

Kamran, S., V. Bates, R. Bakshi et al. 2000. Significance of hyperintense vessels on FLAIR MRI in acute stroke. *Neurology* 55:265–269.

Kidwell, C. S., J. A. Chalela, J. L. Saver et al. 2004. Comparison of MRI and CT for detection of acute intracerebral hemorrhage. *The Journal of the American Medical Association* 292:1823–1830.

Kistler, J. P., R. M. Crowell, K. R. Davis et al. 1983. The relation of cerebral vasospasm to the extent and location of subarachnoid blood visualized by CT scan: A prospective study. *Neurology* 33:424–436.

Kondziolka, D., M. R. McLaughlin, and J. R. Kestle. 1995. Simple risk predictions for arteriovenous malformation hemorrhage. *Neurosurgery* 37:851–855.

Latchaw, R. E., M. J. Alberts, M. H. Lev et al. 2009. Recommendations for imaging of acute ischemic stroke: A scientific statement from the American Heart Association. *Stroke* 40:3646–3678.

Ledezma, C. J., B. L. Hoh, B. S. Carter et al. 2006. Complications of cerebral arteriovenous malformation embolization: Multivariate analysis of predictive factors. *Neurosurgery* 58:602–611; discussion 602–611.

Liscak, R., V. Vladyka, G. Simonova et al. 2007. Arteriovenous malformations after Leksell gamma knife radiosurgery: Rate of obliteration and complications. *Neurosurgery* 60:1005–1014; discussion 1015–1006.

Makkat, S., J. E. Vandevenne, G. Verswijvel et al. 2002. Signs of acute stroke seen on fluid-attenuated inversion recovery MR imaging. *American Journal of Roentgenology* 179:237–243.

Marks, M. P., B. Lane, G. K. Steinberg, and P. J. Chang. 1990. Hemorrhage in intracerebral arteriovenous malformations: Angiographic determinants. *Radiology* 176:807–813.

Maruyama, K., K. Kamada, M. Shin et al. 2005. Integration of three-dimensional corticospinal tractography into treatment planning for gamma knife surgery. *Journal of Neurosurgery* 102:673–677.

Mast, H., W. L. Young, H. C. Koennecke et al. 1997. Risk of spontaneous haemorrhage after diagnosis of cerebral arteriovenous malformation. *Lancet* 350:1065–1068.

Minematsu, K., L. Li, M. Fisher et al. 1992. Diffusion-weighted magnetic resonance imaging: Rapid and quantitative detection of focal brain ischemia. *Neurology* 42:235–240.

Miyasaka, Y., K. Yada, T. Ohwada et al. 1992. An analysis of the venous drainage system as a factor in hemorrhage from arteriovenous malformations. *Journal of Neurosurgery* 76:239–243.

Nicolato, A., F. Lupidi, M. F. Sandri et al. 2006a. Gamma knife radiosurgery for cerebral arteriovenous malformations in children, adolescents and adults. Part I: Differences in epidemiologic, morphologic, and clinical characteristics, permanent complications, and bleeding in the latency period. *International Journal of Radiation Oncology, Biology, Physics* 64:904–913.

Nicolato, A., F. Lupidi, M. F. Sandri et al. 2006b. Gamma Knife radiosurgery for cerebral arteriovenous malformations in children/adolescents and adults. Part II: Differences in obliteration rates, treatment-obliteration intervals, and prognostic factors. *International Journal of Radiation Oncology, Biology, Physics* 64:914–921.

Ogawa, A., E. Mori, K. Minematsu et al. 2007. Randomized trial of intraarterial infusion of urokinase within 6 hours of middle cerebral artery stroke: The middle cerebral artery embolism local fibrinolytic intervention trial (MELT) Japan. *Stroke* 38:2633–2639.

Ohue, S., K. Kohno, K. Kusunoki et al. 1998. Magnetic resonance angiography in patients with acute stroke treated by local thrombolysis. *Neuroradiology* 40:536–540.

Ondra, S. L., H. Troupp, E. D. George, and K. Schwab. 1990. The natural history of symptomatic arteriovenous malformations of the brain: A 24-year follow-up assessment. *Journal of Neurosurgery* 73:387–391.

Ong, B. C. and S. L. Stuckey. 2010. Susceptibility weighted imaging: A pictorial review. *Journal of Medical Imaging and Radiation Oncology* 54:435–449.

Osborn, A. 1994. *Diagnostic Neuroradiology*. St. Louis, MO: CV Mosby.

Pollock, B. E., J. C. Flickinger, L. D. Lunsford, D. J. Bissonette and D. Kondziolka. 1996. Factors that predict the bleeding risk of cerebral arteriovenous malformations. *Stroke* 27:1–6.

Provenzale, J. M., R. Jahan, T. P. Naidich, and A. J. Fox. 2003. Assessment of the patient with hyperacute stroke: Imaging and therapy. *Radiology* 229:347–359.

Rhoten, R. L., Y. G. Comair, D. Shedid, D. Chyatte and M. S. Simonson. 1997. Specific repression of the preproendothelin-1 gene in intracranial arteriovenous malformations. *Journal of Neurosurgery* 86:101–108.

Rowley, H. A., P. E. Grant and T. P. Roberts. 1999. Diffusion MR imaging. Theory and applications. *Neuroimaging Clinics of North America* 9:343–361.

Royal College of Physicians. 2008. *National Collaborating Centre for Chronic Conditions. Stroke: National Clinical Guidelines for Diagnosis and the Initial Management of Acute and Transient Ischemic Attack (TIA)*. London, U.K.: Royal College of Physicians.

Schellinger, P. D., J. B. Fiebach, and W. Hacke. 2003. Imaging-based decision making in thrombolytic therapy for ischemic stroke: Present status. *Stroke* 34:575–583.

Schievink, W. I. 2001. Spontaneous dissection of the carotid and vertebral arteries. *The New England Journal of Medicine* 344:898–906.

Solomon, R. A. 1994. Vascular malformations affecting the nervous system. In *Principles of Neurosurgery*, eds. Rengachary, S. S., and Wilkins, R. H., pp. 12.12–12.15: Chicago, IL: Mosby-Year Book.

Spetzler, R. F. and N. A. Martin. 1986. A proposed grading system for arteriovenous malformations. *Journal of Neurosurgery* 65:476–483.

Stapf, C., H. Mast, R. R. Sciacca et al. 2006. Predictors of hemorrhage in patients with untreated brain arteriovenous malformation. *Neurology* 66:1350–1355.

Stehbens, W. E., C. J. Graf, D. O. Wiebers et al. 1963. Aneurysms and Anatomical Variation of Cerebral Arteries. Prognosis for patients with nonsurgically-treated aneurysms. Analysis of the Cooperative Study of Intracranial Aneurysms and Subarachnoid hemorrhage. Unruptured intracranial aneurysms: Natural history, clinical outcome, and risks of surgical and endovascular treatment. Early management of aneurysmal subarachnoid hemorrhage. A report of the Cooperative Aneurysm Study. Surgical risk as related to time of intervention in the repair of intracranial aneurysms. Relation of cerebral vasospasm to subarachnoid hemorrhage visualized by computerized tomographic scanning. The relation of cerebral vasospasm to the extent and location of subarachnoid blood visualized by CT scan: A prospective study. *Archives of Pathology* 75:45–64.

Chapter 2

Stroke Study Group. 1995. Tissue plasminogen activator for acute ischemic stroke. The National Institute of Neurological Disorders and Stroke rt-PA Stroke Study Group. *The New England Journal of Medicine* 333:1581–1587.

Szikora, I., G. Paal, A. Ugron et al. 2008. Impact of aneurysmal geometry on intraaneurysmal flow: A computerized flow simulation study. *Neuroradiology* 50:411–421.

Thanvi, B., S. K. Munshi, S. L. Dawson and T. G. Robinson. 2005. Carotid and vertebral artery dissection syndromes. *Postgraduate Medical Journal* 81:383–388.

Thomalla, G., C. Schwark, J. Sobesky et al. 2006. Outcome and symptomatic bleeding complications of intravenous thrombolysis within 6 hours in MRI-selected stroke patients: Comparison of a German multicenter study with the pooled data of ATLANTIS, ECASS, and NINDS tPA trials. *Stroke* 37:852–858.

Turjman, F., T. F. Massoud, F. Vinuela et al. 1994. Aneurysms related to cerebral arteriovenous malformations: Superselective angiographic assessment in 58 patients. *American Journal of Neuroradiology* 15:1601–1605.

Valavanis, A., O. Schubiger, and W. Wichmann. 1986. Classification of brain arteriovenous malformation nidus by magnetic resonance imaging. *Acta Radiologica* 369:86–89.

Wardlaw, J. M. and O. Mielke. 2005. Early signs of brain infarction at CT: Observer reliability and outcome after thrombolytic treatment-systematic review. *Radiology* 235:444–453.

Wiebers, D. O., J. P. Whisnant, J. Huston, 3rd et al. 2003. Unruptured intracranial aneurysms: Natural history, clinical outcome, and risks of surgical and endovascular treatment. Early management of aneurysmal subarachnoid hemorrhage. A report of the Cooperative Aneurysm Study Surgical risk as related to time of intervention in the repair of intracranial aneurysms. Relation of cerebral vasospasm to subarachnoid hemorrhage visualized by computerized tomographic scanning. The relation of cerebral vasospasm to the extent and location of subarachnoid blood visualized by CT scan: A prospective study. *Lancet* 362:103–110.

3. Cardiovascular Imaging
State of the Art and Clinical Challenges

Filippo Cademartiri, Cesare Mantini, and Erica Maffei

3.1 Introduction

The study of the heart has always been a diagnostic challenge for physicians, engineers, and physicists because it is a rapidly moving organ, even though in most of the cases it is imaged during breathhold. Nowadays, a further challenge is to evaluate small and tortuous coronary arteries lying on the moving cardiac surface.

Up until the last decade, the main imaging techniques that dealt with heart diseases were

- Radiographic procedures: standard chest x-ray to evaluate gross anatomical changes and invasive tools such as angiography, evaluating the magnitude of x-ray attenuation after contrast medium administration usually through percutaneous access
- Ultrasonic procedures: echocardiography, evaluating the intensity of reflected ultrasonic waves

- Nuclear medical procedures: evaluating the intensity of γ-rays during stress and at rest during injection of radionuclide-tagged pharmaceuticals

Rapid advances in technologies and the development of cardiac synchronization software have allowed the study of the heart with noninvasive and more accurate tools now able to overcome most of the drawbacks of other techniques:

- Computed tomography (CT): able to depict the coronary tree and the whole heart with a high spatial and temporal resolution, analyzing attenuation coefficients of the heart, and with low radiation burden
- Cardiac magnetic resonance (MR): able to evaluate the spatial distribution of proton density in the heart furnishing anatomical and functional information without ionizing radiation dose

Chapter 3

3.2   Coronary Artery Disease (CAD)

Conventional coronary angiography (CCA) has the ability to depict coronary artery stenosis with a high spatial and temporal resolution, and is the gold standard for in vivo coronary stenosis evaluation (Figure 3.1); it permits the avoidance of calcium-related problems such as the blooming effect, which can affect other techniques investigating the coronary tree. However, CCA is not free from drawbacks. Because it is a two-dimensional tool, there are some problems in estimating stenosis at coronary bifurcations, and, furthermore, it is not clear that the remodeled plaque can determine stenosis (Piers et al. 2008).

Currently available cardiac magnetic resonance imaging (MRI) techniques are able to fulfill the aims of imaging in CAD patients evaluating the consequences of CAD to the heart, particularly myocardial perfusion and function and depiction of irreversible myocardial damage (Figure 3.2). Due to its superiority over MRI, cardiac CT is nowadays used for a mere visualization of coronary artery stenoses, cardiac MRI being primarily focused on the assessment of the ischemic consequences of CAD.

Cardiac imaging with multidetector CT has become feasible due to rapid advances in CT technology. With improved temporal resolution (75–200 ms) (Mahesh 2006), submillimeter spatial resolution (<0.75 mm), and cardiac ECG synchronization, the current generation of CT scanners (64–320 row detectors) is able to provide a superb detection of cardiac structures

and function and to identify CAD (Pannu et al. 2003; Flohr et al. 2005).

Since the most quiescent part of the heart cycle is during middiastole, the data obtained in this phase are usually the most readable and important for diagnosis; sometimes, adjunctive image reconstruction is needed for the end-systolic phase to permit a better imaging of the right coronary artery. Image acquisition can occur throughout the R-R interval (retrospective gating) or at a predefined time point (prospective trigger). Myocardial functional data provided with CT are as accurate and reproducible as the functional gold standard, MRI, and nowadays different dose reduction algorithms permit the reduction of dose.

The state of the art of cardiac CT relies on recent literature that claimed this tool was a very accurate technique with a very high negative predictive value and with an important incremental prognostic and further stratifying value over other tools also because it has the privilege of identifying also nonstenotic noncalcified plaque, sometimes referred to as plaque at risk (Mark et al. 2010).

Cardiac CT technology is nowadays spread all over the world, but investments have to be made in human resources because accurate results are provided only by trained personnel with more than a year of full immersion practice with an appropriate expert tutor.

Nuclear imaging of the heart is a valuable and widely used noninvasive procedure revealing information about cardiac perfusion and metabolism. The most widespread imaging technique in nuclear cardiology is single photon emission computerized tomography (SPECT), which is performed during physical or pharmaceutical myocardial perfusion stress and rest tests with isotopes such as technetium-99m labeled isonitriles or diphosphine compounds (Beller 1994). Scintigraphy utilizes gamma cameras to capture emitted radiation from injected radioisotopes to create two-dimensional projection images. The images are three-dimensional reconstructed with SPECT systems and their interpretation can distinguish between infarction and inducible ischemia; large multicenter trials have claimed SPECT as a solid tool to identify patients at augmented risk of future events. However, a lot of false positives and false negatives have been reported during routine clinical practice.

Positron emission tomography (PET) is based on the annihilation of both electron and positron, producing a pair of annihilation (gamma) photons

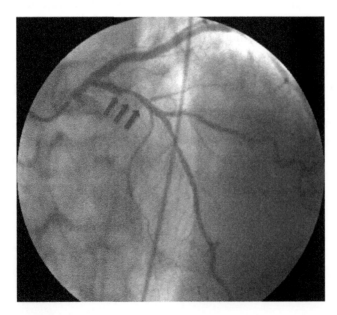

FIGURE 3.1   CCA of right coronary artery.

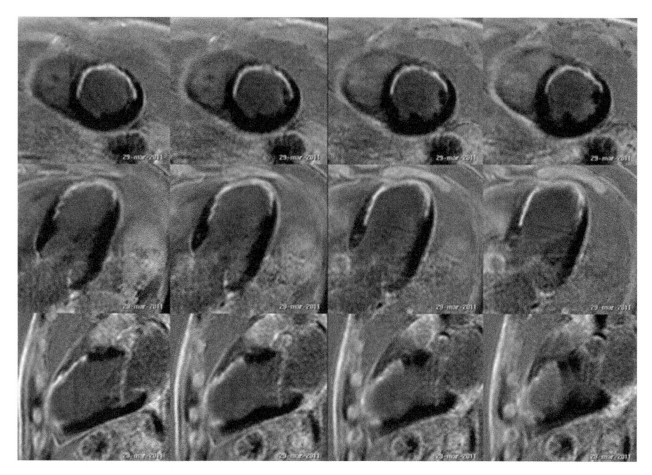

FIGURE 3.2 Contrast enhancement inversion recovery (CE-IR) sequence, acquired in the three planes (short axis, four-chamber, and vertical long axis) till 20 min after contrast medium administration, which, adjusting the inversion time, permits to suppress the healthy myocardium (black) and to highlight the diseased myocardium (white).

moving in approximately opposite directions. PET has a greater sensitivity and specificity than SPECT and can furnish both perfusion and viability information (Schnaiger 1994). Currently, two radiotracers are needed, one for perfusion (e.g., rubidium-82 ornitrogen-13-ammonia) during stress and at rest and one for viability ([18]FDG). The main advantage of PET over SPECT is the possibility to calculate during perfusion the absolute myocardial blood flow per minute per gram of myocardium, overcoming the possibility of a false positive result in three-vessel disease as is common with SPECT; the main disadvantage of PET is the very low half-life of perfusion radiotracers and the subsequent need of having an expensive cyclotron nearby, though commercially available Rb-82 generators can be used.

3.3   Cardiac Function

Since many cardiac diseases have an impact on the performance of the cardiac pump activity, assessment of cardiac performance (cardiac function) in an accurate and reproducible way is crucial to determine the disease severity and evaluate the efficacy of treatment.

Echocardiography utilizes ultrasound to create real-time images of the cardiovascular system in action. A transducer is placed in different projections to obtain different views of the heart. Echocardiographic images are created when the transducer receives the reflected waves due to differences in the acoustic impedance among different cardiac tissues and then converts the ultrasound into electrical signals. Echo signals are amplified electronically and displayed on a monitor using shades of grey (from black to white). Echocardiography is the first-line modality for the noninvasive assessment of cardiac anatomy (Figure 3.3a) and function and may utilize additional

Chapter 3

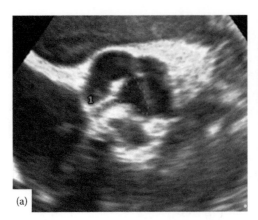 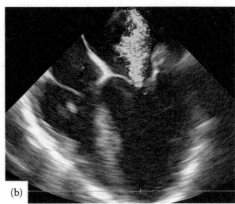

FIGURE 3.3 Transesophageal echocardiography image (a) shows a tricuspid valve with partial fusion of the right coronary and noncoronary cusps. (b) Assessment of mitral regurgitation severity by color Doppler echocardiography.

physical principles, such as the Doppler effect, to add further information useful for diagnosis (Pearlman et al. 1988; Otto 2009) (Figure 3.3b).

Echocardiography is by now a solid tool with many advantages related mainly to its safe noninvasive nature without ionizing radiation exposure; furthermore, it is inexpensive and can be performed at bedside to rapidly evaluate the structure and function of heart. On the other hand, ultrasound imaging has some disadvantages. For example, ultrasound has some difficulty in examining patients with a bad echocardiographic window where the sound cannot penetrate. Also, it can suffer from foreshortening in calculating the exact volume and function of left ventricle, particularly when localized areas of dysfunction are present. Another major drawback of echocardiography is the difficulty in visualizing and quantifying right ventricular volumes and ejection fraction.

Cardiac MRI is the gold standard in vivo for biventricular volumes and systolic functional assessment. Cardiac MRI is based on proton resonance just as it is for MRI of other parts of the human body; however, as previously stated, imaging of the heart has to deal with cardiac movement. Hence, cardiac MRI is acquired during ECG synchronization. Actually, this tool is appropriate for the assessment of a large number of heart diseases, particularly in the evaluation and noninvasive characterization of cardiomyopathy, congenital disease, and infarction. Cardiac MRI enables the avoidance of an invasive procedure and/or improves guidance of subsequent invasive procedures, thereby increasing their accuracy.

The main sequence used currently in cardiac MRI is the cine balanced bright-blood field echo, which allows the important depiction of the anatomy while providing important functional assessment of the moving heart. Another fundamental sequence is contrast enhancement inversion recovery (CE-IR), acquired until 20 min after contrast medium administration that, adjusting the inversion time, permits the suppression of the healthy myocardium and improvement of up to 500% in the contrast resolution difference between healthy and diseased myocardium (Figure 3.2). The main advantage of cardiac MRI relies on the complete absence of ionizing radiation dose and on the minimal administration of paramagnetic contrast medium, considered one of the safest contrast medium (Niendorf et al. 1991; Murphy et al. 1996). The main drawback is related to the lack of qualified personnel (technicians and physicians) and the long training time.

3.4 Nonischemic Myocardial Disease

The myocardium may be involved in a variety of disorders in which the heart muscle is exclusively or preferentially affected (e.g., myocardial ischemia, hypertrophic cardiomyopathy) or in which the myocardial involvement is part of a multiorgan disorder (e.g. hemochromatosis, sarcoidosis), though often no underlying etiology for myocardial dysfunction can be discovered, that is, it is idiopathic. MRI has become part of the investigation of patients with myocardial disorders.

An MRI examination should include morphological analysis of the myocardium, evaluation of the functional impact of the myocardial disorder on ventricular systolic and diastolic function, and valvular function, and detection and/or characterization of pathological myocardium.

For instance, fatty infiltration of the free wall of the right ventricle will be visible as hyperintense intramyocardial spots on T1-weighted SE images, and these abnormalities make the diagnosis of arrhythmogenic right ventricular dysplasia likely. Mature myocardial fibrosis or calcification, on the other hand, appears hypointense on both T1- and T2-weighted sequences and can be found in patients with endomyocardial fibrosis. A hyperintense myocardial area on T2-weighted images in a patient with a recent myocardial infarction is suggestive of increased free-water content due to myocardial edema and/or necrosis. The CE-IR MRI has become the reference technique for myocardial tissue characterization. Myocardial enhancement reflects abnormal myocardium, and the location, distribution pattern, and extent of enhancement is often typical for specific myocardial disorder. For instance, in patients with myocardial infarction, the enhancement typically occurs in a coronary artery distribution territory and involves the subendocardium, with a variable transmural spread. In patients with myocarditis, the enhancement is not related to a coronary artery territory and usually involves the subepicardial part of the myocardium (Mahrholdt et al. 2004). Knowledge of the typical patterns of enhancement is helpful in the differential diagnosis of patients with myocardial diseases. Moreover, concomitant pathology, such as pericardial inflammation or ventricular thrombus formation, can be assessed with CE-IR MRI (Mollet et al. 2002; Bogaert et al. 2004).

3.5   Heart Failure

Definite advantages of echocardiography in the evaluation of changes in the size and shape of the heart in the context of specific drug therapies are the low cost, availability, and its use to perform a bedside examination. Currently, echocardiography offers additional information about cardiac valvular function and diastolic function that is currently unsurpassed by other imaging modalities. Nevertheless, this technique suffers from unpredictable variation in image quality and a number of studies are not interpretable due to patient-related echogenicity. Therefore, the reproducibility and overall accuracy of echocardiography may not be as high as some of the other available techniques.

Cardiac MRI does not have these limitations and is currently the only imaging technique that provides a comprehensive study of the heart, including anatomical evaluation, functional data, and information about myocardial perfusion and viability. An MRI examination is independent of slice orientation or acoustic window, and postprocessing requires no geometric assumptions as in echocardiography to determine ventricular dimensions, which is particularly important in evaluating ventricles of patients with heart failure who may have severely altered morphology as a result of chronic hibernation or infarction and have regional heterogeneity in wall thickness and contractility.

3.6   Pericardial Disease

With the advent of modern noninvasive cardiac imaging modalities such as echocardiography, cardiac CT and cardiac MRI, direct visualization of the pericardium and related pathology has become a reality. Although echocardiography is still the first-line modality used to explore the pericardium, imaging of the pericardium is often suboptimal especially in "nonechogenic" patients such as patients with lung emphysema or chest wall deformities. CT and MRI are second-line imaging modalities that have the ability to overcome, by and large, the limitations of echocardiography. Both are tomographic techniques with excellent spatial and contrast resolution. Since the pericardium, a fibro-fluid structure, is surrounded on both sides by fat (mediastinal and epicardial fat layer), excellent contrast is created, and thus the pericardium is a priori ideally suited for evaluation on CT and MRI (Smith et al. 2001). On routine chest CT examinations, the pericardium and pericardial pathology (fluid, inflammation, calcification, or masses) are very well depicted. The coupling of CT data acquisition with ECG gating enables reduction in cardiac-related motion artifacts and optimal visualization of the pericardium. With MRI accurate information can be obtained not only on the pericardial morphology but also on cardiac function.

Chapter 3

3.7 Cardiac Masses

Before the era of modern cardiac imaging techniques, the diagnosis of cardiac tumors was extremely difficult. Nowadays, the availability of transthoracic and transesophageal echocardiography, cardiac CT, and MRI has greatly facilitated their detection. Accurate differentiation between the cardiac tumors, nontumor masses, and normal variants is a first requirement for each imaging technique. Echocardiography is considered the procedure of choice for the diagnosis of intracardiac tumors. This technique may provide information on the size, mobility, shape, and location of cardiac masses, but it cannot describe histological features. At present, MRI is definitely one of the preferred imaging modalities in the evaluation of patients with suspected cardiac masses. Major advantages of this technique are the excellent spatial resolution; the large field of view; the inherent natural contrast between flowing blood and the surrounding heart chambers, vessel walls, and tumor masses, without the need for contrast agents; the multiplanar imaging capability; and the ability to administer paramagnetic contrast agents in order to obtain a better detection of the tumor borders and the degree of tumor vascularization. These advantages are important in planning the patient's therapy, particularly if surgical intervention is being considered.

3.8 Conclusions

A very accurate study of the heart continues to be a challenge for physicians, but recent advances in technologies have just permitted the medical community to deal with most of the problems that imaging the heart present.

Different imaging tools have advantages and disadvantages, but just because they are based on different physical principles they all can help physicians to obtain complementary information and so all have acquired specific roles in particular situations and diseases. We trust that continued progress will be forthcoming to furnish even more powerful tools to further improve and speed up diagnosis and therapy to patients. The recent introduction of hybrid cardiac imaging combining PET and CT angiography and PET and MRI technology may potentially lead to paradigm shifts in healthcare and improve clinical practice. Hybrid imaging combining PET and multidetector CT angiography allow combining morphologic information about coronary artery stenosis location and degree along with functional information on pathophysiological lesion severity. The new PET-MRI scanners offer attractive possibilities for combining measures of ventricular function, perfusion, viability, and infarct scar for evaluation of ischemic heart disease. It is hoped that further advances will continue as the application of physical principle to improved imaging technology occurs.

References

Beller GA. 1994. Myocardial perfusion imaging with thallium-201. *J Nucl Med* 35(4): 674–680.

Bogaert J, Taylor AM, Van Kerckhove F, Dymarkowski S. 2004. Use of inversion-recovery contrast-enhanced MRI technique for cardiac imaging: Spectrum of diseases. *Am J Roentgenol* 182: 609–615.

Flohr TG, Schaller S, Stierstorfer K. et al. 2005. Multi-detector row CT systems and image-reconstruction techniques. *Radiology* 235: 756–773.

Jan B, Dymarkowski S. 2012. *Clinical Cardiac MRI*. Springer-Verlag, Berlin, Heidelberg.

Mahesh M. 2006. Cardiac imaging: Technical advances in MDCT compared with conventional x-ray angiography. *US Cardiology* 2006; 2(1): 115–118.

Mahrholdt H, Goedecke C, Wagner A. et al. 2004. Cardiovascular magnetic resonance assessment of human myocarditis. A comparison to histology and molecular biology. *Circulation* 109: 1250–1258.

Mark DB, Berman DS, Budoff MJ. 2010. ACCF/ACR/AHA/NASCI/SAIP/SCAI/SCCT 2010 expert consensus document on coronary computed tomographic angiography: A report of the American College of Cardiology Foundation Task Force on Expert Consensus Documents. *J Am Coll Cardiol* 55: 2663–2699.

Mollet NR, Dymarkowski S, Volders W. 2002. Visualization of ventricular thrombi with contrast-enhanced MRI in patients with ischemic heart disease. *Circulation* 106: 2873–2876.

Murphy KJ, Brunberg JA, Cohan RH. 1996. Adverse reactions to gadolinium contrast media: A review of 36 cases. *Am J Roentgenol* 167: 847–849.

Niendorf HP, Haustein J, Cornelius I. et al. 1991. Safety of gadolinium-DTPA: Extended clinical experience. *Magn Reson Med* 22: 222–228.

Otto CM. 2009. *Textbook of Clinical Echocardiography*, 4th edn. Saunders Elsevier, Edinburgh.

Pannu HK, Flohr TG, Corl FM et al. 2003. Current concepts in multi–detector row CT evaluation of the coronary arteries: Principles, techniques, and anatomy. *Radiographics* 23(spec issue): S111–S125.

Pearlman JD, Triulzi MO, King ME et al. 1988. Limits of normal left ventricular dimensions in growth and development: Analysis of dimensions and variance in the two-dimensional echocardiograms of 268 normal healthy subjects. *J Am Coll Cardiol* 12: 1432.

Piers LH, Dikkers R, Willems TP et al. 2008. Computed tomographic angiography or conventional coronary angiography in therapeutic decision-making. *Eur Heart J* 29: 2902–2907.

Schnaiger M. 1994. Myocardial perfusion imaging with PET. *J Nucl Med* 35: 693–98.

Smith WHT, Beacock DJ, Goddard AJP et al. 2001. Magnetic resonance evaluation of the pericardium: A pictorial review. *Br J Radiol* 74: 384–392.

Chapter 3

Physics and Technology

Principal Applications

4. Physics and Technology of X-Ray Angiography

Stephen Rudin and Daniel R. Bednarek

Chapter 4

4.1 Introduction

As discussed in Chapter 1, the goal of angiography involve the visualization of the blood vessels either for diagnostic purposes or for interventions. Historically the oldest angiographic method used x-ray projection imaging, which is still generally the gold standard for angiography because of its unparalleled spatial and temporal resolution capabilities. In this chapter, we will summarize the current state of x-ray angiography with a bit of perspective from its past and hints of where the field is leading.

The use of an injection in a vessel of contrast media, usually an iodine containing compound due to its absorption edge in the middle of the diagnostic x-ray energy range (see Figure 5.4c), has been standard for many years even before the existence of modern digital detectors, when film-screen combination detectors were used. Since the injection only can last a few seconds, it is necessary to take successive exposures to view both the arterial and venous phases and be able to capture near optimal views. Even with film this was achieved using rapid film changers that could mechanically move either cut or roll film between two intensifying screens at a few frames per second. To eliminate the interference in the images of boney structures and soft tissue, analog optical processes were used to subtract the initial frames containing no contrast material but just these structures, from successive frames that contained both contrast and structures. The result, if the process was luckily done properly, might be an image containing primarily only the vasculature that constituted the angiogram. Needless to say this was a tedious, time-consuming procedure with little hope of having information available while the patient was still on the table in the examination room so as to affect the procedure immediately as occurs now with entirely digital systems.

Although the basic technology of detection systems has undergone revolutionary change from analog to digital methods, the other parts of the imaging chain have evolved more slowly. The x-ray sources are still rotating anode x-ray tubes where most of the x-rays are produced using the Bremmstrahlung process with the bombardment of an electron beam onto an anode target to produce a polychromatic spectrum as shown in Figure 4.6. Although high-voltage generator and switching technology have advanced and the antiscatter grids placed between the patient and the image receptor have improved somewhat in efficiency, advancements here have not been near as dramatic as in detector systems.

4.1.1 Visualization of Vessels, Blood Flow, and Endovascular Devices

Most standard x-ray angiography involves the injection of contrast media into an artery via the insertion of a catheter leading into the organ of interest, the brain or the heart in particular. The catheter may be inserted into the femoral artery, radial artery, or even in rare cases into the carotid artery and then, under fluoroscopic guidance, threaded to the location of potential pathology. To visualize the vessel branching, iodine contrast media may be injected during this localization procedure. The amount and concentration of iodine depends upon the vessel, its size and the flow rate, and the distance of the injection origin at the catheter tip to the region of interest (ROI). The determination of the rate at which the contrast media is diluted by the blood to the point where the contrast becomes unable to be detected will depend upon the blood flow and vessel branching structure. Typically, contrast media injection concentrations can be diluted with saline with injection rates that can be carefully controlled using mechanical autoinjectors or by manual injection from a syringe by a clinician. Care must be taken to ensure that the total contrast media burden especially during a long interventional procedure is within safe limits for the patient. Additionally, it may be important for diagnosis or evaluation of interventions to qualitatively assess flow, for example, through a stenotic region of a vessel or inside an aneurysm. Quantitative evaluations of flow using computer fluid dynamic calculations are beginning to be used in the clinic as will be discussed in Chapter 21.

Clearly angiography is not limited to the visualization of vessels because not only is it necessary to visualize catheters during injection of contrast media but there is a whole menagerie of endovascular devices (see Figure 4.1) such as guide wires, stents, retrievers, valves, etc., that must be visualized so as to be able to localize them accurately (Rudin et al. 2008). If the guide wires may be used for gauging distances, or if the catheters are made of x-ray transparent plastic, they may have markers attached made of high atomic number hence high x-ray absorbing material yet inert or biocompatible. Platinum markers are typical. Such markers are also used to denote the ends of stents since the stainless steel or nitinol typically used for stents may not be visualizable especially if the stent struts are just 50–100 μm in diameter.

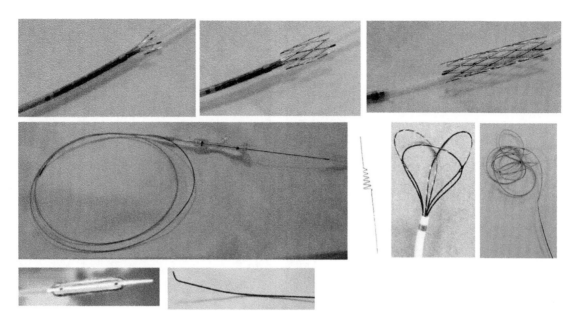

FIGURE 4.1 Examples of endovascular devices: three stages of a stent being deployed out of a catheter, a delivery system, including catheter and guide wire, a clot retrieval device, a device retriever, a coil for treating aneurysms, a balloon at the catheter tip, and the bent end of a guide wire.

4.1.2 Interaction of X-Rays with Matter or What Determines Object Visualization

The physics behind visualization in x-ray imaging involves the basic processes by which x-rays interact with matter. These processes are described in more detail in Section 5.1 but suffice it to say at this point that the two most important processes are photoelectric effect and Compton scattering. The atomic cross section for photoelectric interaction is approximately proportional to Z^4/E^3, (Evans 1955) where Z is the atomic number of the attenuating material and E is the energy of x-ray photon and therefore, the probability of interaction per unit mass or the mass attenuation coefficient for photoelectric effect is proportional to Z^3/E^3 because the atomic weight for materials goes approximately as Z. This would imply that the lower the energy and the higher the Z, the greater would be the chance for attenuating an x-ray; however, if the energy were too low, the x-rays would never penetrate the patient to reach the image receptor. Also if E were too low, there might not be enough energy to remove an inner shell electron in the attenuating material as required for photoelectric effect. For higher energy x-rays and low-Z material such as tissue, Compton scattering, which is less dependent upon energy, becomes the more dominant mode of interaction if E is greater than about 25 keV. Also for higher E, enough energy becomes available to remove an inner shell electron and suddenly the probability for interaction goes way up before it starts to follow the $1/E^3$ relation again. This phenomenon is generally referred to as an absorption edge as depicted in Figure 5.4c. Thus, to visualize vessels, contrast media, and endovascular devices, the proper selection of materials and energy is key.

4.2 Imaging Modes

Modern digital imaging systems allow many kinds of imaging modes to be presented to clinicians as illustrated in Figure 4.2. In what follows is a list of the most common such modes with a brief explanation of their function and utility.

4.2.1 Fluoroscopy

This is the primary mode for image guidance during either a diagnostic or interventional procedure. The image quality need not be the best hence the exposure at the detector per frame is low, of the order of 1–10 μR, resulting in an image with substantial amounts of quantum noise. Toward the lower end of this exposure range, the quantum noise may not even be the dominant noise source in that the instrumentation noise of an image receptor such as a flat panel may become relatively more significant. The exposure at which the quantum noise and the instrumentation noise are equal we (Yadava 2008) have designated the

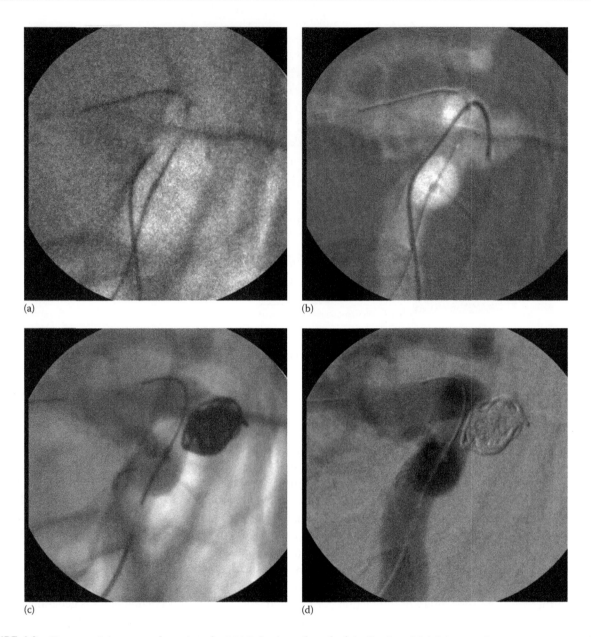

FIGURE 4.2 Pictures giving examples using the MAF detector described in Section 4.3.4.3 in a patient aneurysm procedure. (a) Single frames showing fluoroscopy with increased noise. (b) Roadmapping with catheters within the vessel roadmap. (c) DA with coil filled aneurysm and background as well as vessels. (d) DSA. Although background structure in DSA should be subtracted, this could be incomplete if there is movement of the patient as is indicated in image (d) where the edges of the coil mass and the catheters still appear.

instrumentation noise equivalent exposure (INEE). For exposures above the INEE, a system is quantum noise limited, a desirable state to be in; however, for exposures below the INEE, the detector would be instrumentation noise limited, a state that might be improved with a less noisy detector. Entrance patient exposures during fluoroscopy are limited in the United States by federal device regulations to 10 R/min (air kerma equal to 88 mGy/min) at the compliance point; however, there may be high-dose fluoroscopic modes available up to twice that exposure rate. Long duration procedures even at moderate or low fluoroscopic exposure rates can result in substantial entrance exposures to the patient especially when combined with some of the higher dose angiographic procedures. This will be discussed more fully in Chapters 24, 25, and 27.

Although in older analog systems the frame rate was typically 30 fps in the United States and the exposure was continuous, all new digital systems have pulsed x-rays and allow the frame rates to be changed typically to values such as 7½, 15, 20, and 30 fps. During fluoroscopy, it may be important to maintain a high frame rate up to 30 fps in order to be able to guide whatever procedure is occurring such as placing a guide wire or catheter in the proper artery or localizing an interventional device; however, many clinicians have learned to decrease the frame and pulse rate to save patient exposure. Because the actual viewed image is displayed at a much faster frame rate, which is now independent of the acquired frame rate, there is no perceived flicker that used to occur if lower frame rates were used with older analog systems.

For viewing newer, finer endovascular devices it may be necessary during accurate localization to boost up the exposure so as to reduce the effect of quantum noise. For example, when using a new high-resolution detector, our group has found it beneficial for short but critical periods of time especially during an intervention to use what we designate as "high definition" fluoroscopy (Panse 2010), where the exposure levels can approach those used for digital angiography (DA) mode.

4.2.2   Digital Angiography

DA mode has typically been the mode used in cardiological angiography to record rapid sequence images during an injection of contrast media into the heart's coronary vessels in order to diagnose pathology or evaluate an intervention. When film was used in analog video systems, this mode was referred to as ciné. Although the system functions in DA in a similar way to that for fluoroscopy, the exposure per frame is boosted to reduce the quantum mottle and provide a higher quality image sequence for review. Typical increases in tube current hence exposure may be of the order of 10× so that tens of microroentgens per frame to the detector are used with consequent increases in patient entrance exposure.

4.2.3   Digital Subtraction Angiography (DSA)

In DSA mode, the exposure is boosted yet again to the range of hundreds of microroentgens per image at the detector because now the goal is to visualize the vasculature with the background of bone and soft tissue subtracted. The basic idea is similar to that originally used for film as described previously; however, now the subtraction can be done in real time because all the images are digitized. Thus, a DSA sequence is done by acquiring a few frames before the contrast media appears and then subtracting the average of these or the mask or reference frame from all subsequent frames. The sequence can be done at a few frames per second so as to be able to select the best frame that shows arterial filling or at more rapid frame rates if the goal is to evaluate blood flow features. Since the assumption is that the patient will not move during the whole sequence so that the mask frame will be registered with the subsequent frames, DSA cannot easily be used to record the state of moving cardiac vessels but instead DA is used. This may be satisfactory for cardiac imaging where usually only the major heart vessels are evaluated or treated during interventions; however, for the brain, visualization of small perforating vessels that can be in the range of 50–200 μm in diameter and fine features of aneurysms and bifurcations may be crucial, making the subtraction of background structures extremely important.

The acquired images must be preprocessed before the subtraction by taking the log of the intensity values prior to subtraction. For planar DSA each data value, representing the intensity transmitted as a result of exponential attenuation factors, must be converted by taking the logarithm. The result is a value proportional to the sums of linear attenuation coefficients of the contrast in the vessel plus that of the background that then enables the background from the logarithm of the mask image to be subtracted. Not performing this log processing step will cause unwanted artifacts dependent upon the background features (Balter 1984).

4.2.4   Roadmapping

In order to visualize the vasculature and save injections of contrast media into the patient, it is possible to record an image from a single injection, that is, an image while both contrast media and the background structures are apparent and subtract it from all subsequent images taken during fluoroscopy without contrast. In this case, the background structures will disappear but the vasculature will show up as a white overlay, reversed in grey level from the original black representing absorption by the iodine contrast. This roadmap of the vasculature can be very helpful in guiding the clinician in placing endovascular devices in the patient without the need of an additional injection of contrast media because the paths are all laid out by the vessel roadmap as long as the patient does not move with respect to the original sequence where

Chapter 4

the roadmap was acquired. Should the patient move, then a new roadmap may be required. Even without patient overall movement, vessels may flex so that it might appear during a procedure that a device such as a guide wire or catheter is moving outside the bounds of a vessel. Sometimes, the slight movement, hence misregistration of a device such as a stent, may actually be helpful because the device would not be completely subtracted and its edge enhanced image would be visible and provide a reference feature for the clinician. Likewise, it may be desirable not to completely subtract the background features so that they can be partially made visible without dominating or interfering with the parts of the image that should not be obscured. Some manufacturers have implemented this partial subtraction mode.

There are a variety of ways to acquire roadmaps. One of the common methods is to record the lowest values for every pixel in the field of view (FOV) during a fluoroscopy sequence when contrast is injected. These lowest values are combined into one roadmap image representing the instance when the greatest amount of attenuation is occurring at that pixel. Once enough frames are sampled to obtain a final roadmap, the frame is inverted and is superimposed on subsequent images. Another common approach is to use a frame from a previously acquired DSA run but without the mask subtracted, that is, a raw data frame where both the vessel and the background appear. This again assumes there has been no motion in between runs. The advantages of this method are that the image, done during DSA runs, has less quantum noise since it was taken at higher exposures, and also there is no additional patient exposure since the DSA run has already been recorded in the angiography system's computer memory.

4.2.5   Automatic Control

During fluoroscopic and angiographic procedures, it is important that the clinician not be distracted by the necessity to choose the best technique parameters for imaging. Also, this best setting may change during the procedure as the imaging region moves to different parts of the patient. The clinician should not have to be bothered with such a change in the apparent brightness of the image on the monitor if there were fixed exposure settings during a changing scene. There are various methods used to automatically control the exposure setting during rapid sequence imaging.

This has been referred to by a variety of names such as automatic exposure rate control (AERC), automatic brightness control, and automatic brightness stabilization. In the past with analog systems, the brightness in a preselected region of the FOV was sensed and a feedback signal produced to control the x-ray output. With advent of digital systems, it is now possible to preselect an arbitrarily shaped region located anywhere in the FOV that might be of primary interest to the clinician for forming the control signal. As an example, three possible sampling areas are designated as A, B, and C in Figure 4.3 of a chest to indicate how the image would appear when each of those areas are selected (Yang 1999).

Also for automatically setting parameters for an angiographic sequence, there have been different approaches. Once determined, these DSA parameters should not be changed in order to enable robust subtraction and other quantitative image processing to take place. One method of automatic exposure control is to use a short fluoroscopy sequence without moving the patient during the subsequent DSA run. This way a computer program can be used to calculate what the appropriate x-ray DSA technique parameters should be so as to obtain the desired detector exposure using the measurement of the attenuation of the patient that can be made during the short fluoroscopy sequence. An alternate or additional method is to use a few test shots just before the angiographic sequence is done to provide similar information that can be used to automatically select proper technique parameters for the DSA run to follow.

4.2.6   Display Modes

With the advent of digital image displays, it is possible to provide reference images and image sequences to clinicians, thus saving time and patient exposure. For example, last-frame-hold was one such advance since it allowed the clinician to step off the fluoroscopy footswitch, and have the last image of the patient displayed continuously on the monitor while thinking about the next action to take. It is not uncommon now to have a large bank of flat panel monitors or even one gigantic monitor divided into many images so that the clinician can have either a reference image or a reference sequence looping on one monitor while performing a procedure under the direct real-time guidance of another image or monitor.

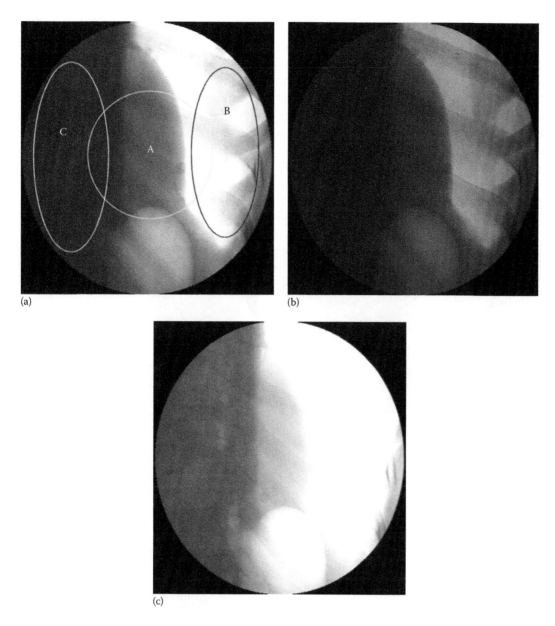

FIGURE 4.3 Images where different sampling areas are used for controlling the AERC system during fluoroscopy. The first image gives outlines of sampling area A, B, and C. Sampling area A is centered on the mediastinum and is used for the first image (a). For the next (b) image, region B is centered on the lung field and finally for the last (c) image region C is centered on the spine. Technique parameters are controlled to optimize viewing in the selected sampling area.

4.2.7   Spot and ROI Fluoroscopy

The basic idea behind ROI and spot fluoroscopy modes is that a high-quality image in the full FOV in many diagnostic and especially in interventional procedures is not really needed. If the clinician can be provided with the best possible image in a smaller part of the FOV in the ROI where the pathology must be imaged, the image in the periphery of the ROI can either act as a static reference in the case of spot fluoroscopy or as a lower quality dynamic reference in the case of ROI fluoroscopy. By not irradiating the patient as much in the periphery, a substantial amount of patient integral or effective dose can be saved, something that may be increasingly important as image guided minimally invasive endovascular interventions become increasingly substituted for invasive surgical interventions and as the image-guided procedures become more complex and potentially longer and with increased patient

Chapter 4

exposure. An additional benefit besides dose reduction is that if only the ROI is irradiated with the full strength of the x-ray beam, then essentially the FOV that is being irradiated at any time, and thus the amount of scatter generated in the patient-irradiated volume, will be substantially reduced and cause less degradation in the ROI image than if the full FOV is fully irradiated.

In spot fluoroscopy (Takahashi 2006), a full FOV reference image is taken and maintained but only in the periphery while the collimation automatically closes to the size of the ROI and real-time fluoroscopy

is taking place in the ROI. If there is no significant movement expected in the periphery, the static reference image should be sufficient for the clinician during the procedure. In contrast, for ROI fluoroscopy (Rudin 1992; Labbe 1994), a partial beam attenuator (Massoumzadeh 1998) is placed over the x-ray source so that all but the ROI is covered. In this way, the beam is reduced in intensity in the periphery but not in the ROI (Figure 4.4). The darker image in the periphery is brightness equalized in the acquisition computer before the full image of equalized brightness is presented to the clinician (Rudin 1995; Rudin 1996;

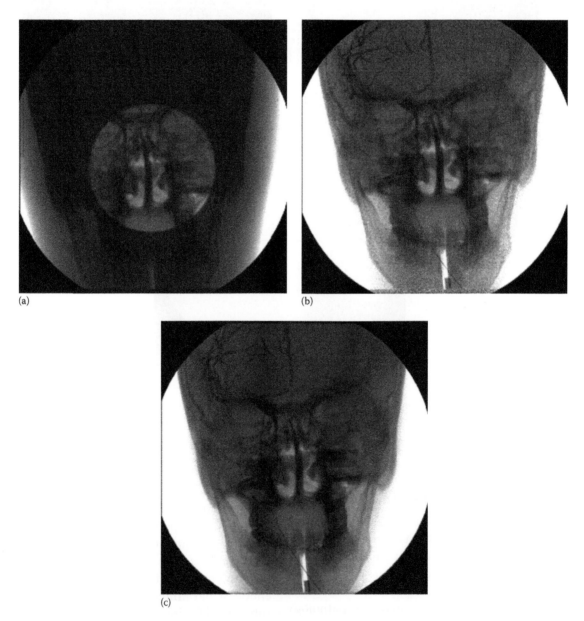

(a)

(b)

(c)

FIGURE 4.4 ROI fluoroscopy. (a) Attenuator in place to lower exposure in the periphery to one-sixth the value in the ROI. (b) Brightness equalized with increased quantum mottle apparent in the periphery. (c) Differential temporal filters applied to smooth out noise.

Kezerashvili 1997; Yang 1998; Rudin 1999). Because the beam is not totally attenuated in the periphery, there is still a real-time image there; however, it is somewhat degraded even with the brightness equalization. This is because there are fewer x-ray photons hence increased quantum noise. Most recently, our group has proposed reducing the apparent noise in the periphery by using two different digital temporal filters one for each region of the image, the ROI and the periphery (Vasan 2011; Vasan 2012b; Vasan 2013). Thus, the peripheral image outside the ROI could have a purposefully increased lag or greater temporal recursive filtering occurring while the image in the ROI has less lag or less temporal recursive filtering. Although the patient dose reduction for ROI fluoroscopy may be slightly smaller than in spot fluoroscopy, the clinician maintains the ability to move the patient around without having to reacquire a new reference image that is necessary in spot fluoroscopy. For additional dose reduction methods, see Chapter 25.

4.3   Angiographic Systems

In this section, there is a review of the variety of angiographic systems that are available as well as a discussion of each major system component including x-ray source, scatter reduction technique, and detector and then some description of our group's experience with workflow and ergonomic design of custom angiographic software.

4.3.1   Gantries

Angiographic gantries can vary from the simplest mobile C-arm to complex robotic arms and biplane systems. For almost all angiographic systems, two degrees of freedom of a C-arm are provided: the rotation around an axis in the plane of the c or "propeller" motion and the other that allows the imaging chain to "slide" in an arc along the c. In addition, for the C-arm that images the A-P plane, there is a degree of freedom that allows the C-arm to rotate around one or more vertical axes one of which is floor mounted while another vertical axis, when used in common with the first, enables translational motion around the patient perimeter so as to orient the C-arm in a variety of configurations. Alternatively, the AP C-arm, or a lateral C-arm for a bi-plane system, may be mounted on a ceiling frame that enables both a degree of freedom around a vertical axes as well as x–y translation in the plane parallel to the floor. Although most C-arm gantries have at least these capabilities, there have been some recent designs that have provided additional axes of rotation. For example, the Toshiba Infinix Multi-Axis Biplane system (Figure 4.5) has an additional vertical axis that enables the flat panel detector (FPD) and the collimator to rotate opposite to the rotation of the C-arm's vertical axis so as to maintain the rectangular FOV steady over the patient during AP gantry movements. The Siemens Artis zeego (see Figure 10.2), which is a C-arm at the end of a floor-mounted robotic arm, appears to have seven rotational degrees of freedom without the slide along the c.

4.3.2   X-Ray Sources for Angiography

Although there have been new technical developments regarding the production of medical x-rays, very little has changed in the fundamental physical mechanisms. Here we review briefly these x-ray production methods and describe the technology as specifically applied to angiography including some new modifications resulting from the effort to provide higher quality and more appropriate clinical image sequences.

4.3.2.1   Principles of X-Ray Production

The use of electrons accelerated typically by a 50–120 kVp potential to bombard a high Z-material target such as tungsten and resulting in the production of a polychromatic spectrum of x-rays via the bremmstrahlung process when the electrons are suddenly decelerated, still is the basic mechanism for x-ray production as it has been virtually since Roentgen's discovery of x-rays. An additional process whereby the bombarding electrons can cause inner shell electrons in the atoms of the target to be removed resulting in a cascade emission of monochromatic characteristic x-rays contributes a smaller amount to the spectrum of x-rays emanating from the focal spot on the target anode. Details of these processes are discussed in enumerable textbooks (Evans 1955; Johns and Cunningham 1983). From an energy-efficiency point of view, the bremmstrahlung process at medical x-ray energies is not very impressive since all but a few percent of the energy in the bombarding electron beam is wasted in heating up the anode target rather than in the actual production of x-rays. Although the intensity of bremmstrahlung

Chapter 4

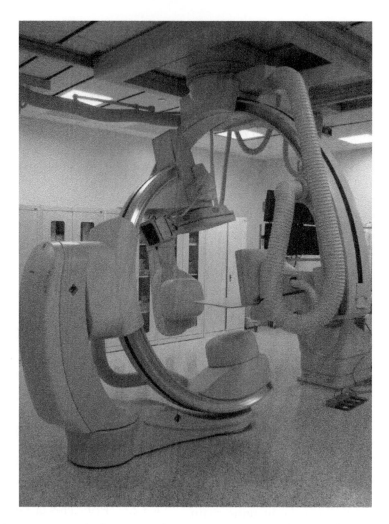

FIGURE 4.5 A 5-axis biplane gantry. For the floor mounted gantry: two axes perpendicular to the floor, one axis parallel to the floor that is the "propeller" degree of freedom, one on the C-arm that is the "slide" degree of freedom, and one axis at the central line of the x-ray beam allowing the detector and collimator to rotate in unison to maintain alignment with the patient's axis while the C-arm rotates. In addition the ceiling mounted C-arm can be translated in the x and y direction as well as having its own propeller and slide degrees of freedom.

x-ray production does improve roughly proportional to ZE^2, where Z is the target material atomic number and E is the energy of the bombarding electrons, the precise shape of the spectrum emitted will depend upon filtration. Inherent filtration in the anode material is due to the x-rays being produced slightly below the target surface hence having to penetrate through the attenuating material of the target itself. As the x-rays leave the vacuum of the tube, they can also be attenuated in the material of the tube envelop and finally aluminum filtration is required to be added so as to purposefully differentially reduce the low-energy spectrum components that would never penetrate the patient to reach the detector anyway, hence eliminating photons that might cause patient dose without providing any image information. Programs and references for generating

simulated x-ray spectrum are readily available (Birch 1979; Tucker 1991; Boone 1997; Poludniowski 2007a; Poludniowski 2007b).

4.3.2.2 X-Ray Tube Specifications and Focal Spot Measurements

The basic production physics may not have changed since the discovery of x-rays but the technology and design of sources has evolved. To dissipate the heat generated in the x-ray tube targets and to enable higher output rates and longer exposure times required for lengthy fluoroscopy and rapid sequence angiographic sequences, rotating anodes are generally used where the heat from the electron beam can be distributed around a circular track of the spinning anode disc. Recent engineering design improvements have contributed to increasing the tube loading through modification of the anode structure and use of liquid metal bearings (Schmidt 2000; Bittl 2006). To minimize the image blur due to the projection of a point in the object or patient by a finite sized focal spot x-ray source at the anode of size F, the magnification, m, must be minimal or close to 1. From simple geometry, one can show that the blurred image of the point in the object is $(m - 1)$ F. To minimize m, the image receptor must be close to the object while the focal spot is relatively far away so as to reduce this blur or geometric unsharpness. Alternatively, the focal spot size, F, can be minimized. This is typically controlled by the size of the filament used to generate the electrons at the cathode and by the so-called "line focus principle," meaning that if the region of the anode emitting x-rays is viewed end-on it appears smaller (Bushberg 2012).

If the cause of image blur is due to the lack of spatial resolution of the detector, then increasing the magnification of the image at the receptor may help in overcoming this resolution limit by making the features in the object appear relatively larger than the blur in the image receptor. The problem with this is that by increasing the magnification usually by moving the image receptor away from the patient while the tube is moved closer, the geometric unsharpness will also consequently increase. Thus, there is a balance between the two causes of image blur that ultimately can only be solved by improving the resolution of the detector or by decreasing the size of the focal spot. This latter reduction in focal spot size usually has negative implications in limiting the current that can bombard a smaller area on the target without causing damage; hence, this lowering of the output can negatively impact the angiographic image quality by increasing the quantum mottle.

4.3.2.3   Design of High Load, Small Focal Spot Tube for High Resolution

One potentially practical method to solve the problem of combining small focal spots with larger tube loading hence higher output is to simply use a smaller anode angle if the imaging FOV can be reduced. Our group has successfully been developing a class of high resolution, ROI detectors specifically for diagnosis and treatment of localized vascular pathology such as aneurysms or stenotic vessel segments. These developments are discussed in detail in Chapter 13. Compatible with these angiography requirements is our group's design illustrated in Figure 4.6 for a reduced anode angle tube of 2° with 4× the length of

the filament for example that can in theory provide 4× the output of a conventional angiography tube with an 8° angle while maintaining the dimensions of the small focal spot of 0.3 mm. To achieve this output, the added aluminum filter is removed (last line of Table 4.1) (Gupta 2012). It can be seen from the figure that the spectrum incident into the patient would be virtually unchanged.

4.3.2.4   Other Specialized Sources

There have been a number of new designs proposed and some implemented to improve the x-ray sources used in angiography. Smaller focal spots can be achieved using electron emission from sharp points of carbon nanotubes (Shan 2012) instead of through thermionic emission used in conventional tubes. Although smaller and more uniform focal spot sizes can be achieved in this way, tube loading at the anode remains limited. Using a scanning electron beam in combination with reverse geometry (Speidel 2006) is another alternative that is discussed in detail in Chapter 11. Multiple focal spots with inverse geometry are also being considered (Baek 2012). To create quasi-monochromatic x-ray beams using synchrotron radiation that in theory should enable improved angiography by selecting x-ray energies just above the absorption edge of iodine contrast agent has also been investigated (Umetani 2009); however, beam strengths and availability and practicality for clinical synchrotron sources have been deterrents. Another new mechanism for creating medical x-rays is using reverse Compton scattering with free electron lasers (Carroll 1990); however, this also appears to be impractical at present. Finally, use of a destructible film or self healing material as the anode along with a laser-produced plasma (LPP) x-ray source has been proposed but not yet practically implemented (Krol 1997). Thus, it appears likely that the evolution in x-ray sources will continue only gradually.

4.3.3   Scatter Reduction Methods

A fundamental problem in x-ray imaging is the contrast reduction caused by Compton scattering originating in the irradiated volume of the object or patient and entering the detector. Scatter radiation entering a specific pixel in the detector and not having come from a scattering source along the line connecting the pixel and the x-ray focal spot erroneously adds to the information in that pixel as well as contributing to the increase in quantum noise. For DSA, there is an

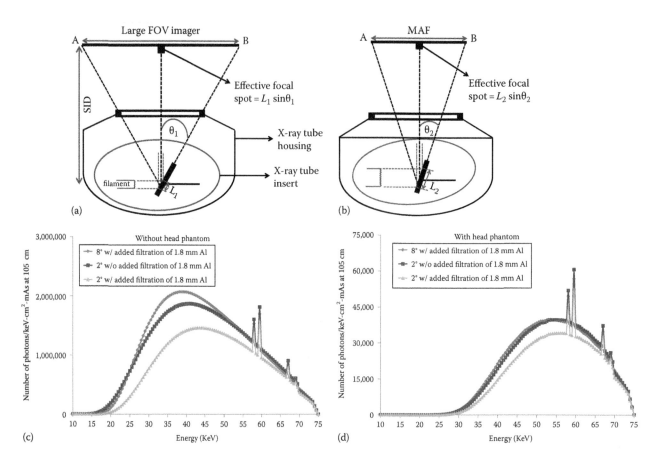

FIGURE 4.6 Proposed design modification of standard x-ray tube with anode angle θ_1 of 8° (a) to a high output x-ray tube with anode angle θ_2 of 2° and 4× elongated filament (b). Resulting spectra comparisons are shown incident to the detector for conditions without (c) and with (d) head phantom in the beam.

Table 4.1 Imaging Parameters at the Image Receptors without Phantom in the Beam

Anode Angle (°)	Relative Beam Intensity (%)	Square Field of View Dimensions (cm × cm)	Length of Filament Projection	Relative Tube Output	Maximum Tube Current (mA)
8	100	29 × 29	L	1.0	160
2	76	7.4 × 7.4	4L	3.0	640
2 (no Al filter)	97	7.4 × 7.4	4L	3.9	640

additional problem that a detection of a scattered x-ray at a given pixel contributes to the intensity before the logarithmic processing of the image occurs; hence, preventing accurate subtraction of the mask image from the subsequent images containing contrast media.

There are three basic methods that have been used for scatter reduction: grids, air gaps, and FOV restrictions such as scanning beams. The most popular method is to place a grid just in front of the image receptor so that the thin lead septa of the grid are focused to the x-ray focal spot to enable primary unscattered radiation from the focal spot to penetrate the grid and reach the detector while scattered radiation from the patient is hopefully absorbed in the septa. Depending upon

the grid ratio and the lead content of the grid, substantial amounts of scatter can be suppressed; however, there is a down side in that a significant amount, typically about 40% of the primary radiation, is absorbed either in the grid interspace material or in the septa. Little or no primary radiation is sacrificed when an air gap is used between the patient and the image receptor; however, for an air gap only the large angle scatter may be suppressed by no longer being able to intersect with the detector and additionally there is the potential loss of spatial resolution due to increased geometric unsharpness. The final method of reducing the volume irradiated may be possible but only if "coning down" with collimation is acceptable. An alternative method of reducing the volume irradiated while not sacrificing FOV is to use scanning beams (Barnes 1979). For this method, a group of parallel slits in an x-ray absorbing material such as lead or tungsten is placed between the patient and the focal spot with corresponding slots between the patient and the image receptor. The slits and slots are maintained in alignment while moving across the FOV during an exposure. In this way, at any instant of time, only a small volume of the patient is exposed so that less scatter is produced and yet the whole FOV can be imaged as the beams created by the slits are scanned across the full FOV. The authors investigated a rapid scanning beam method called rotating aperture wheels that enabled the scanning beams to move across the FOV rapidly enough to enable real-time fluoroscopy and angiography to be done (Rudin 1980a, b; Bednarek 1988). While scanning beams have no interspace material hence no absorption of primary radiation if the slit and slots are maintained in alignment, the fact that only a fraction of the FOV is exposed at any instant implies that the x-ray tube output, hence loading must be increased substantially. These burdens on the x-ray generation and the mechanical complication of moving slit-slot assemblies have prevented rapid scanning beams from being practically applied clinically. The limits of scanning beams and the inadequacy of air gaps have left only grids as the practical scatter reduction method. Thus, truly efficient scatter removal continues to be an unsolved problem. Perhaps in the future, if detectors are able to detect single photons at a time and discriminate with respect to energy as is done in nuclear medical imaging, then maybe the scatter problem in radiography will be solved (Figure 4.7).

4.3.4 Detector Review

After many decades of gradual evolution for dynamic x-ray detectors that are the ones relevant for angiography, dynamic detector design has undergone revolutionary change and is about to undergo another large change. The three designs discussed in this section are the old, reliable x-ray image intensifier (XII), the new FPD, and the newest and not quite available high-resolution CMOS FPDs and ROI detectors. Although there are other detector designs that have been used for static imaging, these three enable fluoroscopy and rapid sequence imaging used for DA and DSA. Also, generally all three of these detector designs have employed an x-ray absorbing convertor phosphor layer of CsI to take some of the energy of the absorbed x-rays

and convert it into light that can be detected further down in the imaging detector chain. These are the so-called indirect detectors because they detect x-rays indirectly by actually detecting the light resulting from the energy deposited in the phosphor by the x-rays. The three detectors can be evaluated by measuring a variety of complex spatial frequency–dependent objective metrics including modulation transfer function (MTF) and detective quantum efficiency (DQE) that will be discussed in a later section. Additionally, they can be evaluated by some simpler metrics such as linearity, dynamic range, and instrumentation noise. In order to do any quantitative image analysis and processing, linear performance is required and all these three detectors are linear in their operational ranges as expected. To satisfy the requirements for the clinical angiographic suite, the detectors must have a large enough dynamic range to be able to do fluoroscopy, DA, and DSA. Finally, for low exposure operation, such as those needed during fluoroscopy, the instrumentation noise must be low enough to have the detector preferably operate in the quantum limited exposure region where the total noise in the image is limited not by instrumentation noise but by the quantum fluctuations. A good metric that our group has been developing to evaluate these noise properties is the INEE (Yadava 2008; Kuhls-Gilcrist 2009). For those exposures below the INEE, the detector is instrumentation noise limited while for exposures above the INEE the detector is the desired quantum noise limited. The instrumentation noise can also be measured in number of electrons in the output noise.

4.3.4.1 XII

The XII is essentially an electron beam intensifier because the electron distribution that is created from the light emitted by the x-ray convertor phosphor layer is accelerated through a high-voltage potential and bombarded into an output phosphor to create a brighter or intensified image. One of the reasons that the XII has lasted so long and dominated the field of dynamic x-ray detectors is its large gain or sensitivity to enable it to change the intensity of the image at the output of the XII. Thus, the XII has the advantage of a very large dynamic range easily enabling it to function over the full exposure range requirements of fluoroscopy, DA, and DSA. This dynamic range has enabled it to provide very low instrumentation noise even for low-exposure values so that the instrumentation noise limits have actually been the result of the noisy pickup television tubes originally used to view the output of

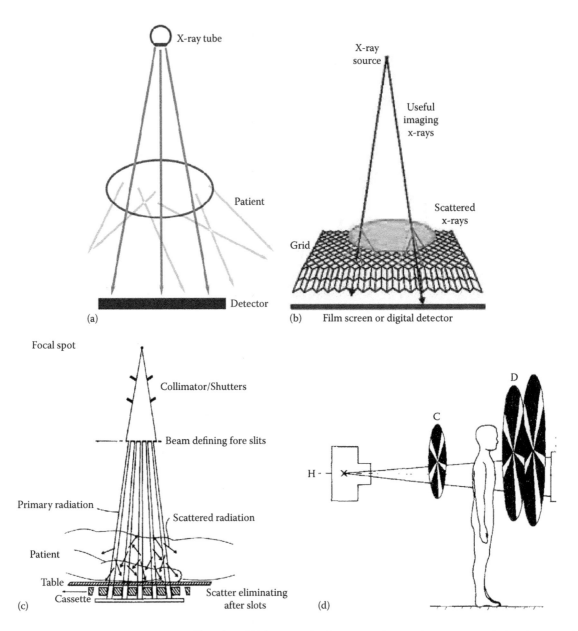

FIGURE 4.7 Antiscatter methods. (a) Air gap, (b) grid, and (c) linear multiple scanning beams. (From Barnes, G.T. and Brezovich, I.A., *Med. Phys.*, 6(3), 197, 1979.) (d) Rotating aperture wheel (RAW) rapid scanning beams.

the XII. Another benefit that XIIs exhibited is the ability to vary the magnification by changes in the electron beam focus. Thus, variation of the FOV and consequent resolution is easily and quickly adjustable.

There are, however, a number of problems associated with XIIs that has led to their decline over the last decade. They are subject to two kinds of image distortion (Figure 4.8). Geometric distortion is caused by the disparity of the front phosphor and photocathode surfaces. That surface is roughly spherically shaped to enable electrons to come off perpendicular to the surface and still be able to be focused; however, the output phosphor is flat to enable light image sensors such as

vidicons and ccds with flat surfaces to image the output images in optical focus. The result is the so-called "pincushion" distortion because a square test grid pattern at the input has an image that appears spread out at the periphery. What is happening is that the effective minification of the flat test pattern is less, that is, the magnification is greater as one goes further away from the center of the XII. Thus, the corners of the test pattern appear pushed out like a pin cushion. The other cause of distortion is the earth's magnetic field causing the electrons in the vacuum of the XII to change path creating an image of a straight-line object to be S-shaped (Rudin 1991; Liu 1999). This distortion that

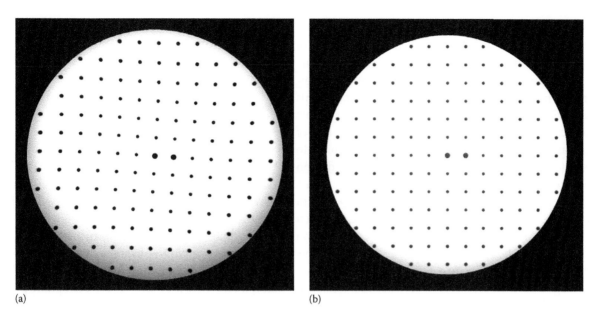

(a)   (b)

FIGURE 4.8 Example of XII image of square test pattern showing pincushion and s-shaped distortion on the (a) with corrected image on the (b).

varies with the position of the XII's gantry with respect to the earth's magnetic field has made it more difficult to perform cone-beam CT with C-arms since all the images each at a different angle have to be processed to correct the distortion. Other negatives of the XII include the so-called "veiling glare" that reduces the contrast of the image and the bulkiness of the large vacuum bottle that is needed to focus and accelerate the electron distribution.

4.3.4.2 Flat Panel Detector

Over the past decade, FPDs have been replacing XIIs as the preferred dynamic image detector. FPDs consist at the backend of a pixilated thin film transistor (TFT) array made of amorphous silicon used to switch charge collected at each pixel. The charge is proportional to the x-ray exposure and is derived from one of two basic types of x-ray converters—direct or indirect. The direct detectors form charge hole-electron pairs in the x-ray absorbing material, usually amorphous Se, which charge is collected and stored until read out at each pixel. Up until recently these direct detectors have not been generally favored because they have exhibited severe ghost and lag properties that are difficult to correct even though their inherent spatial resolution can be superb due to the high electric field used to collect the charge in the x-ray absorber-converter (Rudin 2000; Yamada 2000; Bloomquist 2006; Kabir 2010). Indirect detectors, in contrast, use the same converter phosphors

as XIIs with familiar properties. The light from the phosphor is then sensed by photo-sensors in each pixel and converted to charge, which is then read off the FPD using the amorphous silicon TFT array of switches just as for the direct detectors. The advantage over the direct technology is that any lag appears to be routinely correctable. Additionally, the absorption of medical x-rays in CsI is substantially better for the same thickness of material compared to that of Se. Although the inherent spatial resolution for the CsI phosphors is not as good as it is for Se, the structured crystal growth of the CsI enables the resolution to be generally adequate. An important remaining problem for FPDs, however, has been the instrumentation noise, which is somewhat higher than desirable (Antonuk 2009). The INEE approaches 2 μR, which corresponds to exposures used for low-dose fluoroscopy hence the performance may even appear noisier than such fluoroscopy done with an XII (Roos 2004). Apparently, the only viable approach to reducing the FPD effective readout noise that approaches 2000 electrons is to have an amplifier at each pixel before the readout amplifier. This has been investigated (Karim 2011) but has not become commercially practical as yet. The final problem with FPDs based on amorphous silicon TFTs is the inability to produce pixels smaller than about 150 μm in size without sacrificing active sensing area because of decreasing fill factor. Thus, high-resolution dynamic FPDs based on amorphous silicon TFTs appear out of reach.

4.3.4.3 High-Resolution CMOS and ROI Detectors

Recently, our group has demonstrated the usefulness of very high resolution micro-angiographic detectors during the diagnosis and endovascular interventional treatment of neurovascular pathology such as aneurysms (Kuhls-Gilcrist 2009; Jain 2011). Our microangiographic fluoroscope (MAF), which has an effective pixel size of 35 μm and a large dynamic range, even with the ability to do single photon counting for radiography, is based upon an indirect converter phosphor, a light image intensifier, a fiber-optic taper, and a high-performance CCD camera sensor. The INEE in energy integration mode has been measured to be below 0.2 μR. A more detailed discussion of the MAF can be found in Chapter 13 of this volume. Although limited in FOV to a ROI, the MAF-CCD has encouraged further exploration of detectors one such being a different type of flat panel based upon CMOS technology that we designate the MAF-CMOS.

MAF-CMOS flat panels differ from standard amorphous silicon (aSi) FPDs in the fabrication process and the method for reading out the charge collected in each pixel in that no longer do whole rows at a time have to be read out together. There is more flexibility in the MAF-CMOS FPD design possibilities in that individual MAF-CMOS FPD pixels may be read out. MAF-CMOS flat panels appear to have all the advantages of aSi FPDs without the noise and pixel size limitations. Noise of less than 200 electrons has been reported (Esposito 2011; Konstantinidis 2011) with pixel sizes in the range of 25–100 μm and with multiple sensitivity modes having different full well capacities. Detectors with FOVs that are smaller than conventional FPDs are becoming commercially available. Most recently, a low-noise "crystalline" detector with 160 μm pixels has been shown by a major medical equipment manufacturer. We expect that in the not-too-distance future CMOS-based FPDs may replace TFT-based aSi FPDs.

4.4 Objective Evaluations

Over the years, attempts have been made to obtain a set of objective metrics to enable both absolute and relative evaluations of x-ray image receptors. Presently, manufacturers and imaging medical physicist and engineers have agreed on using linear systems analysis with spatial frequency dependent metrics such as MTF and DQE as the key metrics. Our group and others have generalized these metrics so as to evaluate whole systems including other factors that determine imaging characteristics such as focal spot size and scatter as well as the image receptor.

To illustrate potential benefits of MAF-CMOS FPDs, our group has done simulations of such high-resolution detectors as described in more detail in Section 4.4 (Jain 2013; Loughran 2013).

4.3.5 Workflow and Ergonomic System Design

As image-guided interventional procedures become more complex and longer in duration, innovation in workflow and ergonomic system design will be increasingly important. Where two viewing monitors for a single plane angiography system used to be acceptable, current angiography suites can have upwards of six monitors for a biplane suite with the capability to show a reference loop display for each plane plus road-mapped fluoroscopy in addition to previous study reference cases or computer processed display of metrics not to mention biometric monitors.

In our group's experience with designing a control, acquisition, processing, image display, and storage (CAPIDS) system for the high-resolution MAF, much effort was expended to integrate all these necessary imaging modes into a system that had maximum capability and yet maximum ease of use for the operator (Wang 2010; Vasan 2012a). Combining data integrity, reliability, accuracy, yet operator flexibility and upgradability, is no trivial task. As time goes on with increasing performance demands, the system designer will have to start using labor-saving devices such as voice command, control through operator eye tracking, (Guez 2012) and even perhaps smarter control computers with machine learning capability. For example, such a system might automatically determine what the best C-arm orientations might be to minimize vessel foreshortening and proceed to position C-arms with the prior knowledge of all occupants of the room to eliminate potential collisions and accidents.

4.4.1 Detector Evaluation with Frequency Dependant Metrics: MTF and DQE

Since most practical medical imaging receptors are designed to be linear in their response to x-ray energy deposited in each pixel, linear systems analysis can be used. There are a number of good detailed descriptions of the analytic methods used to develop the MTF and DQE metrics (Dobbins 1995; Cunningham 2000). We will only summarize the definition of MTF as being the Fourier transform (FT) of the point spread function of a detector. Physically this means that if an infinitely small point test object is imaged, its image will be blurred into a spread function. Since a point object or delta function has all spatial frequency components in it, its blurred image or the point spread function will have decreased magnitude of the higher frequency components. Thus, the MTF indicates how well each spatial frequency component of the signal is imaged or literally transferred from the object to the image. To measure the MTF, the assumption is made that a sufficiently large signal is recorded so that noise is relatively negligible. Although the two-dimensional MTF is the FT of the point spread function, it is often sufficient to use a slit or an edge object to determine the one-dimensional MTF from the line or edge spread function to characterize an image receptor especially if its resolution is independent of direction. A simplified and efficient method of determining the 2D MTF which members of our group have reported does not require either an edge or a slit. The so-called NR (noise response) method obtains the MTF from simple measurement of the noise pattern in the flat field images (Kuhls-Gilcrist 2010a; Kuhls-Gilcrist 2010b).

The DQE is defined as the output signal to noise ratio squared compared to that of the input for each spatial frequency. It can be shown (Cunningham 2000) that the DQE is determined from measurable quantities as indicated in the following equation:

$$\mathrm{DQE}(u,v,X) = \frac{\mathrm{MTF}^2(u,v,X)}{\phi_{in}(X)\mathrm{NNPS}(u,v)} \tag{4.1}$$

where
u and v are spatial frequencies
X is the exposure to the detector
NNPS is the normalized noise power spectrum
ϕ_{in} is the input x-ray fluence

As an example of the value of these metrics, we present results of a simulation study of a new MAF-CMOS FPD design that should be possible to implement in comparison with other detectors (Figure 4.9).

4.4.2 Generalized System Evaluation Metrics: GMTF and GDQE

To determine the imaging performance of the whole system, one must include the focal spot size and its effect on geometric unsharpness, and the scatter as well as the image receptor characteristics. Equation 4.2 provides the quantitative description of the generalized MTF designated GMTF where the spatial frequency is now evaluated in a specific plane in the object such that m is the magnification of images of features in that plane (Kyprianou et al. 2005). In the second term of the bracketed part of the equation, the scatter MTF$_S$ is weighted by the scatter fraction ρ and in the first term the focal spot MTF$_F$ is weighted by the primary fraction or $1 - \rho$. The term multiplying the bracketed sum provides for the image detector characteristics as it views both the magnified image and the scatter hence the inclusion of the MTF$_D$ factor.

$$\mathrm{GMTF}(u,v,\rho,m) = \left[(1-\rho)\mathrm{MTF}_F\left(\left(\frac{m-1}{m}\right)(u,v)\right) \right. $$
$$\left. + \rho\mathrm{MTF}_S\left(\frac{(u,v)}{m}\right) \right]\mathrm{MTF}_D\left(\frac{(u,v)}{m}\right) \tag{4.2}$$

Following the detector-only formulas of the previous section, the GDQE is given by (Kyprianou 2005; Jain 2010):

$$\mathrm{GDQE}(u,v,\rho,X,m) = \frac{\mathrm{GNEQ}(u,v,\rho,X,m)}{m^2\phi_{in}(X,m)} \tag{4.3}$$

Similar generalized metrics have been used by other groups (Samei 2009). Examples of the use of GMTFs and GDQEs are provided in Section 13.7.1 and shown in Figure 13.10.

4.4.3 Relative Object Detectability

To evaluate a system's ability to detect an object compared to that of an ideal system of the same type, one can use a new metric we designate the fractional relative object detectability (F-ROD) as indicated in the following equation:

$$\mathrm{F\text{-}ROD}(\rho,X,m) = \frac{\iint |\mathrm{OBJ}(u,v)|^2 \mathrm{DQE}(u,v,\rho,X,m)\,dudv}{\iint |\mathrm{OBJ}(u,v)|^2\,dudv} \tag{4.4}$$

where OBJ(u, v) is the object's spatial frequency distribution. The F-ROD will obviously be dependent on the specific object so that a comparison of the F-ROD for two different objects could give a quantitative comparative assessment for a given detector design. Also, one

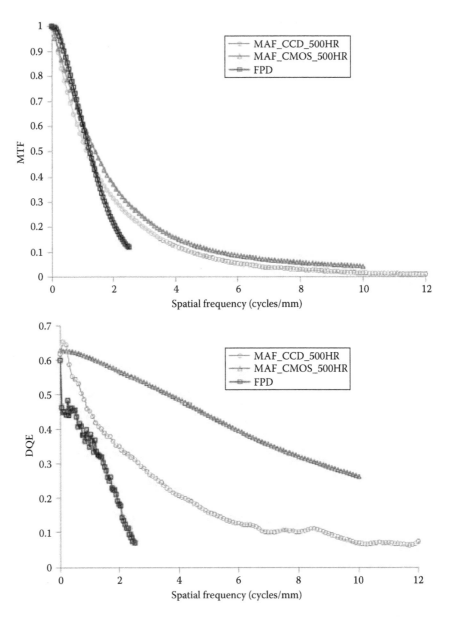

FIGURE 4.9 Linear systems analysis metrics of MTF and DQE for two high-resolution detectors (the MAF-CCD and the MAF-CMOS) compared with those of a standard amorphous Si FPD. The MAF-CMOS was simulated while the MAF-CCD and FPD without a grid were measured. Both MAFs had 500 μm thick CsI phosphors with no reflective layer. RQA5 spectrum with the same collimation and exposure of 40 μR was used. The MAF-CMOS was assumed to have a 50 μm pixel with 100 electrons rms instrumentation noise and light conversion efficiency of 50% with 80% light coupling efficiency. Although aliasing was not considered here, it would have to be for a practical implementation. (From Jain, A. et al., Experimental and theoretical performance analysis for a CMOS-based high resolution image detector, in *Proceedings from Medical Imaging 2014: Physics of Medical Imaging*, paper 9033-132, vol. 9033, SPIE, San Diego, CA, 2014.)

can keep the object constant and compare the relative performance of two different detectors by substituting for the denominator in the equation an expression similar to the numerator but with a DQE for a second detector rather than the ideal detector where the DQE equals 1. This is designated the ROD (Singh 2014a). Further generalization of this new metric concept can be obtained by substituting GDQE for DQE resulting in the FG-ROD for a given detector system or

for comparing two different detector systems G-ROD (Singh 2014b) as stated in Equation 4.5:

$$
\begin{aligned}
&\text{G-ROD}(\rho, X, m) \\
&= \frac{\iint |\text{OBJ}(u,v)|^2 \text{GDQE}_2(u,v,\rho,X,m)dudv}{\iint |\text{OBJ}(u,v)|^2 \text{GDQE}_1(u,v,\rho,X,m)dudv}
\end{aligned}
\tag{4.5}
$$

G-ROD can be used for simulated objects where the object function's FT can be calculated and where the two detector system DQEs are known either through simulation or measurement to derive their GDQEs; however, for actual radiographed objects, Equation 4.6 is more appropriate since the FT of the measured object function obtained from images $OBJ(u, v)_{meas}$ is already convolved with the MTF of the detector and the NNPS of each detector is also measured. This new metric is thus designated the generalized measured ROD (GM-ROD) where detector 2 is the MAF and 1 is the FPD in Equation 4.6.

$$\text{GM-ROD} = \frac{\left[\iint \frac{\left|OBJ(u,v)_{meas}\right|^2}{\text{NNPS}(u,v)} \, du \, dv\right]_{MAF}}{\left[\iint \frac{\left|OBJ(u,v)_{meas}\right|^2}{\text{NNPS}(u,v)} \, du \, dv\right]_{FPD}} \quad (4.6)$$

4.4.4 Phantoms

Various phantoms have been used to evaluate angiographic imaging systems (Figure 4.10). There are

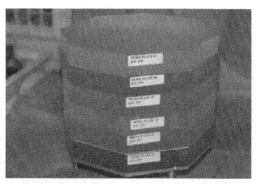
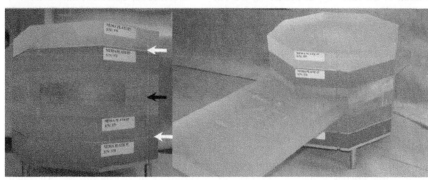
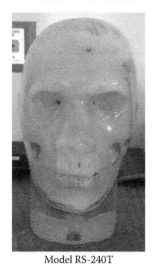
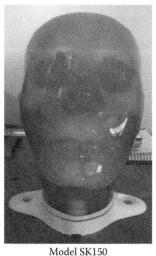

Model RS-240T   Model SK150   PBU-50

FIGURE 4.10 NEMA modified phantom made of flat slabs and some head phantoms.

anthropomorphic phantoms (Ionita 2012) some containing iodinated structures that simulate vessel trees such as the Model RS-240T head angiography phantom (Radiology Support Devices Inc., Long Beach, CA 90810). On other anthropomorphic phantoms such as the Kyoto Kagaku Model PBU-50 (Kyoto Kagaku Co. Ltd., Fushimi-ku Kyoto, Japan 612-8388) or the Model SK150 Sectional Head Phantom with Upper Cervical Vertebrae (The Phantom Laboratory, Salem, NY 12865), vessels can be simulated using tubes fastened appropriately. Some phantoms are regular solid acrylic slabs

with simulation vessels embedded such at the Stenosis/ Aneurysm Artery Block Model 76-705 vascular phantom (Nuclear Associates, Fluke). These slabs can be placed within other uniform phantoms such as the NEMA cardiac phantom better to simulate realistic scatter and magnification conditions (Ionita 2009). To simulate the lateral head angiographic set-up, the AAPM has recommended a standard ANSI phantom; however, our group has found such a head phantom must be modified and increased in attenuator thickness for it to be properly used for the PA head view (Ionita 2012).

4.5 Summary and Conclusion

In this chapter, angiographic operational modes, system components, and evaluation methods were introduced and summarized. While later chapters provide discussions of other and sometimes competing methods for cardio- and neurovascular imaging, x-ray angiography remains the gold standard for angiography because of its high-resolution, real-time imaging capabilities.

References

(Antonuk 2009) Antonuk L. E., Zhao Q., El-Mohri Y., Du H., Wang Y., Street R. A., Ho J., Weisfield R., and Yao W. 2009. An investigation of signal performance enhancements achieved through innovative pixel design across several generations of indirect detection, active matrix, flat-panel arrays. *Med. Phys.*, 36, 3322–3339.

(Baek 2012) Baek J., Pelc N. J., Deman B., Uribe J., Harrison D., Reynolds J., Neculaes B., Inzinna L., and Caiafa A. 2012. Initial results with a multi-source inverse-geometry CT system. SPIE vol. 8313-46. In: *Proceedings from Medical Imaging 2012: Physics of Medical Imaging*, San Diego, CA, paper 83131A:1-7.

(Balter 1984) Balter S., Ergun D., Tscholl E., Buchmann F., and Verhoeven L. 1984. Digital subtraction angiography: Fundamental noise characteristics. *Radiology*, 152, 195–198.

(Barnes 1979) Barnes G.T. and Brezovich I.A. 1979. The design and performance of a scanning multiple slit assembly. *Med. Phys.*, 6(3), 197–204.

(Bednarek 1988) Bednarek D. R., Rudinger G., and Rudin S. 1988. Slit design considerations for rotating aperture scanning beam radiography. *Med. Phys.*, 15(1), 29–35.

(Birch 1979) Birch R. and Marshall M. 1979. Computation of bremsstrahlung x-ray spectra and comparison with spectra measured with a Ge(Li) detector. *Phys. Med. Biol.*, 24, 505–513, IPEM Report 78.

(Bloomquist 2006) Bloomquist A. K., Yaffe M. J., Mawdsley G. E., Hunter D. M., and Beideck D. J. 2006. Lag and ghosting in a clinical flat-panel selenium digital mammography system. *Med. Phys.*, 33, 2998.

(Boone 1997) Boone J. M. and Seibert J. A. 1997. An accurate method for computer-generating tungsten anode x-ray spectra from 30 to 140 kV. *Med. Phys.*, 24, 1661–1670.

(Bushberg 2012) Jerrold T., Bushberg J., Anthony S., Edwin M. L., Jr., and John M. B. 2012. *The Essential Physics of Medical Imaging*, 3rd edn., p. 182. Lippincott Williams & Wilkins.

(Carroll 1990) Carroll F. E., Waters J. W., Price R. R., Brau C. A., Roos C. F., Tolk N. H., Pickens D. R., and Stephens W. H. 1990. Near-monochromatic x-ray beams produced by the free electron laser (FEL) and Compton backscatter. *Investi. Radiol.*, 25(5), 465–471.

(Cunningham 2000) Cunningham I. A. 2000. Applied linear systems theory. In *Handbook of Medical Imaging*, J. Beutel, H. L. Kundel, and R. L. Van Metter Eds., Vol. 1, Chapter 2, pp. 79–162, SPIE Press, Bellingham, WA.

(Dobbins 1995) Dobbins J. T., Ergun D. L., Rutz L., Hinshaw D. A., Blume H., and Clark D. C. 1995. DQE(f) of four generations of computed radiography acquisition devices. *Med. Phys.*, 22(10), 1581–1593.

(Esposito 2011) Esposito M., Anaxagoras T., Fant A., Wells K., Konstantinidis A., Osmond J. P. F., Evans P. M., Speller R. D., and Allinson N. M. 2011. DynAMITe: A wafer scale sensor for biomedical applications, *J. Instrum.*, 6, C12064.

(Evans 1955) Evans R. 1955. *The Atomic Nucleus*, New York, McGraw-Hill Book Co., p. 698, eqn. 1.2.

(Bittl 2006) Bittl H. 2006. X-ray tube with liquid-metal fluid bearing. US Patent 7050541 B2. Assigned to Siemens Corp.

(Guez 2012) Guez A. July 26, 2012. Radiation control and minimization system and method. US Patent Application Publication No. US 2012/0187312 A1.

(Gupta 2012) Gupta S., Jain A., Bednarek D. R., and Rudin S. 2012. Overcoming x-ray tube small focal spot output limitations for high-resolution region of interest imaging. SPIE, vol. 8313-192. In: *Proceedings from Medical Imaging 2012: Physics of Medical Imaging*, San Diego, CA, paper 831358:1-11. NIHMSID 391962. http://www.ncbi.nlm.nih.gov/pmc/articles/PMC3419545/.

(Ionita 2009) Ionita C. N., Dohatcu A., Jain A., Keleshis C. M., Hoffmann K. R., Bednarek D. R., and Rudin S. 2009. Modification of the NEMA XR21-2000 cardiac phantom for testing of imaging systems used in endovascular image guided interventions. SPIE, vol. 7258. In: *Proceedings from Medical Imaging 2009*: *Physics of Medical Imaging*, Orlando, FL, paper 7258-176, 72584R:1–9. http://www. pubmedcentral.gov/articlerender.fcgi?artid=2743264.

(Ionita 2012) Ionita C. N., Loughran B., Jain A., Swetadri Vasan S. N., Bednarek D. R., and Rudin S. 2012. New head equivalent phantom for task and image performance evaluation representative for neurovascular procedures occurring in the Circle of Willis. SPIE, vol. 8313-25. In: *Proceedings from Medical Imaging 2012*: *Physics of Medical Imaging*, San Diego, CA, paper 83130Q:1-12. NIHMSID 391963. http://www.ncbi.nlm.nih.gov/pmc/articles/PMC3767006/.

(Jain 2010) Jain A., Kuhls-Gilcrist A. T., Bednarek D. R., and Rudin S. 2010. Generalized two-dimensional (2D) linear system analysis metrics (GMTF, GDQE) for digital radiography systems including the effect of focal spot, magnification, scatter, and detector characteristics. SPIE, vol. 7622. In: *Proceedings from Medical Imaging 2010*: *Physics of Medical Imaging*, San Diego CA, paper 7622-19, 76220K:1–10. http://www.pubmedcentral.gov/articlerender.fcgi?artid=3021385.

(Jain 2011) Jain A., Bednarek D. R., and Rudin S. July 2011. A theoretical and experimental evaluation of the micro-angiographic fluoroscope (MAF): A high-resolution region-of-interest x-ray imager. *Med. Phys.*, 38(7), 4112–4126. http://www.ncbi.nlm.nih.gov/pubmed/21859012.

(Jain 2013) Jain A., Bednarek D. R., and Rudin S. 2013. Theoretical performance analysis for CMOS based high resolution detectors. SPIE, vol. 8668-212. In: *Proceedings from Medical Imaging 2013*: *Physics of Medical Imaging*, Orlando, FL, paper 86685U:1-9. NIHMSID # 483443. http://www.ncbi.nlm.nih.gov/pmc/articles/PMC3864964/.

(Jain 2014) Jain A., Bednarek D. R., and Rudin S. 2014. Experimental and theoretical performance analysis for a CMOS-based high resolution image detector. SPIE, vol. 9033. In: *Proceedings from Medical Imaging 2014*: *Physics of Medical Imaging*, San Diego, CA, paper 9033-132.

(Johns and Cunningham1983) Johns, H. E. and Cunningham, J. R. 1983. *The Physics of Radiology*, pp. 59–68. Charles C. Thomas.

(Kabir 2010) Kabir M. Z., Chowdhury L., DeCrescenzo G., Tousignant O., Kasap S. O., and Rowlands J. A. 2010. Effect of repeated x-ray exposure on the resolution of amorphous selenium based x-ray imagers. *Med. Phys.* 37, 1339.

(Karim 2011) Karim K. S., Taghibakhsh F., and Izadi M. H. August 9, 2011. High gain digital imaging system. US Patent 7,995,113,B2.

(Karim 2012) Karim K. S. and Taghibakhsh F. June 12, 2012. Device and pixel architecture for high resolution digital. US Patent 8,199,236 B2.

(Kezerashvili 1997) Kezerashvili M., Rudin S., and Bednarek D. R. 1997. Automatic filter placement device for Region of Interest (ROI) Fluoroscopy. *Health Phys*, 72(1), 141–146.

(Konstantinidis 2011) Konstantinidis A. C. 2011. Evaluation of digital x-ray detectors for medical imaging applications. PhD thesis, University College, London, U.K.

(Krol 1997) Krol A., Ikhlef A., Kieffer J. C., Bassano D. A., Chamberlain C. C., Jiang Z., Pépin H., and Prasad, S. C. 1997. Laser-based microfocused x-ray source for mammography: Feasibility study. *Med. Phys.*, 24(5), 725–732.

(Kuhls-Gilcrist 2009) Kuhls-Gilcrist A. T., Bednarek D. R., and Rudin S. 2009. Component analysis of a new solid state x-ray image intensifier (SSXII) using photon transfer. SPIE,

vol. 7258. In: *Proceedings from Medical Imaging 2009*: *Physics of Medical Imaging*, Orlando, FL, paper 7258-42, 725817:1–10. http://www.pubmedcentral.gov/articlerender.fcgi?artid=2745170.

(Kuhls-Gilcrist 2010a) Kuhls-Gilcrist A. T., Jain A., Bednarek D. R., Hoffmann K. R., and Rudin S. 2010. Measurement of the MTF of digital radiographic image receptors using the noise response. *Med. Phys.*, 37(2), 724–735. http://www.ncbi.nlm.nih.gov/pubmed/20229882.

(Kuhls-Gilcrist 2010) Kuhls-Gilcrist A., Bednarek D. R., and Rudin S. 2010. A method for the determination of the two-dimensional MTF of digital radiography systems using only the noise response. SPIE, vol. 7622. In: *Proceedings from Medical Imaging 2010*: *Physics of Medical Imaging*, San Diego CA, paper 7622-178, 76224W:1–9. http://www.pubmedcentral.gov/articlerender.fcgi?artid=3003440.

(Kyprianou 2005) Kyprianou I., Rudin S., Bednarek D. R., and Hoffmann K. R. 2005. Generalizing the MTF and DQE to include x-ray scatter and focal spot unsharpness: Application to a new micro-angiographic system for clinical use. *Med. Phys.*, 32(2), 613–626. http://www.ncbi.nlm.nih.gov/pubmed/15789608.

(Labbe 1994) Labbe M. S., Chiu M. Y., Rzeszotarski M. S., Bani-Hashemi A. R., and Wilson D. L. 1994. The x-ray fovea, a device for reducing x-ray dose in fluoroscopy. *Med. Phys.*, 21(3), 471–481.

(Liu 1999) Liu R., Rudin S., and Bednarek D. R. 1999. Super-global distortion correction for a rotational C-arm image intensifier. *Med. Phys.*, 26(9), 1802–1810.

(Loughran 2013) Loughran B., Vasan S. N. S., Singh V., Ionita C. N., Jain A., Bednarek D. R., Titus A. H., and Rudin S. 2013. Design considerations for a new, high resolution Micro-Angiographic Fluoroscope based on a CMOS sensor. SPIE, vol. 8668-4. In: *Proceedings from Medical Imaging 2013*: *Physics of Medical Imaging*, Orlando, FL, paper 866806:1-9. NIHMSID #483430. http://www.ncbi.nlm.nih.gov/pmc/articles/PMC3864963/.

(Massoumzadeh 1998) Massoumzadeh P., Rudin S., and Bednarek D. R. February 1998. Filter material selection for region of interest radiologic imaging. *Med. Phys.*, 25(2), 161–171.

(Panse 2010) Panse A., Ionita C. N., Wang W., Jain A., Bednarek D. R., and Rudin S. October 30–November 6, 2010. The micro-angiographic fluoroscope (MAF) in high definition (HD) mode for improved contrast-to-noise ratio and resolution in fluoroscopy and roadmapping. In *IEEE Nuclear Science Symposium and Medical Imaging Conference*, Knoxville, TN, paper M18-204, NSS Conference Record, p. 3217–3220. NIHMSID 303301 http://www.pubmedcentral.gov/articlerender.fcgi?artid=3137370.

(Poludniowski 2007a) Poludniowski, G. G. and Evans P. M. 2007. Calculation of x-ray spectra emerging from an x-ray tube: Part I. Electron characteristics in x-ray targets. *Med. Phys.*, 34(6), 2164–2174.

(Poludniowski 2007b) Poludniowski, G. G. 2007. Calculation of x-ray spectra emerging from an x-ray tube: Part II. X-ray production and filtration in x-ray targets. *Med. Phys.*, 34, 2175–2186.

(Roo 2004) Roos P. G., Colbeth R. E., Mollov I., Munro P., Pavkovich J., Seppi E. J., Shapiro E. G. et al., 2004. Multiple gain ranging readout method to extend the dynamic range of amorphous silicon flat panel imagers. Physics of Medical Imaging. *Proc. SPIE*, 5368, 139–149.

(Rudin 1980a) Rudin S. 1980. Rotating aperture wheel (RAW) device for improving radiographic contrast. *Opt. Eng.*, 19(1), 132–138.

(Rudin 1980b) Rudin S. and Bednarek D. R. 1980. Improving contrast in special procedures using a rotating aperture wheel (RAW) device. *Radiology*, 137(2), 505–510.

(Rudin 1992) Rudin S. and Bednarek D. R. September/October 1992. Region of interest fluoroscopy. *Med. Phys.*, 19(5), 1183–1189.

(Rudin 1995) Rudin S., Bednarek D. R., Guterman L. R., Wakhloo A., Hopkins L. N., Fletcher L., and Massoumzadeh P. November 1995. Implementation of region of interest fluoroscopy using the road mapping mode of a real-time digital radiographic unit. *Radiographics*, 15(6), 1465–1470.

(Rudin 1996) Rudin S., Guterman L. R., Granger W., Bednarek D. R., and Hopkins L. N. 1996. Neuro-interventional radiologic application of region of interest (ROI) imaging techniques. *Radiology*, 199, 870–873.

(Rudin 1999) Rudin S., Bednarek D. R., and Yang C. Y. , July 1999. Real-time equalization of region of interest fluoroscopic images using binary masks. *Med. Phys.*, 26(7), 1359–1364.

(Rudin 1991) Rudin S., Bednarek D. R., and Wong R. 1991. Accurate characterization of image intensifier distortion. *Med. Phys.*, 18(6), 1145–1151.

(Rudin 2000) Rudin S., Yang C. J., Wang Z., Gopal A., Chattopadhyay A., Bednarek D. R., Hoffmann K. et al., September 2000. The clinical potential of x-ray flat panel detectors for dynamic images B the evaluation of animal angiographic studies using a direct conversion type flat panel detector. *Toshiba Med. Rev.*, (73), 1–12.

(Rudin 2008) Rudin S., Bednarek D. R., and Hoffmann K. R. 2008. Endovascular image guided interventions (EIGI). Invited Vision 20/20 paper. *Med. Phys.*, 35(1), 301–309. http://www.pubmedcentral.gov/articlerender.fcgi?artid=2669303.

(Samei 2009) Samei E., Ranger N. T., MacKenzie A., Honey I. D., Dobbins J. T., and Ravin C. E. 2009. Effective DQE (eDQE) and speed of digital radiographic systems: An experimental methodology. *Med. Phys.*, 36, 3806.

(Schmidt 2000) Schmidt T. and Behling R. November 2000. MRC: A successful platform for future x-ray tube development MEDICA MUNDI 44/2, pp. 50–55. Philips Medical Systems Corp. http://www.healthcare.philips.com/main/about/News/Publications/MedicaMundi/v44200.wpd.

(Shan 2012) Shan J, Zhou O, and Lu J. Anode thermal analysis of high power micro-focus CNT X-ray tubes for in-vivo small animal imaging. 2012. SPIE, vol. 8313-23. In: *Proceedings from Medical Imaging 2012: Physics of Medical Imaging*, Orlando, FL, paper 83130O:1-9.

(Singh 2014a) Singh V., Jain A., Bednarek D. R., and Rudin S. 2014. Relative object detectability (ROD): A new metric for comparing x-ray detector performance for a specified imaging task. SPIE, vol. 9033-203. In: *Proceedings from Medical Imaging 2014: Physics of Medical Imaging*, San Diego, CA, paper 90335I:1-6. NIHMSID #619227. http://www.ncbi.nlm.nih.gov/pmc/articles/PMC4188352/.

(Singh 2014b) Singh V., Jain A., Bednarek D. R., and Rudin S. June 2014. Generalized relative object detectability (G-ROD) as a metric for comparing x-ray imaging systems (abstract). *Med. Phys.*, 41(6), June 2014, AAPM Annual Meeting Program, p. 136, SU-E-I-29.

(Speidel 2006) Speidel M. A., Wilfley B. P., Star-Lack J. M., Heanue J. A., and Van Lysel M. S. 2006. Scanning-beam digital x-ray (SBDX) technology for interventional and diagnostic cardiac angiography. *Med. Phys.*, 33, 2714.

(Takahashi 2006) Takahashi T. and Kurihara T. October 3, 2006. Diagnostic x-ray system. US Patent 7,116,752 B2.

(Tucker 1991) Tucker D. M., Barnes G. T., and Chakraborty D. P. 1991. Semiempirical model for generating tungsten target x-ray spectra. *Med. Phys.*, 18, 211–218.

(Umetani 2009) Umetani K., Uesugi K., Kobatake M., Yamamoto A., Yamashita T., and Imai S. 2009. Synchrotron radiation microimaging in rabbit models of cancer for preclinical testing. *Nucl. Instr. Methods Phys. Res.*, A609, 38–49.

(Vasan 2011) Vasan S. N. S., Sharma P., Ionita C. N., Titus A. H., Cartwright A. N., Bednarek D. R., and Rudin S. August 30–September 3, 2011. Spatially different real-time temporal filtering and dose reduction for dynamic image guidance during neurovascular interventions. In *Proceedings of IEEE-EMBS 2011 Annual Meeting*, Boston, MA, paper FrP25.3 (poster), X-Ray, CT, PET, SPECT. NIHMSID 391928. PBID 22255753.

(Vasan 2012a) Vasan S. N., Ionita C. N., Titus A. H., Cartwright A. N., Bednarek D. R., and Rudin S. 2012. Graphics processing unit (GPU) implementation of image processing algorithms to improve system performance of the control acquisition, processing and image display system (CAPIDS) of the micro-angiographic fluoroscope (MAF). SPIE, vol. 8313-159. In: *Proceedings from Medical Imaging 2012: Physics of Medical Imaging*, San Diego, CA, paper 83134C:1-8. NIHMSID 391955. http://www.ncbi.nlm.nih.gov/pmc/articles/PMC3767001/.

(Vasan 2012b) Vasan S. N., Panse A., Jain A., Sharma P., Ionita C. N., Titus A. H., Cartwright A. N., Bednarek D. R., and Rudin S. 2012. Dose reduction technique using a combination of a region of interest (ROI) material x-ray attenuator and spatially different temporal filtering for fluoroscopic interventions. SPIE vol. 8313-191. In: *Proceedings from Medical Imaging 2012: Physics of Medical Imaging*, San Diego, CA, paper 831357:1-11. NIHMSID 391952. http://www.ncbi.nlm.nih.gov/pmc/articles/PMC3766980/.

(Vasan 2013) Vasan S. N. S., Pope L., Ionita C. N., Titus A. H., Cartwright A. N., Bednarek D. R., and Rudin S. 2013. Dose reduction in fluoroscopic interventions using a combination of a region of interest (ROI) x-ray attenuator and spatially-different, temporally-variable temporal filtering. SPIE vol. 8668-142. In: *Proceedings from Medical Imaging 2013: Physics of Medical Imaging*, Orlando, FL, paper 86683Y:1-8. NIHMSID # 483436.

(Wang 2010) Wang W., Ionita C. N., Keleshis C., Kuhls-Gilcrist A. T., Jain A., Bednarek D. R., and Rudin S. 2010. Progress in the development of a new angiography suite including the high resolution micro-angiographic fluoroscope (MAF), a control, acquisition, processing, and image display system (CAPIDS), and a new detector changer integrated into a commercial C-Arm angiography unit to enable clinical use. SPIE vol. 7622. In: *Proceedings from Medical Imaging 2010: Physics of Medical Imaging*, San Diego CA, paper 7622-200, 76225I:1-10. https://www.pubmedcentral.gov/articlerender.fcgi?artid=3021378.

(Yadava 2008) Yadava G., Rudin S., Kuhls-Gilcrist A. T., Patel V., Hoffmann K. R., and Bednarek D. R. September 21, 2008. A practical x-ray exposure-equivalent metric for instrumentation noise in high-sensitivity and fluoroscopic x-ray imagers.

Phys. Med. Biol., 53(18), 5107–5121. PMCID2562256 http://www.pubmedcentral.nih.gov/articlerender.fcgi?tool=pubmed&pubmedid=18723932.

(Yamada 2000) Yamada S., Takahashi A., Umazaki H., Honda M., Shiraishi K., Rudin S., Yang C-Y., and Bednarek D. R. 2000. Image quality evaluation of a new selenium-based flat panel digital X-ray detector system based on animal studies. SPIE vol. 3977, pp. 429–436. In: *Proceedings from Medical Imaging 2000: Physics of Medical Imaging*, San Diego, CA. paper #46.

(Yang 1998) Yang C. J., Rudin S., and Bednarek D. R. 1998. Image acquisition and real-time processing in region of interest fluoroscopy with variation of source to image distance. SPIE vol. 3336, pp. 660–667. In: *Proceedings from Medical Imaging 1998: Physics of Medical Imaging*, San Diego, CA.

(Yang 1999) Yang C. J., Rudin S., and Bednarek D. R. 1999. Variable sampling area for automatic brightness control in digital fluoroscopy. SPIE vol. 3659, pp. 826–832. In: *Proceedings from Medical Imaging 1999: Physics of Medical Imaging*, San Diego, CA. paper #88.

Chapter 4

5. Physics and Technology of CT Angiography

Michael D. Silver

Computed tomography angiography (CTA) uses CT as a tool to assess the blood vessels and blood flow through organs seen with the aid of a contrast agent, also known as a dye, injected into the blood stream. Diagnostic CT scanners give exquisite 3D anatomic images of humans at spatial resolutions of the order of 0.5 mm and contrast differences of the order of 1% for diagnostic CT scanners. An example of a modern, high-end CT system is shown in Figure 5.1. A clinical goal of these scanners is to diagnose the vasculature, including the aorta and coronary and neurological arteries, for stenoses (blockages) and aneurysms (a bulge in a weakened vessel wall). Such diagnoses use the ability of a CT scanner to faithfully render the anatomy of the patient. Another clinical goal is to assess organ function mostly through dynamic scanning by measuring the blood flow through the organ and cardiac function. This helps the physician decide on the treatment for the affected organ, for example, deciding how aggressive to intervene in the case of a stroke. Still another clinical goal is to aid interventional radiologists to understand complicated vascular structure, especially in the brain, guide the catheter with its treatment devices such as stents and coils to the target, and to assess placement

Chapter 5

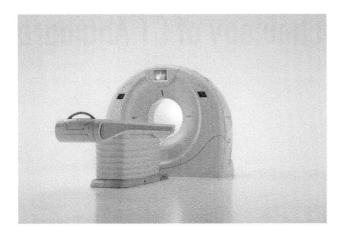

FIGURE 5.1   A high-end, modern diagnostic CT scanner.

of such devices and vessel function before completing the interventional procedure in as minimally invasive manner as possible with a minimal disruption of the patient workflow. Most interventions are done with the patient on the table of x-ray C-arm imaging system, which has the option of CT imaging. An example is shown in Figure 5.2; Chapters 10 and 22 cover this aspect of x-ray angiography. Spatial resolutions can be about three times finer for these C-arm-based cone-beam CT scanners than with diagnostic scanners although contrast discrimination is less. Readers interested in trade-offs between spatial and contrast resolutions along with dose considerations are directed to the literature, for example, Chapter 5 of Reference 1 and the references therein.

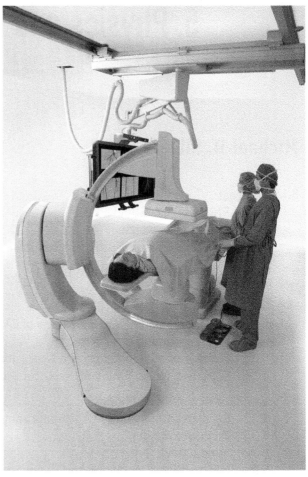

FIGURE 5.2   A C-arm, x-ray imaging system designed for interventional radiography and surgery with CT-capabilities.

5.1   Contrast

All imaging requires contrast between features of interest and the background. In computed tomography (CT) and x-ray imaging in general, contrast comes from the ability of different tissues (and objects) to absorb or scatter x-rays differently out of a narrow pencil beam of x-rays. The transmitted intensity, which is measured, tells us how much is left in the beam. We will return to the basic mathematics of what this means in the next section. In this section, we explain the physical mechanisms behind the absorption and scattering. In general, absorption and scattering remove x-rays from the beam; this is called attenuation. The property of the tissue or material to do this is called the *linear attenuation coefficient*, usually designated by the Greek letter μ and has units of inverse length such as cm^{-1}. A closely related concept is the *mass attenuation coefficient*, usually designated by *m*:

$$\mu(E) = \rho \times m(E) \qquad (5.1)$$

where
  ρ is the physical density of the material (g/cm^3)
  The unit for *m* is typically cm^2/g

Both the linear and mass attenuation coefficients are functions of the x-ray energy.

Several physical processes contribute to the attenuation coefficients. At medical diagnostic energies (from about 20 to 140 keV), the two most important processes are

1. Photoelectric absorption
2. Compton scatter

The attenuation coefficient of an arbitrary material is the sum of these two processes:

$$\mu = \mu_P + \mu_C = \rho \times m = \rho \times (m_P + m_C), \qquad (5.2)$$

where the subscripts stand for the two processes and each has its own dependence on energy. If the elemental formula is known for the material such as H_2O, the mass attenuation coefficient of water can be found by

$$m = \frac{\sum_i N_i A_i m_i}{\sum_k N_k A_k},$$

(5.3)

where the sum is over the elements and N_i and A_i are the number of atoms of the ith element and its atomic weight. If the fractional weight of a tissue is known, such as fat with 12% hydrogen, 77% carbon, and 11% oxygen [2,3], then the mass attenuation coefficient is given by

$$m = \sum_i w_i m_i$$

(5.4)

with w_i the fractional weight of element i or one of several composite materials that make up the tissue. Values for m_i for the elements and composite materials can be found or calculated on the NIST website [4].

CT reconstructs maps of the linear attenuation coefficients. The CT numbers in the CT image are given by the Hounsfield scale (HU) and related to the attenuation coefficient by

$$\text{CT-Number}(\text{HU}) = 1000\frac{\mu - \mu_{H_2O}}{\mu_{H_2O}}$$

(5.5)

where
 μ_{H_2O} is the linear attenuation of water
 μ is the linear attenuation of some locale in the CT image

In angiography, the μ of interest is often an iodinated contrast agent that has been injected into the patient and diluted by the blood. There are some problems with the HU, which we will return to in Section 5.4. Before we look at the attenuation curves as a function of energy, let us understand the two processes that make up the attenuation coefficient.

5.1.1 Photoelectric Effect

Photoelectric effect is the absorption of an x-ray by an atom with the ejection of a bound electron carrying away the energy leftover after overcoming the binding energy of the electron (see Figure 5.3) [5]. Even if the same or other electrons cascade back down to fill

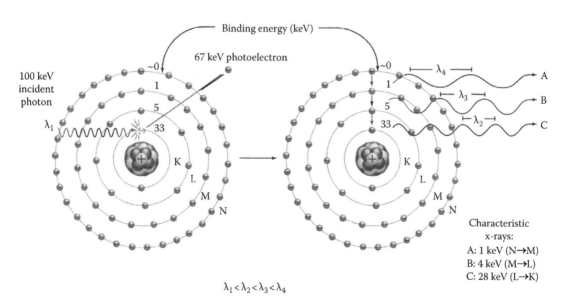

FIGURE 5.3 Photoelectric absorption. Left: A 100 keV incident x-ray undergoes photoelectric absorption by ejecting a K-shell electron and thus ionizing the atom. Because the binding energy of a K-shell electron for an iodine atom is 33 keV, the photoelectron is emitted with the kinetic energy of the difference, 67 keV for this example. Right: The vacancy in the K-shell causes a cascade of x-rays with energies that correspond to the different atomic shells, as shown in the figure. It is unlikely that these secondary x-rays contribute to the transmitted beam both because they are emitted equally likely in all directions and these characteristic x-rays have a much lower energy and are thus highly likely to undergo further absorptions. (From Bushberg, J.T. et al., *The Essential Physics of Medical Imaging*, 3rd edn., Lippincott Williams & Wilkins, 2012;Figure 3.9.)

the missing electron, the x-ray energies emitted tend to have much less energy than the incident x-ray and are emitted equally likely in any direction, thus no longer along the original transmitted beam. Because of the shell structure of the atom [6], at certain energy thresholds (e.g., K-shell), new interaction channels become possible and the attenuation coefficient takes an abrupt jump as seen in Figure 5.3 for iodine. Even above the K-edge, the photoelectric absorption has a complex dependence on atomic number Z and energy E.

Some texts [7] reduce these dependences to this simple power law for the mass attenuation coefficient

$$m_P \propto \frac{Z^3}{E^3}. \tag{5.6}$$

The strong Z-dependence is even stronger when one considers the linear attenuation coefficient with an extra power of Z from the density. As we see from Figure 5.4, the strong difference of attenuation coefficient from

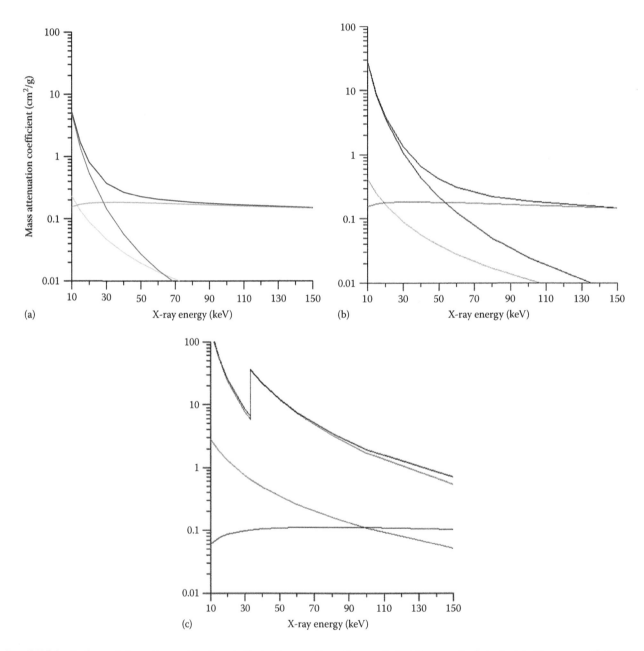

FIGURE 5.4   Each panel shows the contributions to the total mass attenuation coefficient (top curve): photoelectric (the curve with the biggest contribution at low energies), Compton (the curve with flattest energy dependence), and Rayleigh (the curve with lightest shade of gray). Panel (a) is for water, panel (b) for cortical bone, and panel (c) is for iodine. Multiply by their densities to get the linear attenuation coefficients. All three panels are shown to the same scale to make comparisons visually direct. Notice how similar the Compton contributions are.

soft tissue (water-like) and a vascular contrast agent based on iodine. This is what drives the contrast in angiography, more so at lower energies where the photoelectric effect dominates over scattering.

5.1.2 Compton Scattering

Compton scattering, also called incoherent scatter, is a loss of energy of the incident x-ray while being deflected by electrons in the atom; the greater the deflection, the more the energy loss as given by

$$E' = \frac{E_0}{1 + \gamma(1 - \cos\theta)}, \tag{5.7}$$

where

E' is the energy of the scattered x-ray

E_0 is its incident energy

$\gamma = E_0/m_e c^2$ where the denominator is the rest energy of the electron, 511 keV

θ is the scattering angle

The Compton interaction is shown in Figure 5.5 and the scattered x-ray energy in Figure 5.6. If the scattering is considered as off free electrons (a reasonable

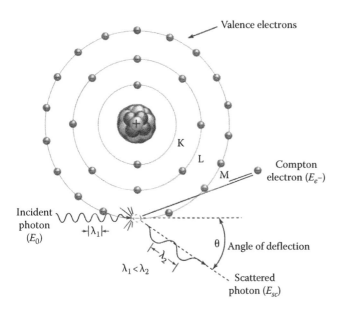

FIGURE 5.5 Compton scattering. The incident x-ray, energy E_0 (or wavelength 11), interacts with a valence shell electron knocking it free of the atom and thus ionizing the atom. The incident x-ray is scattered with an energy loss related to the scattering angle as given by Equation 5.7 in the text. (From Bushberg, J.T. et al., *The Essential Physics of Medical Imaging*, 3rd edn., Lippincott Williams & Wilkins, 2012; Figure 3.7.)

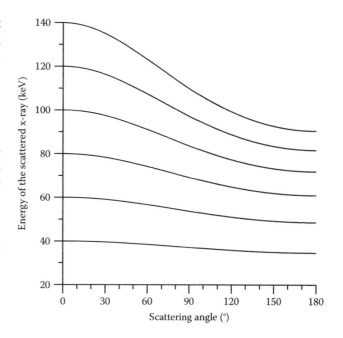

FIGURE 5.6 Energy of scattered x-rays as a function of scattering angle. From top to bottom: 140, 120, 100, 80, 60, and 40 keV incident x-ray energy.

approximation), then the Klein–Nishina formula applies. Its energy and atomic number dependence go as [8]

$$m_C \propto \left(1 - 2\gamma + \tfrac{26}{5}\gamma^2 + \cdots\right) \quad \gamma \ll 1. \tag{5.8}$$

Remember that in medical imaging, the x-ray energy is below 140 keV so that $\gamma \ll 1$ is weakly satisfied. Note that there is no Z-dependence and the energy dependence is weak. A linear Z-dependence comes in from the density when we look at the linear attenuation coefficient. The Z-dependence and energy dependence differences between photoelectric and Compton are exploited in spectral imaging (see Chapter 12).

5.1.3 Relationship of Z-Number, Attenuation, Contrast, and Signal-to-Noise Ratio

Because Z is so important to the attenuation coefficient, the reader might ask how to define an effective Z, Z_{eff}, for a compound. This turns out to have a rather complicated answer [9] but a commonly used approximation uses the following model:

$$Z_{eff}^{2.94} = \sum_i \left(\frac{N_i Z_i^{2.94}}{\sum_k N_k} \right). \tag{5.9}$$

The reader might suggest using a higher Z-based contrast agent given the strong dependence of contrast on Z_{eff}. However, heavy metals are usually toxic and require chelating for safety. And this text is interested in vascular imaging and so the agent should stay within the blood stream. Nonetheless, experimental, high Z agents are being developed and tested, for example, gadolinium (Gd, Z = 64, or gold nanoparticles, Au, Z = 79), although their use might not be restricted to vascular imaging [10].

We have discussed the need for contrast in medical imaging and how iodine in the contrast agent in angiographic imaging is used to enhance that contrast in the vasculature. We also discussed that attenuation is a reduction of the transmitted signal due to absorption and scattering. Is there a level of transmission that is best; for example, is tuning the scan parameters to get 50% transmission optimal? This is a question of signal-to-noise ratio (*SNR*). Because we need a little of the formalism of CT imaging to best answer this question, we shift to that formalism by asking the question, How do we measure contrast in CT in the next section before returning to *SNR* in Section 5.2.6.

5.2 How Do We Measure Contrast in CT?

5.2.1 Beer's Law

Consider a narrow beam of x-rays. As the x-rays pass through a thin slab of material, some are absorbed, some scattered, and the rest transmitted. The more x-rays in the beam, the more some are absorbed and scattered. Thus, the loss of intensity in the beam should be proportional to the strength of the beam where the proportionality constant is a property of the material of the slab as given by

$$-\frac{dI}{dx} = I\mu, \tag{5.10}$$

where we put the minus in explicitly because this represents a loss of x-rays from the beam and the linear attenuation coefficient μ is the proportionality constant. This integrates to

$$I = I_0 e^{-\mu t}, \tag{5.11}$$

where I_0 is the incident beam intensity before entering the material with linear attenuation coefficient μ of thickness *t*. Equation 5.11 is known as Beer's law [11] and it represents a very simple model of x-ray transmission through matter. More accurately, we have to account for the fact that the incident beam is polychromatic up to the potential difference between the anode and the cathode of the x-ray tube. Some examples are shown in Figure 5.7 [12]. Also recall that the attenuation coefficients are energy dependent. Finally, any transmitted path for a narrow beam of x-rays could pass through many tissues.

Thus, Equation 5.11 generalizes to a more physically correct equation:

$$I(\ell) = \int_0^{kV} S(E) \exp\left[-\int_\ell \mu(E,x,y,z) d\ell \right] \times \left\{ 1 - \exp\left[-\mu_d(E) t_d \right] \right\} E dE \tag{5.12}$$

where
$I(\ell)$ is the intensity absorbed in the detector
ℓ is the integration path (a line integral)

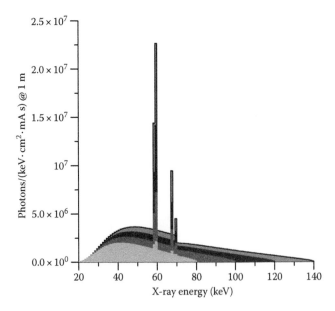

FIGURE 5.7 Output spectra for an x-ray tube with a 7° anode, 1 mm Be window, and 5 mm Al filtration at 140 (gray), 120 (black), 100 (dark gray), and 80 (light gray) kV across the tube. The areas under each spectrum are the photon counts at 1 m/mA/s for a 1 cm² cross section at the center of the beam. Note that the higher the operating voltage, the more efficient the conversion to x-rays.

$S(E)$ is the spectrum of the x-ray tube with voltage kV; the term in curly brackets represents the stopping efficiency of the detector material with linear attenuation coefficient μ_d and thickness t_d, and the extra factor of E is because in clinical CT systems currently on the market, whether diagnostic or C-arm, the detectors integrate over the energy deposited in each sensor element. In the future, photon counting detectors might become practical and then this factor of E should be stricken from Equation 5.12.

5.2.2   Radon Transform

In CT imaging, we want a map of $\mu(x,y,z)$. Let us stay simple with a monochromatic x-ray source of energy $\bar{E}$ and a perfect detector. Then

$$I(\ell) = I_0 \exp\left[-\int_\ell \mu(\bar{E}, x, y, z)d\ell\right]. \qquad (5.13)$$

By taking the logarithm, we can isolate the line integral:

$$p(\ell) = \ln\frac{I_0}{I(\ell)} = \int_\ell \mu(\bar{E}, x, y, z)d\ell, \qquad (5.14)$$

where p is known as the projection or ray-sum along path ℓ. A simple example of this equation is shown in Figure 5.8 for a given orientation of parallel paths ℓ.

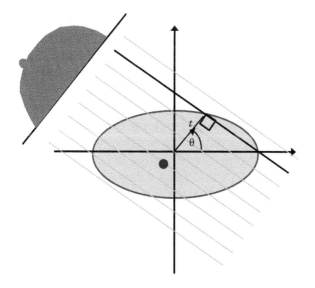

FIGURE 5.8   Parallel-ray geometry can be achieved with either first- or second-generation CT designs. Here each parallel family of ray-sums is labeled by θ and each ray-sum within the family, often called a view, labeled by t. The upper left of the drawing shows the projection values for this view with a high-density inclusion added in the object.

The solution of how to invert the line integral results (in a plane) to get at the integrand $\mu(\bar{E}, x, y)$ was found by Czech, German, and/or Austrian (all three claim him as a native son) mathematician Johan Radon in 1917 [13] and rediscovered by co–Nobel Prize winner Allan Cormack in 1963 [14]. Equations of the form of Equation 5.14 are called the Radon transform when in two dimensions $(\mu(\bar{E}, x, y)$ with ℓ restricted to the xy-plane$)$. Equations of the form of Equation 5.13 are called the x-ray transform.

5.2.3   Radon Space and a Brief History of CT [15]

Radon space is the collection of all projections (or ray-sums or line integrals in 2D). In planar CT, this is also known as the sinogram because each voxel of the scanned object (think of a small feature) traces a sinusoidal trajectory through Radon space as shown in Figure 5.9. In higher dimensions, Radon space is one less dimension than the object being reconstruction. Thus in 3D imaging, it is the collection of all planar integrals through the object. CT systems have evolved through several generations of design since the first clinical prototype scanner (by EMI) in 1971 [16]. The first commercial medical CT scanner (by EMI) that went on sale in 1973 was a two-slice scanner of first-generation design [17]. All current clinical CT scanners, whether diagnostic or C-arm, are of third-generation design, although from the mid-1970s through the early 1980s, second-generation models were common and, from the late 1970s through the end of the 1990s, fourth-generation scanners were offered by some equipment vendors [18]. Acquisition geometries are shown in Figure 5.10. Two other CT designs are worthy of notice because of their role in opening up the clinical possibilities of cardiac scanning:

1. The dynamic spatial reconstructor developed at Mayo Clinic funded by American taxpayers was designed with 28 imaging chains (although perhaps only operational with 14) consisting of pairs of x-ray tubes and opposing x-ray cameras arrayed along the ring of the gantry so that a quick, small motion of the gantry gave complete angular coverage of the cardiac region [19].
2. Electron-beam CT due to Douglas Boyd of University of California, San Francisco (UCSF) and Imatron, sometimes called a fifth-generation design, was used from the mid-1980s into the early 2000s.

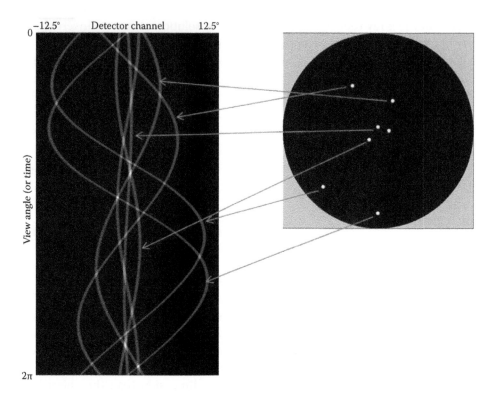

FIGURE 5.9 This figure with the sinogram, left, of a "bed-of-nails" phantom, right, shows the relation between image space and data (sinogram) space via the Radon transform. Because third-generation geometry (diverging fan beam) was used in this simulation, the traces in data space are slightly distorted from a true sinusoidal. If first- or second-generation geometry (parallel beam) were used, then the traces would be exactly sinusoidal.

It had no moving parts as an electron gun in a large vacuum shell electromagnetically steered the electrons to hit strips of tungsten formed as an arc to produce x-rays opposite arcs of detector elements to give temporal resolutions down to 50 ms [20].

Although the first CT system had two slices (i.e., two contiguous imaging planes through the patient by two detector elements stacked atop of one another), all other clinical scanners were single slice until Elscint introduced a two-slice scanner in the mid-1990s. GE's introduction of a four-slice scanner in late 1998 led to the "slice wars" among the vendors leading to Toshiba's 320-slice scanner in 2007. Currently for diagnostic CT, the market is supporting 16-slice and above CT systems (see Figure 5.11) [21]. Slice number refers to the number of rows of detector elements in the 2D array of x-ray-sensing elements, although marketing departments of vendors with a longitudinal flying focal spot cite double the number of detector rows for the slice count.*

Good coverage of multislice CT can be found in the books by Kalender [22] and Mahesh [23]. C-arm CT systems were experimented with early in the history of CT using an x-ray image intensifier with camera as the x-ray detector [24]. C-arm CT development took off with the introduction of flat panel detectors [25]. These come in a variety of sizes and readout configurations (see Chapter 10).

The development of slip ring technology by Varian and EMI (then EMI spin-off Bio-Imaging Research) introduced continuous rotation of the CT gantry, fast rotations (current record holder: a Philips' gantry at 0.27 s), and large volume scanning by helical acquisition (table translates while the gantry is spinning) within a few seconds [26–28]. It was these parallel developments of multislice and helical CT made possible by slip ring technology that open the door to angiographic imaging by diagnostic CT. On the other hand, C-arms have winding cables—not slip rings—and, because they are open to medical personnel attending to the patient, with restrictive speeds. Depending on vendor and model, C-arm gantries can cover from 200° to over 360° before needing to unwind with speeds typically less than 60°/s.

---

* A flying spot means that the focal spot on the anode of the x-ray tube rapidly oscillates on the order of a few 1000 Hz. The purpose is to increase sampling density—to satisfy the Nyquist criterion—and thus reduce aliasing artifacts (see Section 5.4).

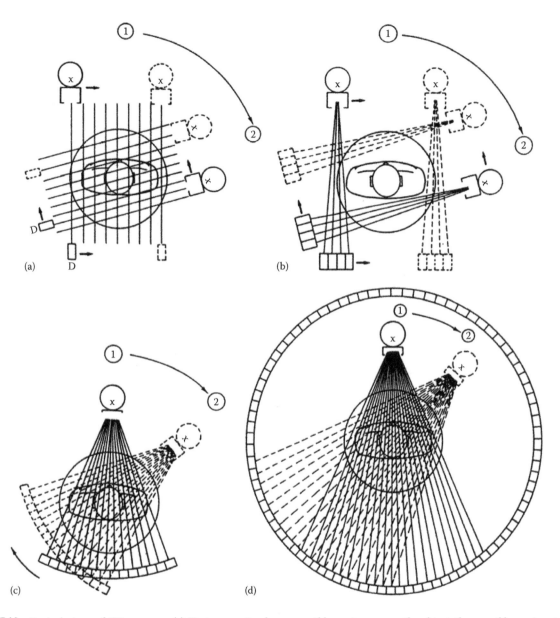

FIGURE 5.10 Basic designs of CT scanners. (a) First generation has a pencil beam transverse the object; the pencil beam is rotated for another pass. Repeat until sufficient data are collected. This is an example of translate–rotate, parallel beam geometry. (b) Second generation is another example of translate–rotate, parallel beam geometry. The number of rotations is 180° divided by the fan angle formed by the detector array. (c) Third generation is also known as rotate–rotate, fan beam geometry, as both the x-ray source and detector rotate together around the object. The detector array is large enough so that the fan encompasses the object. Shown here is a float detector such as found on C-arm gantries. Diagnostic scanners have the detector elements on an arc whose radius is the source-to-detector distance. (d) Fourth generation is known as stationary–rotate. The x-ray source rotates within a stationary ring of detectors. This can be shown to also be of fan beam geometry. An alternative design placed the detector ring inside the orbit of the x-ray tube where the detector ring nutated to clear the fan beam prepatient.

5.2.4   Inverse Radon Transform

There are several mathematically equivalent formulations of the inverse Radon transform, the inversion of Equation 5.14. Most commercial CT systems use a solution based on what is commonly called convolution backprojection or filtered backprojection (FBP).

This section provides a sense of where FBP comes from and then provides a sense of what it means. We derive FBP from the Fourier slice theorem [29], which demonstrates the deep connection between Radon transforms and Fourier transforms, as emphasized in Dean's book [13]. Take the parallel-ray projection geometry (Figure 5.8), where each ray-sum is denoted by θ,t.

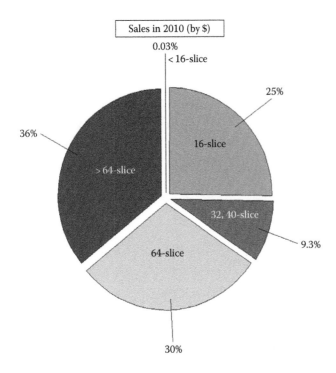

0.03%
< 16-slice

25%

16-slice

36%

> 64-slice

32, 40-slice

64-slice

9.3%

30%

FIGURE 5.11 Sales of diagnostic CT systems in the United States in 2010. The pie size is by dollars: the more slices, the more expensive the system. (Courtesy of Medical Imaging and Technology Alliance [MITA], a division of the National Electrical Manufacturers Association, Arlington, VA.)

θ represents the projection angle of a family of ray-sums indexed by t. Then the Fourier slice theorem (2D version) states that the 1D Fourier transform of a projection for fixed θ is equal to the 2D Fourier transform of our object function (the linear attenuation map $\mu(x,y)$) evaluated along the projection angle θ (see Figure 5.12

and Section 3.2 of Reference 30 or Section 3.3 of Reference 31 for detailed proof):

$$S(\omega,\theta)$$

$$= \int p(\theta,t)e^{-2\pi i\omega t}dt$$

(1D Fourier transform of a projection at fixed θ

$$= \iint \mu(x,y)e^{-2\pi i\omega(x\cos\theta + y\sin\theta)}dxdy$$

(2D Fourier transform of object function),

(5.15)

where ω is the spatial frequency associated with $t, i = \sqrt{-1}$.

To get to FBP, express the object function as a 2D inverse Fourier transform

$$\mu(x,y) = \iint \tilde{\mu}(u,v)e^{2\pi i(ux+vy)}dudv. \qquad (5.16)$$

Substitute

$$u = \omega\cos\theta$$

$$v = \omega\sin\theta$$

$$dudv = \omega d\omega d\theta$$

(5.17)

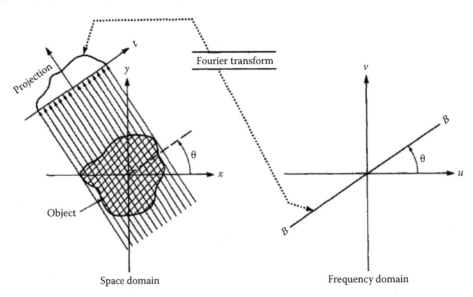

FIGURE 5.12 The Fourier slice theorem relates the Fourier transform of a projection to the Fourier transform of the object along a radial line. (From Pan, S.X. and Kak, A.C., *IEEE Trans. Acoust. Speech Signal Processing*, 31, 1262, 1983.)

and use symmetry properties to show [30,31]

$$\mu(x,y) = \int_0^\pi \left[\int_{-\infty}^\infty \tilde{\mu}(\omega,\theta)|\omega|e^{2\pi i\omega t}\,d\omega \right] d\theta. \quad (5.18)$$

And then apply the Fourier slice theorem to arrive at

$$\mu(x,y) = \int_0^\pi \left\{ \int_{-\infty}^\infty \left[p(\theta,t')e^{-2\pi i\omega t'} \right]|\omega|e^{2\pi i\omega t}\,d\omega \right\} d\theta. \quad (5.19)$$

The integral in the curly brackets {…} is an example of the convolution theorem, where $|\omega|$ is the 1D Fourier transform of some spatial domain filter; that is, the spatial domain (Figure 5.13) version of the ramp function $|\omega|$ is convolved with the projection $p(\theta,t)$ for each θ. The outer integral over θ is backprojection, hence FBP. Figures 5.14 and 5.15 (adapted from Deans [13]) summarize and underscore the tight connection between the Radon transform pair and the Fourier transform pair. More formal mathematical presentations can be found in the books by Natterer [31,34].

In clinical practice, the data are not parallel beam but rather diverging fan beam (2D) or cone beam (3D).

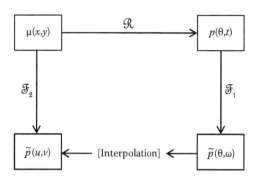

FIGURE 5.14 Another view of the Fourier slice theorem. $\mathcal{R}$ is the Radon transform, $\mathcal{F}_1$ is the 1D Fourier transform, and $\mathcal{F}_2$ is the 2D Fourier transform. The interpolation is from polar coordinates to Cartesian coordinates (in the frequency domain) using the first two equations of [17]. (Adapted from Deans, S.R., *The Radon Transform and Some of Its Applications*, Wiley-Interscience, 1983, Figure 6.1.)

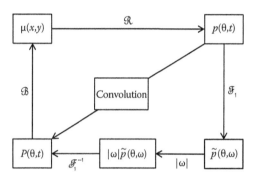

FIGURE 5.15 Diagram of FBP. The lower right–hand triangle is the convolution theorem. The CT scanner plus logarithmic conversion step does $\mathcal{R}$. Going around the rest of the perimeter does the inverse Radon transform, that is FBP. The last step $\mathcal{B}$ is backprojection. (Adapted from Deans, S.R., *The Radon Transform and Some of its Applications*, Wiley-Interscience, 1983, Figure 6.6.)

Either the acquired data can be resampled to parallel beam or a change from Cartesian to polar coordinates on Equation 5.19 can be applied to give similar equations for these geometries [30,34]; they are still of FBP structure with a ramp filter but include additional factors due to the diverging geometry. Also in clinical practice, the pure ramp filter $|\omega|$ is rarely used. Rather, modifications are made to suppress some frequencies and emphasize others. Because stochastic noise in CT tends to be blue, often the ramp is rolled off at higher frequencies to suppress the noise especially for thick and large body cross sections such as adult abdomens. On the other hand, angiographic vascular imaging of the lungs uses a filter that boosts mid-to-high frequencies to emphasize the fine blood vessels (Figure 5.16).

Let us return to backprojection. Earlier in Equations 5.12 through 5.14, we used ℓ to represent a ray-sum.

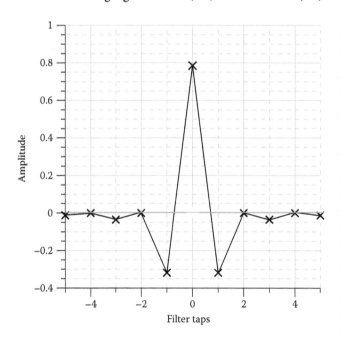

FIGURE 5.13 The center few taps of the spatial domain ramp kernel [32]. The kernel points continue to ±Nyquist frequency. Only the center tap is positive; the other even values are 0 while the odd values are all negative whose envelope falls as one over square of the sampling pitch within a projection. Other kernel designs, such as that due to Shepp–Logan [33], do not have this odd/even oscillation. The sum of the kernel values nearly equals zero; see text.

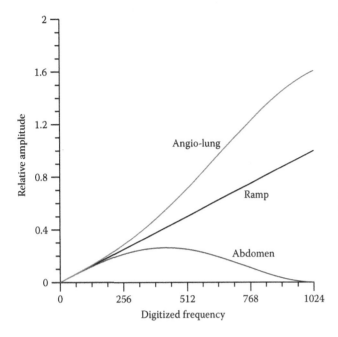

FIGURE 5.16 Two of the three reconstruction kernels are examples of tuning the kernel for the clinical application. The third is the theoretical ramp kernel shown for comparison. The kernel marked angio-lung has a mid-to-high frequency boost to bring out better the small vessels in the lung, which are high-contrast features above a very-low-density background. The kernel marked abdomen has a high-frequency rollover to suppress noise to bring out the low-contrast tumors in the liver and other organs in the abdomen. The kernels are shown in the frequency domain. Note that all kernels are the same at low frequencies in order to preserve the mean values of the reconstructed organs and tissues. These kernels are not exact representations of any vendors' kernels.

In 2D, parallel beam geometry, ℓ, is parameterized by θ,t. Therefore, we can write Equation 5.14 as

$$p(\theta,t) = \int \mu(\overline{E},x,y)\delta(x\cos\theta + y\sin\theta - t)dxdy. \quad (5.20)$$

The delta function expresses the relation between the Cartesian coordinates of the object function x,y and the coordinates of the measured data θ,t. Thus, as we calculate the contribution to the image $\mu(x,y)$ at x,y from projection θ, we use the delta function to tell us which ray-sum to backproject, namely, the one at

$$t = x\cos\theta + y\sin\theta. \quad (5.21)$$

The image is built up by looping through all x,y and summing the contributions from the ray-sum satisfying (5.21) from each projected view θ. If we ignore the convolution step, then backprojection gives a blurred result, as shown in Figure 5.17 for simple pin objects. The ramp filter is a type of high-pass filter that gives just the negative slide lobs needed to remove the blur when all views are integrated. For readers who like to think in the frequency domain, for the parallel geometry case, each integral of the ray-sums at any fixed θ is the same; it represents the DC component of the data. Because there are an infinite number of views (here we are still using continuous mathematics rather than discrete samples of the real CT system), the DC component is severely oversampled.

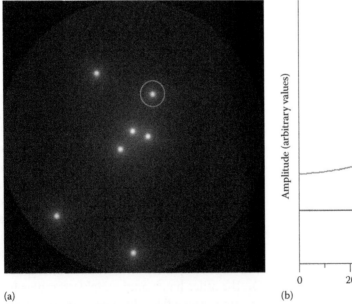

(a)

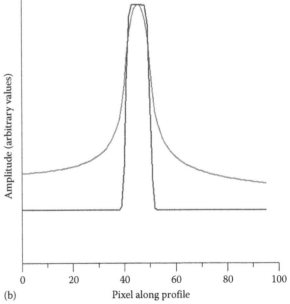

(b)

FIGURE 5.17 (a) Bed-of-nails phantom of Figure 5.9 with backprojection only, no convolution. Notice how blurry the "nails" appear. (b) Profiles through the circled "nail" comparing with (rectangular profile) and without (smoother profile) convolution. Some of the contribution to the red profile is from the other six "nails." Convolution with a ramp-like kernel is required to recover the object.

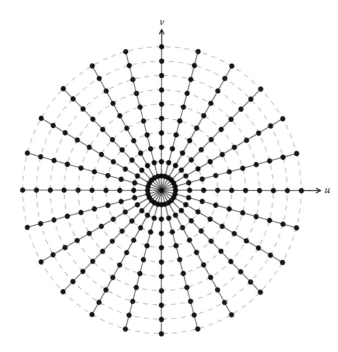

FIGURE 5.18 The lower right-hand corners of Figures 5.14 and 5.15 are the Fourier transforms of the projection data. This figure represents how it is sampled (when we leave the continuous mathematical model for discrete sampling) along radial lines. Each line samples DC while the high frequencies are sparsely sampled. Furthermore, to apply direct Fourier reconstruction (using the Fourier slice theorem in Figure 5.12), we must use interpolation to be able to use FFTs. The interpolation is the weakness of this approach yielding high-frequency artifacts, which is why FBP is the preferred approach in CT.

The Fourier slice theorem (Figure 5.12) implies that the high-frequency components of the Radon space are undersampled as shown in Figure 5.18. Hence, a ramp filter brings into balance the low- and high-frequency components of the data. Once we discretize and band-limit the CT system, the Gordian knot of infinite number of views times a zero for the DC from the ramp filter is cut to give the appropriate average value for the image (Figure 3.16 in Reference 30 and in Reference 1, especially Figure 18 and surrounding discussion).

When we go to 3D with a circular acquisition (stationary table), the most common reconstruction methods are variations of the Feldkamp algorithm [35]. This treats the filtering step as a set of 1D convolutions, typically along the detector rows, followed by true 3D backprojection. For helical scanning, if the nominal aperture (detector row pitch multiplied by the number of detector rows) is small, extensions of the single-slice helical reconstruction algorithm [36] work very well [37,38]. Once the scanners started to have eight or more slices, helical extensions of the Feldkamp algorithm were introduced [39,40]. These families of solutions are not mathematically exact but often good enough in clinical practice. An exact solution to helical multislice geometry was derived by Katsevich [41], where helical pitch is the ratio of the table travel to the full-detector aperture projected at system center. The long list of references in Chapter 10 of Reference 1 gives an idea of the many variations of reconstruction methods.

Other solutions to image reconstruction from its projections exist, some mathematically equivalent and others that converge to the nominal solution. Examples of the first case include reconstruction by direct Fourier methods as indicated by Figure 5.14, and Hilbert filtering methods where differentiation between views is followed by filtering with Hilbert kernels (instead of the ramp) and then backprojection [42,43]. Backprojection can even go before the filtering with a different type of filtering as the second step [13,44]. Recently, iterative reconstruction methods, used in the early days of CT and for the past decade or two in PET and SPECT, have been reintroduced to CT [45–47]. The basic idea is to reproject through an estimate of the image, compare with measured data, and update the image estimate. Repeat until convergence or image appears clinically useful. Often this is cast as an optimization problem to minimize the difference with a penalty term added to suppress any noise buildup while maintaining features and image sharpness. Another way of looking at iterative reconstruction is as a blind deconvolution problem with a spatially variant blur function. The penalty term is needed to counter the noise amplification caused by the deconvolution process. The advocates of iterative reconstruction—and there are as many variants as there are researchers on iterative reconstruction—argue that the noise/spatial resolution trade-off common with linear methods can be torn asunder with good noise suppression without sacrificing and perhaps improving small image features, low-contrast targets, and sharpness. The evidence is still out whether only iterative approaches can achieve these worthy goals. Noise suppression can be equated with the potential for dose reduction. A good introduction to dose reduction in CT can be found in a special report from the Summit on Management of Radiation Dose in CT held in February 2011 [48].

5.2.5 Completeness Condition and Acquisition Modes

The completeness condition in 2D requires that any line (ray-sum) through the region of interest must intercept the source (x-ray tube) trajectory at least once. If this is not true, the data are incomplete and exact reconstruction is impossible (but possibly good enough if the violation is small). If a ray-sum is measured more than once, then the redundancy must be taken into account [49,50]. A simple example is 360° acquisition. Every ray-sum is measured twice, whether parallel beam or fan beam; hence, redundancy weights are all ½. The completeness requirement can be recast as every point in the reconstructable image must be irradiated by at least a 180° transit of the source as seen from that point.

A popular acquisition mode in both diagnostic and interventional CT is half scan, due to the need for temporal resolution in CTA for the former and mechanical restrictions of gantry motion for the latter. Figure 5.19 shows the reconstructionable region for a detector array that spans 2Γ in third-generation, fan beam geometry. The trajectory of the source must cover at least $\pi + 2\Gamma$ (180° + fan angle) so that the image voxel at A is irradiated through 180°; however, all other image voxels get contributions from ray-sums from over more than 180° but less than 180° + 2Γ but, as seen from each voxel in the image, requiring redundancy weighting [49,50]. Displayed as a sinogram (Figure 5.20), there are pairs of ray-sums that measure the same path:

$$p(\beta,\gamma) = p(\beta + \pi \pm 2\gamma, -\gamma), \qquad (5.22)$$

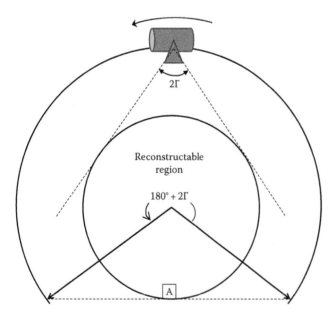

FIGURE 5.19 Half-scan geometry. The outer arc is the trajectory of the x-ray tube. For the voxel at position A in the reconstructable region to receive radiation from at least 180° from its point of view, the source must travel 180° plus full fan angle. But now all other voxels get redundant information.

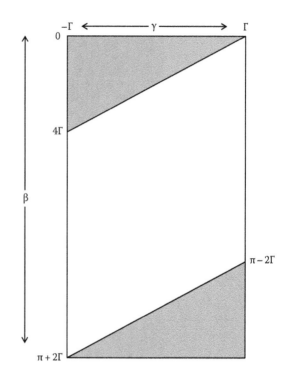

FIGURE 5.20 Half-scan sinogram. The areas in the two-shaded triangles are reflections of each other as given by Equation 22. The weights for each pair of redundant data must sum to 1; the weights in the white region are all 1. The transition of the weights must be smooth. β is angular position of the source and γ is the angle between a detector element and the line from the source through the isocenter of the gantry. Thus, γ labels the ray-sums for a given source position (β or view) with $\gamma = 0$ called the central ray.

where the ± ambiguity is related to how the positive directions are defined for β and γ. Derivations of examples for the redundancy weights are given in Refs. [36,49] (Figure 5.20).

The completeness condition in 3D requires that any plane through the region of interest must cut the source trajectory at least once [51,52]. Notice that for a circular trajectory, Radon space is not complete (Figure 5.21). Take the plane of the trajectory: all planes parallel to the trajectory plane but are not the trajectory plane fail to intersect the trajectory plane. The further the plane is off the trajectory plane, the more severe the missing data problem, which can lead to artifacts (see Section 5.4). It is the completeness condition that limits the helical pitch for multislice helical acquisitions [54].

5.2.6 Signal-to-Noise Ratio and Protocol Optimization

We now return to the question poised at the end of Section 1.4: what percentage transmission maximizes signal-to-noise? We take a very simple model of a

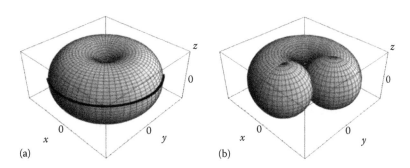

FIGURE 5.21 (a) 3D Radon space for a circular, full-scan cone-beam acquisition. The doughnut "hole" represents missing data. Note that the "hole" does not go all the way through the center. The equatorial plane represents the complete data for fan beam (2D) acquisitions. As we try to reconstruct image slices further and further off the equatorial plane, there is more and more missing data and thus increased likelihood of artifact. (b) 3D Radon space for a circular, half-scan cone-beam acquisition.

monochromatic x-ray source with a water cylinder representing the patient and the mean for all ray-sums are the same assuming an appropriate bow-tie filter.* The signal in the image is the linear attenuation coefficient $\mu(x,y,z)$. It is reconstructed from its projection data by

$$\tilde{\mu} = \Re^{-1}\tilde{p}, \tag{5.23}$$

where

The tilde indicates that these are random variables
$\Re^{-1}$ is the reconstruction operator, the inverse Radon transform

The variance of the signal is

$$\mathrm{Var}\{\tilde{\mu}\} = \left[\Re^{-1}\right]^2 \mathrm{Var}\{\tilde{p}\}. \tag{5.24}$$

Because $\Re^{-1}$ is a linear operator. Using Equation 5.14, the variance of the projection data is given by

$$\mathrm{Var}\{\tilde{p}\} = \frac{1}{\bar{I}^2}\mathrm{Var}\{\tilde{I}\} + \frac{1}{\bar{I}_0^2}\mathrm{Var}\{\tilde{I}_0\}, \tag{5.25}$$

where the bar indicates the mean values of $\tilde{I}, \tilde{I}_0$. The generation of x-rays is a stochastic process, which implies that x-rays obey Poisson statistics:

$$\mathrm{Var}\{\tilde{I}\} = \bar{I}, \quad \mathrm{Var}\{\tilde{I}_0\} = \bar{I}_0. \tag{5.26}$$

The unattenuated photon count is much higher than the attenuated count so that $\bar{I}_0 \gg \bar{I}$, then Equation 5.25 reduces to $\mathrm{Var}\{\tilde{p}\} \approx 1/I$ and Equation 5.24 becomes

$$\mathrm{Var}\{\tilde{\mu}\} = \left[\Re^{-1}\right]^2 \frac{1}{\bar{I}}. \tag{5.27}$$

The square of the *signal-to-noise* is

$$\mathrm{SNR}^2 \equiv \frac{\bar{\mu}^2}{\mathrm{Var}\{\tilde{\mu}\}} = \frac{\bar{\mu}^2}{\left[\Re^{-1}\right]^2 \frac{1}{\bar{I}}} = \frac{\bar{\mu}^2 \bar{I}}{\left[\Re^{-1}\right]^2}. \tag{5.28}$$

We want to maximize the SNR as a function of attenuation represented by μt, where t is the object's characteristic size (cylinder's diameter):

$$\frac{d\mathrm{SNR}^2}{d\mu t} = \frac{d\mathrm{SNR}^2}{d\mu}\frac{d\mu}{d\mu t} = \frac{1}{t}\frac{d\mathrm{SNR}^2}{d\mu} = 0. \tag{5.29}$$

The last step is valid because μ and t are not related: μ is a local property of the "patient" and t its overall size. Substituting in Equation 5.28 and dropping the constants

$$\frac{1}{t\left[\Re^{-1}\right]^2}\frac{d\left(\bar{\mu}^2\bar{I}\right)}{d\mu} = 0, \tag{5.30}$$

which leads to

$$2\bar{\mu}\bar{I} + \bar{\mu}^2\frac{d\bar{I}}{d\mu} = 0,$$

and using $\bar{I} = \bar{I}_0 e^{-\bar{\mu}t}$ and doing the derivative, we find that

$$\bar{\mu}t \approx 2, \tag{5.31}$$

---

* A bow-tie filter is a shaped piece typically of aluminum that is near and rotates with the x-ray source. Its purpose is to equalize both the intensity and beam quality (hardness of the x-rays) across the detector array. These conditions are only roughly satisfied as the humans do not have perfect cylindrical symmetry, and even if so, it is impossible to exactly satisfy both conditions with any design for the bow tie.

which implies that a transmission of $e^{-2} = 13.5\%$ maximizes the *SNR* [55]. The linear attenuation coefficient of water is 0.020 mm^{-1} at $\bar{E} = 64$ keV. SNR maximizes for water-like cross sections of the patient when the characteristic diameter of the patient is $t = 2/\mu = 100$ mm. Only limbs, smaller breasts, and babies satisfy this size characteristic. Hence, the overwhelming majority of CT scans take place far from the optimum. In the diagnostic energy range, attenuation coefficients are decreasing functions of E. Thus, you might think increasing the operating tube voltage of the scanner would help us get closer to maximizing SNR.

On the other hand, contrast between tissues decreases, with increasing energy including for iodine contrast from a soft tissue background. This is a difficult optimization problem to improve conspicuity while using the lowest dose practical, considering modern reconstruction methods that use adaptive filtration on data and images designed to reduce noise while leaving contrast differences intact, whether in FBP or in iterative reconstruction. Therefore, the rule of thumb is to use the lowest potential on the x-ray tube possible consistent with enough penetration through the patient.

5.3 Vascular Acquisitions Methods

Angiographic imaging involves the injection of a contrast agent into the blood stream (the vasculature). Nearly all contrast agents for angiography are based on iodine. In diagnostic CT imaging, this is called CTA and is usually done by venous injection. The contrast is generally injected through the antecubital vein in the arm via a flexible catheter, returns to the heart for further mixing with the blood, and then enters during the arterial phase into the vasculature. Acquisition is performed at the peak HU enhancement for arterial imaging. The delay from injection to start of scan is a matter of using known average transit times from the injection site to the region of interest or can be monitored and then acquired at peak HU enhancement.

A review of CTA techniques after the introduction of helical, single-slice CT but before multislice was given by Kalender [56]. An updated review in the early days of helical, multislice CT was given by Kalender and Prokop [57]. A review of coronary cardiac angiography appeared a few years ago [58]. The rest of this section gives a flavor of recent developments.

There are four generic types of vascular applications:

1. Cardiac/coronary imaging (cCTA) for coronary artery disease
2. Arterial imaging
3. Functional CTA
4. Interventional guidance and assessment

5.3.1 Cardiac and Coronary CT Angiography

With the advent of 4-slice and especially with 16-slice and then 64-slice CT systems, cCTA of the whole heart became practical. The main clinical purpose is to search the coronary tree for stenoses and to assess their severity. The original acquisition techniques were retrospectively gated helical. This means the patient's ECG is taken concurrently and recorded with the data. After the data are acquired, we use only those views needed for reconstruction that appear in a certain cardiac phase window, typically the quietest phase, usually the diastolic cardiac phase. The x-ray tube is on while the scanner slowly covers the heart such that each slice of the heart is seen by enough contiguous views in the diastolic phase [58 and Figure 5.19, bottom, except that one gray rectangle covers the width of the figure]. This implied a relatively slow helical pitch of around 0.2 (ratio of the table motion per rotation of the gantry to the nominal aperture of the detector). The main drawbacks of this approach are high radiation dose (around 20 mSv) and high-contrast dosage to ensure opacification throughout the entire acquisition. Image quality is dependent on heart rate: the slower and more steady the heart rate, the better the image quality.

This led to several countermeasures, one is drug-related and the others are technology related. Beta-blockers are often used to lower and steady the heart rate, where under 60 bpm is the goal. They are administered orally an hour before the scan or through an IV minutes before the scan [60]. For patients whose heart rates cannot be lowered or the patient has contraindications, then there is a technique called "segmented reconstruction." Here the views needed for reconstruction come from two or even more heartbeats and rotations of the gantry and are stitched together to make a complete data set [61 and Figure 5.22]. This technique lowers the effective temporal resolution of the scanner; however, it assumes that the heartbeat is perfectly repeatable.

Prospective gating greatly reduced the dose to the patient to the 3–5 mSv range for 64-slice CT. Here the ECG signal is used to predict the next diastolic

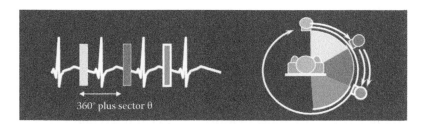

FIGURE 5.22 Data acquisition for segmented cardiac scans. Such scans can be done for circular cone-beam, step-and-shoot, or helical modes. The segments can be disjoint if we use the reflection condition in Equation 22 [61]. Now the acquisition time window narrows (gray rectangles in Figure 5.23 are narrower) and neighboring heartbeats are used to make an image slice putting greater emphasis on the repeatability of the heartbeat and on making sure the heartbeat is not resonant with scanner rotation or else each acquisition covers the same angular views. This example shows three sectors. (From Edyvean, S., Technical Aspects of Cardiac CT, Harefield Cardiac Course from ImPACT, available for download at http://www.impactscan.org/presentations/Technical_Aspects_of_Cardiac_CT_Harefield2011.pdf.)

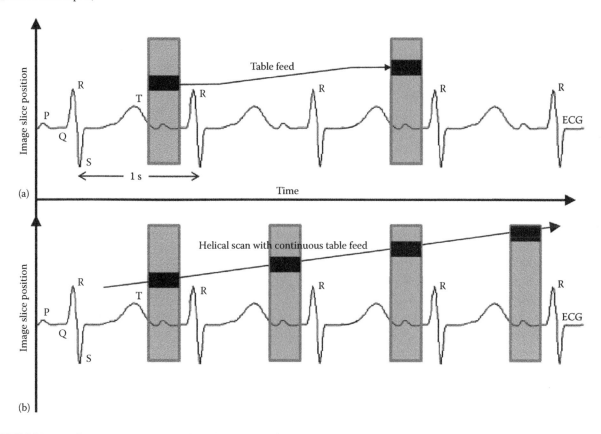

FIGURE 5.23 Cardiac acquisition modes for half scan. This figure shows a 60 bpm heart rate with acquisition at 75% of the R–R wave near end diastolic for a gantry with 0.35 s rotation period. Remember that half scan means that data are acquired for $(180 + 2\Gamma)/360 \times 0.35$ s ≈ 0.25 s. Gray rectangles represent data used in reconstruction. Black bands represent the z-range of image volume reconstructed. For retrospective gating, the patient is continuously irradiated. For prospective gating, the acquisition rectangle is nearly the same as the data used in reconstruction. (a) Step-and-shoot. Typically a heartbeat is skipped while the table is repositioned. (b) Helical acquisition. Data are used from every heartbeat (assuming no arrhythmia), but the helical pitch must be set to minimize image slice overlap while at the same time have no gaps in the coverage. (Adapted from Dewey, M., *Coronary CT Angiography*, Springer-Verlag, 2009, Fig. 2.7.)

phase and radiation is delivered only during the time necessary to reconstruct a volume (with a little margin to account for small variations in the heart rate). The different vendors took different paths for their current top-of-the-line solutions to cCTA. GE uses a step-and-shoot approach with a 40 mm aperture [63]

that requires several steps and a different heartbeat for each step to cover the full heart volume (Figure 5.23). This implies that banding artifacts are a potential problem, meaning that different volume slabs of the heart either have slightly different contrast (due to dynamic changes in the contrast concentration) or are slightly shifted (due to nonrepeatability of the heartbeats). Philips has the fastest gantry at 0.27 s per rotation with an 80 mm aperture. Toshiba developed a wide-coverage scanner (Aquilion ONE™) to acquire the whole heart with a 160 mm aperture in a single scan [64]. Siemens developed a gantry (Definition Flash™) with two imaging chains at right angles with very high helical pitch (~3.2) so that the entire heart is acquired within a heartbeat even with a 40 mm aperture [65]. These latest approaches make sub-mSv scanning possible.

The temporal resolution τ of an image slice can be estimated by

$$\tau \geq \frac{T}{2 \times N \times M} \tag{5.32}$$

where

    T is the rotation period; the factor of 2 is because of half scan (Section 2.5)

    N is the number of imaging chains (1 and 2 for the Siemens solution)

    M is the number of segments in segmented reconstruction

But because of potential resonance between the rotation period and the heart rate, this is really a lower limit of temporal resolution when $M = 2$ or higher. For the Toshiba solution, all images in the volume have the temporal resolution window centered at the same time point; other solutions either change gradually with image slice location (Siemens) or have one (Philips) or more (GE) discontinuous jumps.

Often a precursor to cCTA is a coronary artery calcium scoring exam, also known as an Agatston score, but strictly speaking, this is not an angiographic exam since the vasculature is not observed nor is a dye injected. An estimate of the amount of calcification in the coronaries is made by adding the area of each calcium speck within a 3 mm thick slice weighted by a step function correlated to the HU of the Ca, summed over all the slices that cover the heart [66]. The cCTA exam would then give the location of each Ca speck as well as the severity of the blockage caused by the Ca (hard plaque) or by soft plaque. One complication is when the Ca specks have a similar HU value as the iodine contrast agent. Dual-energy CT has the potential to solve this problem and characterize the plaque but requires spatial and temporal registration of the low- and high-energy data (see Chapter 12).

A more advanced—and controversial—extension to cCTA is the triple rule-out study [67,68]. Its purpose is to exclude three potential causes of chest pain in a single scan:

- Coronary artery disease
- Aortic dissection
- Pulmonary embolism

One study of emergency room patients with chest pains at low and intermediate risk from acute coronary syndrome found that "TRO CT can safely eliminate the need for further diagnostic testing in over 75% of patients" [69]. On the other hand, it uses higher dose from scanning a third larger volume than standard cCTAs, there is difficulty in administering the protocol, particularly the contrast because peak aortic enhancement is about 10 s after pulmonary vessel enhancement, and there is a lack of studies showing improved outcomes or downstream cost savings. Elliot Fishman concludes: "So in the right patient, where you really up the wall, you can't figure out what's going on, but you think it is cardiac-related, it is a reasonable study to do, particularly in patients over 40. … the numbers suggest that triple rule-out is not a routine [emergency room] study,…it's not for everyone, and before you order one, you really need to think about whether you need one" [70].

5.3.2 Arterial Imaging

There are several clinical applications covered by arterial imaging. Great vessel inspection includes the aorta from the arch down through the thorax to the bifurcation into the common iliac arteries and further into the left and right external and internal iliacs. We can go further, on down to the toes as well. The imaging volume follows a bolus of contrast. The tricky part is to be in synchronization with the bolus travel: do not get ahead of it or too far behind. Clinical reasons include to inspect the vascular system for stenoses and/or aneurysms, and to plan surgeries and navigation for vascular repair such as stent placement. Head and neck arterial imaging, especially of the carotids, is very similar except the bolus is followed in the other direction from the heart. Other arteries of interest are those that feed specific organs such as the renal arteries for the kidneys, hepatic artery for the liver, and the superior mesenteric artery and its branches that supply the intestines.

5.3.3 Functional CTA

Functional imaging goes beyond an examination of the anatomy. Perhaps the simplest example is measuring the ejection fraction (*EF*). Here, the fraction of the blood volume of the left ventricle is compared at end systole (*ESV*) to end diastole (*EDV*):

$$EF = \frac{EDV - ESV}{EDV}. \qquad (5.33)$$

In CT, images reconstructed at end diastole typically have better image quality than those at end systole due to the greater motion within the acquisition window at end systole. This could mean that extracting the volume from the end systole is somewhat more complicated. By extension, images could be reconstructed throughout the cardiac cycle and a movie made of the beating heart.

The most common functional technique with CTA is perfusion tomography (CTp). The goal of perfusion imaging is to assess the blood flow through the organ under investigation as a surrogate for its condition. It gives the doctor the information needed on how aggressively to treat victims of stroke in neuroperfusion and blockages in other organs or microvascular diseases. It seems to be one of those ideas in medical imaging that has come up several times in the past, but not catching on until the technology had advanced enough to make it practical [71–74]. Contrast is injected and image volumes are collected starting just before contrast reaches the organ of interest through intake and washout. Typically an organ could be imaged over 1 to even 2 or more minutes, with a sampling of around every 2 s during intake and perhaps at a lower frequency toward the end of the study. Because CTp can involve 20–30 time points, low-dose techniques are necessary and radiation overdose problems can result if a wrong protocol is used [75]. What is called a time density curve (TDC) for each voxel of the image is recorded as the HU value as a function of time. TDCs are analyzed by a variety of algorithms to give maps of perfusion parameters such as blood volume, blood flow, and mean transit time [76]. Because perfusion is done during take-up phase at 2 s intervals, CTp is done by diagnostic scanners although there is some research in the use of C-arm systems [77]. A more detailed discussion of TDC appears in Chapter 17 and of neuro-CTp in Chapter 18.

Neuroperfusion of the brain tissue is the most common target for CTp. Organ perfusion of the liver to assess blood flow, of the kidneys for blood flow and permeability, of the heart muscle to replace or supplement nuclear medicine studies of myocardial defect, and even of tumors have all been investigated [78].

The aforementioned examples of functional imaging in CT require dynamic scanning: examining the same tissue volume over some period of time. Recent innovations combine CT imaging with computational flow dynamics (CFD) modeling to derive functional information. One example is to derive fractional flow reserve (FFR) from a static CT image. FFR is the ratio of the mean pressure distal to a stenosis to the mean aortic pressure during maximal flow. It has been found to more accurately assess the severity of the stenosis than cCTA [79]. Traditional FFR is an invasive technique involving the insertion of a pressure wire during invasive coronary angiography. The new technique, FFT from CT + CFD (FFR_{CT}), was evaluated by the DeFACTO study, recently reporting its encouraging results in Reference 79, although long times are required to solve the CFD equations.

Another example of CT + CFD takes the static CTA image and also applies CFD to solve the Navier–Stokes equation for the wall shear stress along a portion of the vasculature. The goal here is to calculate the probability of atherogenesis in coronary arteries [80] and also to the probability of aneurysms developing along the portion of the vasculature imaged and to predict the likelihood of hemorrhage for existing aneurysms. Chapter 21 covers this topic in greater detail.

5.3.4 Interventional Guidance and Assessment

Minimal invasive interventional surgery typically uses C-arm imagers because of their open access for the surgeon as he or she guides a treatment device to the point of disease. Many of these applications use the CT option that is available of several C-arm models. These applications are covered in Chapters 10 and 22. One recent use of diagnostic CT (although C-arm CT could be used as well) is planning for transcatheter aortic valve replacement. A CT scan is done prior to the procedure, perhaps up to a month, to provide detailed information about the aortic root noninvasively, to aid in the selection of the correct size for the valve, and to judge the best angle for subsequent fluoroscopy guidance from the C-arm imager [81].

Chapter 5

Assessment can be immediate, such as checking aneurysm coils and stent placements done with C-arm systems, or over a period of time to assess treatment such as the shrinkage of tumors. The later example may or may not use CTA techniques and often requires quantitative measures [82].

Another example of CTA for surgical planning is organ transplantations. These rely on CTA exams of the donor and the recipient to plan the surgical reconnections of both the arteries and veins to ensure restore blood flow [83,84].

5.3.5 Contrast Agent Administration

Kyongtae Bae wrote a state of the art review of contrast medium administration for intravenous injection [85]. Total injection amount and rates vary depending on patient, clinical application, and scan parameters. Patient factors include mainly weight, but also height, gender, age, and cardiac and renal function. Scan factors include number of detector rows, scan duration, tube voltage, and scan direction. Bae's Table 1 gives typical contrast and scanning parameters for common clinical protocols. A simplified and shortened version appears here as Table 5.1 for 64-slice CT. Table 5.2 is the analogous table for a 320-slice CT scanner.

Some organs such as the liver are scanned at various phases of the contrast: precontrast, arterial phase, portal venous, or postcontrast (70–90 s after the arterial phase contrast scan), and delayed phase some 5–10 min later. Different kinds of lesions can appear with more contrast in one phase over the other due to differential uptake and washout rates.

The strength of the iodine in contrast media, also called dyes, is given in micrograms of iodine per milliliter (mgI/mL) of solution. Commercial dyes range from 400 mgI/mL down to 270 mgI/mL with dilution always possible if required by the application. Injection is often by automated power injectors. We see from Figure 5.4 that pure iodine gives a huge contrast jump over soft tissue (represented by water). However, the injected iodine is in solution so we must use Equations 5.3 and 5.4 to give the clinical agent's true attenuation and thus contrast from background. Table 5.3 compares the attenuation values and other physical properties of water and iodine at 64 keV. Let us use the tools developed earlier in this chapter to estimate the iodine difference in the vascular between venous injection used in CTA from a diagnostic scanner and arterial injection used in 3D-DSA, where 3D means rotational acquisition from the C-arm gantry and DSA means

Table 5.1 Key Contrast Injection and Scan Parameters for Common Clinical Protocols for a 70 kg Patient [Typical Values for 64-slice CT]

Examination	Total Contrast	Injection Rate	Injection Duration	Scan Duration	Saline Flush
Brain parenchyma	80 mL of 300 mg/mL	1 mL/s	1–2 min	Variable	Not essential
Neck soft tissue	100 mL of 300 mg/mL	2 mL/s	50 s	Variable	Not essential
Neck and brain CTA	75 mL of 350 mg/mL	4.5 mL/s	17 s	5–10 s	Essential
Brain perfusion	50 mL of 350 mg/mL	4–10 mL/s	<10 s	<60 s	Essential
Routine chest	70 mL of 300–350 mg/mL	2–3 mL/s	5–10 s	40–60 s	Not essential
Pulmonary CTA	100 mL of 350 mg/mL	4–5 mL/s	20–25 s	5–10 s	Essential
cCTA	75 mL of 350 mg/mL	4.5–5 mL/s	15 s	5–10 s	Essential
Peripheral runoff CTA	125–140 mL of 350 mg/mL	3.5–4 mL/s	35 s	40 s	Useful
Liver and routine abdomen	100 mL of 350 mg/mL	4 mL/s (dual phase) 2–3 mL/s (hepatic phase only)	25–30 s (dual) 35–50 s (hepatic)	5–10 s	Essential (dual) Not essential (hepatic)
Pancreas	100–140 mL of 350 mg/mL depending on phases	2.5–5 mL/s depending on phases	25–30 s (dual) 50 s (single)	5–10 s	Essential (dual) Not essential (single)
Kidney	100–120 mL of 350 mg/mL	2.5–4 mL/s	25–30 s (dual) 3–50 s (single)	5–10 s	Essential

Source: Kyongtae, T.B., *Radiology*, 256, 32, 2010.

Table 5.2 Key Contrast Injection and Scan Parameters for Common Clinical Protocols for a 70 kg Patient [Typical Values for Volume CT Scanner]

Examination	Total Contrast (Fixed)	Injection Rate	Contrast Injection Duration	Fixed Delay after Start of Injection	Scan Duration	Saline Flush (50 mL)
Brain parenchyma	50 mL of 320 mg/mL	2 mL/s	25 s	2 min	0.5 s	Not essential
Neck soft tissue	75 mL of 320 mg/mL	3–4 mL/s	18–25 s	40 s	7 s	Not essential
Neck and brain CTA	60 mL of 350 mg/mL	4–5 mL/s	12–15 s	Bolus tracking or ~20 s	6 s	Essential
Brain perfusion	50 mL of 350 mg/mL	6 mL/s	8 s	5 s	43 s	Essential
Routine chest	70 mL of 320 mg/mL	3 mL/s	23 s	Bolus tracking or ~35 s	3.1 s	Not essential
Pulmonary CTA	70 mL of 350 mg/mL	4–5 mL/s	14–17 s	Bolus tracking or ~10 s	2.8 s	Essential
cCTA	70mL of 350 mg/mL	4–5 mL/s	14–17 s	Bolus tracking or ~25 s	0.35 s	Essential
cCTA (Biphasic mix)	P1–60mL of 350 mg/mL P2–30 mL 30% contrast 70% saline	4–5 mL/s for both phases	14–17 s	Bolus tracking or ~25 s	0.35 s	Essential
Peripheral runoff CTA (biphasic)	100 mL of 350 mg/mL	30 mL at 5 mL/s 70 mL at 4 mL/s	23 s	Bolus tracking or ~35 s	15 s	Useful
Abdominal aorta	70 mL of 350 mg/mL	4–5 mL/s	14–17 s	Bolus tracking or ~25 s	4 s	Useful
Liver (hypovascular) and routine abdomen	70 mL of 320 mg/mL	2–4 mL/s	23–35 s	~65–70 s	2.7 s (liver) 4.0 s (abdomen)	Essential (dual) Not essential (hepatic)
Liver (hypervascular) and routine abdomen	70 mL of 320 mg/mL	4 mL/s	23 s	Bolus tracking or ~30 s	2.7 s (liver) 4.0 s (abdomen)	Essential (dual) Not essential (hepatic)
Pancreas	70 mL of 320 mg/mL depending on phases	2–4 mL/s depending on phases	23–35 s	~65–70 s	0.5 s (pancreas) 4 s (abdomen)	Essential (dual) Not essential (single)
Kidney	70 mL of 350 mg/mL	3–4 mL/s	17–23 s	Corticomedullary 40 s Nephrographic 100 s	0.5 s (kidney) 4 s (abdomen)	Essential

Source: Data in Table 4.2 courtesy of Schultz, K., Toshiba Medical Research Institute, Vernon Hills, IL.

Table 5.3 Attenuation at 64 keV along with Physical Properties

	Mass Attn Coeff. (cm²/g)	Linear Attn Coeff. (cm⁻¹)	Density (g/cm³)	Effective Atomic Number (Z_{eff})
Water	0.2	0.2	1	7.42
Iodine	6.375	31.46	4.935	53

Table 5.4 Properties of the Contrast Agent in the Vasculature

	CTA	Interventional
Injection	Venous (*v*)	Arterial (*a*)
HU	500	5000
$\mu_{v,a}$	0.3 cm⁻¹	1.2 cm⁻¹
Fractional weight of I	1.4%	10%
$\rho_{v,a}$	1.05 g/cm³	1.4 g/cm³
Iodine concen.	14.6 mg/cm³	140 mg/cm³

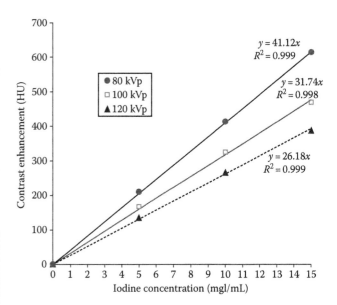

FIGURE 5.24 Graph showing the relationship between iodine concentration and CT enhancement at three voltage settings. For a given voltage, the proportionality of contrast enhancement to iodine concentration is near constant. An increase in concentration by 1 mg of iodine per milliliter yields contrast enhancement of 41.12 HU at 80 kVp, 31.74 HU at 100 kVp, and 26.18 HU at 120 kVp. Thus, use of lower voltage results in stronger contrast enhancement per iodine concentration. (From K.T. Bae., *Radiology*, 256, 32, 2010, Fig. 1.)

digital subtraction angiography: subtraction of a precontrast mask from the angiographic scan. We will assume anything not iodine is water, whether blood or diluting solution in the contrast agent. All calculations are at 64 keV where $\mu_{H_2O} = 0.2$ cm⁻¹. A typical value for contrast in CTA is 500 HU and closer to 5000 HU in interventional. Using Equation 5.5 and then 5.4, along with the constraint that the density of the contrast agent in the vasculature is the sum of the fractional weights (by weight, fixed volume) of its components, we can fill in Table 5.4. If we assume iodine takes up very little of the volume in the contrast agent, then typical agent with 350 mgI/mL is about (350/1350 × 100%) = 26% iodine by weight. Arterial injection reduces this to about 10% and venous injection to 1.4%. Of course, these are very rough estimates to give a sense of how much (or how little) iodine is in the blood stream during angiographic imaging. These results compare well with Figure 1 from Ref. [85], reproduced here as Figure 5.24.

5.4 Artifacts

Artifacts are features in the CT image that do not correspond to patient anatomy or physiology. They arise because the physical CT scanner does not match the mathematical purity of reconstruction theory from projections (the Radon transfer and its inverse). For example, the mathematics assumes continuous, square-integral functions. Reality is sampled data, each datum the result not of a line integral but rather a varying tube of volumetric integration. This leads to aliasing artifacts, especially off high-contrast boundaries such as bone and soft tissue, observed as streaks in the image. The mathematics also assumes data completeness, which as we saw in Section 2.5 is not always satisfied. It is beyond the scope of this introductory chapter to discuss all causes of artifacts and their countermeasures. Peter Joseph made an early list of the causes and potential cures for artifacts, in Chapter 114 in the Big Red Book [16], and Jiang Hsieh and Thorten Buzug have a more complete and modern list of causes and cures in Chapter 7 in the former [1] and Chapter 9 in the latter [87]. However, there are two types of artifacts that especially affect angiographic imaging that deserve coverage here.

5.4.1 Artifacts due to Motion

The mathematical model assumes a stationary scan object, at least during the acquisition of each set of data going to make a volume or set of image slices within the volume. Dynamic changes in the patient can be a good thing, such as in perfusion studies as we watch the contrast enter and then washout of a tissue of interest. Here the change is slow compared to acquisition time; on the other hand, for coronary imaging, motion of a coronary artery for some patients can be substantial during an acquisition leading to blurred or arterial cross sections of strange shapes. This is why most cCTA imaging is acquired during the diastolic heart phase when the heart is at its quietest, and betablockers are often administered to reduce and make more regular the patient's heartbeat. An area of active research is a motion compensated reconstruction method with no consensus solution yet. A session on this topic at a recent CT meeting gives a flavor of current efforts as does Chapter 16 of this text [88,89].

5.4.2 Beam Hardening

The mathematical model assumes monochromatic imaging; it is the only way to isolate the line integral and satisfy the Radon transform as shown in Equation 5.13. However, reality is Equation 5.12. No longer does taking the logarithm of the transmission measurements give a line integral. This unfortunate fact is ignored, which has two major consequences:

1. There is no quantitative and unique HU value for each tissue type. Equation 5.5, which defines the HU scale, does not tell us at what energy the linear attenuation should be evaluated. Scanning with different tube voltages give different HU values, except that water and air should be near 0 and −1000, respectively. Because the x-ray source is polychromatic, we do not have an analytic expression for the attenuation coefficient at any energy.

2. Artifacts arise by the "misuse" of the inverse Radon transform. Two examples are the dishing artifact and dark streaks between boney masses both shown in the simple simulation in Figure 5.25. These artifacts can be particularly bothersome in head imaging. They are called beam hardening artifacts because as the x-ray beam goes through the patient, the less energetic (softer) photons are preferentially absorbed or scattered leaving the more energetic (harder) to carry on. Because the harder x-rays are attenuated less, the beam hardened measurements are more air-like than expected by the mathematical model, giving rise to the lower CT values along hardened ray paths. All tissue hardens the beam; bone more so than soft tissue per unit length. But iodinated contrast is also a strong agent of beam hardening. For example, myocardial perfusion results depend on good HU values, and there could be shadows cast across the myocardium by beam hardening because of contrast filling a heart chamber. We now introduce several measures taken to counteract beam hardening artifacts.

The bow-tie filter is a shaped piece of aluminum (perhaps with other materials as well) such that bow-tie attenuation is weak through the center of the patient and stronger through his peripheral regions. Not only does this tend to equalize signal (and reduce dynamic range requirements on the detector system) but it also tends to equalize the beam hardening for an idealized patient, namely, a water cylinder. Patients are not water cylinders, and equalizing signal and beam hardening together is impossible because Z_{eff} of the bow tie differs from that for patients.

Water cylinder calibration is a way to attempt to cancel beam hardening. Two common correction schemes are as follows: (1) reconstruct against a water cylinder [90]. Then whatever systematic errors are in the patient are also in the water cylinder and thus tend to cancel. But again, patients are not water cylinders and this method only imperfectly corrects the beam hardening. (2) Correct all projection readings as if they were water. That is, from theory (or calibration) we know the expected projection value through an arbitrary path of water [91]. This means any polychromatic reading can be replaced by its monochromatic equivalent and thus solves the problem of a patient's shape not matching a cylinder. However, any tissue that is not water-like, especially bone or contrast, along the path is improperly corrected.

Second-pass reconstruction was introduced early in CT for head imaging (Chapter 114 in [16]). Here the first-pass reconstruction, while imperfect, does differentiate readily between bone and soft tissue. The image is segmented by simple thresholding into bone-like and water-like components, which are reprojected into two data sets where each pair of data values (for each ray-sum) serves as entries into a 2D

Chapter 5

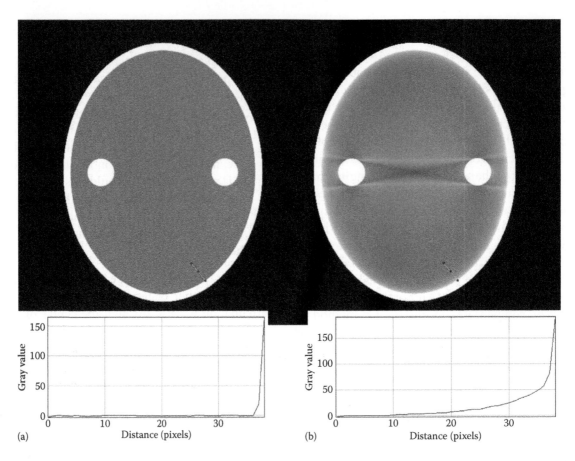

FIGURE 5.25 Beam hardening artifacts are seen by comparing a monochromatic simulation (a) with a polychromatic simulation (b) without any beam hardening correction. The gray tissue was simulated as water and the skull-like feature and two rods as cortical bone. The images are displayed at a level of 0 HU and width of 200 HU. The dishing artifact is an increase of the water HU near the bone, as shown in the profile inserts at 5 o'clock of the simulations.

lookup table containing the correction value. These values are derived from theory or calibration and are then reconstructed into a correction image that is added to the original reconstruction. If there is substantial contrast, such as in angiography, then the segmentation needs to include a separate contrast-like component so that the lookup table becomes 3D.

We should not forget that the bow-tie filter gives a different incident spectrum to each source–detector element combination.

Dual-energy methods can completely eliminate beam hardening if the two (or more) data sets are registered in space and time [92,93]. Chapter 12 covers spectral CT methods in angiography.

5.5   Concluding Remarks

As we approach the 40th anniversary of the clinical availability of Godfrey Hounsfield's wonderful invention, we must marvel at the innovations from industry, academia, and the clinic that made diagnostic medicine so much easier and more accurate and precise than in the (literally) dark ages that proceeded. Patients have been the beneficiaries of this unique collaboration among these three contributing groups in medical imaging. Since the advent of rapid volume scanning by CT around the year 2000,

clinical angiographic applications have become an important diagnostic tool. In this chapter, we presented the physical and mathematical background to CT with an emphasis on contrast differences within the images. Angiography is a high-contrast imaging task due to the use of iodine dye in the blood vessels making it useful not only to see the vasculature but also to derive functional information from observing blood flow over time. We gave a brief introduction of the many angiographic applications and the protocols to obtain these images. Other chapters in this volume describe some of the more interesting clinical applications by CT and other chapters deal with other modalities and the similar or complementary information from those types of scans.

Finally, we mentioned some problems in making perfect images in CT and some of the countermeasures being investigated. Forty years has seen revolutionary developments; let us see what the next 40 years brings.

References

1. J. Hsieh, *Computed Tomography: Principles, Design, Artifacts, and Recent Advances*, 2nd edn., SPIE Press (Wiley-Interscience), 2009.
2. ICRP, Report of the Task Group on Reference Man, ICRP Publication 23, Pergamon Press, Oxford, U.K., 1975.
3. Table 106. The Hospital Physicists' Association (Radiotherapy Topic Group), *Phantom Materials for Photons and Electrons*, Scientific Report Series-20, The Hospital Physicists' Association, London, U.K., 1977.
4. http://physics.nist.gov/PhysRefData/Xcom/Text/.
5. J. T. Bushberg, J. A. Seibert, E. M. Leidholdt, Jr., and J. M. Boone, *The Essential Physics of Medical Imaging*, 3rd edn., Lippincott Williams & Wilkins, 2012.
6. F. K. Richtmyer, E. H. Kennard, and T. Lauritsen, *Introduction to Modern Physics*, 5th edn., McGraw-Hill, 1955.
7. R. D. Evans, *The Atomic Nucleus*, McGraw-Hill, 1955.
8. A. C. Melissinos, *Experiments in Modern Physics*, Academic Press, 1966.
9. M. L. Taylor, R. L. Smith, F. Dossing, and P. D. Franich, Robust calculation of effective atomic numbers: The Auto-Zeff software, *Medical Physics*, **39**, 1769–1778, 2012.
10. D. P. Cormode, Nanoparticle contrast agents for CT: Their potential and the challenges that lie ahead, *International Symposium on Multidetector Row CT*, San Francisco, CA, June 17–20, 2012. A summary appears on the Aunt Minnie web-site by Eric Barnes, "CT nanoparticle contrast: Good as gold?" http://www.auntminnie.com/index.aspx?sec=sup&sub=cto&pag=dis&ItemID=99929, posted July 12, 2012.
11. Beer's law is also known as Beer-Lambert law and sometimes is also associated with Bouguer. August Beer, 1825–1863, was a German physicist, Johann Heinrich Lambert, 1728–1777, was a Swiss polymath, and Pierre Bouguer, 1698–1758, was a French mathematician, geophysicist, astronomer, and naval architect. All studied the passage of light through materials. Wikipedia. Accessed on April 7, 2015.
12. G. G. Poludniowski and P. M. Evans, Calculation of x-ray spectra emerging from an x-ray tube: Part I. Electron characteristics in x-ray targets, *Medical Physics*, **34**, 2164–2174, 2007; G. G. Poludniowski, Calculation of x-ray spectra emerging from an x-ray tube: Part II. X-ray production and filtration in x-ray targets, *Medical Physics*, **34**, 2175–2186, 2007.
13. S. R. Deans, *The Radon Transform and Some of its Applications*, Wiley-Interscience, 1983.
14. A. M. Cormack, Representation of a function by its line integrals, with some radiological applications, *Journal of Applied Physics*, **34**, 2722–2727, 1963.
15. A detailed history of the development of CT can be found in S. Webb, *From the Watching of Shadows: The Origins of Radiological Tomography*, Adam Hilger, Bristol, U.K., 1990.
16. J. Bull, History of computed tomography, Chapter 108 in T. H. Newton and D. G. Potts, eds., *Radiology of the Skull and Brain: Technical Aspects of Computed Tomography*, The C. V. Mosby Company, 1981.
17. G. N. Hounsfield, Computerized transverse axial scanning (tomography): Part 1. Description of system, *British Journal of Radiology*, **46**, 1016–1022, 1973; J. Ambrose, Computerized transverse axial scanning (tomography): Part 2. Clinical application, *British Journal of Radiology*, **46**, 1023–1047, 1973.
18. E. C. McCullough, Industrial corner: X-ray transmission CT scanner survey, *Journal of Computer Assisted Tomography*, **6**, 423–428, January 1982.
19. R. A. Robb, E. A. Hoffman, L. J. Sinak, L. D. Harris, and E. L. Ritman, High-speed three-dimensional x-ray computed tomography: The dynamic spatial reconstructor, *Proceedings of the IEEE*, **71**, 308–319, 1983.
20. D. P. Boyd and M. J. Lipton, Cardiac computed tomography, *Proceedings of the IEEE*, **71**, 298–307, 1983.
21. Medical Imaging and Technology Alliance (MITA), a division of the National Electrical Manufacturers Association, Arlington, VA.
22. W. A. Kalender, *Computed Tomography*, Publicis MCD Verlag, 2000.
23. M. Mahesh, *MDCT Physics: The Basics Technology, Image Quality and Radiation Dose*, Lippincott Williams & Wilkins, 2009.
24. N. A. Baily, Video techniques for x-ray imaging and data extraction from roentgenographic and fluoroscopic presentations, *Medical Physics*, **7**, 472–491, 1980.
25. L. E. Antonuk, K.-W. Jee, Y. El-Mohri, M. Maolinbay, S. Nassif, X. Rong, Q. Zhao, J. H. Siewerdsen, R. A. Street, and K. S. Shah, Strategies to improve the signal and noise performance of active matrix, flat-panel imagers for diagnostic x-ray applications, *Medical Physics*, **27**, 289–306, 2000.
26. I. Mori, Computerized tomographic apparatus utilizing a radiation source, U.S. Patent 4630202, December 16, 1986.
27. W. A. Kalender, W. Seissler, E. Klotz, and P. Vock, Spiral volumetric CT with a single-breathhold technique, continuous transport, and continuous scanner rotation, *Radiology*, **176**, 181–182, 1990.
28. W. A. Kalender and A. Polacin, Physical performance characteristics of spiral CT scanning, *Medical Physics*, **18**, 910–915, 1991.
29. R. N. Bracewell, Strip integration in radio astronomy, *Australian Journal of Physics*, **9**, 198–217, 1956.
30. A. C. Kak and M. Slaney, *Principles of Computerized Tomographic Imaging*, IEEE Press, 1988.
31. F. Natterer, *The Mathematics of Computerized Tomography*, John Wiley & Sons, 1986.
32. A. V. Lakshimarayanan, Reconstruction from divergent ray data, Technical Report 92, Department of Computer Science, State University of New York at Buffalo, Buffalo, NY, 1975.

33. L. A. Shepp and B. F. Logan, The Fourier reconstruction of a head section, *IEEE Transactions on Nuclear Science*, **21**, 21–43, 1974.

34. F. Natterer and F. Wübbeling, *Mathematical Methods in Image Reconstruction*, SIAM, 2001.

35. L. A. Feldkamp, L. C. Davis, and J. W. Kress, Practical cone-beam algorithm, *Journal of Optical Society of America A*, **1**, 612–619, 1984.

36. C. R. Crawford and K. F. King, Computed tomography scanning with simultaneous patient translation, *Medical Physics*, **17**, 967–982, 1990.

37. K. Taguchi and H. Aradate, Algorithm for image reconstruction in multi-slice helical CT, *Medical Physics*, **25**, 550–561, 1998.

38. H. Hu, Multi-slice helical CT: Scan and reconstruction, *Medical Physics*, **26**, 1–18, 1999.

39. H. Kudo and T. Saito, Three-dimensional helical-scan computed tomography using cone-beam projections, *Journal of the Electronics, Information, and Communication Society of Japan*, **J74-D-II**, 1108–1114, 1991.

40. M. D. Silver, K. Taguchi, and K. S. Han, Field-of-view dependent helical pitch in multi-slice CT, in L. E. Antonuk and M. J. Yaffe, eds., *Proceedings of the SPIE*, **4320**, 839–850, 2001.

41. A. Katsevich, An improved exact filtered backprojection algorithm for spiral computed tomography, *Advanced in Applied Mathematics*, **32**, 681–697, 2004.

42. G. T. Herman, *Image Reconstruction from Projections: The Fundamentals of Computerized Tomography*, Academic Press, 1980.

43. A. A. Zamyatin, K. Taguchi, and M. D. Silver, Practical hybrid convolution algorithm for helical CT reconstruction, *IEEE Transactions on Medical Imaging*, **53**, 167–174, 2006.

44. Y. Zou, X. Pan, and E. Y. Sidky, Theory and algorithms for image reconstruction on chords and within regions of interest, *Journal of the Optical Society of America A*, **22**, 2372–2384, 2005.

45. X. Pan, E. Y. Sidky, and M. Vannier, Why do commercial CT scanners still employ traditional, filtered back-projection for image reconstruction? *Inverse Problems*, **25**, 1–36, 2009.

46. J. A. Fessler, Statistical image reconstruction methods for transmission tomography, in M. Sonka and J. Michael Fitzpatrick, eds., *Handbook of Medical Imaging*, Vol. 2: Medical Image Processing and Analysis, SPIE, pp. 1–70, 2000.

47. B. DeMan and J. A. Fessler, Statistical iterative reconstruction for x-ray computed tomography, in Y. Censor, M. Jiang, and G. Wang, eds., *Biomedical Mathematics: Promising Directions in Imaging, Therapy Planning and Inverse Problems*, Medical Physics Publishing, pp. 113–140, 2010.

48. C. H. McCollough, G. H. Chen, W. Kalender, S. Leng, E. Samei, K. Taguchi, G. Wang, L. Yu, and R. I. Pettigrew, Achieving routine Submillisievert CT scanning, *Radiology*, **264**, 567–580, 2012.

49. D. L. Parker, Optimal short scan convolution reconstruction for fan-beam CT, *Medical Physics*, **9**, 254–257, 1982.

50. M. D. Silver, A method for including redundant data in computed tomography, *Medical Physics*, **27**, 773–774, 2000.

51. H. K. Tuy, An inversion formula for cone-beam reconstruction, *SIAM Journal of Applied Mathematics*, **3**, 546–552, 1983.

52. B. D. Smith, Image reconstruction from cone-beam projections: Necessary and sufficient conditions and reconstruction methods, *IEEE Transactions on Medical Imaging*, **4**, 14–25, 1985.

53. K. Taguchi, Temporal resolution and the evaluation of candidate algorithms for four-dimensional CT, *Medical Physics*, **30**, 640–650, 2003.

54. M. D. Silver, High-helical-pitch, cone-beam computed tomography, *Physics in Medicine and Biology*, **43**, 847–855, 1998.

55. I first read this argument in a paper by Phil Engler. But I no longer can find his paper or him. An alternate derivation of Equation 4.28 can be found in Harrison H. Barrett and William Swindel, *Radiological Imaging: The Theory of Image Formation, Detection, and Processing*, Academic Press, 1981, their Equation 10.193. However, they did not then try to maximize the *SNR* to optimize the scan protocol. I would like to thank Yu Zou for help with this section.

56. W. A. Kalender, Spiral CT angiography, in L. W. Goldman and J. B. Fowlkes, eds., *Medical CT and Ultrasound: Current Technology and Applications*, AAPM, 1995.

57. W. A. Kalender and M. Prokop, 3D CT angiography, *Critical Reviews in Diagnostic Imaging*, **42**, 1–28, 2001.

58. B. M. Ohnesorge, C. R. Becker, T. G. Flohr, and M. F. Reiser, *Multi-Slice CT in Cardiac Imaging*, Springer-Verlag, 2002.

59. M. Dewey, *Coronary CT Angiography*, Springer-Verlag, 2009.

60. H. K. Pannu, W. Alverez, Jr., and E. K. Fishman, *American Journal of Roentgenology*, **186**, S341–S345, 2006. http://www.ajronline.org/content/186/6_Supplement_2/S341.full. Accessed on April 7, 2015.

61. K. Taguchi, B. S. Chiang, and I. A. Hein, Direct cone-beam cardiac reconstruction algorithm with cardiac banding artifact correction, *Medical Physics*, **33**, 521–539, 2006.

62. S. Edyvean, Technical aspects of cardiac CT, Harefield Cardiac Course from ImPACT, available for download at http://www.impactscan.org/presentations/Technical_Aspects_of_Cardiac_CT_Harefield2011.pdf. Accessed on April 7, 2015.

63. J. Hsieh, J. Londt, M. Vass, J. Li, X. Tang, and D. Okerlund, Step-and-shoot data acquisition and reconstruction for cardiac x-ray computed tomography, *Medical Physics*, **33**, 4236–4248, 2006.

64. M. Dewey, E. Zimmermann, F. Deissenrieder, M. Laule, H.-P. Dübel, P. Schlattmann, F. Knebel, W. Rutsch, and B. Hamm, Noninvasive coronary angiography by 320-row computed tomography with lower radiation exposure and maintained diagnostic accuracy: Comparison of results with cardiac catheterization in a head-to-head pilot investigation, *Circulation*, **120**, 867–875, 2009.

65. T. G. Flohr, S. Leng, L. Yu, T. Allmendinger, H. Bruder, M. Petersilka, C. D. Eusemann, K. Stierstofer, B. Schmidt, and C. H. McCollough, Dual-source spiral CT with pitch up to 3.2 and 75 ms temporal resolution: Image reconstruction and assessment of image quality, *Medical Physics*, **36**, 5641–5653, 2009.

66. A. S. Agatston, W. R. Janowitz, F. J. Hildner, N. R. Zusmer, M. Viamonte, and R. Detrano, Quantification of coronary artery calcium using ultrafast computed tomography, *Journal of American College of Cardiology*, **15**, 827–832, 1990.

67. C. White, Triple rule-out CT in the emergency department, Suppl. *Applied Radiology*, 35–44, November 2007.

68. M. J. Gallagher and G, L. Raff, Use of multislice CT for the evaluation of emergency room patients with chest pain: The so-called triple rule-out, *Catheterization and Cardiovascular Interventions*, **71**, 92–99, 2008.

69. E. J. Halpern, Triple-rule-out CT angiography for evaluation of acute chest pain and possible acute coronary syndrome, *Radiology*, **252**, 332–345, 2009.

70. E. Fishman, http://www.medscope.com/viewarticle/760544. Accessed on October 18, 2012.

71. G. Ladurner, E. Zilkha, L. D. Iliff, G. H. du Boulay, and J. Marshall, Measurement of regional cerebral blood volume by computerized axial tomography, *Journal of Neurology, Neurosurgery, and Psychiatry*, **39**, 152–158, 1976.

72. L. Axel, Cerebral blood flow determination by rapid-sequence computed tomography, *Radiology*, **137**, 679–686, 1980.

73. A. A. Konstas, G. V. Goldmakher, T.-Y. Lee, and M. H. Lev, Theoretic basis and technical implementations of CT perfusion in acute ischemic stroke, Part 1: Theoretic basis, *American Journal of Neuroradiology*, **30**, 662–668, 2009.

74. A. A. Konstas, G. V. Goldmakher, T.-Y. Lee, and M. H. Lev, Theoretic basis and technical implementations of CT perfusion in acute ischemic stroke, Part 2: Technical implementations, *American Journal of Neuroradiology*, **30**, 885–892, 2009.

75. W. Bogdanich, Radiation overdoses point up dangers of CT scans, *The New York Times*, October 15, 2009.

76. R. E. Latchaw et. al., Guidelines and recommendations for perfusion imaging in cerebral ischemia: A scientific statement for healthcare professionals by the writing group on perfusion imaging, From the Council on Cardiovascular Radiology of the American Heart Association, *Stroke*, **34**, 1084–1104, 2003.

77. A. Ganguly, A. Fieselmann, M. Marks, J. Rosenberg, J. Boese, Y. Deuerling-Zhang, M. Straka, G. Zaharchuk, R. Bammer, and R. Fahrig, Cerebral CT perfusion using an interventional C-arm imaging system: Cerebral blood flow measurements, *American Journal of Neuroradiology*, **32**, 1525–1431, 2011.

78. K. A. Miles and M. R. Griffiths, Perfusion CT: A worthwhile enhancement? *The British Journal of Radiology*, **76**, 220–231, 2003.

79. J. K. Min, J. Leipsic, M. J. Pencina, D. S. Berman, B.-K. Koo, C. V. Mieghem, A. Erglis et al., Diagnostic accuracy of fractional flow reserve from anatomic CT angiography, *JAMA*, **308**(12), 1237–1245, 2012.

80. S. Jin, Y. Yang, J. Oshinski, A. Tannebaum, J. Gruden, and D. Giddens, Flow patterns and wall shear stress distributions at atherosclerotic-prone sites in a human left coronary artery—An exploration using combined methods of CT and computational fluid dynamic, *Proceedings of the 26th Annual International Conference of the IEEE EMBS*, San Francisco, CA, September 1–5, 2004.

81. L. F. Tops, D. A. Wood, V. Delgado, J. D. Schuijf, J. R. Mayo, S. Pasupati, F. P. L. Lamers et al., Noninvasive evaluation of the aortic root with multislice computed tomography, *Journal of American College of Cardiology Imaging*, **1**, 321–330, 2008.

82. Quantitative imaging from several modalities is being spearheaded by an RSNA initiative called QIBA: Quantitative Imaging Biomarkers Alliance, https://www.rsna.org/QIBA.aspx. Accessed on April 7, 2015.

83. B. Pomahac et al., Three patients with full facial transplantation, *New England Journal of Medicine*, **366**, 715–722, 2012.

84. S. Soga et al. Surgical planning for composite tissue allo-transplantation of the face using 320-detector row computed tomography, *Journal of Computed Assisted Tomography*, **34**, 766–769, 2010.

85. K. T. Bae, Intravenous contrast medium administration and scan timing at CT: Considerations and approaches, *Radiology*, **256**, 32–61, 2010.

86. Data in Table 4.2 courtesy of Kurt Schultz, Toshiba Medical Research Institute, Vernon Hills, IL.

87. T. M. Buzug, *Computed Tomography: From Photon Statistics to Modern Cone-Beam CT*, Springer, 2008.

88. J. D. Pack and B. Claus, An analysis of motion artifacts in CT and implications for motion compensation, *The Second International Conference on Image Formation in X-Ray Computed Tomography*, Salt Lake City, UT, June 24–27, pp. 322–325, 2012.

89. A. Katsevich, A. Zamyatin, and M. Silver, A novel motion estimation scheme, *The Second International Conference on Image Formation in X-Ray Computed Tomography*, Salt Lake City, UT, June 24–27, pp. 326–329, 2012.

90. W. D. McDavid, R. G. Waggoner, W. H. Payne, and M. J. Dennis, Correction for spectral artifacts in cross-sectional reconstruction from x-rays, *Medical Physics*, **4**, 54–57, 1977.

91. P. M. Joseph and R. D. Spital, A method for correcting bone induced artifacts in computed tomography scanners, *Journal of Computed Assisted Tomography*, **2**, 100–108, 1978.

92. R. E. Alvarez and A. Macovski, Energy-selective reconstruction in x-ray computerized reconstruction, *Physics in Medicine and Biology*, **21**, 733–744, 1976.

93. Y. Zou and M. D. Silver, Analysis of fast kV-switching in dual energy CT using a pre-construction decomposition technique, in J. Hsieh and E. Samei, eds., *Physics of Medical Imaging Proceedings of the SPIE*, **6913**, 13-1–13-12, 2008.

94. S.X. Pan and A.C. Kak, *IEEE Trans. Acoust. Speech Signal Processing*, **ASSP-31**, 1262-1275, 1983.

Chapter 5

6. Physics and Technology of MR Angiography

Octavia Bane, Parmede Vakil and Timothy J. Carroll

Chapter 6

6.1 Introduction

In this chapter, we present the basic physics and technology of magnetic resonance imaging (MRI)-based angiography (MRA). In the time since its inception by Dr. Martin Prince [1], MRA has moved from a research tool to a widely used clinical tool. Its widespread acceptance is due in part to the minimally invasive nature of the procedure when compared to the reference standard, catheter x-ray angiography. Although MRA lacks the spatial resolution and speed of x-ray techniques, it provides useful feedback to clinicians on the status of the vasculature, sufficient to perform diagnosis. In this chapter, we review the basic physical principles of widely used MRA technologies. We will review the basic physics of MRI and in particular those relevant to angiography. Both noncontrast MRA and contrast-enhanced techniques will be described. We highlighted included sections on "Clinical Relevance" to underscore the importance of the concepts to practicing clinicians.

6.2 Physics of Magnetization of Nuclear Spins

Magnetic resonance methods that depict images of the vasculature comprise the field of MRA. These techniques image the nuclei of atoms that have an odd number protons or neutrons. Such atoms can be described as possessing a property called "spin" due to the spinning of their protons around an axis in the presence of a magnetic field. The spinning of charged particles induces a small magnetic moment, which is the basis of the nuclear magnetic resonance, or NMR phenomenon. Hydrogen ^{1}H atoms, which are the most abundant in the body, exhibit this property and are routinely imaged in MRI.

Spins of hydrogen atoms are distributed randomly throughout the body. However, in the presence of an external magnetic field, B_0, they orient either 54°44′ to the direction of B_0 (parallel) or 54°44′ to the opposite direction B_0 (antiparallel).

The ratio of the number of spins that are parallel ($N_\uparrow$) to those that are antiparallel ($N_\downarrow$) is defined by the Boltzmann distribution:

$$\frac{N_\uparrow}{N_\downarrow} = \exp\left(\frac{\Delta E}{KT_S}\right) \tag{6.1}$$

where
ΔE is the energy difference between the spins states
K is the Boltzmann constant
T_S is the absolute temperature of the spin system

The spins then precess around the B_0 direction with the angular frequency given by

$$\omega_0 = \gamma B_0 \tag{6.2}$$

where

γ is a constant known as gyromagnetic ratio
For H atoms, $\gamma = 267.54 \times 10^6$ rad/s/T
The ω_0 is called Larmor frequency, and for H atoms, it is $\gamma/2\pi = 42.58$ Hz/T

Then the many precessing parallel and antiparallel spins are summed up to a vector to form a net magnetization M_0 in the z direction at equilibrium.

6.3   Excitation and Relaxation: Signal Formation in MRI

Measurement of the magnetic signal occurs through excitation and relaxation of an imaging volume of interest. Spins must first be excited in from their alignment along the z-axis into the transverse plane in order to measure their net signal. Excitation is achieved by an RF pulse.

When discussing excitation and relaxation, the spins are often considered in rotating frame of reference in the xy plane using the symbols x' and y'. The reference from rotates at the Larmor frequency with the spins around the z-axis, so all the events occur relative to ω_0, which greatly simplifies the visualization of the phenomenon.

6.3.1   Excitation

When a radiofrequency (RF) pulse with a same frequency as the Larmor frequency, called the B_1 field, is applied in the xy plane, the equilibrium magnetization is tipped toward the xy plane, while continuing to precess at ω_0. This is the "resonance" part of MRI. On the rotating reference frame, this simply appears as the M_0 vector rotating at an angle α about the x'-axis since the xy plane also spins at ω_0. The angle α is called the flip angle and is defined by

$$\alpha = \int_0^{\tau_p} \gamma B_1(t)dt \tag{6.3}$$

where τ_p is the duration of excitation. In other words, the flip angle depends on the area of the RF pulse multiplied by the gyromagnetic ratio. Immediately after excitation at arbitrary time 0, $x'y'$ and z' components $M_{x'y'}$ and $M_{z'}$ of net magnetization arise, where $M_{x'y'}(0_+) = M_{z'}(0_-)\sin(\alpha) + M_{x'y'}(0_-)\cos(\alpha)$ and $M_{z'}(0_+) = M_{z'}(0_-)\cos(\alpha) - M_{x'y'}(0_-)\sin(\alpha)$, where 0_- and 0_+ are times immediately before and after excitation. If the system was at equilibrium, $M_{z'}(0_-) = M_0$ and $M_{x'y'}(0_-) = 0$. $M_{x'y'}$ signal induces changes in magnetic flux of through the receiver coils, called free induction decay (FID), and current is detected.

6.3.2   Specific Absorption Rate

During RF excitation, heat energy is deposited in the body. For long scans, or those with frequent excitation pulses, this can become a safety concern as the human tissue can only absorb a certain amount of RF energy before body temperatures begin to raise. The energy of the RF excitation field is determined by its wavelength, which is quite long for 1.5 T magnets and lower, and as such, very rarely are critical limits reached at 1.5 T. However, with increasing usage of higher field, 3.0 T and 7.0 T, clinical and research scanners, governmental guidelines set limits on the specific absorption rates (SARs—usually W/kg) of heat radiation that is deposited into the body from an MRI scan. A SAR of 1 W/kg applied for an hour will increase the temperature by 1°C. The SAR criteria will vary for specific anatomy but specific guidelines require a whole body SAR of less than 4 W/kg. In addition, no pulse sequence should raise the body temperature by more than 1°C. SAR is proportional to the square of the main magnetic field, B_0, the square of the flip angle θ, and the bandwidth of the RF pulse, f, and inversely proportional to repetition time (TR), described in the following:

$$SAR \sim \frac{B_0^2 \theta^2 \Delta f}{TR} \tag{6.4}$$

As a result SAR can be a limiting factor for certain pulse sequences. For example, in gradient-echo sequences, the TR may need to be extended for a given flip angle.

6.3.3   Relaxation

After spins are excited, the tipped magnetization vector begins to return to the equilibrium state, or "relax." The $M_{z'}$ and $M_{x'y'}$ components undergo relaxation independently. The relaxation of the z and the xy components are called longitudinal and transverse,

Chapter 6

respectively. The longitudinal component exponentially gets restored to the equilibrium value M_0, while the transverse magnetization exponentially decays to the equilibrium value of 0 according to the following equations:

$$M_{x'y'}(t) = M_{x'y'}(0_+)\exp(-t/T_2) \tag{6.5}$$

$$M_{z'}(t) = M_0[1 - \exp(-t/T_1)] + M_{z'}(0_+)\exp(-t/T_1) \tag{6.6}$$

The rate of the relaxation for longitudinal and transverse magnetization depends on constant parameters T_1 and T_2, respectively, which are different for each type of biological tissue. T_1 relaxation, also known as spin–lattice relaxation, is due to exchange of energy between the spins and surrounding tissue. T_2 spin–spin or transverse relaxation, the time in which the transverse magnetization decays to zero, occurs due to field fluctuations similar to T_1 relaxation. In addition, loss of phase coherence among adjacent spins also contributes to the transverse relaxation. Therefore, T_1 is always greater than T_2. Typical T_1 and T_2 values for different types of tissues are listed in Table 6.1. Note that T_1 values depend on the main

Table 6.1 Approximate T_1 and T_2 Values

Tissue	T_2 (ms)	T_1(ms) (at 1.5 T)
Gray matter	100	920
White matter	92	780
Muscle	47	890
Fat	85	260
Kidney	58	650
Liver	4	490

field strength B_0, but T_2 values are independent of B_0. The signal received by the receiver coils depend on $M_{x'y'}$ and $M_{z'}$ values, and those values depend on T_1 and T_2. Therefore, T_1 and T_2 values provide contrast between tissues. In reality, the B_0 field is not perfectly homogeneous. This inhomogeneity causes transverse relaxation according to the constant T_2'. Combined with T_2 relaxation, the overall transverse relaxation is constant is T_2^*, given by

$$\frac{1}{T_2^*} = \frac{1}{T_2} + \frac{1}{T_2'} \tag{6.7}$$

However, the effect of T_2' can be recovered in spin-echo imaging sequence using a 180° RF pulse.

6.4 Mathematical Description of MRI: The Bloch Equation

The behavior of the net magnetization of protons due to excitation and relaxation can be summarized by the Bloch equation [2]:

$$\frac{d\mathbf{M}}{dt} = \mathbf{M} \times \gamma\mathbf{B} - \frac{M_x\mathbf{i} + M_y\mathbf{j}}{T_2} - \frac{(M_z - M_0)\mathbf{k}}{T_1}, \tag{6.8}$$

where
  M is the net magnetization
  B is the applied magnetic fields (including B_0, B_1, and gradient fields)
  i, **j**, and **k** are unit vectors in the x, y, and z directions, respectively

The description shown earlier has three terms: The first is related to the excitation of the magnetization by the RF pulse provided by the scanner. The spins simultaneously return their equilibrium position with a rate, T_1, as described by the third term. The rate at which the group of spins returns to equilibrium is known as the relaxation rates. The second term governs the dephasing in the xy or "transverse" plane as T_2 is therefore known as the transverse relaxation rate. The rate at which the spins reestablish themselves along the z-axis, T_1, is referred to as the longitudinal relaxation rate.

6.4.1 Clinical Relevance

The T_1 and T_2 relaxation rates are a specific to each tissue (see Table 6.1).

MRI pulse sequences are designed to be sensitive to T_1 or T_2 changes, consequently, to make specific anatomic structures appear more (or less) bright in an MRI image. The ability to "tune" the MRI scanner to make pathologic tissue (i.e., tumors) to appear bright is a strength that is unique to MRI. When images are referred to as "T_1-weighted" or "T_2-weighted," it indicates which tissue property (T_1 or T_2) will determine its brightness in an image.

Note that the behavior depends on both T_1 and T_2, which results in contrast between tissues with different T_1 and T_2 values. Solving this governing equation for **M** with time-varying gradient fields results in the signal equations for different pulse sequences.

6.5 Formation of MR Images: Fourier Encoding and Magnetic Gradients

Now that there is signal and contrast between tissues, this section discusses how signal can be spatially encoded to localize the signals. MR images are acquired in the frequency domain, also known as the k-space, which makes MRI less intuitive. After the data are acquired in k-space, a Fourier transform is performed to obtain the values in the image space. The spatial encoding is achieved by magnetic gradients varying in x, y, and z directions. In order to localize signals from a 3D object, all three x, y, and z dimensions must be encoded.

6.5.1 Selection of a Slice through the Body: 2D Slice Selection

A slice can be selected by turning on a gradient in the z direction of the imaging volume to cause linearly and spatially varying Larmor frequencies along the z direction. Since the RF pulse only excites protons precessing at a certain Larmor frequency, only a slice with the frequency that falls inside the bandwidth of the RF pulse will be excited. The thickness of the slice can be varied by the strength of the gradient. A stronger gradient will cause greater differences in precession frequency, so a thinner slice will be selected if the same RF pulse is used. A smaller gradient amplitude will select thicker slices. In multislice imaging, RF frequency can be varied to select a different slice location.

6.5.2 Encoding the Position of Emitted Signal: Phase Encoding and Readout

After the slice is selected, a gradient is applied in the y direction. When the gradient is on, the spins precess at slightly different rates. When the gradient is turned off, the spins return to same Larmor frequency but have accrued different phases depending on the position in the y-axis. This is called phase encoding, and the gradient is known as the phase-encoding gradient. Then, another gradient in the x direction is turned on, and the signal is sampled simultaneously. The x gradient in this case is called the frequency-encoding gradient, because during acquisition the spins precess at different frequencies. This acquisition acquires one line of k-space, and the process is repeated with different values of phase-encoding gradients until the entire k-space is acquired. In essence, the amplitude of the phase-encoding gradient determines which k_y line to acquire. The gradients encode the spins so that values at different spatial frequencies are collected in k-space. Each slice is then Fourier transformed to image space.

6.5.3 Looking at the Whole Body: 3D Imaging

To image 3D volumes, the imaging can be repeated with different z-gradient and/or RF pulse frequency combinations to select different slices. This acquires images slice by slice and results in a stack of 2D images. However, RF pulses in general are not perfect, and there may not be uniform excitation for a given slice, especially around the edges. To fix this problem, the entire slab of the 3D volume can be excited and Fourier encoding can be used to encode the slice direction. The reconstruction then requires another Fourier transform in the slice direction.

6.5.4 Detection and Recording of the MRI Signal: k-Space Sampling

The k-space is the Fourier domain in 2D. The k-space, instead of spatial coordinates, has coordinates of spatial frequencies of the image. Most of the signal are concentrated around the origin of the k-space, in the lower spatial frequencies. The Fourier encoding encodes the image using frequency and phase is equivalent to sampling data in k-space.

The idea of encoding and localizing spins can be thought of as simply traversing and sampling parts of the k-space. The values of the gradients can be thought of as knobs of an "etch-a-sketch" toy. For example, negative x gradient values draw imaginary lines in the $-k_x$ direction, and positive x gradient values move the line in the $+k_x$ direction. The scheme described in Section 6.5.2 would traverse the k-space from left to right, with a different k_y value for each TR, eventually filling the entire k-space. This is the most intuitive and conventional trajectory, the Cartesian sampling.

The density of phase-encoding lines and readout samples required is determined by the Nyquist sampling criterion:

$$\Delta t \le \frac{2\pi}{\gamma|G_x|FOV_x} \tag{6.9}$$

$$\Delta G_y \le \frac{2\pi}{\gamma T_{PE}FOV_y} \tag{6.10}$$

where

Δt is the readout sampling interval
G_x is the frequency-encoding gradient amplitude
FOV_x and FOV_y are field of view sizes in x and y directions
ΔG_y is the phase-encoding gradient step size
T_{PE} is the phase-encoding time interval

6.6 Basic Pulse Sequences

6.6.1 Spin Echo

When spins are excited, the transverse signal starts to decay immediately because of dephasing. In order to sample the data when the phase of all spins is coherent, "echoes" are created. In spin-echo imaging, this is accomplished by adding a 180° RF pulse to flip the spins back into phase. This is analogous to having runners with different speeds start running from the same starting point, and after some time having all runners turn around and run back. The runners will return to the start at the same time even though they all have different speeds.

Figure 6.1a illustrates a basic spin-echo sequence. A slice is first selected using an RF pulse (the α-pulse) and a z-gradient. Immediately after, another gradient with opposite polarity and half the area of the slice select gradient is applied to rephase the out-of-phase spins from the gradient. Then the phase-encoding gradient is applied in the y direction. At the same time, a prephaser gradient with half the area of the readout gradient is applied in the x direction so that there are no unwanted phase differences during readout. A 180° pulse is followed to produce an echo, and FID is sampled. There are a few things to note in this method. The echo time (TE), the time

If this requirement is not met, the reconstructed images will have aliasing artifacts, where copies of images will be superimposed.

There are different ways to traverse the k-space using different combinations of gradient values. Instead of using Cartesian coordinates, the k-space can be sampled radially, each radial line going through the center of the k-space in equally spaced angles. Spiral trajectories also exist, where the k-space is sampled starting from the center and spiraling outward. Other trajectories such as vastly undersampled isotropic projection reconstruction (VIPR) [3] kooshball-like trajectories and sampling pattern that resemble the propellers [4] of an airplane are also under development.

6.5.5 Image Contrast and MRA

Unlike other imaging modalities, MRI has "adjustable" contrast. The scanner can run pulse sequences that are designed to make blood appear bright or dark. Bright blood techniques are those most often used to image blood vessels.

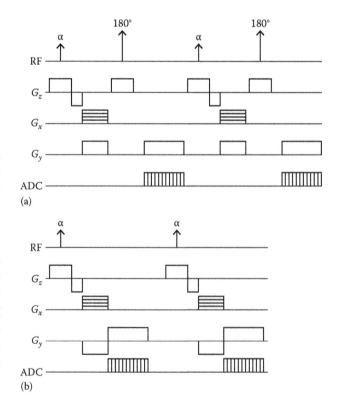

FIGURE 6.1 Diagram of the (a) spin-echo and (b) gradient-echo sequence.

between the α-pulse and echo, is exactly twice the time between the α-pulse and 180° pulse. Also, the prephaser has the same polarity as the readout gradient because the phases are negated by the 180° pulse. More echoes can be produced by applying more 180° pulses, but each subsequent echo will have an exponential amplitude decrease from T_2 relaxation characterized by $\exp(-t/T_2)$. The entire process is repeated every *TR*.

6.6.1.1 Clinical Relevance

Flowing fluid, like blood or cerebrospinal fluid, is inherently dark on spin echo (SE) images. The dark-blood property has been exploited to image myocardial thickening in heart failure or the build-up of plaque within the layers of the arterial wall in atherosclerosis.

6.6.2 Gradient Echo

Converse to spin-echo images, gradient refocused echo (GRE) is inherently bright blood, and therefore a logical tool in angiography.

The spin-echo imaging has the disadvantage that 180° pulse is always necessary. This makes the acquisition time longer, and increases the amount of RF energy applied to the subject, which may be harmful.

The gradient-echo sequence produces echoes by inverting the gradients. When the first gradient is applied, the spins at different spatial locations acquire different phases. Immediately after the first gradient,

another gradient with opposite polarity follows. When the gradient equals 0, the dephasing due to gradients has reversed and an echo is produced. Subsequent echoes can be produced by applying another gradient followed by an opposite gradient. However, this does not reverse the dephasing from B_0 inhomogeneity, so the peak amplitude of echoes decays by $\exp(-t/T_2^*)$. The gradient-echo sequence is illustrated in Figure 6.1b.

Gradient echo is used with small α-angles for fast imaging. In fact, the basic sequence that is used for fast dynamic MRA is the spoiled gradient-echo (SPGR) sequence. The SPGR is a gradient-echo sequence with low flip angle and very short TR, plus a spoiler gradient to remove any transverse magnetization. With the gradient-echo sequence with a short TR, spins are excited before it returns to equilibrium from the previous repetition, so $M_z \neq M_0$ and $M_{xy} \neq 0$. Therefore, a strong gradient is applied in the slice direction to remove or "spoil" the residual transverse magnetization by dephasing it. The amplitude of the echo A_E is characterized by the following equation:

$$A_E = \frac{M_z^0\left(1 - e^{-TR/T_1}\right)}{1 - \cos\alpha \, e^{-TR/T_1}} \sin\alpha \, e^{-TE/T_2^*} \qquad (6.11)$$

With short *TR* and *TE*, The result is a fast gradient-echo sequence with T_1 weighting.

6.7 Noncontrast MRA

MRA of the vessels without exogenous contrast agents exploits the signal differences between flowing blood and stationary tissues (i.e., flow-related enhancement). Noncontrast MRA sequences manipulate the magnitude of the magnetization, such that a larger signal is emitted by moving blood and a lesser signal by stationary tissue.

6.7.1 Time-of-Flight Techniques

6.7.1.1 Time-of-Flight Effect

Time of flight (TOF) is the oldest flow-related enhancement technique, still used for intracranial MRA. In TOF, the amplitude of the signal from flowing blood changes as the blood moves through the imaging volume. The flow of blood during applied gradients also results in a change in phase due to motion. These two effects, on signal magnitude and phase, are known as

TOF effects and can be used to distinguish flowing fluid from stationary tissue.

In TOF, contrast between flowing blood and tissue depends on the pulse sequence used. With spin-echo sequences, the signal from flowing spins is diminished (black blood TOF). However, SPGR (SPGRE, fast low-angle shot [FLASH]) sequences in which signal decreases with TR/T_1 yield high signal from flowing blood and low signal from background tissue (white blood TOF). The enhancement of blood is due to the specific contrast mechanism of SPGRE sequences (see section on FLASH), but foremost to the saturation of stationary tissue by repeated RF excitation pulses. Blood flowing in perpendicularly to the imaging slab experiences fewer saturation pulses, and thus appears brighter than static tissue. The saturation achieved is also called partial saturation, to distinguish it from the state in which the magnetization is zero.

Chapter 6

There are three main acquisition modes for TOF imaging: a series of sequentially acquired thin slices (2D), a single 3D volume, and multiple overlapping thin-slab acquisition (MOTSA).

6.7.1.2 Image Contrast and Artifacts

The signal intensity of moving blood in white blood TOF depends on the number of RF saturation pulses experienced by blood spins before and during their passage through the imaging slice. Assuming blood plug flow perpendicular to the imaging slice and uniform velocity *v*, the blood spins will move a distance $dz = v\ TR$ during the time between RF pulses. If *dz* is greater than or equal to the chosen slice thickness *z*, then the signal of the vessel portion included in the slice will be replenished by the unsaturated spins in the inflowing blood. If *dz* is less than *z*, then there will be sections of the imaging slice in which the blood spins will experience a different number of RF pulses (see Figure 6.2). Thus, the vessel segment included in the slice will contain bands of length *dz* in which blood is partially saturated, and contrast of the blood will vary accordingly. This variation of contrast with partial saturation may cause a vessel narrowing artifact. This artifact can be avoided when slice thickness *z* and *TR* are chosen so that their ratio is equal, or smaller, than blood flow velocity in a vessel perpendicular to the slice. The ratio of slice thickness and *TR* is also known as the critical velocity, V_c:

$$V_c = \frac{z}{TR} \qquad (6.12)$$

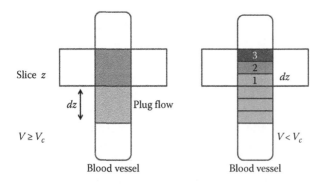

FIGURE 6.2 TOF contrast dependence of imaging parameters slice thickness *z* and *TR*, expressed as critical velocity $V_c = z/TR$. The color scale on the vessel increases with the number of RF pulses experienced by moving spins and is inversely proportional to white blood image contrast. In the case of $V < V_c$ (right), sections *dz* of plug flow will experience different numbers of RF pulses, shown in Arabic numerals.

For 3D TOF and MOTSA, the relationship is expressed in terms of slab thickness (number of slices × slice thickness).

6.7.2 Phase-Contrast Techniques

A basic physics principle of MRI is that the phase of the rotating spin (relative to reference) is used to determine the position of the spin. Phase-contrast (PC) MRI uses this fact to encode the change in position, or velocity of moving spins. In PC-MRA the moving spins are blood.

Among MRI methods, phase PC-MRA allows noninvasive visualization of blood vessels, as well as computation of flow, velocity field, and pressure. Although PC-MRA was one of the first applications of MRI, it was largely abandoned in favor of contrast-enhanced MRA because it compared poorly with the latter in terms of acquisition times, image artifacts, and diagnostic accuracy. However, concerns about the toxicity of gadolinium (Gd)-based contrast agents, technical improvements in MRI pulse sequences, sampling and image reconstruction algorithms, as well as PC's suitability for quantifying blood velocity fields have motivated the resurgence of PC angiography studies.

6.7.3 Development of Basic PC–MRA Pulse Sequences

PC angiography appeared as a result of the observation of Grant and Black in 1982 that MR images of flowing liquid through a tube showed additional dephasing when flow was in the direction of a magnetic field gradient [5]. Because spins moving in the direction of the field gradient experience a higher magnetic field, they will precess faster around the reference axis than spins that remain stationary with respect to the same field. Thus, the net phase accumulation of moving spins is proportional to displacement. Although Grant and Black stated in their conclusions that this property could be used to distinguish between different types of flow and to measure velocity, Moran described the flow-encoding technique most widely used in PC-MRA [6].

Moran's flow-encoding bipolar gradients (a pair of opposite sign magnetic field gradients) were added to a standard spin-echo sequence to tag spins as a function of their mean velocity, similar to spatial location tagging [6]. Since phase shift is proportional to gradient amplitude [7] (Equation 6.13), the bipolar gradients do not add net phase shift to stationary spins;

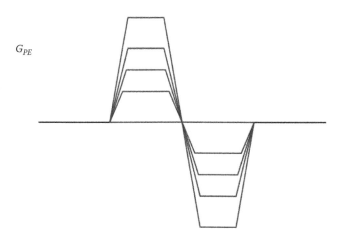

G_{PE}

FIGURE 6.3 Fourier velocity encoding by stepped gradients.

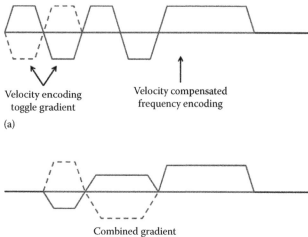

Velocity encoding toggle gradient

Velocity compensated frequency encoding

(a)

Combined gradient

(b)

FIGURE 6.4 Toggling of bipolar flow-encoding gradient: bipolar gradient (a) and combined bipolar and frequency-encoding gradient (b).

moving spins, on the other hand, accumulate a net phase shift proportional to their velocity, the gradient amplitude and duration, as well as the time T between the opposite gradient lobes (Equation 6.14). The amplitude of flow-encoding gradients is incremented, like in the classical 2-DFT phase encoding (Figure 6.3), so that either the phase or frequency-encoding direction is used as a scale of velocity. Moran's method, also known as Fourier velocity encoding [7], can be used to create 2D (one velocity-encoded, one spatially encoded direction) or 3D images (by adding two spatially encoded directions to the velocity-encoded direction):

$$\varphi(t) = \int x(t)G(t)dt \tag{6.13}$$

$$\varphi = \gamma V G \tau T \tag{6.14}$$

It was not until 1987 when Dumoulin and Hart of GE added a version of Moran's bipolar flow-encoding gradients to an ECG-gated GRE sequence, to obtain the pulse sequence most commonly used up to this day in PC angiography [8]. To obtain 2D projection angiograms, they acquired two sets of images by changing polarity ("toggling" [7]) of the flow-encoding bipolar gradients applied first along the phase-encoding direction and then along the frequency-encoding direction (Figure 6.4 top). The *k*-space data for the two toggles of the gradient were subtracted, thus preserving the velocity-dependent phase shift. In each direction, the toggled bipolar gradient was followed by a velocity-compensated frequency (or phase)-encoding gradient, forming a five-lobe waveform. The waveform can be shortened to three lobes (Figure 6.4 bottom), by combining the flow-encoding lobes with

the frequency-encoding lobes, thus shortening the minimum TE and consequently decreasing TR and dephasing.

Dumoulin and colleagues expanded their technique to 3D in 1989 [9] by adding toggled bipolar gradients to each gradient direction, for a total of six images [5]. The toggling of bipolar gradients is highly effective at eliminating phase shifts of stationary spins due to field inhomogeneities and eddy currents, so that the net phase shift in stationary objects is zero [10]. The same effect, however, can be achieved by subtracting an initial image with no flow-encoding from one acquired with a flow-encoding gradient [10]. In 3D, this method reduces the six image acquisitions to four, but is more vulnerable to misregistration due to subjects' breathing and peristalsis [10].

6.7.4 Image Reconstruction and Types of PC Angiograms

Regardless of the acquisition modality used, phase shift due to flow velocity can be displayed in different ways, to create several types of angiograms suitable to different clinical applications. Image reconstruction can be performed by either complex difference or phase difference.

Complex difference reconstruction was presented in the early PC papers of Moran and Dumoulin [6,11]. Images are obtained by subtracting the *k*-space data obtained by opposite polarity flow-encoding gradients and then taking the inverse Fourier transform of the resulting magnitude image. Because Fourier transforms are linear, the same result is achieved by complex (vector) subtraction in the image domain. The complex difference is a relationship between the magnitudes of the signal in the subtracted pixels and a sine or cosine of the phase difference. If the toggling modality is used, the signal magnitude of pixels in the two images can be assumed to be the same, so the complex difference becomes [7]

$$CD = \sqrt{2}M\sqrt{1-\cos(\Delta\varphi)} = 2M\left|\sin\left(\frac{\Delta\varphi}{2}\right)\right| \qquad (6.15)$$

The average velocity value over a pixel, for flow in one direction can be calculated from Equations 6.2 and 6.3. Images sensitive to flow in all three directions (Dumoulin [9]), or speed displays, can be obtained by taking the square root of the sum of squared complex differences for flow encodings applied along each of the three directions [7]. However, in speed displays, phase does not have a linear relationship to velocity.

Complex difference images in general are not suitable for velocity quantification, as one cannot obtain a vector field–like map of velocities, which is why they are mostly used to generate projection angiograms and in 2D scout or localizer sequences [10]. However, complex difference reconstruction has the advantage of being unaffected by partial-volume effects, by which stationary tissue contributes signal to the phase difference, as it often happens at the low-velocity boundary between blood and vessel walls [7].

Phase difference images are obtained by subtracting the phases of the two images encoded by toggled bipolar gradients, or the flow-encoded image from the flow-compensated one, if this modality is used. In either modality, flow is encoded in one direction at a time, and signal intensity is proportional to phase, and thus to velocity, according to Equation 6.13. Thus, high-velocity flow in the "positive" axis direction appears bright, while flow in the opposite direction appears dark [10].

Because phase difference preserves both flow direction and magnitude, it is the technique of choice for velocity mapping and flow quantization [7,10]. A particular application is cine PC, in which images are acquired with and without bipolar gradients at different times during the cardiac cycle, with ECG triggering [10]. It is also possible to acquire a velocity map in 3D, without ECG gating, in which case the flow velocity is an average over one cardiac cycle [7,10].

Before the late 1990s, when 3D PC acquisition times were impractically long, imaging the direction of flow was performed with thick-slab 2D PC. While partial-volume effects are kept under control by setting the "thick slab" to smaller thickness, flow cannot be distinguished when two vessels cross within the same slice, as the flow through each vessel contributes to the measured phase [10]. Although the scan times have since been brought to more manageable values by parallel imaging, it is still difficult to measure 3D flow velocity vectors with these acquisition times because of breathing artifacts.

6.7.5 Choosing Pulse Sequence Parameters

In addition to choice of acquisition and reconstruction method, choice of area and spacing of bipolar gradient lobes allows tailoring of PC imaging to specific applications. In PC imaging, two parameters summarize the characteristics of the gradient waveform: gradient moment and velocity-encoding parameter (VENC).

The relationship between gradient moments and phase is easily obtained by expanding the displacement $x(t)$ in Equation 6.15 into a Taylor series around 0 (Equation 6.16). Writing Equation 6.15 this way makes it readily apparent that the phase is a weighted sum of spins' initial position, as well as higher time derivatives of displacement (Equation 6.16). The weights are the gradient moments, functions of gradient amplitude, and time (Equation 6.17):

$$x(t) = x_0 + v_0 t + \frac{1}{2}a_0 t^2 + \dots \qquad (6.16)$$

$$\varphi(t) = \int x(t)G(t)dt = \gamma(m_0 x_0$$
$$+ m_1 v_0 + \dots + \frac{1}{n!}m_n\left(\frac{d^n x}{d^n t}\right)_{t=0} + \dots \qquad (6.17)$$

$$m_n = \int_0^t G(t')t'^n dt' \qquad (6.18)$$

To encode particular derivatives of displacement (velocity, or acceleration) along one of the Cartesian axes into the phase, one has to set the lower order gradient moments to zero [7]. For example, to obtain a mapping of fluid acceleration, as necessary for local pressure mapping from flow, $m_0 = m_1 = 0$, nonzero m_2 is encoded. Usually, $m_0 = 0$, because the area under the bipolar gradient that stationary spins experience is null. Errors in calculation of velocity derivatives from phase arise when higher gradient moments are not null, because of a large *TE* during acquisition [7].

The most prominent errors in velocity estimation from phase are caused by aliasing. Since the phase angle (radians) has a finite range of $[-\pi, \pi]$ and is periodic with period $2k\pi$, where k integer, values of phase outside this range will be aliased, and so will be their corresponding velocity values. The velocity that corresponds to an absolute phase value of π is known as the aliasing velocity or VENC. In a toggle gradient sequence, VENC is inversely proportional to the change in the first moment [7]:

$$VENC = \frac{\pi}{\gamma \, |\Delta m_1|} \qquad (6.19)$$

6.8 True-FISP

Balanced steady-state free precession (bSSFP) (*true fast imaging with steady-state free precession* [true-FISP], fast imaging employing steady-state excitation [FIESTA], balanced fast field echoes [FFE]) is, unlike PC-MRA, a method for performing flow-independent MRA. In bSSFP, image acquisition is sped up by recycling, instead of spoiling, the transverse magnetization after each TR. There are two conditions for achieving steady-state free precession: First, the RF pulses over several TR intervals have to have the same phase in the rotating frame, or else a simple phase cycle such as sign alternation [7], and second, the phase accumulated by the transverse magnetization must be the same in each TR interval. To avoid dephasing of the transverse magnetization, the area under the gradient waveforms must be zero (gradients are completely balanced; see Figure 6.5).

The signal equation when steady state has been achieved is given by the following equation:

$$M_{SS} = \frac{M_0 \sin\theta \left(1 - E_1\right) e^{-TE/T_2}}{1 - \left(E_1 - E_2\right) \cos\theta - E_1 E_2} \qquad (6.20)$$

where
$E_1 = \exp(-TR/T_1)$
$E_2 = \exp(-TR/T_2)$

The signal equation illustrates the T_2/T_1 contrast specific to bSSP imaging. Thus, media with large T_2/T_1,

If VENC is lower than the maximum velocity in a blood vessel, signal intensity wrapping artifacts occur: positive velocities higher than VENC would appear smaller and of opposite direction, or as corresponding lower signal intensities next to high intensities for velocities close to, but lower than VENC. Just as setting the VENC too low can cause aliasing, setting it too high brings about a penalty in signal-to-noise ratio of phase measurements, which is inversely proportional to VENC [7]. Thus, the optimum VENC should be set slightly higher than the known maximum velocity in the vessels to be imaged. Doppler ultrasound measurements provide basic knowledge of what velocity limits one should expect in blood vessels of interest; however, VENC choices often depend on the imaging method (ECG gated or not) and reconstruction mode (complex difference or phase difference). A more computationally intensive way to work around fine-tuning the VENC is to use phase unwrapping postprocessing algorithms [7].

such as blood and fluids, will appear brighter in bSSFP images, which makes bSSFP naturally suitable for imaging blood vessel. The T_2/T_1 contrast can be further enhanced by choice of the optimal flip angle, $\alpha_{opt} = \alpha \cos(T_1/T_2 - 1)/(T_1/T_2 - 1)$, with expected T_1 and T_2 of the tissue to be imaged. At the optimal flip angle, the signal amplitude is proportional to the square root of T_2/T_1:

$$M_{SS} = \frac{1}{2} M_0 \sqrt{\frac{T_2}{T_1}} \qquad (6.21)$$

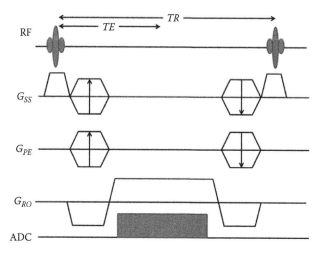

FIGURE 6.5 Pulse sequence diagram for a bSSFP pulse sequence.

However, the same contrast mechanism makes other tissues with large T2/T1, such as fat, appear hyperintense, which requires the use of fat suppression methods during bSSFP imaging.

6.8.1 QISS

Quiescent interval single shot (QISS) [12] is a bSSFP-based, cardiac-gated pulse sequence form MRA that takes advantage of the bright blood contrast in bSSFP while suppressing fat, venous inflow, and background tissue. Following the ECG R-wave, 90° slice selective saturation pulses are played to null the longitudinal magnetization of tissues within the slice. Another 90° tracking pulse is played to saturate the venous spins. The quiescent interval corresponds to peak systolic flow, which allows the signal to be replenished by the inflowing spins of the arterial blood. The quiescent period is followed by a chemical shift–selective fat saturation RF pulse and an RF $\alpha/2$ catalyzation pulse that forces the magnetization into steady-state free precession. bSSFP readout follows these preparation pulses. A two-center trial in 53 patients referred for lower extremity MRA found QISS to have equivalent diagnostic accuracy as contrast-enhanced MRA (CE-MRA) and DSA [1].

6.9 Contrast-Enhanced MRA

Gadolinium (Gd) is a highly paramagnetic element with maximum number of unpaired electrons in the outer shell. This property is causes shortening of T_1 relaxation of protons. The T_1 shortening is described by the following equation:

$$\frac{1}{T_1} = \frac{1}{T_{1_0}} + R_1 \times [\text{Gd}] \tag{6.22}$$

where

T_{1_0} is the longitudinal relaxation time of blood without Gd

T_1 is the shortened longitudinal relaxation time of blood to presence of Gd

r_1 is the relaxivity of Gd, which is multiplied by the concentration of [Gd]

Since the gadolinium itself is highly toxic, it must be chelated before it can be injected into the vessels. Currently, there are several different chelates widely used in CE-MRA, which include Gd-DTPA (Magnevist), gadodiamide (Omniscan), gadobenate (Multihance), and gadobutrol (Gadovist). These compounds are permeable through the blood vessel walls, enhancing tissue signals as well. Chelates that are not permeable through the vessel walls, called intravascular or blood pool agents, also exist (gadofosveset trisodium).

6.9.1 Contrast Agent Administration

Contrast agent is injected as a bolus during the scan using a power injector, which delivers with accuracy in dose and rate. Typically, 0.1 mmol/kg (considered single dose) is injected at 0.5–5 mL/s. Saline flush immediately follows to clear the venous access and push the contrast agent further into the circulation.

Timing of injection used to be an important issue, since it takes some time for the peak of the bolus to go from the venous system to the heart and then to the vessels of interest. With conventional MRA, the imaging and injection was carefully timed so that the peak of the bolus occurs at acquisition of the center of k-space for maximum signal. However, with radial acquisition with sliding window reconstruction, no timing of the injection is necessary.

6.9.1.1 Bolus Timing and Venous Overlay

Typically, an injection volume of contrast agent of 20 mL is injected into the veins over a short time interval (5–10 s). The goal of CE-MRA arterial imaging is to image the maximum arterial enhancement, achieved when the injection bolus first passes through the artery of interest at its peak concentration. Therefore, acquisition speed is the main design criterion for first-pass CE-MRA pulse sequences. Usually, CE-MRA pulse sequences are SPGR pulse sequences that minimize TE and TR and very often use partial k-space acquisitions in two or three directions. CE-MRA venography does not have this time constraint, which permits the acquisition of high SNR and spatial resolution images after the veins enhance.

Single-phase CE-MRA sequences use the entire acquisition time to obtain data for one set of 3D images in order to increase spatial resolution and coverage. This method requires timing of the bolus arrival, by timing a test bolus or fluoroscopic triggering.

Time-resolved MRA images the volume of interest at different moments in time, which allows visualization of of the various stages of signal enhancement in the vessels. Time-resolved MRA frees the operator from having to measure the bolus circulation time (time from injection to peak arterial enhancement in the vessel of interest). However, since multiple acquisitions of the imaging volume have to be performed into a time interval comparable to the circulation time of the bolus, acquisition speed is even more important than in single-phase CE-MRA. To decrease acquisition time, parallel imaging (SENSE) or partial k-space updating methods such as keyhole imaging or TRICKS are used. The keyhole method updates the center of k-space periodically, while TRICKS reacquires high-spatial-frequency data that contain the imaging information on small vessels.

6.9.2 Blood Pool Contrast Agents

MS-325 (gadofosveset trisodium; Ablavar, Lantheus Medical Imaging; AngioMark, Vasovist; EPIX Pharmaceuticals, Schering AG) is a gadolinium-based, large-molecule, T_1-shortening contrast agent that, unlike other gadolinium (Gd) chelates, binds reversibly to serum albumin, in a proportion of 80%–90% for humans [13]. In 2010, it has obtained FDA approval for MRA in the United States and has been used for over a decade in Europe for angiographic applications. Albumin binding slows down the leakage of gadofosveset into extravascular space and increases its half-life, so that the T_1 shortening effect is observed up to 4 h postinjection and interpretable steady-state images can be obtained up to an hour postinjection [13]. The reversible binding allows excretion through the kidneys, or uptake by hepatocytes [14]. Excretion of unbound agent by the kidneys makes repeat injections feasible, unlike with other intravascular contrast agents [15]. Previous versions of albumin-bound Gd-DTPA, the prototype compounds HSA-(Gd-DTPA)$_{30}$, and polylysine-Gd-DTPA, did not extravasate and displayed high signal enhancement, but could not be excreted through the kidneys, which made them unsuitable for human use.

The albumin bond also slows down the rotation rate of the complex, which enhances its relaxivity (T_1-shortening effect, in mmol/L/s) 6–10 times compared to other nonbinding Gd chelates [13] and 4–10-fold compared to unbound gadofosveset [14]. As consequence of the high relaxivity/relaxation rate $1/T_1$ and of the long half-life in the vessel, the gadofosveset dosage necessary for quality perfusion imaging is lower than that of Gd-DTPA (0.03 mmol/kg

vs. 0.1 mmol/kg). With lower doses, T_2^* effects, as well as patient reaction to Gd, are of lesser concern. Gadofosveset has been characterized as intravascular in the literature [13,14,16], so leakage can be assumed nonexistent. However, we may find that gadofosveset violates the assumption of a truly intravascular agent in animal models [17], due to its variable and moderate binding affinity to albumin and plasma proteins across species, which results in variable T_1 relaxation rates and possibly some leakage [14]. Toxicity studies in animals [18], as well as trials in healthy volunteers, have shown gadofosveset to have the same tolerability as conventional contrast agents [13].

Gadofosveset is particularly advantageous to MRA because images can be acquired with the first pass of the contrast agent bolus, as well as in steady state (1–5 min to up to 1 h postinjection) [19]. Steady-state MRA with gadofosveset allows the acquisition of bright, high-resolution images of the smaller branches of the vasculature [20]. Multicenter studies in Europe have shown comparative efficacy of gadofosveset MRA to conventional x-ray angiography [21,22]. In the past 2 years, gadofosveset MRA has expanded from peripheral vessel imaging to thoracic imaging [23] and the imaging of aortic endoleaks [24] and abdominal perforators [25].

6.9.2.1 Ferumoxytol

Ferumoxytol (Feraheme, AMAG Pharmaceuticals, Lexington, MA) is an ultrasmall paramagnetic iron oxide (USPIO) that has obtained FDA approval as an injectable iron supplement for the treatment of severe anemia. It has been used off-label as an MRA blood pool agent, in the imaging of the carotid arteries, thoracic and abdominal aorta, and peripheral vessels. Initial studies have shown the feasibility of first-pass and steady-state MRA in humans, with the initial concentration of 537.2 µmol elemental Fe/mL diluted four- (134.3 µmol Fe/mL) or eightfold (67.1 µmol Fe/mL). The highest dose for MRA was 71.6 µmol/kg, injected at a rate of 1 mL/s. Dilution of the contrast agent minimizes its marked T_2^* shortening effects, which can lead to signal loss [26,27].

6.9.3 FLASH Imaging

SPGR pulse sequences are the most common choice for CE-MRA since they provide fast acquisition of T_1-weighted contrast images. These sequences are known by different names, depending on the vendor: SPGR, T_1 fast field echo (T_1-FFE), and FLASH. FLASH achieves most spatially and temporally uniform spoiling of the transverse magnetization by phase cycling the RF

Chapter 6

excitation pulses according to a preset schedule, in addition to a phase-encoding rewinder gradient. In each TR, the RF excitation converts the longitudinal magnetization into transverse magnetization, which is rephased by a gradient echo. If perfect spoiling is achieved, the transverse magnetization is zero before the new TR begins (there is no residual transverse magnetization from previous TR intervals). After a sufficient number of excitation pulses are applied, the longitudinal magnetization reaches steady state, and the FLASH signal at steady state is dependent on T_1, T_2^* and the flip angle:

$$S_{SS} = \frac{M_0 \sin\theta (1 - E_1)e^{-TE/T_2^*}}{1 - \cos\theta E_1} \tag{6.23}$$

6.10   Non-Cartesian Sampling

Most MRI data are sampled on a rectilinear grid [28] using spin-warp imaging. This Cartesian method of sampling was adopted due to its reliability against inhomogeneity-inducing scanner imperfections in the static magnetic field. With recent advances in scanner hardware and consequent improvements in static field homogeneities, clinical and research protocols utilizing non-Cartesian sampling schemes have been increasingly explored. In the following paragraphs, we will address a very popular non-Cartesian sampling technique called radial sampling.

6.10.1   General Considerations

6.10.1.1   Nyquist Criteria

Although radial trajectories have recently become a popular focus in MRI research, it is in fact a relatively old technique, developed and used by Paul Lauterbur in his first MRI acquisitions [29]. Instead of acquiring k-space lines orthogonally in Cartesian k-space, radial spokes are acquired in equal angular spaces, as shown in Figure 6.6.

Every radial spoke, or projection, acquires N_r read out points in accordance with Nyquist criteria. As a result, the radial trajectory oversamples the center of k-space and under samples the edges. The oversampling of low-frequency central k-space has known advantages in time-resolved imaging [26]; however, larger numbers of projections are required to meet the minimum Nyquist criteria. Those criteria state that the largest spacing between adjacent k-space samples must be no larger than 1/FOV. For points on adjacent spokes at the edge of k-space where the spacing is largest, we have the following:

The FLASH signal is maximized at a flip angle called the Ernst angle:

$$\theta_E = \cos^{-1}\left(e^{-TR/T_1}\right) \tag{6.24}$$

Since the Ernst angle increases monotonically as the ratio TR/T_1 increases, TR values much shorter than T_1 are an advantage, not a trade-off, to signal intensity. In SPGR, much shorter TR values ($TR \ll T_1$) are possible than in spin echo, since no lengthy period of time is required for T_1 recovery. Contrast is also dependent on T_2^*, not on T_2, so $TE \ll T_2$. These properties of the sequence make the fast ($TR \sim 2$–50 ms) acquisition needed in CE-MRA possible.

According to the Nyquist theory, we must have $\Delta k_{\max} < 1/\text{FOV}$ and $\Delta k_r = 1/\text{FOV}$. The number of projections required to meet Nyquist criteria within the FOV is therefore

$$\Delta k_{\max} = \frac{N_r \pi}{2N_p}\Delta k_r \tag{6.25}$$

According to the Nyquist theory, we must have $\Delta k_{\max} < 1/\text{FOV}$ and $\Delta k_r = 1/\text{FOV}$. The number of projections required to meet Nyquist criteria within the FOV is therefore

$$N_p \geq \frac{\pi}{2}N_r \tag{6.26}$$

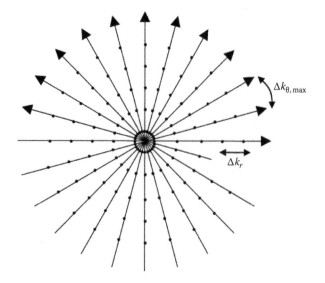

FIGURE 6.6   Radial k-space trajectory.

Acquiring fewer than $N_{p,opt}$ results in a reduction of the artifact-free FOV (CITE Scheffler reduced FOV paper).

6.10.1.2 Spoiling

Spoiling in gradient recalled echo pulse sequences is addressed differently in non-Cartesian scans as the readout direction is changing with each projection acquisition. Spoiling residual transverse magnetizations equally at the end of each read out interval requires balancing of gradient moments. A simple method of accomplishing this may be to first rewind all gradients to the origin and subsequently wind the magnetization to the same point in k-space along the k_x or k_y axis. Two methods for doing this are shown in Figure 6.7. This ensures the net magnetization at the end of each TR interval is at the same point in k-space [27].

6.10.2 Benefits for Radial MRA

6.10.2.1 Reduced Sampling Region

Two popular strategies for sampling 3D k-space with radial trajectories include cylindrical sampling with a radial stack of stars or spherical sampling using the true radial VIPR sequence (Figure 6.8). In each scheme, there is a reduction in the total amount of sampled space relative to a cube sampled on a 3D Cartesian grid. For example, a cylindrical sampling scheme acquires 21.5% less sampling and spherical sampling 47.6% less sampling relative to a cube.

Radial trajectories sample the center of k-space often and at regular time intervals. This has an averaging effect, which serves to reduce motion artifacts evident in Cartesian sampling schemes. In addition, studies have shown that Cartesian acquisitions can distort the bolus profile during the vascular time course depending on the direction of travel relative to the gradient read out direction, while radial acquisitions produce more uniform bolus profiles regardless of bolus direction.[6]

6.10.3 Reconstruction of Non-Cartesian Data

Data acquired on Cartesian rectilinear grids are easily reconstructed into images using a simple inverse Fourier transformation. However, data sampled by radial trajectories must first be resampled onto a Cartesian grid in order for the Fourier transform to be utilized. Acquired non-Cartesian data are first interpolated by convolution with a smoothing function and then resampled onto a Cartesian grid. The entire

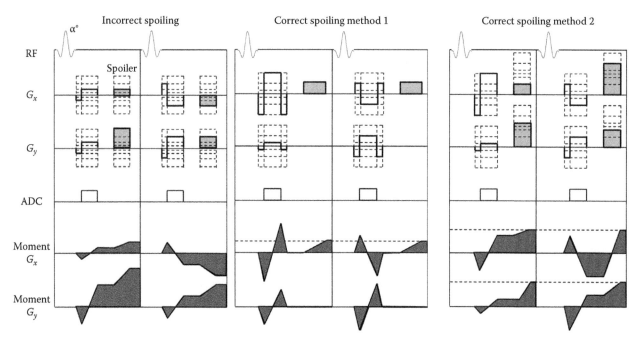

FIGURE 6.7 Pulse sequence diagrams of consecutive projections in pseudorandom radial FLASH acquisitions showing examples of incorrect spoiling and two correct spoiling methods. Incorrect spoiling results in unbalanced gradient moments visualized in the blue-shaded moment plots. The initial projection has a net positive moment accrual along G_x, while the second projection has a net negative moment accrual along G_x. In each of the corrected spoiling methods, the magnetization is winded out to the same point in k-space resulting in balanced gradient moments among consecutive projections. In method 1, the G_x and G_y gradients are rewound and a constant spoiler is played along G_x. In method 2, the spoiler moment is adjusted in accordance with the projection angle, ensuring equal positive moments along G_x and G_y at the end of the TR interval.

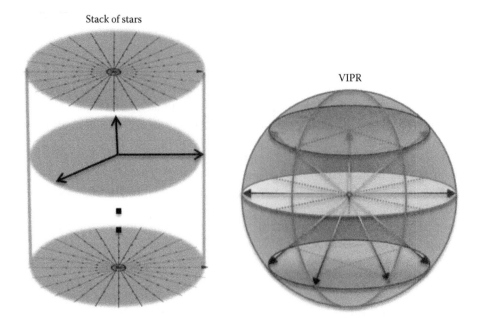

FIGURE 6.8 3D radial *k*-space sampling schemes include stack of stars or cylindrical sampling and "true radial" or VIPR sampling.

process including the inverse Fourier Transform is called gridding. Radially sampled data are sometimes reconstructed using filtered backprojection, similar to the reconstruction of computed tomography data; however, this method is considerably less commonly used than gridding and will not be addressed here.

The interpolative process of gridding uses a convolution kernel and accounts for differences in sampling density. Since *k*-space data are nonzero for a finite space, sampling theory informs us that any point can be sampled accurately with SINC interpolation at or above Nyquist sampling frequency. This results in multiplication by a RECT function in the image domain and hence suppresses aliasing. However, since the SINC function has infinite extent, the calculation of each new *k*-space point requires the multiplication of all points with the SINC function, a process that is time-intensive. For that reason, a convolution kernel that has finite extent is utilized. However, the corresponding Fourier transform is not finite, potentially resulting in image aliasing. The choice of interpolation kernels therefore trades off speed and accuracy. Finally, convolution in *k*-space amounts to multiplication in the image domain resulting in shading in the reconstructed image. For this reason, the final image is typically divided by the Fourier transform of the convolution kernel.

6.10.3.1 Kaiser–Bessel Kernel

Considerable efforts have undergone the identification of the best convolution kernel in terms of speed and

accuracy. Many agree that the Kaiser–Bessel window is the most precise for *k*-space convolution in terms of accuracy of the reconstructed image, determined with comparison to SINC interpolation using least-squares metrics [30]. The 1D Kaiser–Besel function is given by the equation

$$KB(k) = \frac{1}{w} I_0 \left(b \left[1 - \left(2k/w \right)^2 \right] \right) RECT(2k/w) \quad (6.27)$$

where
w is the width of the kernel
b a scaling parameter
I_0, a zero-order modified Bessel function of the first kind and $RECT(x)$ a rectangle function defined by

$$RECT\left(x\right) = \begin{cases} 1, |x| \le 1 \\ 0, |x| > 1 \end{cases}$$

The width parameter w is typically chosen to be 2–4 *k*-space samples. Increasing w suppresses aliasing artifacts in the image domain, but also increases the computation time.

6.10.3.2 Sampling Density

Since the data are denser as it gets closer to the origin, filters such as Ram-Lak are used to weight radial points accordingly during the regridding process. Then 2D Fourier transform is performed on the regridded data. There are a number of algorithms for calculating the sampling density with the most popular being an iterative approach proposed by Pipe [31].

6.10.3.3 Sliding Window Reconstruction

Even with previously discussed methods in section, the acquisition time is still lengthy. With partial Fourier in all three dimensions, a typical MRA would take about 10 s to acquire one 3D volume. This results in a frame rate of 0.1 frame/s, whereas x-ray DSA uses 3–6 frames/s to capture the hemodynamics.

A technique to achieve higher frame rate is to use sliding window reconstruction, first developed by Riederer [32]. This method reconstructs intermediate frames between the actual measurements by combining echoes from two consecutive volumes. After the first measurement is reconstructed, it is updated by replacing the "oldest" k-space lines of the image with the corresponding lines from the next measurement. After the updated volume is reconstructed, it is again updated by replacing the next "oldest" k-space lines, which are now the "oldest" in the latest frame. When all the lines are replaced, the image will be equal to the second measurement. This process is repeated with the second and third measurements.

This technique is suitable for radial acquisition since each line has information about the center of k-space, so each update is equivalent. The slice loop needs to be inside the angular loop of course, to update projections. For Cartesian acquisitions, each phase-encoding line contains different spatial frequencies. Therefore, it would not capture motion smoothly since the center of k-space is not sampled as often. Figure 6.9 shows a diagram of the radial sliding reconstruction.

Each update could be as small as one projection, which could result in a frame rate of about 12 frames/s for an acquisition with 128 projections per image. However, although the image is *updated* more often, the update rate is not the true temporal resolution. Each image still contains about 10 s of dynamic information, whereas in x-ray DSA, each frame is acquired instantaneously. The direct result of having 10 s of dynamic information is temporal blurring of the images. In order to increase the true temporal resolution, acquisition time needs to be shortened.

One approach to mitigating the effects of temporal blurring is sliding mask subtraction that can be used in addition to the sliding window reconstruction. Conventional subtraction in CE-MRA datasets uses an initial precontrast acquisition as the "mask" volume, which is subtracted from every subsequent contrast acquisition. While the imaging acquisition continues the temporal separation between the mask and the imaging slice increases. In pulse sequences with long imaging times, this can result in blurring of arterial

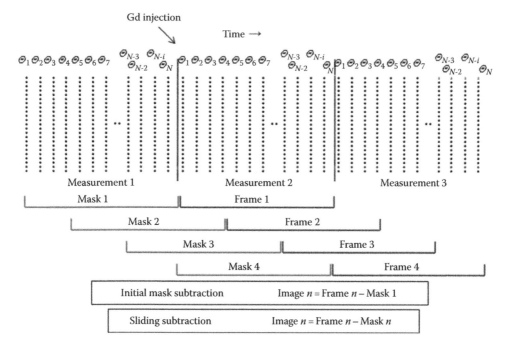

FIGURE 6.9 Radial sliding window reconstruction with sliding mask. The N radial acquisitions $\theta_1,\dots\theta_N$ make up one measurement repetition. Intermediate frames are reconstructed by sharing echoes from two consecutive measurements that fall in the sliding time window. Sliding mask subtraction uses a subtraction mask reconstructed in a similar manner that follows the frame. Conventional subtraction, or initial mask, uses Mask 1 for all frames.

and venous phases. Utilizing a mask that advances synchronously in time with the imaging acquisition improves the arterial and venous separation [26]. This is especially useful for diagnosis of vascular diseases that feature high-flow pathology such as arteriovenous malformations and fistulas.

6.10.4 Radial Imaging Errors

There are errors associated with radial imaging. One error is from gradient delays. To sample echoes radially, the x and y gradients need to turn on at the same time. But there exists some delay in turning on the gradients in the order of few microseconds. Moreover, the delays are different for each gradient coil and the model of the MR scanner. These delays cause the radial lines to be centered incorrectly, offset from the origin. Plus, the radial lines in different projection angles do not intersect at a single point. These errors appear as streaks at the edges and uneven shading on the images. For Cartesian trajectories, the gradient delays do not result in any artifacts.

The gradient delay can be corrected by several different ways. One method involves measuring gradient delay for each of the scanner and adjusting the gradient amplitude during the scan [33]. If the delays are known, gradient amplitudes can be calculated to make adjustments so that the projection intersects the origin. Image-based postprocessing methods are also available, where the algorithm iteratively corrects the images based on user-defined constraints [34].

Another type of artifacts in radial imaging is clock-shift errors. The phases of the numeric crystal oscillator of the scanner sometimes may not align with the phase of the echo. This results in no visible errors for Cartesian imaging, since readout direction is always the same and because phase shift is constant. But for radial imaging, the readout direction changes for each scan. The varying phase shifts for each readout cause artifacts resembling halos around the objects.

Similar to gradient delays, the phase shifts can also be measured and corrections can be applied prior to scanning.

6.11 Conclusions

We have presented an overview of noninvasive, high-resolution, and high-frame-rate MRA techniques that make noninvasive visualization of arteries and veins possible. We chose to give equal consideration to exogenous and endogenous contrast methods, in the context of concerns about nephrotoxicity and nephrogenic systemic fibrosis associated with gadolinium-based contrast agents. Knowledge of both contrast and noncontrast MRA would allow clinicians to choose the best imaging method for their patients on a case-by-case basis.

References

1. Prince MR, Yucel EK, Kaufman JA, Harrison DC, Geller SC. Dynamic gadolinium-enhanced three-dimensional abdominal MR arteriography. *J Magn Reson Imaging* 1993;3(6):877–881.
2. Nishimura DG. *Principles of Magnetic Resonance Imaging*; Stanford University Press, Stanford, CA, 1996.
3. Barger AV, Block WF, Toropov Y, Grist TM, Mistretta CA. Time-resolved contrast-enhanced imaging with isotropic resolution and broad coverage using an undersampled 3D projection trajectory. *Magn Reson Med* 2002;48(2):297–305.
4. Pipe JG. Motion correction with PROPELLER MRI: Application to head motion and free-breathing cardiac imaging. *Magn Reson Med* 1999;42(5):963–969.
5. Graves MJ. Magnetic resonance angiography. *Br J Radiol* 1997;70:6–28.
6. Moran PR. A flow velocity zeugmatographic interlace for NMR imaging in humans. *Magn Reson Imaging* 1982;1(4):197–203.
7. Bernstein M, King K, Zhou X. *Handbook of MRI Pulse Sequences*; Elsevier Academic Press, Amsterdam, the Netherlands, 2004.
8. Dumoulin CL, Souza SP, Hart HR. Rapid scan magnetic resonance angiography. *Magn Reson Med* 1987;5(3):238–245.
9. Dumoulin CL, Souza SP, Walker MF, Wagle W. Three-dimensional phase contrast angiography. *Magn Reson Med* 1989;9(1):139–149.
10. Edelman RR, Hesselink J, Zlatkin M. Chapter 27. *Clinical Magnetic Resonance Imaging*, 3rd ed.; Saunders-Elsevier, Philadelphia, PA, 2005, pp. 696–711.
11. Dumoulin CL, Hart HR, Jr. Magnetic resonance angiography. *Radiology* 1986;161(3):717–720.
12. Edelman RR, Sheehan JJ, Dunkle E, Schindler N, Carr J, Koktzoglou I. Quiescent-interval single-shot unenhanced magnetic resonance angiography of peripheral vascular disease: Technical considerations and clinical feasibility. *Magn Reson Med* 2010;63(4):951–958.
13. Henness S, Keating GM. Gadofosveset. *Drugs* 2006;66(6):851–857.
14. Eldredge HB, Spiller M, Chasse JM, Greenwood MT, Caravan P. Species dependence on plasma protein binding and relaxivity of the gadolinium-based MRI contrast agent MS-325. *Invest Radiol* 2006;41(3):229–243.
15. Kraitchman DL, Chin BB, Heldman AW, Solaiyappan M, Bluemke DA. MRI detection of myocardial perfusion defects due to coronary artery stenosis with MS-325. *J Magn Reson Imaging* 2002;15(2):149–158.

16. Kroft LJ, de Roos A. Blood pool contrast agents for cardiovascular MR imaging. *J Magn Reson Imaging* 1999;10(3):395–403.

17. Corot C, Violas X, Robert P, Gagneur G, Port M. Comparison of different types of blood pool agents (P792, MS325, USPIO) in a rabbit MR angiography-like protocol. *Invest Radiol* 2003;38(6):311–319.

18. Steger-Hartmann T, Graham PB, Muller S, Schweinfurth H. Preclinical safety assessment of Vasovist (Gadofosveset trisodium), a new magnetic resonance imaging contrast agent for angiography. *Invest Radiol* 2006;41(5):449–459.

19. Goyen M. Gadofosveset-enhanced magnetic resonance angiography. *Vasc Health Risk Manag* 2008;4(1):1–9.

20. Grist TM, Korosec FR, Peters DC, Witte S, Walovitch RC, Dolan RP, Bridson WE, Yucel EK, Mistretta CA. Steady-state and dynamic MR angiography with MS-325: Initial experience in humans. *Radiology* 1998;207(2):539–544.

21. Goyen M, Edelman M, Perreault P, O'Riordan E, Bertoni H, Taylor J, Siragusa D et al. MR angiography of aortoiliac occlusive disease: A phase III study of the safety and effectiveness of the blood-pool contrast agent MS-325. *Radiology* 2005;236(3):825–833.

22. Rapp JH, Wolff SD, Quinn SF, Soto JA, Meranze SG, Muluk S, Blebea J et al. Aortoiliac occlusive disease in patients with known or suspected peripheral vascular disease: Safety and efficacy of gadofosveset-enhanced MR angiography—Multicenter comparative phase III study. *Radiology* 2005;236(1):71–78.

23. Frydrychowicz A, Russe MF, Bock J, Stalder AF, Bley TA, Harloff A, Markl M. Comparison of gadofosveset trisodium and gadobenate dimeglumine during time-resolved thoracic MR angiography at 3T. *Acad Radiol* 2010;17(11):1394–1400.

24. Wieners G, Meyer F, Halloul Z, Peters N, Ruhl R, Dudeck O, Tautenhahn J, Ricke J, Pech M. Detection of type II endoleak after endovascular aortic repair: Comparison between magnetic resonance angiography and blood-pool contrast agent and dual-phase computed tomography angiography. *Cardiovasc Intervent Radiol* 2010;33(6):1135–1142.

25. Zou Z, Kate Lee H, Levine JL, Greenspun DT, Allen RJ, Vasile J, Rohde C, Prince MR. Gadofosveset trisodium-enhanced abdominal perforator MRA. *J Magn Reson Imaging* 2012;35(3):711–716.

26. Cashen TA, Jeong H, Shah MK, Bhatt HM, Shin W, Carr JC, Walker MT, Batjer HH, Carroll TJ. 4D radial contrast-enhanced MR angiography with sliding subtraction. *Magn Reson Med* 2007;58(5):962–972.

27. Vakil P, Ansari SA, Hurley MC, Bhat H, Batjer HH, Bendok BR, Eddleman CS, Carroll TJ. Magnetization spoiling in radial FLASH contrast-enhanced MR digital subtraction angiography. *J Magn Reson Imaging* 2012;36(1):249–258.

28. Edelstein WA, Hutchison JM, Johnson G, Redpath T. Spin warp NMR imaging and applications to human whole-body imaging. *Phys Med Biol* 1980;25(4):751–756.

29. Lauterbur PC. Progress in n.m.r. zeugmatography imaging. *Philos Trans R Soc Lond B Biol Sci* 1980;289(1037):483–487.

30. Jackson JI, Meyer CH, Nishimura DG, Macovski A. Selection of a convolution function for Fourier inversion using gridding [computerised tomography application]. *IEEE Trans Med Imaging* 1991;10(3):473–478.

31. Zwart NR, Johnson KO, Pipe JG. Efficient sample density estimation by combining gridding and an optimized kernel. *Magn Reson Med* 2012;67(3):701–710.

32. Riederer SJ, Tasciyan T, Farzaneh F, Lee JN, Wright RC, Herfkens RJ. MR fluoroscopy: Technical feasibility. *Magn Reson Med* 1988;8(1):1–15.

33. Peters DC, Derbyshire JA, McVeigh ER. Centering the projection reconstruction trajectory: Reducing gradient delay errors. *Magn Reson Med* 2003;50(1):1–6.

34. Lee KJ, Paley MN, Griffiths PD, Wild JM. Method of generalized projections algorithm for image-based reduction of artifacts in radial imaging. *Magn Reson Med* 2005;54(1):246–250.

Chapter 6

7. Carotid Ultrasound Imaging
Physics, Technology, and Applications

Kai Erik Thomenius

7.1   Introduction

Imaging of the carotid arteries with ultrasound is a mature field that continues to grow and expand at a steady pace both due to technological improvements and newer imaging applications with somewhat different clinical roles. This clinical area is one of the oldest applications of ultrasound with publications going back to the early 1960s with the work of Robert Rushmer, Eugene Strandness, and colleagues.[1,2] The clinical focus of these early years was that of locating flow abnormalities associated with hemodynamically significant plaques in the common and internal carotid arteries. Initially, techniques such as Doppler ultrasound were used blindly without any imaging support. Duplex techniques combined these two technologies and permitted image guidance for the identification of critical sites for flow measurement. By the mid-1980s, technology had advanced to the point that 2D Doppler interrogation of blood flow became possible; this significantly speeded up clinical examinations. Subsequently, additional waves of advancement have occurred and are ongoing. It is a key goal of this chapter to review the most important of these.

This chapter is organized in the following manner. Several sections are dedicated toward a review of the physics and technology behind ultrasonic scanners aimed mainly at imaging the carotid artery. Some material is included that covers the basics of ultrasonic imaging; the goal is not to be comprehensive but rather give enough information and references for the reader to pursue topics of interest as desired. Typical specifications of these scanners are described in relation to the carotid anatomy being imaged. The following sections cover the carotid anatomy and pathologies often diagnosed by ultrasound and the role of the imagers in these areas. Finally, a section is dedicated to the several advanced topics, which are currently being researched but which may have not yet reached the clinic.

Chapter 7

7.2   Physics and Technology of Medical Ultrasound

Looking at it broadly, the use of ultrasound in clinical applications is based on a pulse-echo methodology much like radar and sonar. Briefly, a burst of acoustic energy is transmitted into tissue from a transducer array. These acoustic waves encounter variations in mechanical properties (density and compressibility) of tissue and generate echoes, which are received by the same transducer array and converted into electrical signals. This process is described in the block diagram in Figure 7.1. The transmitted acoustic energy is usually focused in a desired direction and a focal depth. This process is repeated as needed to form a 2D image of the region of interest (see Figure 7.2). In newer scanners, the array can be moved mechanically to form a 3D image. It is likely these mechanical 3D scanners will be replaced by fully electronic acquisition of volumetric image data in the near future. Processors within the scanner take the echo information and process it to form an image of the scanned tissue.[3,6] The following sections elaborate on these process steps.

Echo generation in tissue, as hinted earlier, depends on the local acoustic impedance, Z_{AC}, usually given in units referred to as Rayls (for Lord Rayleigh). For most tissues, the values lie in the vicinity of 1.5 MegaRayls. This parameter is usually defined as follows:

$$Z_{AC} = \rho_o c = \sqrt{\frac{\rho_o}{\kappa}},$$

where

ρ_o is the density in kg/m^3

c is the local speed of sound in m/s

κ is the tissue compressibility or bulk modulus in pascals

Variations in these quantities at any tissue bed or pathology site define the amount of acoustic energy scattered from such target and hence play a key role in establishing the conspicuity of that target. Thus, the diagnostic potential of ultrasound depends on the interaction of a propagating acoustic wave with variations in the local acoustic impedance. Sufficient differences do exist with major organs (e.g., liver vs. kidneys) and their boundaries that permit easy anatomic identification. In B-mode (B for brightness) imaging, the amplitude of the echoes from the scattering sites is displayed on a spatial grid defined by the assumed speed of sound and the direction of the acoustic beam as in Figure 7.3. This figure shows a fetal face in amniotic fluid; at the top of the image is the maternal placenta. Most fluid-filled targets such as blood vessels, cardiac chambers, regions with amniotic fluid, or cystic fluids are readily detectable due to the lack of obvious scattering in B-mode imaging within such chambers or vessels. A good example of this is shown in Figure 7.4, where the dark lumens associated with the carotid artery bifurcation are clearly visible. It should be noted that targets such as red blood cells, while not visible in such images as Figure 7.4, do scatter enough acoustic energy to permit detection and analysis of blood flow by Doppler techniques. These techniques form an important aspect of carotid artery imaging by ultrasound.

In addition to major sources of backscatter such as organ or blood vessel boundaries, most tissues have density and compressibility variations that are nonresolvable by clinically suitable transducer frequencies. Analogous to laser speckle, these sources generate a

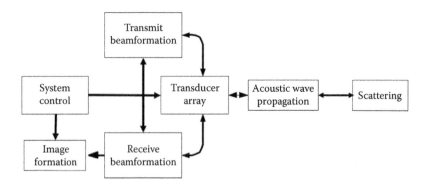

FIGURE 7.1   Typical scanner block diagram. The transmit beamformer applies appropriately delayed electrical signals to an array of transducer elements that convert those signals to acoustic form. The acoustic energy propagates in tissue and scatters back echoes during this transit. These echoes are converted back to electrical form by the same transducer array and are eventually delayed to form an electrical signal related to the acoustic impedance mismatches encountered by the transmitted beam. This last signal will be used to form one line of the resultant ultrasound image.

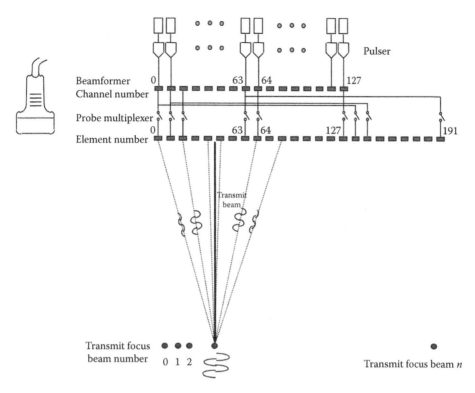

FIGURE 7.2 Linear array element selection. The displayed scheme shows how a linear array is used to form an acoustic beam out of a subset of the entire array of elements. In the case shown, 66 elements are participating in forming a beam aimed at the transmit focus. If we wish to get acoustic data from adjacent areas, we merely need to turn off element 0 and turn on element 66 and the centroid of the acoustic beam has shifted by one element pitch. In addition to this geometric beam selection, one can also use electronic delays, as suggested in Figure 7.5.

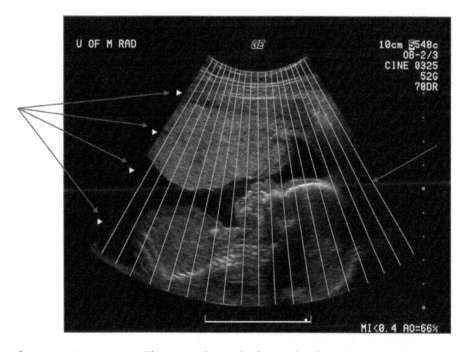

FIGURE 7.3 Image data acquisition sequence. This image shows a background with an ultrasound image of a fetal head and part of the torso. The bright lines indicate the data acquisition pattern typical of this type of an ultrasound scanner. For the depth shown in the image, about 10 cm, the round trip travel time for an ultrasound beam is close to 130 μs. One such image might have some 100 lines giving a frame rate of 13 ms. However, as suggested by the arrows on the left, one can use multiple transmit focal locations; this will, of course, slow down the frame rate proportionately.

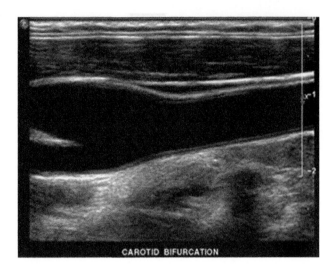

FIGURE 7.4 Normal carotid artery bifurcation. The upper branch is the internal carotid; the lower or thinner branch is the external. The thin line on the far wall corresponds to echoes from the intimal lining of the carotid. The brighter band of echoes immediately below it is due to the adventitial layer. The distance between them is the intima–media thickness or the c-IMT. The arrows indicate the outer boundary of the intima layer. (Courtesy of http://www.specialistvascularclinic.com.au/carotid-interventions.html.)

speckle pattern that carries some useful information about the underlying tissues. This pattern is composed of "speckle cells," which represent areas of constructive and destructive interferences from these unresolved scatterers.

Much of image formation in medical ultrasound depends on the assumption of uniformity of speed of sound in tissue. The displayed depth of acoustic scatterers and the beamformation process during transmit-and-receive operation are based on it (see Figure 7.3 where the position of the fetal features is based on

the speed of sound). Delay calculations for Figure 7.5 assume a speed of sound value. This condition is clearly not met in human tissues where the speed of sound varies from values as low as 1400 m/s in adipose tissue to over 3000 m/s and higher in certain types of bone. However, in most cases with soft tissues, the range is from about 1400 to 1600 m/s. The great majority of scanners assume a constant speed of sound of 1540 m/s, a representative value over a wide variety of tissues, and this appears to be a sufficiently accurate first-order approximation. Increasingly new scanners give a user an option to vary this nominal assumption; this is particularly important in certain imaging areas such as the breast where the speed of sound can average as low as 1430 m/s.

Spatial resolution in medical ultrasound is largely dependent on the acoustic frequency and spectral bandwidth of the transmitted signal and the size of the aperture being used during the transmit-and-receive operations. Spatial resolution typically is not isotropic and has clear lateral, elevational, and axial characteristics. Here lateral is used to refer to the resolution dimension, which is within the image plane and perpendicular to the acoustic beam. Elevational spatial resolution refers to the slice thickness at any point in the ultrasound image. Unlike some other imaging modalities (e.g., MRI), the slice thickness with ultrasound images is variable and depends on how the acoustic energy is focused in the elevation direction. Finally, axial resolution refers to the resolution along the acoustic beam. The operating frequencies of the large majority of scanners fall into the decade from 1.5 to 15 MHz. For carotid artery imaging, the upper

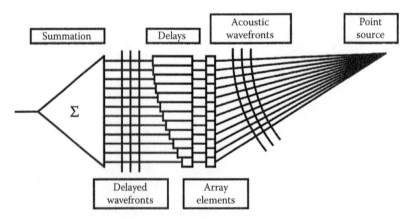

FIGURE 7.5 Beamformation process for a steered and focused beam from a linear array. A point source, after receiving an acoustic wavefront, will return echoes that form a spherical wavefront as shown. These echoes are converted to electrical form by the transducer array. Each of the signals from the array elements undergoes an electrical delay calculated to compensate for the travel time differences from the desired focal point to each of the array elements. The echoes from the point source will be completely coherent after the beamforming delays and will add up constructively during the summation step. The end result, shown as the output of the summer, is referred to as the beamsum signal.

end of this range if often used with transducers ranges from 7 to 12 MHz. If we assume a speed of sound of 1540 m/s, this will translate to a wavelength of around 125–200 μm. As a rough estimate, laboratory measurements of lateral resolution are less than 500 μm, typically around 300 μm. The elevational resolution tends to have a wide range over the entire image and ranges from submillimeter values to possibly as high as 1 or 2 mm. Axial resolution is fairly uniform throughout the image and usually corresponds to one or two cycles of the pressure wave. In actual imaging, the resolution values given become somewhat worse due to the aforementioned speed of sound variations and the errors that these introduce to the beamformation process. In addition, the attenuation suffered by propagating acoustic waves is frequency dependent roughly to the first power in frequency. Over larger depths, this preferentially attenuates the higher frequencies, causing widening of axial resolution by as much as 10%–20%.

7.2.1 Transducer Arrays

The huge majority of ultrasound scanners today use some form of piezoelectricity for this electrical/acoustic and acoustic/electrical transduction. The traditional material used in these arrays has been some form of piezoceramic such as lead zirconate titanate (PZT), although in the last 5 years the commercialization of single crystal materials has been gaining strength due to their broader bandwidths and greater sensitivities. There is considerable research ongoing into the use of microelectromechanical (MEMS) devices,[4,5] although the commercialization of such technologies has been quite slow.

The great majority of arrays used today in carotid imaging are 1D linear arrays composed of 128–192 elements (see Figure 7.2). A typical element in such an array might be about 4–6 mm tall and 0.2 mm wide for a total length dimension in the neighborhood of 4 cm. With such an array, one can image some 4 cm of the long axis of the carotid artery or get a full view of the cross section of the vessel in the short axis.

The last decade has seen a number of studies involving 3D imaging of the carotids. This is usually accomplished with a mechanically swept linear array (such as the one described previously). A number of new types of diagnostics will become available with tools such as plaque volume estimation. The main challenge here is the rather slow speed of sound in tissue, which limits the data acquisition rate to 5–15 volumes per second.

7.2.2 Transmit-and-Receive Beamformation

Figure 7.1 shows an overall block diagram for a conventional ultrasound scanner. The basic sequence of operations involves the initiation of the data acquisition along a single line (see Figure 7.3) of the region to be imaged. A system controller, typically a processor, sends the scan acquisition information to transmit-and-receive beamformers and then initiates the transmit-and-receive operations. The transmit beamformer generates timing signals that cause high-voltage (ca. 100 V) pulses to be sent to the transducer array. The timing signals are designed to compensate the travel time differences from each of the elements to the desired focal point. The transducer elements convert these timing signals to acoustic pulses that propagate along the desired acoustic ray and also focus on the desired depth location along that ray. During this transit, the acoustic beam encounters scattering sites of differing acoustic impedance; these cause echoes to be sent back to the array, which then reconverts these back to electrical signals. These echoes are then amplified, delayed to compensate for travel time differences, and summed to form a single echo signal representative of the density and compressibility variations of the tissue along the acoustic ray. This process is shown in Figure 7.5 for a linear array. During receive beamformation, the delay experienced by the echo signal received by each element is varied continuously to maintain "dynamic" focus on the propagating wavefront. This echo data can be envelope-detected by various means and eventually converted to a raster scan suitable for display by any of various interpolation algorithms. One can also retain the phase information from the signal by use of in-phase/quadrature (I/Q) conversion that will be critical in Doppler signal processing.

The process just described is an example of what is now referred to as delay-and-sum (DAS) beamformation.[6] In most scanners today, this is implemented with dedicated application-specific integrated circuits, that is, with dedicated special purpose devices. In the last several years, with the availability of higher speed and greater circuit density in processors, this type of implementation is being supplanted by software beamformers. A further benefit of such beamformers is the use of non-DAS designs such as plane wave beamformers.

7.2.3 Doppler Signal Processing

Information about the velocity of scatterers can be gained by Doppler processing of the echoes. As noted earlier, echoes from individual red blood cells within the focal region of an ultrasound beam form a summed echo signal, which retains its shape over multiple transmit repetitions (as long as the red blood cells within the ultrasound beam retain their relative positions). The goal of Doppler processing is to measure the phase (or time) shift between successive transmissions and thereby get an estimate of the flow velocity. These will change with turbulence as well as flow reversals during a cardiac cycle; however, typically, one gets enough information to be able to measure the temporal shift of such echoes. Two often used methods for determining this shift are by measuring the phase shift between transmits for narrow band transmits and by RF correlation for wider band transmits. Given a scatterer moving at a velocity v in a medium with the speed of sound of c, the Doppler shift is given by

$$f_D = f_t - f_o = \frac{2v\cos\theta}{c - v\cos\theta} f_o \qquad (7.1)$$

where

f_D is the Doppler shift
f_t is the frequency as seen by the transducer during reception
f_o is the transmitted frequency
v is the velocity of the scatterer
θ is the angle between the velocity of the scatterer and the direction of the acoustic beam
c is the speed of sound within the medium

The last equation can be simplified by recognizing that the speed of sound in the medium, c, will always be significantly larger than the blood flow velocity, thereby resulting in the following commonly used form:

$$f_D = \frac{2v\cos\theta}{c}. \qquad (7.2)$$

Equation 7.2 illustrates the processing steps used in generating the Doppler shift signal; this approach is commonly referred to as quadrature modulation. Let us assume that we are transmitting a sinusoid with a frequency of $f_o = \omega_o/(2\pi)$. Thus,

$$transmit = \cos(\omega_o t).$$

If we now assume that the moving scatterers impart a frequency shift of f_D, the received signal will be

$$receive = A\cos((\omega_o + \omega_D)t).$$

The resulting signals, referred to as in-phase and quadrature signals, will be, after some manipulation and taking advantage of trigonometric identities,

$$I(t) = LPF\{A\cos[(\omega_o + \omega_D)t]\cos(\omega_o t)\} = \frac{A}{2}\cos(\omega_D t)$$

$$Q(t) = LPF\{A\cos[(\omega_o + \omega_D)t]\sin(\omega_o t)\} = \frac{A}{2}\sin(\omega_D t)$$

In these expressions, *LPF* refers to low-pass filtering used to eliminate the frequency content at twice the fundamental frequency; it arises as part of the product of the two cosines. With the further processing, such as the spectral analysis, the in-phase and quadrature components are combined as a single complex signal, with the in-phase being the real part and the quadrature the complex (Figure 7.6). A typical Doppler spectral display may use 64 samples from a single-user selectable range gate, acquired at the rate of the pulse repetition frequency. These data are Fourier-transformed, and the resulting spectrum is displayed in gray scale as one vertical line in a 2D image. As new data comes in, new spectra are formed to complete the spectral image.

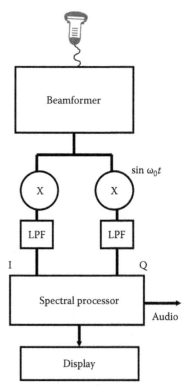

FIGURE 7.6 Generation of in-phase and quadrature components for the Doppler signal followed by the spectral analysis of the resulting complex I/Q signal. The resulting spectral data are displayed as shown at the bottom of Figure 7.8. In that display, time is on x-axis, Doppler frequency is on the y-axis, and the displayed brightness gives the amount of energy in the frequency bins.

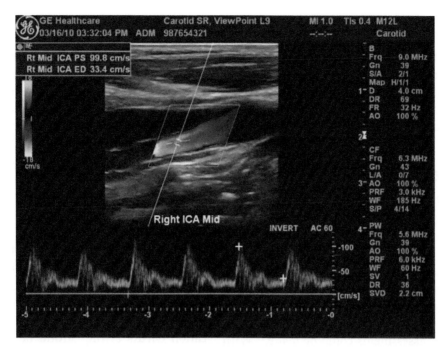

FIGURE 7.7   Display of the anatomy and flow within the carotid artery. This format is sometimes referred to as the triplex mode since it combines B-mode, spectral Doppler, and the 2D color flow Doppler (CFD). In this image, the user has placed a cursor within the color flow region of the internal carotid to study the flow velocities there. Based on the systolic cursor, the maximum flow velocity at peak systole is about 100 cm/s.

The most basic Doppler display can be seen at the bottom of Figure 7.7. In this Doppler spectral graph, the x-axis corresponds to time, the y-axis to Doppler frequency, and the trace brightness to the amount of energy at that frequency. The Doppler signal used comes from a single range gate. The user may select any point of interest using a selection tool (e.g., track ball) to move the cursor on the B-mode image. In Figure 7.7, the range gate is placed in the middle of the internal carotid artery in the middle of the color segment to be described shortly. The Doppler spectrum is usually converted to a Doppler velocity (as opposed to frequency) spectrum, largely since blood flow velocities are far more understandable from a clinical decision-making point of view. There are, however, several caveats to the use of the Doppler spectral information. The most critical of these is the uncertainty of the angle between the acoustic beam and the blood flow. As shown in Equation 7.2, the $\cos(\theta)$ term has a direct impact on the velocity

magnitudes measured and for large Doppler angles, this could introduce a sizeable error. Most systems today have tools to estimate the Doppler angle and supply a degree of correction.

In addition to the spectral analysis of Doppler echoes, signal-processing techniques have been developed to estimate the mean Doppler shift for the displayed pixels in a 2D region of interest. These techniques, usually referred to as color flow mapping or color flow Doppler (CFD), are based on the phase changes of the in-phase and quadrature samples for each of the pixels on that grid. Typically, one might take 8–12 pairs of I/Q samples and average the phase changes between the successive transmissions needed to generate those samples.[7,8] CFD has turned out to be a very useful and helpful tool for vascular as well as cardiac studies.[9] Figure 7.7 shows a CFD display of the flow in the internal carotid artery; this is the bright yellow and orange region within the artery.

7.3   Ultrasound Imaging Technology and Carotid Arteries

Among the various imaging targets for ultrasound, carotid arteries fall into a class of relatively near-field (e.g., less than 5 cm) targets. Typical transducers used at these depths range in frequency from about 7 to 12 MHz. Due to its location in the neck, there are few anatomical obstructions, thereby permitting the use of linear arrays whose image plane dimension may be in the range of 40–50 mm, although individual cases may present difficulties in gaining access. This is in direct contrast with applications such as echocardiography where the ribs limit ultrasonic access to less than 25 mm. A typical array used today will be around 40 mm long by 4 or 6 mm in height. Such an array might have 192 array elements used to form the individual acoustic beams needed for image formation. In the image slice thickness direction, today's scanners mostly have a fixed mechanical lens in the short dimension of the array. Alternative designs have multiple elements in the shorter elevation dimension to permit electronic focusing in that direction. With latter designs, the slice thickness of a typical carotid scan will be 1 mm or less.

7.4 Clinical Roles of Carotid Ultrasound

The earliest applications of ultrasound in the diagnosis of carotid pathologies involved the use of Doppler techniques by which the velocity of red blood cells was measured by the phase shifts introduced by successive interrogations.[10] The impacts of plaques in the common and internal branches of the carotid were typical targets. With the introduction of duplex ultrasound scanners, simultaneous B-mode imaging and Doppler flow assessments could be made.[11] With further development of imaging techniques, the anatomical appearance and the presence of plaques can be readily studied.

7.4.1 Carotid Anatomy

The left common carotid artery breaks off from the aorta at its arch, while the right common branches break off from the brachiocephalic trunk. They both travel up on their respective sides of the neck eventually to branch into their internal and external branches, as shown in the anatomic sketch in Figure 7.8. This bifurcation occurs at the carotid bulb; a key anatomical region for several reasons. Due to the flow division and the high blood flow velocities and associated wall shear stresses, this region is associated with considerable

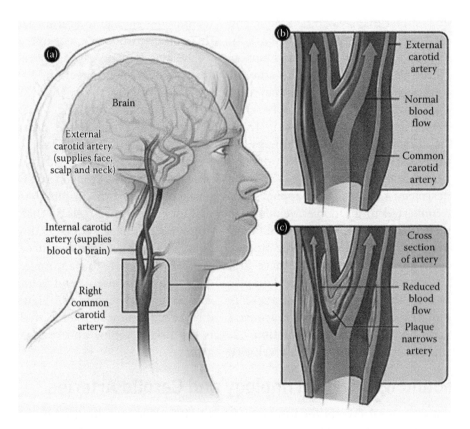

FIGURE 7.8 Carotid artery location and anatomy (a) both for a normal case (b) and with advanced plaque (c). (Image courtesy of NHLBI's Health Topics web site, http://www.nhlbi.nih.gov/health/health-topics/topics/cu/.)

vessel wall pathology. Figure 7.4 shows an ultrasound image of a normal carotid artery bifurcation. The lower branch is the internal carotid; the upper branch is the external.

The carotid artery wall, much like other similarly situated arteries, is composed of multiple layers usually referred to as the intima, media, and adventitia layers. These layers have the following compositions[12]:

- *Intimal layer (or tunica media)*: A thin layer of endothelial cells along with smooth muscle cells. This layer is usually not visualized transcutaneously with most current imaging devices.
- *Medial layer (or tunica media)*: This layer is composed of smooth muscle cells and elastin tissue. These two inner layers of the common carotid define the intima–media layer; typical dimensions of this layer range from say 0.5 mm to greater than 1 mm depending on age or disease progress.
- *Adventitial layer (or tunica adventitia)*: Traditionally, the adventitia has been viewed as a collagen-rich connective tissue containing fibroblasts and nerves; however, more recently, a far more dynamic role has been recognized[13]

The slow development of atherosclerosis, or stiffening of the carotid wall, has increased attention to the walls of the carotid arteries and the roles of the different layers just described. One key parameter measured by ultrasound scanners is the carotid intima–media thickness.[14,15]

As we have continued to learn more about cardiovascular disease, the key role played by blood vessel walls has become increasingly apparent. The complex interplay by low-density lipoproteins and high-density lipoproteins (LDLs and HDLs, respectively) in the process of plaque formation has been extensively discussed in the following.[16,17] Most basic role of ultrasound is in the assessment of carotid plaques and their likely impact on cerebral circulation.

7.4.2 Diagnostic Roles of Ultrasonic Imaging of the Carotid Artery

The importance of the carotid arises from two main factors: (1) its role as a key supplier of nutrients to the cerebral circulation and (2) its similarity to coronary arteries in the arterial tree and hence a potential role as a surrogate. It might be noted that the carotids are much more accessible to ultrasonic imaging than the coronaries themselves. The clinical relevance is magnified by the availability of highly miniaturized low-cost scanners today.[27] While the use of diagnostic surrogates is somewhat controversial, there is a fairly strong set of literature supporting the idea of using the carotid as a coronary surrogate and hence the justification of carotid ultrasound as a possible indicator of coronary pathologies.[18]

As noted, the carotids play a key role with the cerebral circulation. The main tool in this role is duplex Doppler (B-mode and spectral Doppler) or triplex Doppler (duplex plus color flow mapping). Because of the relative success of these tools, the standard of care for the initial diagnosis of the carotid artery diseases is duplex or triplex Doppler.[19] With these tools, one can rapidly evaluate the lumens of the common carotid as well as those of the internal and external branches. The main target will be detection of plaque; if found, one can then analyze the impact of the plaque on the surrounding blood flow velocities and the degree of turbulence present.

In principle, this sounds straightforward. However, there are a good number of pitfalls and challenges one often runs into. Among these are instrumentation and data-acquisition-related issues such as the presence of acoustic noise in the lumen of the artery. These are typically caused by strong out-of-plane scatterers or, more specifically at the anterior wall, reverberations due to the echoes from that wall. Due to the relatively planar surface of the anterior wall, a strong echo will be returned to the transducer array. This strong echo will reflect off the transducer surface and will immediately encounter the acoustic impedance variations at the skin and the various tissue layers in the first few millimeters below the skin. In the final B-mode image, these reverberant echoes will be displayed on what should be a perfectly echo-free lumen. The presence of those echoes will make the detection of plaques more difficult especially for the fresher plaques whose acoustic impedance will not be that different from that of the blood. In such cases, use of Doppler techniques will be helpful as will be the use of contrast agents (microbubbles) as will be discussed later. Doppler will be much more robust to static acoustic noise (such as the reverberations) and hence will have a better chance of determining the lumen size than will B-mode.

Perhaps a more serious challenge in carotid assessments is the presence of calcified plaque in the anterior wall of the artery. Calcium is usually present in older and possibly more stable plaques. Due to their very large acoustic impedance, the calcifications will scatter the acoustic energy strongly and allow relatively little penetration into the artery itself. Thus, it is difficult to image the lumen or measure the velocities in these shadowed areas. One may be able to manipulate the probe to an angle that may get around the obstructions; however, this is by no means certain.

The Doppler and B-mode tools give a good amount of useful clinical information about the plaque and its impact. As a rule, however, estimation of the vulnerability of plaque based on transcutaneous ultrasound has not been very successful. In the next section, the use of ultrasound contrast in this context will be discussed.

7.4.3 Potential New Roles for Carotid Ultrasound

More recently, ultrasound has started to play new roles beyond what has been discussed so far. These relate more closely to progress of atherosclerosis and to characterization of observed plaques. Two of these will now be discussed.

As the use of ultrasound elastography has increased in organs such as the liver or the breast, increasing numbers of researchers have sought to apply it to the carotid artery. The rationale of this is based on gaining a measure of the dimensional changes of the artery wall in response either to the arterial pressure wave or to an extrinsic source such as the acoustic radiation force impulse (ARFI).[20] Techniques based on strain imaging have been successful when combined with multiangle B-mode imaging data sets.[21,22]

Ultrasound contrast agents are microbubbles usually made with either phospholipid or albumin shells and contain a perfluorocarbon-type insoluble gas. Their dimensions are in the neighborhood of a few

microns and thereby will readily pass through capillaries. On the other hand, they are much too large to get through the endothelium and thus remain as blood pool agents. In the United States, the approved use is for left ventricular opacification, that is, to improve the delineation between the walls of the left ventricle and the blood pool. In other countries (e.g., Japan), the use of such agents for imaging the liver has been approved mainly to help detect lesions associated with hepatocellular carcinoma. The use of these agents with arteries has been somewhat slower to evolve, but clear new and valuable applications are being developed.

As has already been mentioned, B-mode carotid ultrasound is being used for measurements of the thickness of the intima–media layers. Most often this is being done with the far wall of the carotid (when viewed from the skin). The near or anterior wall is often made more difficult to see by acoustic noise as noted previously. The use of contrast improves this substantially.

Another interesting new use of contrast is in the study of the vulnerability of carotid plaques.[23] During the development of plaques, the inflammatory processes generate ischemia within the plaque itself. Neovascularization of the vasa vasorum develops in response to this need. Recently Hellings et al. have followed up carotid endarterectomy patients, analyzing their clinical outcomes and relating these to their plaque characteristics.[24] They have shown that two specific characteristics of the excised plaques were markedly associated with these events, namely, intraplaque hemorrhage and vessel formation. Feinstein has shown that microbubbles can identify areas of neovascularization[25] and other researchers, for example, Hoogi et al. have associated contrast-derived measures with histopathology.[26] While this is still a prospective area needing more research, this appears to be one of the more intriguing ways of assessing plaque vulnerability.

7.5 Concluding Comments

Carotid ultrasound has had an interesting history and, perhaps more importantly, a future with excellent opportunities. It is clearly the first line of defense in cases with suspected cerebrovascular impairment. In view of the dramatically decreasing cost and size of ultrasound scanners[27] and the relative technical

simplicity of a carotid exam, its utilization is likely to move closer to the primary care physician or even earlier. The newer roles in assessing arterial vessel wall health and the characterization of plaques are likely to have a broader impact well beyond the current role of carotid ultrasound.

References

1. Strandness, D. E., Jr., Schultz, R. D., Sumner, D. S., and Rushmer, R. F. Ultrasonic flow detection: A useful technic in the evaluation of peripheral vascular disease. *Am. J. Surg.* 113: 311–320, 1967.

2. Strandness Jr, D. E., McCutcheon, E. P., and Rushmer, R. F. Application of transcutaneous Doppler flowmeter in evaluation of occlusive vascular disease. *Surg. Gynecol. Obstet.* 122: 1039–1045, 1967.

3. Szabo, T. L. *Diagnostic Ultrasound Imaging: Inside Out.* Access Online via Elsevier, Amsterdam, the Netherlands, 2004.

4. Ladabaum, I. et al. Surface micromachined capacitive ultrasonic transducers. *IEEE Trans. Ultrason. Ferroelectr. Freq. Control* 45(3): 678–690, 1998.

5. Thomenius, K. et al. Mosaic arrays using micromachined ultrasound transducers. U.S. Patent No. 6,865,140, March 8, 2005.

6. Thomenius, K. E. Evolution of ultrasound beamformers. *Proc. IEEE Ultrason. Symp.* 2: 1615–1622, 1996.

7. Kasai, C., Namekawa, K., Koyano, A., and Omoto, R. Real-time two-dimensional blood flow imaging using an autocorrelation technique. *IEEE Trans. Sonics Ultrason.* 32(3): 458–464, 1985.

8. Omoto, R. and Kasai, C. Physics and instrumentation of Doppler color flow mapping. *Echocardiography* 4(6): 467–483, 1987.

9. Grant, E. G., Benson, C. B., Moneta, G. L., Alexandrov, A. V., Baker, J. D., Bluth, E. I. et al. Carotid artery stenosis: Gray-scale and Doppler US diagnosis—Society of Radiologists in Ultrasound Consensus Conference. *Radiology* 229(2): 340–346, 2003.

10. Baker, D. W. Pulsed ultrasonic Doppler blood-flow sensing. *IEEE Trans. Sonics Ultrason.* 17(3): 170–184, 1970.

11. Barber, F. E. et al. Ultrasonic duplex echo-Doppler scanner. *IEEE Trans Biomed. Eng.* 2: 109–113, 1974.

12. Nicolaides, A., Beach, K. W., Kyriacou, E., and Pattichis, C. *Ultrasound and Carotid Bifurcation Atherosclerosis.* Springer, New York, 2012.

13. Majesky, M. W. et al. The adventitia: A dynamic interface containing resident progenitor cells. *Arterioscler. Thromb. Vasc. Biol.* 31: 1530–1539, 2011.

14. Chahal, N. S., Lim, T. K., Jain, P., Chambers, J. C., Kooner, J. S., and Senior, R. The distinct relationships of carotid plaque disease and carotid intima-media thickness with left ventricular function. *J. Am. Soc. Echocardiogr.* 23(12): 1303–1309, 2010.

15. Rossi, A. C., Brands, P. J., and Hoeks, A. P. G. Automatic localization of intimal and adventitial carotid artery layers with noninvasive ultrasound: A novel algorithm providing scan quality control. *Ultrasound Med. Biol.* 36(3): 467–479, 2010.

16. Libby, P. Managing the risk of atherosclerosis: The role of high-density lipoprotein. *Am. J. Cardiol.* 88(12): 3–8, 2001.

17. Libby, P., Ridker, P. M., and Maseri, A. Inflammation and atherosclerosis. *Circulation* 105(9): 1135–1143, 2002.

18. Nambi, V., Chambless, L., Folsom, A. et al. Carotid intima-media thickness and presence or absence of plaque improves prediction of coronary heart disease risk: The ARIC (Atherosclerosis Risk in Communities) study, *J. Am. Coll. Cardiol.* 55: 1600–1607, 2010.

19. Clevert, D. A., Paprottka, P., Sommer, W. H., Helck, A., Reiser, M. F., and Zengel, P. The role of contrast-enhanced ultrasound in imaging carotid arterial diseases. *Sem. Ultrasound CT MR* 34(3): 204–212, 2013.

20. Reneman, R. S., Meinders, J. M., and Hoeks, A. P. G. Non-invasive ultrasound in arterial wall dynamics in humans: What have we learned and what remains to be solved. *Eur. Heart J.* 26(10): 960–966, 2005.

21. Hansen, H. H. G., Lopata, R. G. P., and de Korte, C. L. Noninvasive carotid strain imaging using angular compounding at large beam steered angles: Validation in vessel phantoms. *IEEE Trans. Med. Imaging,* 28(6): 872–880, 2009.

22. Idzenga, T., Hansen, H., Lopata, R., and De Korte, C. Non-invasive assessment of shear strain in the carotid arterial wall based on ultrasound radiofrequency data. *Ultrasonics Symposium (IUS), 2009 IEEE International,* pp. 2000–2459, 2009.

23. Feinstein, S. B. The powerful microbubble: From bench to bedside, from intravascular indicator to therapeutic delivery system, and beyond. *Am. J. Physiol. Heart Circ. Physiol.* 287: H450–H457, 2004.

24. Hellings, W. E., Peeters, W., Moll, F. L. et al. Composition of carotid atherosclerotic plaque is associated with cardiovascular outcome. *Circulation* 121:1941–1950, 2010.

25. Feinstein, S. B. Contrast ultrasound imaging of the carotid artery vasa vasorum and atherosclerotic plaque neovascularization. *J. Am. Coll. Cardiol.* 48(2): 236–243, July 18, 2006.

26. Hoogi, A., Adam, D. et al. Carotid plaque vulnerability: Quantification of neovascularization on contrast-enhanced ultrasound with histopathologic correlation. *Am. J. Roentgenol.* 196: 431–436, 2011.

27. Thomenius, K. E. Miniaturization of ultrasound scanners. *Ultrasound Clin.* 4(3): 385–389, 2009.

References

8. Cardiovascular Imaging with Nuclear Medicine Techniques

Piotr J. Slomka, Robert deKemp, and Guido Germano

8.1 Introduction

Nuclear medicine techniques are used widely in mainstream clinical practice for detection and evaluation of patients with coronary artery disease. In the United States alone, approximately 9 million myocardial perfusion scans are performed annually [1]. In comparison to other modalities, the primary advantage of the nuclear imaging techniques is the objective quantitative evaluation of stress and rest perfusion and function. The most prevalent nuclear technique used clinically is myocardial perfusion SPECT (MPS). New types of SPECT devices and software methods for image reconstruction have recently been introduced, dramatically increasing the speed of MPS imaging. Myocardial perfusion PET is performed less frequently due to higher cost of the equipment, but offers some advantages over SPECT, such as routine attenuation correction (AC), ability to quantify stress/rest myocardial blood flow (MBF), and perfusion flow reserve. This chapter provides a general overview of the instrumentation and image acquisition methods used in cardiovascular imaging with SPECT and PET.

8.2 SPECT

8.2.1 Conventional Anger Camera SPECT

Three-dimensional SPECT cardiac imaging is performed using conventional gamma cameras or Anger cameras rotating around the patient. In Figure 8.1, we illustrate a generation of a single-projection image by one detector of a gamma camera. In standard myocardial perfusion imaging, two detectors are typically used. In this example, a ^{99m}Tc-based radiopharmaceutical is injected into a patient, taken up by the heart muscle (myocardium), and radioactivity is emitted in the form of photons (gamma rays) having energy of 140 keV. Some of those photons are absorbed by a scintillation detector positioned with its plane parallel to the patient's long axis, the "allowed" angle of impact being constrained reasonably close to 90° through the use of a collimator. The scintillation detector is made of high-density material such as thallium-activated sodium iodide or NaI(Tl). When radiation penetrates it, energy dissipates by conversion into light photons, which are in turn converted into electrons and amplified into an electrical current by a set of photomultiplier tubes (PMTs) directly coupled to the scintillating

crystal. The current is proportional to the amount of energy dissipated in the detector, which ideally should be equal to that of the original gamma ray, but is considered acceptable if it lies within a 15%–20% window around the expected 140 keV value. For each "hit" by a photon, the point of impact (X, Y) on the detector is derived by analysis of the fraction of light photons collected by different PMTs; if the associated energy Z is within the window, energy discrimination circuitry generates a signal that allows the event to be accumulated into a computer matrix, or "projection image."

8.2.2 Gated SPECT

In a SPECT acquisition, the camera detector(s) rotate around the long axis of the patient, acquiring one projection image at each of many, evenly spaced angular locations (steps) along the acquisition orbit. If the acquisition is gated, several (8 or 16) projection images are acquired at each projection angle, with each image corresponding to a specific phase of the cardiac cycle termed "interval" or "frame" (Figure 8.2). All projection images for a given interval can be reconstructed

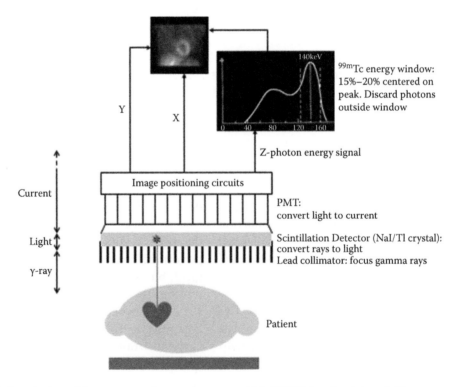

FIGURE 8.1 Basic functioning of the gamma or Anger camera, on which SPECT imaging is based. Gamma ray scintillation within the sodium iodide crystal indicated by asterisk (*). The diagram in the upper right illustrates the energy spectrum of technetium 99m (^{99m}TC). The dashed vertical lines indicate the 20% energy window surrounding the 140 keV primary photopeak of ^{99m}TC.

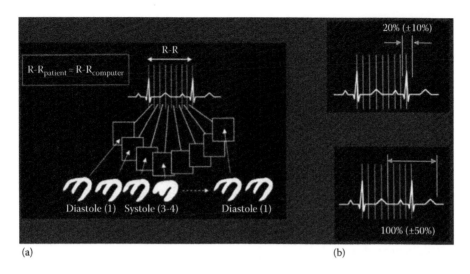

FIGURE 8.2 Principles of gated SPECT acquisition: (a) selection of gating window and (b) for beat acceptance.

into a SPECT or tomographic image volume using filtered back projection or iterative reconstruction techniques. Image volumes from the various gated SPECT intervals can be displayed in 4D format (x, y, z, and time), allowing for the assessment of dynamic cardiac function. In addition, summing all individual intervals' projections at each angle before reconstruction produces an "ungated" or "summed gated" SPECT image volume. Thus, a gated SPECT acquisition results in a standard SPECT data set ("summed" SPECT), from which perfusion is assessed, and a larger "gated" SPECT data set, from which contractile function is evaluated. The strong appeal of gated SPECT imaging is a direct consequence of the ease and modest expense with which perfusion assessment is "upgraded" to perfusion + function assessment, and accounted for the growth of the technique over the past decade. As long as adequate count statistics are achieved, there is no limitation as to the specific perfusion agent that can be imaged with the gated SPECT technique. No matter what radiopharmaceutical is used, the quality of the gated SPECT study will be closely and directly related to the number of counts in its individual frames. Count statistics are influenced by numerous factors, including injected activity, acquisition time, patient size, camera configuration and sensitivity, collimation, number of frames and energy window acceptance criteria.

8.2.3   Attenuation Correction in SPECT

To determine the accurate distribution of the radio-pharmaceutical, detected counts need to be corrected

for attenuation losses. To determine the spatial extent of attenuation, a map of the attenuation coefficient (μ-map) needs to be determined. Although these systems have not been in wide clinical use in nuclear cardiology [2], incremental value has been shown improving the diagnostic accuracy of MPS imaging. Gated SPECT images, unlike PET images (see Section 8.3), are typically not corrected for attenuation.

Various strategies have been developed for AC for nuclear cardiology cameras. The conventional approach is to obtain transmission maps using the gamma camera and an external radiation source such as line source. One example of such system utilizes a scanning gadolinium-153 line source to create a transmission image of the patient while simultaneously acquiring an emission image from a myocardial perfusion study using either technetium-99m or thallium-201 isotopes. The collimated line sources are positioned 180° across from each detector, and are mechanically scanned across the field of view at each angle, which projects a transmission image onto the opposing detector (Figure 8.3). A sliding electronic window, which is synchronized to the moving position of the line source, allows for the simultaneous acquisition of the transmission data with the collection of the emission data in a different energy window. Concurrent acquisition of the transmission and emission scans eliminates problems of image registration and does not require additional patient scan time. The source is collimated with a narrow slit aperture running the length of the source. Additional copper shielding is present initially to slightly reduce the radioactive

Chapter 8

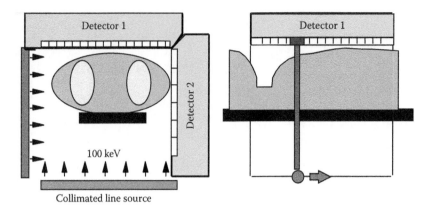

FIGURE 8.3 Example of the system for attenuation correction of SPECT with an external transmission (collimated line source) source on a dual-headed SPECT detector system.

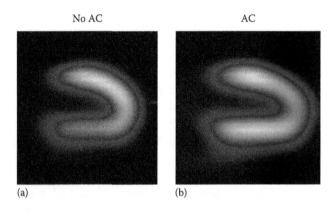

FIGURE 8.4 Example of noncorrected (a) and attenuation-corrected vertical long axis myocardial perfusion SPECT images (b) illustrating inferior wall count reduction in images when not corrected for photon attenuation.

emissions from a new source, but this shielding is later removed as decay occurs and the activity of the transmission source decreases. Attenuation maps can be created from the transmission images using filtered back projection or iterative reconstruction (see reconstruction section). Other designs have also been proposed with the transmission source scanning in different geometries. In Figure 8.4, we show an example of attenuation-corrected SPECT demonstrating improved homogeneity of the myocardial perfusion in a normal subject after AC is applied.

Another approach to AC is to utilize the x-ray CT scans obtained on hybrid SPECT/CT systems. Examples of such systems include gamma cameras coupled with slow-rotation CT or faster diagnostic CT systems [3]. Some researchers proposed the use of coregistered CT data obtained on stand-alone systems [4], but this approach has not yet been adopted clinically.

AC can be associated with a small, but additional radiation dose, especially if CT is used for this purpose in SPECT/CT scanners. The typical stress and rest dose from such CT–AC scans is in the order of 0.3–1.3 mSv [5] and therefore it is not negligible when compared to the dose from radioisotopes, especially for new low-dose protocols. To reduce the AC radiation dose, various new designs have been developed for SPECT scanners. One vendor has developed an integrated system with 5 μSv effective dose in which photons from x-ray source are detected by solid-state detectors with a fan beam collimator operating in high-counting-rate mode [6]. The scan time for the acquisition of the AC maps in this mode is 60 s. Another example of low-dose AC technology is the flat-panel x-ray detector system, providing low-dose (0.12 mSv) CT images [7]. A transmission image of the entire heart can be obtained from a single 60 s rotation of the gantry while the patient is breathing normally. By acquiring the transmission image during tidal respiration, the attenuation data are averaged over multiple respiratory cycles to match the position of the heart during the SPECT acquisition. Furthermore, since the x-ray detector and the SPECT detectors are located on the same rotatable gantry, the emission and transmission images can be acquired with little or no table translation between scans, which also reduces misalignment between emission and transmission data.

8.2.4 SPECT Image Quality

The quality of the reconstructed and reoriented SPECT images depends on that of the projection images. The projection images, in turn, can be compromised by a host of technical and physical factors, including

Compton scatter, photon absorption, variable resolution for noncircular acquisition orbits, and patient or organ motion. Figure 8.5 presents a schematic representation of the first two phenomena, which is particularly important in cardiac imaging because of the close proximity of the heart to other organs of much different densities and activities.

Specifically, scatter may cause myocardial counts to be incorrectly assigned to other areas, reducing the apparent myocardial uptake (case 2a in Figure 8.5). Extracardiac counts may be a interpreted as coming from the myocardium, with consequent potential loss of image resolution and defect contrast (case 2b in Figure 8.5).

Additionally, photons may be absorbed causing attenuation artifacts. Typical artifacts associated with photon attenuation are apparent perfusion defects in the anterior and/or lateral myocardial wall (breast attenuation), and defects in the inferior wall from diaphragmatic attenuation (Figure 8.4). Scatter and AC schemes are used to compensate for these effects.

Patient motion is believed to affect as many as 10%–20% of conventional cardiac SPECT studies. It causes inconsistencies in SPECT projections obtained from different angles. Generally speaking, it can be divided into organ motion (e.g., "upward creep"), and patient motion ("bouncing" translations and rotations of the returning or nonreturning type). "Upward creep" of the heart is often found when imaging is started too soon following exercise stress [8] and reflects the gradual return of the diaphragm to its pre-exercise location in a patient's chest as the rate and depth of respiration decreases. "Bouncing" refers to an up-and-down, oscillating pattern of motion caused by breathing or other factors, usually along the

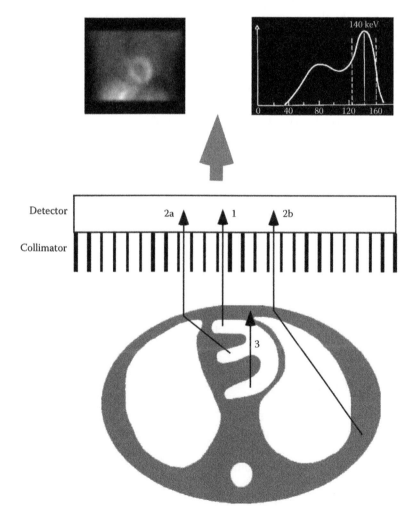

FIGURE 8.5 Different categories of photons in nuclear imaging: primary photons, which are uncontaminated and account for the photopeak in the energy spectrum (top right) (1); Compton-scattered photons, which have lower energy and carry incorrect positional information (2a–2b); absorbed photons, which are not collected in forming the projection image (top left) (3). (Modified and reproduced from Germano, G., *J. Nucl. Med.*, 42, 1499, October 2001. With permission.)

Chapter 8

anterior–posterior direction. Returning and nonreturning translations by the patient may occur along the vertical direction, horizontal direction, or both.

The effects of motion on a SPECT acquisition depend on the type and extent of the motion, time at which it occurs, and the number and configuration of camera detectors employed, as investigated in detail by Matsumoto et al. [9] Artifacts will range from subtle to severe deformations of the LV cavity. Although motion correction software is widely available and reasonably effective in correcting various types of translational motion, its effectiveness is limited with rotational motion [9]. For this and other reasons, it is advisable to use motion compensation techniques only in cases where there is a clear need for it, and not as a matter of routine. The best solution for minimizing the effect of patient motion is to prevent it from occurring during SPECT acquisition, for example, by reducing the total acquisition time as is allowed by the new fast cameras.

Another phenomenon in cardiac SPECT imaging is partial volume (PV) averaging. The problematic effects of PV averaging are divided into two categories: count spill-in and count spill-out of the organ of interest. Both are problems for cardiac SPECT imaging of the myocardium. Additionally, the thickness of the LV wall is usually on the order of 8–12 mm, while a typical clinical image has a reconstructed resolution in the range of 10–15 mm, resulting in an underestimation of the true myocardial activity due to spill-out. These problems are compounded by the facts that the myocardial wall thickness is not constant over the entire LV and that some parts of the myocardium are more susceptible to motion than others. For example, the activity of the apex, is commonly underestimated because the apical myocardium is thinner and undergoes more motion than the basal wall of the LV myocardium. If left uncorrected, these effects can lead to inaccurate quantification and interpretation.

8.2.5 New Designs for SPECT Cameras

Although MPS imaging using Anger camera technology is widely used in current clinical practice, it suffers from some fundamental limitations including long image acquisition, low image resolution, and high radiation doses. In the last two decades, MPS was performed commonly by dual-head scintillation cameras with parallel-hole collimators, typically configured in a 90° detector geometry and image reconstruction based on standard filtered back

projection algorithms. The scan times were as long as 15–20 min for both stress and rest MPS acquisition to provide adequate imaging statistics, resulting in longer test times and frequent artifacts caused by patient motion during the scan, as well as compromised patient comfort. Recently, it has become very important to address these limitations, since MPS has new competitors in the noninvasive imaging area. The most notable new modality introduced is coronary CT angiography (CCTA), which allows diagnostic imaging to be acquired within a very short time. In addition, the clinical practice of combining MPS with other modalities such as CCTA for better diagnostic certainty has intensified concerns regarding the total radiation dose delivered to the patient, making it more imperative to reduce the radiation dose with MPS. [10].

There have been significant efforts recently by both industry and academia to develop new imaging systems with increased sensitivity and new methods of image reconstruction for optimizing image quality, which will simultaneously allow higher photon sensitivity and improve both image quality and resolution. These efforts address the main limitations of MPS by combining several approaches, such as changing the detector geometry and optimal tomographic sampling of the field of view for myocardial imaging, improving the detector material and collimator design, and optimizing the image reconstruction algorithms. In particular, innovative designs of the gantry and new detectors have been developed, which allow increased sampling of the myocardial region, and thus allow better local sensitivity. These systems combine improvements in both spatial resolution and sensitivity. With faster imaging times due to increased sensitivity, and eliminating the standard positioning of the patient's arms above the head (by imaging in an upright or reclining position), patient comfort is dramatically improved. As a result of faster imaging times and more comfortable patient positioning, these systems also have the additional benefit of reducing patient motion during a scan. Furthermore, claustrophobic effects are reduced and the floor space requirements are more flexible since the new detectors and the associated mechanical requirements are significantly smaller in comparison to standard MPS equipment.

8.2.5.1 New Detectors

Cadmium zinc telluride (CZT) detectors for SPECT: At least two vendors now have introduced cameras with full CZT solid-state detectors which have replaced the Anger camera design presented in Figure 8.1, with simpler and lighter detectors where photons are directly detected (Figure 8.6). While CZT detectors are higher in cost, they have advantages of superior isotope energy resolution (by a factor of approximately 1.7 at 140 keV) and are compact in size, as compared to the combination of NaI(Tl) with PMTs of the conventional Anger camera. In one design, pixelated CZT detector arrays are mounted in nine vertical columns, and placed in 90° gantry geometry (Figure 8.7). Each column consists of an array of 1024 CZT elements (2.46 × 2.46 × 5 mm thick), arranged in a 16 × 64 element array with an approximate size of 40 × 160 mm. In another design, a round CZT crystal is used with optimized geometry for cardiac imaging, and coupled with pinhole collimators (see next section). It has been shown that for fixed energy acceptance windows, the asymmetric CZT energy response shape leads to a 30% reduction of the scatter component in measured data [11]. Preliminary work has shown that simultaneous dual-isotope MPS with CZT is feasible using Tl-201 and Tc-99m, taking advantage of the improved energy resolution of CZT [12].

Indirect photon detection with solid-state technology: It is also possible to use indirect, solid-state detectors consisting of pixelated CsI(Tl) and photodiodes to configure detector heads that are more compact than conventional cameras equipped with PMTs (Figure 8.8) [13]. This design has been used in a 2- or 3-detector configuration. Each detector head contains an array of 768 6.1 × 6.1 × 6 mm thick CsI(Tl) crystals, coupled to individual silicon photodiodes, which are used to convert the light output of the crystals to electrical pulses. Digital logic and software are both used to process the signals instead of analog Anger positioning circuits.

8.2.5.2 New Acquisition Geometry and Collimator Designs

In addition to solid-state detectors, novel collimator designs have been recently proposed to maximize the detection of photons for cardiac imaging geometry. The collimator is the key nuclear scanner component responsible for the loss of over 99.9% of photons in the traditional gamma camera with a high-resolution collimator [14]. Therefore, improvements of the collimator sensitivity without sacrificing image resolution are highly desirable. In one approach, several small high-sensitivity collimators on separate detector columns can be positioned close to the patient's body and moved independently (Figure 8.9). The photons from a

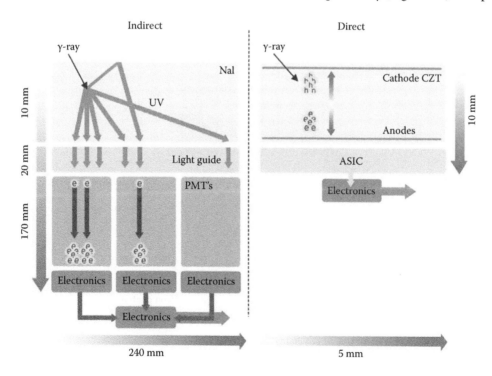

FIGURE 8.6 Comparison of indirect/Anger camera (left) and direct solid-state CZT camera (right) photon detection principles (images courtesy of Aharon Peretz, GE Healthcare). ASIC, application-specific integrated circuits.

Chapter 8

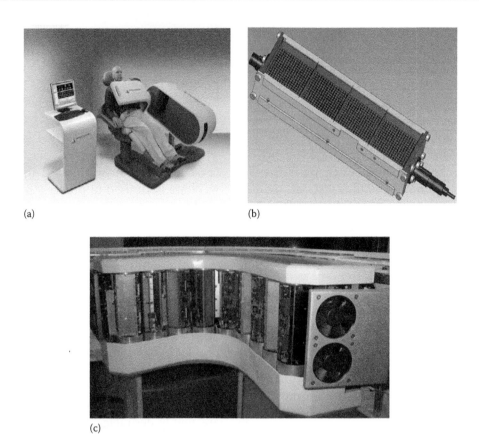

FIGURE 8.7 New detector design. (a) Photograph of the camera showing patient position, (b) a diagram of a single detector column of the detector, and (c) a photograph of a 9-detector column configuration. (Images Courtesy of Nathaniel Roth [Spectrum Dynamics, Caesarea, Israel].)

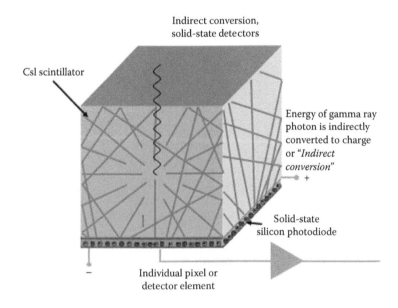

FIGURE 8.8 Indirect conversion of photon energy for the solid-state detector. (Reproduced from Bai, C. et al., *J. Nucl. Cardiol.*, 17, 459, June 2010. With permission.)

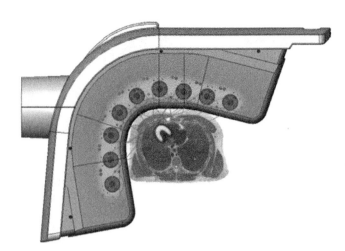

FIGURE 8.9 The ROI-centric collimation technique to optimize data collection from the myocardium. (Images Courtesy of Nathaniel Roth [Spectrum Dynamics, Caesarea, Israel].)

given location are detected at multiple angles by multiple columns as the fields of view of the detectors are swept through the region of interest. Each detector column is fitted with a square, parallel-hole, high-sensitivity collimators, such that the dimensions of each hole are matched to the size of a single detector element. The collimators have a larger effective diameter than conventional high-resolution collimators used with scintillation cameras, yielding a significant gain in their geometric efficiency. The compensation for the loss in geometric spatial resolution that results from this design is accomplished by the use of CZT with its superior energy resolution and digital compensation methods. The overall system resolution is 5 mm in line

source experiments, superior to that of the standard Anger camera systems. The photon sensitivity of such a system for the centrally located point has been reported to demonstrate a sensitivity of >1400 counts/μCi/min compared to the 160–240 counts/μCi/min range generally observed with standard cameras [15]. In a cardiac imaging study, the myocardial count rate (with the same injection of the isotope) was 7–8 times higher than the Anger camera (Figure 8.10) [16].

Cardiofocal collimators: Another way to increase sensitivity is to map the organ of interest (the myocardium) to a larger portion of the detector by exploiting the intrinsic magnification properties of a fan beam or cone beam collimator. One vendor utilizes astigmatic collimator with optimized organ-of-interest centered acquisition and iterative reconstruction [17]. The collimator is designed so that the center of the field of view magnifies the heart both in axial as well as in transaxial direction, while the edges sample the entire body to avoid truncation artifacts common to single focal collimators when imaging the torso. With an appropriate orbit, this variable-focus collimator increases the number of detected events from the heart by more than a factor of two in each direction compared to that of a parallel-hole collimator with equivalent resolution, and magnifying the heart while imaging the rest of the torso under traditional conditions [17]. The principle of image acquisition with these collimators is shown in Figure 8.11. The reported image acquisition time of this system can be as short as 4 min. These dedicated cardiac imaging collimators can be added to the conventional Anger cameras [18].

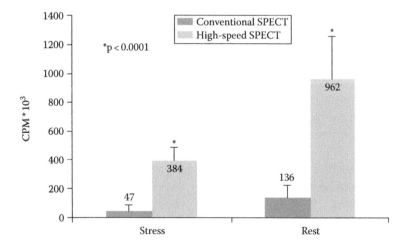

FIGURE 8.10 The higher system sensitivity of the dedicated cardiac camera system with high-sensitivity collimators is demonstrated by a significantly higher myocardial count rate (7–8 times), compared with conventional SPECT stress and rest images. CPM: counts/min. (Reproduced from Sharir, T. et al., *J. Am. Coll. Cardiol. Cardiovasc. Imaging* 1, 156, 2008. With permission.)

Chapter 8

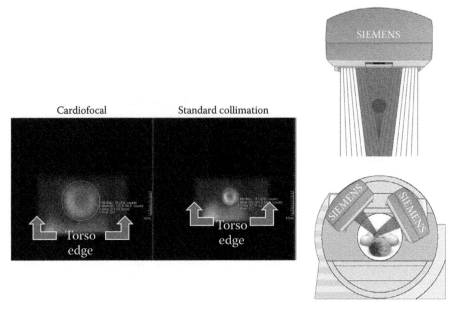

FIGURE 8.11   Principle of operation for the cardiofocal collimator. The holes of the collimator are focused on the heart, but outside of the heart region, the collimator is diverging to avoid image truncation. Thus, only the heart region is magnified, but the entire body is covered. The astigmatic collimator is designed to achieve a two-time magnification of the heart in all directions (thus a four-time sensitivity increase for the heart region without truncation of the torso in a specified orbit). In the specified orbit, the detector heads are positioned at 76°, keeping a fixed radius of 28 cm about the "center of rotation," which now is located in the heart region. The scan range is 208°, combining views from both detectors. (Images courtesy of Hans Vija Siemens Medical Solutions USA, Inc., Malvern, PA; Molecular Imaging, Hoffman Estates, IL.)

Multipinhole collimation: Multipinhole collimation provides an alternative approach to parallel-hole collimators. The multipinhole approach allows many views to be acquired simultaneously throughout the entire image acquisition period without moving the detector, collimator, or the patient. This capability allows image acquisition to be accomplished without the need for any electromechanical hardware, potentially reducing the manufacturing and servicing costs. In addition to the increase in detection sensitivity, the use of stationary detectors equipped with multipinhole collimation provides coincident sets of raw images, eliminating view-to-view inconsistencies, and thereby reducing artifacts induced by patient motion. Therefore, by the multipinhole approach, all views are acquired simultaneously for the entire acquisition period, providing a compatible data set for input to iterative SPECT reconstruction algorithms.

A standard dual-detector SPECT camera for imaging the human heart can be adapted to the multipinhole SPECT technique. The performance characteristics for a multipinhole SPECT system applicable to cardiac imaging were reported using a 9-pinhole collimator design [19]. The detection efficiency was increased 10-fold at a 30% loss of spatial resolution as compared to standard collimator. These data predict a fivefold increase in sensitivity with comparable resolution to that of a standard gamma camera.

One issue in multipinhole imaging concerns the minimal number of views and the optimal viewing geometry required for clinical cardiac SPECT. There are both advantages and disadvantages to having fewer views. As the number of views in the SPECT data set is decreased, the geometric appearance of the heart is visibly altered. This aspect must be weighed against the fact that the increased statistical content and simultaneous views improve the comparability of the stress versus rest data sets acquired on the same patient. The improved statistical count-density of multipinhole SPECT images is also a key factor in supporting the ability to image multiple isotopes simultaneously.

One vendor recently introduced a cardiac imaging SPECT camera with multipinhole collimation system combined with an array of solid-state CZT pixilated detectors. The design of the camera is shown in Figure 8.12. The use of simultaneously acquired views improves the overall sensitivity and gives complete and consistent angular data needed for both dynamic studies and for the reduction of motion artifacts. In this system, the detectors and collimators do not move during acquisition and all lines of response are acquired simultaneously through a multipinhole collimator with a large number of pinholes. Patients are imaged in a supine position with their arms placed over their heads. It has been shown that the combination of CZT with a pinhole

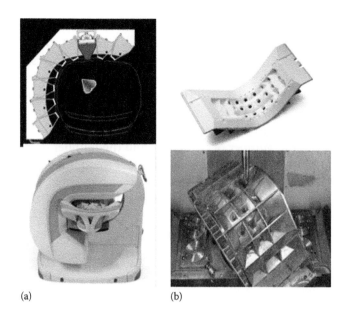

(a)

(b)

FIGURE 8.12 Design of multipinhole collimator SPECT system. Patient position (a) and images of the curved multipinhole collimator (b). (Images courtesy of Aharon Peretz, GE Healthcare, Buckinghamshire, U.K.)

collimator is seen to further enhance the improved energy resolution available as compared to from CZT alone [20], which may facilitate new applications such as simultaneous dual-isotope imaging. In a recent preliminary report, compared to the standard, state-of-the-art SPECT Anger camera, this design has demonstrated improvements of 1.65-fold in energy resolution, 1.7- to 2.5-fold in spatial resolution and 5- to 7-fold in sensitivity with energy resolution of 5.70% and spatial resolution in the 4.3–4.9 mm range [21].

8.2.5.3 Reconstruction Algorithms

Filtered back projection: Traditionally filtered back projection reconstruction has been used for cardiac imaging [22] (Figure 8.13). The 2D projection images yield a series of "activity profiles" (also termed "count profiles" or "scan profiles") for each transaxial plane. Each profile represents the integrated sum of the activity subtended by the detector along a given angle in that particular plane, and all profiles are used to reconstruct the image representing the activity in that plane (transaxial image). Since no information is available concerning the depth of the activity responsible for a peak in a profile, the simplest approach is to assume that it was uniformly distributed. Thus, the counts in each profile are uniformly redistributed (back-projected) onto the transaxial image, following the linear superposition of back projection scheme. A drawback of this approach is that it results in loss of resolution, loss of contrast, and creation of the characteristic "star artifact." To alleviate this problem, count profiles can be altered (filtered) before reconstruction using an oscillating function, which has both positive and negative side-lobe values, so as to cancel out the star's "rays." This latter approach is termed "filtered back projection" and represents the traditional method used for image reconstruction in SPECT.

Image filtering: Nuclear cardiac images are relatively count-poor due to limitations in the injected activity, the use of physical collimation and the attenuation introduced by the thorax, so projection images are affected by Poisson-distributed noise. Noise by

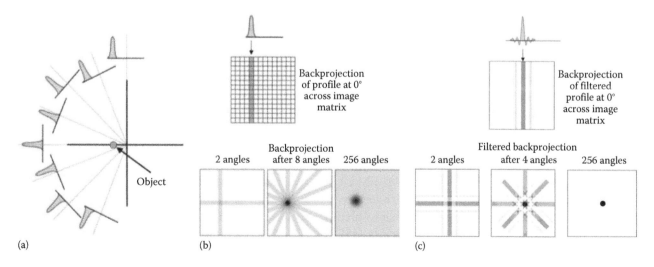

(a)

Object

2 angles   Backprojection after 8 angles   256 angles

Backprojection of profile at 0° across image matrix

(b)

2 angles   Filtered backprojection after 4 angles   256 angles

Backprojection of filtered profile at 0° across image matrix

(c)

FIGURE 8.13 Projection samples are measured at discrete angles around the object (a). Tomographic reconstruction is illustrated (b) using linear superposition of simple back projections and (c) filtered back projections. The ramp filter is used to restore the high frequencies in the measured projection data, but also amplifies noise. (Modified and reproduced from Physics in Nuclear Medicine. Cherry SR, Sorenson J, Phelps ME [22]. With permission.)

Chapter 8

definition involves high frequencies, and as such it can be reduced by smoothing with low-pass filters. The most popular low-pass filters used in nuclear cardiology belong to the Butterworth family, with other less widespread options including the Hanning filter and the Metz and Wiener adaptive filters. Low-pass filtering is usually accomplished on the projection data before reconstruction (or during reconstruction, modifying the ramp filter by a "low-pass window"). These approaches should be equivalent to filtering the reconstructed images (two-dimensionally or three-dimensionally), given the theoretical linearity of the process. In practice, however, nonlinearities are introduced by some camera manufacturers by "clipping" negative pixels resulting from filtering, and it is generally accepted that pre-reconstruction filtering is the most desirable way to achieve noise reduction in nuclear cardiology imaging.

8.2.5.4 Iterative Reconstruction Techniques

More recently, iterative reconstruction methods have been utilized extensively in cardiac SPECT. Nuclear cardiac imaging is significantly affected by Poisson noise, scatter, AC, and variable image resolution [23]. The iterative methods allow incorporation of accurate corrections for these degrading factors into the reconstruction process so that the reconstructed image is a better representation of the true object being imaged. Iterative maximum likelihood expectation minimization (MLEM) reconstruction methods allow the geometry of the acquisition to vary for each projection, greatly enhancing the flexibility in modeling the physical parameters.

Currently, the most widely used iterative technique is based on the ordered subset expectation maximization (OSEM) approach, which is an accelerated version of the MLEM. This technique groups projection data into an ordered sequence of subsets for efficient computation. One iteration of the OSEM algorithm is defined as a single pass through all of the subsets [24]. Typically 2–4 projections per subset are used, with 4–12 iterations that are computationally less demanding than 1 iteration of standard MLEM algorithm (assuming 64 projections). Even with one iteration of OSEM and 32 subsets, it is possible to obtain a reasonable initial reconstruction. Typically OSEM results in an order of magnitude decrease of computing time without measurable loss of image quality [24]. Computational efficiency of OSEM allows for incorporation of more complicated modeling during the reconstruction process. In OSEM, reconstruction

image data are updated for each subset during each iteration. Therefore, the number of updates is the product of iterations times the projection subsets. As the number of updates increases, the spatial resolution increases due to the recovery of higher frequencies; however, the increasing noise necessitates an optimization process where the noise smoothing filter, the number of iterations, and the number of subsets are properly balanced in order to obtain optimal image quality. Therefore, most current algorithms utilize various forms of noise suppression or "regularization" during iterative reconstruction.

Scatter and attenuation compensation can be integrated within iterative reconstruction. In addition, nonuniform resolution recovery techniques can be incorporated to correct for losses in spatial resolution due to image blurring by the collimator [25]. The current algorithms simultaneously address these problems by modeling the instrumentation and imaging parameters used for a specific application in order to eliminate the degrading physical effects and suppress noise in the image reconstruction process. The resolution recovery aspects of these algorithms can be emphasized to provide significant improvements in spatial resolution and MPS image quality, and the noise suppression aspects can be emphasized to decrease imaging times.

8.2.6 Advances in SPECT Reconstruction

To maximize the capabilities of the new hardware and to compensate for the resolution loss due to high sensitivity, collimation vendors of optimized solid-state systems have developed various versions of iterative reconstruction with resolution recovery and various compensations to improve image quality [26,27]. Additionally, the use of cardiac shape priors has been developed as an alternate form of regularization [26].

Standard Anger cameras can be also retrofitted with new reconstruction techniques, allowing them to reduce imaging time on standard equipment. Cardiofocal collimation had been coupled with iterative fast OSEM reconstruction with 3D resolution recovery, 3D collimator and detector response correction, and attenuation and scatter compensation [28]. Acquisition time can be reduced to between 33% and 50%, as compared to the standard acquisition protocols with FBP reconstruction with this combination [29].

It is possible to correct for the variations in spatial resolution by measuring the changes in spatial resolution with distance from the collimator and incorporating this collimator information both into the back projection and the forward projection parts of the reconstruction. To avoid amplification of statistical noise during the reconstruction process, noise can be controlled by smoothing both the estimated projection data and the measured projection data internally during the reconstruction process [30,31]. This method also allowed reduction of imaging time by 50% [32].

Another approach includes modeling of the integrated collimator and detector response function in an iterative reconstruction algorithm and performs image resolution recovery [33] based on these parameters [34]. This compensation technique is accomplished by convolving the projected photon ray with the corresponding line spread function during iterative projection and back projection [33,35,36]. The following parameters are accounted for and compensated: collimator hole and septa dimensions, intrinsic detector resolution, crystal thickness, collimator to detector gap, and projection-angle-specific center of rotation to collimator face distances. These collimator-specific data are embedded in the software in the form of look-up tables. Some of the relevant acquisition parameters (such as object to collimator distance) are obtained directly from the raw projection data.

Advanced reconstruction methods modeling the physics and geometry of the emission and detection processes can also be offered as a stand-alone workstation that can reconstruct data from most existing gamma cameras with standard collimators [37]. The resolution-recovery reconstruction may also require the information about the detector's distance from the patient. This distance can be obtained automatically on new cameras, but also can be obtained by image processing techniques and definition of the 3D patient body contour from standard images [37]. Equivalent image quality and defect characterization has been demonstrated with simulated fast imaging with times as low as 1/4th of standard imaging time as compared to full-time standard reconstruction [38].

A related technical development resulting in improved MPS image quality is the "motion-frozen" processing of gated cardiac images, which eliminates blurring of perfusion images due to cardiac motion [39]. This technique applies a nonlinear, thin-plate-spline warping algorithm and shifts counts from the whole cardiac cycle into the end diastolic position (Figure 8.14). The "motion-frozen" images have the

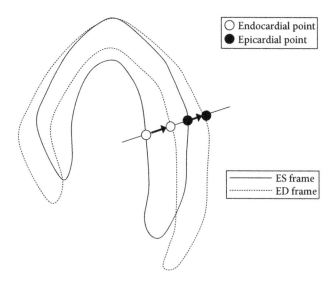

FIGURE 8.14 Principle of "motion-frozen" reconstruction to eliminate the effect of cardiac motion on static perfusion imaging. Counts from ES frame are warped by nonlinear image alignment to the ED frame position. (Reproduced from Slomka, P.J. et al., *J. Nucl. Med.*, 45, 1128, July 2004. With permission.)

appearance of diastolic frames but are significantly less noisy since the counts from the entire cardiac cycle are used. The spatial resolution of such images is higher than that of summed gated images. Recently, diagnostic improvement in specificity has been demonstrated by "motion-frozen" technique in MPS scans of obese patients [40]. The combination of such advanced approaches dedicated to cardiac imaging and the general advances in image reconstruction described earlier could result in further gains in image quality.

8.2.7 Radiation Dose Reduction in SPECT

To date, the new cardiac scanners and image reconstruction software technologies have been put into clinical practice primarily to reduce acquisition times since fast imaging provides immediate benefits in terms of patient throughput and patient comfort [16]. However, equivalent radiation dose reduction is also possible by performing standard-time acquisitions with lower injected activities. In view of the recent sensitivity toward reduced patient radiation dose, a few clinical studies have been conducted to evaluate the new equipment specifically with the reduced-dose protocols [41]. These studies are summarized in Table 8.1. Low-dose stress-only protocols allowed the patient radiation dose as low as a 4.2 mSv for a 12.5 mCi 5 min stress-only Tc-99m scan [42]. A low-dose rest-stress Tc-99m protocol has been reported resulting in 5.1–6.1 mSv exposure to patients [43,44].

Table 8.1 Recent Fast Clinical SPECT Studies with Reduced Dosing and New Instrumentation

Study	N	Protocol	Injected Dose	Total Radiation Dose
Husmann et al. [64]	100	Stress only/prospective gated CTA	Stress, 8.1 mCi	(5.4 ± 0.8) mSv
Duvall et al. [42]	209	Stress only ^{99m}Tc sestamibi	Stress, 12.5 mCi	4.2 mSv
DePuey et al. [45]	160	Tc-99m sestamibi	R, 5.8 ± 0.6 S, 17.5 ± 2.5 mCi	6.8 mSv
Nkoulou et al. [46]	50	1-d ^{99m}Tc tetrofosmin adenosine stress–rest	Stress/rest, 8.6/17.3 mCi 5 min/5 min	4.3 mSv
Duvall et al. [43]	131	Tc-99m sestamibi 5–8 min rest, 3–5 min stress	Rest, 5 mCi Stress, 15 mCi	5.8 mSv
Gimelli et al. [44]	137	Tc-99m tetrofosmin	Stress, 5–6 mCi Rest, 10–12 mCi	5.10 mSv (men), 6.12 mSv (women)*

Low-dose imaging was also demonstrated on a standard SPECT camera, utilizing an iterative reconstruction technique with standard acquisition times [45]. In another study, the patient dose was reduced to 4.3 mSv for stress-rest protocols when background activity subtraction was applied [46]. One problem with conducting such prospective low-dose studies is that one needs to be conservative in dose reduction since the quality of the images could potentially suffer significantly if the injected dose is too low. From the diagnostic standpoint, this would be especially critical for stress studies.

Gradual lowering of the dose was simulated by several reconstructions obtained from the list mode data [47]. It has been demonstrated that low-dose protocols producing as low as 1.0 million myocardial count images appear to be sufficient to maintain excellent agreement in quantitative perfusion and function parameters as compared to those determined from 8.0 million count images. With a solid-state camera, these images could be obtained over 10 min with an effective radiation dose of less than 1 mSv without a significant sacrifice in accuracy.

8.3 PET

Although PET MPI is currently not as widely utilized compared to SPECT, it offers significant potential advantages because in addition to relative perfusion, it allows routine AC, and measurements of (stress/rest MBF) and perfusion flow reserve [48]. New PET systems registering all photon events in list-mode allow reconstruction of static and ECG-gated perfusion, and blood flow quantification from one scan. PET is based on the coincidence detection of photons emitted from the decay of radiolabeled tracers introduced into the body to map a particular physiological process. Like SPECT imaging, PET is used to produce 3D images of radionuclide distribution in organs of interest, such as the heart. In PET, the radionuclide decays via positron emission, and subsequent annihilation with a free electron results in the production of 2 colinear 511 keV gamma-rays. These coincident annihilation gamma-rays are measured to form projections that can be reconstructed using FBP or OSEM methods to create a 3D image. The PET camera consists of stationary, 360°, detector rings allowing all projections through the heart to be obtained simultaneously. In contrast, SPECT uses radionuclides that decay via single gamma-ray emission.

8.3.1 PET Radiopharmaceuticals

Several PET radiopharmaceuticals are used in cardiac imaging (Table 8.2). All of these tracers have fairly short half-lives ranging from just over a minute to just under 2 h. Rb-82 is a perfusion imaging agent (analogous to Tl-201 used with SPECT) commonly used in sites without access to a cyclotron since it can also be distributed via a Sr-82 generator system (placed in the imaging room), which has a half-life of 25.5 days and a practical shelf life of 1–2 months. F-18 with a longer 110 min half-life can also be used without an on-site cyclotron for viability imaging with F-18 fluorodeoxyglucose, and more recently for perfusion imaging with F-18 flurpiridaz [49]. The very short half-lives of other isotopes such as C-11, N-13, and O-15 mean that

Table 8.2   Properties of Common PET Radioisotopes

	Half-Life (min)	Maximum Positron Energy (keV)	Positron Range in Water[a] (mm)	Production
F-18	110	635	0.23	Cyclotron
C-11	20	960	0.42	Cyclotron
N-13	10	1190	0.57	Cyclotron
O-15	2.04	1720	1.02	Cyclotron
Rb-82	1.25	3350	2.60	Sr-82 generator

[a] The ranges given here correspond to the root-mean-square distance travelled by the positron from the site of decay [65].

it is only practical to use pharmaceuticals labeled with these isotopes if there is a cyclotron or radiochemistry facilities onsite that permit local production.

An important property of PET radioisotopes is the positron range. The range refers to the distance that the positron travels before it slows down enough to annihilate with an electron, and is dependent on the kinetic energy of the emitted positron. As positron decay entails the emission of both a positron and a neutrino (a particle that is not detected), the kinetic energy of the positron forms a distribution of values up to some maximum value that depends on the radioisotope. In addition, like the electron, the positron is very light and can undergo large changes in different direction when it scatters off the nuclei of surrounding tissues. The combined effects of variable initial energies and the tortuous random paths of the positron result in an "inverse-cusp" distribution of the distance travelled from the site of decay. The range of the positron is thus a statistical measure of the distance travelled by all positrons emitted by a particular isotope, and is often calculated as the root-mean-square distance. The largest impact that the positron range has on PET imaging is as a reduction of spatial resolution. However, for most of the isotopes used in cardiac imaging, the range is small enough that the loss in resolution for human imaging is very small. The exception to this is Rb-82 whose range is on the same order as some other components of scanner resolution and can significantly affect the final image resolution.

8.3.2   Coincidence Detection

Positron emission tomography relies on the technique of simultaneous photon detection in "coincidence" to generate tomographic images of radiolabeled tracer distributions within the body. The positron emitted from the decay of the injected isotope travels a short distance in the body before annihilating with a nearby

electron, resulting in the simultaneous emission of two 511 keV photons in almost exactly opposite directions, unlike SPECT where only one photon is emitted (Figure 8.15). If both photons are detected within a short time, known as the coincidence timing window, a prompt coincidence count is recorded. The coincidence window is typically on the order of 2–10 ns, accounting for the finite time needed for photons to reach the detectors, and the limited precision of the electronic timing and detection systems. Coincidence photons recorded within this window indicate that a positron decay event occurred somewhere along the line between the two detectors, referred to as a line of response (LOR). A PET scanner consists of detectors arranged in multiple rings, resulting in numerous transaxial planes in which coincidences can be recorded, as shown in Figure 8.16. By detecting the coincidence events from a large number of LORs, a volumetric image of the activity distribution within the field of view can be reconstructed. The typical geometry of a PET imaging system is shown with multiple axial rings of detectors, as shown in Figure 8.17.

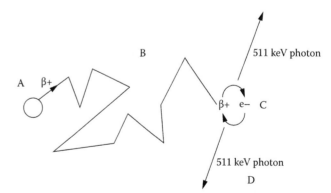

FIGURE 8.15   Positron (β+) decay results in emission from the nucleus (A), loss of kinetic energy over some maximum range (B), annihilation with a free electron (C), and production of two 511 keV photons travelling in opposite directions (D). (Reproduced with permission from Wells, R.G. et al., Positron emission tomography, in *Nuclear Cardiology Technical Applications* G.V. Heller, A. Mann, and R. Hendel (Eds.), McGraw-Hill, New York, 2009.)

Chapter 8

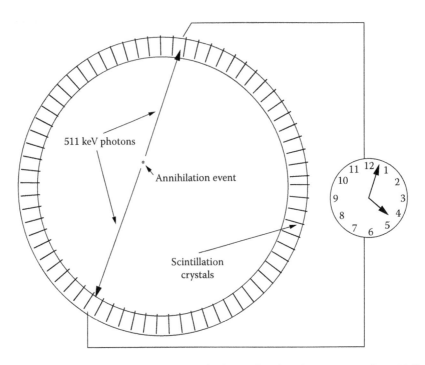

FIGURE 8.16   Illustration of PET coincidence detection principle. (Reproduced with permission from Wells, R.G. et al., Positron emission tomography, in *Nuclear Cardiology Technical Applications* G.V. Heller, A. Mann, and R. Hendel (Eds.), McGraw-Hill, New York, 2009.)

FIGURE 8.17   Typical geometry of a multiring PET system.

There are three (or four) types of coincidences recorded in the prompt event window: true, scattered, and random (as well as prompt gamma). True coincidences do not undergo any scatter or absorption in the body tissues. Scattered coincidences are recorded when one or both of the photons from a single annihilation event are scattered in the body and subsequently mispositioned (Figure 8.18). Random coincidences occur when two photons from separate annihilation events are detected by chance or accidentally within the coincidence window (Figure 8.19). Some radionuclides such

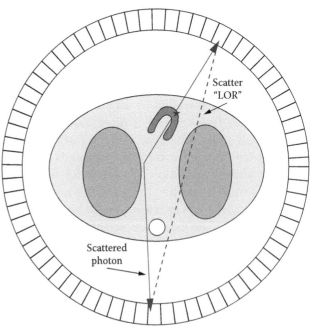

FIGURE 8.18   Scattered coincidence event recorded along incorrect line of response (LOR). Prompt gammas can also result in coincidence events recorded along incorrect LOR. (Modified and reproduced with permission from Wells, R.G. et al., Positron emission tomography, in *Nuclear Cardiology Technical Applications* G.V. Heller, A. Mann, and R. Hendel (Eds.), McGraw-Hill, New York, 2009.)

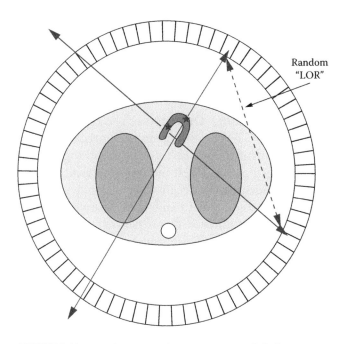

FIGURE 8.19 Random coincidence event recorded along incorrect line of response. (Reproduced with permission from Wells, R.G. et al., Positron emission tomography, in *Nuclear Cardiology Technical Applications* G.V. Heller, A. Mann, and R. Hendel (Eds.), McGraw-Hill, New York, 2009.)

as Rb-82 also produce a fraction of "prompt-gamma rays" together with the emitted positron. Depending on the energy, these gamma rays can be detected in coincidence with one of the annihilation photons, resulting in a prompt-gamma coincidence that is also mispositioned. Random and prompt-gamma rays versus scattered coincidences add relatively uniform background activities, respectively, which increase image noise and decrease contrast. They must be corrected in order to obtain accurate measures of the true activity distribution.

8.3.3   PET Detectors

Annihilation photons are detected and converted into visible light photons by scintillation crystals forming the detectors similar to conventional SPECT detectors. As in SPECT, the visible photons are also detected by PMTs where they are converted into an electrical pulse that is later processed into an image. The imaging performance depends on the number of light photons detected, which is determined by the scintillator characteristics. Commonly used PET scintillator materials and their properties at 511 keV are listed in Table 8.3.

The ability of a scintillator to stop and detect photons is described by the linear attenuation coefficient μ (cm^{-1}) reflecting the fraction of photons absorbed per cm. The higher the μ value, the better the detection sensitivity. The light output determines the precision of the arrival time and energy measurements. Therefore, with higher light output, the coincidence time window can be narrowed to reduce randoms, and the energy window can be narrowed to reduce scatter. The light decay time determines the PMT electronic pulse integration time, and therefore the detector dead time. So-called "fast" detectors have high light output and low dead time; therefore higher count rates can be achieved without saturating the detectors. This is of particular importance for dynamic cardiac imaging of short-lived isotopes such as ^{82}Rb and ^{15}O-water, where higher activities are typically administered compared to the longer lived PET isotopes.

8.3.4   PET Image Acquisition

Coincidence counts recorded along the LORs can be organized into transaxial sinograms or angular projections as in SPECT, where each bin corresponds to a unique axial, radial, and angular location in the field of view. The sinogram format is typically used with 2D reconstruction to form an image. On the most current PET systems, however, events can also be stored in list-mode format where the LOR coordinates for each count are stored individually along with the corresponding arrival time (and possibly the difference time of flight [TOF]). While this mode requires more storage space for large data sets, it provides the flexibility to re-bin the data retrospectively into projections. This is highly advantageous, as it allows reconstruction of static, cardiac-gated, respiratory-gated and dynamic

Table 8.3   Properties of Common PET Detector Scintillator Materials at 511 keV

	Scintillator	Density (g/cm³)	μ (cm^{-1})	Light Decay Time (ns)	Relative Light Output (%)
NaI	Sodium iodide	3.67	0.34	230	100
BGO	Bismuth germanate	7.13	0.95	300	21
LSO	Lutetium oxyorthosilicate	7.4	0.88	40	68
LYSO	Lutetium yttrium oxyorthosilicate	7.1	0.83	41	75
GSO	Gadolinium oxyorthosilicate	6.71	0.70	60	36

Chapter 8

images, all from a single injection without lengthening the time of the imaging session.

Static perfusion imaging: Static images are created by summing the list-mode data acquired over the uptake phase of the scan, usually starting at 2–5 min after the injection of the tracer up until the end of the acquisition, thus providing images of a *relative* tracer distribution. For perfusion tracers such as ^{82}Rb, and ^{13}N-ammonia, this technique is used for conventional myocardial perfusion imaging. For ^{18}F-FDG and similar metabolic tracers, viability, and standard uptake values (SUVs, in units of g/mL) can be evaluated by normalizing the PET images for activity distribution (Bq/mL or µCi/mL) to the injected activity (MBq or mCi) and the patient weight (kg). An SUV of 1.0 corresponds to a uniform distribution in the body, whereas an SUV > 1.0 is observed with selective tracer retention and an SUV < 1.0 is observed with relative tracer clearance.

ECG gating: PET images, as in SPECT, are used to obtain information about the contractile function of the beating heart. An ECG trigger allows data to be sorted into the different phases of the cardiac cycle. Each phase is reconstructed into a separate image (gate), creating a cine-loop that can be played to visually evaluate wall motion. Analysis of end systolic and diastolic volumes, wall thickening and left-ventricular ejection fraction can also be performed. To obtain adequate image quality, a greater number of total counts must be obtained

to compensate for the division of counts over multiple gates. This can be achieved by lengthening the scan time or injecting more tracer activity, but this must be balanced with the tracer half-life, detector dead-time losses, total imaging time, and patient radiation dose.

Respiratory gating: One approach to correct for blurring and CT–AC misregistration artifacts (caused by movement of the heart during breathing) is to use respiratory-gating. Chest motion is determined either with a video camera registering the movement of a marker on the patient's abdomen, or by an inductive respiration monitor with an elasticized belt around the patient's chest. These signals are used as triggers to divide the list-mode data into different respiratory phases before image reconstruction. This method can also be useful in correcting for the upward creep of the heart during exercise-stress imaging as the heart rate slows down and the body relaxes over time. Dual respiratory and ECG gating is possible; however, as with ECG gating alone, higher total counts are needed to maintain good image quality [50].

Dynamic imaging: Dynamic images provide an evolving picture of the distribution of the tracer in the heart over time. By reconstructing the data into a series of short time frames following tracer injection and subsequent clearance from the blood into the tissues (Figure 8.20), quantitatively accurate measurements of myocardial physiology and biology can be obtained. Dynamic imaging is essential

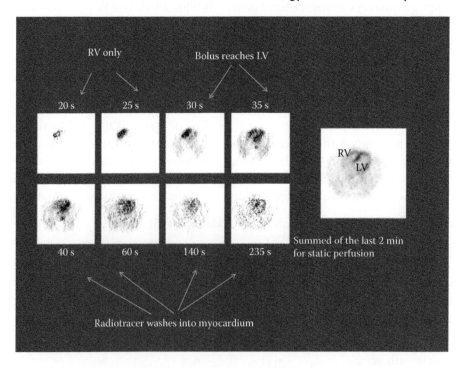

FIGURE 8.20   Sequential transaxial images obtained dynamically for the Rb-82 images.

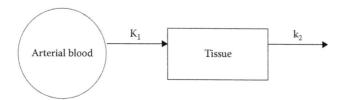

FIGURE 8.21 MBF measurement. A simple 1-tissue compartment model can be used to model MBF with tracers such as Rb-82 and N-13 ammonia. In this model, the rate of the exchange of the tracer from the arterial blood (red) into the myocardial tissue (blue) K_1 is related to MBF in mL/min/g. The tracer clearance rate from the myocardium is described by k_2/min.

for evaluating MBF, an important measure for diagnosing cases of multivessel and microvascular diseases that cannot be identified using standard relative quantification of the perfusion in the LV myocardium. To obtain MBF values, dynamic image-derived time activity curves from the arterial blood and myocardial tissue regions are used as input to a tracer kinetic model. This model describes the exchange of tracer between the blood and the tissue compartments over time. The rate of tracer uptake into the tissue provides an estimate of MBF in absolute units of mL/min/g (Figure 8.21).

PET 2D versus 3D acquisition: Acquiring oblique LORs with a multiring configuration greatly increases the geometric sensitivity of modern PET scanners, that is, the fraction of the annihilation photons that it is capable of detecting. Unfortunately, it also greatly increases the number of random events detected (because the random rate increases as the square of the associated detector singles rates) and the amount of scatter that is detected. One solution to this problem is to insert thin absorbing tungsten septa between the detector planes (Figure 8.22). Scanners configured with interplane septa acquire

data in what is referred to as a *2D acquisition* mode. The length of the septa and the axial width of the crystal planes determine the maximum accepted angle of the LOR and hence the maximum possible ring difference. The ring difference might be further restricted by electronically rejecting those LORs with larger-than-desired ring differences. In 2D imaging, the maximum ring difference is typically 3 or 4 rings.

An alternative solution is *3D acquisition,* for which the septa are not used and any ring difference is possible (Figure 8.22). The principal advantage of the 3D PET acquisition (accepting photons from all angles) over the 2D PET with parallel septa is much higher sensitivity (4–6 times) for photon detection, albeit at the cost of increased random and scatter fraction, and more complex corrections needed as part of the image reconstruction [51]. 3D ^{82}Rb PET acquisition has presented image quality problems due to increased random events when older 2D-3D BGO scanners were used [52]. Most literature in cardiac PET imaging describes studies performed in 2D mode. However, currently, the users of new PET–CT equipment are unable to acquire images in 2D, since almost all new

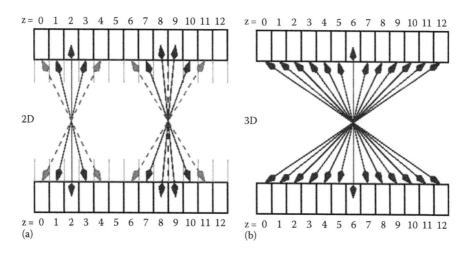

FIGURE 8.22 Two-dimensional (a) versus 3D (b) PET image acquisition. Short lines extending from the detectors represent septa used with 2D imaging. Blue coincidences are detected and red coincidences are rejected. (Reproduced with permission from Wells, R.G. et al., Positron emission tomography, in *Nuclear Cardiology Technical Applications* G.V. Heller, A. Mann, and R. Hendel (Eds.), McGraw-Hill, New York, 2009.)

Table 8.4 Summary of PET–CT System Performance

System	Resolution (mm)	Sensitivity (cps/kBq)	Reference
Siemens BioGraph HiRez	4.6	4.92	Brambilla et al. [66]
GE ST	5.2 (6.0)	9.2 (1.95)	Mawlawi et al. [67]
GE RX	5.1 (5.0)	7.3 (1.7)	Kemp et al. [68]
Philips Gemini TF	4.8	6.6	Surti et al. [69]

scanners operate in 3D mode only and come without septa. Recently, PET scanners have been optimized to cope with the high count rates encountered in first-pass Rb-82 imaging for MBF measurements [53], and with additional Rb-82 gamma-prompt coincidences affecting 3D imaging [54]. Although the initial application of the new PET technology was in the field of oncology, 3D acquisition and new software methods offer a potential for dose reduction in cardiac imaging, beyond the existing low-radiation-dose levels. Key performance characteristics for several currently available 3D PET–CT scanners are shown in Table 8.4.

8.3.5 PET Image Reconstruction and Corrections

Tomographic image reconstruction has traditionally been performed using FBP reconstruction; however, iterative techniques are becoming standard in PET similar to cardiac SPECT imaging. These algorithms provide images with decreased background noise in comparison to FBP (see SPECT section). Alternatively, image quality can be maintained comparable to FBP and the injected activity reduced with iterative reconstruction. The disadvantage of this technique is that the algorithm convergence and spatial resolution may vary locally within the image. The final image contrast can change depending on the local count statistics and should be carefully evaluated for quantitative imaging. Iterative reconstruction facilitates highly accurate corrections to compensate for detector efficiency, detector response profiles, photon attenuation, and random, scattered, and prompt-gamma coincidences.

Recent technical advances include developments of new high-definition iterative reconstruction techniques based on scanner-specific 3D modeling of detector response functions with effective tomographic resolution as low as 2 mm [55]. Preliminary reports have been published demonstrating the application of these improved PET reconstruction methods to cardiac imaging [56]. Further image quality improvements have been proposed [57] by combining new reconstruction with

cardiac motion-frozen techniques to reduce blurring due to cardiac motion [39] (Figure 8.23).

Another type of reconstruction, specific to PET that is possible with list-mode data and picosecond timing resolution from fast coincidence electronics and detectors is TOF [58]. With this technique, the approximate location of the coincidence event is determined using the difference in arrival times of the two annihilation photons. This allows the possible locations of individual events to be constrained to a shorter distance along the LOR during reconstruction, resulting in higher signal-to-noise ratios for the same number of recorded counts. Initial applications of TOF imaging are expected to demonstrate some improvements in cardiac image quality, particularly in obese patients.

Detector corrections: Unlike in SPECT acquired data must be corrected for detector dead-time losses, to obtain reconstructed images that are proportional to the absolute activity concentration in Bq/mL. To ensure optimal detection sensitivity, daily quality assurance should be performed to check that the detector response has not drifted in comparison to an established reference scan. Lost counts due to the dead-time effect can be accounted for using a model of the camera's response to a range of injected activities acquired prior to the imaging session.

AC: As annihilation photons travel through the tissue they can be absorbed or Compton scattered, decreasing the likelihood that they will actually exit the body and strike a detector. Although energy of the 511 keV photons is higher than in SPECT (typically 140 keV), the requirement for coincidence counting results in an attenuation effect even more pronounced than in SPECT (Figure 8.24). To correct for this effect, the acquired data is scaled using a map of the tissue attenuation coefficients in the FOV. This "μ-map" is obtained via an isotope (e.g., ^{68}Ge/Ga) transmission scan on dedicated PET systems or a coregistered x-ray CT scan on hybrid PET–CT systems. Unlike in SPECT, AC in PET is universally applied due to the simple and accurate

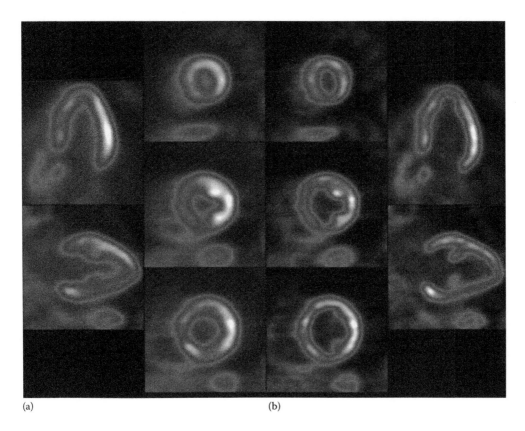

(a)   (b)

FIGURE 8.23 Standard PET imaging (a) and improvement image quality obtained with dual motion frozen imaging (b) for the F-18 fluorodeoxyglucose cardiac imaging with latest generation PET/CT scanners.

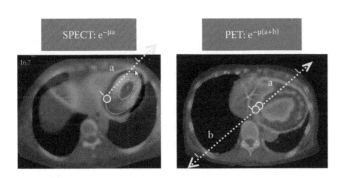

FIGURE 8.24 Photon attenuation in SPECT (left) and in PET (right). The effective photon path is shorter in SPECT (a) than in PET (a + b).

physics of annihilation photon coincidence detection. MR-based AC on combined PET-MR machines is also under development, but has the added complexity that MR image contrast is not linearly related to tissue density and attenuation [59].

The CT attenuation (CT-AC) image is typically measured using a fast (2–20 s), low-dose (0.1–1.0 mSv) scan (Figure 8.25). CT-AC is acquired ideally before a resting scan and after the stress scan, or in between the two scans, to minimize the chance of misalignment with the PET image due to patient movement or a shift in the heart position during stress. Careful review of the fused PET–CT images should be done prior to reconstruction, with the CT being shifted to align with the PET image if necessary. Since the CT is much faster than transmission imaging, it can produce respiratory motion artifacts in the corrected PET image, as the CT is only acquired at one point in the breathing cycle and the PET is obtained over many cycles. To reduce this problem, patients could be instructed to shallow breathe or hold their breath at end-expiration for the duration of the CT scan [60].

It should be noted that there is a small additional radiation dose associated with CT–AC scans for PET imaging. Considering the dose reducing efforts undertaken by vendors for SPECT-AC the PET-CT-AC method remain relatively dose inefficient. Further efforts to reduce the radiation dose associated with PET AC are therefore warranted, especially in the view of very low dose incurred from the injection of the PET radiotracers.

The hybrid PET–CT systems also have the potential for obtaining CTA and coronary artery calcium (CAC) measurements [48]. CTA allows the fusion of MPI or MBF

Chapter 8

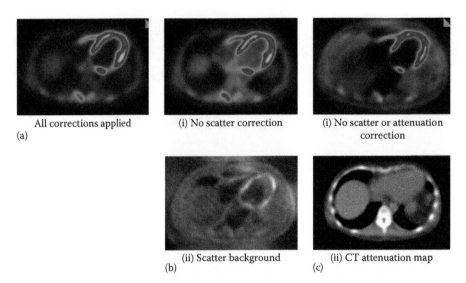

FIGURE 8.25 Image corrections. When all corrections are enabled, excellent image quality is achieved in this transverse FDG image, quantified in absolute units of activity concentration (Bq/cc) (a). Without scatter correction, an increase in background noise is observed, particularly in the blood cavity due to scatter from the surrounding myocardium with high tracer uptake (b.i). This is also evident in the corresponding subtracted scatter background (b.ii) where the myocardium demonstrates the highest scatter levels. If neither scatter or attenuation correction is applied, the myocardial and cavity activities are reduced (c.i). The apparent background activity increases in the lungs and periphery, while the cavity activity appears to decrease due to higher photon attenuation in that area, as evidenced on the CT attenuation map (c.ii). (Reproduced with permission from *Handbook of Nuclear Cardiology: Cardiac SPECT and Cardiac PET*, G.V. Heller, R.C. Hendel (Eds.), Springer, London, U.K., 2013.)

images with the coronary anatomy. Software alignment is nevertheless required since CTA images are acquired at "breath-hold" unlike PET images [61]. CAC scans can be used for both AC and calcium scoring [62].

Random and scatter corrections: The annihilation photon interactions within the body must also be compensated precisely to generate quantitatively accurate images. Corrections for random, scattered and prompt-gamma coincidences are typically performed by subtraction of their measured or modeled event rates from the total measured prompts when using FBP reconstruction. Alternatively, the measured or modeled event rates can be added in the forward projection step when using iterative reconstruction methods, to better model the Poisson noise effects in the raw projection data. The qualitative effects of these corrections on the reconstructed images are shown in Figure 8.25.

Decay correction: For dynamic imaging especially for short-lived isotopes like Rb-82, it is also necessary to correct the images for the effect of isotope decay, typically back to the start time of the acquisition, taking into account the exponential decrease in activity over time.

8.3.6 PET Image Quality

Several other physical factors outlined below affect PET image quality. The distance the positron travels from the nucleus before annihilation is determined by the initial positron energy; the greater the energy the larger the distance. This can significantly degrade image quality. Thus, high-energy positron emitters such as ^{82}Rb generally result in lower resolution images as compared to lower-energy isotopes, such as ^{18}F. In a combined PET-MR system, the positron range can actually be reduced by the spiral movement in the magnetic field, potentially resulting in images with better in-plane or transaxial resolution, at the expense of axial resolution [63]. Organs that are adjacent to the heart, such as the liver, lungs, and stomach wall, can take up activity in significant concentrations, resulting in spillover into the myocardium. This can complicate CT–AC alignment and mask both MPI and MBF defects, and should be taken into account when interpreting images.

PV effect due to the finite spatial resolution of the PET scanner is another limiting factor similar to SPECT imaging (see SPECT quality factors). As additional complication in PET, activity from the arterial blood pool in the cavity of the LV spills over onto the myocardium, resulting in an overestimation of activity, particularly in the first-pass transit through the heart. Patient and organ motion from the respiratory and cardiac cycles, results in blurring of PET images, leading to a mixing of activity from adjacent pixels. This is different than in SPECT where motion may cause severe artifacts due to the inconsistency of the projections.

8.4   Summary

We have reviewed the instrumentation acquisition and reconstruction methods for cardiac SPECT and PET. The most prevalent nuclear technique used clinically is MPS. New types of SPECT devices and software methods for image reconstruction have been recently introduced dramatically increasing the speed of SPECT images and lowering the radiation dose to the patient. Myocardial perfusion PET is performed less frequently due to higher cost of the equipment and of the radiotracer, but offers some advantages over SPECT, such as higher spatial and temporal resolution, ability to quantify MBF and perfusion flow reserve. New reconstruction methods in PET and SPECT compensate for artifacts resulting from various physical and biological effects, recover image resolution and consequently allow further reduction of radiation dose and imaging time.

References

1. A. J. Einstein, Medical imaging: The radiation issue, *Nat Rev Cardiol*, 6, 436–438, June 2009.
2. G. Germano et al., Attenuation correction in cardiac SPECT: The boy who cried wolf?, *J Nucl Cardiol*, 14, 25–35, 2007.
3. R. G. Wells et al., Comparing slow-versus high-speed CT for attenuation correction of cardiac SPECT perfusion studies, *J Nucl Cardiol*, 19, 719–726, August 2012.
4. T. Schepis et al., Use of coronary calcium score scans from stand-alone multislice computed tomography for attenuation correction of myocardial perfusion SPECT, *Eur J Nucl Med Mol Imaging*, 34, 11–19, January 2007.
5. A. J. Einstein et al., Agreement of visual estimation of coronary artery calcium from low-dose CT attenuation correction scans in hybrid PET/CT and SPECT/CT with standard agatston score, *J Am Coll Cardiol*, 56, 1914–1921, 2010.
6. C. Bai et al., Phantom evaluation of a cardiac SPECT/VCT system that uses a common set of solid-state detectors for both emission and transmission scans, *J Nucl Cardiol*, 17, 459–469, June 2010.
7. D. Sowards-Emmerd et al., CBCT-subsystem performance of the multi-modality Brightview XCT system (M09–26), in *2009 IEEE Nuclear Science Symposium Conference Record (NSS/MIC)*, Orlando, FL, 2009, pp. 3053–3058.
8. J. Friedman et al., "Upward creep" of the heart: A frequent source of false-positive reversible defects during thallium-201 stress-redistribution SPECT, *J Nucl Med*, 30, 1718–1722, 1989.
9. N. Matsumoto et al., Quantitative assessment of motion artifacts and validation of a new motion-correction program for myocardial perfusion SPECT, *J Nucl Med*, 42, 687–694, May 2001.
10. A. J. Einstein et al., Estimating risk of cancer associated with radiation exposure from 64-slice computed tomography coronary angiography, *JAMA*, 298, 317–323, July 18, 2007.
11. L. Volokh et al., Effect of detector energy response on image quality of myocardial perfusion SPECT, in *IEEE Nuclear Science Symposium and Medical Imaging Conference*, Dresden, Germany, 2008.
12. S. Ben-Haim et al., Simultaneous dual isotope myocardial perfusion scintigraphy (DI MPS)–Initial experience with fast D-SPECT [Abstract], *J Nucl Med*, 49(Suppl. 1), 72, 2008.
13. H. Babla et al., A triple-head solid state camera for cardiac single photon emission tomography (SPECT), in L. A. Franks et al. (Eds.), *Proc SPIE*, 6319, 63190M1–63190M5, 2006.

14. D. L. Gunter, Gamma camera collimator characteristics and design, in *Nuclear Medicine*, vol. 1, R. E. Henkin (Ed.), Philadelphia, PA: Elsevier, 2006, pp. 107–126.
15. J. Patton et al., D-SPECT: A new solid state camera for high speed molecular imaging, *Soc Nucl Med*, 47, 189–189, 2006.
16. T. Sharir et al., High-speed myocardial perfusion imaging initial clinical comparison with conventional dual detector anger camera imaging, *J Am Coll Cardiol Cardiovasc Imaging* 1, 156–163, 2008.
17. P. C. Hawman and E. J. Haines, The cardiofocal collimator: A variable focus collimator for cardiac SPECT, *Phys Med Biol*, 39, 439, 1994.
18. H. Vija et al., IQ SPECT technology white paper, Siemens Medical Solutions USA, *Mol Imaging*, 2008, 1–7.
19. T. Funk et al., A novel approach to multipinhole SPECT for myocardial perfusion imaging, *J Nucl Med*, 47, 595–602, 2006.
20. I. Blevis et al., CZT Gamma camera with pinhole collimator: Spectral measurements, in *IEEE 2008 Nuclear Science and Medical Imaging Conference*, Dresden, Germany, 2008.
21. E. V. Garcia et al., 2.05: A new solid state, ultra fast cardiac multi-detector SPECT system, *J Nucl Cardiol*, 15, S3, 2008.
22. J. A. Sorenson and M. A. Phelps, The anger camera: Performance characteristics, in *Physics in Nuclear Medicine*, 2nd ed., J. A. Sorenson and M. A. Phelps (Eds.), Philadelphia, PA: WB Saunders Company, 1987, pp. 318–345.
23. G. El Fakhri et al., Relative impact of scatter, collimator response, attenuation, and finite spatial resolution corrections in cardiac SPECT, *J Nucl Med*, 41, 1400–1408, August 1, 2000.
24. H. M. Hudson and R. S. Larkin, Accelerated image reconstruction using ordered subsets of projection data, *IEEE Trans Med Imaging*, 13, 601–609, 1994.
25. C. E. Metz, The geometric transfer function component for scintillation camera collimators with straight parallel holes, *Phys Med Biol*, 25, 1059–1070, 1980.
26. B. Rousso and M. Nagler, Multi-dimensional image reconstruction, US Patent 7176466, February 13, 2007.
27. J. Maddahi et al., Prospective multicenter evaluation of rapid, gated SPECT myocardial perfusion upright imaging, *J Nucl Cardiol*, 16, 351–357, May–June 2009.
28. A. H. Vija et al., Development of rapid SPECT acquisition protocol for myocardial perfusion imaging, *IEEE Nucl Sci Symp Conf Record*, 3, 1811–1816, 2006.
29. E. P. Ficaro et al., 15.34: Effect of reconstruction parameters and acquisition times on myocardial perfusion distribution in normals, *J Nucl Cardiol*, 15, S20, 2008.

Chapter 8

30. J. Ye et al., Iterative SPECT reconstruction using matched filtering for improved image quality, *IEEE Nucl Sci Symp Conf Record*, 4, 2285–2287, 2006.

31. J. Ye et al., Iterative reconstruction with enhanced noise control filtering, WO Patent WO/2007/034,342, 2007.

32. T. M. Bateman et al., Multicenter investigation comparing a highly efficient half-time stress-only attenuation correction approach against standard rest-stress Tc-99m SPECT imaging, *J Nucl Cardiol*, 16, 726–735, September–October 2009.

33. B. M. W. Tsui et al., Implementation of simultaneous attenuation and detector response correction in SPECT, *IEEE Trans Nucl Sci*, 35, 778–783, February 1988.

34. E. DePuey et al., OSEM and wide beam reconstruction (WBR) "half-time" gated myocardial perfusion SPECT functional imaging: A comparison to "full-time" filtered back projection, *J Nucl Cardiol*, 15(4), 547–563, 2008.

35. B. M. W. Tsui et al., The importance and implementation of accurate 3D compensation methods for quantitative SPECT, *Phys Med Biol*, 39, 509–530, March 1994.

36. B. M. W. Tsui and G. T. Gullberg, The geometric transfer-function for cone and fan beam collimators, *Phys Med Biol*, 35, 81–93, 1990.

37. S. Borges-Neto et al., Clinical results of a novel wide beam reconstruction method for shortening scan time of Tc-99m cardiac SPECT perfusion studies, *J Nucl Cardiol*, 14, 555–565, 2007.

38. E. G. DePuey et al., 2.01: Quarter-time myocardial perfusion SPECT wide beam reconstruction, *J Nucl Cardiol*, 15, S2, 2008.

39. P. J. Slomka et al., "Motion-frozen" display and quantification of myocardial perfusion, *J Nucl Med*, 45, 1128–1134, July 2004.

40. Y. Suzuki et al., Motion-frozen myocardial perfusion SPECT improves detection of coronary artery disease in obese patients, *J Nucl Med*, 49, 1075–1079, 2008.

41. M. Henzlova and W. Duvall, The future of SPECT MPI: Time and dose reduction, *J Nucl Cardiol*, 18, 580–587, 2011.

42. W. Duvall et al., Reduced isotope dose with rapid SPECT MPI imaging: Initial experience with a CZT SPECT camera, *J Nucl Cardiol*, 17, 1009–1014, 2010.

43. W. Duvall et al., Reduced isotope dose and imaging time with a high-efficiency CZT SPECT camera, *J Nucl Cardiol*, 18, 847–857, 2011.

44. A. Gimelli et al., High diagnostic accuracy of low-dose gated-SPECT with solid-state ultrafast detectors: Preliminary clinical results, *Eur J Nucl Med Mol Imaging*, 39, 83–90, September 2, 2011.

45. E. DePuey et al., A comparison of the image quality of full-time myocardial perfusion SPECT vs wide beam reconstruction half-time and half-dose SPECT, *J Nucl Cardiol*, 18, 273–280, 2011.

46. R. Nkoulou et al., Semiconductor detectors allow low-dose-low-dose 1-day SPECT myocardial perfusion imaging, *J Nucl Med*, 52, 1204–1209, August 1, 2011.

47. R. Nakazato et al., Myocardial perfusion imaging with a solid state camera: Simulation of a very low dose imaging protocol, *J Nucl Med*, 54, 373–379, 2012.

48. M. F. Di Carli et al., Clinical myocardial perfusion PET/CT, *J Nucl Med*, 48, 783–793, May 2007.

49. D. S. Berman et al., Improvement in PET myocardial perfusion image quality and quantification with flurpiridaz F 18, *J Nucl Cardiol*, 19(Suppl. 1), S38–S45, February 2012.

50. F. Buther et al., List mode-driven cardiac and respiratory gating in PET, *J Nucl Med*, 50, 674–681, May 2009.

51. K. Knešaurek et al., Comparison of 2-dimensional and 3-dimensional 82Rb myocardial perfusion PET imaging, *J Nucl Med*, 44, 1350–1356, August 1, 2003.

52. J. R. Votaw and M. White, Comparison of 2-dimensional and 3-dimensional cardiac 82Rb PET studies, *J Nucl Med*, 42, 701–706, May 2001.

53. R. Klein et al., Quantification of myocardial blood flow and flow reserve: Technical aspects, *J Nucl Cardiol*, 17, 555–570, 2010.

54. F. Esteves et al., Prompt-gamma compensation in Rb-82 myocardial perfusion 3D PET/CT, *J Nucl Cardiol*, 17, 247–253, 2010.

55. V. Y. Panin et al., Fully 3-D PET reconstruction with system matrix derived from point source measurements, *IEEE Trans Med Imaging*, 25, 907–921, 2006.

56. L. Le Meunier et al., Enhanced definition PET for cardiac imaging, *J Nucl Cardiol*, 17, 414–426, 2010.

57. L. Le Meunier et al., Motion frozen (18)F-FDG cardiac PET, *J Nucl Cardiol*, 18, 259–266, Epub 2010, December 16, 2011.

58. B. W. Jakoby et al., Physical and clinical performance of the mCT time-of-flight PET/CT scanner, *Phys Med Biol*, 56, 2375–2389, Epub 2011, March 22, 2011.

59. V. Keereman et al., Challenges and current methods for attenuation correction in PET/MR, *MAGMA*, 26, 81–98, August 9, 2012.

60. R. A. Cook et al., Respiration-averaged CT for attenuation correction in canine cardiac PET/CT, *J Nucl Med*, 48, 811–818, May 2007.

61. R. Nakazato et al., Automatic alignment of myocardial perfusion PET and 64-slice coronary CT angiography on hybrid PET/CT, *J Nucl Cardiol*, 19, 482–491, June 2012.

62. I. Mylonas et al., Measuring coronary artery calcification using positron emission tomography-computed tomography attenuation correction images, *Eur Heart J Cardiovasc Imaging*, 13, 786–792, September 2012.

63. B. E. Hammer et al., Use of a magnetic field to increase the spatial resolution of positron emission tomography, *Med Phys*, 21, 1917–1920, December 1994.

64. L. Husmann et al., Diagnostic accuracy of computed tomography coronary angiography and evaluation of stress-only single-photon emission computed tomography/computed tomography hybrid imaging: Comparison of prospective electrocardiogram-triggering vs. retrospective gating, *Eur Heart J*, 30, 600, 2009.

65. R. Lecomte, Technology challenges in small animal PET imaging, *Nucl Instr Methods Phys Res A*, 527, 157–165, 2004.

66. M. Brambilla et al., Performance characteristics obtained for a new 3-dimensional lutetium oxyorthosilicate-based whole-body PET/CT scanner with the National Electrical Manufacturers Association NU 2-2001 standard, *J Nucl Med*, 46, 2083–2091, December 2005.

67. O. Mawlawi et al., Performance characteristics of a newly developed PET/CT scanner using NEMA standards in 2D and 3D modes, *J Nucl Med*, 45, 1734–1742, October 2004.

68. B. J. Kemp et al., NEMA NU 2-2001 performance measurements of an LYSO-based PET/CT system in 2D and 3D acquisition modes, *J Nucl Med*, 47, 1960–1967, December 2006.

69. S. Surti et al., Performance of Philips Gemini TF PET/CT scanner with special consideration for its time-of-flight imaging capabilities, *J Nucl Med*, 48, 471–480, March 2007.

70. G. Germano, Technical aspects of myocardial SPECT imaging, *J Nucl Med*, 42, 1499–507, October 2001.

71. R. G. Wells et al., Positron emission tomography, in *Nuclear Cardiology Technical Applications*, G.V. Heller, A. Mann, and R. Hendel (Eds.), New York: McGraw-Hill, 2009.

72. Renaud et al., PET Instrumentation. *Handbook of Nuclear Cardiology: Cardiac SPECT and Cardiac PET*, G.V. Heller, R.C. Hendel (Eds.), London, U.K.: Springer, pp. 127–139, 2013.

Chapter 8

9. Morphological Imaging of the Heart

George C. Kagadis, Stavros Spiliopoulos, Konstantinos Katsanos, Periklis Davlouros, and Dimitris Karnabatidis

9.1   Introduction

Imaging of the heart is in principle challenging due to its continuous motion. However, the continuous and innovative technological advances in the field of radiology radically improved both the morphological and functional imaging of the cardiovascular system, making novel imaging modalities indispensable for the evaluation of cardiac pathology. More specifically, modern diagnosis of congenital and acquired cardiac diseases is made possible with the use of magnetic resonance imaging (MRI), multidetector computed tomography (CT), Ultrasound (U/S), digital subtraction angiography (DSA), optical coherence tomography, and intravascular ultrasound (IVUS). This chapter focuses on the role of these modern imaging modalities for the morphological assessment of the heart and its vessels.

Chapter 9

9.2 Magnetic Resonance Imaging (MRI)

9.2.1 General Considerations

Today MRI is considered a first-line diagnostic tool for the morphological imaging of the heart. Its non-invasive, radiation-free, nature has established cardiac MRI in everyday clinical practice for the assessment of a wide variety of cardiac diseases, while its diagnostic accuracy and clinical applications are continuously increasing. Cardiac MRI is typically implied for the diagnosis, postsurgical evaluation, and follow-up of congenital heart disease (CHD), whereas for the diagnosis and clinical management of a great variety of acquired cardiac pathologies, the morphological assessment of the heart using MRI is considered the gold standard.

MRI of the heart necessitates specific imaging sequences, while the majority of the dedicated cardiac MRI techniques require rapid data acquisition with gradient systems that permit short repetition and echo time. Although cardiac imaging can be performed using 1.0 T field, short data-sampling times are achieved with 1.5 T and new 3.0 T systems that allow high signal-to-noise ratio and therefore faster and more accurate imaging techniques (Wen et al. 1997; Wintersperger et al. 2006). Rapid image acquisition, within a fraction of the cardiac cycle, minimizes motion artifacts and improves image quality. However, even rapid data sampling is subjective to the heart's pulsatile motion and to the respiratory motion artifacts, and therefore, synchronization of the cardiac motion should be performed using prospective ECG triggering or retrospective ECG gating techniques (Ridgway 2010; Barnwell et al. 2012).

For the morphologic cardiac assessment, typical transverse, sagittal, and coronal images from the arch to the diaphragm should be obtained, while various imaging packages are available from each company that include the necessary multiple sequence techniques and postprocessing possibilities (Yorimitsu et al. 2012).

Accurate triggering can be performed by tracing the ECG signal using an MR-integrated monitoring unit or by an external ECG monitor connected to the scanner (Figure 9.1). These techniques allow data sampling either within systole or diastole according to the clinical indication. As far as the elimination of respiratory artifacts is concerned, this can be attained

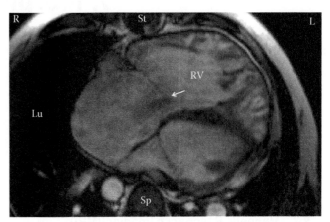

FIGURE 9.1 Cardiac magnetic resonance (CMR): This imaging method is based on classical MRI, using specially designed MRI sequences and chest coils, to acquire images of the heart. Due to the motion of the heart during the cardiac cycle, special software allowing gating with the electrocardiogram (ECG) is required. This image of the heart was acquired using a gradient-echo MRI sequence, in which blood is shown white and the myocardium is shown gray. The four chambers of the heart are shown, with the right ventricle (RV) of this patient being severely dilated and hypertrophied. Note the hypertrophic trabeculations of the heart muscle. The black signal shown with the arrow corresponds to regurgitant blood through the tricuspid valve from the RV to the right atrium (RA). CMR is noninvasive and provides excellent images of the heart due to the high contrast between blood and myocardium. Sp: spine, St: sternum, Lu: lung, L: left, R: right.

using various techniques as breath-hold imaging data averaging, respiratory gating, and real-time imaging (Reiser et al. 2008).

9.2.2 Clinical Applications

MRI is used for the evaluation of nearly all types of cardiac pathology, such as CHDs, cardiac involvement in various systemic diseases (amyloidosis, sarcoidosis, iron-overload cardiomyopathy, etc.), cardiomyopathies, assessment of cardiac masses, morphological imaging of the cardiac valves, and cardiac inflammatory disease (pericarditis, myocarditis), while the use of MR angiography (MRA) techniques enable the evaluation of ischemic heart disease and postbypass graft surveillance. For the evaluation of CHD, cardiac mass and inflammatory disease segmented FSE/TSE with black blood preparation can be utilized as it provides best soft-tissue differentiation. The determination of the situs and the morphology of cardiac chambers are essential for the comprehension of the histologic tissue derivation of the cardiac structures as to distinguish characteristic CHD aberrations from normal anatomy. Contrast media must be used when the differential diagnosis includes a cardiac mass and inflammation, while the combination with MRA is imperative for the assessment of various CHDs involving the great vessels (e.g., tetralogy of Fallot, transposition of the great arteries, coarctation of the aorta, pulmonary atresia and agenesis, of the pulmonary artery, lusory artery, etc.) (deRoos and Roest 2000; Constantine et al. 2004). Delayed enhancement techniques are also appropriate for the appraisal of cardiac inflammatory disease, cardiomyopathies, and the detection of a cardiac mass (Reiser et al. 2008). Another very useful MR imaging technique for the evaluation of ischemic disease is myocardial perfusion imaging. Perfusion functional imaging is an innovative modality that has recently established its value in clinical practice for the assessment of reversible perfusion insufficiency and the detection of viable tissue. Myocardial perfusion is performed using contrast media, and images are obtained at rest and following pharmacologic stress testing induced with adenosine or dipyridamole (Nassenstein et al. 2008). Delayed enhancement imaging is another

accurate method for the assessment of myocardial viability as it delineates diverse contrast agent concentrations within the myocardium, detecting myocardial infarction and/or fibrosis as areas of higher contrast concentration (Hunold et al. 2005).

Finally, MRA and contrast-enhanced (CE) MRA techniques have significantly evolved in past years in established diagnostic tests for the assessment of pulmonary artery disease (thromboembolism and pulmonary hypertension), while its implication in ischemic coronary artery disease is under continuous development (Ley et al. 2003; Nael et al. 2005). Various MR protocols, including time of flight MRA (TOF-MRA), time-resolved noncontrast-enhanced MRA (Tagging), and 3D CE MRA, are noninvasive imaging modalities, which provide high spatial resolution of the 3D datasets combined with morphologic and functional information in a single examination. Moreover, paramagnetic agents (when needed) demonstrate lower allergic reactions and nephrotoxicity rates compared to iodinated contrast media used in other imaging apparatus (Prince et al. 1996; Fujisaki et al. 2011; Amene et al. 2012).

Some disadvantages of cardiac MRI include the long examination time and the noncompatibility of the method with various metallic implants and medical devices (Kanal and Shellock 1994).

9.3   Computed Tomography (CT)

9.3.1   General Considerations

Another modern, noninvasive, imaging modality of the heart is CT. ECG-synchronized rapid CT scanning combines speed and high spatial resolution, allowing the precise imaging of the cardiac morphology. Of note, over the past years, CT has evolved in a particularly valuable noninvasive imaging tool for the assessment of ischemic heart disease using various protocols like CT colonography and calcium score (Abdulla et al. 2011; Ghadri et al. 2012). This was made possible following the extraordinary progress of CT technology that enabled faster data acquisition within few milliseconds (50–100 ms using electron beam technology and 125–500 ms with helical imaging). Consequent to the increased temporal resolution, CT is mildly influenced by physiologic cardiac motion, and in synergy to the improved spatial resolution (identification of

0.5 mm^3 isotropic voxels), even small cardiac structures can be accurately assessed. Moreover, a variety of novel postprocessing possibilities like multiplanar reformation (MPR) and 3D reconstruction have enabled the visualization of the true short and longitudinal anatomical axes, resulting in a detailed imaging of the cardiac anatomy (Pontone et al. 2012).

9.3.2   Clinical Applications

Nonenhanced thin-slice CT techniques have been widely used for the assessment of coronary calcifications. Using CT, an accurate identification and quantification (calcium score) of vessel-wall calcifications of the coronary arteries can be performed and this validates the presence of early-stage, subclinical, and atherosclerotic coronary disease (Becker et al. 2000; Ohnesorge et al. 2002; Yoon et al. 2002).

Chapter 9

The differentiation between various cardiac tissues such as the pericardium and the endocardium is also possible due to their diverse CT attenuation, especially when an underlying pathology, for instance, a pericardial effusion, is present. This diverse attenuation is enhanced by the additional use of iodinated contrast material. CE multidetector CT also allows the imaging of cardiac valve morphology and it is comparable to other established modalities like surgery and echocardiography. It is very precise in evaluating valve calcification and the measurement of the diameter of the aortic valve annulus (Willmann et al. 2002).

Various CE CT protocols demonstrate high sensitivity and specificity rates in detecting intracardiac filling defects such as thrombi and tumors as well as defects of the pericardium, while pulmonary artery CT pulmonary artery angiography (CTPA) is considered today as the gold-standard imaging modality in the diagnosis of pulmonary embolism. Although the imaging method of choice for cardiac masses is 2D echocardiography in patients presenting various restrictions to U/S such as obesity and pulmonary emphysema, CE MDCT cross-sectional techniques constitute a very useful alternative for the diagnosis of suspected cardiac and paracardiac tumors (Wintersperger et al. 2000).

In addition, novel ECG-gated multislice CT angiography with ultrathin-slice acquisition and gantry rotation times less than 0.5 s that allows the complete scanning of the organ during a single breath-hold has been recently established in everyday clinical practice for the evaluation of coronary stenosis or occlusions, the characterization of coronary atherosclerotic plaque, bypass planning, and postbypass graft surveillance (Lee et al. 2011; Pregowski et al. 2011). Myocardial viability following infarction can also be performed using CT perfusion protocols, but this is considered as functional imaging and is beyond the scope of this chapter.

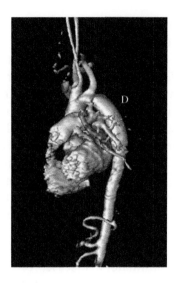

FIGURE 9.2 Noninvasive 3D aortography: This image depicts an aortography of a patient with a dacron graft (G) placed in the descending aorta during repair of a ruptured aneurysm.

Finally, 3D reformatting following CTA (Figure 9.2) is considered an important additional diagnostic imaging approach for the evaluation of anomalous extracardiac anatomy present in a variety of CHD like aortic coarctation, tetralogy of Fallot, anomalous pulmonary venous return, and double aortic arch. Noninvasive CT angiography using modern CT scanners can furnish important and precise information concerning the presence, morphology, dimensions, and route of the cardiac vessels, as well as any possible anomalous anastomosis or collaterals, contributing in the overall therapeutic plan of these pathologies (Westra et al. 1999; Kawano et al. 2000; van Ooijen et al. 2007).

Some disadvantages of even the most advanced CT scanners include the inability to completely obliterate motion artifacts, use of a significant amount of ionized radiation received by the patient, and various contraindications to contrast administration, which is usually mandatory, such as declined renal function and history of severe allergic reaction to the contrast agent.

9.4 Echocardiography

9.4.1 General Considerations

Echocardiography is still considered as a first-line imaging modality for the morphological assessment of the heart in everyday clinical care as it is a fast, radiation-free, and cost-effective imaging method that provides very valuable clinical information regarding the cardiac integrity and function. Although standard cardiac ultrasound (Figure 9.3) is performed with the transducer (or probe) placed on the chest wall of the patient (transthoracic echocardiography [TTE]), several diverse

ICE is performed using novel low-frequency, steerable phased array catheters and provides a real-time, direct visualization of the cardiac structures. The development of ultrasound technology established these techniques in the era of modern noninvasive cardiac imaging, while the appropriate use of each modality is based on well-defined clinical indications (Douglas et al. 2011).

9.4.2 Clinical Applications

The indications for the morphological evaluation of the heart using the transthoracic examination include the assessment of known or suspected adult CHD, suspected cardiomyopathy, the evaluation of cardiac masses (tumor or thrombus), the evaluation of myocardial ischemia and infarction-related complications, the initial assessment of valvular or structural heart disease, and the visualization of prosthetic valves as postplacement baseline imaging. Moreover, TTE is highly accurate in the evaluation of various pericardial conditions as pericardial effusions and constrictive pericarditis. Finally, TTE enables the morphological evaluation of the proximal aortic root and the mitral valve and is also used for the identification and follow-up of Marfan syndrome. The limitations of TTE include the decrease of diagnostic accuracy in obese patients, cases of chronic obstructive pulmonary disease, chest-wall deformities, and noncooperative patients (Ommen et al. 2000).

TEE is mainly utilized in the evaluation of suspected acute aortic pathology (dissection/transaction), the diagnosis of endocarditis with a moderate or high pretest probability (bacteremia, especially staph bacteremia or fungemia), assessment of persistent fever in patients with intracardiac device, and the evaluation of atrial fibrillation/flutter in order to assist in the therapeutic plan regarding anticoagulation and/or cardioversion and/or radiofrequency ablation (Ommen et al. 2000). ICE is today considered the imaging modality of choice during cardiac catheterization due to its many advantages over fluoroscopy, TTE, and TEE in various interventional cardiology procedures. The diagnostic indications of this modality include the diagnosis of intracardiac masses, the evaluation of intracardiac thrombus, atrial appendage occlusion, and the visualization of

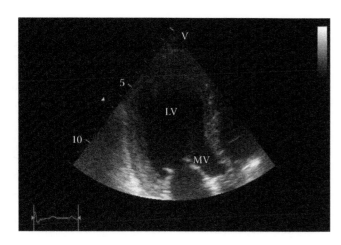

FIGURE 9.3 Echocardiogram: This image depicts a long axis echocardiographic view of the left ventricle (LV) of the heart. Blood is dark, whereas the wall of the ventricle is gray. The leaflets of the open mitral valve (MV) are shown.

imaging techniques are employed to further assess the cardiac pathology as 3D TTE, transesophageal echocardiography (TEE), and intracardiac echocardiography (ICE).

TEE is performed using special probes, positioned on a flexible endoscope, guided through the patient's mouth cavity into the esophagus, providing an accurate and direct morphological estimation of the heart and its chambers without the interference from the chest wall and ribs (Figure 9.4).

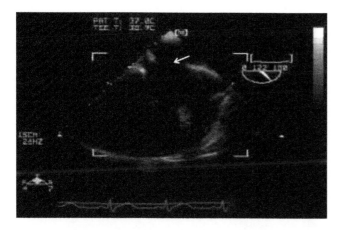

FIGURE 9.4 TOE: TOE follows the same principle with classical echocardiography, with the ultrasound transducer being mounted on an endoscopic catheter similar to that used for gastroscopy. The endoscopic catheter is advanced into the esophagus behind the heart, and images at multiple planes are acquired. This image shows a communication between two chambers of the heart at the atrial level (atrial septal defect, white arrow).

Chapter 9

coronary sinus, while at the same time enabling the performance of biopsy and other therapeutic interventions (Oishi et al. 2000; Ali et al. 2011).

Finally, the aforementioned ultrasound techniques are already incorporated into patient management algorithms regarding the assessment of cardiac morphology in emergency clinical settings. The technological development of various ultrasound systems such as ultraminiaturized pocket ultrasound devices broadened the spectrum of the indications of emergency intraoperative echocardiography and echocardiography in the emergency room in acute coronary syndromes, cardiac tamponade, acute native and prosthetic valve disease, acute diseases of the great vessels, cardiac-related embolism, cardiac arrest, chest trauma, and cardiogenic shock, identification of ventricular enlargement/hypertrophy, and estimation of the size and respiratory changes of the inferior vena cava (IVC) diameter (Biais et al. 2012) (Figure 9.5).

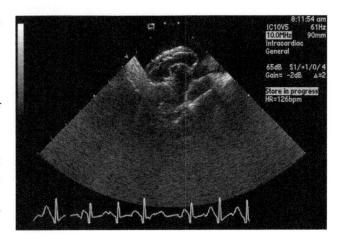

FIGURE 9.5 Invasive ICE: Invasive echocardiography follows the same principle with classical echocardiography, with the ultrasound transducer being mounted on a catheter, which is introduced into the heart from a peripheral vein or artery. This image shows an umbrella device used to seal a communication between two chambers of the heart (atrial septal defect). This kind of technology is mainly used to guide invasive therapeutic procedures.

9.5 Morphological Imaging of the Cardiac Vessels

The aforementioned imaging modalities are able to assess both cardiac morphology and its vessels. The following invasive modalities are used to evaluate only the cardiac vasculature.

9.5.1 Selective Coronary Angiography

Selective coronary angiography has been first introduced in clinical practice by Dr. Mason Sones in 1958 as the imaging modality of choice for the assessment of coronary arteries (Schoenhagen et al. 2003). Coronary angiography is an invasive imaging test typically performed by obtaining femoral, brachial, or ulnar arterial access and the advancement of a special diagnostic catheter that enables the selective administration of contrast media for the detailed angiographic evaluation of the coronary bed (Figure 9.6).

Coronary angiography offers the possibility to detect, quantify, and treat coronary artery steno-occlusive disease using established percutaneous catheter-based methods. Moreover, cardiac catheterization using femoral or jugular venous access permits the endovascular treatment of pulmonary embolism using selective pulmonary artery thrombolysis and/or mechanical thrombectomy (Siablis et al. 2005).

9.5.2 Intravascular Ultrasound (IVUS)

IVUS is another invasive imaging modality, which uses a specially designed catheter with a miniaturized

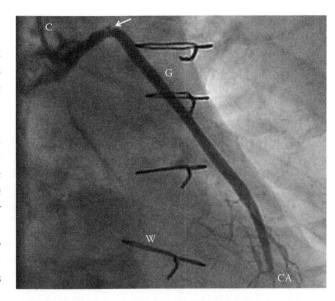

FIGURE 9.6 Classical selective angiography: This image shows a selective invasive angiography of a saphenous vein graft (G), which provides blood to a native coronary artery (CA). A contrast is injected through a catheter (C) into the graft and an ulcerated atheromatous plaque is depicted at its proximal part (arrow). The radiopaque wires of the sternotomy are also seen (W).

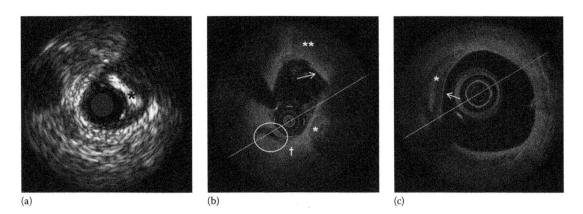

FIGURE 9.7 Invasive imaging of coronary arteries with (a) IVUS and (b, c) optical coherence tomography (OCT). IVUS is based on emission of ultrasound 360° around a catheter introduced into an artery. Panel (a) shows the catheter as a gray circle at the center of the artery. Blood is shown dark around the catheter, inside the lumen of a severely stenosed coronary artery. An atheromatous plaque between 12 and 5 o'clock is shown, with significant reflection of ultrasound creating a bright white image due to calcification (asterisk) and an echolucent wedge area behind calcium due to ultrasound attenuation. Panels (b) and (c) show similar atheromatous plaques using OCT. The latter is based on emission of infrared light. Calcium deposits inside an atheromatous plaque are shown as sharply demarcated areas of light attenuation (white single * in b and c). Fibrous tissue inside the coronary plaque is shown as a bright dense area († in b), whereas the area of diffuse light attenuation is considered to represent lipid accumulation (** in b). The microchannel within the white circle in (b) represents neovascularization inside the atheroma. Note the bright thin rim of intimal cap of the atheroma (arrow in b), an appearance that may be due to dense macrophage accumulation. Thin intima is also lining the calcific deposit in (c). Obviously, OCT has a much higher axial resolution (10 µm), compared to IVUS (100 µm), allowing a better visualization and characterization of the vessels' details.

ultrasound probe attached to its distal end, while the proximal end is connected to the ultrasound main unit. Again like in selective angiography, arterial access is necessary and therefore, in the majority of the cases, IVUS is performed during selective coronary angiography as an adjunct to the evaluation of the coronary disease (Figure 9.7). The main advantage of IVUS is the ability to evaluate the diameter and morphology of the vessel wall as well as the detailed characterization of the atheromatic plaque (Kawasaki et al. 2010; Ohshima et al. 2012).

9.5.3 Optical Coherence Tomography

Optical coherence tomography (OCT) is an innovative, invasive, intravascular imaging modality, which allows the acquisition of high-resolution in vivo imaging of the coronary arteries and veins. OCT presents a much higher image resolution and penetration compared to IVUS, resulting in superior imaging of the vessel wall (Figure 9.7) (Kawasaki et al. 2006). OCT is today considered as an established method for the evaluation of coronary atherosclerotic plaque, as its safety, efficiency, and reproducibility have been widely reported (Kawasaki et al. 2006). A variety of atherosclerotic histological features, such as lipid-rich neointima, calcifications, and neovascularization, as well as features of vulnerability like thin cap fibroatheroma (TCFA), lipid/necrotic core, thrombus, and plaque rupture, have been characterized using OCT (Kang et al. 2011; Nakazawa et al. 2011). Moreover, data regarding the OCT evaluation of stent struts coverage, neoatherosclerosis, stent strut malapposition, as well as quantitative data on neointimal hyperplasia and their correlation with in stent thrombosis, are constantly amassing (Kang et al. 2011; Mandelias et al. 2012; Tsantis et al. 2012).

9.6 Plain X-Ray Films

Finally, despite the development of all the aforementioned sophisticated imaging technologies, plain x-ray films remain among the most commonly requested diagnostic tests performed in everyday clinical practice, for the initial evaluation of the cardiac morphology. Standard posteroanterior and lateral chest films obtained in inspiration are performed in order to assess the cardiac size and contour, as well as data regarding the basic cardiac anatomy, its great vessels, diagnosis of numerous pathological conditions including both congenital and acquired heart disease, and the evaluation of the position of various cardiac implants (Burney et al. 2007; Khosangruang and Chitranonth 2008; Mahnken et al. 2009).

9.7 Summary

In summary, the technological advance of already existing imaging modalities such as MRI, CT, and echocardiography, as well as the development of various novel imaging technologies such as IVUS and OCT, has enabled the accurate evaluation of the cardiac morphology and has innovated the majority of medical protocols to endorse these methods in the diagnosis and treatment of cardiac disease.

References

Abdulla, J., C. Asferg et al. (2011). Prognostic value of absence or presence of coronary artery disease determined by 64-slice computed tomography coronary angiography a systematic review and meta-analysis. *Int J Card Imag* **27**(3): 413–420.

Ali, S., L. K. George et al. (2011). Intracardiac echocardiography: Clinical utility and application. *Echocardiography* **28**(5): 582–590.

Amene, C., L. A. Yeh-Nayre et al. (2012). Incidental MRI findings of acute gadolinium hypersensitivity. *Case Rep Neurol* **4**(1): 68–70.

Barnwell, J. D., J. L. Klein et al. (2012). Image-guided optimization of the ECG trace in cardiac MRI. *Int J Card Imag* **28**(3): 587–593.

Becker, C. R., T. F. Jakobs et al. (2000). Helical and single-slice conventional CT versus electron beam CT for the quantification of coronary artery calcification. *Am J Roentgenol* **174**(2): 543–547.

Biais, M., C. Carrie et al. (2012). Evaluation of a new pocket echoscopic device for focused cardiac ultrasonography in an emergency setting. *Crit Care* **16**(3): R82.

Burney, K., N. Thayur et al. (2007). Imaging of implants on chest radiographs: A radiological perspective. *Clin Radiol* **62**(3): 204–212.

Constantine, G., K. Shan et al. (2004). Role of MRI in clinical cardiology. *Lancet* **363**(9427): 2162–2171.

de Roos, A. and A. A. Roest (2000). Evaluation of congenital heart disease by magnetic resonance imaging. *Eur Radiol* **10**(1): 2–6.

Douglas, P. S., M. J. Garcia et al. (2011). ACCF/ASE/AHA/ASNC/HFSA/HRS/SCAI/SCCM/SCCT/SCMR 2011 appropriate use criteria for echocardiography. A report of the American College of Cardiology Foundation appropriate use criteria task force, American Society of Echocardiography, American Heart Association, American Society of Nuclear Cardiology, Heart Failure Society of America, Heart Rhythm Society, Society for Cardiovascular Angiography and Interventions, Society of Critical Care Medicine, Society of Cardiovascular Computed Tomography, Society for Cardiovascular Magnetic Resonance American College of Chest Physicians. *J Am Soc Echocardiogr* **24**(3): 229–267.

Fujisaki, K., A. Ono-Fujisaki et al. (2011). Rapid deterioration of renal insufficiency after magnetic resonance imaging with gadolinium-based contrast agent. *Clin Nephrol* **75**(3): 251–254.

Ghadri, J. R., M. Fiechter et al. (2012). The value of coronary calcium score in daily clinical routine, a case series of patients with extensive coronary calcifications. *Int J Cardiol* **162**(2): e47–e49.

Hunold, P., T. Schlosser et al. (2005). Myocardial late enhancement in contrast-enhanced cardiac MRI: Distinction between infarction scar and non-infarction-related disease. *Am J Roentgenol* **184**(5): 1420–1426.

Kanal, E. and F. G. Shellock (1994). The value of published data on MR compatibility of metallic implants and devices. *Am J Neuroradiol* **15**(7): 1394–1396.

Kang, S. J., G. S. Mintz et al. (2011). Optical coherence tomographic analysis of in-stent neoatherosclerosis after drug-eluting stent implantation. *Circulation* **123**(25): 2954–2963.

Kawano, T., M. Ishii et al. (2000). Three-dimensional helical computed tomographic angiography in neonates and infants with complex congenital heart disease. *Am Heart J* **139**(4): 654–660.

Kawasaki, M., B. E. Bouma et al. (2006). Diagnostic accuracy of optical coherence tomography and integrated backscatter intravascular ultrasound images for tissue characterization of human coronary plaques. *J Am Coll Cardiol* **48**(1): 81–88.

Kawasaki, M., A. Hattori et al. (2010). Tissue characterization of coronary plaques and assessment of thickness of fibrous cap using integrated backscatter intravascular ultrasound. Comparison with histology and optical coherence tomography. *Circ J* **74**(12): 2641–2648.

Khosangruang, O. and C. Chitranonth (2008). Radiographic findings of screening chest films in Priest Hospital. *J Med Assoc Thai* **91**(Suppl. 1): S21–S23.

Lee, C. K., Y. M. Kim et al. (2011). The detection of pulmonary embolisms after a coronary artery bypass graft surgery by the use of 64-slice multidetector CT. *Int J Card Imag* **27**(5): 639–645.

Ley, S., H. U. Kauczor et al. (2003). Value of contrast-enhanced MR angiography and helical CT angiography in chronic thromboembolic pulmonary hypertension. *Eur Radiol* **13**(10): 2365–2371.

Mahnken, A. H., B. B. Wein et al. (2009). Value of conventional chest radiography for the detection of coronary calcifications: Comparison with MSCT. *Eur J Radiol* **69**(3): 510–516.

Mandelias, K., S. Tsantis et al. (2012). SU-E-I-90: Fast and robust algorithm towards vessel lumen and stent strut detection in optical coherence tomography. *Med Phys* **39**(6): 3645–3646.

Nael, K., G. Laub et al. (2005). Three-dimensional contrast-enhanced MR angiography of the thoraco-abdominal vessels. *Magn Reson Imag Clin N Am* **13**(2): 359–380.

Nakazawa, G., F. Otsuka et al. (2011). The pathology of neoatherosclerosis in human coronary implants bare-metal and drug-eluting stents. *J Am Coll Cardiol* **57**(11): 1314–1322.

Nassenstein, K., K. U. Waltering et al. (2008). Magnetic resonance coronary angiography with Vasovist: In-vivo T1 estimation to improve image quality of navigator and breath-hold techniques. *Eur Radiol* **18**(1): 103–109.

Ohnesorge, B., T. Flohr et al. (2002). Reproducibility of coronary calcium quantification in repeat examinations with retrospectively ECG-gated multisection spiral CT. *Eur Radiol* **12**(6): 1532–1540.

Ohshima, K., S. Ikeda et al. (2012). Impact of culprit plaque volume and composition on myocardial microcirculation following primary angioplasty in patients with ST-segment elevation myocardial infarction: Virtual histology intravascular ultrasound analysis. *Int J Cardiol* **167**(3): 1000–1005.

Oishi, Y., M. Okamoto et al. (2000). Cardiac tumor biopsy under the guidance of intracardiac echocardiography. *Jpn Circ J* **64**(8): 638–640.

Ommen, S. R., R. A. Nishimura et al. (2000). Clinical utility of Doppler echocardiography and tissue Doppler imaging in the estimation of left ventricular filling pressures: A comparative simultaneous Doppler-catheterization study. *Circulation* **102**(15): 1788–1794.

Pontone, G., D. Andreini et al. (2012). Feasibility and diagnostic accuracy of a low radiation exposure protocol for prospective ECG-triggering coronary MDCT angiography. *Clin Radiol* **67**(3): 207–215.

Pregowski, J., C. Kepka et al. (2011). Comparison of usefulness of percutaneous coronary intervention guided by angiography plus computed tomography versus angiography alone using intravascular ultrasound end points. *Am J Cardiol* **108**(12): 1728–1734.

Prince, M. R., C. Arnoldus et al. (1996). Nephrotoxicity of high-dose gadolinium compared with iodinated contrast. *J Magn Reson Imag* **6**(1): 162–166.

Reiser, M. F., W. Semmler et al. (2008). *Magnetic Resonance Tomography*. Berlin, Germany: Springer-Verlag.

Ridgway, J. P. (2010). Cardiovascular magnetic resonance physics for clinicians: Part I. *J Cardiovasc Magn Reson* **12**: 71.

Schoenhagen, P., R. D. White et al. (2003). Coronary imaging: Angiography shows the stenosis, but IVUS, CT, and MRI show the plaque. *Cleve Clin J Med* **70**(8): 713–719.

Siablis, D., D. Karnabatidis et al. (2005). AngioJet rheolytic thrombectomy versus local intrapulmonary thrombolysis in massive pulmonary embolism: A retrospective data analysis. *J Endovasc Ther* **12**(2): 206–214.

Tsantis, S., G. C. Kagadis et al. (2012). Automatic vessel lumen segmentation and stent strut detection in intravascular optical coherence tomography. *Med Phys* **39**(1): 503–513.

van Ooijen, P. M., G. de Jonge et al. (2007). Coronary fly-through or virtual angioscopy using dual-source MDCT data. *Eur Radiol* **17**(11): 2852–2859.

Wen, H., T. J. Denison et al. (1997). The intrinsic signal-to-noise ratio in human cardiac imaging at 1.5, 3, and 4 T. *J Magn Reson* **125**(1): 65–71.

Westra, S. J., J. Hurteau et al. (1999). Cardiac electron-beam CT in children undergoing surgical repair for pulmonary atresia. *Radiology* **213**(2): 502–512.

Willmann, J. K., D. Weishaupt et al. (2002). Electrocardiographically gated multi-detector row CT for assessment of valvular morphology and calcification in aortic stenosis. *Radiology* **225**(1): 120–128.

Wintersperger, B. J., K. Bauner et al. (2006). Cardiac steady-state free precession CINE magnetic resonance imaging at 3.0 tesla: Impact of parallel imaging acceleration on volumetric accuracy and signal parameters. *Invest Radiol* **41**(2): 141–147.

Wintersperger, B. J., C. R. Becker et al. (2000). Tumors of the cardiac valves: Imaging findings in magnetic resonance imaging, electron beam computed tomography, and echocardiography. *Eur Radiol* **10**(3): 443–449.

Yoon, H. C., A. M. Emerick et al. (2002). Calcium begets calcium: Progression of coronary artery calcification in asymptomatic subjects. *Radiology* **224**(1): 236–241.

Yorimitsu, M., K. Yokoyama et al. (2012). Whole-heart 3D late gadolinium-enhanced MR imaging: Investigation of optimal scan parameters and clinical usefulness. *Magn Reson Med Sci* **11**(1): 9–16.

Chapter 9

Focused Applications and Dedicated Technology

Geometries, Sources, Detectors, Advanced Image Reconstruction, and Quantitative Analysis

10. Cone-Beam CT for Vascular Imaging

Lei Zhu and Rebecca Fahrig

10.1 Introduction

Conventional vascular imaging techniques, including digital subtraction angiography (DSA) and standard fluoroscopy, attempt to visualize three-dimensional (3D) vascular structures from two-dimensional (2D) projection images. These methods well serve the needs of many clinical applications, and have the advantage of low radiation dose as compared to other 3D imaging modalities. However, as devices and angiographic interventions become more complicated, an accurate 3D geometry of vessels and neighboring structures often provides additional critical information. The lost dimension in the 2D images results in localization ambiguities in the projection direction, which significantly hinders the performance of these interventions. Many efforts have been made to develop imaging systems for generating 3D datasets for use during interventional radiology procedures. To obtain the lost information in the conventional 2D vascular images, one viable option is to acquire more projection data from a sweep of gantry angles. 3D tomographic images are then reconstructed

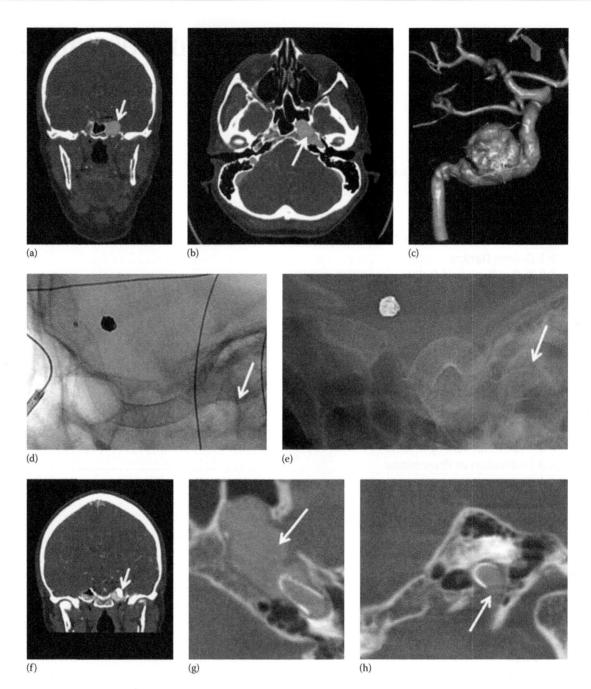

FIGURE 10.1   Use of C-arm CT in the reconstruction of a dysplastic, aneurysmally dilated, petrocavernous segment of the left internal carotid artery with a pipeline embolization device (PED). Panels (a) and (b) show two views from a conventional CT Angiogram (CTA) prior to the intervention. Panel (c) demonstrates a high-resolution 3D vascular reconstruction obtained in the interventional suite prior to intervention; measurements of the aneurysm neck and vessel diameter are used to choose the appropriate device size. Panel (d) shows a projection image of the aneurysm (arrow) following placement of overlapping PEDs in native mode. Note that this lateral image also shows a coil mass (left of center of image) from a prior intervention of the right anterior circulation. Panel (f) shows a follow-up conventional CTA 6 months post intervention, demonstrating contrast within the device and within the nonthrombosed aneurysm (arrow). Panel (e) shows a projection image acquired in the interventional suite (contrast reversed) where there is new prolapse of the PED into the aneurysm and fore-shortening of the proximal end of the device (arrow). Because of the lack of aneurysm thrombosis, the interval change in the configuration of the proximal PED, and the fact that conventional CTA images do not allow an accurate distinction of an endoleak (blood flow around the ends of the device), a 3D C-arm CT was acquired in the interventional suite (Siemens AX, DynaCT, injecting 10% contrast at 1.5 cc/s). Use of dilute contrast permits confirmation of apposition of the PED to the normal vessel wall proximal and distal to the aneurysm neck. Panel (h) shows the proximal end of the PED in good apposition to the vessel wall, excluding endoleak. Panel (g) shows an oblique slice through contrast-filled aneurysm, due presumably to incomplete neointimal coverage of the PED as an endoleak has been excluded. (Case study courtesy of Dr. R. McTaggart and Dr. M. Marks, Dept. of Radiology, Stanford University, Stanford, CA.)

using the principles of computed tomography (CT) imaging. Figure 10.1 shows one clinical example of aneurysm treatment with interventional guidance.

Interventional CT imaging is used at various stages of the treatment procedures to provide more accurate patient geometry.

10.2   CBCT System Hardware and Data Acquisition

The x-ray physics and mathematical reconstruction theories implemented on 3D x-ray tomographic systems for vascular imaging are the same as those on clinical CT scanners. The system geometries, however, are distinct due to different design considerations for interventional use. Angiographic interventions require more space beside the patient couch and more flexibility in choice of the x-ray projection angle. 3D vascular imaging systems therefore adopt an open-gantry configuration. The x-ray flat-panel detector has a large active detection area, and each x-ray projection covers a volume of cone beam on the patient. These systems are referred to as cone-beam CT (CBCT) systems. Figure 10.2 shows the Artis zeego® robotic platform (Siemens Healthcare, Forchheim, Germany). The system is often called C-arm CT, since the gantry has a "C" shape. In this section, we provide an overview of C-arm CBCT with a focus on its main differences from clinical CT scanners used for diagnosis. Several commercial vendors offer C-arm CT imaging in the angiographic suite; this imaging mode is known as DynaCT (Siemens AG), Innova CT (GE Healthcare), XperCT (Philips Healthcare), and 3D-LCI (Toshiba Medical Systems). All of these implementations have similar system geometries and data acquisition approaches.

10.2.1   C-Arm Gantry

The C-arm gantry supports the x-ray tube and the flat-panel detector. By moving the robot arm, the system is able to acquire fluoroscopic projection images from almost all view angles, as long as the patient bench and/or patient do not obstruct the gantry. The gantry can also rotate around the patient with a circular or a more sophisticated trajectory to acquire data for 3D tomographic reconstruction. In this chapter, we discuss C-arm CT scans with a circular trajectory on a plane, the standard CBCT imaging protocol used on current C-arm systems.

10.2.2   X-Ray Tube and Collimator

The C-arm CBCT system uses a rotating-anode x-ray tube for x-ray generation, with the same basic design of the x-ray tube on clinical CT scanners but slightly different ranges of parameters. High voltage is applied on the cathode and the anode inside the vacuum tube. Accelerated electron beams hit a high atomic number target (e.g., tungsten) and generate x-rays due to the Bremsstrahlung effect. The x-ray intensity is controlled by the electron beam current and the applied voltage. The output x-ray has a polyenergetic spectrum with a range from 0 to the peak energy determined by the tube voltage (kVp). Readers are referred to Chapter 4 or Reference 1 for detailed physics and to Reference 1 for tube diagram. The x-ray tube also includes an inherent metal filter (i.e., aluminum) to remove low-energy x-rays that cannot penetrate through the patient and mainly contribute to patient skin dose. The beam filtration increases the mean x-ray energy and therefore reduces image contrast of soft tissues. To improve the quality of fluoroscopic imaging, the C-arm x-ray tube uses relatively weak beam filtration as compared to clinical CT scanners. This configuration, however,

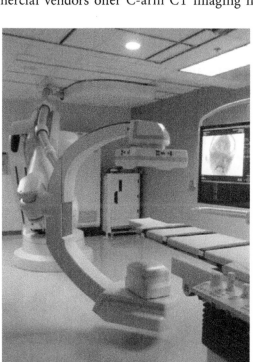

FIGURE 10.2 Artis zeego® system (Siemens Healthcare, Forchheim, Germany). In the gantry position shown here, the x-ray flat-panel detector is on the ceiling side, and the x-ray tube is located close to the floor.

FIGURE 10.3   X-ray beam collimator.

leads to more beam hardening and scatter artifacts when the system operates in the CBCT mode. An x-ray beam collimator is mounted outside the x-ray tube, as shown in Figure 10.3. The collimator blades control the illumination field size in two orthogonal directions such that the radiation dose is delivered only to the volume of interest. To further optimize the radiation dose of CBCT imaging, the x-ray imaging system uses automatic exposure control (AEC), where the tube current is automatically adjusted based on the measured photon intensity on the detector. Lower exposure is used when the imaged object has weak attenuation (e.g., in the anteroposterior (AP) view on a human torso). The x-ray tube is switched to a higher exposure rate when more photons are needed for sufficient measurement accuracy (e.g., in the lateral view on a human torso).

10.2.3   Flat–Panel Detector and Antiscatter Grid

Flat-panel detectors are used in fluoroscopic imaging to provide large-area projection images of vasculatures. Since C-arm hardware does not have a slip ring, the standard clinical approach using spiral or helical CT to provide *z*-axis coverage (superior–inferior direction along the patient) cannot be implemented. Therefore, to acquire enough patient data for timely treatment monitoring, the detector needs to cover a large patient *z*-extent in each view. It is therefore a natural choice to use the full *z*-length of the detector during a sweep, which results in a cone-beam geometry.

The design of flat-panel detectors is constrained by the need for high-resolution projection images of

the vasculature, as well as by considerations of manufacturing cost and detector dose efficiency. Current commercial flat plat-panel detectors use an indirect x-ray measurement approach. X-rays escaping from the subject being imaged first strike a scintillator layer on the detector, which converts the x-rays into visible light photons. The scintillator is made of good x-ray absorbers, such as gadolinium oxysulfide or cesium iodide (CsI), as used in standard screen/film radiography and fluoroscopy. CsI is the most commonly used scintillator material due to its high x-ray absorption efficiency.

Directly behind the scintillator is an amorphous silicon layer, which contains an array of photodiodes. Each photodiode generates an electrical signal proportional to the detected light photon intensity. The signal from the photodiode is then amplified and encoded by additional electronics positioned at the corner of or behind each detector element. The recorded data are extracted via massive data lines controlled by readout chips and driver chips. As a typical example shown in Figure 10.4, the Varian PaxScan CB4030 flat-panel detector has an effective detector area of 40-by-30 cm², which contains a maximum of 2048-by-1536 pixels with a pixel pitch of 194 μm. Due to the large number of pixels, the flat-panel detector requires an enormous integrated circuit. The manufacturing process is very similar to that used to make LCD televisions and computer monitors. The circuit contains an array of thin-film transistors (TFTs) patterned in amorphous silicon on a glass substrate.

In front of the flat-panel detector is an antiscatter grid specially designed to "screen" or remove scattered

FIGURE 10.4   Varian PaxScan CB4030 CsI-aSi flat-panel detector and its command processor.

FIGURE 10.5   Antiscatter grid on a C-arm system.

photons in order to improve imaging accuracy (see Figure 10.5). The details of the grid design are presented in the following text in the context of scatter signal and its correction methods.

10.2.4   Bed-Side Control Panel and Monitors

A typical bed-side control panel is shown in Figure 10.6 for an Artis zeego system (Siemens Healthcare, Forchheim, Germany). The main goal of the C-arm hardware is to provide maximum flexibility to the

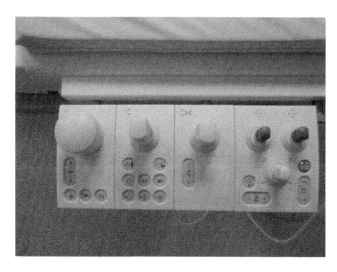

FIGURE 10.6   Bed-side control panel on a C-arm system. Proceeding from left to right across the photograph: the left-most panel controls table motion (up and down, lateral and axial translation); the next panel controls C-arm position (cranial/caudal and left/right angulation and rotation) and provides access to pre-programmed C-arm positions and geometries; the third panel is unique to the Siemens zeego® system and provides control over the C-arm position in Cartesian space; the fourth panel controls detector zoom, height, and collimation.

interventional radiologist/cardiologist with respect to image angulation, quality, and location. The procedures typically start with access at the groin by placement of venous or arterial sheaths. The guidewires, catheters, stents, and other devices are then manipulated through the vasculature to the location of interest. Throughout this process, the table and the C-arm are moved in order to keep the operating end of the device in the middle of the field of view (FOV). Tableside displays provide feedback on the system geometry, and display real-time projection images and reconstructed 3D volumes during interventions.

10.2.5   Data Acquisition

To generate a C-arm CBCT volume, the simplest trajectory that acquires adequate information for satisfactory image reconstruction is provided by moving the source and detector on a circle through about 200° (i.e., the short scan mode, see Section 10.3.1). The imaging protocol starts with first positioning the patient so that the region of interest (ROI) is centered on the isocenter of the rotational trajectory. This is usually done by acquiring fluoroscopic images with the x-ray tube and the detector oriented so as to acquire images in first the AP view, and then in the lateral view. Table height, left/right position, and cranial/caudal position are adjusted to maximize coverage of the ROI. The C-arm is then moved to the trajectory end position, and a safety run at slow speed is carried out, moving the C-arm to the trajectory start position, to ensure that collision cannot occur between the patient and the C-arm. After initializing the AEC system, which automatically chooses the x-ray tube energy and current (i.e., kVp and mAs) for the rotational acquisition according to the requested image quality parameters and the size of the patient, the rotational run is initiated with or without the injection of contrast. A digital subtraction acquisition acquires two sweeps (or rotations): a first mask run with no injection of iodinated contrast followed by a fill run. Sometimes a delay between mask and fill runs is necessary to ensure complete filling of the vasculature of interest. Following data acquisition, the projection images are corrected for gain, offset, and other nonidealities (see Section 10.4 for further details), and then reconstructed using an appropriate algorithm. Reconstructions of a 256^3 volume of voxels are now routinely supplied within 1 min following data acquisition, and secondary reconstructions at higher resolution over smaller FOVs are often carried out to provide enhanced visualization of small vessels, aneurysm necks, stent struts, etc.

10.3 Image Reconstruction and Processing

A major difference between the data-processing chains of planar vascular imaging (i.e., fluoroscopy imaging) and CBCT imaging is that the latter involves intensive mathematical calculations. The data acquisition of CBCT imaging uses x-ray projection, in which the object information is inherently lost in the x-ray incident direction. As such, tomographic reconstruction techniques are implemented to indirectly recover the missing dimension. 3D CBCT images contain large data volumes that cannot be easily displayed on 2D monitors. Image-processing methods are often used such that soft tissue contrast and vessels can be examined more clearly.

10.3.1 CBCT Reconstruction

Despite the differences in hardware configurations, CBCT uses the same underlying physical principle as that of clinical CT imaging. The same reconstruction algorithms are applied to generate CBCT image from line integral measurements.

X-ray projection images can be expressed as

$$I \approx I_0 e^{-\int \mu(\vec{x})d\vec{x}}, \tag{10.1}$$

where

I_0 is the incident x-ray photon intensity
μ is the spatially varying attenuation coefficient of the object, the target of CBCT imaging

The integration is taken along the projection line from the x-ray source to the detector element. Note that Equation 10.1 is an approximation. Other nonlinear factors, including scatter and beam-hardening effects, are ignored in the basic CT reconstruction algorithm. These effects will be discussed in detail in a later section. From Equation 10.1, one obtains the line integral of object attenuation:

$$q = \int \mu(\vec{x})d\vec{x} \approx \ln\left(\frac{I_0}{I}\right) \tag{10.2}$$

The goal of CT reconstruction is to recover the spatial distribution of μ from its line integrals measured at different projection angles and at different displacements from the rotation center.

Over the 40 years since the invention of CT, significant research has been conducted on the development of CT reconstruction theories. The field still remains active at the time of this writing. Reconstruction methods that use some form of filtered back projection (FBP) are predominantly used on current CBCT imaging due to its mathematical simplicity, computational efficiency, and low memory size requirement. Since FBP applied to a circular trajectory is an approximate 3D reconstruction algorithm on a circular trajectory, cone-beam artifacts become more pronounced in slices far from the central slice along the z-axis (see Section 10.4.6 for more detail). Studies have been carried out to take advantage of the high flexibility of a C-arm system in data acquisition and acquire more projections from different scanning trajectories to ensure sufficient measurements for exact reconstruction. For these complicated imaging geometries, general algorithms of exact CT reconstruction (e.g., the PI-line method [2] and the Kasetvich's algorithm [3]) need to be implemented to obtain improved image quality. Recent advances in iterative algorithms have shown successes in accurate CT reconstruction from very noisy and/or angularly sparse projection data [4,5]. The potential of significant dose reduction makes iterative CT reconstruction one of the future directions of advanced CBCT imaging. However, these methods are considered impractical for clinical use with the current computer technology, due to their intensive computational complexities and resulting long calculation time.

An exhaustive review of existing CT algorithms is beyond the scope of this book. In this chapter, we present an overview of the FBP reconstruction methods implemented on commercial CBCT systems. Readers can find details of the method derivation in References 8 and 9. Other advanced reconstruction techniques are discussed in Chapters 5 and 14 and their references.

We first define the geometry of a CBCT scan. The x-ray source has a circular trajectory on a plane, which is often referred to as the midplane. Other planes parallel to the midplane are called off-planes. As shown in Figure 10.7, during data acquisition, the x-ray source S rotates around the z axis in the x–y plane, with a fixed distance D to the center of rotation O. In the Cartesian coordinate system with the origin at O, the source position is expressed as

$$\vec{s}(\beta) = (D\cos\beta, D\sin\beta, 0), \tag{10.3}$$

where β is the view angle.

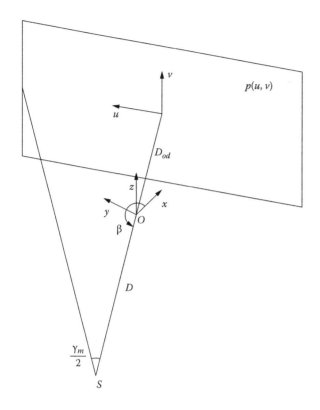

FIGURE 10.7   CBCT geometry.

The projection image at angle β on the detector is a set of half line integrals, written as

$$p(u,v,\beta) = \int_0^\infty \mu(\vec{s}(\beta) + l\vec{r_0}(u,v,\beta))dl, \qquad (10.4)$$

where
  u and v are the horizontal and the vertical coordinate axes of the detector, with the origin defined as the projection of the rotational center on the detector
  Vector $\vec{r_0}$ is the unit vector of the line integral direction connecting the x-ray source and the detector element

The most popular CBCT reconstruction algorithm on a circular trajectory is the FDK method, developed by Feldkamp, Davis, and Kress [9]. The method uses 1D shift-invariant filtering and 3D back-projection, a structure that allows axial truncation of the projection images and is also desirable for parallel data processing. The FDK algorithm was originally derived as a heuristic extension of the exact FBP reconstruction for 2D CT. FDK treats the transverse lines on the projection images with the same value of axial coordinate v as

fan-beam projection lines of a 2D object. The exact fan-beam reconstruction is then applied to each group of the "fan-beam sinograms." Different weighting factors are used for each group of data to compensate for the impact of oblique projection angles. Mathematically, the FDK reconstruction can be written as two steps: a filtering step to properly "weight" the contribution to the final reconstructed image from each projection, and a back-projection step to assemble all the contributions from different projections. The algorithm is summarized as follows:

  The FDK algorithm for CBCT with a full scan

1. Filtering of each weighted projection image:

$$q_F(u,v,\beta) = \int_{-\infty}^{\infty} \frac{D}{\sqrt{\bar{u}^2 + v^2 + D^2}}$$
$$p\left(\bar{u}\frac{D+D_{od}}{D}, v\frac{D+D_{od}}{D}, \beta\right) g(u-\bar{u})d\bar{u}, \qquad (10.5)$$

where
  D_{od} is the distance from the detector to the rotational center
  g is the ramp filter kernel, defined as

$$g(u) = \int_{-W}^{W} |\omega| e^{j2\pi\omega u} d\omega \qquad (10.6)$$

W is the cut off frequency that depends on the resolution requirement of the reconstruction.

2. Weighted back-projection to generate the reconstructed volume:

$$\hat{\mu}_{FDK}(x,y,z)$$
$$= \frac{1}{2}\int_0^{2\pi} \left(\frac{D}{x+D}\right)_\beta^2 q_F\left(\left(\frac{Dy}{x+D}\right)_\beta, \left(\frac{Dz}{x+D}\right)_\beta, \beta\right)d\beta \qquad (10.7)$$

where the subscript β stands for the coordinate transformation of rotation about the z axis by β. This operation on a 3D function $k(x, y, z)$ is defined as

$$(k(x,y,z))_\beta = k(x\cos\beta - y\sin\beta, x\sin\beta + y\cos\beta, z). \qquad (10.8)$$

Although derived heuristically, the FDK method has been shown to be the optimal if the effects of noise are not considered and if the contributions from the unmeasured line integrals of the imaged object are assumed to be zeros [10,11]. Nevertheless, the algorithm presented earlier (Equations 10.5 through 10.8) is for reconstruction of the projection data acquired from a full rotation. On C-arm CBCT systems, a scan with a rotation of slightly larger than 180° is preferred due to its short data acquisition, or may be the only scan mode available due to the limitations of the gantry mechanics and cable management system.

In the case of single-slice CT imaging, a scan through a full rotation acquires more than enough projection data for accurate image reconstruction, since every line integral is measured twice from two opposite directions. The projection data have a uniform redundancy of 2. It can be shown that a scan with a minimum angular coverage of 180° plus the angular coverage of the divergent beam, referred to as a short scan, guarantees an exact reconstruction on the central slice. On a fan-beam short scan, the data redundancy is not uniform over the whole measured data set. Weighting factors are needed to compensate for the nonuniform redundancy. If the redundant data are assigned equal weights in the FBP reconstruction, the weighting function has a piece-wise constant shape across the projection image. Severe artifacts appear in the reconstructed images due to filtering across the discontinuities. To suppress these artifacts, Parker developed a fan-beam reconstruction algorithm using a weighting function that is smooth between regions with different data redundancy (i.e., the first-order derivatives equal zero) [12]. Using the same fan-beam to cone-beam extension as in the derivation of the FDK algorithm on a full scan, the FDK algorithm with Parker's weighting (P-FDK) is predominantly used on CBCT short scans of clinical C-arm systems.

Mathematically, the P-FDK algorithm can be written in the FDK structure with an additional weighting on the projection images before the filtering and the backprojection:

The P-FDK algorithm for CBCT with a short scan

1. Filtering of each weighted projection image:

$$q_p(u,v,\beta) = \int_{-\infty}^{\infty} w_p(\bar{u},v,\beta) \frac{D}{\sqrt{\bar{u}^2 + v^2 + D^2}}$$

$$p\left(\bar{u}\frac{D+D_{od}}{D}, v\frac{D+D_{od}}{D}, \beta\right) g(u-\bar{u})d\bar{u},$$

$$(10.9)$$

where w_p is Parker's weighting, defined as

$$w_p(u,v,\beta)$$

$$= \begin{cases} \sin^2\left(\dfrac{\pi}{4}\dfrac{\beta}{\gamma_m/2 - \gamma}\right), & 0 \leq \beta \leq \gamma_m - 2\gamma \\ 1, & \gamma_m - 2\gamma \leq \beta \leq \pi - 2\gamma \\ \sin^2\left(\dfrac{\pi}{4}\dfrac{\pi + \gamma_m - \beta}{\gamma_m/2 + \gamma}\right), & \pi - 2\gamma \leq \beta \leq \pi + \gamma_m \end{cases}$$

$$(10.10)$$

γ_m is the full fan angle determined by the size of the detector (see Figure 10.7). Parameter γ is the angle between the projection line and the central vertical plane perpendicular to the detector, calculated as

$$\gamma = \arctan\left(\frac{u}{\sqrt{v^2 + (D + D_{od})^2}}\right) \qquad (10.11)$$

2. Weighted backprojection to generate the reconstructed volume:

$$\hat{\mu}_{P-FDK}(x,y,z)$$

$$= \int_0^{\pi+\gamma_m} \left(\frac{D}{x+D}\right)_\beta^2 q_p\left(\left(\frac{Dy}{x+D}\right)_\beta, \left(\frac{Dz}{x+D}\right)_\beta, \beta\right) d\beta$$

$$(10.12)$$

where the subscript β stands for the coordinate transformation of rotation about the z axis by β as defined in Equation 10.8.

10.3.2 Image Operations in Vascular Imaging

Compared to projection imaging, CBCT acquires 3D tomographic images with much higher contrast and less geometric ambiguity. Although the enriched patient information potentially improves the accuracy of vessel visualization in angiographic interventions, it poses new difficulties in image presentation. Treatment guidance from CBCT is less convenient in a clinical setting, as it is time consuming to scrutinize the complete CBCT volume on a 2D display monitor. Image-processing techniques are therefore needed to facilitate this process.

Volume rendering and maximum intensity projection (MIP) are the two techniques commonly used to convert the 3D patient geometry to a 2D image while preserving critical structural information. To render a 2D view of the 3D dataset, one defines the camera position and the opacity and color of every sample voxel. With proper settings, the rendered image shows the surface of the target structure whose voxel values are within a certain range of CT values. Figure 10.1c is one example. Different from direct volume rendering, MIP projects on the 2D image only the voxels with the maximum intensity along the ray path. This technique is useful to make high-intensity structures more separable, but loses the depth information of the original 3D data. To gain the sense of 3D, it is a common practice to generate a series of MIP images with monotonically changing view angles. These images are then displayed continuously to create the illusion of rotation.

Some objects are not easily differentiable on CBCT images, since they may have similar x-ray attenuation and therefore CT values. One typical example in vascular imaging is that the injected contrast agent leads to high voxel intensities comparable to the CT values of bones. Bone structures are not separable from vessels on a rendered view or an MIP image. It is therefore helpful to segment and remove organs and structures to improve the angiographic image presentation. Manual segmentation is cumbersome in a busy clinical setting. To reduce the burden, algorithms have been developed to provide automatic image segmentation. Basic segmentation algorithms, including thresholding and edge detection, are usually available as standard tools in the image-processing software of a C-arm system.

In the studies involving more than two CBCT volumes (e.g., images on the same patient but acquired at different times), it is common to investigate the CT value difference for the same structures. One example is the perfusion CT study where blood flow is monitored continuously. Another example is DSA where the subtraction of two CBCT volumes is expected to cancel the background signals and highlight the vascular structures containing contrast agent. In these applications, creation of a voxel-to-voxel mapping between the CBCT volumes is needed before algorithmic operations, a procedure known as image registration. In the registration process, one CBCT volume is selected as a fixed frame. Rigid or deformable transformations are then applied to other CBCT volumes, referred to as moving frames, to match the fixed frame. The matching is quantified by designed objectives, such as the mean square of the image difference and/or the mutual information. Iterative algorithms are then used to tune the transformation and minimize the objective for the best matching. The optimized transformation is finally applied to the moving frames to register all the CBCT volumes. Similar to segmentation, basic image registration is usually included in commercial image-processing software.

10.4 Image Artifacts and Correction Methods

CBCT reconstruction theory is built on the line integral model described in Equation 10.1. In reality, this model is only an approximation. The line integral measurements from x-ray projections contain errors from various sources. Furthermore, a CBCT scan may not acquire enough projection data for accurate reconstruction, leading to CBCT image artifacts even if there are no measurement errors.

10.4.1 Scatter Contamination

One prominent error source in CBCT imaging is scatter contamination. Inside patients, x-rays can be not only completely absorbed but also scattered into a different direction with reduced energies. If detected, these scatter photons cause large errors of line integral measurements, as they do not travel along a straight line connecting the detector element and the x-ray source.

For a quantitative estimation of errors induced by scatter, we start from first principles. We assume a monoenergetic x-ray source and study the errors from scatter only. The detected photons contributing to the correct line integral measurement are referred to as primary signals. Denote I_P and I_S as the measured primary and scatter signals at the detector, that is, $I = I_P + I_S$. From Equation 10.2, simple derivation gives

$$q = \ln(I_0/I_P) - \ln(1 + SPR) \qquad (10.13)$$

where SPR is the scatter to primary ratio, defined as $SPR = I_S/I_P$.

Equation 10.14 shows that the measured line integral is too small by $\ln(1 + SPR)$ as compared to the scatter-free value $\ln(I_0/I_P)$. After signal processing and reconstruction, these errors distribute across the whole

CBCT image. Scatter artifacts are difficult to estimate, since they not only depend on the SPR for each projection view but also have a complicated relationship with the projection errors due to the reconstruction process. In general, two types of scatter artifacts are common in CBCT imaging. For a uniform cylindrical object, the SPR is identical for different views. Global cupping artifacts therefore appear in the CT image with an error proportional to $\ln(1 + SPR)$. For a nonuniform object, besides the cupping artifacts, along rays where the primary signals are small and SPR is large (e.g., around dense objects), the line integrals are significantly underestimated and local streaks will result.

The scatter artifacts are well observed in the image comparison of Figure 10.8. These images are generated using a tabletop CBCT system, whose major imaging components match those of a C-arm system. A scan of a fan-beam equivalent geometry is carried out on the phantom with the collimator narrowly opened in the longitudinal direction to inherently suppress scatter. The resultant CT image is used as a scatter-free reference (Figure 10.8b). The CBCT image (Figure 10.8a) contains severe scatter artifacts, shown as global shading and distortions around the high-contrast rods. The measured SPR is around 2 at the center of projections, leading to a CT number error of up to more than 180 Hounsfield Unit (HU) on the CBCT image.

A clinically relevant image quality metric is the contrast-to-noise ratio (CNR), which is defined as the ratio of image contrast to the average background noise. Scatter not only induces CT number errors but also significantly reduces image CNR. If CNR_s and CNR_0 are denoted as the CNRs of CBCT images with and without the presence of scatter, we have the following approximation:

$$CNR_S \approx CNR_0 \cdot (1 - \ln(1 + SPR)) \cdot \sqrt{1 + SPR} \qquad (10.14)$$

where

The term $(1 - \ln(1 + SPR))$ is due to the contrast loss as shown in Equation 10.13

The term $\sqrt{1 + SPR}$ accounts for the decrease in noise, since scatter photons improve the statistics of the detected photon signal

Using Taylor's expansion at $SPR = 0$, for small SPR, we simplify the approximation as shown in Reference 13:

$$CNR_S \approx CNR_0 \cdot (1 - SPR/2) \qquad (10.15)$$

Many methods have been proposed for scatter correction with different strengths and weaknesses. They can be divided into two major categories: preprocessing methods [14–17], and postprocessing methods [18–27]. The preprocessing methods correct for scatter signals by preventing the scatter photons from reaching the detector, and thus, scatter signals are suppressed in the projections. Two typical examples are increasing the air gap (i.e., the distance between the patient and the detector) [14] and/or using an antiscatter grid [13,15,16] between the object and the detector. As the air gap increases, the detection rate of the diffusive scatter photons decreases, while the primary signals are not affected. It is difficult to further increase the air

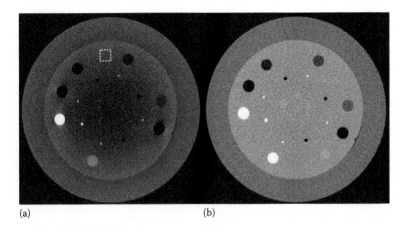

(a)   (b)

FIGURE 10.8 CT axial images of an evaluation phantom (Catphan©600, The Phantom Laboratory, Salem, NY). The experiments are performed on a tabletop CBCT system (a) using a CBCT geometry; (b) using a fan-beam geometry where the collimator is narrowly opened to inherently suppress scatter. Display window [−200, 300] HU. Inside the ROI (indicated by dash square in (a)), the CT number difference is 180 HU.

gap on C-arm systems, since the method is limited by the detector size and the clinical room size. The anti-scatter grid is used as a standard method on C-arm systems to suppress scatter signals (see Figure 10.5). The grid consists of lead strips focused at the x-ray focal spot, and reject scattered photons with oblique incident angles. The current commercial grids provide a typical SPR reduction by a factor of ~3, which does not guarantee a satisfactory image quality (SPR in CBCT can be >8 behind dense structures, e.g., bones [22–28]). Furthermore, extra dose is required to compensate for the inevitable primary attenuation due to the grid [29].

The imaging software of the C-arm system includes a postprocessing step to correct for errors in scatter-contaminated projection images after they are acquired in a CBCT scan. The method estimates scatter from the measured projection based on a scatter model obtained from system calibration. However, scatter has a complicated dependence on the imaged object. Approximations have to be applied in the model to make it computationally practical. These simplifications degrade the algorithm performance. The method usually works well only on simple geometries, such as uniform water-equivalent objects.

In the literature, numerous techniques have been developed for scatter correction on volumetric CT in general. Besides the scatter model method, other examples include Monte Carlo (MC) simulation [30], scatter measurement [31], and modulation-based approaches [28]. An optimal solution to this problem is yet to be found.

10.4.2   Beam-Hardening Effects

Even in the absence of x-ray scatter, the approximation of Equation 10.1 is inaccurate when the x-ray source is polyenergetic. Although many methods have been developed to generate monoenergetic x-ray beams (e.g., based on Compton effects [32,33]), the x-ray tube used on current CBCT systems provides by far the most effective production of x-rays. The tube is based on Bremsstrahlung effects and generates polyenergetic x-ray spectra. The penetration ability of x-rays increases as the energy increases. As a result, the x-ray spectrum is shaped toward a higher mean energy as the beam travels through a human body, that is, the beam hardens. The dependence of material attenuation coefficient on x-ray energy destroys the linearity relationship shown in Equation 10.2 and leads to beam-hardening artifacts in CBCT images.

Errors due to the beam-hardening effects can be analyzed following the method in Reference 34. We start from the error analysis of a homogeneous object with a simple geometry (e.g., water-equivalent elliptical cylinder). When a polyenergetic source is used and scatter is not considered, the measured line integral can be approximated on such an object as

$$q_S \approx \sum_{i=1}^{N} \alpha_i \ln^i (I_0/I_p). \tag{10.16}$$

where

N represents the order of polynomials used to model the beam-hardening effects

α_i depend on both the object material and the incident photon spectrum

The errors in q_s from the true line integrals lead to global cupping artifacts, which can be well compensated for by measurement of α_i in the system calibration and correction on the projection data with a polynomial fitting or a lookup table. The method is implemented on current CBCT systems, known as water precorrection or uniformity correction.

Challenges to beam-hardening correction are presented by inhomogeneous objects. When two different materials are present, for example, tissue and bone, the measured integral becomes

$$q_m \approx \sum_{i=1}^{N} \alpha_i \ln^i (I_0/I_{p'}) + \sum_{i=1}^{N} \beta_i \ln^i (I_{p'}/I_p) \tag{10.17}$$

where

β_i are parameters of the second high-attenuation material (e.g., bone) similar to α_i

$I_{p'}$ is the transmitted beam intensity after the first low-attenuation (e.g., tissue) material

Since the errors on q_s can be well suppressed by system calibration, the errors in CBCT images mainly stem from the difference between q_s and q_m, that is,

$$\Delta q = q_s - q_m \approx (\alpha_1 - \beta_1)\xi + (\alpha_2 - \beta_2)\xi^2 \tag{10.18}$$

where $\xi = \ln(I_{p'}/I_p)$ is the line integral of the high-intensity object. Higher order terms in ξ are ignored in the approximation. In Equation 10.18, the linear error term causes a low-frequency CT number shift in the CT image, and the nonlinear term produces streaking and shading artifacts. Data correction of q_m is difficult, since both β_i and ξ are dependent on the high-attenuation material distribution, which is unknown at the time of imaging. Beam-hardening errors for an object with more than two different materials can be analyzed similarly.

On CBCT systems, the CT image errors from beam-hardening effects are typically smaller than those from scatter, but are still significant especially around dense objects (e.g., bones). To better view the scatter and beam-hardening artifacts separately, we carry out a comparison study on an anthropomorphic head phantom, shown in Figure 10.9. The experiments are performed on the same tabletop CBCT system as used in Figure 10.8. With scatter signal inherently suppressed, the image of fan-beam CT (i.e., using a narrowly opened collimator) (Figure 10.9b) is considered to contain beam-hardening artifacts only. To obtain a reference image with small beam-hardening effects, we use the same fan-beam geometry and a 39 mm Al layer to heavily filter the x-ray beam in a third scan (Figure 10.9c). The incident x-rays are close to monoenergetic, but significantly attenuated. The scan is therefore repeated 20 times and averaged to make the image noise matchable to that in Figure 10.9a and b. The comparison indicates that the beam-hardening artifacts are secondary to scatter errors, but still reach ~70 HU around high-intensity areas.

One way to alleviate the beam-hardening problem is to prefilter the x-ray beam, which increases the mean x-ray beam energy and narrows the x-ray output spectrum. This physics-based approach is not favored on C-arm systems, since a higher x-ray beam energy reduces the contrast of fluoroscopic images. Besides the uniformity correction by system calibration described earlier, many advanced algorithms have been proposed to combat beam-hardening effects. For example, by estimating the distribution of high-attenuation structures, iterative algorithms achieve improved beam-hardening correction when two or more materials are present in the object [34–37].

A more elegant solution to the beam-hardening problem in CT imaging is based on the dual-energy theory [36–38]. In the diagnostic energy range where Rayleigh scattering can be ignored, the linear attenuation coefficient of any material is a weighted summation of two universal energy-dependent basis functions accounting for photoelectric and Compton interactions. From measured projections with two different x-ray spectra (e.g., with low and high tube kVp energies), the weights of the two basis functions, which are energy independent, are uniquely determined and their spatial distributions are reconstructed using the same CT principle. A beam-hardening free image is finally synthesized from the two weight images for any given energy. The practical value of dual-energy imaging, however, is constrained by the requirement of having two separate projection measurements at each angle with different x-ray energies. The method is not available on current C-arm systems.

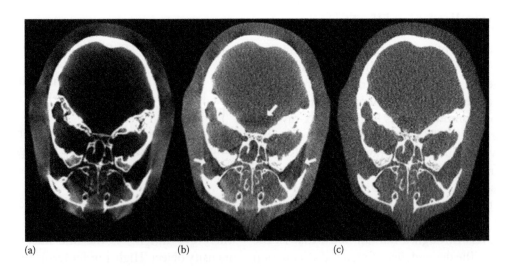

(a)   (b)   (c)

FIGURE 10.9 CT images of an anthropomorphic head phantom. Display window: [–100,200] HU. (a) CBCT; (b) fan-beam CT (with a narrowly open collimator to suppress scatter), with arrows highlighting the beam-hardening artifacts; (c) fan-beam CT with a 39 mm Al beam filter to suppress beam-hardening effects.

Chapter 10

10.4.3   Truncation of Projections

Besides errors from physical processes of x-ray photon interaction with the patient, limitations on CBCT system hardware lead to considerable artifacts as well. The manufacturing cost of a flat-panel detector increases dramatically with the effective detector area. The current C-arm system uses a flat-panel detector of 40-by-30 cm^2, which in general is not large enough to cover a human torso, especially in the AP view. As a result, the maximum FOV available for reconstruction may not cover the whole imaged object.

A somewhat counterintuitive fact is that, in FBP reconstruction, the missing projection signals outside the detector lead to image errors even if the reconstructed volume is inside the FOV. This effect arises, because the ramp filter kernel used in FBP reconstruction (see Equation 10.6) has an infinite data support. After the ramp filtering, the local missing signals generate global errors at detector locations even far from the edge. The backprojection step distributes these errors across the whole reconstructed volume, typically resulting in cupping artifacts (referred to as truncation artifacts). Figure 10.10 shows the truncation artifact in a CBCT image of an in vivo porcine thorax.

Compared to scatter and beam-hardening errors, truncation artifacts are of a smaller concern in vascular imaging, since they cause negligible image contrast loss within the center of the FOV. Extrapolations can be implemented outside the detected area to suppress the artifacts at the edge of the FOV and even to extend the FOV with simulated data [41]. Application of one such data extrapolation approach is shown in Figure 10.10 [42]. Advanced reconstruction algorithms, including derivative backprojection filtration [43] and iterative reconstruction [44], are able to obtain a theoretically exact CT image inside the FOV in certain scenarios. The truncation errors can also be avoided by measuring more projection lines with a modified imaging geometry. For example, the flat-panel detector can be shifted in the lateral direction to cover only half of the object. The missing half of the projection is measured from the opposite direction when the x-ray source rotates by 180°. Such a half-fan geometry enlarges the radius of FOV by a factor of approximately 2 without increasing the detector size. However, a scan of full rotation is required for sufficient data acquisition.

10.4.4   Detector Lag

Lag is signal present in detector frames following the frame in which it was actually generated. The primary cause of detector lag is charge trapping in the a-Si layer of the detector [45–47], although scintillator afterglow and incomplete charge readout also contribute to the lag signal. Traps are inherent in a-Si because of the lack of long-range crystalline order and dangling bonds within the material. Trapping in the a-Si also leads to a gain change of the photodiode, since when traps are filled, subsequently generated charge is collected by the readout electronics instead of falling into the trap states [48]. Because of the broad spectrum of trap states that exist in a-Si, lag effects can be seen at a wide range of panel frame rates [49–50]. At higher frame rates, measured lag can appear to decrease

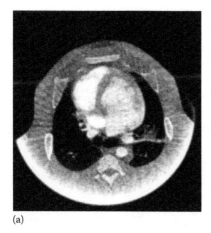

(a)

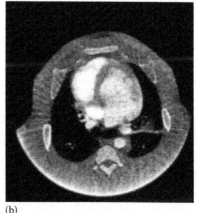

(b)

FIGURE 10.10   Retrospectively gated CBCT image of the thorax of an in vivo porcine model acquired using a clinical C-arm CT system (Siemens Axiom Artis dTA). Note the presence of significant streak artifact throughout the volume due to view starvation and contrast dynamics. (a) Image without truncation correction. (b) Image with truncation correction applied. (Images courtesy of Dr. Andreas Maier, U. Erlangen-Nurnberg, Erlangen, Germany.)

because of shorter integration times, but the lag then persists for a longer number of frames.

Typically, the first frame lag after a significant length of irradiation is ~2%–10% of the irradiating signal [46]. While the lag signal decays exponentially, a significant amount of signal remains for many seconds and could easily equal the signal behind a large object. In CT reconstructions, lag can lead to a range of image artifacts, such as streaks, comet-tails off of high contrast objects, or blurring and shading artifacts for noncircular or off-center objects [46,51,52]. The shading artifacts can be quite significant, depending on the geometry and detector state, especially near the object edge. Figure 10.11 shows a lag artifact of 49 HU for a Varian 4030CB panel used to image a large pelvic phantom.

Several software [51,53] and hardware methods to deal with lag in flat-panel imagers exist. One effective hardware method uses LEDs built into the panels to backlight the photodiodes, which saturates the traps. [45] Forward biasing to push current through the diode can be used to achieve the same goal [45,54,55]. Note that neither of these approaches correct for scintillator afterglow. Current software methods typically model an impulse response function for the panel by fitting a suitable model, such as a multiexponential [56] or power function [57], to the lag decay. A single impulse response function can be used to describe the entire panel, or an independent function can be fit for each individual pixel, which allows for variation in lag across the detector. The accuracy of the linear time-invariant model depends on the size of the object (i.e., on the dynamic range seen by individual pixels) and on the exposure at which the calibration of lag coefficients and lag rates is carried out [58]. This nonlinear behavior can be captured using an exposure-dependent calibration and can be corrected by applying the principle of charge conservation, which further reduces the artifact in CBCT reconstructions [59].

10.4.5 Motion Artifacts

CBCT imaging on a C-arm system has a low temporal resolution. The minimum data acquisition of projections takes 4–6 s on a rotation of about 200°. Even if the patient holds still, both respiratory and cardiac motions cause significant image errors. The resultant CBCT is the average of the patient at different phases, with blurring on the boundaries of moving structures and streak artifact. Example images showing motion artifacts can be found in Section 10.6.2. Figure 10.13 more specifically shows the impact of cardiac motion combined with the long imaging time of C-arm CT (see column labeled "Ungated C-arm CT").

Dynamic imaging is a general challenge for x-ray CT. Faster CBCT by increasing the C-arm gantry rotation speed generates more concerns for patient safety and system mechanical stability. To suppress motion artifacts without changing the CBCT machine settings, one of the popular techniques is by gating [60]. If the motion of the object is assumed to be periodic, gating methods first continuously measure projections for multiple rotations. Surrogate signals are also recorded simultaneously to provide phase indices. For example, electrocardiogram (ECG) signals are used to indicate the phase of cardiac motion. The projections are then sorted into different phases, and the 3D volume of each phase is reconstructed individually. Sufficient projections are needed for each phase that is reconstructed in order to avoid view aliasing artifacts. The number of rotations of data acquisition increases with the number of phases, resulting in a dose increase by the same factor.

10.4.6 Circular CBCT Artifacts from Approximate Reconstruction

Based on CT reconstruction theory [10], it can be easily verified that a circular CBCT scan does not acquire

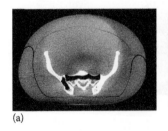 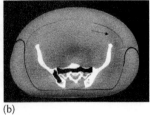 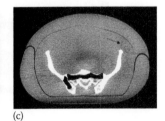

(a)　　　　　　　　　(b)　　　　　　　　　(c)

FIGURE 10.11 CBCT image of an anthropomorphic pelvis phantom, showing a detector lag artifact before (a) and after correction using a linear time-invariant model (b) and a nonlinear time-invariant model (c). The improvement in uniformity is seen particularly with respect to the so-called radar artifact, as indicated by the arrows.

sufficient projection data for exact reconstruction of the whole volume. The FDK algorithm provides an approximate reconstruction with an implicit assumption in its derivation that the projection lines outside the plane of the source trajectory (i.e., the midplane) can be treated as if they were measured through planes parallel to the mid-plane. This approximation becomes less accurate in both full and short scans when the reconstructed voxel moves away from the midplane, resulting in streaks around object edges, image blurs, and intensity drop errors [11,61].

The CBCT errors from approximate reconstruction are small around the midplane, and reach tens of HU at the two ends of the reconstructed volume along the z-axis. The image quality can be improved by using more accurate approximations in the reconstruction algorithm design, at the price of increased computational complexity. Exact reconstruction of the whole reconstructed volume becomes possible if sufficient projection lines are measured using more sophisticated scan trajectories. Advanced reconstruction techniques beyond the FDK algorithm need to be implemented on these datasets to ensure CBCT image qualities.

10.4.7 Metal Artifacts

If no effective beam-hardening correction is implemented, streak artifacts appear around dense objects (e.g., bones) in the CBCT image. The error becomes more severe as the attenuation of the object increases. In the extreme case when the incident x-rays are totally attenuated (or blocked) by the object (e.g., metal implants or dental fillings), the image errors are referred to as metal artifacts. In vascular CBCT imaging, metal artifacts are commonly seen in patients with coil-filled aneurysms, skull clips due to transcranial surgery, implanted defibrillators, and many other high-density implants.

Distinct from other CBCT errors discussed earlier, it is theoretically impossible to perfectly remove metal artifacts by system calibration or by more projection measurements. The rays passing through the metal are lost in the sinogram, which voids the data sufficiency condition for exact reconstruction of the whole volume. The artifacts, however, can be suppressed by approximations (e.g., interpolation) to the missing data when the metal size is small [62] or by iterative reconstruction with prior knowledge of the object [63,64].

10.5 Radiation Dose

The imaging chain components in C-arm systems have been optimized to provide high-resolution, and short exposure time per frame during vascular interventional procedures. The x-ray source kVp and filtration, and detector converter and electronics achieve excellent image qualities when the imaging task is the visualization of iodine-filled vessels in native or digitally subtracted projection images. Optimal image quality is ensured through use of the AEC system, which is designed to balance image signal-to-noise ratio (usually set at a level chosen by the operator) and patient dose. However, these settings/design choices are not necessarily optimized for CBCT imaging. For example, one difference is in the choice of applied filtration. Because the C-arm systems have lower-power x-ray tubes than clinical CT systems, the maximum mAs that can be applied during multiimage acquisitions is lower, and additional filtration would further reduce the x-ray fluence contributing to the images. In general, the x-ray spectra provided by fluoroscopy systems have less added filtration and is therefore "softer" than that of clinical CT systems [42]. This difference increases dose per mAs, as well as making C-arm CBCT images more susceptible to beam-hardening artifact as described earlier. A second dose

penalty arises when the system is used for C-arm CT of larger body regions such as abdomen or pelvis. To achieve sufficient penetration through long path lengths in the lateral direction (not typically used for projection imaging, but required for CT reconstruction), the AEC increases the kVp of the x-ray beam to maintain a constant exposure at the detector. Because the CsI flat panel is optimized for high-resolution imaging at lower kVp, the quantum detection efficiency drops from about 75% to less than 50% as the energy of the incident x-ray beam increases from 70 to 120 kVp. This also lowers the dose efficiency of C-arm CBCT.

The rather poor detector optimization is exacerbated by the low detector frame rates (as compared to clinical CT scanners), which limits the number of projections used to form each 3D CT reconstruction. Depending on the acquisition protocol (total scan time, DSA vs. native, detector frame rate), the number of projections contributing to a volume ranges from ~150 to ~500 as compared to at least 1000 projections acquired during a typical clinical CT examination. Note that the reduced number of projections contributing to a reconstruction leads to view aliasing artifact, creating streaks around the periphery of the reconstructed volume.

Chapter 10

The estimation of dose for a particular acquisition is complicated by the AEC, as well as by the use of a short-scan acquisition with gantry rotation through only 200°. The x-ray tube travels in an arc under the patient bed, and skin dose is higher on the tube side of the patient as compared to the detector side by at least a factor of 2 for head imaging [42]. Taking 2/3 of the average skin dose plus 1/3 of the central dose measured in a standard CT dose index (CTDI) phantom [42,65], the CTDI for a medium-dose protocol at 80 kVp and 543 views was measured to be 37 mGy, which is below the American College of Radiology (ACR) reference value of 60 mGy for a clinical CT scan [66]. Again, the presence of the AEC complicates dose estimation, although it is quite analogous to mA (and kVp) modulation in clinical CT, and depends on the body mass index. Some measurements of effective dose have been carried out using thermoluminescent dosimeters (TLDs) in an anthropomorphic phantom on a Siemens C-arm CBCT system. An abdominal scan at 90 kVp and 397 frames (dose request at detector of 0.36 μGy/frame) delivered a dose of 6.71 mSv, while a 70 kVp, 248 frame cardiac protocol (dose request at detector of 0.54 μGy/frame) delivered

a dose of 3.31 mSv. Relevant clinical CT doses for abdominal and heart scans are 4.3 and 8.3 mSv, respectively [67]. Such measurements should be repeated for a range of patient sizes in order to fully characterize the behavior of the AEC system and the resulting impact on dose. Head imaging is more straightforward, with less variability in size across the population. Similar TLD-based measurements in a head phantom showed that dose changed by an order of magnitude depending on the application of interest [68]. A vascular DSA-type study (0.36 μGy/frame, 5 s scan, 266 frames) delivered a dose of 0.3 mSv, while a study aimed at detecting bleeds or other low-contrast signals (1.2 μGy/frame, 20 s scan, 496 frames) delivered a dose of 2 mSv.

Several software techniques have shown promise on reduction of CBCT dose with negligible image quality loss. Based on statistical models, optimization algorithms suppress projection noise from low mAs while preserving image edges [4,69]. Iterative reconstruction obtains accurate CBCT volumes even if the projection view number is small [5,6,7]. The CBCT dose can be further reduced if prior knowledge of patient images is used in the iterative reconstruction [70].

10.6 Clinical Applications

In the early days of x-ray image intensifier-based C-arm CBCT, the 3D images were primarily of the intracranial vasculature. These 3D reconstructions provided additional guidance particularly during complicated procedures such as aneurysm coiling, when the exact geometry of the aneurysm neck and the relationship to the parent artery was key to successful and complete filling of the aneurysm. The intracranial application was ideal, since the head and its internal vasculature could be held reasonably still during the long imaging times associated with slow C-arm rotation speeds. The nonidealities of the x-ray image intensifier, such as veiling glare and distortion, did not degrade the ability to visualize high-contrast, high-resolution structures. With the advent of digital flat-panel detectors, the image quality of C-arm CBCT has improved dramatically, providing low-contrast visibility that is starting to approach that of clinical CT. Some fundamental limits on image quality remain, since the imaging chain (limited dynamic range of the detector, relatively high noise floor) and imaging geometry (cone beam, scatter) are not ideally suited to CT; however, increased detector readout rates in concert with fast rotation speeds have expanded the range of applications to include guidance of abdominal interventions

as well as cardiac imaging. Applications requiring multiple sequential sweeps around the patient are also under investigation, including ECG-gated imaging of the heart, and quantitative measurement of blood perfusion in the brain. In the following, we describe in more detail the newer applications of single and dual sweep (DSA) C-arm CBCT, as well as two multisweep vascular applications that are rapidly entering the clinic. Perhaps the fastest-growing area of clinical use for C-arm CBCT is in concert with navigation systems. Intraprocedural 3D C-arm CT volumes have been fused with electromagnetic tracking, 2D fluoroscopy, prior clinical CT volumes and intravascular and intracardiac ultrasound, magnetic resonance imaging data and/or PET images. Examples of this approach include electromagnetic navigation fused with C-arm CBCT for radiofrequency ablation (RFA) of atrial fibrillation, and use of 2D fluoroscopy overlaid on C-arm CBCT to guide transcatheter aortic-valve replacement (TAVR).

10.6.1 Intracranial Applications

Therapy for ischemic stroke caused by large vessel occlusions is slowly migrating from systemic application of thrombolytic drugs to local drug application, or local mechanical thrombectomy. Administration of such therapies occurs in the interventional suite after a baseline CT or MR scan has ruled out hemorrhage, and a volume of salvageable tissue has been identified. One of the potentially most useful applications of native reconstructions from a C-arm CT system would be the ability to distinguish between ischemic and hemorrhagic stroke, and then to provide an evaluation of the severity of the ischemic insult and an estimate of the extent of salvageable tissue by measuring cerebral blood volume (CBV) and cerebral blood flow. Providing this information in the interventional suite would shorten the time to therapy while providing access to all of the available therapeutic techniques. Preliminary studies have demonstrated the ability of C-arm CT to detect intracranial hemorrhages and have highlighted other applications for low-contrast intracranial imaging [71–74]. A recent study by Ahmed et al. [75] has shown that clinically relevant measures of CBV can be obtained, with values that compare well with clinical CT measures of blood volume in a healthy canine model.

The primary limitation of the C-arm CBCT system with respect to blood flow imaging is the slow rotation speed of the C-arm, which is a minimum of 4–5 s for one rotation through ~200°. In addition, a turnaround time of 1.5–2 s is required between rotations. Thus, each consecutive 3D volume acquired during continuous C-arm rotation represents an average of at least 4 s, with temporal spacing between volumes of at least 5.5 s. This sampling results in sparse time-density curves, and the arterial input function required for quantitative evaluation of blood flow is severely undersampled. Several new approaches have been proposed to increase the temporal sampling and provide accurate blood flow maps in the interventional suite. One approach provides improved sampling by interleaving two or three six-sweep acquisitions, with changing offsets between iodinated contrast injection and start of each multisweep run, combined with block interpolation reconstruction techniques. Accurate results have been shown in an healthy porcine model when compared against clinical CT [76] and are described in more detail in Chapter 18. Neukirchen et al. [77] have proposed an iterative method that uses a decomposition model-based approach (rather than interleaved scans) in an attempt to increase the temporal sampling of data acquired using a C-arm system. The goal of the algorithm is to estimate, on a voxel-by-voxel basis, the time-attenuation curves caused by the temporal variation of blood contrast concentration in vessels and

tissue (see also Chapter 17). A parametric representation, $\tilde{X}_d(y)$, is used to describe the spatial and temporal characteristics of the time-attenuation curves in the object. The model parameters y should have a sufficient degree of freedom to reflect all possible variations in the spatial and temporal properties of the time-attenuation curves, while having limited complexity for computation reasons. This balance is achieved by using prior knowledge about expected features, defining a set of N basis functions $b_n(t)$ that cover the time interval of dynamic acquisitions. Temporal basis functions could be, for example, time-shifted Gaussian functions or gamma-variate functions. The accuracy of the algorithm was investigated for a six-sweep, backward/forward C-arm acquisition using noisy data simulated from a CT perfusion dataset. Normalized error in the arterial time-attenuation curve was between 12% and 16%, while error in normal and hypoperfused tissue was between 4% and 6%. Image maps of cerebral blood flow, CBV and mean transit time reproduced the main features of the clinical CT data well, although streak artifact due to the turning points in the trajectory led to a partial masking of the infarct area.

Rotation speeds of C-arm systems continue to increase. The recently released zeego® robotic C-arm platform (Siemens Healthcare, Forchheim, Germany) is capable of very high-rotation speeds up to and beyond 60°/s, and therefore, a single volume can be acquired in under 3 s. Alternatively, the two arms of a biplane system could be rotated in tandem to cut the image acquisition time in half. Both of these approaches reduce the need for complicated sampling or data-fitting routines. Intraprocedural monitoring of blood flow perfusion will be a more straightforward application of approaches already optimized for clinical CT perfusion imaging.

10.6.2 Cardiac Applications

One important application for single-sweep cardiac imaging is in the area of procedural planning and improved image guidance during RFA procedures. RFA is becoming a first-line treatment for many cardiac arrhythmias, particularly for atrial fibrillation and ventricular tachycardia. The success of RFA depends on the extent of the lesion and the creation of lesions at specific anatomic locations and/or in continuous lines. It is currently possible to register preprocedural CT or MR images with electroanatomical mapping systems, but a more seamless integration and automatic registration of intraprocedural

Chapter 10

3D C-arm CT cardiac images would be of benefit. Intraprocedural C-arm CT images more accurately represent the hemodynamic loading conditions and heart rhythm of the patient over preprocedural CT or MR images and may also help to visualize surrounding structures, such as the esophagus [78,79], which are vulnerable to damage. Single-sweep C-arm CT images provide good image quality in 3D volumes of the left atrium and pulmonary vein anatomy, as well as the left ventricle [80]. Additionally, adenosine-induced asystole and rapid ventricular pacing have been investigated to improve the image quality of single-sweep acquisitions [82–86]. An example of registration between an intraprocedural C-arm CT image of the left atrium following segmentation and a prototype 2D-image-based mapping system applied to guidance of cryoablation is shown in Figure 10.12.

ECG-gated C-arm CT, introduced by Lauritsch et al. [58], provides more accurate images of the cardiac chamber, reconstruction at any time within the cardiac phase, and improved soft tissue contrast

due to decreased motion artifact [87,88]. A typical imaging protocol acquires images during four 5-s bidirectional sweeps, and requires a single breath-hold of ~25 s. Iodinated contrast is injected into the inferior vena cava (6 f pigtail catheter, 75%–25% iodinated contrast-saline, at 5 cc/s, total volume injected of 185 mL) throughout the acquisition for cardiac chamber and pulmonary vein visualization. For myocardial visualization, contrast injection ends just prior to the start of imaging in order to reduce artifact from rapidly changing contrast concentration in the cardiac chambers.

The multisweep approach combines data from several cardiac cycles to reconstruct the desired cardiac phase. The challenge is to ensure that complementary coverage is obtained from each sweep so that a minimal ϕ plus fan angle set of data is acquired. This approach is analogous to multisegment reconstruction in cardiac spiral CT. An example of the image quality that can be achieved using the multisweep protocol is shown in Figure 10.13. A study comparing

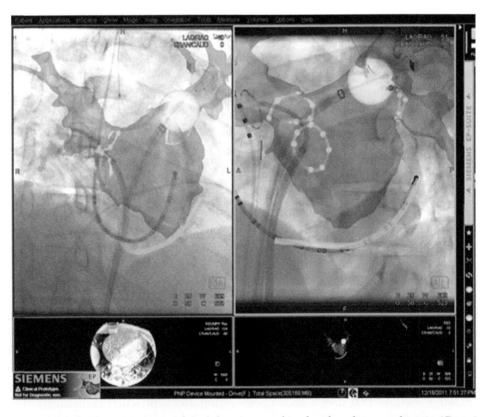

FIGURE 10.12 Intraprocedural C-arm CT images of the left atrium combined with catheter tracking in 2D projection images for guidance of cryoablation procedures. (bottom left) Multiplanar reformatted C-arm CT image of the contrast-filled left atrium, (bottom right) left atrium after segmentation, (upper left) 2D–3D overlay prior to registration between the two images, (upper right) 2D–3D overlay following registration. The current location of the cryoballoon is indicated by the white sphere, and the rings targeted for ablation within the ostia of the pulmonary veins are indicated in green, turquoise, and yellow. The excellent overlap of balloon and target is indicated by the presence of a single pink dot just showing on the boundary of the white sphere in the upper right panel. (Images courtesy of Dr. Amin Al-Ahmad, Stanford University Medical Center, Stanford, CA.)

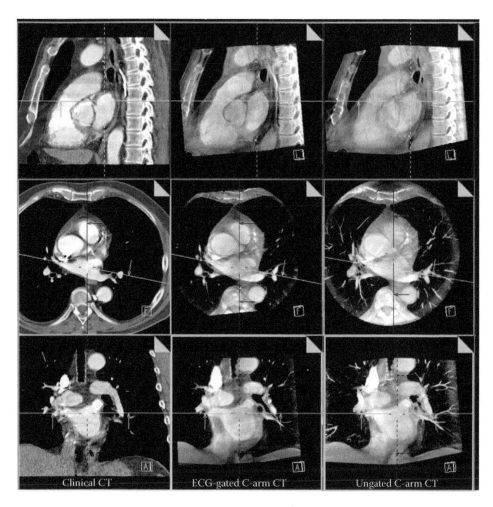

FIGURE 10.13 Comparison of clinical CT, ECG-gated C-arm CT, and ungated C-arm CT images obtained in the same patient. Image quality of ECG-gated C-arm CT permits delineation of the walls of the cardiac chamber, which are blurred in the single-sweep reconstruction. Efforts to improve image quality include cardiac motion correction algorithms, and reduction of total imaging time to reduce breathing-related motion.

gated and nongated C-arm CT against clinical CT in seven patients showed that measurements of pulmonary vein dimensions more closely agreed with those of CT when gating was applied (correlation between 0.329 and 0.899 for ungated and between 0.732 and 0.962 for gated, with correlation coefficients that are highly dependent on which vessel was measured). Applications in low-contrast imaging beyond visualization of the cardiac chambers, such as intraprocedural visualization of RFA lesions [89,90] and regions of myocardial infarction [91], are now under investigation. While ECG gating requires multiple sweeps of the C-arm and therefore a higher radiation dose, it is possible to further improve the signal-to-noise ratio (SNR) and low-contrast imaging within the myocardium using motion-corrected reconstructions [92]. Cardiac C-arm CBCT is an imaging modality that brings together the current 3D anatomy of the patient with real-time fluoroscopic guidance in a system that permits open access to the patient [83,93,94]. Future integration with a catheter guidance system will permit more minimally invasive procedures in the cardiac catheterization lab that would otherwise prove too complicated [95].

10.7 Summary

In this chapter, we introduce the C-arm system for vascular imaging and its CBCT imaging mode that is being increasingly used before, during, and after inventions for treatment guidance and verification. As developed on the same underlying principle, CBCT systems employ the same mathematical model,

Chapter 10

reconstruction algorithms, and data correction methods as those implemented on diagnostic CT scanners. Nevertheless, a C-arm gantry is used for its flexibility in interventional applications, resulting in inferior qualities of current CBCT imaging. For example, high scatter contamination due to the large size of x-ray illumination in one projection causes severe shading artifacts. The relatively slow gantry motion blurs the CBCT image on a moving object. Several clinical applications are discussed as typical examples of CBCT-based interventional radiology procedures.

Sophisticated algorithms are being continuously developed to further improve CBCT imaging, with the goal of approaching the image quality of clinical CT. New flat-panel detector technologies that are under development may also enhance the dose performance of CBCT such that multiple volumes can be acquired during interventions to track the progress. These advances in both software and hardware promote the performance of C-arm CBCT to a next stage, and may even extend its clinic use beyond interventional applications.

References

1. Hsieh, J. *Computed Tomography: Principles, Design, Artifacts, and Recent Advances.* 2009. SPIE, Bellingham, WA.
2. Zou, Y. and X. Pan, Exact image reconstruction on PI-lines from minimum data in helical cone-beam CT. *Physics in Medicine and Biology*, 2004. **49**(6): 941.
3. Katsevich, A., A general scheme for constructing inversion algorithms for cone beam CT. *International Journal of Mathematics and Mathematical Sciences*, 2003. **2003**(21): 1305–1321.
4. Wang, J. et al., Penalized weighted least-squares approach to sinogram noise reduction and image reconstruction for low-dose X-ray computed tomography. *IEEE Transactions on Medical Imaging*, 2006. **25**(10): 1272–1283.
5. Sidky, E.Y. and X. Pan, Image reconstruction in circular cone-beam computed tomography by constrained, total-variation minimization. *Physics in Medicine and Biology*, 2008. **53**(17): 4777.
6. Niu, T. and Zhu, L. Accelerated barrier optimization compressed sensing (ABOCS) reconstruction for cone-beam CT: phantom studies. *Medical physics*, 2012. **39**(7), 4588–4598.
7. Niu, T. et al., Accelerated barrier optimization compressed sensing (ABOCS) for CT reconstruction with improved convergence. *Physics in medicine and biology*, 2014. **59**(7), 1801.
8. Kak, A.C. and M. Slaney, *Principles of Computerized Tomographic Imaging*, Vol. 33. 1988. IEEE Press, New York.
9. Feldkamp, L., L. Davis, and J. Kress, Practical cone-beam algorithm. *Journal of the Optical Society of America A*, 1984. **1**(6): 612–619.
10. Grangeat, P., Mathematical framework of cone beam 3D reconstruction via the first derivative of the Radon transform, in *Mathematical Methods in Tomography*, Gabor T. Herman, Alfred K. Louis and Frank Natterer, Eds. 1991. Springer, Berlin, Germany. pp. 66–97.
11. Zhu, L., S. Yoon, and R. Fahrig, A short-scan reconstruction for cone-beam CT using shift-invariant FBP and equal weighting. *Medical Physics*, 2007. **34**: 4422.
12. Parker, D.L., Optimal short scan convolution reconstruction for fan beam CT. *Medical Physics*, 1982. **9**: 254.
13. Endo, M. et al., Magnitude and effects of x-ray scatter in a 256-slice CT scanner. *Medical Physics*, 2006. **33**(9): 3359–3368.
14. Siewerdsen, J.H. and D.A. Jaffray, Optimization of x-ray imaging geometry (with specific application to flat-panel cone-beam computed tomography). *Medical Physics*, 2000. **27**(8): 1903–1914.
15. Wiegert, J. et al., Performance of standard fluoroscopy anti-scatter grids in flat detector based cone beam CT. *Proceedings of the SPIE—The International Society for Optical Engineering*, 2004. **5368**: 67–78.
16. Siewerdsen, J.H. et al., The influence of antiscatter grids on soft-tissue detectability in cone-beam computed tomography with flat-panel detectors. *Medical Physics*, 2004. **31**(12): 3506–3520.
17. Neitzel, U., Grids or air gaps for scatter reduction in digital radiography: A model calculation. *Medical Physics*, 1992. **19**(2): 475–481.
18. Boone, J.M. and J.A. Seibert, An analytical model of the scattered radiation distribution in diagnostic radiology. *Medical Physics*, 1988. **15**(5): 721–725.
19. Seibert, J.A. and J.M. Boone, X-ray scatter removal by deconvolution. *Medical Physics*, 1988. **15**(4): 567–575.
20. Spies, L. et al., Correction of scatter in megavoltage cone-beam CT. *Physics in Medicine and Biology*, 2001. **46**(3): 821–833.
21. Jarry, G. et al., Characterization of scattered radiation in kV CBCT images using Monte Carlo simulations. *Medical Physics*, 2006. **33**(11): 4320–4329.
22. Ning, R., X. Tang, and D. Conover, X-ray scatter correction algorithm for cone beam CT imaging. *Medical Physics*, 2004. **31**(5): 1195–1202.
23. Zhu, L., N. Strobel, and R. Fahrig, X-ray scatter correction for cone-beam CT using moving blocker array. *Proceedings of the SPIE—The International Society for Optical Engineering*, 2005. **5745**(1): 251–258.
24. Boone, J.M. and J.A. Seibert, Monte Carlo simulation of the scattered radiation distribution in diagnostic radiology. *Medical Physics*, 1988. **15**(5): 713–720.
25. Seibert, J.A. and J.M. Boone, Medical image scatter suppression by inverse filtering. *Proceedings of the SPIE—The International Society for Optical Engineering*, 1988. **914**: 742–750.
26. Bertram, M., J. Wiegert, and G. Rose, Potential of software-based scatter corrections in cone-beam volume CT. *Proceedings of the SPIE: Progress in Biomedical Optics and Imaging*, 2005. **5745**(I): 259–270.
27. Bani-Hashemi, A. et al., Cone beam x-ray scatter removal via image frequency modulation and filtering. *Medical Physics*, 2005. **32**(6): 2093–2093.
28. Zhu, L., N.R. Bennett, and R. Fahrig, Scatter correction method for X-ray CT using primary modulation: Theory and preliminary results. *IEEE Transaction on Medical Imaging*, 2006. **25**(12): 1573–1587.

29. Kyriakou, Y. and W. Kalender, Efficiency of antiscatter grids for flat-detector CT. *Physics in Medicine and Biology*, 2007. **52**(20): 6275–6293.

30. Poludniowski, G. et al., An efficient Monte Carlo-based algorithm for scatter correction in keV cone-beam CT. *Physics in Medicine and Biology*, 2009. **54**(12): 3847–3864.

31. Zhu, L. et al., Scatter correction for cone-beam CT in radiation therapy. *Medical Physics*, 2009. **36**(6): 2258–2268.

32. Hao, X. et al., Terawatt femtosecond laser storage cavity with cholesteric liquid crystals for an x-ray source based on Compton scattering. *Optics Letters*, 2010. **35**(9): 1361–1363.

33. Beer, G.P.D., A compton scatterer as a source of mono-energetic gamma rays. *Nuclear Instruments and Methods*, 1970. **78**(1): 13–18.

34. Hsieh, J. et al., An iterative approach to the beam hardening correction in cone beam CT. *Medical Physics*, 2000. **27**(1): 23–29.

35. Kyriakou, Y. et al., Empirical beam hardening correction (EBHC) for CT. *Medical Physics*, 2010. **37**(10): 5179–5187.

36. Reitz, I. et al., Enhancement of image quality with a fast iterative scatter and beam hardening correction method for kV CBCT. *Zeitschrift für Medizinische Physik*, 2009. **19**(3): 158–172.

37. Stenner, P. et al., Dynamic iterative beam hardening correction (DIBHC) in myocardial perfusion imaging using contrast-enhanced computed tomography. *Investigative Radiology*, 2010. **45**(6): 314–323.

38. Alvarez, R.E. and A. Macovski, Energy-selective reconstructions in x-ray computerized tomography. *Physics in Medicine and Biology*, 1976. **21**(5): 733–744.

39. Stonestrom, J.P., R.E. Alvarez, and A. Macovski, A framework for spectral artifact corrections in x-ray CT. *IEEE Transaction on Biomedical Engineering*, 1981. **28**(2): 128–141.

40. Coleman, A.J. and M. Sinclair, A beam-hardening correction using dual-energy computed tomography. *Physics in Medicine and Biology*, 1985. **30**(11): 1251–1256.

41. Hsieh, J. et al., A novel reconstruction algorithm to extend the CT scan field-of-view. *Medical Physics*, 2004. **31**: 2385.

42. Ohnesorge, B., et al., Efficient correction for CT image artifacts caused by objects extending outside the scan field of view. *Medical Physics*, 2000. **27**(1): 39–46.

43. Noo, F., R. Clackdoyle, and J.D. Pack, A two-step Hilbert transform method for 2D image reconstruction. *Physics in Medicine and Biology*, 2004. **49**(17): 3903.

42. Yu, H. and G. Wang, Compressed sensing based interior tomography. *Physics in Medicine and Biology*, 2009. **54**(9): 2791.

45. Overdick, M., T. Solf, and H.A. Wischmann, Temporal artefacts in flat dynamic x-ray detectors. *Proceedings of the SPIE—The International Society for Optical Engineering*, 2001. **4320**: 47–58.

46. Siewerdsen, J.H. and D.A. Jaffray, Cone-beam computed tomography with a flat-panel imager: Effects of image lag. *Medical Physics*, 1999. **26**(12): 2635–2647.

47. Street, R.A., *Hydrogenated Amorphous Silicon*. 1991. Cambridge University Press, Cambridge, U.K.

48. Roberts, D.A. et al., Charge trapping at high doses in an active matrix flat panel dosimeter. *IEEE Transactions on Nuclear Science*, 2004. **51**(4): 1427–1433.

49. Edward, G.S. et al., Multidetector-row CT with a 64-row amorphous silicon flat panel detector. *Proceedings of SPIE*, 2007. **6510**: 65103X.

50. Siewerdsen, J.H. and D.A. Jaffray, A ghost story: Spatio-temporal response characteristics of an indirect-detection flat-panel imager. *Medical Physics*, 1999. **26**(8): 1624–1641.

51. Mail, N. et al., An empirical method for lag correction in cone-beam CT. *Medical Physics*, 2008. **35**(11): 5187–5196.

52. Starman, J. et al., TH-E-330A-04: Investigation into the cause of a new artifact in cone beam CT reconstructions on a flat panel imager. *Medical Physics*, 2006. **33**(6): 2288–2288.

53. Hsieh, J., O.E. Gurmen, and K.F. King. Recursive correction algorithm for detector decay characteristics in CT. In *Proceedings of the SPIE on Medical Imaging 2000: Physics of Medical Imaging*. 2000. San Diego, CA.

54. Mollov, I., C. Tognina, and R. Colbeth, Photodiode forward bias to reduce temporal effects in a-Si based flat panel detectors. *Proceedings of the SPIE—The International Society for Optical Engineering*, 2008. **6913**: 69133S-1–69133S-9.

55. Starman, J. et al., Parameter investigation and first results from a digital flat panel detector with forward bias capability. *Proceedings of the SPIE—The International Society for Optical Engineering*, 2008. **6913**: 69130L-1–69130L-9.

56. Yang, K., A.L. Kwan, and J.M. Boone, Computer modeling of the spatial resolution properties of a dedicated breast CT system. *Medical Physics*, 2007. **34**(6): 2059–2069.

57. Weisfield, R.L. et al., High-performance amorphous silicon image sensor for x-ray diagnostic medical imaging applications. In *Conference on Physics of Medical Imaging*. 1999. San Diego, CA. pp. 307–317.

58. Starman, J. et al., Investigation into the optimal linear time-invariant lag correction for radar artifact removal. *Medical Physics*, 2011. **38**(5): 2398–2411.

59. Starman, J. et al., A forward bias method for lag correction of an a-Si flat panel detector. *Medical Physics*, 2012. **39**(1): 18–27.

60. Lauritsch, G. et al., Towards cardiac C-arm computed tomography. *IEEE Transactions on Medical Imaging*, 2006. **25**(7): 922–934.

61. Zhu, L., J. Starman, and R. Fahrig, An efficient estimation method for reducing the axial intensity drop in circular cone-beam CT. *International Journal of Biomedical Imaging*, 2008. **2008**: 242841.

62. Veldkamp, W.J. et al., Development and validation of segmentation and interpolation techniques in sinograms for metal artifact suppression in CT. *Medical Physics*, 2010. **37**: 620.

63. Wang, G., T. Frei, and M.W. Vannier, Fast iterative algorithm for metal artifact reduction in x-ray CT. *Academic Radiology*, 2000. **7**(8): 607–614.

64. Wang, G. et al., Iterative deblurring for CT metal artifact reduction. *IEEE Transactions on Medical Imaging*, 1996. **15**(5): 657–664.

65. Kroon, J.N., 3-Dimensional rotational X-ray imaging, 3D-RX: Image quality and patient dose simulation for optimisation studies. *Radiation Protection Dosimetry*, 2005. **114**(1–3): 341–349.

66. Gray, J.E. et al., Reference values for diagnostic radiology: Application and impact. *Radiology*, 2005. **235**(2): 354–358.

67. Das, M. et al., Angiographic CT: Measurement of patient dose. In *Radiological Society of North America*. 2008. Chicago, IL. p. SSE25-06.

68. Strobel, N. et al., 3D imaging with flat-detector C-arm systems, in *Multislice CT*, 3rd edn., M.F.E.A. Reiser, Ed. 2009, Springer, Berlin, Germany, pp. 33–51.

Chapter 10

69. Zhu, L., J. Wang, and L. Xing, Noise suppression in scatter correction for cone-beam CT. *Medical Physics*, 2009. **36**(3): 741–752.

70. Chen, G.-H., J. Tang, and S. Leng, Prior image constrained compressed sensing (PICCS): A method to accurately reconstruct dynamic CT images from highly undersampled projection data sets. *Medical Physics*, 2008. **35**: 660.

71. Arakawa, H. et al., Experimental study of intracranial hematoma detection with flat panel detector C-arm CT. *American Journal of Neuroradiology*, 2008. **29**(4): 766–772.

72. Zellerhoff, M. et al., Low contrast 3D reconstruction from C-arm data. *Proceedings of the SPIE—The International Society for Optical Engineering*, 2005. **5745**(1): 646–655.

73. Orth, R.C., M.J. Wallace, and M.D. Kuo, C-arm cone-beam CT: General principles and technical considerations for use in interventional radiology. *Journal of Vascular Interventional Radiology*, 2008. **19**(6): 814–820.

74. Wallace, M.J. et al., Three-dimensional C-arm cone-beam CT: Applications in the interventional suite. *Journal of Vascular Interventional Radiology*, 2009. **20**(7 Suppl): S523–S537.

75. Ahmed, A.S. et al., C-arm CT measurement of cerebral blood volume: An experimental study in canines. *American Journal of Neuroradiology*, 2009. **30**(5): 917–922.

76. Ganguly, A. et al., Cerebral CT perfusion using an interventional C-arm imaging system: Cerebral blood flow measurements. *American Journal of Neuroradiology*, 2011. **32**(8): 1525–1531.

77. Neukirchen, C., M. Giordano, and S. Wiesner, An iterative method for tomographic x-ray perfusion estimation in a decomposition model-based approach. *Medical Physics*, 2010. **37**(12): 6125–6141.

78. Orlov, M.V. et al., Three-dimensional rotational angiography of the left atrium and esophagus—A virtual computed tomography scan in the electrophysiology lab? *Heart Rhythm: The Official Journal of the Heart Rhythm Society*, 2007. **4**: 37–43.

79. Nölker, G. et al., Three-dimensional left atrial and esophagus reconstruction using cardiac C-arm computed tomography with image integration into fluoroscopic views for ablation of atrial fibrillation: Accuracy of a novel modality in comparison with multislice computed tomography. *Heart Rhythm: The Official Journal of the Heart Rhythm Society*, 2008. **5**(12): 1651–1657.

80. Thiagalingam, A. et al., Intraprocedural volume imaging of the left atrium and pulmonary veins with rotational X-ray angiography: Implications for catheter ablation of atrial fibrillation. *Journal of Cardiovascular Electrophysiology*, 2008. **19**(3): 293–300.

81. Bartolac, S. et al., A local shift-variant Fourier model and experimental validation of circular cone-beam computed tomography artifacts. *Medical Physics*, 2009. **36**(2): 500–512.

82. Gerds-Li, J. et al., Rapid ventricular pacing to optimize rotational angiography in atrial fibrillation ablation. *Journal of Interventional Cardiac Electrophysiology: An International Journal of Arrhythmias and Pacing*, 2009. **26**(2): 101–107.

83. Kriatselis, C. et al., A new approach for contrast-enhanced X-ray imaging of the left atrium and pulmonary veins for atrial fibrillation ablation: Rotational angiography during adenosine-induced asystole. *Europace: European Pacing, Arrhythmias, and Cardiac Electrophysiology: Journal of the Working Groups on Cardiac Pacing, Arrhythmias, and Cardiac Cellular Electrophysiology of the European Society of Cardiology*, 2009. **11**(1): 35–41.

84. Tang, M. et al., Reconstructing and registering three-dimensional rotational angiogram of left atrium during ablation of atrial fibrillation. *Pacing and Clinical Electrophysiology*, 2009. **32**(11): 1407–1416.

85. Müller, K. et al. Evaluation of interpolation methods for surface-based motion compensated tomographic reconstruction for cardiac angiographic C-arm data. *Medical Physics*, 2013. **40**(3): 031107.

86. Müller, K. et al. Interventional heart wall motion analysis with cardiac C-arm CT systems. *Physics in Medical Biology*, 2014. **59**(9), 2265–2284.

87. Al-Ahmad, A. et al., Time-resolved three-dimensional imaging of the left atrium and pulmonary veins in the interventional suite—A comparison between multisweep gated rotational three-dimensional reconstructed fluoroscopy and multislice computed tomography. *Heart Rhythm: The Official Journal of the Heart Rhythm Society*, 2008. **5**(4): 513–519.

88. Fahrig, R. et al., Comparison of gated vs. non-gated cardiac C-arm CT (DYNA CT) of the left atrium and pulmonary veins in humans. *Journal of Interventional Cardiac Electrophysiology*, 2008. **21**: 169–174.

89. Girard-Hughes, E. et al., Visualization and enhancement patterns of radiofrequency ablation lesions with iodine contrast-enhanced cardiac C-arm CT. *Proceedings of SPIE*, 2009. **7262**: 72621N

90. Girard, E.E. et al., Contrast-enhanced C-arm CT evaluation of radiofrequency ablation lesions in the left ventricle. *JACC: Cardiovasc Imaging*, 2011. **4**(3): 259–268.

91. Girard, E.E. et al., Contrast-enhanced C-arm computed tomography imaging of myocardial infarction in the interventional suite. *Investigative Radiology*, 2015. [Epub ahead of print].

92. Prümmer, M. et al., Cardiac C-arm CT: 4D non-model based heart motion estimation and its application. *Proceedings of SPIE*, 2007. **6510**: 651015.

93. Li, J.H. et al., Segmentation and registration of three-dimensional rotational angiogram on live fluoroscopy to guide atrial fibrillation ablation: A new online imaging tool. *Heart Rhythm: The Official Journal of the Heart Rhythm Society*, 2009. **6**(2): 231–237.

94. Ector, J. et al., Biplane three-dimensional augmented fluoroscopy as single navigation tool for ablation of atrial fibrillation: Accuracy and clinical value. *Heart Rhythm: The Official Journal of the Heart Rhythm Society*, 2008. **5**(7): 957–964.

95. Wallace, M.J. et al., Three-dimensional C-arm cone-beam CT: Applications in the interventional suite. *Journal of Vascular Interventional Radiology*, 2008. **19**(6): 799–813.

11. Digital Techniques with a Nonstandard Beam Geometry

Michael Van Lysel and Michael Speidel

The design of an x-ray imaging system starts with the fact of x-ray beam divergence. It directly affects the x-ray fluence level at the detector and the achievable anatomic coverage. Spatial resolution is also affected as the designer balances the size of detector elements, focal spot intensity distribution, and the placement of the patient along the source–detector axis. The configuration used by the conventional fluoroscopic system is a wide-area detector and fixed-point x-ray source, with the patient placed directly in front of the detector. This geometry achieves both a high fluence level and good spatial resolution but at the expense of high levels of detected x-ray scatter, degrading image quality, and high dose to the patient and the interventionalist. In search of alternative solutions, many investigators have explored the concept of the scanned system (Barnes 1998). Motivations for scanning include scatter rejection, exposure equalization, tomography, and extended field-of-view (FOV) imaging with a small detector.

Scanning systems, which can take on several forms, most commonly employ a fixed source with moving collimator(s) fore, and sometimes aft, of the patient. Due to mechanical considerations, most scanned systems are limited to low frame rates although rapid scanning beam systems with complex mechanisms have been investigated (Rudin and Bednarek 1980, Rudin et al. 1982). High-speed scanning can be achieved by the use of spatially distributed x-ray sources. Carbon nanotube field emission x-ray sources are under investigation for volumetric and micro-CT (Cao et al. 2010, Xu et al. 2011). The dynamic spatial reconstructor was an early system using multiple, conventional x-ray tubes to perform CT imaging of the heart (Robb et al. 1983). A multisource inverse-geometry CT (MS-IGCT) system is currently under development for the purpose of volumetric scanning without cone-beam artifacts (Frutschy et al. 2010, Baek et al. 2012). Additional potential advantages of MS-IGCT include lower x-ray scatter and patient dose. An alternative method to relax the physical limitations imposed upon mechanically scanned systems is electromagnetic scanning of the x-ray source electron beam. The electron beam CT scanner made use of this technique to acquire axial images in as little as 50–33 ms (Boyd and Lipton 1983).

Electronic scanning greatly relaxes the physical limitations imposed upon mechanically scanned systems. Modern CT scanners are engineering marvels that achieve impressive scanning rates, but they exploit the advantage

Chapter 11

of circular motion. This chapter describes a fluoroscopic system based upon electromagnetic scanning of the x-ray source. High frame rate imaging supports interventional procedures while the distributed array of source positions enables tomosynthesis, stereoscopic imaging, and regionally adaptive exposure equalization. The x-ray field is comprised a collection of narrow beamlets converging to the detector. The inverse, narrow-beam geometry results in significant dose reduction for both the patient and interventional staff.

11.1   System Description

11.1.1   X-Ray Source

The scanning beam digital x-ray (SBDX) system accomplishes high-frame-rate scanning by electromagnetic scanning of a small, focused electron beam. The SBDX source (Figure 11.1) resembles the basic design of a cathode ray tube (CRT) television monitor. The source has a large 2D array of focal spot positions within a single vacuum vessel. Electrons are produced by thermionic emission from a dispenser cathode and accelerated across a variable potential (70–120 kVp). Coils in the neck of the tube focus and scan the electron beam magnetically across a tungsten transmission target, from which the usable x-rays are transmitted out the opposite surface. The target is approximately 15 μm thick supported by a 5 mm thick beryllium substrate. Cooling water flowing across the exit surface of the beryllium plate allows the tube to be operated continuously at full power.

A multihole collimator restricts the x-rays to a beamlet (1.6 × 3.2 cm at isocenter) directed at the detector. Each rectangular collimator hole is individually focused on the detector located 150 cm from the target. The electron beam dwells behind a collimator hole for 1.04 μs, then moves to the next hole (0.24 μs transit time). The detector image is read out at the end of each dwell and sent to the reconstruction engine, where full FOV images are reconstructed. After one frame period, data from the entire field of view has been collected. SBDX has frame rates of 10, 15, and 30 fps, as well as single frame acquisition.

Figure 11.2 shows the geometry of the scanning process as it relates to images formed by individual collimator holes and the anatomic FOV. The collimator is an array of 100 × 100 holes on a pitch of 2.3 mm. The collected set of beamlets transmitted by this array yield a reconstructed image 16 × 16 cm at the isocenter of the system, similar in coverage to a conventional cardiac system. Figure 11.2 shows a scan utilizing the central 71 × 71 holes, emulating a magnification mode on a conventional system.

11.1.2   X-Ray Detector

The SBDX detector is a small direct-conversion high speed imager. It consists of 2 mm thick CdTe crystals bump bonded to underlying application-specific integrated circuit (ASIC) read-out chips. The detector is constructed from 32 CdTe/ASIC hybrids covering an area of 5.3 × 10.6 cm (Figure 11.3). Each ASIC is divided into a 40 × 40 array of detector elements (dels) on a pitch of 0.33 mm, producing a total array

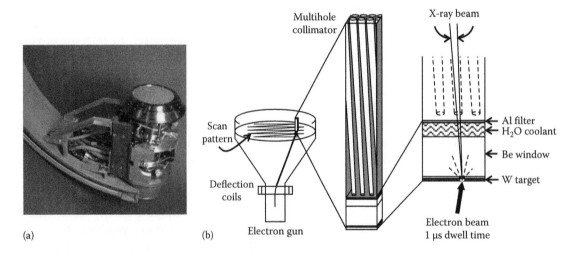

(a)   (b)

FIGURE 11.1   (a) SBDX source with exterior covers removed. (b) Source schematic with scale cutaway view of the multihole collimator and details of the target, exit window, cooling, and filtration.

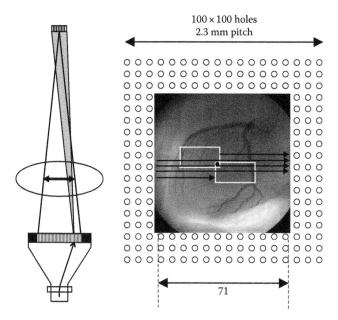

FIGURE 11.2 Relationship between the coverage by x-ray beamlets from individual collimator holes and the total FOV. The white rectangles indicate the first and last x-ray beamlets that overlap a given point in the object (black dot) as the scan proceeds from left-to-right and top-to-bottom. The temporal window for motion blur opens with the first illumination of an object and closes with the last illumination.

FIGURE 11.3 SBDX photon-counting detector. The detector is constructed of 32.2 mm thick CdTe crystals bonded to underlying readout chips. The array is stepped to facilitate electrical connections.

size of 160 × 320, though the usual read-out mode uses 2 × 2 binning of the 0.33 mm del (80 × 160 detector array). This detector is a recent upgrade of the 5.5 × 5.5 cm detector described in a previous report (Speidel et al. 2006a).

The mean photon fluence incident on the detector during the 1 μs capture period is typically less than 1 photon per detector element, low enough that the array is operated in photon-counting mode. To perform high-speed photon counting, each del is subdivided into a 2 × 2 array of sub-dels. Each sub-del is connected to a preamplifier and threshold discriminator, producing a binary output (0 or 1). A del sample is the sum of its binary sub-del outputs. Ideally, each photon that interacts with the CdTe x-ray converter via photoelectric absorption produces one sub-del count. However, the stochastic detection of the subsequent k-fluorescence photon in a neighboring sub-del ("double counting") represents a noise process. To address this, neighboring sub-del outputs are connected to an anticoincidence network, the purpose of which is to convert double counts back into a single count. At the end of each 1 μs read-out period, the array of detected photon counts is sent to the reconstruction engine.

11.1.3 Image Reconstruction

As the acquisition of an image frame proceeds, the detector images are reconstructed with a shift-and-add algorithm into a full FOV image in real-time using dedicated hardware. Since any particular point in the imaging volume is irradiated by several different source positions during the course of the scan, the reconstructed image is planar tomographic in nature. That is, for a given set of reconstruction parameters, there is a plane in the imaging volume (the *focal plane*) that is rendered without tomographic blur. Objects outside of the focal plane are blurred in proportion to their distance from the focal plane (Figure 11.4a).

Because the detector images are collected individually, this data can be used to perform digital tomosynthesis (Dobbins and Godfrey 2003). Any desired plane can be reconstructed from the acquired data in a frame period, but each reconstruction will contain blurred representations of out-of-plane structures. The usefulness of this image presentation for real-time image guidance is limited. Figure 11.5 shows a hardware implementation of an algorithm to reconstruct multiple planes (16 in the figure) and select for display, on a pixel-by-pixel basis, the plane(s) that contain objects of interest. The result is a multiplane composite reconstruction with objects from multiple planes in focus in the same display. Figure 11.4b shows the output of the multiplane composite reconstruction, with all numbers in focus. Multiplane coronary angiographic images are seen in Figures 11.12, 11.13, and 11.16.

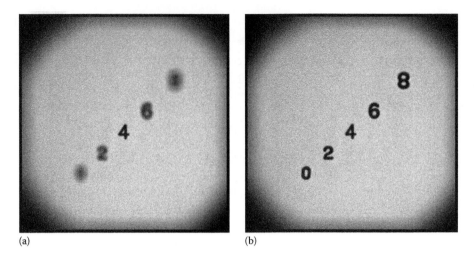

FIGURE 11.4 Phantom demonstration of the SBDX multiplane composite display. The phantom consists of lead numbers spaced at 2 in. intervals along the source–detector axis (number "0" is closest to the source). (a) A single-plane reconstruction of the plane located 4 mm above the number "4." (b) A multiplane composite generated from single-plane images with 12 mm spacing.

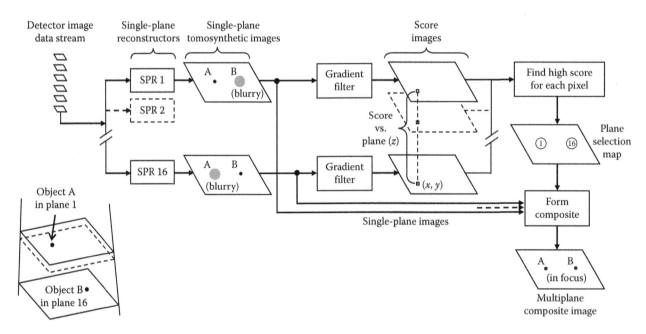

FIGURE 11.5 Method of real-time multiplane composite reconstruction (example geometry shown at left). Tomosynthetic images are simultaneously reconstructed at 16 planes spaced throughout the imaging volume. Each pixel in a reconstructed image is scored to reflect the degree of local sharpness. The score images are searched at fixed pixel positions (x, y) to find the plane with the highest score. The plane selection map indicates, for each pixel position, which single-plane image to draw a pixel value from when forming the composite.

11.2 System Performance

11.2.1 Dose Efficiency

The pairing of a scanned focal spot with a multihole collimator that directs each beamlet toward a small distant detector creates inverse-geometry beam scanning, a nonmechanical technique that carries the benefits of scatter rejection and that inverts the normal fluence versus distance dependence. Figure 11.6 compares the geometry of a conventional system with the inverse-geometry of SBDX. While each individual SBDX beamlet exhibits normal divergence from the focal spot, the fluence from an entire frame ensemble

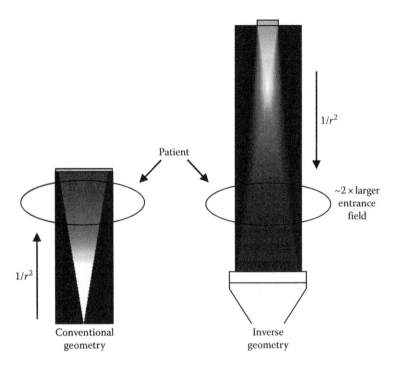

FIGURE 11.6 Comparison of the fluence-versus-distance dependence for conventional and SBDX systems. For SBDX, the patient is located in a lower intensity portion of the field, resulting in a skin-sparing reduction in entrance dose.

exhibits $1/r^2$ dependence as measured from the *detector*. The decreasing field size allows for the use of a small detector that in turn makes it economically viable to use a high-Z, crystalline direct conversion detector. Direct conversion detectors do not have the absorption/resolution trade off inherent in indirect conversion CsI phosphors. The ability to use a thick converter is a distinct advantage when imaging large patients at high energy (Figure 11.7a).

The narrow x-ray beam and large air gap provides for significant scatter rejection (Figure 11.7b) without the use of an antiscatter grid. The area of the x-ray field as it enters the patient is approximately twice that of a conventional system, thus decreasing skin dose for a given number of photons entering the patient. Skin dose reduction is a significant consideration for fluoroscopically guided interventional procedures. The most important factors controlling the entrance

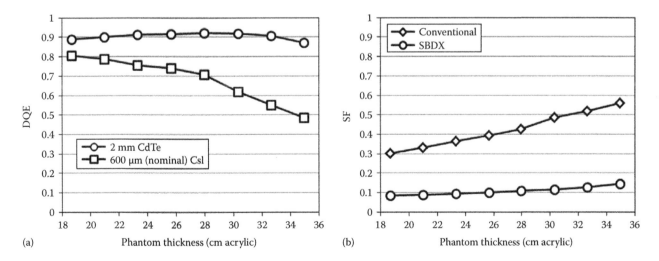

FIGURE 11.7 (a) Modeled DQE of CdTe and CsI x-ray converters as a function of phantom thickness. X-ray beam energy versus phantom thickness from Table II of Speidel et al. (2006b). CsI DQE is based on data from Zhao et al. (2004) and the effective energy of the phantom exit beam. The nominal CsI thickness is 600 μm. (b) Measured scatter fraction for conventional and SBDX systems. The SBDX fraction is the sum of x-ray scatter and off focus radiation from the source (roughly equal contributions).

Chapter 11

dose required of the two systems to achieve a given signal-to-noise (SNR) ratio are as follows:

$$\frac{D^{SBDX}}{D^{conv}} \approx \left(\frac{1 - SF^{conv}}{1 - SF^{SBDX}} \right) \times \left(\frac{DQE^{conv}}{DQE^{SBDX}} \right) \times \left(\frac{A^{conv}}{A^{SBDX}} \right)$$

where

D is skin dose

SF is detected scatter fraction

DQE is the detective quantum efficiency of the detector

A is the area of the x-ray field where it enters the patient

Other factors include the use of an anti-scatter grid (conventional system) and nonideal beam collimation (SBDX). A careful consideration of all factors pertaining to the dose efficiency of these two systems shows SBDX skin dose to be three to seven times less than a conventional system for the same SNR (Speidel et al. 2006b). The greatest dose savings obtains for the largest patients, where dose is highest. The dose savings to the interventional staff is expected to be half that of the patient, because all factors related to patient dose are also related to the number of photons scattered into the procedure room except the area of the entrance beam.

11.2.2   Signal-to-Noise and Resolution

Optimization of source output is critical for scanned systems. In the case of SBDX, x-ray output is limited by the constraints imposed by the narrow beamlets (only a small fraction of the x-rays generated at the target illuminate the FOV at any instant in time) and focal spot size. Conventional sources make use of target angulation to increase the available power by a factor of five or more at a given effective focal spot size. This strategy is not available with a transmission target (Figure 11.8). In addition, the high geometric magnification of the system makes

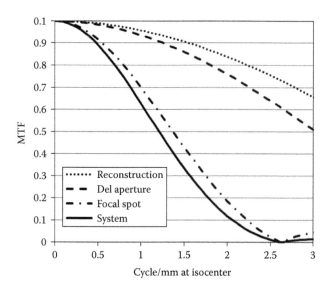

FIGURE 11.9 Measured modulation transfer function (MTF) at the isocenter plane. The system MTF is the product of the focal spot, 0.66 mm detector aperture, and image reconstruction MTFs, each scaled to isocenter. The reconstruction MTF describes the sampling of backprojected detector element values onto the regular pixel grid; a nearest-neighbor binning scheme is shown.

SBDX resolution very sensitive to the focal spot intensity distribution (the collimator-hole size is not a limiting factor). Figure 11.9 shows the modulation transfer function (MTF) at isocenter. The current nominal focal spot is a 0.6 mm diameter pillbox at full power. Smaller spot size could be used at lower power, such as during fluoroscopy, but this feature has not yet been implemented.

A significant difference between a conventional source and SBDX is the beam on-time per frame. While motion blurring limits the pulse width of a conventional system to about 10 ms (15% on-time at 15 fps), the SBDX duty cycle is 63% for the same frame rate. Temporal blurring is limited to a similar amount as a conventional system because blurring of an object is limited to the period between first and last illuminations of that object during the frame period (Figure 11.2, Table 11.1).

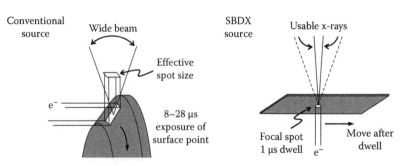

FIGURE 11.8 Comparison of the x-ray beam solid angle, anode geometry, electron beam spatial distribution, and exposure time for a fixed point on the anode.

Table 11.1   SBDX Scanning Techniques

Focal Spot Positions (Mode)	Frame Rate (Hz)	Passes (Dwells per Hole)	Beam-On Time/ Frame (ms)	Effective Pulse Width (ms)
71 × 71	30	4	21.0	4.7
(7″ mode)	15	8	41.9	9.4
	10	12	62.9	14.1
100 × 100	30	2	20.8	3.0
(10″ mode)	15	4	41.6	6.1

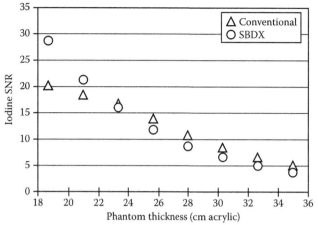

FIGURE 11.11   Measured iodine (35 mg/cm²) SNR versus phantom thickness for a conventional flat panel system (16 μR/frame detector exposure) and SBDX. Flat panel system under AEC control. SBDX operated at 120 kVp, 24 kWp, and 15 frames/s.

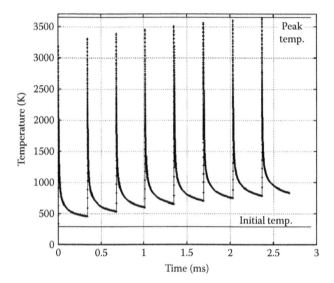

FIGURE 11.10   Target surface temperature model for an 8-pass scan (71 × 71 mode, 15 frames/s, 120 keV electrons, 36 kWp).

A multiple-pass scanning strategy in which each collimator hole is visited several times per image frame is used to boost tube output. Figure 11.10 shows the temperature of the target behind a single hole for the 71 × 71 hole, 15 fps scanning mode in which each hole receives eight passes of the electron beam. Target temperature returns almost to baseline between passes, increasing the total fluence derived from the hole position for much the same reason that rotating anodes allow increased x-ray output for conventional sources. This strategy supports a peak electron beam current density several times that of a conventional source

(e.g., 700 mA/mm² at 120 kVp). Thus, x-ray output constraints imposed by beam collimation and focal spot geometry are mitigated by an increase in duty factor and current density.

Figure 11.11 compares the measured SNR of angiographic contrast agent (i.e., iodine) for a conventional flat detector and the current SBDX system. The conventional system was operated at 15 frames/s under automatic exposure control (AEC) with 16 μR/frame incident on the detector. SBDX was operated at 120 kVp, 24 kWp, 15 frames/s for all thicknesses. The conventional system raises kVp as thickness increases yet SNR drops because of decreasing primary contrast, decreasing detector DQE, and increasing scatter fraction. SBDX SNR drops primarily due to patient attenuation alone. Perhaps paradoxically, as the conventional system is forced to operate at higher power and kVp for larger patients it becomes easier for SBDX to match conventional performance. Figure 11.12 shows a porcine study in which the conventional system was operated under AEC control and SBDX was operated at a subject entrance exposure 4× less than the conventional system.

11.3   Applications in Support of Interventional Procedures

11.3.1   X-Ray Beam Equalization

One of the difficulties with fluoroscopic imaging is the large scene dynamic range in some projections. Conventional systems make use of nonlinear contrast

curves and image processing techniques to flatten the displayed dynamic range. X-ray beam equalization has been investigated as an alternative method (Molloi et al. 2001). A related technique, region-of-interest fluoroscopy, is advanced as a method to reduce x-ray dose

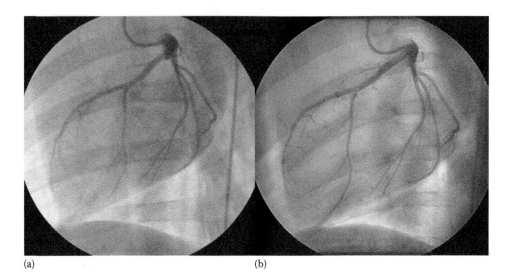

(a) (b)

FIGURE 11.12 Porcine coronary angiograms on a flat panel system (a) and SBDX (b). Images are phase-matched for comparison. Entrance exposure for SBDX was 23% of the conventional system.

(Rudin et al. 1999). Because the contribution of each collimator hole to an image frame is accumulated over multiple visits of the electron beam, the implementation of regionally adaptive x-ray exposure control is relatively straightforward on the SBDX system. In Figure 11.13a each hole is visited eight times per frame, producing the typical homogeneous patient entrance beam. In Figure 11.13b, the detector signal from a single beam pass has been used to produce a map of the number of hole illuminations (from 1 to 8) necessary to produce a uniform patient exit fluence. This map is used to either enable or blank the electron beam on subsequent visits to each hole. The result is a desirable decrease in scene dynamic range (without image processing) and a further reduction in patient dose. The scene-specific reduction in dose-area-product was roughly a factor of two in preliminary studies (Burion et al. 2013).

11.3.2 Three-Dimensional Analysis of Device Location and Vessel Size

The real-time production of an SBDX image in which all of the objects across the desired depth of field are

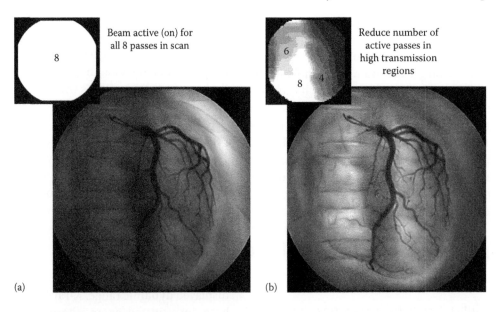

(a) (b)

FIGURE 11.13 SBDX images of anthropomorphic phantom without (a) and with (b) regionally adaptive exposure equalization. Insets show number of scans at each source point.

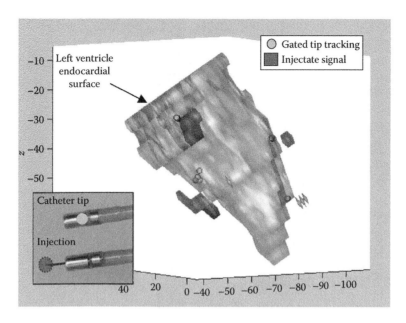

FIGURE 11.14 In vivo SBDX 3D tracking of an injection catheter (light circles) in the left ventricle, compared to MRI-derived positions of the Ferex injectate (dark regions).

in focus is challenging but necessary for real-time fluoroscopic guidance. However, meeting that challenge provides three-dimensional information from an ostensibly 2D imaging system. It follows that if the SBDX reconstruction hardware correctly identifies the plane on which an object resides, that information can be reported out as the z-coordinate of the object. 3D object location is potentially valuable in a host of interventional procedures, such as radiofrequency (RF) catheter ablation procedures or fluoroscopically guided injections during stem cell therapy. Using a reconstructed plane separation of 12 mm, the z-coordinate of the metallic electrodes of radiofrequency ablation catheters have been localized and tracked in 3D with approximately 1 mm accuracy (Speidel et al. 2010).

Figure 11.14 shows the results of an in vivo tracking study in which SBDX fluoroscopic guidance was used to inject iron oxide particles (Ferex) at four locations on the endocardial surface of the left ventricle of a pig. Subsequent to the injections cardiac gated MRI imaging of the pig was performed. SBDX-to-MRI coordinate registration was performed using external multimodality fiducials. Figure 11.14 compares 3D tracking of the catheter position to the locations of the injected Ferex signal. The SBDX tip positions were located near the endocardial surface and at the expected sites of the Ferex signal. The expected difference between catheter coordinate and Ferex was 7 mm, equal to ½ of the 6 mm tip length plus 4 mm

needle extension into the myocardium. The mean distances for the four sites were 7.0, 7.5, 9.8, and 10.7 mm, with an average frame-to-frame deviation of ±1.0 mm.

In addition to device tracking, the z-plane location returned from the tomosynthetic multiplane compositing algorithm can be used in support of quantitative analysis of vessels and lesions. The primary application is vessel sizing for device selection (e.g., balloons and stents). As outlined in Figure 11.15, the algorithm determines the 3D position of points along the vessel centerline and borders. The value of the z-coordinate yields the magnification of the vessel, allowing the

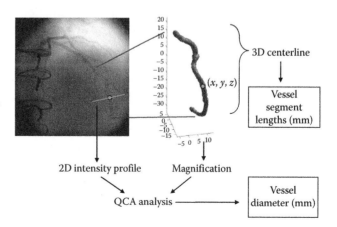

FIGURE 11.15 Workflow of a tomosynthetically based vessel-sizing algorithm. Tomosynthetic multiplane compositing was used to produce the 3D surface–rendered presentation to aid in visualization.

determination of vessel diameter in absolute units (e.g., millimeters) using standard quantitative coronary angiography (QCA) methods, but without the need for a calibration object such as a reference catheter. Furthermore, the 3D centerline tracing enables the measurement of vessel centerline length (e.g., lesion length) even when the vessel appears foreshortened in the 2D image. In phantom measurements, errors in vessel diameter averaged less than 0.15 mm for all magnification studies, while errors in vessel length averaged less than 1% over a range of foreshortening angles 0°–75° (Tomkowiak et al. 2011). In vivo studies are currently underway.

11.3.3 Monoplane Stereoscopic Imaging

While 3D position information is potentially valuable, if it cannot be provided to the interventionalist during the procedure, it is of limited use. For transcatheter interventions taking place inside cardiac chambers, such as left atrial RF ablation or left ventricular myocardial stem cell injections, devices have freedom to move in three dimensions. Stereoscopic visualization of device depth and orientation during these procedures may provide valuable guidance. The multiple viewing angles provided by the extended SBDX source provide the views necessary to create a stereo display. This is achieved by dividing the detector data into left and right halves, from which separate multiplane composite images are made (Tomkowiak et al. 2013).

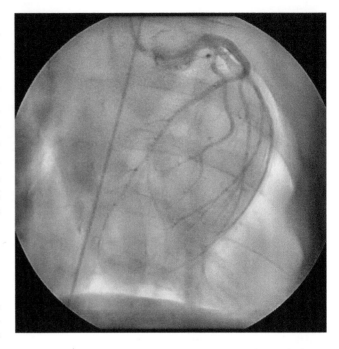

FIGURE 11.16 Red/cyan stereoscopic anaglyph of a single frame from a 15 frame/s SBDX porcine coronary angiogram.

This display is produced without gantry motion and does not introduce a temporal disparity between the images of the stereo pair, which is critical for imaging the beating heart. In principle, an SBDX stereoscopic display can be generated in real time, although the current reconstruction engine was not designed with this in mind. Figure 11.16 shows a frame from a stereo angiographic run.

11.4 Summary

Fluoroscopic x-ray systems provide a combination of high spatial and temporal resolution, real-time visualization, ease of use, and device compatibility that is highly desirable in the cardiac catheterization lab. However, these benefits must be tempered with the fact that fluoroscopically guided interventional procedures have the potential to deliver unhealthy levels of x-ray exposure to both the patient and the interventional staff. Unfortunately, further advancements in interventional techniques may exacerbate the safety problem as increasing complexity leads to increased procedure duration.

Another limitation of conventional fluoroscopic imaging technology is that the 2D projection format does a poor job of portraying the 3D cardiac anatomy and its relationship to the size and position of interventional devices. Scanning systems address this limitation by adding a tomographic component to image formation, but rotational scanning can be cumbersome in the cath lab. The SBDX system allows the interventionalist to visualize and measure the 3D location of anatomy and devices without the need of physical gantry motion. Importantly, this new information can be provided at a significantly lower x-ray dose to both the patient and the medical staff. As we seek to place new tools in the hands of the interventionalist, it is our responsibility to explore safer alternatives to current technology.

Chapter 1

References

Baek J., Pelc N.J., Deman B. et al. 2012. Initial results with a multi-source inverse-geometry CT system. *Proc. SPIE* 8313:1A1-1A7.

Barnes G.T. 1998. Scatter control in imaging recording. In Taveras J.M. and Ferrucci J.T. (Eds.) *Radiology: Diagnosis, Imaging, Intervention*, Vol. 1. Philadelphia, PA: Lippincott-Raven Publishers.

Burion S, Speidel M.A., and Funk T. 2013. A real-time regional adaptive exposure method for saving dose in x-ray fluoroscopy. *Med. Phys.* 40(5):051911.

Boyd D.P. and Lipton M.J. 1983. Cardiac computed tomography. *Proc. IEEE* 71:298–307.

Cao G., Burk L.M., Lee Y.Z. et al. 2010. Prospective-gated cardiac micro-CT imaging of free-breathing mice using carbon nanotube field emission x-ray. *Med. Phys.* 37:5306–5312.

Dobbins J.T. III and Godfrey D.J. 2003. Digital x-ray tomosynthesis: Current state of the art and clinical potential. *Phys. Med. Biol.* 48:R65–R106.

Frutschy K., Neculaes B., Inzinna L. et al. 2010. High power distributed x-ray source. *Proc. SPIE* 7622:1H1-1H11.

Molloi S., Van Drie A., and Wang F. 2001. X-ray beam equalization: Feasibility and performance of an automated prototype system in a phantom and swine. *Radiology* 221:668–675.

Robb R.A., Hoffman E.A., Sinak L.J., Harris L.D., and Ritman E.L. 1983. High-speed three-dimensional x-ray computed tomography: The dynamic spatial reconstructor. *Proc. IEEE* 71:308–328.

Rudin S. and Bednarek D.R. November 1980. Improving contrast in special procedures using a rotating aperture wheel (RAW) device. *Radiology* 137(2):505–510.

Rudin S., Bednarek D.R., and Wong R. May/June 1982. Design of rotating aperture cones for radiographic scatter reduction. *Med. Phys.* 9(3):385–393.

Rudin S., Bednarek D.R., and Yang C.-Y.J. 1999. Real-time equalization of region-of-interest fluoroscopic images using binary masks. *Med. Phys.* 26:1359–1364.

Speidel M.A., Tomkowiak M.T., Raval A.N., and Van Lysel M.S. 2010. Three-dimensional tracking of cardiac catheters using an inverse geometry x-ray fluoroscopy system. *Med. Phys.* 37:6377–6389.

Speidel M.A., Wilfley B.P., Star-Lack J.M., Heanue J.A., and Van Lysel M.S. 2006a. Scanning-beam digital x-ray (SBDX) technology for interventional and diagnostic cardiac angiography. *Med. Phys.* 33:2714–2727.

Speidel M.A., Wilfley B.P., Star-Lack J.M., Heanue J.A., Betts T.D., and Van Lysel M.S. 2006b. Comparison of entrance exposure and signal-to-noise ratio between an SBDX prototype and a wide-beam cardiac angiographic system. *Med. Phys.* 33:2728–2743.

Tomkowiak M.T., Speidel M.A., Raval A.N., and Van Lysel M.S. 2011. Calibration-free device sizing using an inverse geometry x-ray system. *Med. Phys.* 38:283–293.

Tomkowiak M.T., Van Lysel M.S., and Speidel M.A. 2013. Monoplane stereoscopic imaging methods for inverse geometry x-ray fluoroscopy. *Proc. SPIE* 8669:2W1-2W10.

Xu X., Kim J., Laganis P., Schulze D., Liang Y., and Zhang T. 2011. A tetrahedron beam computed tomography benchtop system with a multiple pixel field emission x-ray tube. *Med. Phys.* 38:5500–5508.

Zhao W., Ristic G., and Rowlands J.A. 2004. X-ray imaging performance of structured cesium iodide scintillators. *Med. Phys.* 31:2594–2605.

Chapter 11

12. Dual-Energy and Multienergy Techniques in Vascular Imaging

Adam S. Wang, Scott S. Hsieh, and Norbert J. Pelc

The underlying physical parameter reflected in a computed tomography (CT) image is the x-ray attenuation coefficient, which depends on the energy of the x-rays used. Soon after the advent of CT, the basic principles of this dependence were understood, and it was found that x-ray attenuation reflected two physical quantities—the effective atomic number and the electron density—and that measurement with at least two distinct energies (hence the name dual energy) could be used to estimate these physical quantities and to distinctly identify materials in cross-sectional images.[1,2] These theoretical concepts of dual-energy CT (DECT) were quickly tested on phantoms and in tissue,[3,4] but early experiments were plagued by several practical difficulties and initial implementations of DECT were poor, suffering from drawbacks that prevented long-term clinical adoption.[5,6] Nonetheless, the major reconstruction concepts and several potential applications were seen early on, including the measurement of bone mineral density in the vertebrae for the diagnosis of osteoporosis.[7–10] With the recent availability of modern commercial implementations of dual-energy CT, many new applications are being adopted or actively investigated for clinical practice. While we will focus on vascular imaging in this chapter, other applications of dual energy exist, ranging from the classification of kidney stones to the diagnosis of gout to performing lung ventilation studies.

This chapter begins with an overview of DECT applications in vascular imaging, followed by an in-depth explanation of the physics and mathematical algorithms used to form images from dual-energy measurements. We also discuss the recent commercial implementations of DECT that have served to popularize the technique in recent years, along with some of their strengths and drawbacks.

12.1 Clinical Applications

Dual-energy imaging has long been used in wide-ranging applications such as chest radiographs, dual-energy x-ray absorptiometry (DEXA), security screening, and nondestructive evaluation. Most clinical applications of DECT can be traced back to its ability to perform material-specific imaging from two well-registered (spatially and temporally) scans acquired at two different x-ray energies. Three of the most common presentations of dual-energy scans are images showing only a certain material (such as iodine contrast), images with

Chapter 12

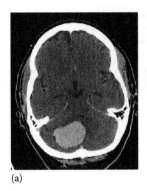 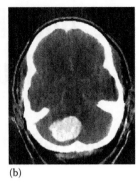 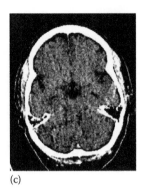

(a)   (b)   (c)

FIGURE 12.1   DECT taken 5 min post contrast injection to rule out active bleeding provides (a) a low-noise, monoenergetic 70 keV image resembling conventional CT and allows for decomposition into (b) an iodine map and (c) an iodine-subtracted image (VNC). Lack of blood in the VNC image (the tumor appears isodense rather than a hyperdense acute hemorrhage in the VNC and only enhances in the iodine image) indicates a vascular tumor and confirms no active bleed. (Images Courtesy of Lior Molvin, Stanford Hospital & Clinics, Stanford, CA.)

a certain material virtually removed, and equivalent monoenergetic images. Equivalent monoenergetic images are designed to mimic CT with images made with photons of only a single energy. In the same way that the x-ray tube voltage can be tuned to an optimal kVp level to maximize dose efficiency, the energy of the equivalent monoenergetic image (in keV) can be tuned to optimize contrast and noise characteristics. One key advantage of DECT is that the energy used for the equivalent monoenergetic image can be freely adjusted after the scan.

In vascular imaging, many of the primary applications of DECT relate to its ability to produce iodine-specific images, leveraging the prevalent use of iodine contrast in such studies or its ability to remove the effects of bone. Conventional contrast studies typically require a non-contrast scan and post-contrast arterial and/or venous phase images, but a single DECT scan following contrast administration can reduce the need for the other scans by producing an iodine-specific image and an iodine-subtracted image known as a virtual non-contrast (VNC) image. The iodine image is an approximation of the difference image between the contrast and non-contrast scans, just as the VNC image offers an approximation to the true non-contrast (TNC) study with the additional benefits of near-perfect image registration, operational efficiency, and potential for reducing radiation dose of the entire study. However, it is important to note that the iodine and VNC images will generally have higher levels of noise, and bony structures will appear in both images, though at lower contrast in the VNC than in a TNC since they will be partially removed along with the iodine. Rather than decomposing an object into its iodine and non-iodine components, the dual-energy

scans can also be combined to form a low-noise image such as a virtual monochromatic image that resembles a conventional CT image with minimal beam hardening artifacts. Examples of these images are shown in Figure 12.1, where the DECT study was used to rule out suspicion of active bleeding. Other applications of iodine-specific imaging include the monitoring of stents for endoleaks following endovascular repair of aortic aneurysms,[11–13] diagnosing coronary artery stenosis, and evaluating myocardial ischemia.[14,15] Modern DECT scanners have enough volumetric coverage and temporal resolution to freeze heart motion and obtain high-resolution anatomical information as well as the needed spectral information to visualize blockage of iodine contrast by stenoses. Iodine-specific images also prove beneficial for perfusion studies and, unlike single-energy scans, do not rely on CT numbers alone and can correct for beam hardening effects.

Iodine can also be differentiated from calcium in bone or calcified plaque using DECT, allowing for the evaluation of arterial stenosis, cerebral aneurysms, and other vascular diseases. For example, the vasculature near bony structures, especially at the base of the skull and near the cervical vertebrae, is often difficult to visualize, since segmentation and the removal of bony structures based on a single-energy CT scan is prone to failure, especially in the regions near bone or calcified plaque. Subtraction of pre- and post-contrast scans can remove bone, leaving only the iodinated vessels, but any motion, even swallowing, leads to imperfect subtraction. On the other hand, DECT typically employs well-registered low- and high-energy scans to differentiate between iodine and bone, thereby providing bone-removed images that potentially improve the assessment of the vasculature for problems such as vessel lumen

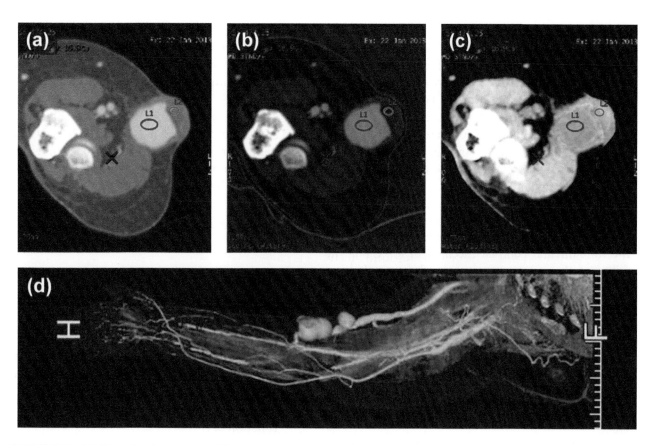

FIGURE 12.2   DECT used to diagnose a malfunctioning arteriovenous fistula. (a) Monoenergetic image, with the energy (40 keV) optimized to maximize the contrast-to-noise ratio (CNR) between the iodinated vessel (L1) and soft tissue (L2); (b) iodine image; (c) virtual non-contrast; (d) 3D rendering of patient arm, with the fistula clearly visualized. (Images Courtesy of Lior Molvin, Stanford Hospital & Clinics, Stanford, CA.)

narrowing or aneurysms. Additionally, some studies have indicated that DECT protocols may achieve this at lower dose than conventional bone-subtraction CT angiography (CTA).[16,17] Similar material-specific techniques can be applied to peripheral artery imaging, such as for the assessment of arteriovenous fistulas (Figure 12.2) or stenoses in the femoral arteries, where automated DE-CTA can be used to remove bone and calcifications to enable clear visualization of the lumen.[18-20]

DECT can also be used for imaging materials other than iodine. For example, when quantifying calcified plaque, DECT has been shown to improve accuracy in determining calcification size and reduce blooming and beam hardening artifacts.[21] Because atherosclerotic plaque can result in different clinical outcomes depending not only on the narrowing of the lumen but

also on the composition of the plaque, DECT has been investigated for its ability to discriminate between iodinated lumen and calcified, fibrous, and fatty arterial plaque, with the latter two proving the most difficult to differentiate.[22] The large differences in effective atomic number Z_{eff} for materials such as iodine, calcium, and soft tissue enable DECT to identify them from another with relative ease. Conversely, the small separation in Z_{eff} between fibrous and fatty tissue is challenging for DECT and discrimination between such materials relies more on mass density, much like single-energy CT. Therefore, for this task, it may be beneficial to use monoenergetic images to form a single anatomic image with high contrast-to-noise ratio (CNR) rather than a material-specific image for plaque characterization.[23]

12.2   Principles

The underlying principles of DECT are best understood by first examining the physical mechanisms of x-ray contrast. An important fact is that there are (in

most circumstances) only two mechanisms for x-ray contrast. DECT can be thought of as measuring these two fundamental contrast mechanisms in the image.

Chapter 12

These contrast measurements can equivalently be interpreted as measurements of *basis materials*. The nature of these basis materials lies at the heart of DECT, and they are often chosen so that they have physical and clinical significance. For example, when iodine is chosen as a basis material, the resulting image showing the distribution of iodinated contrast media has significant utility in an angiographicor perfusion study, and the image of the other basis material will be free of iodine signal.

12.2.1 Physics

The fundamental contrast mechanism for CT is the ability of tissue to scatter or absorb x-ray photons. The behavior of this attenuation is governed by the Beer–Lambert law. For an x-ray photon with energy E, the total amount of attenuation per unit length of matter is characterized by the linear attenuation coefficient $\mu(E)$, which leads to the relationship

$$I(E) = I_0(E)e^{-\mu(E)t}, \tag{12.1}$$

where
  $I(E)$ is the number of transmitted photons
  $I_0(E)$ is the number of incident photons on a homogeneous object of linear attenuation $\mu(E)$
  t is the thickness

The x-rays employed in medical imaging generally have energies that range from 20 to 150 keV. As was discussed in Chapter 5, in this diagnostic energy range, the interactions between matter and x-rays predominantly arise from two mechanisms: photoelectric absorption and Compton scattering.[24] In photoelectric absorption, the x-ray photon transfers all its energy to an atom, ionizing an electron. In Compton scattering, the x-ray photon transfers some of its energy to an electron and then propagates at reduced energy in a different direction, with the electron and photon following the laws of energy and momentum conservation. A third mechanism, Rayleigh scattering, will be ignored since it plays a much more minor role.

The two physical mechanisms for attenuation, photoelectric absorption and Compton scattering, depend not only on the x-ray energy but also on the object. The object-dependent parameters are the effective atomic number Z_{eff} and the electron density ρ_e. These two properties are the driving force behind the difference in attenuation between two different materials such as bone and tissue. In fact, the probability of Compton scattering is proportional to the electron density (electrons per unit volume, which, with the exception of hydrogen, is roughly proportional to the mass density) and photoelectric absorption increases with higher Z_{eff} and mass density ρ. Therefore, calcium ($Z = 20$) will be more attenuating than aluminum ($Z = 13$) at the same mass density. In particular, the attenuation $\mu(E)$ is the sum of the photoelectric $\mu_P(E)$ and Compton $\mu_C(E)$ components, which can be approximated by

$$\mu_P(E) = \rho_e C_P \frac{Z_{\text{eff}}^m}{E^n} \tag{12.2}$$

$$\mu_C(E) = \rho_e f_{KN}(E), \tag{12.3}$$

where
  C_P is a constant
  Exponents m and n have been empirically determined as $m = 3.8$, $n = 3.2$ and commonly approximated as $m, n = 3$,[24-26]
  The Klein–Nishina function $f_{KN}(E)$ is a decreasing function of E

The photoelectric absorption, which is dominant at low energies and for higher Z_{eff} materials, decreases quickly with increasing energy while the Compton attenuation falls off much slower and is more important at higher energies and for lower Z_{eff} materials (Figure 12.3a). The photoelectric absorption (Equation 12.2) expression does not account for K-edges, which will be discussed later.

Let us call $E^{-3.2}$ and $f_{KN}(E)$ the photoelectric and Compton basis functions, respectively. It is clear that any material's attenuation as a function of energy is a linear combination of these basis functions, where the weight of each basis function depends on the electron density ρ_e and effective atomic number Z_{eff}. Given the preceding equations, it should be clear that if an object's attenuation can be measured with at least two different energies, then the object-specific parameters ρ_e and Z_{eff} can be determined by *decomposing* the material's attenuation into the amount of each physical basis function. For example, if the object is measured with two different monoenergetic beams, the preceding equations will give us a system of equations from which ρ_e and Z_{eff} can be easily found. These two parameters can be very useful in identifying an object's composition, as first reported in the seminal paper by Alvarez and Macovski.[1]

In addition to representing a material as a linear combination of the physical basis functions, any material can be represented as a linear combination of two other (different) materials since these other materials are themselves a linear combination of the physical basis functions. For example, the attenuation of water and calcium can be written as a unique linear combination of the photoelectric and Compton basis functions. Aluminum could also be written as a linear combination of the photoelectric and Compton basis functions, but it could be equally well described as being a linear combination of water and calcium simply by solving a system of linear equations. In fact, in terms of the mass attenuation (linear attenuation divided by the mass density) we have that

$$\left(\frac{\mu}{\rho}\right)_{Al}(E) = a_1 \cdot \left(\frac{\mu}{\rho}\right)_{H_2O}(E) + a_2 \cdot \left(\frac{\mu}{\rho}\right)_{Ca}(E), \qquad (12.4)$$

where
$$a_1 = 0.678$$
$$a_2 = 0.211$$

In other words, an area density of 1 g/cm^2 of Al is indiscernible from an area density mixture of 0.678 g/cm^2 water and 0.211 g/cm^2 calcium by x-rays of any energy. This illustrates that the attenuation of an unknown material can be decomposed into two basis materials (such as water and calcium). Although nearly any two basis materials will work, it is often useful to choose basis materials that have physical relevance. For example, when imaging the human body, water and calcium may be a good choice—soft tissue will be decomposed

mostly into water since its Z_{eff} is close to water, while bone will have a substantial calcium component. One limitation of dual energy is that some ambiguity still remains. The aluminum previously described remains indiscernible from the mixture of water and calcium that it can be decomposed into. A hypothetical extension to "triple energy" imaging would not be helpful in this regard since the attenuation of all these materials are fundamentally well described by only two basis functions. Therefore, basis materials should be carefully chosen so that they are suitable for the imaging task.

The vast majority of the human body is composed of atomic elements of $Z \le 30$. However, a contrast agent such as iodine may be introduced into the vasculature to increase the local attenuation wherever it travels to, in large part due to the high atomic number of iodine ($Z = 53$). Besides simply increasing the effective atomic number Z_{eff}, however, the attenuation function of iodine also features a large and sudden increase at 33.2 keV, known as the K-edge (Figure 12.3b). The K-edge energy is the binding energy of the K-shell (innermost) electron, and only photons exceeding this energy can eject these K-shell electrons, as indicated by the discontinuity in attenuation at the K-edge. Iodine will much more strongly attenuate photons with energies just exceeding 33.2 keV than photons just under 33.2 keV. The K-edge energy increases with atomic number Z—for instance, the K-edge of gadolinium ($Z = 64$) is 50.2 keV and gold ($Z = 79$) is 80.7 keV. For elements with $Z \le 30$, the K-edge energy lies well below the diagnostic energy range and can therefore be neglected. Clearly, the K-edge introduces a new basis function since the attenuation curve cannot be represented by

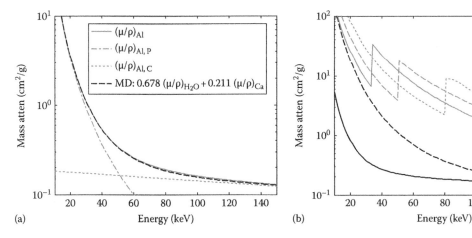

FIGURE 12.3 (a) The mass attenuations of several materials are plotted as a function of energy. (b) The contributions from photoelectric absorption and Compton scatter are shown for aluminum, as well as the material decomposition into water and calcium that overlays the aluminum curve. Note the prominent K-edge at 33.2 keV for iodine, as well as the K-edges for gadolinium and gold, which increase in energy with increasing atomic number.

the continuous Compton and photoelectric basis functions alone. However, for vascular imaging, the vast majority of the transmitted x-rays have energy greater than the K-edge of iodine since their average energy is well above 33 keV and photons with energy less than 33 keV do not penetrate the body or head easily. In this case, for energies above the K-edge, the attenuation of iodine *can* be modeled by the Compton and photoelectric basis functions. Therefore, while vascular imaging does not generally leverage iodine's K-edge, objects that are sufficiently thin, such as in preclinical or breast imaging, can utilize an incident x-ray spectrum containing a significant number of photons with energy less than 33 keV to uniquely identify iodine.

12.2.2 Implementation

As the name "dual energy" suggests, it is necessary to image with x-rays of two different effective energies. Viable implementations of dual-energy CT must satisfy two requirements. First, the effective energies of the spectra should have a sufficiently large separation so that the difference in attenuation between the energies is significant. When the energy separation is small, the noise of the material decomposition becomes large and the images become unusable. Second, the time between the two acquisitions at different energies must be short. For stationary objects, dual-energy imaging could be implemented simply by taking two sequential scans at different energies, but in clinical applications the errors resulting from motion or contrast uptake/washout could be problematic. The recent, widespread adoption of dual energy into clinical CT scanners is largely due to implementations of DECT that satisfy both these requirements, as well as the other capabilities of modern CT scanners, such as high temporal and spatial resolution and volume coverage. There are four primary system strategies used for dual-energy imaging: dual source, tube voltage switching, dual-layer detectors, and energy-resolving photon counting detectors. We will describe the first three here and discuss photon counting later.

Both dual source and tube voltage switching utilize measurements with two different peak x-ray tube voltages (kVp) and are both therefore considered dual kVp approaches. The low and high kVp beams have different effective energies, where the need for photons with intensity and energy high enough for the object to be penetrated places a lower limit on kVp, and tube limits provide an upper bound to the tube voltage. A dual-source system utilizes two source-detector pairs in the gantry (Figure 12.4a). One source is operated at a lower voltage (e.g., 80 kVp), while the other source is operated at a higher voltage (e.g., 140 kVp). Because the source-detector pairs are approximately in orthogonal directions, they can be operated simultaneously. After one rotation, a complete CT data set is acquired at both the low and high energies. Because both tubes are operated independently, it is easy to apply different filtrations, and a beam hardening filter such as tin on high-energy x-ray source can greatly increase the energy separation between the two spectra.[27,28] However, there are drawbacks to the system design including the increase in hardware cost and complexity due to the extra source-detector pair. Motion during the acquisition can result in collected data that is not perfectly registered between the low- and high-energy scans. Limited gantry space may require that the second source-detector pair have a smaller effective field of view than the first source-detector pair. Finally, cross-scatter from one source to the other detector can increase the overall scatter when the sources are operated simultaneously since they are not typically pulsed due to the increased complexity of high-speed pulsing.[29]

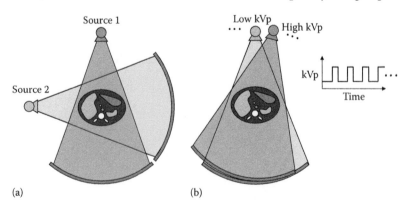

FIGURE 12.4 Dual kVp geometries. (a) Two source-detector pairs operate simultaneously as they rotate around the patient. In DECT mode, one source is operated at a lower kVp than the other (e.g., 80 and 140 kVp). (b) The tube voltage rapidly oscillates between the low and high kVp as the source rotates about the patient, and these interleaved projections are captured separately by the detector.

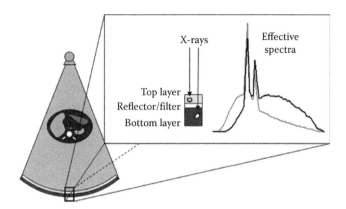

FIGURE 12.5 Dual-layer geometry. The x-ray tube is operated at a single kVp, and energy-specific information is captured by the layered detector. The effective spectrum for each layer demonstrates that the top layer is more sensitive to lower energies, while the bottom layer captures more of the higher energies.

Another approach to dual kVp is to use a single source that switches between tube voltages (Figure 12.4b). While it is possible to scan once at low energy followed by another scan at high energy (or vice-versa), this sequential scan approach may suffer from patient motion or different contrast distributions between the scans. Therefore, rapid kVp switching has also been implemented, whereby the source alternates very rapidly between low and high kVp as the source-detector pair rotates about the object.[6,30] The difference in time between successive low- and high-energy projections is very small, providing well-registered and motion-free projection pairs. The switching frequency is high enough that a sufficient number of projections equally spaced in angle are acquired with each spectrum. There are a number of technical challenges with this approach, most notably the kHz-rate changing of x-ray tube voltage from 80 to 140 kVp that is needed for sub-second gantry rotation. Ideally, the voltage would change instantaneously, but in practice x-ray sources have limited voltage slew rate that may reduce the spectral separation.[31] Changing the tube current in synchrony with the rapid kVp switching is also very difficult, so differences in mAs between the low- and high-energy projections can instead be accomplished with control of the dwell time at each voltage. Finally, although in principle a synchronized switching filter can be used to provide different filtration to each spectrum, mechanical constraints may preclude this.[32] It is important to note that in each of these dual kVp techniques, the dose is not necessarily increased; instead it is commonly split between the two energies and can still provide the additional spectral information for comparable dose to a single kVp acquisition.[33,34]

The other class of systems used for dual-energy systems are energy-sensitive detectors, including dual

layer and photon counting detectors, which generally utilize a single kVp spectrum. A dual-layer, or sandwich, detector is comprised of a thin top scintillator and a thicker bottom scintillator (Figure 12.5). To provide spectral separation, the top scintillator is chosen to be more sensitive to lower energies, while high energy photons are more likely to penetrate the top layer and be absorbed in the bottom layer. The spectral separation can be further increased by placing an absorbing filter between the two layers, with the filter preferentially absorbing lower energy photons. While this decreases the quantum efficiency in single-energy CT tasks, the increase in spectral separation can (perhaps surprisingly) increase dose efficiency for spectral imaging tasks.[35,36] The effective spectrum for each layer is also shown in Figure 12.5. Dual-layer detectors have the advantage of perfect spatial registration between the low- and high-energy projections, but generally suffer from poor energy separation, even with the selection of optimal materials and thicknesses for the top layer, bottom layer, and filter in between.[37]

12.2.3 Estimation

The simplest view of the material decomposition estimation problem is to solve for two unknown parameters (the basis material lengths) from two measurements (the dual-energy data). There are two domains in which the estimation can be done: image space or projection space. In the image-space approach, the dual-energy scans are first reconstructed independently to form low- and high-energy images. These images are then used to estimate material-specific images. In projection space, each line integral is decomposed into a line integral through each basis material. The resulting material sinograms can then be reconstructed to form the basis material images.

Image-space decomposition is especially suited for unregistered projection data or if there is motion between the two images. The dual-energy data could be acquired from two entirely different trajectories, but if the same location of the object is reconstructed, the images can be paired together. Consider a cylindrical acrylic (PMMA) phantom filled with solutions of different iodine concentrations scanned at 80 kVp and separately at 140 kVp (Figure 12.6). Due to the different effective energies of the spectra, the iodine contrast is markedly different. A weighted subtraction of the images can recover the iodine image or (with different weights) the water image by solving the set of linear equations for density images $\rho_{H_2O}(\mathbf{x})$, $\rho_I(\mathbf{x})$ from the measured attenuation at the low and high energies:

$$\mu(\mathbf{x}; E_{low}) = \rho_{H_2O}(\mathbf{x}) \cdot \left(\frac{\mu}{\rho}\right)_{H_2O} (E_{low}) + \rho_I(\mathbf{x}) \cdot \left(\frac{\mu}{\rho}\right)_I (E_{low})$$

$$\mu(\mathbf{x}; E_{high}) = \rho_{H_2O}(\mathbf{x}) \cdot \left(\frac{\mu}{\rho}\right)_{H_2O} (E_{high}) + \rho_I(\mathbf{x}) \cdot \left(\frac{\mu}{\rho}\right)_I (E_{high})$$

(12.5)

However, this is the case only if the two images are equivalent to monoenergetic images; otherwise, the material decomposition will be imperfect due to beam hardening. Nonetheless, linear (or even nonlinear) combinations of the images, variously known as mixed, weighted average, or blended images, can be formed in such a way that the contrast-to-noise ratio (CNR) is maximized for a particular imaging task,[33,38] or so that a material (like iodine) is fully canceled. Moreover, extension to three-material decomposition has been demonstrated with methods that partition the 2D space of low- and high-energy attenuation using classification techniques to identify various materials such as soft tissue, bone, and iodine,[39] or by adding another constraint such as mass conservation.[40] While these techniques often perform well in practice, there can be scenarios that lead to ambiguous three-material decompositions.

Alternatively, material decomposition can be done in projection space by decomposing each pair of ray measurements into line integrals of the basis materials. The resulting basis material sinograms can then be reconstructed to form basis material images (Figure 12.7). Projection-based methods can, in principle, be

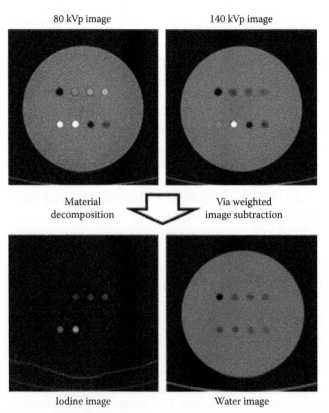

80 kVp image 140 kVp image

Material decomposition Via weighted image subtraction

Iodine image Water image

FIGURE 12.6 Image-space decomposition. The 80 and 140 kVp images of this phantom are used to form an iodine and water image. The iodine is clearly separated from the rest of the object and quantitatively matches the iodine concentration of the solutions. Note that the patient table appears in both images since it is neither water nor iodine.

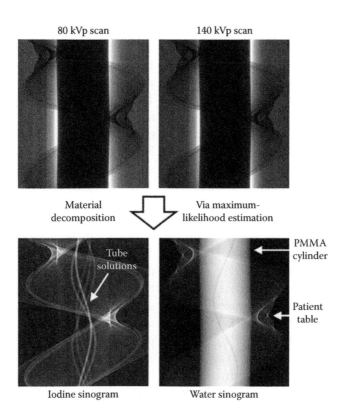

80 kVp scan   140 kVp scan

Material   Via maximum-
decomposition   likelihood estimation

Tube   PMMA
solutions   cylinder

                              Patient
                              table

Iodine sinogram   Water sinogram

FIGURE 12.7 Projection-space decomposition. The raw data, shown as a sinogram, is used to estimate the line integrals of iodine and water. These material decomposition sinograms can then be reconstructed separately.

made immune to beam hardening artifacts. In conventional, single-energy CT, beam hardening artifacts are corrected by assuming that the object being measured is water like. Objects that are not water like, such as bone, create residual beam hardening artifacts. However, in DECT, both water and bone basis material lengths can be estimated, which allows for a more accurate beam hardening artifact correction. However, this technique generally requires the data to be well registered or else the errors will be quite large—different low- and high-energy trajectories make projection space decomposition impossible, as do motion or data inconsistencies. Algorithms for estimating the decomposition include maximum-likelihood estimation (MLE), linearized (weighted) least squares, or a polynomial fit to calibration data. MLE solves for the decomposition $t = (t_1, t_2)$ that most likely produced the low- and high-energy measurements $d = (d_l, d_h)$. The estimated decomposition $\hat{t}$ is mathematically described as

$$\hat{t} = \arg\max_{t} f(d \mid t), \qquad (12.6)$$

where $f(d \mid t)$ is the probability density function of measurement d from an object with decomposition line integral t and incorporates a polyenergetic model of the incident spectrum, x-ray attenuation, and detector response. Because MLE models the physical and statistical processes of how the data were acquired, it is also well suited for the overdetermined case when more than two energy-sensitive measurements are acquired, as is possible with photon counting detectors discussed in the next section.[41-43] Likewise, a linearized weighted least squares approach is a computationally faster approximation of MLE that also handles the overdetermined case well by weighting each measurement depending on its statistics.[44] Finally, the polynomial fit method is well suited as an empirical technique, using calibration data to fit a polynomial function that maps measured data to the material thickness, but may not optimally handle the statistics of the measurements.[45]

In both image- and projection-space methods, noise in the dual-energy measurements leads to noise in the material decomposition. For measurements with uncorrelated noise (commonly the case), the decomposition noise is negatively correlated. When performing material decomposition from dual-energy measurements, the noise is approximately the same with all decomposition algorithms. The material decompositions can be combined to meet task-specific objectives, such as the task of forming the highest iodine CNR image from a weighted sum of the material decomposition images. In particular, when the weights are the mass attenuation of each basis material at energy E', a monoenergetic image is produced that represents the object as if it were scanned with a monoenergetic beam of energy E' and is similar in appearance to the weighted average images previously mentioned. Generally, monoenergetic images in the range of 60–75 keV have the lowest noise as they combine the material decompositions such that the negatively correlated noise tends to cancel, while monoenergetic images at lower energy have higher contrast, suggesting that an optimal energy can be found to maximize CNR. The material-specific images are themselves useful— for example, the iodine image can be viewed by itself or as a color overlay on a grayscale anatomic image. Furthermore, for a water/iodine decomposition, the water image is inherently free of iodine and therefore provides a virtual non-contrast image.

12.3   Future Directions

Many of the clinical applications in the first section are still being actively investigated, and DECT technology is continuing to evolve. Methods for more dose efficient dual-energy techniques, estimation, and noise reduction are also active areas of investigation. DECT is a rich area for research, and in particular we will briefly touch on some promising technical developments.

In addition to the dual-energy implementations of existing clinical systems, multienergy imaging using energy-resolving photon counting detectors forms another approach that is a very active area of research. These detectors are designed to count individual photons as they arrive at the detector and additionally discern the energy of each photon so that spectral information about the transmitted x-rays can be obtained from a single polyenergetic exposure. While significant technical challenges remain, especially in the limited speeds of such detectors, a number of key advantages can be exploited. Ideal photon counting detectors are known to be more dose efficient than energy integrating detectors because of the additional energy information made available,[46,47] which is usually recorded as the number of counts within different energy windows or bins.

A minimum of two energy bins are necessary to obtain dual-energy information, but if three or more energy bins can be used, the traditional two-basis material set can be extended to include K-edge contrast agents such as iodine, gadolinium, or gold. This exciting use of photon counting detectors to uniquely identify contrast agents has been recently demonstrated preclinically and has the potential to better determine atherosclerotic plaque composition.[48,49]

Lastly, one novel application of dual-energy imaging, known as synthetic CT, could be used for CT protocol optimization, such as determining lower-dose protocols for studies with relatively high levels of radiation dose. With synthetic CT, users can retrospectively synthesize clinical images from arbitrary low-dose CT protocols, including changes to tube voltage, current, and filtration.[50,51] The synthesized images appear as if they were acquired with the alternate protocol and accurately reflect the contrast and noise of this protocol. With such a tool, it is possible to study the effect of protocol selection on image quality and dose distribution,[52] enabling CT users to perform task-dependent protocol optimization that lead to lower-dose techniques.

12.4   Conclusions

For materials normally found in vivo and within the diagnostic x-ray spectrum, the fundamental x-ray physics suggests that there are two primary mechanisms for x-ray attenuation, allowing for measurements at two energies to estimate object-specific properties related to its composition. This material-specific imaging is quantitative and leads to increased diagnostic confidence based on the distribution of a material of interest, such as iodine contrast. Dual-energy imaging is an exciting area of growth and research, with a number of active areas of development. It has many benefits in vascular imaging and will continue to see increased use as more CT systems deploy this technology.

Acknowledgments

The authors thank Lior Molvin and Jiang Hsieh for their assistance in acquiring the phantom data and acknowledge the financial support of GE Healthcare and the Lucas Foundation.

References

1. R.E. Alvarez, A. Macovski, Energy-selective reconstructions in x-ray computerised tomography, *Physics in Medicine and Biology* **21**, 733–744 (1976).

2. R. Rutherford, B. Pullan, I. Isherwood, Measurement of effective atomic number and electron density using an EMI scanner, *Neuroradiology* **11**(1), 15–21 (1976).

3. G. Di Chiro, R.A. Brooks, R.M. Kessler, G.S. Johnston, A.E. Jones, J.R. Herdt, W.T. Sheridan, Tissue signatures with dual-energy computed tomography, *Radiology* **131**(2), 521 (1979).

4. M.R. Millner, W.D. McDavid, R.G. Waggener, M.J. Dennis, W.H. Payne, V.J. Sank, Extraction of information from CT scans at different energies, *Medical Physics* **6**, 70 (1979).

5. R. Ritchings, B. Pullan, A technique for simultaneous dual energy scanning, *Journal of Computer Assisted Tomography* **3**(6), 842 (1979).

6. W. Kalender, W. Perman, J. Vetter, E. Klotz, Evaluation of a prototype dual-energy computed tomographic apparatus. I. Phantom studies, *Medical Physics* **13**, 334 (1986).

7. H.K. Genant, D. Boyd et al., Quantitative bone mineral analysis using dual energy computed tomography, *Investigative Radiology* **12**(6), 545 (1977).

8. J. Adams, S. Chen, P. Adams, I. Isherwood, others, Measurement of trabecular bone mineral by dual energy computed tomography, *Journal of Computer Assisted Tomography* **6**(3), 601 (1982).

9. J. Vetter, W. Perman, W. Kalender, R. Mazess, J. Holden, Evaluation of a prototype dual-energy computed tomographic apparatus. II. Determination of vertebral bone mineral content, *Medical Physics* **13**, 340 (1986).

10. W.A. Kalender, E. Klotz, C. Suess, Vertebral bone mineral analysis: An integrated approach with CT, *Radiology* **164**(2), 419–423 (1987).

11. P. Stolzmann, T. Frauenfelder, T. Pfammatter, N. Peter, H. Scheffel, M. Lachat, B. Schmidt, B. Marincek, H. Alkadhi, T. Schertler, Endoleaks after endovascular abdominal aortic aneurysm repair: Detection with dual-energy dual-source CT, *Radiology* **249**(2), 682 (2008).

12. H. Chandarana, M.C.B. Godoy, I. Vlahos, A. Graser, J. Babb, C. Leidecker, M. Macari, Abdominal aorta: Evaluation with dual-source dual-energy multidetector CT after endovascular repair of aneurysms—Initial observations, *Radiology* **249**(2), 692 (2008).

13. W.H. Sommer, A. Graser, C.R. Becker, D.A. Clevert, M.F. Reiser, K. Nikolaou, T.R.C. Johnson, Image quality of virtual noncontrast images derived from dual-energy CT angiography after endovascular aneurysm repair, *Journal of Vascular and Interventional Radiology* **21**(3), 315–321 (2010).

14. B. Ruzsics, H. Lee, P.L. Zwerner, M. Gebregziabher, P. Costello, U.J. Schoepf, Dual-energy CT of the heart for diagnosing coronary artery stenosis and myocardial ischemia-initial experience, *European Radiology* **18**(11), 2414–2424 (2008).

15. S.M. Ko, J.W. Choi, M.G. Song, J.K. Shin, H.K. Chee, H.W. Chung, D.H. Kim, Myocardial perfusion imaging using adenosine-induced stress dual-energy computed tomography of the heart: Comparison with cardiac magnetic resonance imaging and conventional coronary angiography, *European Radiology*, **21**(1), 26–35 (2011).

16. K. Deng, C. Liu, R. Ma, C. Sun, X. Wang, Z. Ma, X. Sun, Clinical evaluation of dual-energy bone removal in CT angiography of the head and neck: Comparison with conventional bone-subtraction CT angiography, *Clinical Radiology* **64**(5), 534–541 (2009).

17. K. Uotani, Y. Watanabe, M. Higashi, T. Nakazawa, A.K. Kono, Y. Hori, T. Fukuda, S. Kanzaki, N. Yamada, T. Itoh, others, Dual-energy CT head bone and hard plaque removal for quantification of calcified carotid stenosis: Utility and comparison with digital subtraction angiography, *European Radiology* **19**(8), 2060–2065 (2009).

18. B. Meyer, T. Werncke, W. Hopfenmüller, H. Raatschen, K.J. Wolf, T. Albrecht, Dual energy CT of peripheral arteries: Effect of automatic bone and plaque removal on image quality and grading of stenoses, *European Journal of Radiology* **68**(3), 414–422 (2008).

19. C. Brockmann, S. Jochum, M. Sadick, K. Huck, P. Ziegler, C. Fink, S.O. Schoenberg, S.J. Diehl, Dual-energy CT angiography in peripheral arterial occlusive disease, *Cardiovascular and Interventional Radiology* **32**(4), 630–637 (2009).

20. D.N. Tran, M. Straka, J.E. Roos, S. Napel, D. Fleischmann, Dual-energy CT discrimination of iodine and calcium: Experimental results and implications for lower extremity CT angiography, *Academic Radiology* **16**(2), 160–171 (2009).

21. D.T. Boll, E.M. Merkle, E.K. Paulson, R.A. Mirza, T.R. Fleiter, Calcified vascular plaque specimens: Assessment with cardiac dual-energy multidetector CT in anthropomorphically moving heart phantom, *Radiology* **249**(1), 119–126 (2008).

22. M. Barreto, P. Schoenhagen, A. Nair, S. Amatangelo, M. Milite, N.A. Obuchowski, M.L. Lieber, S.S. Halliburton, Potential of dual-energy computed tomography to characterize atherosclerotic plaque: Ex vivo assessment of human coronary arteries in comparison to histology, *Journal of Cardiovascular Computed Tomography* **2**(4), 234–242 (2008).

23. F.F. Behrendt, B. Schmidt, C. Plumhans, S. Keil, S.G. Woodruff, D. Ackermann, G. Mühlenbruch, T. Flohr, R.W. Günther, A.H. Mahnken, Image fusion in dual energy computed tomography: Effect on contrast enhancement, signal-to-noise ratio and image quality in computed tomography angiography, *Investigative Radiology* **44**(1), 1–6 (2009).

24. A. Macovski, *Medical Imaging Systems*. (Prentice-Hall, Englewood Cliffs, NJ, 1983).

25. J.T. Bushberg, J.A. Seibert, E.M. Leidholdt, J.M. Boone, *The Essential Physics of Medical Imaging*, 2nd ed. (Lippincott Williams & Wilkins, Philadelphia, PA, 2002).

26. J. Hsieh, *Computed Tomography: Principles, Design, Artifacts, and Recent Advances*. (SPIE Press, Bellingham, WA, 2003).

27. A. Primak, J.C.R. Giraldo, X. Liu, L. Yu, C. McCollough, Improved dual-energy material discrimination for dual-source CT by means of additional spectral filtration" *Medical Physics* **36**(4), 1359–1369 (2009).

28. A.N. Primak, J.C.R. Giraldo, C.D. Eusemann, B. Schmidt, B. Kantor, J.G. Fletcher, C.H. McCollough, Dual-source dual-energy CT With additional tin filtration: Dose and image quality evaluation in phantoms and in vivo, *American Journal of Roentgenology* **195**(5), 1164 (2010).

Chapter 12

29. M. Petersilka, K. Stierstorfer, H. Bruder, T. Flohr, Strategies for scatter correction in dual source CT, *Medical Physics* **37**, 5971 (2010).

30. J.E. Tkaczyk, D. Langan, X. Wu, D. Xu, T. Benson, J.D. Pack, A. Schmitz, A. Hara, W. Palicek, P. Licato, Quantization of liver tissue in dual kVp computed tomography using linear discriminant analysis, *SPIE Medical Imaging*, **72512**, 72580G (2009).

31. D. Xu, D.A. Langan, X. Wu, J.D. Pack, T.M. Benson, J.E. Tkaczky, A.M. Schmitz, Dual energy CT via fast kVp switching spectrum estimation, *Proceedings of SPIE*, **7258** 72583T (2009).

32. M.M. Goodsitt, E.G. Christodoulou, S.C. Larson, Accuracies of the synthesized monochromatic CT numbers and effective atomic numbers obtained with a rapid kVp switching dual energy CT scanner, *Medical Physics* **38**, 2222 (2011).

33. L. Yu, A.N. Primak, X. Liu, C.H. McCollough, Image quality optimization and evaluation of linearly mixed images in dual-source, dual-energy CT, *Medical Physics* **36**, 1019 (2009).

34. J.C. Schenzle, W.H. Sommer, K. Neumaier, G. Michalski, U. Lechel, K. Nikolaou, C.R. Becker, M.F. Reiser, T.R.C. Johnson, Dual energy CT of the chest: How about the dose?, *Investigative Radiology* **45**(6), 347 (2010).

35. G.T. Barnes, R.A. Sones, M.M. Tesic, D.R. Morgan, J.N. Sanders, Detector for dual-energy digital radiography, *Radiology* **156**(2), 537–540 (1985).

36. D.M. Gauntt, G.T. Barnes, X-ray tube potential, filtration, and detector considerations in dual-energy chest radiography, *Medical Physics* **21**, 203 (1994).

37. R.E. Alvarez, J.A. Seibert, S.K. Thompson, Comparison of dual energy detector system performance, *Medical Physics* **31**, 556 (2004).

38. D.R. Holmes III, J.G. Fletcher, A. Apel, J.E. Huprich, H. Siddiki, D.M. Hough, B. Schmidt et al., Evaluation of non-linear blending in dual-energy computed tomography, *European Journal of Radiology* **68**(3), 409–413 (2008).

39. M. Petersilka, H. Bruder, B. Krauss, K. Stierstorfer, T.G. Flohr, Technical principles of dual source CT, *European Journal of Radiology* **68**(3), 362–368 (2008).

40. X. Liu, L. Yu, A.N. Primak, C.H. McCollough, Quantitative imaging of element composition and mass fraction using dual-energy CT: Three-material decomposition, *Medical Physics* **36**, 1602 (2009).

41. J.A. Fessler, I. Elbakri, P. Sukovic, N.H. Clinthorne, Maximum-likelihood dual-energy tomographic image reconstruction, *Proceedings of SPIE: Medical Imaging 2002: Image Processing* **4684**, 38–49 (2002).

42. J. Schlomka, E. Roessl, R. Dorscheid, S. Dill, G. Martens, T. Istel, C. Bäumer et al., Experimental feasibility of multi-energy photon-counting K-edge imaging in pre-clinical computed tomography, *Physics in Medicine and Biology* **53**, 4031–4047 (2008).

43. E. Roessl, C. Herrmann, Cramér-Rao lower bound of basis image noise in multiple-energy x-ray imaging, *Physics in Medicine and Biology* **54**, 1307–1318 (2009).

44. R.E. Alvarez, Estimator for photon counting energy selective x-ray imaging with multibin pulse height analysis, *Medical Physics* **38**, 2324 (2011).

45. H.N. Cardinal, A. Fenster, An accurate method for direct dual-energy calibration and decomposition, *Medical Physics* **17**, 327 (1990).

46. M.J. Tapiovaara, R. Wagner, SNR and DQE analysis of broad spectrum x-ray imaging, *Physics in Medicine and Biology* **30**, 519 (1985).

47. A.S. Wang, N.J. Pelc, Sufficient statistics as a generalization of binning in spectral X-ray imaging, *IEEE Transactions on Medical Imaging* **30**(1), 84–93 (2011).

48. D.P. Cormode, E. Roessl, A. Thran, T. Skajaa, R.E. Gordon, J.P. Schlomka, V. Fuster et al., Atherosclerotic plaque composition: Analysis with multicolor CT and targeted gold nanoparticles, *Radiology* **256**(3), 774 (2010).

49. A.M. Alessio, L.R. MacDonald, Quantitative material characterization from multi-energy photon counting CT, *Medical Physics* **40**, 031108 (2013).

50. A.S. Wang, N.J. Pelc, Synthetic CT: Simulating low dose single and dual energy protocols from a dual energy scan, *Medical Physics* **38**, 5551 (2011).

51. A.S. Wang, N.J. Pelc, Synthetic CT: Simulating arbitrary low dose single and dual energy protocols, *Proceedings of SPIE* **7961**, 79611R (2011).

52. M. Bazalova, J.F. Carrier, L. Beaulieu, F. Verhaegen, Dual-energy CT-based material extraction for tissue segmentation in Monte Carlo dose calculations, *Physics in Medicine and Biology* **53**, 2439–2456 (2008).

13. Special Detectors for Digital Angiography, Micro-Angiography, and Micro ROI CBCT

Stephen Rudin, Amit Jain, and Daniel R. Bednarek

13.1   Introduction

Vascular disease seems to be all pervasive. Cardiovascular disease is the number one killer in developed countries and neurovascular disease can result in stroke, which annually affects 795,000 Americans (AHA 2010). About 140,000 will die and many will be left permanently disabled or neurologically impaired, stroke being the leading cause of adult disability. The recent evolution of treatment for cardiovascular and neurovascular diseases has been away from invasive surgery toward minimally invasive endovascular interventions. Of all the medical imaging modalities, x-ray imaging remains the dominant mode for guiding vascular and other

Chapter 13

interventions and for investigating rapidly moving phenomenon such as vascular flow. No other modality has the combination of high image resolution, capabilities for high speed, and minimal encumbrance on the patient treatment space. Yet rapid sequence x-ray imaging is far from achieving its potential in part due to nonoptimal detectors; however, as the accuracy of treatments have become more demanding with the use of smaller and more complex endovascular devices (Nesbit 2004, Rudin 2008), the need for higher resolution both for diagnosis and interventional guidance has become more crucial. New detector developments of such a higher-resolution, region-of-interest, micro-angiographic fluoroscope or MAF may fulfill this need and thus have a large positive impact on improving the standard of care for the diagnosis and minimally invasive treatment of patients with vascular and specifically for neurovascular disease and potentially also for the rapidly expanding field of trans-aortic valve replacement (TAVR).

The underlying idea behind using ROI detectors is that x-ray photons incident on a patient should be best distributed where they are most essential in forming a diagnosis or guiding an intervention rather than exposing a patient with a uniform field. In the past, this basic idea involved using a material filter to suppress the intensity at the periphery of a central region of interest (ROI)

while the full photon fluence is preserved within the ROI. The resulting image was digitally corrected so that the displayed field was essentially uniform with a noisier peripheral field due to increased quantum mottle. We designated this technique as "ROI fluoroscopy" (Rudin 1992) and colleagues in another group (Labbe 1994) designated it the x-ray "fovea" reminiscent of the way the eye functions. An example of the most modern version of ROI Fluoroscopy was shown in Chapter 4 where Figure 4.4 displayed how differential temporal filtering can improve the image quality in both the ROI and the periphery. Initially, the primary attraction of this technique was for dose reduction with perhaps some increased image quality within the ROI due to scatter reduction; however, the immediate need for ROI techniques became less urgent with the introduction of all digital imaging systems when dose reduction for the whole field-of-view (FOV) became possible using reduced frame rates and image processing techniques. Nevertheless, if the basic idea of using x-rays where they are most needed could be used to radically improve the image within an ROI, then not only could integral or effective dose be reduced as discussed in detail in Chapters 4 and 25, but also the requirements outlined earlier for vascular imaging could also be fulfilled. As discussed in the next section, no such detector with the required capability existed until our group developed the MAF.

13.2　Conventional Dynamic X-Ray Detectors: X-Ray Image Intensifiers and Flat Panel Devices

As endovascular devices such as stents, heart valves, delivery catheters, and clot protection devices become more complex and yet finer, they are more difficult to visualize with standard x-ray detectors. Also to reduce mortality and morbidity during vascular procedures and improve their outcomes, it is increasingly important to visualize the detailed movement of heart valves during deployment and the detailed structure of vessels in the brain as well as small perforator vessels that may not have redundancy in feeding regions of brain tissue and that when damaged can have severe consequences. These perforators are most often not even visualized at all by current standard dynamic x-ray detectors, that is, x-ray image intensifiers (XIIs) and

flat panel devices (FPD) that have typically 200 µm pixels and generally are limited to visualizing two to three line pairs per millimeter. Additionally, FPDs have degraded performance because of lag and ghosting (Siewerdsen 1999) and during fluoroscopy because of their excessive readout noise while XIIs are limited by geometric and magnetic-induced distortion and physical bulkiness as discussed in Sections 4.3.4.1 and 4.3.4.2. As will be described in what follows, to overcome these weaknesses and maintain dynamic imaging capabilities, detectors such as the MAF provide the higher resolution needed at least over a ROI in keeping with the basic idea of using the x-ray exposure only where it is most needed.

13.3　Specialized Detector Design Review

In Section 4.3.4, a brief review of dynamic x-ray detectors including XIIs, FPDs, and MAFs was provided. Although amorphous silicon thin film transistor (TFT)

-based FPDs demonstrated some improvement over XIIs, the new generation of dynamic x-ray detectors must not only be able to have the larger dynamic range to be used in fluoroscopy and angiography, they must satisfy a list of features that may be difficult to achieve in one detector design. Such additional features include substantially higher spatial resolution, lower noise to enable improved signal-to-noise ratio, high speed, physical size constraints, expandability of FOV, cost constraints, preferably reduced complexity or fewer components, and of course improved objective performance metrics such as modulation transfer function (MTF) and detective quantum efficiency (DQE). Various technologies have only been demonstrated in early prototypes and are too immature to project to early clinical application. These include the use of amplifying stages in each pixel for FPDs (Wu 2010, Karim 2012) and the use of polysilicon enabling FPDs to exhibit improved noise performance (Antonuk 2009). Also direct detectors using

cadmium zinc telluride (CZT) such as the modules from the Medipix project based at CERN has reported on a photon counting device; however, the FOV is still quite small (14 mm × 14 mm) with a matrix size of only 256 × 256 of 55 μm pixels. Even an array of these (Koenig 2010, Zwerger 2010a) is limited to 28 mm in one direction since it is only three-side buttable. The present 2 × 3 configuration as recently reported is also at this time quite slow and only available binned 3:1, hence resulting in pixels of 156 μm pixels (Zwerger 2010b) that are too large for the high-resolution ROI vascular application. In what follows, we will restrict the discussion to high-resolution indirect detectors that at this time have a restricted FOV and are designated micro-angiographic fluoroscopes (MAFs) because they are capable of both angiography and fluoroscopy.

13.4 Evolution of Micro-Angiographic Detector Systems

The micro-angiographic detectors that our own group at the University at Buffalo has developed over the years (Rudin 2001) have progressively improved in physical characteristics. Figure 13.1 gives a comparison of the various designs that were explored and are discussed in the following text.

13.4.1 MA

The first ROI detector built by the authors' group (Ganguly 2003a,b) obtained the improvement in resolution by using a thin 200 μm CsI(Tl) phosphor on a fiber optic plate (FOP) and a fiber optic taper (FOT) to

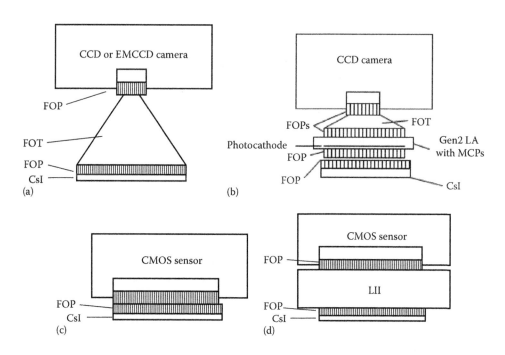

FIGURE 13.1 Schematics of designs for ROI micro-angiographic detectors where the electron multiplying charge coupled device (EMCCD) and LII designs have built-in gain. (a) The upper left was the MA that had no pixel gain or the SSXII with EMCCD for gain. (b) The upper right was the MAF-CCD that had a LII for gain and charge coupled device (CCD) as the light sensor. (c) The lower left is the MAF-CMOS eliminating the FOT by replacing the CCD with a CMOS sensor. (d) The lower right is the proposed MAF-LII-CMOS with the LII to enable variable gain.

(a)   (b)

FIGURE 13.2   Images of an undeployed coronary stent on a 2.7 F catheter taken with (a) an XII with highest magnification mode of 4.5 in., (b) taken with the MA camera.

funnel the light to a CCD camera resulting in an effective pixel size at the phosphor of 50 μm. It is clear from Figure 13.2 that the improvement in spatial resolution enabled detailed features of the crimped coronary stent such as the individual struts to be visualized, whereas these features could not be seen using a standard XII operating in its highest magnification mode. Because of the lack of built-in gain and the substantial light losses in the FOT, the MA was only able to be used for angiography but not during fluoroscopy exposure levels.

13.4.2   MAF-CCD

To increase the gain of the ROI detectors, for the MAF, a light image amplifier (LIA) or intensifier (LII) was placed between the x-ray converter phosphor and the input to the FOT (Figure 13.1b) to compensate for any losses in the FOT and to increase the signal above the noise level of the readout CCD sensor (Jain 2011). The LII operates somewhat like night vision goggles by maintaining the light distribution of an image only increasing the brightness at the output. Light coming from the x-ray converter phosphor is absorbed on the LII's input photocathode and converted to an electron

distribution that is in turn incident on micro-channel plate (MCP) devices built into the LII. These MCPs act as charge current amplifiers through successive collisions within the micro-channels when the electrons are accelerated through the high voltage applied to the MCP. The image in the form of the electron distribution, however, is maintained as is the spatial resolution due to the small size of the channels, but the electrons are multiplied and caused to bombard the output phosphor creating a brighter image through the output FOP and onto the FOT. By controlling the voltage applied to the MCPs, the gain can be adjusted: higher for fluoroscopy and lower for angiography.

13.4.3   Solid State X-Ray Image Intensifier

Although the MAF-CCD is serving successfully in initial clinical testing as will be discussed in the following text, there are practical limits on the diameter, hence field of view available. Currently, the largest practical diameter is 4 cm. To increase the FOV, the University at Buffalo (UB) group has been exploring the use of an alternative modular method to obtain variable gain enabling in theory arbitrary FOV increase. This alternative is designated the solid state x-ray image intensifier (SSXII) based on electron multiplying CCDs (EMCCDs) viewing a CsI x-ray convertor phosphor through a fiber optic taper in a similar fashion to the CCD in the MA as illustrated schematically in Figure 13.1a. An EMCCD operates similar to a CCD with the addition, prior to readout, of a multielement register where the charge packets are transported through a region of slightly higher voltage. This voltage of around 50V is sufficient to cause a small amount of charge multiplication up to a few percent due to impact ionization at each element. Successive multiplication of even a small amount in about 600 elements can result in a total charge multiplication of up to 2000× thus amplifying the signal prior to the addition of noise in the readout amplifier; however, problems of stability, temperature control, and FOT losses still remain to be fully resolved. While progressing (Kuhls-Gilcrist 2008, Kuhls-Gilcrist 2009, Kuhls-Gilcrist 2010a, Qu 2010a,b, Huang 2010, Huang 2013), the SSXII is still some time away from having a practical clinical use.

13.4.4   MAF-LII-CMOS and MAF-CMOS

With larger-area complementary metal oxide semiconductor (CMOS) sensors becoming available in recent years, two new MAF designs have become possible as illustrated in Figure 13.1c and d. In Figure 13.1d, the CCD is replaced by the CMOS sensor while maintaining the LII for large variable gain. Any degradation due to the FOT is also eliminated because if the CMOS sensor has a large enough FOV, the FOT becomes expendable. In Figure 13.1c, the LII is eliminated if the CMOS sensor can be made to have a sufficient dynamic range and low noise to enable the MAF-CMOS to operate over the full range of angiography and fluoroscopy. Commercial dynamic CMOS FPDs are beginning to become available from a number of manufacturers that have pixel sizes in the range of 75–100 μm. CMOS detectors do have substantially lower noise than commercial aSi FPDs and appear to be on the verge of replacing them in standard imaging applications; however, it is too early to conclude whether their combination of pixel size, electronic noise, dynamic range, and speed will be sufficient to surpass the overall performance of the MAF-CCD for high-resolution (HR) imaging. A comparison in Table 13.1 gives the features of proposed 50 μm pixel MAF-CMOS detectors with the MAF-CCD and the standard FPD.

Although clinical testing with the MAF has begun, a number of improvements will be forthcoming including in the selection of component features such as for the phosphor and the light sensor, new system features such as noise reduction, application to biplane capability, and development of improved x-ray sources as was reviewed in Chapter 4.

13.4.5 MAF Component Selection and MAF Comparison Metrics

Improvement in at least two major components, the phosphor and image sensor, appears possible. Our group recently reported on a comparative study of various phosphor types and thicknesses (Kuhls-Gilcrist 2010b). As can be seen from Figure 13.3, although there are some losses in resolution for thicker phosphors, by selecting HR type phosphor with less sensitivity but better MTF rather than high light (HL) (which has more light output through the window per absorbed x-ray photon but has poorer resolution), a better overall DQE at higher spatial frequencies is obtained.

Also if the CCD camera FOT combination could be replaced by a flat CMOS sensor with 50 μm pixels, improved DQE as well as smaller overall package would be possible. A comparison for a 500 μm

Table 13.1 Comparison of Detectors

Specs/Detector	MAF-LII-CCD	MAF-LII-CMOS (Simulation)	MAF-CMOS (Simulation)	FPD
Pixel size	35 μm	50 μm	50 μm	194 μm
Scintillator	CsI (Tl)	CsI (Tl)	CsI (Tl)	CsI (Tl)
Thickness/type	500 μm/HR	500 μm/HR	500 μm/HR	600 μm/HL
Gain boost	Yes/variable	Yes/variable	No	No
Dyn. Range	Huge (>14 bit)	Huge (>14 bit)	~14 bit	~14 bit (dual)
Total noise	~0–50 e⁻	~?–100+ e⁻	~100–1500 e⁻	~1000–2000e⁻

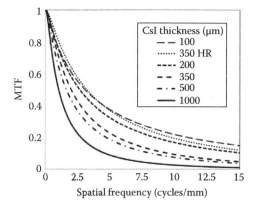
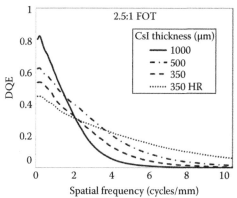

FIGURE 13.3 MTF for various CsI modules at RQA5. The DQE measurement includes a FOT with 2.5:1 ratio.

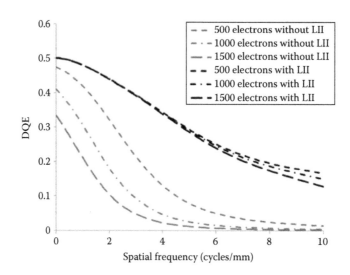

FIGURE 13.4 DQE plots for MAF-CMOS detectors with 500, 1000, and 1500 electrons for a 50 μR detector exposure with and without an LII.

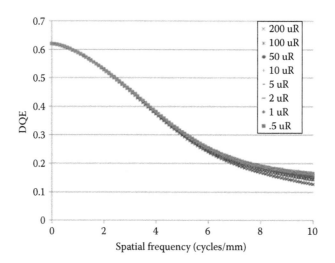

FIGURE 13.5 MAF-LII-CMOS where gain is adjusted to provide constant light illumination to the CMOS sensor with 500 electrons noise, for various exposures.

HR phosphor layer with such a CMOS readout sensor is shown in Figure 4.9, indicating clear improvement once the FOT degradation is eliminated. In that figure, the MAF-CMOS was assumed to have a low noise level of 100 electrons. If the noise is substantially higher as it is for some commercial CMOS sensors, then substantial degradation can occur without an LII as is illustrated in Figure 13.4 (Jain 2013).

Although the LII adds complication and expense to an MAF detector, it also has the benefit of enabling a larger dynamic range and to some extent independence from instrumentation noise for arbitrary exposures as both Figures 13.4 and 13.5 demonstrate for an assumed noise of 500–1500 electrons.

13.4.6 MAF Noise Degradation Reduction

Because of the smaller pixels (~35 μm in the case of the MAF-CCD and 50 μm for the simulated MAF-CMOS), a lower number of quanta per pixel and higher resolution of the MAF hence increased visualization of high-frequency noise components, images, although quantum limited, can still appear too noisy unless there is provision for noise reduction. This is especially true in fluoroscopy mode when standard x-ray parameters based upon the characteristics of conventional FPD or XII detectors are selected for use with the MAF. Additionally in MAF angiography modes, the strong preference for using the smallest focal spot to minimize geometric unsharpness limits the x-ray tube current, again, potentially implying increased quantum noise. One partial solution to these problems that our group has implemented is a simple real-time temporal recursive filter with operator controlled weighting factor, changeable for both acquisition and playback display to help reduce the apparent image noise. Another method we have implemented during fluoroscopy is to increase the x-ray output by using the digital angiography (DA) mode normally controlled by a separate footswitch from the fluoroscopy footswitch while the MAF continues to acquire in its fluoroscopy mode. We designate this as the HD or high definition mode for the MAF (Panse 2010b). The automatic gain control of the LII within the MAF enables the operator to go quickly back and forth between standard fluoroscopy where the MAF images are somewhat noisy and the HD mode where there is considerably less noise by simply going from one footswitch to the other, depending upon the quality of image needed during the intervention. The reason this is possible is that in normal fluoroscopy, for example, the mA may be limited to up to 50 mA and the pulse duration limited to under 12 ms even though the medium focal spot is usually the default. For HD mode, which uses the DA mode of the commercial unit, the limits for the small focal spot may go up to 125 mA and 25 ms at 15 and 30 fps.

For digital subtraction angiography (DSA), normally done at 3 fps using the standard detectors, the requirement for the MAF that the small focal spot be used prevents the exposure per frame for the MAF from being even as large as that for the FPD or XII where normally medium focal spots with larger mAs capability is used. To address this problem, we have done preliminary efforts that appear to indicate that if we increase the frame rate, for example, to 15 fps we can achieve a satisfactory increase in the effective mAs using a temporal weighting factor of 4–6 as well. The equation for a basic temporal recursion filter is given in the following text (Vasan 2012):

$$y(t) = \frac{1}{k} \times x(t) + \left(1 - \frac{1}{k}\right) \times y(t-1)$$

where

  $x(t)$ is the current input signal at time or frame number t

  $y(t-1)$ is the previous output

  $y(t)$ is the present output of the filter

  k is the filter weight (lag or memory) greater than or equal to 1

The output of the filter is the weighted sum of its current input and previous outputs.

13.4.7 MAF Biplane Considerations and X-Ray Tube Requirements

Other improvements and extensions in ROI imaging with the MAF may include use in biplane gantries and the development of improved high output x-ray sources. One of the potential problems in the use of a smaller FOV detector especially during a vascular intervention may be the inability for the clinician to visualize potential hazards suddenly occurring outside the limited FOV of a ROI detector. A solution to this problem is to use the larger FOV standard detector on the other plane of a biplane system. For example, if the MAF is being used in the AP plane, the larger FPD or XII on the lateral plane can be used to monitor the larger region around the intervention. If an MAF is mounted on both planes then the lateral MAF could be used while the AP standard detector is used as an overall monitor.

Another improvement involves the minimization of geometric unsharpness by always using a small x-ray tube focal spot and overcoming tube loading limitations. A method for increasing x-ray tube output while maintaining a small focal spot was described in Chapter 4 where an option to use a longer filament combined with a smaller anode angle would enable larger tube loads to occur without increasing the size of the effective focal spot. Such a tube design would enable reduced quantum noise in MAF image sequences. Although the dose over the small FOV could be increased, the overall integral or effective dose might actually be decreased due to the reduced FOV and any reductions in procedure time due to improved image quality. (See Chapters 4 and 25 for further discussion of patient dose reduction considerations for ROI imaging and Chapter 24 for dose tracking.)

13.4.8 MAF Setup

For evaluation and clinical testing of the ROI micro-angiography detectors, a complete self-contained system had to be developed. For example, although the MAF has been installed on a changer mounted on a commercial angiography C-arm and made able to be deployed when needed or parked when not in use (Figure 13.6), the image sequences from the MAF and

FIGURE 13.6   The MAF shown mounted on a commercial angiography C-arm in parked and deployed positions.

FIGURE 13.7   The control console running CAPIDS in the control room of a clinical suite placed between the angio control console on left and AP display monitor on right. The in-room AP monitor is switched between the XII or MAF video signals.

the standard modes such as DA, DSA, and roadmapping had to be independently provided.

Such a computer control acquisition, image processing, display and storage (CAPIDS) system was developed (Wang 2010) such that controls could be accessed conveniently adjacent to the controls for the commercial system in the clinical suite as shown in Figure 13.7.

13.5   Noise Evaluation of Detectors Using the Instrumentation Noise Equivalent Exposure Concept

As mentioned earlier in Section 13.4.6, instrumentation or read-out noise is an important limiting feature in dynamic detectors especially during fluoroscopy procedures. Although typically it is more convenient for detector developers and engineers to parameterize the noise in the number of electrons per pixel per frame, these quantities are difficult or nearly impossible for clinical medical physicists and field engineers to measure directly in a closed commercial system. An indirect measure of noise designated the instrumentation noise equivalent exposure (Yadava 2008b) or INEE is much more accessible to measurements of exposure and image

noise obtained on systems operating in the field. If we assume a simple model that the image noise variance is composed of a combination of random quantum noise variance plus constant read-out noise variance, then by taking data from flat field (uniform) images for various exposures and plotting the total noise variance versus exposure, a straight line should be obtained whose value at zero exposure is the read-out noise in units of exposure rather than in electrons. This INEE determination will provide an indication as to whether the detector system is working in a quantum limited region of exposures that is desirable or is working in a region where the read-out noise dominates. We compared the INEE for standard detectors with that for the MAF and the SSXII (Kuhls-Gilcrist 2008, 2009) and found as did others (Roo 2004) that the INEE for FPDs was about equal to typical fluoroscopy exposure values. Thus, for exposures below this value (a few μR per frame), no longer are FPDs quantum limited because the instrumentation noise contributes increasingly more than the quantum noise. Because the MAF due to the multichannel plate light image intensifier (see Figure 13.4) and the SSXII

due to the EMCCD have inherently adjustable large gain at each pixel before any instrumentation noise is added, the INEE is extremely low and the instrumentation noise is always negligible causing the MAF and SSXII to be always quantum limited. Due to this high gain capability, the MAF and the SSXII as will be discussed in more detail in the next section have even been shown to be usable for single photon counting in x-ray and gamma ray imaging (Kuhls-Gilcrist 2010a, Panse 2010a, Jain 2012).

13.6 Single Photon Counting Mode

13.6.1 Introduction to CT, Nuclear Med, and Medipix Single Photon Counting

Up until recently, most radiographic imaging is in the energy integrating (EI) mode where the sum of energy of all incident x-ray photons on a detector pixel per frame acquired results in a recorded signal. In nuclear medical imaging, each gamma ray photon is detected, assessed for energy individually, and counted with the total resultant count representing the signal in each pixel per frame. Because x-ray photons impinge on detectors with much less energy and much more rapidly than is typical in nuclear medicine imaging, the detector systems until recently have been unable to accomplish single photon counting (SPC) during radiographic imaging. Recently, SPC methods have begun to be attempted in some areas of radiography. In computed tomography (CT) where there are discrete detector elements, there have been encouraging reports of attempts to have elements each with associated electronic amplifiers and counters able to do SPC although pulse pileup due to the high count

rates must still be considered (Taguchi 2011). A consortium of institutions centered at CERN has developed a small 2D imaging sensor, the Medipix, which has amplifiers and counters associated with each element. Attempts to join six of these small devices have resulted in a 42 mm × 28 mm detector; however, the frame rates and resolution have been limited. Aside from energy differentiation, there are also potential advantages of SPC toward improved spatial resolution as will be demonstrated in the next section where the MAF was used in SPC mode.

13.6.2 MAF in SPC Mode

To demonstrate the potential advantage of SPC over energy integration (EI) mode, the MAF with its huge range of gains was used to perform EI and SPC on a thin slit to obtain the line spread function and then from its Fourier transform, the MTF (Jain 2012). Figure 13.8 shows the large potential advantage that SPC may have over conventional EI.

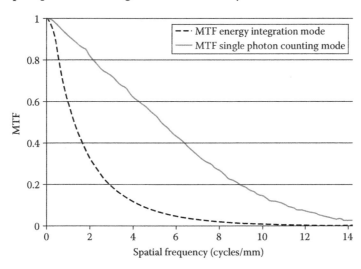

FIGURE 13.8 Comparison of MTFs for the MAF in EI and SPC modes.

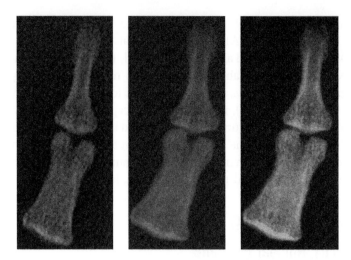

FIGURE 13.9 Comparison radiograph of distal and intermediate phalanges bones taken with an FPD, MAF in EI mode, and MAF in SPC mode.

To achieve SPC with the MAF, the exposure per frame had to be reduced to impractically low levels so that individual x-ray photons would result in detectable individual scintillations just as occurs in nuclear medicine cameras. In fact subsequently the MAF was used in another study to demonstrate its capability to image a radionuclide distribution as well. To achieve the improved resolution, the center of mass of each detected x-ray's photon scintillation distribution was determined and substituted for the scintillation. Then after thresholding, each event was counted in the particular pixel that it coincided with. When comparing the same data treated in each mode, EI and SPC, the improvement in displayed detail can be easily seen in Figure 13.9 and also compared with the additional degradation occurring when a standard lower resolution FPD is used and false details with a mottled background appear.

While it is much too early for practical application of SPC to the severe demands of dynamic vascular imaging, this demonstration illustrates the vast future potential of such advances.

13.7 Generalized and Relative Evaluation Metrics for Micro-Angiographic Detector Systems

13.7.1 Detector System Comparisons with GMTF and GDQE Versus MTF and DQE

While comparison of linear systems metrics of MTF and DQE are essential for detectors such as those shown earlier in Figure 4.9, they do not give the complete information regarding how successful the detectors will operate in a realistic clinical environment where geometric unsharpness due to focal spot size and where scatter generated in the patient can have substantial impact on patient quality. It is for these reasons that our group and others have developed generalized metrics described in more detail in Section 4.4.2. Although we have shown already that the MAF has far better metrics than standard detectors as seen in Figure 4.9 for an RQA5 spectrum,

Figure 13.10 compares both plane detector MTF and DQE and the generalized linear systems parameters of GMTF and GDQE (Kyprianou 2005, Yadava 2005, Yadava 2008a, Jain 2010, Gupta 2011) for an FPD and the MAF with two different phosphors. The generalized functions were calculated assuming a standard head-equivalent phantom where the spatial frequency was considered for a central plane, hence the magnification factor was 1.11. The field size was that of the MAF, hence the low scatter fraction (SF) of 0.33. The zero frequency drops are due to the inclusion of scatter in the generalized metrics. The Hamamatsu HR phosphor for the MAF, designated MAF-HR, although having reduced light yield compared to the HL phosphor has the best spatial resolution and maintains GDQE superiority as well.

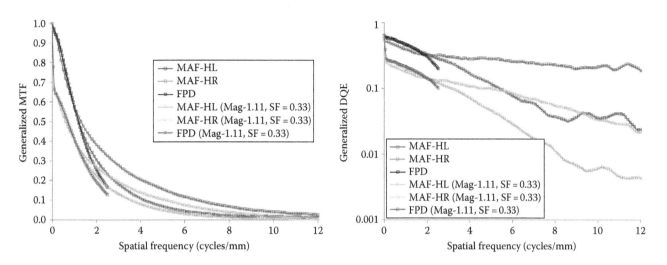

FIGURE 13.10 GMTF and GDQE of FPD without grid and with ~600 μm CsI versus MAF with 300 μm HL and HR CsI(Tl). (From Gupta, S. et al., Evaluation and comparison of high-resolution (HR) and high-light (HL) phosphors in the micro-angiographic fluoroscope (MAF) using generalized linear systems analyses (GMTF, GDQE) that include the effect of scatter, magnification and detector characteristics, in *SPIE: Proceedings from Medical Imaging 2011: Physics of Medical Imaging*, Orlando, FL, vol. 7961–173, pp. 79614S:1–10, 2011, http://www.pubmedcentral.gov/articlerender.fcgi?artid=3134255.) FPD Nyquist is only 2.6 cycles/mm.

13.7.2 Relative Object Detectability for Detector Comparisons

Although generalized metrics of GMTF and GDQE may be more helpful in describing actual detector systems, they still may not provide a realistic comparison of detector systems when encountering actual objects to visualize. The concept of relative object detectability (ROD) (Singh 2014a) and its generalizations in G-ROD and GM-ROD were defined in Section 4.4.3, to attempt to more adequately make evaluations of different systems when viewing specific objects in specific geometries. As an example of

a specific comparison between the MAF and the FPD, the ROD and the G-ROD are plotted in Figure 13.11 for the same x-ray system with small focal spot, no scatter, and for various sizes of spherical lesions and magnifications (Singh 2014b). As the diameter of the lesion decreases the ROD (where the magnification is 1), and the G-RODs demonstrate substantial benefit for the MAF over the FPD, however, this benefit is decreased when there is more geometric unsharpness due to increased magnification. Further specific comparisons could be made by measurements of images of relevant clinical objects or features such as stents or catheter markers so as to calculate GM-RODs.

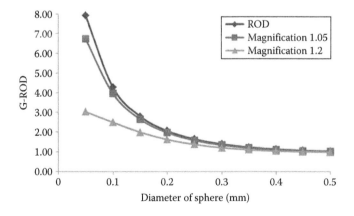

FIGURE 13.11 G-ROD for the MAF versus the FPD as a function of diameter of spherical object being imaged for three different magnifications: 1 (equivalent to the ROD), 1.05, and 1.2. Small focal spot is used with no scatter.

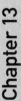

Chapter 13

13.8 Micro ROI Cone-Beam Computed Tomography

Just as cone-beam computed tomography (CBCT) can be achieved using the rotation of an angiography C-arm using a flat panel detector (FPD), high resolution ROI CBCT can be achieved by inserting an MAF detector in front of the standard FPD and rotating the C-arm during serial image acquisition. Nonuniformities in the mechanical gantry motion, however, have to be corrected in the reconstruction (Farig 2000).

Figure 13.12 indicates the starting position of such a C-arm at the beginning of its rotation when the FPD is in place and next when the MAF is deployed. A comparison of the projection images obtained with each detector is given in Figure 13.13.

To reduce potential truncation artifacts resulting from the ROI reduced FOV, the projection images from both acquisition with MAF and FPD separately can be blended prior to reconstruction (Chityala 2005, Patel 2008). A resulting high-resolution 3D rendering of a stent is shown in Figure 13.14.

Consideration has been given to what happens if the ROI is not at the center of rotation (COR) of the C-arm. A complete set of projections must be obtained either by moving the patient's ROI so that it is in the COR or by moving the ROI detector so that it is always viewing the patient's ROI for every projection image (Chityala 2006).

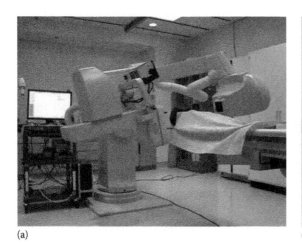
(a)

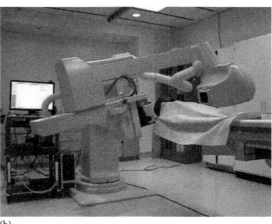
(b)

FIGURE 13.12 CBCT acquisition using angiography C-arm with (a) FPD and (b) MAF.

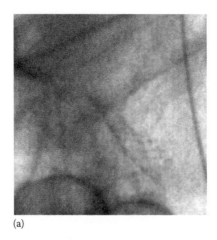
(a)

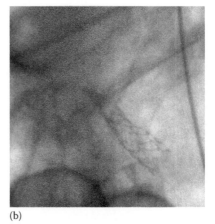
(b)

FIGURE 13.13 Stent deployed in a rabbit carotid vessel showing constriction toward the distal (top) part of the vessel. (a) FPD and (b) MAF.

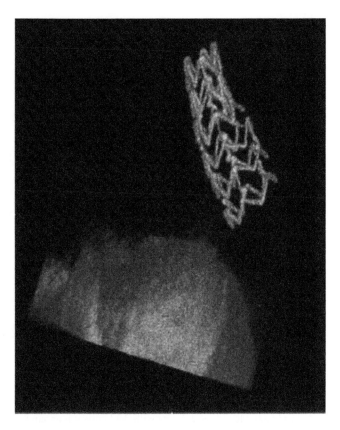

FIGURE 13.14 A view of a 3D rendered CBCT taken with the MAF-FPD combination demonstrating ability to visualize individual stent struts as well as the distal constriction.

13.9   Examples of MAF-CCD Application

Ultimately, the significance of improved high resolution detector development is to provide an improved standard of care for image guided interventional procedures. By exploring the use of the MAF, for example, in various specific neurovascular pathologies, improvements may be achieved in such applications as aneurysm coiling, stent and distal protection device placement, determination and treatment of arteriovenous malformation niduses, the study of blood flow and flow modifying devices, visualization (hence preservation) of small but very important perforator vessels, guidance of thrombolysis and deployment of clot removal devices for acute stroke, determination of the morphology of complex pathology, and all with a reduction in effective dose due to the decreased exposed FOV (Gill 2011).

To illustrate the MAF's potential significance, we present examples of stent and coil deployments. The MAF detector was mounted on a specially designed changer system containing safety features such as mechanical touch sensor and a shock absorber as shown in Figure 13.6.

13.9.1   Stent Deployment Guided by the MAF

A pipeline stent was imaged as it was released from its delivery catheter and with the MAF the many densely packed thin struts not usually seen with a standard FPD or XII were visualized. This stent is designed for flow modification but is difficult to see ordinarily because it has no markers attached at the ends (Figure 13.15). The MAF not only visualized all the struts but also showed how this stent shortens appreciably as it expands out of the catheter making it hard to localize with standard FPDs compared to the MAF, which provides an imaging sequence with superior acuity (Wang 2011).

The MAF was also used in the pre-clinical deployment of a number of enterprise stents (Codman &

Chapter 13

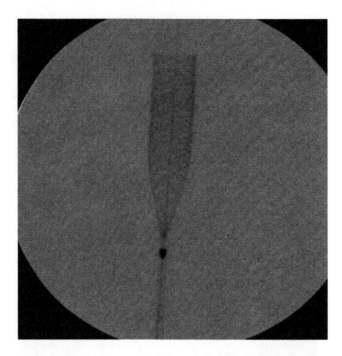

FIGURE 13.15   Pipeline stent, with guide wire at the center, being released from its delivery catheter (marker at lower end). Video was presented at the SPIE 2011 meeting. (From Wang, W. et al., Study of stent deployment mechanics using a high-resolution x-ray imaging detector, in *SPIE: Proceedings from Medical Imaging 2011: Biomedical Applications: Molecular, Structural, and Functional Imaging*, Orlando, FL, vol. 7965–51, 2011, pp. 79651G:1–8, http://www.pubmedcentral.gov/articlerender.fcgi?artid=3144509.) Tiny but densely packed struts can be seen and the change in axial length of the stent with deployment is apparent using the MAF.

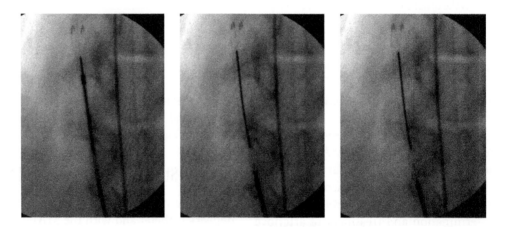

FIGURE 13.16   Enterprise stent deployment in rabbit. Notice bullet on guide wire marking final release point (see text).

Shurtleff, Div. of Johnson and Johnson, Inc.) in animal models. This stent has the desirable property that for most of the deployed extent it can be retrieved back into the delivery catheter; however, only with the MAF images was the clinician able to identify the bullet-shaped marker on the delivery guide wire that captures the proximal group of four stent markers next to the distal end of the thicker portion of the guide wire. Once the end of the catheter overlaps and passes the bullet marker, the stent can no longer be retrieved (Figure 13.16). Only with the MAF images was this precise juxtaposition of features seen, hence contributing to the improved control of this stent delivery system. Once these features are experienced by clinicians, they prefer that all future such stent deliveries be done under MAF image guidance. Thus, even before complete clinical testing can be done, the MAF is becoming a standard tool in the pre-clinical animal lab available to any researcher.

13.9.2 Clinical Aneurysm Coiling

The MAF was recently used in a round of patient studies mostly to guide the coiling of neurovascular aneurysms (Kan 2012) with overwhelming acceptance by interventionalists (Wang 2012). In one of these early cases with the MAF, the clinician was able to save the patient a stent deployment during the coiling procedure. With the usual x-ray receptor image, the clinician indicated that he could not have determined completely whether the coils had herniated into the aneurysm feeding vessel and hence he would have played it safe and deployed a stent with consequent anti-platelet therapy. With the MAF image, he indicated that he was certain that the last coils remained in the aneurysm and had not herniated into the vessel with the potential for follow-on stroke, and hence a stent was not needed. From just the first few cases (see Figure 13.17), clinicians in general stated that "…the smallest movements of the micro-catheters and coils can be seen, allowing the operator to visualize kickback of the catheter before it actually kicks out of the aneurysm. The MAF was useful in evaluating for residual filling to determine if further coiling is required and in identifying small pockets within the aneurysm to reposition the micro-catheter for more complete coiling. The micro-catheter can be repositioned with great precision, which is especially helpful when repositioning the micro-catheter within small intra-aneurysmal pockets" (Binning 2011). As the MAF is applied to more cases, we expect further significant impact such that the minimally invasive treatment of neurovascular aneurysms should become more accurate, with reduced risk of perforation and hemorrhage, reduced likelihood of recurrence, and the possibility of treatment for wide-necked, large or giant and fusiform aneurysms that are presently untreatable. Because of the high spatial resolution, flow modification should have reduced risk to small but crucial perforator vessels unique to the cerebrovasculature. Also, the duration of treatment and patient discomfort may be reduced since improved images should lead to greater accuracy in an intervention.

13.10 Summary and Conclusion

As minimally invasive image guided endovascular interventions increasingly replace invasive surgery, the demands to improve image quality will continue. We have shown that current dynamic detectors can be vastly improved both in spatial resolution and in sensitivity. We have demonstrated that this superior

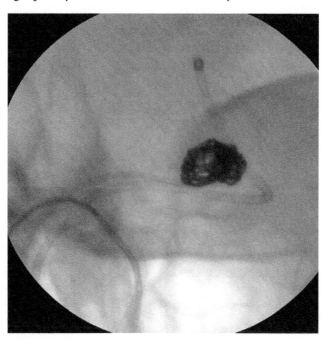

FIGURE 13.17 MAF fluoroscopy for guidance of human aneurysm coiling. Details of the actual coil structure (Pt coated envelope), coil delivery micro-catheter windings, and stent delivery catheter (stent not used here) windings are apparent with the MAF. Other frames show interactive strain on coil wire, and catheter tip placement that can only be seen on the uncompressed video sequence accompanying the publication. (From Binning, M.J. et al., *Neurosurgery*, 69(5), 1131, 2011, (post acceptance epub June 18, 2011, doi:10.1227/ NEU.0b013e3182299814) NIHMSID 391939, http://www.ncbi.nlm.nih.gov/pubmed/21694658 Videos with voice-over description for this paper is available at: http://www.youtube.com/watch?v=fY-OpedS1H0 and at http://www.youtube.com/watch?v=dT6-tReQCcg.)

image guidance can be achieved over a smaller but sufficient field of view with new detector designs such as that of the micro-angiographic fluoroscope that not only has about three times the spatial resolution of current commercial detectors but also has such a large range of sensitivities that single photon counting for x-rays is possible. In order to better measure and compare the performance of the new detectors when embedded in realistic systems where scatter and geometric unsharpness due to finite focal spot size are included, new generalized systems metrics such as the GMTF, GDQE, F-ROD, G-ROD, and GM-ROD and noise metrics such as the INEE were developed. It was also shown that the improved detectors could be used not only for image guidance during an intervention but also to acquire 3D images using standard C-arms rotations with ROI cone-beam CT reconstruction algorithms. While the new detector technologies have been demonstrated in pre-clinical experiments and the potential for substantial procedure improvement was seen in a limited number of actual neurovascular patient procedures, the prospect for wide distribution can be projected to be plausible in the near future once these new capabilities are integrated into standard angiography and interventional suites.

References

(AHA 2010) American Heart Association. 2010. *Heart Disease and Stroke Statistics-2010 Update*, Dallas, TX. http://www.americanheart.org/presenter.jhtml?identifier=3000090.

(Antonuk 2009) Antonuk L. E., Koniczek M., El-Mohri Y., Zhao Q. 2009. Active pixel and photon counting imagers based on poly-Si TFTs—Rewriting the rule book on large area, flat panel x-ray devices. In *SPIE: Proceedings from Medical Imaging 2009: Physics of Medical Imaging*, (ed.) E. Samei, J. Hsieh, Lake Buena Vista, FL, Vol. 7258, 725814pp.

(Binning 2011) Binning M. J., Orion D., Yashar P., Webb S., Ionita C. N., Jain A., Rudin S., Hopkins L. N., Siddiqui A. H., Levy E. I. November 2011. The use of the micro-angiographic fluoroscope for coiling of intracranial aneurysms. *Neurosurgery*, 69(5): 1131–1138. (post acceptance epub June 18, 2011. doi:10.1227/NEU.0b013e3182299814) NIHMSID 391939. http://www.ncbi.nlm.nih.gov/pubmed/21694658 Videos with voice-over description for this paper is available at: http://www.youtube.com/watch?v=fY-OpedS1H0 and at http://www.youtube.com/watch?v=dT6-tReQCcg.

(Chityala 2005) Chityala R., Hoffmann K. R., Bednarek D. R., Rudin S. 2005. Region of interest (ROI) computed tomography (CT): Comparison with full field of view (FFOV) and truncated CT for a human head phantom. In *SPIE: Proceedings from Medical Imaging 2005: Physics of Medical Imaging*, San Diego, CA, vol. 5745, pp. 583–590, paper #64. NIHMSID # 267843. http://www.ncbi.nlm.nih.gov/pmc/articles/PMC3035320/.

(Chityala 2006) Chityala R. N., Patel V., Hoffmann K. R., Rudin S., Bednarek D. R. 2006. Region of interest (ROI) cone-beam CT using dual resolution, dual detector image acquisition: Evaluation of off-centered ROIs (abstract). In *Program of 92nd Scientific Assembly and Annual Meeting of RSNA*, Chicago, IL, November 26–December 1, 2006, scientific presentation LL-PH4194-H03, p. 923.

(Farig 2000) Fahrig R., Holdsworth D. 2000. Three-dimensional computed tomographic reconstruction using a C-arm mounted XRII: Image-based correction of gantry motion nonidealities. *Medical Physics*, 27: 30–38.

(Ganguly 2003a) Ganguly A., Rudin S., Bednarek D. R., Hoffmann K. R., Kyprianou I. 2003a. Micro angiography for neuro vascular imaging, Part 1: Experimental measurements and feasibility. *Medical Physics*, 30(11): 3018–3028. http://www.ncbi.nlm.nih.gov/pubmed/14655949.

(Ganguly 2003b) Ganguly A., Rudin S., Bednarek D. R., Hoffmann K. R. 2003b. Micro angiography for neuro vascular imaging, Part 2: Cascade model analysis. *Medical Physics*, 30(11): 3029–3039. http://www.ncbi.nlm.nih.gov/pubmed/14655950.

(Gill 2011) Gill K., Ionita C. N., Bednarek D. R., Rudin S. 2011. Effective-dose rate comparison between the micro-angiographic fluoroscope (MAF) and the X-ray image intensifier (XII) used during neuro-endovascular device deployment procedures (abstract). *Medical Physics*, 38(6): 3440, SU-E-I-191.

(Gupta 2011) Gupta S., Jain A., Bednarek D. R., Rudin S. 2011. Evaluation and comparison of high-resolution (HR) and high-light (HL) phosphors in the micro-angiographic fluoroscope (MAF) using generalized linear systems analyses (GMTF, GDQE) that include the effect of scatter, magnification and detector characteristics. In *SPIE: Proceedings from Medical Imaging 2011: Physics of Medical Imaging*, Orlando, FL, vol. 7961–173, paper 79614S:1–10. http://www.pubmedcentral.gov/articlerender.fcgi?artid=3134255.

(Huang 2010) Huang Y., Qu B., Sharma P., Kuhls-Glicrist A. T., Wang W., Titus A. H., Cartwright A. N., Bednarek D. R., Rudin S. 2010. Component level modular design of a solid state x-ray image intensifier for an M × N Array. In *2010 IEEE Nuclear Science Symposium and Medical Imaging Conference*, Knoxville, TN, October 30–November 6, 2010, paper M14-3, *NSS Conference Record*, p. 2714–2717. http://www.ncbi.nlm.nih.gov/pmc/articles/PMC3596890/.

(Huang 2013) Huang Y., Qu B., Jain A., Kuhls-Gilcrist A. T., Titus A. H., Cartwright A. N., Bednarek D. R., Rudin S. February 2013. Design, characterization and modeling for a modular high resolution solid state x-ray image intensifier (SSXII). *IEEE Transaction on Nuclear Science*, 60(1): 20–29. 10.1109/TNS.2012.2226913. http://ieeexplore.ieee.org/stamp/stamp.jsp?tp=&arnumber=6409968.

(Jain 2010) Jain A., Kuhls-Gilcrist A. T., Bednarek D. R., Rudin S. 2010. Generalized two-dimensional (2D) linear system analysis metrics (GMTF, GDQE) for digital radiography systems including the effect of focal spot, magnification, scatter, and detector characteristics. In *SPIE: Proceedings from Medical Imaging 2010: Physics of Medical Imaging*, San Diego CA, vol. 7622, paper 7622-19, 76220K:1–10. http://www.pubmedcentral.gov/articlerender.fcgi?artid=3021385.

(Jain 2011) Jain A., Bednarek D. R., Rudin S. July 2011. A theoretical and experimental evaluation of the micro-angiographic fluoroscope (MAF): A high-resolution region-of-interest x-ray imager. *Medical Physics*, 38(7): 4112–4126. http://www.ncbi.nlm.nih.gov/pubmed/21859012

(Jain 2012) Jain A., Panse A., Bednarek D. R., Yao R., Rudin S. October 2012. Evaluation of the micro-angiographic fluoroscope as a high resolution, single-photon counting and energy integrating imager for transmission and emission imaging. *IEEE Transaction on Nuclear Science*, 59(5): 1927–1933. http://ieeexplore.ieee.org/stamp/stamp.jsp?tp=&arnumber=6213164.

(Jain 2013) Jain A., Bednarek D. R., Rudin S. 2013. Theoretical performance analysis for CMOS based high resolution detectors. In *SPIE: Proceedings from Medical Imaging 2013: Physics of Medical Imaging*, Orlando, FL, vol. 8668–212, paper 86685U:1–9. http://www.ncbi.nlm.nih.gov/pmc/articles/PMC3864964/.

(Kan 2012) Kan P., Yashar P., Ionita C. N., Jain A., Rudin S., Levy E. I., Siddiqui A. H. 2012. Endovascular coil embolization of a very small ruptured aneurysm using a novel micro-angiographic technique: Technical note. Accepted for publication in *Journal of NeuroInterventional Surgery (JNIS)*, published online January 21, 2012 in advance of print journal, doi:10.1136/neurintsurg-2011-010154. NIHMSID 391936. PMID 22266790. http://jnis.bmj.com/content/early/2012/01/21/neurintsurg-2011-010154.full.html.

(Karim 2012) Karim, K. S., Taghibakhsh, F. 2012. Device and pixel architecture for high resolution digital. US Patent 8,199,236 B2, June 12,2012.

(Koenig 2010) Koenig T., Zwerger A., Schuenke P., Zuber M., Steinke M., Nill S., Fauler A., Fiederle M., Oelfke U. 2010. Properties of a CdTe Medipix Hexa detector designated for small animal imaging. *Proceedings of 2010 IEEE Nuclear Science Symposium and Medical Imaging Conference*, Knoxville, TN, October 30–November 6, 2010, paper M09-181.

(Kuhls-Gilcrist 2008) Kuhls-Gilcrist A. T., Yadava G. K., Patel V., Jain A., Bednarek D. R., Rudin S. 2008. The solid-state x-ray image intensifier (SSXII): An EMCCD-based x-ray detector. In *SPIE: Proceedings from Medical Imaging 2008: Physics of Medical Imaging*, San Diego, CA, vol. 6913, paper 6913-19. PMCID2557100 http://www.pubmedcentral.nih.gov/articlerender.fcgi?tool=pubmed&pubmedid=18836568

(Kuhls-Gilcrist 2009) Kuhls-Gilcrist A. T., Bednarek D. R., Rudin S. 2009. Component analysis of a new solid state x-ray image intensifier (SSXII) using photon transfer. In *SPIE: Proceedings from Medical Imaging 2009: Physics of Medical Imaging*, Orlando, FL, vol. 7258, paper 7258-42, 725817:1–10. http://www.pubmedcentral.gov/articlerender.fcgi?artid=2745170.

(Kuhls-Gilcrist 2010a) Kuhls-Gilcrist A. T., Jain A., Bednarek D. R., Rudin S. 2010a. The solid state x-ray image intensifier (SSXII) in single photon counting (SPC) mode. In *SPIE: Proceedings from Medical Imaging 2010: Physics of Medical Imaging*, San Diego CA, vol. 7622, paper 7622-60, 76221P:1–9. http://www.pubmedcentral.gov/articlerender.fcgi?artid=3021377.

(Kuhls-Gilcrist 2010b) Kuhls-Gilcrist A. T., Jain A., Bednarek D. R., Rudin S. 2010b. Performance trade-off analysis comparing different front-end configurations in a digital x-ray imager. In *2010 IEEE Nuclear Science Symposium and Medical Imaging Conference*, Knoxville, TN, October 30–November 6, 2010, paper M13-2, *NSS Conference Record*, pp. 2491–2494. http://www.pubmedcentral.gov/articlerender.fcgi?artid=3127232.

(Kyprianou 2005) Kyprianou I., Rudin S., Bednarek D. R., Hoffmann K. R. 2005. Generalizing the MTF and DQE to include x-ray scatter and focal spot unsharpness: Application to a new micro-angiographic system for clinical use. *Medical Physics*, 32(2): 613–626, 2005. http://www.ncbi.nlm.nih.gov/pubmed/15789608.

(Labbe 1994) Labbe, M. S., Chiu, M. Y., Rzeszotarski, M. S., Bani-Hashemi, A. R., Wilson, D. L. 1994. The x-ray fovea, a device for reducing x-ray dose in fluoroscopy. *Medical Physics*, 21(3): 471–481.

(Nesbit 2004) Nesbit, G. M., Luh, G., Tien, R., and Barnwell, S. L. 2004. New and future endovascular treatment strategies for acute ischemic stroke. *Journal of Vascular and Interventional Radiology*, 15: S103–S110.

(Panse 2010a) Panse A., Jain A., Wang W., Yao R., Bednarek D. R., Rudin S. 2010a. High resolution emission and transmission imaging using the same detector. In *2010 IEEE Nuclear Science Symposium and Medical Imaging Conference*, Knoxville, TN, October 30–November 6, 2010, paper M19-65, *NSS Conference Record*, pp. 3372–3375.

(Panse 2010b) Panse A., Ionita C. N., Wang W., Jain A., Bednarek D. R., Rudin S. 2010b. The micro-angiographic fluoroscope (MAF) in high definition (HD) mode for improved contrast-to-noise ratio and resolution in fluoroscopy and roadmapping. In *2010 IEEE Nuclear Science Symposium and Medical Imaging Conference*, Knoxville, TN, October 30–November 6, 2010, paper M18-204, *NSS Conference Record*, pp. 3217–3220. http://www.pubmedcentral.gov/articlerender.fcgi?artid=3137370.

(Patel 2008) Patel V., Hoffmann K. R., Ionita C. N., Keleshis C., Bednarek D. R., Rudin S. October 2008. Rotational micro-CT using a clinical rotational angiography gantry. *Medical Physics Letter, Medical Physics*, 35(10): 4757–4764. PMCID: PMC2663391; PMID: 18975720 http://www.pubmedcentral.gov/articlerender.fcgi?artid=2663391.

(Qu 2010a) Qu B., Huang Y., Wang W., Sharma P., Kuhls-Gilcrist A. T., Cartwright A. N., Titus A. H., Bednarek D. R., Rudin S. 2010a. Optical demonstration of a medical imaging system with an EMCCD-sensor array for use in a high resolution dynamic x-ray imager. In *2010 IEEE Nuclear Science Symposium and Medical Imaging Conference*, Knoxville, TN, October 30–November 6, 2010, paper M13-207, *NSS Conference Record*, pp. 2607–2609. http://www.ncbi.nlm.nih.gov/pmc/articles/PMC3596892/

(Qu 2010b) Qu B., Kuhls-Gilcrist A. T., Huang Y., Wang W., Cartwright A. N., Titus A. H., Bednarek D. R., Rudin S. 2010b. Quantum performance analysis of an EMCCD-based x-ray detector using the photon transfer technique. In *2010 IEEE Nuclear Science Symposium and Medical Imaging Conference*, Knoxville, TN, October 30–November 6, 2010, paper M19-170, *NSS Conference Record*, pp. 3438–3441. http://www.pubmedcentral.gov/articlerender.fcgi?artid=3127248.

(Roo 2004) Roos P. G., Colbeth R. E., Mollov I., Munro P., Pavkovich J., Seppi E. J., Shapiro E. G. et al. 2004. Multiple gain ranging readout method to extend the dynamic range of amorphous silicon flat panel imagers. *Proceedings of the SPIE*, 5368: 139–149.

(Rudin 1992) Rudin, S., Bednarek D. R. 1992. Region of interest fluoroscopy. *Medical Physics*, 19(5): 1183–1189.

(Rudin 2001) Rudin S., Bednarek D. R., Wakhloo A. J., Lieber B. B. 2001. Region of interest micro-angiography. US Patent No. 6,285,739, September 4, 2001.

Chapter 13

(Rudin 2008) Rudin, S., Bednarek, D. R., Hoffmann, K. R. 2008. Endovascular image guided interventions (EIGI). Invited vision 20/20 paper. *Medical Physics*, 35(1): 301–309. http://www.pubmedcentral.gov/articlerender.fcgi?artid=2669303.

(Siewerdsen 1999) Siewerdsen, J. H. and Jaffray, D. A. 1999. A ghost story: Spatio-temporal response characteristics of an indirect-detection flat-panel imager. *Medical Physics*, 26: 1624.

(Singh 2014a) Singh V., Jain A., Bednarek D. R., Rudin S. 2014a. Relative object detectability (ROD): A new metric for comparing x-ray detector performance for a specified imaging task. In *SPIE: Proceedings from Medical Imaging 2014: Physics of Medical Imaging*, San Diego, CA, vol. 9033–9203, Paper 903351:1-6. NIHMSID #619227. http://www.ncbi.nlm.nih.gov/pmc/articles/PMC4188352/

(Singh 2014b) Singh V., Jain A., Bednarek D. R., Rudin S. June 2014b. Generalized relative object detectability (G-ROD) as a metric for comparing x-ray imaging systems (abstract). *Medical Physics*, 41(6). *AAPM Annual Meeting Program*, SU-E-I-29, Paper 903362:1-9. NIHMSID #619230. http://www.ncbi.nlm.nih.gov/pmc/articles/PMC4189125/

(Taguchi 2011) Taguchi K., Zhang M., Frey E. C., Wang X., Iwanczyk J. S., Nygard E., Hartsough N. E., Tsui B. M. W., Barber W. C. 2011. Modeling the performance of a photon counting x-ray detector for CT: Energy response and pulse pileup effects. *Medical Physics*, 38: 1089.

(Vasan 2012) Swetadri Vasan S. N., Panse A., Jain A., Sharma P., Ionita C. N., Titus A. H., Cartwright A. N., Bednarek D. R., Rudin S. 2012. Dose reduction technique using a combination of a region of interest (ROI) Material x-ray attenuator and spatially different temporal filtering for fluoroscopic interventions. In *SPIE: Proceedings from Medical Imaging 2012: Physics of Medical Imaging*, San Diego, CA, vol. 8313–191, paper 831357:1–11. http://www.ncbi.nlm.nih.gov/pmc/articles/PMC3766980/.

(Wang 2010) Wang W., Ionita C. N., Keleshis C., Kuhls-Gilcrist A. T., Jain A., Bednarek D. R, Rudin S. 2010. Progress in the development of a new angiography suite including the high resolution micro-angiographic fluoroscope (MAF), a control, acquisition, processing, and image display system (CAPIDS), and a new detector changer integrated into a commercial C-arm angiography unit to enable clinical use. In *SPIE: Proceedings from Medical Imaging 2010: Physics of Medical Imaging*, San Diego CA, vol. 7622, paper 7622-200, 76225I:1–10. https://www.pubmedcentral.gov/articlerender.fcgi?artid=3021378.

(Wang 2011) Wang W., Ionita C. N., Bednarek D. R., Rudin S. 2011. Study of stent deployment mechanics using a high-resolution x-ray imaging detector. In *SPIE: Proceedings from Medical Imaging 2011: Biomedical Applications: Molecular, Structural, and Functional Imaging*, Orlando, FL, vol. 7965-51, paper 79651G:1–8. http://www.pubmedcentral.gov/articlerender.fcgi?artid=3144509.

(Wang 2012) Wang W., Ionita C. N., Huang Y., Qu B., Panse A., Jain A., Bednarek D. R., Rudin S. 2012. Region-of-interest micro-angiographic fluoroscope detector used in aneurysm and artery stenosis diagnoses and treatment. In *SPIE: Proceedings from Medical Imaging 2012: Physics of Medical Imaging*, San Diego, CA, vol. 8313–8343, paper 831317:1–9. http://www.ncbi.nlm.nih.gov/pmc/articles/PMC3877313/.

(Wu 2010) Wu, D., Safavian, N., Yazdandoost, M. Y., Izadi, M. H., Karim, K. S. 2010. Electronic noise comparison of amorphous silicon current mode and voltage mode active pixel sensors for large area digital x-ray imaging. In *SPIE: Proceedings from Medical Imaging 2010: Physics of Medical Imaging*, San Diego CA, vol. 7622, paper 7622-134, 76223Q.

(Yadava 2005) Yadava G. K., Kyprianou I. S., Rudin S., Bednarek D. R., Hoffmann K. R. 2005. Generalized performance evaluation of x-ray image intensifier compared with a microangiographic system. In *SPIE: Proceedings from Medical Imaging 2005: Physics of Medical Imaging*, San Diego, CA, vol. 5745, pp. 419–429, paper #49.

(Yadava 2008a) Yadava G. K., Rudin S., Kuhls-Gilcrist A. T., Bednarek D. R. 2008. Generalized objective performance assessment of a new high-sensitivity microangiographic fluoroscopic (HSMAF) imaging system. In *SPIE: Proceedings from Medical Imaging 2008: Physics of Medical Imaging*, San Diego, CA, vol. 6913, paper 6913-29. http://www.pubmedcentral.nih.gov/articlerender.fcgi?tool=pubmed&pubmedid=18836567.

(Yadava 2008b) Yadava G., Rudin S., Kuhls-Gilcrist A. T., Patel V., Hoffmann K. R., Bednarek D. R. September 21, 2008. A practical x-ray exposure-equivalent metric for instrumentation noise in high-sensitivity and fluoroscopic x-ray imagers. *Physics in Medicine and Biology*, 53(18): 5107–5121. PMCID2562256 http://www.pubmedcentral.nih.gov/articlerender.fcgi?tool=pubmed&pubmedid=18723932.

(Zwerger 2010a) Zwerger A., Fauler A., Fiederle M. 2010a. Large 12 cm^2 monolithic CdTe pixel sensors with medipix readout. In *Proceeding of 2010 IEEE Nuclear Science Symposium and Medical Imaging Conference*, Knoxville, TN, October 30–November 6, 2010, paper R18-2.

(Zwerger 2010b) Zwerger A. 2010b. Personal communication at IEEE-MIC2010.

14. Advanced Image Reconstruction in Cardiovascular Imaging

Pascal Thériault-Lauzier and Guang-Hong Chen

14.1 Physical Principles of X-Ray Computed Tomography

In order to understand how images can be generated from a beam of x-rays, one should have some knowledge of the physical processes that take place when radiation interacts with matter. X-rays are essentially a form of electromagnetic (EM) radiation (Attix, 2004). For the purpose of this discussion, the wave-like nature of x-rays is neglected to focus on their particle-like behavior. Photons are quanta, finite packets, of EM energy. The energy of x-rays used in diagnostic medical imaging varies between 10 and 150 keV. At such energy levels, photons interact with matter using two main mechanisms: Compton and photoelectric effects (Attix, 2004).

The Compton effect consists in the scattering of photons incident on a free or loosely bound electron. It results in a change of direction of the photon with a loss in energy. Photoelectric effect is an interaction with the electron cloud surrounding an atom. The photon is absorbed by the atom, which adopts an excited state and then returns to the ground state with the emission of a fluorescent photon (Attix, 2004).

The combination of both Compton scattering and photoelectric absorption is termed attenuation. Quantitatively, this phenomenon is described by the attenuation coefficient (μ), which gives the rate of

Chapter 14

relative attenuation of a beam of photons with respect to the distance of propagation. For a monochromatic beam, one may write (Kak and Slaney, 1988):

$$\frac{1}{n}\frac{dn}{dx} = \mu(x) \tag{14.1}$$

where $n(x)$ is the mean number of photons having propagated a distance x through the material without being attenuated. Given that $n(0) = f$, this differential equation can be solved by simple integration to yield:

$$n(x) = f \exp\left(-\int_0^x \mu(s)ds\right) \tag{14.2}$$

where s is an integration variable. A detector can be placed in the beam to measure n. The object under study can then be positioned between the x-ray source and the detector. The attenuation of the beam is a function of the material's atomic number and electron density, in addition to the energy of the x-ray beam. This relation explains the contrast one can observe between different x-ray projection measurements depending on the physical properties of the object. We refer to this principle as the x-ray imaging contrast mechanism.

In projection x-ray imaging, every point of the image is a function of a line integral through the object. The contrast of a particular structure is thus diminished by the presence of overlaid anatomy. For example, if one attempts to see the heart on a chest x-ray, one will notice that it is obscured by ribs, lungs, and other superimposed tissues. This lack of conspicuity is what tomography aims to improve. The imaging equation of x-ray CT is the following:

$$\ln\left(\frac{f}{n}\right) = \int_0^x \mu(s)ds. \tag{14.3}$$

It is customary to define the projection datum as $\int_0^x \mu(s)ds \equiv y$. The left-hand side of the equation corresponds to the measurement (y), while the right-hand side involves the object's property to be reconstructed (μ).

A parallel beam scanning geometry assumes a nondivergent beam of x-rays, which is incident on a linear array of detectors oriented perpendicular to the direction of propagation of the beam. A projection view angle is defined as the set of measurements recorded by all detectors at a specific beam angle. A complete projection dataset is comprised of a set of projection view angles acquired by rotating the beam over 180° around an object.

In a 2D case, the spatial distribution of an object's attenuation coefficient can be defined as $\mu : \mathbb{R}^2 \to \mathbb{R}$. We define $\Omega \subset \mathbb{R}^2$ as the spatial support of the object. The attenuation coefficient outside of the object, that is, in the set $\mathbb{R}^2 \backslash \Omega$, is defined as zero. A projection dataset is also defined as function $\lambda : \Gamma \to \mathbb{R}$, where $\Gamma = [\rho_{min}, \rho_{max}] \times [0, \pi]$ is the sinogram space. The position along the detector array is described by variable $\rho \in [\rho_{min}, \rho_{max}]$ while the angle of the x-ray beam is given by variable $\theta \in [0, \pi]$. The spatial position is described by variables r_1 and r_2. If one assumes that the projection that includes point $(r_1, r_2) = (0, 0)$ is located at $\rho = 0$ for all θ, we can write:

$$\lambda(\rho, \theta) = \int_{-\infty}^{\infty}\int_{-\infty}^{\infty} dr_1 dr_2 \mu(r_1, r_2)\delta(\rho - r_1\cos(\theta) - r_2\sin(\theta))$$

$$\tag{14.4}$$

where the Dirac δ distribution is used according to its customary definition. In essence, this relationship defines nothing but a set of line integrals over the object. It is known widely as the Radon transform.

14.1.1 Filtered Backprojection

Filtered backprojection (FBP) is the most widely used reconstruction algorithm in commercial CT scanners today. FBP is typically described as an *analytical* reconstruction algorithm, which means that a closed-form formula can be derived to solve the inverse Radon transform.

It can be shown that given λ one can reconstruct μ using the following formula:

$$\mu(r_1, r_2) = \int_0^{\pi} d\theta \int_{\rho_{min}}^{\rho_{max}} d\rho\, h(r_1\cos(\theta) + r_2\sin(\theta) - \rho)\lambda(\theta, \rho),$$

$$\tag{14.5}$$

where h is the ramp filter defined as $h(\rho) = \int_{-\infty}^{\infty} d\xi\, |\xi| \exp(2\pi i \xi \rho)$. Typically, the integral over ρ is called *filtration*, while that over θ is called *backprojection*. This explains the name of filtered backprojection for this reconstruction formula.

In current generation CT scanners, x-ray projections are measured using an array of detectors arranged in a plane or a cylinder. The source and the detector array are mounted on rigid gantry. It is possible to derive FBP reconstruction formuli specially adapted to these fan-beam and cone-beam geometries (Kak and Slaney, 1988). In fan-beam geometry, data must be acquired over an angular range of at least 180° plus the fan angle in order to be used with an FBP reconstruction algorithm. This range is called a *short scan*. It is important to note that the temporal resolution achievable in this case is approximately half of the scan time.

Due to its closed form, FBP has the advantage of being computationally efficient. It is also simple to study using linear systems theory. However, it has several limitations described in the next section.

14.2 Limitations of Filtered Backprojection

14.2.1 Undersampling

The question of sampling has been discussed since the inception of CT. Allan McLeod Cormack, co-recipient of the Nobel Prize in Medicine and Physiology for the development of CT, put the issue this way:

> In practice one can make only a finite number of measurements with beams of finite width, and the question which arises is how many observations should be made, and how should they be related to each other in order to reconstruct the object.

Cormack (1978)

A typical approach to derive the number of view angles (N_a) and detector elements (N_d) required for CT imaging is to use the Nyquist–Shannon sampling theorem. In parallel beam geometry, it can be shown that the following requirement should be approximately met (Kak and Slaney, 1988):

$$\frac{N_a}{N_d} \approx \frac{\pi}{2}. \tag{14.6}$$

A similar, yet much more complicated, relation was also derived for the fan-beam geometry (Natterer, 1993). In practice, both relationships dictate that the number of view angles and detector elements be roughly equal (Kak and Slaney, 1988). For example, on General Electric's state of the art scanner, 984 view angles are acquired each composed of 888 detectors arranged in 128 rows.

A possible approach to increase the temporal resolution or decrease the dose of radiation received by patients during a CT scan is to acquire only a subset of the view angles normally measured. This is referred to as undersampling. However, in doing so, the Nyquist–Shannon sampling theorem is violated; the cost to pay is the introduction of artifacts in the image (Kak and Slaney, 1988) if FBP is used as a reconstruction algorithm. These are referred to as undersampling or aliasing artifacts. They typically manifest themselves in the form of streaks parallel to the projections that were present in the dataset. An example is shown in Figure 14.1. Several reconstruction methods have been proposed to mitigate aliasing artifacts and enable the reconstruction of high-quality images from highly undersampled projection datasets (McKinnon and Bates, 1981; Garden and Robb, 1986; Sidky et al., 2006;

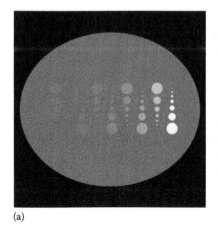 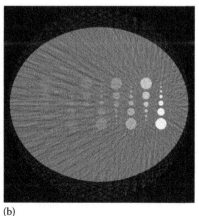

(a)   (b)

FIGURE 14.1   Illustration of the impact of undersampling on image quality of FBP reconstructions of a numerical phantom. (a) Shows a reconstruction from a fully sampled dataset. (b) Shows the same object reconstructed this time from an undersampled dataset. Aliasing artifacts (streaks) are clearly visible in (b).

Chapter 14

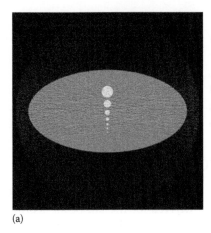
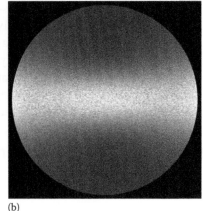

(a)   (b)

FIGURE 14.2 Illustration of the impact of Poisson distributed noise on a CT image. (a) Shows the reconstruction of a numerical phantom with noise. (b) Shows a map of the noise standard deviation obtained by Monte Carlo methods where grayscale values increase with noise level.

Chen et al., 2008; Leng et al., 2008b; Supanich et al., 2009). Some of these are presented later in this chapter.

14.2.2 Decrease in X-Ray Tube Current

The dose of ionizing radiation received by a patient during a CT scan is proportional to the number of photons emitted by the x-ray source. The total number of photons itself has a linear dependence on the scan time and the tube current. Therefore, by reducing the tube current, one can reduce the dose proportionally. Of course, there is a price to pay for such a decrease. In effect, when the tube current is reduced, the image noise increases. This is a consequence of the stochastic nature of photon detection. It is well understood that the detection of x-rays is essentially a counting process and that as such, it is governed by Poisson statistics (Ter-Pogossian, 1967; Kak and Slaney, 1988). The Poisson distribution has a variance equal to its mean. If an average count of n photons is measured, one expects a signal-to-noise ratio (SNR) of $\sqrt{n}$. Roughly speaking, this means that when the dose is reduced by a factor of 4, the SNR of projection data is cut by half.

The dependence of the CT image noise on the tube current is more complicated than that of projection data. In this case, the noise is a function of the structure of the object being imaged. The noise level at a particular position is typically dominated by the intersecting x-ray projections with the highest noise level. This phenomenon is illustrated in Figure 14.2. It is possible to observe that the noise has a somewhat horizontal structure. One can also notice an increase in the noise level along the long axis of the phantom.

Both of these phenomena are explained by the reduced number of photons reaching the detector along that direction. When using FBP, projection data are given weights that do not account for the relative amount of noise between different projections.

14.2.3 Decrease in the Size of the Scanner Field of View

The scanner field of view (SFOV) is defined as the region of space that is covered by the x-ray beam from all view angles. In single-slice geometries, this region is typically a circle, while in cone-beam geometry, it is approximately cylindrical. Figure 14.3 illustrates the SFOV in relation to the x-ray source and beam.

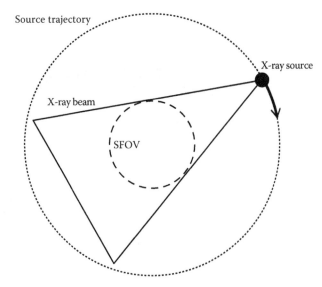

FIGURE 14.3 Illustration of the scanner field of view (SFOV) for a fan-beam geometry.

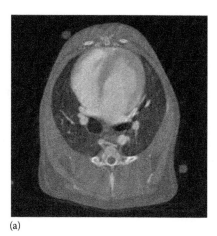
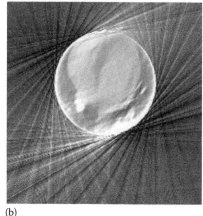

(a)   (b)

FIGURE 14.4 Illustration of the impact of lateral projection data truncation on a CT image. (a) Presents a reconstruction without lateral data truncation. (b) Shows a reconstruction of the same object but with bilateral data truncation of all view angles. In the latter case, a severe cupping artifact is observed.

If the object is not fully contained within the SFOV during the scan, some of the projection view angles will be laterally truncated, which means that they do not extend to a region where the attenuation coefficient is zero. A projection dataset is said to be fully truncated when the SFOV is entirely contained within the support of the object. Truncation can have the advantage of reducing the dose of radiation imparted to the patient. However, it introduces artifacts in images reconstructed using FBP.

In effect, the filtration step of FBP is a nonlocal operation, which means that a missing segment at one location can cause an error in the whole filtered projection. After backprojection, the image is plagued with artifacts as illustrated in Figure 14.4.

Several schemes have been proposed to enable the reconstruction of a volume of interest from truncated projections. Some of these methods are reviewed later in this chapter.

14.3 Iterative Reconstruction Algorithms

The last section outlined three situations where FBP does not perform adequately: (1) when the projection data is undersampled; (2) when the noise level in the projection data is too high; and (3) when projection data are laterally truncated. It has been demonstrated that iterative reconstruction algorithms can outperform FBP in these—and other—situations. This section introduces some mathematical formalism, as well as several such iterative methods.

In order to discuss iterative CT reconstruction methods, it is necessary to lay down a framework that describes the discrete numerical modeling of tomographic systems. In a 3D case, the distribution of x-ray attenuation coefficients of an object can be defined as a compactly supported function $\mu : \Omega \to \mathbb{R}$, where $\Omega \subset \mathbb{R}^3$ is the spatial support of the object. We define the x-ray projection measurements vector $\bar{\mathbf{y}} \in \mathbb{R}^{N_{proj}}$, with $[\bar{\mathbf{y}}]_i = \bar{y}_i$, storing a set

of line integrals over the lines $\{\ell_i \subset \mathbb{R}^3 : i \in [1, N_{proj}]\}$ with $\mathbf{r} \in \mathbb{R}^3$ such that

$$\bar{y}_i = \int_{\ell_i \cap \Omega} ds \mu(\mathbf{r}). \tag{14.7}$$

$\bar{\mathbf{y}}$ is essentially the discrete representation of function λ from the continuous representation. The bar notation indicates a mean value, which is necessary since the measurement of projections is a stochastic process. Furthermore, we define the discrete representation of the attenuation coefficient distribution as $\mathbf{x} \in \mathbb{R}^{MNP}$ with $[\mathbf{x}]_j = x_j$—this corresponds to an $M \times N \times P$ image volume. We assume a voxelized discretization defined using basis functions $B_j : \mathbb{R}^3 \to \mathbb{R}$. Each function $B_j(\mathbf{r})$ has a value of 1 for position $\mathbf{r}$ inside voxel j and a value of 0 when $\mathbf{r}$ is outside of that voxel. We write the approximation as:

$$\mu(\mathbf{r}) \approx \sum_{j=1}^{MNP} x_j B_j(\mathbf{r}). \tag{14.8}$$

In this representation, the approximate x-ray projections—typically called forward projection and denoted by the tilde here—become

$$\tilde{y}_i = \int_{\ell_i \cap \Omega} ds \sum_{j=1}^{MNP} x_j B_j(\mathbf{r})$$

$$= \sum_{j=1}^{MNP} x_j \int_{\ell_i \cap \Omega} ds B_j(\mathbf{r})$$

$$= \sum_{j=1}^{MNP} A_{ij} x_j, \tag{14.9}$$

where $\mathbf{A} \in \mathbb{R}^{Nproj \times MNP}$ with $[\mathbf{A}]_{i,j} = A_{ij}$ is the intersection length of line i with voxel j. We thus write $\tilde{\mathbf{y}} = \mathbf{Ax}$.

As mentioned earlier, x-ray detectors do not directly measure $\bar{\mathbf{y}}$ but rather measure numbers of photons $\mathbf{n} \in \mathbb{N}^{Nproj}$ with $[\mathbf{n}]_i = n_i$. We assume that a photon counting detector is used for convenience; energy integrating detectors are installed in the majority of the modern CT systems. If we model only counting statistics, each n_i is distributed according to a Poisson distribution with mean $\bar{n}_i \in \mathbb{R}$ such that

$$\bar{n}_i = f_i \exp(-\bar{y}_i), \tag{14.10}$$

where f_i is the x-ray fluence incident on the object. We also define the stochastic projection variables

$$y_i = \log\left(\frac{f_i}{n_i}\right). \tag{14.11}$$

Generally, in medical CT, the values f_i are not equal to each other due to the bow-tie compensation applied to the x-ray beam (Bushberg, 2002).

The goal of image reconstruction can thus be understood as the solution of a linear system. Indeed, one would like to determine $\mathbf{x}$ that solves

$$\mathbf{Ax} = \mathbf{y}. \tag{14.12}$$

This formulation of the reconstruction problem was used by Sir Godfrey Hounsfield, the second co-recipient of the Nobel Prize in Medicine and Physiology for the development of CT. In fact, the first scanner built by Hounsfield at the EMI company used the algebraic reconstruction technique (ART) (Gordon et al., 1970), which is based on a linear algebra method developed by Kaczmarz to solve large systems of equations (Natterer, 2001; Herman, 2009). ART is an iterative algorithm that preceded even the use of FBP reconstruction. However, FBP gained widespread acceptance because of its computational efficiency.

14.3.1 Statistical Image Reconstruction

In the current section, we discuss a reconstruction method based on the statistical nature of the x-ray detection process. The conditional probability for measuring count values $\mathbf{n}$ given an object with attenuation coefficients $\mathbf{x}$ is

$$P(\mathbf{n} \mid \mathbf{x}) = \prod_{i=1}^{Nproj} \frac{\bar{n}_i^{n_i} \exp(\bar{n}_i)}{n_i!}. \tag{14.13}$$

This relation is simply the product of Poisson distributions, one for each projection. The log likelihood can be shown (Sauer and Bouman, 1993; Erdogan and Fessler, 1999; Whiting, 2002) to simplify to

$$L(\mathbf{n} \mid \mathbf{x}) = -\sum_{i=1}^{Nproj} [n_i \tilde{y}_i + f_i \exp(-\tilde{y}_i)]. \tag{14.14}$$

Using Equation 14.11, we can write

$$L(\mathbf{n} \mid \mathbf{x}) = -\sum_{i=1}^{Nproj} [n_i \tilde{y}_i + n_i \exp(y_i - \tilde{y}_i)], \tag{14.15}$$

which is possible to approximate using a Taylor expansion (Sauer and Bouman, 1993; Tang et al., 2009) to yield

$$L(\mathbf{n} \mid \mathbf{x}) = -\frac{1}{2} \sum_{i=1}^{Nproj} [(y_i - \tilde{y}_i)^2 n_i] \tag{14.16}$$

$$= -\frac{1}{2}(\mathbf{y} - \mathbf{Ax})^T \mathbf{D}(\mathbf{y} - \mathbf{Ax}), \tag{14.17}$$

where $\mathbf{D} \in \mathbb{R}^{N_{proj} \times N_{proj}}$ is a diagonal matrix with the elements of vector $\mathbf{n}$ on its diagonal. The notation $(\cdot)^{\mathrm{T}}$ indicates a matrix transpose.

The vector that maximizes $L(\mathbf{n}|\mathbf{x})$ is called the maximum likelihood (ML) estimate (Rockmore and Macovski, 1977). However, this result is often unstable since it does not incorporate prior information about the image to be reconstructed. It is possible to include this information via a regularizing field $R : \mathbb{R}^{MNP} \to \mathbb{R}$. The vector that solves

$$\hat{x} = \underset{\mathbf{x} \in \mathbb{R}^{MNP}}{\operatorname{argmax}}[L(\mathbf{n} \mid \mathbf{x}) + R(\mathbf{x})] \qquad (14.18)$$

is called the maximum a posteriori (MAP) estimate. The function $R(\mathbf{x})$ is sometimes called a roughness penalty. In the Bayesian terminology, this function plays the role of prior distribution. It depends only on the reconstructed image and enables one to include information known a priori about its structure. Typically, this function is simple and favors smooth images (Bouman and Sauer, 1993; Thibault et al., 2007).

It has been demonstrated that a statistical reconstruction framework can enable the reconstruction of images with a lower noise level (Bouman and Sauer, 1993; Sauer and Bouman, 1993; Erdogan and Fessler, 1999; Whiting, 2002; Thibault et al., 2007; Tang et al., 2009). Algorithms that decrease image noise can alternatively be used to enable low x-ray tube current acquisitions. Thus, statistical image reconstruction methods have the potential to decrease the dose of radiation imparted to patients during CT scans.

14.3.2 Compressed Sensing

Compressed sensing (CS)—also known as compressive sampling—is a signal processing theory that enables the exact recovery of a signal from very few samples, given that the sampling system and the signal possess certain characteristics. Specifically, CS stipulates that if a signal is *sparse* in a known domain and that this domain is *incoherent* with the sampling basis, then the signal can be reconstructed exactly from few samples (Candès et al., 2006a,b; Donoho, 2006). As this definition is somewhat cryptic, let us discuss the basic concepts of CS before describing its application in imaging.

The concept of sparsity can be considered a measure of the information content of a signal. If an invertible transformation can be applied to a signal such that most of the elements of the transformed signal are zero, then we say that it is sparse. It means that all the

information of the original signal is now concentrated in a few nonzero elements. In other words, the signal is compressible. This concept is interesting in the context of reconstruction because it provides a priori information about the signal to be reconstructed from samples. In particular, we are interested in undetermined linear systems such as $\mathbf{Ax} = \mathbf{y}$ where $\mathbf{A} \in \mathbb{R}^{M \times N}$ with $M < N$, $\mathbf{x} \in \mathbb{R}^{N}$ is the signal to be reconstructed given data samples $\mathbf{y} \in \mathbb{R}^{M}$. It is well known that such systems have an infinite number of solutions. However, if we know a priori that $\mathbf{x}$ is sparse under a transformation Ψ, then we can pick the sparsest solution out of all of those compatible with the measurements. This is the idea of CS reconstruction.

The previous discussion begs the following question: how can one *pick* the sparsest solution? The answer comes from vector norms. The ℓ_2-norm—defined as $\|\mathbf{x}\|_2 = \sqrt{\sum_i (x_i)^2}$—yields the Euclidean distance between two points or the common concept of *length* of a vector. The ℓ_2-norm is rotationally invariant, which means that a vector will have the same *length* even if it is rotated. The ℓ_1-norm—defined as $\|\mathbf{x}\|_1 = \sum_i |x_i|$—is not rotationally invariant. A vector of a particular *length* will have a minimal ℓ_1-norm when it lies parallel to basis vectors. In other words, this means that a vector with many zero-valued elements would have a smaller ℓ_1-norm than a vector with the same ℓ_2-norm but only a few zero elements. The ℓ_1-norm offers a metric of sparsity. The reconstruction problem thus becomes one of optimization (Candès et al., 2006a,b; Donoho, 2006; Candes and Romberg, 2007).

In practice, the signal is reconstructed by minimizing an objective function that includes a sparsity-promoting norm such as the ℓ_1-norm:

$$\hat{\mathbf{x}} = \underset{\mathbf{x}}{\operatorname{argmin}} \|\Psi \mathbf{x}\|_1 \quad \text{such that } \mathbf{Ax} = \mathbf{y}. \qquad (14.19)$$

Note that the optimization is constrained in the sense that it enforces that the reconstructed signal be consistent with the data.

In addition to sparsity, the concept of incoherence is also important in the context of CS. It is related to the question of how many samples are required to ensure that a signal can be exactly reconstructed from an undersampled acquisition. Imagine a signal that is sampled in a space where it is sparse. For example, take a discrete signal of length N with only one nonzero element. If one were to sample this signal by measuring the value of one of its elements at a time, at least N measurements would be required to have a high probability of recording the nonzero element. In this case, no undersampling would be possible. In this simple example, the sparsity domain is *coherent* with the sampling basis. For undersampling to be achievable, the sampling matrix $\mathbf{A}$ and the sparsifying transform Ψ must be mutually *incoherent*. The level of incoherence can be quantified and, based on this metric, it is possible to derive the number of samples required for a sparse signal to be exactly reconstructed (Candes and Romberg, 2007). Note that in the case of CT imaging, the sampling basis is predetermined by the imaging equation and the scanner geometry, while the domain of sparsity is a function of the object's structure.

Medical imaging is intrinsically linked with the history of CS. In fact, Emmanuel Candès, the mathematician who is credited with the discovery of CS, was experimenting with the well-known Shepp–Logan phantom in 2004 when he discovered that he could obtain near-perfect reconstructions from highly undersampled datasets. Quickly, ℓ_1-norm minimization was applied to magnetic resonance imaging (MRI) (Lustig et al., 2007). CS is particularly interesting since it could enable a reduction in the number of projection view angles to be acquired during a scan. As discussed in Section 14.2.1, this could reduce the dose of ionization radiation imparted to patients.

14.3.3 Prior Image Constrained Compressed Sensing

In medical imaging applications, an image similar to that to be reconstructed is often available. This image is referred to as the prior image and its origin depends on the specific clinical application. An appropriate prior image should have a high SNR and a low level of artifacts. However, it may differ from the target image, which is to be reconstructed; it may differ in temporal resolution, noise level, spatial resolution, and/or the object's structure. The goal of prior image constrained

compressed sensing (PICCS) is to integrate this prior image into the CS signal recovery procedure. To this end, the following constrained optimization problem was proposed (Chen et al., 2008):

$$\hat{\mathbf{x}} = \underset{\mathbf{x} \in \mathbb{R}^{MNP}}{\operatorname{argmin}} \left[\alpha \left\| \Psi_1 (\mathbf{x} - \mathbf{x}_p) \right\|_1 + (1-\alpha) \left\| \Psi_2 \mathbf{x} \right\|_1 \right] \text{ s.t. } \mathbf{A}\mathbf{x} = \mathbf{y},$$

(14.20)

where
$\mathbf{x}_p \in \mathbb{R}^{MNP}$ is the prior image
Ψ_1 and Ψ_2 are sparsifying transforms

$\alpha \in [0, 1]$ is a scalar that controls the relative weight to be allocated to the prior image term and to the original CS term from Equation 14.19.

PICCS has been applied to many medical imaging problems since its inception. In the interventional imaging field, it has been employed to offer 3D surgical guidance on a C-arm system from a tomosynthetic acquisition (Nett et al., 2008). In the context of image-guided radiation therapy, PICCS was used to reconstruct streak-free 4D cone-beam CT images from retrospectively gated respiratory datasets (Leng et al., 2008a; Qi and Chen, 2011a,b). In diagnostic CT imaging, it has been shown that PICCS enables an improvement in temporal resolution for cardiac CT imaging (Chen et al., 2009a; Tang et al., 2010) and also allows one to reduce the radiation dose imparted during multidetector CT scans (Lubner et al., 2011; Tang et al., 2011). In neuro-interventional imaging, PICCS has been shown to reduce the radiation dose for 3D angiography on a C-arm system (Nett et al., 2009). In dual energy CT imaging, the application of the PICCS algorithm reduces hardware constraints in kVp switching data acquisition (Szczykutowicz and Chen, 2010). PICCS was also formulated as a nonconvex objective function to moderately improve the potential undersampling factor (Ramirez-Giraldo et al., 2010, 2011).

14.4 Applications in Diagnostic Cardiovascular CT

14.4.1 Temporal Resolution Improvement in Coronary CTA

Motion artifacts are one of the main challenges when imaging the heart. In CT imaging, the temporal resolution of an image is limited by the acquisition time of

a complete projection dataset. For a short-scan angular range, the temporal resolution is approximately $t_{rot}/2$, where t_{rot} is the 360° rotation time of the scanner (Kalender, 2002; Hsieh, 2003). Contemporary CT scanners have rotation speeds of several revolutions per second. Such high speeds result in a substantial

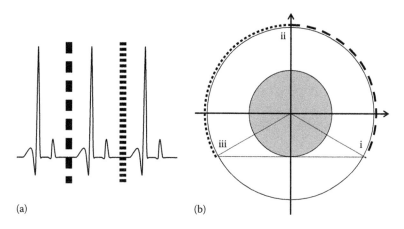

(a)   (b)

FIGURE 14.5 Schematic illustration of the ECG-gated multisegment method. (a) Presents an ECG signal and (b) presents the x-ray source position during the acquisition. The short-scan angular range (i–iii) consists of the union of the angular ranges from two heartbeats. In the first heartbeat, data acquisition takes place at the selected cardiac phase (long-dashed line) and the source travels from (i) to (ii). Ideally, the acquisition at the following heartbeat (short-dashed line) completes the projection dataset between (ii) and (iii). As the gantry continues its rotation between each segment, the temporal resolution achievable is highly dependent on the match between the heart rate and gantry rotation speed.

centrifugal force on the gantry and thus limit the minimum rotation time achievable. To circumvent this problem, scanners with multiple x-ray sources have also been developed to acquire projections simultaneously and reduce the total scan time (Robb et al., 1983; Hsieh and Senzig, 2002; Flohr et al., 2006; Kachelriess et al., 2006). Current dual-source scanners can provide an improvement by a factor of 2 in temporal resolution. These methods provide temporal resolutions between 75 and 200 ms. However, it has been demonstrated that temporal resolutions below 20 ms may be required to obtain artifact-free images at any cardiac phase (Ritchie et al., 1992). Many advanced reconstruction algorithms have been developed to complement hardware methods.

14.4.1.1 ECG-Gated Multisegment Acquisition and Reconstruction Methods

A possibility is to use multisegment acquisitions and reconstruction methods (Kachelriess et al., 2000; Ohnesorge et al., 2000a; Taguchi and Anno, 2000; Flohr and Ohnesorge, 2001). These methods are also called *multiphasic, multisector, snapshot burst,* and *adaptive cardio volume reconstruction* depending on the manufacturer. The essential idea is to improve the effective temporal resolution by acquiring incomplete projection data segments from several heartbeats at the same cardiac phase. All segments are then assembled to produce a complete dataset using electrocardiogram (ECG) gating. The ideal temporal resolution thus achieved is $t_{rot}/2N_{seg}$, where N_{seg} is the number of

incomplete segments into which the complete acquisition is divided. Figure 14.5 illustrates the method in the case of two segments reconstruction ($N_{seg} = 2$).

Multisegment methods were shown to be beneficial in terms of image quality for multislice CT at heart rates above 65 bpm (Halliburton et al., 2003; Herzog et al., 2007). However, the diagnostic accuracy of coronary CTA is not significantly different between single- and two-segment image reconstruction (Herzog et al., 2007). While these methods do provide improvements in image quality, they are limited in nonideal conditions. In particular, if the heart contractions are inappropriately synchronized with the scanner rotation, there could exist an extensive overlap between the view angles measured in two segments. Such cases are not unusual and thus the temporal resolution improvement is limited.

14.4.1.2 Motion Compensation

Another method to improve image quality is motion correction by estimation of a motion field (Ritchie et al., 1996; Grangeat et al., 2002; Pack and Noo, 2004; Roux et al., 2004; Li et al., 2005; Taguchi and Kudo, 2007, 2008; Isola et al., 2008, 2010a,b; van Stevendaal et al., 2008). In this case, a motion vector field is estimated for all positions in the image based on a priori information such as projection data or images. A correction algorithm based on the motion field is then applied to correct motion artifacts. These methods may improve image quality in coronary CTA but due to their novelty, their clinical impact remains to be demonstrated.

Chapter 14

14.4.1.3 Compressed Sensing Methods

Compressed sensing can also be used to improve temporal resolution in coronary CTA. It has been demonstrated that a method called temporal resolution improvement using PICCS (TRI-PICCS) can improve temporal resolution by up to a factor of 2 with respect to FBP in an ideal situation (Chen et al., 2009a; Tang et al., 2010). The idea of the method is illustrated in Figure 14.6. In essence, a short-scan acquisition is used to generate a prior image. Then, the projection data from different view angles are separated in two segments, each with a shorter temporal window. A PICCS-based reconstruction algorithm is then applied to each projection dataset. The resulting images are consistent with the projection datasets from half of the short-scan range. Figure 14.7 demonstrates the application of the TRI-PICCS reconstruction framework in coronary CTA in a swine. Motion artifacts present in the FBP reconstruction are corrected in the TRI-PICCS image due to an improvement temporal resolution.

14.4.2 Myocardial Perfusion Imaging

Cardiac CT has the potential to offer angiography (CTA) simultaneously with myocardial perfusion imaging (MPI), thus enabling an assessment of both coronary anatomy and function from a single scan (Henneman et al., 2006; Kachenoura et al., 2009). In CT MPI, many time frames are acquired while an iodinated contrast agent is administered through intravenous injection. A sample should be recorded about every second to adequately sample the time attenuation curve (TAC) (Wintermark et al., 2004). The reconstructed image series can then be analyzed to yield quantitative measurements of myocardial perfusion such as the myocardial blood flow (MBF), myocardial blood volume (MBV), and mean transit time (MTT) (George et al., 2007; Bamberg et al., 2011). Radiation dose is an important concern due to the repeated scans that need to be obtained over the same anatomical region. One mean of dose reduction is to decrease the x-ray tube current. However, this increases the noise level in images reconstructed using FBP.

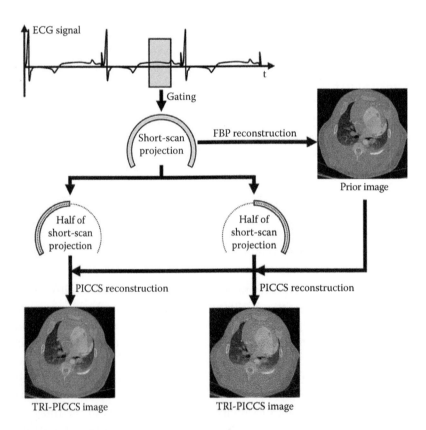

FIGURE 14.6 Work flow of the TRI-PICCS method. The projection data acquired within a short-scan cardiac window were used to reconstruct a low temporal resolution prior image using the conventional FBP method. The short-scan cardiac window is then divided into two contiguous subwindows. The projection data from each of the subwindows are used together with the low temporal resolution prior image to reconstruct shading artifact-free and higher temporal resolution cardiac images. The two subcardiac windows correspond two new cardiac phases with half of the temporal window width. (This figure is used with permission from Tang, J. et al., *Med. Phys.*, 37(8), 4377–4388, 2010. Copyright © 2010 American Association of Physicists in Medicine.)

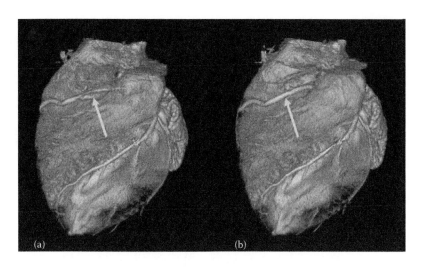

FIGURE 14.7   Comparison of performance between single-source FBP reconstruction (a) and TRI-PICCS reconstruction (b). Note that the motion artifacts in the RCA are removed after applying TRI-PICCS. (This figure is used with permission from Chen, G.-H. et al., *Med. Phys.*, 36(6), 2130, 2009a. Copyright © 2009 American Association of Physicists in Medicine.)

Furthermore, the selected projection data must be acquired within the same relatively static cardiac phase, which is often selected at the end of diastole (Ohnesorge et al., 2000a; Flohr and Ohnesorge, 2001; Hsieh et al., 2006). Due to the need for a high temporal resolution, rapid sequential scanning is necessary, and empirically, the start angle of the short-scan range is generally not constant between time frames. In effect, the nonsynchronicity between the gantry rotation and the cardiac contraction causes a shift in the short-scan central angle over the duration of the entire MPI data acquisition. In images reconstructed using FBP, the noise spatial distribution depends on the source trajectory from frame to frame and on the structure of the object being imaged. Consequently, a region that had a low noise for a certain rotation range can have a much greater noise level once a shift has occurred. When the tube current is reduced to very low levels, noise streaks are often observed in images. While the presence of these streaks depends mostly on high attenuation structures in the object, their orientation and severity depends also on the scan central angle. These effects could be particularly significant if one attempts to obtain a quantitative perfusion map.

As was described in Section 14.3.1, the expected noise present in the projection data is integrated into statistical image reconstruction (SIR) algorithms. In the context of CT MPI, a simple SIR method was shown to mitigate the structured gradient observed in the noise distribution of FBP images, as well as variations in noise level between time frames. It was also demonstrated that the accuracy of quantitative perfusion measurements may be improved by using an SIR approach of Thériault Lauzier et al. (2012b). Figure 14.8 illustrates the noise structure of images reconstructed using FBP and SIR from low tube current projection data.

14.5   Applications in Interventional Cardiovascular CT

14.5.1   View Angle Undersampling in Slow-Acquisition Cone-Beam CT

Recently, an interesting data acquisition scheme (Lauritsch et al., 2006) was explored as an attempt to address the need for tomographic imaging during cardiac interventions. In this scheme, multiple back-and-forth gantry rotations were combined with a multisegment cardiac image reconstruction method to enable cardiac imaging. The preliminary results are promising. However, the prolonged data acquisition increases the likelihood of losing synchronization between the gantry rotation and the cardiac motion. As a result, the quality of images reconstructed could be degraded. In order to combat data inconsistency, a motion compensation method was also proposed for this acquisition method (Prummer et al., 2006, 2009).

An alternative cardiac C-arm cone-beam CT (CBCT) imaging method was also recently proposed (Chen et al., 2012). Instead of multiple sweeps of the C-arm gantry, cone-beam data are acquired using a single slow rotation over a short-scan angular range.

Chapter 14

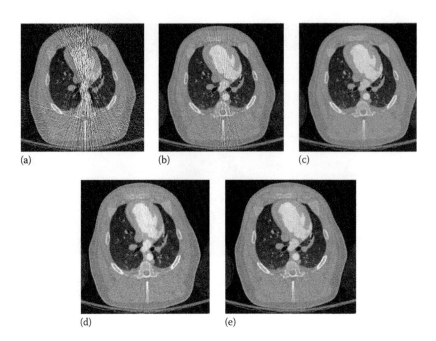

FIGURE 14.8 Illustration of the structured noise obtained in CT MPI from low-current CT protocols. Images (a–c) were reconstructed using FBP and the tube currents used in the acquisitions were, respectively, 25, 50, and 500 mA. Images (d) and (e) were reconstructed using SIR from the same projection data as (a) and (b), respectively. It is possible to notice high noise regions within images reconstructed using FBP at 25 and 50 mA. When SIR was used, high noise projections received a low weight in the reconstruction and their deleterious effect on the image was mitigated. Images reconstructed using SIR have a higher contrast to noise ratio (CNR). (This figure is used with permission from Thériault Lauzier, P. et al., *Med. Phys.*, 39(7), 4079, 2012b. Copyright © 2012 American Association of Physicists in Medicine.)

This method has the potential to allow the user to select a narrow temporal gating window. Such a choice would achieve the highest temporal resolution possible at a given detector frame rate. However, it results in highly undersampled projection datasets once ECG gating has been applied. The dataset associated with each phase has about the same number of view angles as the number of heartbeats that occurred during the scan, between 15 and 25. This high level of undersampling would result in severe artifacts if the datasets were reconstructed using an analytical reconstruction algorithm such as filtered backprojection. In order to mitigate aliasing artifacts, the data acquisition method is coupled with prior image constrained compressed sensing (PICCS) (Chen et al., 2008). The prior image is reconstructed from the whole nongated projection dataset using the classical Feldkamp–Davis–Kress (FDK) algorithm (Feldkamp et al., 1984). FDK is a heuristic FBP-type algorithm adapted to cone-beam geometries. The prior image does not suffer from undersampling artifacts but it contains information from all cardiac phases. PICCS is then used to recover an image for each cardiac phase separately. The work flow of the reconstruction procedure is summarized in Figure 14.9.

To illustrate this method experimentally, a canine was scanned using a clinical C-arm interventional suite. An image volume was reconstructed using the FDK algorithm without gating and thus without temporal information or undersampling artifacts. This image volume, shown in Figure 14.10, was used as a prior image in the work flow presented earlier. After gating, only 14 view angles were available for each of 20 cardiac phases. The FDK and PICCS reconstructions at the end of systole and of diastole are also presented in Figure 14.10. The FDK reconstructions were plagued with a high level of streaking artifacts due to undersampling that interfered with the visualization of the cardiac anatomy. The PICCS reconstructions did not show such artifacts and allowed proper appreciation of the cardiac contraction dynamics. The deformation of the myocardium and of the heart chambers was clearly observed. The edges of the ventricles were reconstructed with high spatial resolution.

14.5.2 Lateral Projection Data Truncation in Cone-Beam CT

The flat panel detector used with most cardiac intervention suites is typically only about 20 cm across.

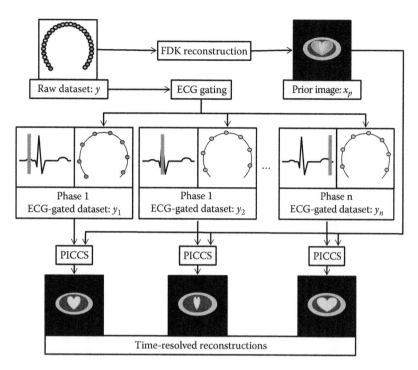

FIGURE 14.9 Flowchart of the cardiac CBCT PICCS algorithm. The dots in the raw dataset represent x-ray source positions where projections were acquired. (This figure is used with permission from Chen, G.-H., Thériault Lauzier, P., Tang, J., Nett, B., Zambelli, J., Qi, Z., Bevins, N. et al., Time-resolved interventional cardiac C-arm cone-beam CT: An application of the PICCS algorithm, *IEEE Transactions on Medical Imaging* 31(4), 907–923. © 2012 IEEE.)

This results in a small scanner field of view (SFOV), which is often too small to cover the entire width of adult patients. Consequently, such systems are prone to lateral data truncation artifacts if FBP methods are used. These artifacts severely limit the quality of the cardiac C-arm CBCT images.

The problem of image reconstruction from truncated projections has been studied extensively (Natterer, 1986). It is well known that it results in severe cupping artifacts. Several exact and approximate methods have been suggested to improve image quality. The projection data extrapolation method aims to approximate the missing projection segments using some a priori information (Herman and Lewitt, 1981; Ogawa et al., 1984; Ohnesorge et al., 2000b; Ruchala et al., 2002; Hsieh et al., 2004; Sourbelle et al., 2005; Starman et al., 2005; Wiegert et al., 2005; Anoop and Rajgopal, 2007; Maltz et al., 2007; Zamyatin and Nakanishi, 2007; Kolditz et al., 2010, 2011; Zhao et al., 2011). It is also possible to increase the spatial and angular sampling to mitigate cupping artifacts moderately. Such an increase minimizes the impact of missing segments in the center of the field of view.

The idea of differentiated backprojection (DBP) has been shown to enable exact reconstruction in cases of partial data truncation (Chen, 2003; Noo et al., 2004; Zhuang et al., 2004; Zou and Pan, 2004; Pan et al., 2005; Defrise et al., 2006). The DBP can be used to exactly reconstruct images from fully truncated projection data using a projection onto convex sets (POCS) algorithm provided that accurate information about the image object is known for a small region of interest inside the object (Ye et al., 2007a,b, 2008; Kudo et al., 2008; Yu et al., 2008; Li et al., 2009).

Recently, the reconstruction of piecewise constant image objects from truncated projections has been shown to be possible using total variation–based compressed sensing (TVCS) (Han et al., 2009; Yu and Wang, 2009; Yu et al., 2009). In these publications, TVCS was applied to fully sampled projection datasets. Higher-order TVCS has also been shown to enable the exact reconstruction of piecewise polynomial objects (Yang et al., 2010). A method combining the TVCS and the POCS approaches has been suggested (Taguchi et al., 2011). Hardware solutions such as multiresolution (Maass et al., 2011) acquisitions and x-ray filtration (Schafer et al., 2010) have also been investigated.

An adapted version of PICCS has also been developed in the context of time-resolved cardiac CBCT

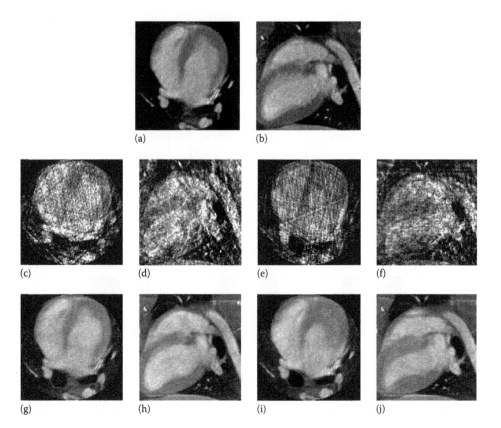

FIGURE 14.10 Reconstructions obtained using PICCS in cardiac interventional imaging in an in vivo animal model. Two slices through the prior image reconstructed using the FDK algorithm are shown in (a) and (b). The rest of the images were reconstructed from gated dataset with 14 view angles. (c–f) are FDK reconstructions, while (g–j) are PICCS reconstructions. Two cardiac phases are shown: the end of the diastole (c), (d), (g), and (h), and the end of the systole (e), (f), (i), and (j). The images are presented in two orthogonal views. (This figure is used with permission from Chen, G.-H., Thériault Lauzier, P., Tang, J., Nett, B., Zambelli, J., Qi, Z., Bevins, N. et al., Time-resolved interventional cardiac C-arm cone-beam CT: An application of the PICCS algorithm, *IEEE Trans. Med. Imaging*, 31(4), 907–923. © 2012 IEEE.)

(Chen et al., 2009b, 2012; Theriault-Lauzier 2012). The prior image is reconstructed using the conventional FDK algorithm combined with a projection data extrapolation scheme—called E-FDK here. The PICCS objective function is also slightly modified to account for truncated projections.

To illustrate this method, a canine was scanned without lateral data truncation using a clinical interventional C-arm system with a detector size of 40×40 cm^2. Data truncation was then introduced to simulate a detector size of 15×15 cm^2. The images reconstructed using FDK, E-FDK, TVCS, and PICCS are presented in Figure 14.11. The level and nature of artifacts vary between images reconstructed using different algorithms. The FDK reconstructions are plagued by a high level of truncation (cupping) and undersampling (streaks) artifacts. The latter artifacts also contaminate the E-FDK images.

However, truncation artifacts are mostly corrected on these E-FDK images. The TVCS method mitigated both cupping and streaks. However, the object studied is only approximately piecewise constant. The use of TVCS results in images that contain a large amount of patchy artifacts. Small-scale and low-contrast structures are particularly degraded by this algorithm. The images reconstructed using PICCS have mitigated both undersampling and truncation artifacts. Furthermore, the algorithm accurately reconstructs both small-scale and low-contrast structures and does not result in obvious patchy artifacts.

The reconstructions with various levels of truncation are presented in Figure 14.12. In each case, the reconstruction is only expected to be accurate within the SFOV denoted by the blue dashed line. Except for the reduction in SFOV size, no qualitative difference is observed within the SFOV.

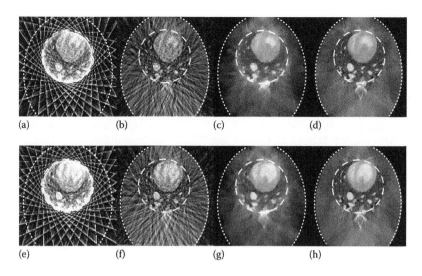

FIGURE 14.11 Axial slice through the reconstructions from truncated projection dataset simulating a 15 × 15 cm² detector size for the four reconstruction algorithms studied: (a, e) FDK, (b, f) E-FDK, (c, g) TVCS, and (g, h) PICCS. Images of the first row (a–d) were acquired at the end of systole, while images of the second row (e–h) were acquired at the end of diastole. The long-dashed line encircles the scanner field of view and the short-dashed line encircles the assumed support of the object. (This figure is used with permission from Thériault Lauzier, P. et al., *Phys. Med. Biol.*, 57(9), 2461, 2012a. Copyright © 2012 Institute of Physics and Engineering in Medicine.)

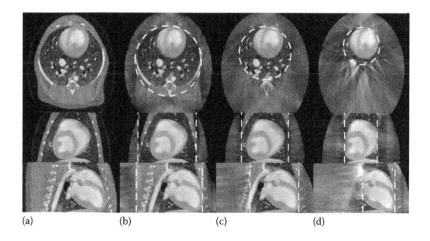

FIGURE 14.12 (a-d) Images reconstructed using PICCS with varying levels of truncation. Three orthogonal slices are presented: axial (top), coronal (middle), and sagittal (bottom). The dashed line encircles the SFOV in each case. (This figure is used with permission from Thériault Lauzier, P. et al., *Phys. Med. Biol.*, 57(9), 2461, 2012a. Copyright © 2012 Institute of Physics and Engineering in Medicine.)

14.6   Conclusion

This chapter demonstrates that despite CT image reconstruction being a mature field, it remains the topic of a considerable research effort. New mathematical theories such as CS may soon have an impact on patient care, by reducing the dose of radiation and improving image quality. Innovations such as PICCS may even enable CBCT to be used in the operating room. The steady increase in affordable computing power may make iterative reconstruction the next frontier in cardiovascular imaging.

Chapter 14

References

Anoop, K.P. and K. Rajgopal. Estimation of missing data using windowed linear prediction in laterally truncated projections in cone-beam CT. In *IEEE EMBS*. Cit Internationale, Lyon, France, 2007, pp. 2903–2906.

Attix, F.H. *Introduction to Radiological Physics and Radiation Dosimetry*. Weinheim, Germany: Wiley-VCH, 2004.

Bamberg, F., A. Becker, F. Schwarz, R.P. Marcus, M. Greif, F. von Ziegler, R. Blankstein et al. Detection of hemodynamically significant coronary artery stenosis: Incremental diagnostic value of dynamic CT-Based myocardial perfusion imaging. *Radiology* 260, 3: (2011) 689–698.

Bouman, C. and K. Sauer. A generalized Gaussian image model for edge-preserving MAP estimation. *IEEE Transactions on Image Processing* 2, 3: (1993) 296–310.

Bushberg, J.T. *The Essential Physics of Medical Imaging*. Philadelphia, PA: Williams & Wilkins, 2002.

Candès, E.J., J. Romberg, and T. Tao. Robust uncertainty principles: Exact signal reconstruction from highly incomplete frequency information. *IEEE Transactions on Information Theory* 52, 2: (2006a) 489.

Candes, E.J. and J.K. Romberg. Sparsity and incoherence in compressive sampling. *Inverse Problems* 23: (2007) 969.

Candès, E.J., J.K. Romberg, and T. Tao. Stable signal recovery from incomplete and inaccurate measurements. *Communications on Pure and Applied Mathematics* 59, 8: (2006b) 1207–1223.

Chen, G.-H. A new framework of image reconstruction from fan beam projections. *Medical Physics* 30, 6: (2003) 1151–1161.

Chen, G.-H., J. Tang, and S. Leng. Prior image constrained compressed sensing (PICCS): A method to accurately reconstruct dynamic CT images from highly undersampled projection data sets. *Medical Physics* 35, 2: (2008) 660.

Chen, G.-H., J. Tang, and J. Hsieh. Temporal resolution improvement using PICCS in MDCT cardiac imaging. *Medical Physics* 36, 6: (2009a) 2130.

Chen, G.-H., J. Tang, B. Nett, S. Leng, J. Zambelli, Z. Qi, N. Bevins, S. Reeder, and H. Rowley. High temporal resolution cardiac cone-beam CT using a slowly rotating C-arm gantry. In *Proceedings of SPIE*. 2009b, volume 7258, 72580C. Orlando, FL.

Chen, G.-H., P. Thériault Lauzier, J. Tang, B. Nett, J. Zambelli, Z. Qi, N. Bevins et al. Time-resolved interventional cardiac C-arm cone-beam CT: An application of the PICCS algorithm. *IEEE Transactions on Medical Imaging* 31, 4: (2012) 907–923.

Cormack, A.M. Sampling the Radon transform with beams of finite width. *Physics in Medicine and Biology* 23, 6: (1978) 1141.

Defrise, M., F. Noo, R. Clackdoyle, and H. Kudo. Truncated Hilbert transform and image reconstruction from limited tomographic data. *Inverse Problems* 22, 3: (2006) 1037.

Donoho, D.L. Compressed sensing. *IEEE Transactions on Information Theory* 52, 4: (2006) 1289–1306.

Erdogan, H. and J.A. Fessler. Ordered subsets algorithms for transmission tomography. *Physics in Medicine and Biology* 44, 11: (1999) 2835.

Feldkamp, L.A., L.C. Davis, and J.W. Kress. Practical cone-beam algorithm. *Journal of the Optical Society of America A* 1, 6: (1984) 612–619.

Flohr, T. and B. Ohnesorge. Heart rate adaptive optimization of spatial and temporal resolution for electrocardiogram-gated multislice spiral CT of the heart. *Journal of Computer Assisted Tomography* 25, 6: (2001) 907–923.

Flohr, T.G., C.H. McCollough, H. Bruder, M. Petersilka, K. Gruber, C. Süss, M. Grasruck et al. First performance evaluation of a dual-source CT (DSCT) system. *European Radiology* 16, 2: (2006) 256–268.

Garden, K.L. and R.A. Robb. 3-D reconstruction of the heart from few projections: A practical implementation of the McKinnon-Bates algorithm. *IEEE Transactions on Medical Imaging* 5, 4: (1986) 233–239.

George, R.T., M. Jerosch-Herold, C. Silva, K. Kitagawa, D.A. Bluemke, J.A.C. Lima, and A.C. Lardo. Quantification of myocardial perfusion using dynamic 64-detector computed tomography. *Investigative Radiology* 42, 12: (2007) 815–822.

Gordon, R., R. Bender, and G.T. Herman. Algebraic reconstruction techniques (ART) for three-dimensional electron microscopy and x-ray photography. *Journal of Theoretical Biology* 29, 3: (1970) 471–481.

Grangeat, P., A. Koenig, T. Rodet, and S. Bonnet. Theoretical framework for a dynamic cone-beam reconstruction algorithm based on a dynamic particle model. *Physics in Medicine and Biology* 47, 15: (2002) 2611.

Halliburton, S.S., A.E. Stillman, T. Flohr, B. Ohnesorge, N. Obuchowski, M. Lieber, W. Karim, S.A. Kuzmiak, J.M. Kasper, and R.D. White. Do segmented reconstruction algorithms for cardiac multi-slice computed tomography improve image quality? *Herz* 28, 1: (2003) 20–31.

Han, W., H. Yu, and G. Wang. A general total variation minimization theorem for compressed sensing based interior tomography. *International Journal Biomedical Imaging* 2009: (2009) 1–3.

Henneman, M.M., J.D. Schuijf, J.W. Jukema, H.J. Lamb, A. De Roos, P. Dibbets, M.P. Stokkel, E.E. Van Der Wall, and J.J. Bax. Comprehensive cardiac assessment with multislice computed tomography: Evaluation of left ventricular function and perfusion in addition to coronary anatomy in patients with previous myocardial infarction. *Heart* 92, 12: (2006) 1779–1783.

Herman, G.T. *Fundamentals of Computerized Tomography: Image Reconstruction from Projections*. Dordrecht, the Netherlands: Springer Verlag, 2009.

Herman, G.T. and R.M. Lewitt. Evaluation of a preprocessing algorithm for truncated CT projections. *Journal Computer Assisted Tomography* 5, 1: (1981) 127–35.

Herzog, C., S.A. Nguyen, G. Savino, P.L. Zwerner, J. Doll, C.D. Nielsen, T.G. Flohr, T.J. Vogl, P. Costello, and U.J. Schoepf. Does two-segment image reconstruction at 64-section CT coronary angiography improve image quality and diagnostic accuracy? 1. *Radiology* 244, 1: (2007) 121–129.

Hsieh, J. *Computed Tomography: Principles, Design, Artifacts, and Recent Advances*. Bellingham, WA: SPIE, 2003.

Hsieh, J., E. Chao, J. Thibault, B. Grekowicz, A. Horst, S. McOlash, and T.J. Myers. A novel reconstruction algorithm to extend the CT scan field-of-view. *Medical Physics* 31, 9: (2004) 2385–2391.

Hsieh, J., J. Londt, M. Vass, J. Li, X. Tang, and D. Okerlund. Step-and-shoot data acquisition and reconstruction for cardiac x-ray computed tomography. *Medical Physics* 33: (2006) 4236–4248.

Hsieh, J. and R. Senzig. Dual cardiac CT scanner. U.S. Patent No. 6,421,412 (July 16, 2002).

Isola, A.A., M. Grass, and W.J. Niessen. Fully automatic nonrigid registration-based local motion estimation for motion-corrected iterative cardiac CT reconstruction. *Medical Physics* 37, 3: (2010a) 1093–1109.

Isola, A.A., A. Ziegler, T. Koehler, W.J. Niessen, and M. Grass. Motion-compensated iterative cone-beam CT image reconstruction with adapted blobs as basis functions. *Physics in Medicine and Biology* 53: (2008) 6777–6797.

Isola, A.A., A. Ziegler, D. Schäfer, T. Köhler, W.J. Niessen, and M. Grass. Motion compensated iterative reconstruction of a region of interest in cardiac cone-beam CT. *Computerized Medical Imaging and Graphics* 34, 2: (2010b) 149–159.

Kachelriess, M., M. Knaup, and W.A. Kalender. Multithreaded cardiac CT. *Medical physics* 33, 7: (2006) 2435–2447.

Kachelriess, M., S. Ulzheimer, and W.A. Kalender. ECG-correlated imaging of the heart with subsecond multislice spiral CT. *IEEE Transactions on Medical Imaging* 19, 9: (2000) 888–901.

Kachenoura, N., T. Gaspar, J.A. Lodato, D.M.E. Bardo, B. Newby, S. Gips, N. Peled, R.M. Lang, and V. Mor-Avi. Combined assessment of coronary anatomy and myocardial perfusion using multidetector computed tomography for the evaluation of coronary artery disease. *The American Journal of Cardiology* 103, 11: (2009) 1487–1494.

Kak, A. and M. Slaney. *Principles of Computerized Tomographic Imaging*. Philadelphia, PA: SIAM, 1988.

Kalender, W. *Computed Tomography: Fundamentals, System Technology, Image Quality, Applications.* Munich, Germany: MCD Verlag, 2002.

Kolditz, D., Y. Kyriakou, and W.A. Kalender. Volume-of-interest (VOI) imaging in C-arm flat-detector CT for high image quality at reduced dose. *Medical Physics* 37, 6: (2010) 2719–30.

Kolditz, D., M. Meyer, Y. Kyriakou, and W.A. Kalender. Comparison of extended field-of-view reconstructions in C-arm flat-detector CT using patient size, shape or attenuation information. *Physics in Medicine and Biology* 56, 1: (2011) 39.

Kudo, H., M. Courdurier, F. Noo, and M. Defrise. Tiny a priori knowledge solves the interior problem in computed tomography. *Physics in Medicine and Biology* 53, 9: (2008) 2207.

Lauritsch, G., J. Boese, L. Wigstrom, H. Kemeth, and R. Fahrig. Towards cardiac C-arm computed tomography. *IEEE Transactions on Medical Imaging* 25, 7: (2006) 922–934.

Leng, S., J. Tang, J. Zambelli, B. Nett, R. Tolakanahalli, and G.-H. Chen. High temporal resolution and streak-free four-dimensional cone-beam computed tomography. *Physics in Medicine and Biology* 53, 20: (2008a) 5653.

Leng, S., J. Zambelli, R. Tolakanahalli, B. Nett, P. Munro, J. Star-Lack, B. Paliwal, and G.-H. Chen. Streaking artifacts reduction in four-dimensional cone-beam computed tomography. *Medical Physics* 35, 10: (2008b) 4649–4659.

Li, L., K. Kang, Z. Chen, L. Zhang, and Y. Xing. A general region-of-interest image reconstruction approach with truncated Hilbert transform. *Journal of Xray Science and Technology* 17, 2: (2009) 135–152.

Li, T., E. Schreibmann, Y. Yang, and L. Xing. Motion correction for improved target localization with on-board cone-beam computed tomography. *Physics in Medicine and Biology* 51, 2: (2005) 253.

Lubner, M.G., P.J. Pickhardt, J. Tang, and G.-H. Chen. Reduced image noise at low-dose multidetector CT of the abdomen with prior image constrained compressed sensing algorithm. *Radiology* 260, 1: (2011) 248–256.

Lustig, M., D. Donoho, and J.M. Pauly. Sparse MRI: The application of compressed sensing for rapid MR imaging. *Magnetic Resonance in Medicine* 58, 6: (2007) 1182–1195.

Maass, C., M. Knaup, and M. Kachelriess. New approaches to region of interest computed tomography. *Medical Physics* 38, 6: (2011) 2868–2878.

Maltz, J.S., S. Bose, H.P. Shukla, and A.R. Bani-Hashemi. CT truncation artifact removal using water-equivalent thicknesses derived from truncated projection data. In *IEEE EMBS.* Cit Internationale, Lyon, France, 2007, vol. 2007, pp. 2907–2911.

McKinnon, G.C. and R.H. Bates. Towards imaging the beating heart usefully with a conventional CT scanner. *IEEE Transactions on Biomedical Engineering* 28, 2: (1981) 123–127.

Natterer, F. *The Mathematics of Computerized Tomography.* Philadelphia, PA: Society for Industrial and Applied Mathematics, 1986.

Natterer, F. Sampling in fan beam tomography. *SIAM Journal on Applied Mathematics* 53, 2: (1993) 358–380.

Natterer, F. *The Mathematics of Computerized Tomography.* Philadelphia, PA: Society for Industrial and Applied Mathematics, 2001.

Nett, B., J. Tang, B. Aagaard-Kienitz, H. Rowley, and G.-H. Chen. Low radiation dose C-arm cone-beam CT based on prior image constrained compressed sensing (PICCS): Including compensation for image volume mismatch between multiple data acquisitions. In *Proceedings of SPIE.* 2009, vol. 7258, p. 725803. Orlando, FL.

Nett, B., J. Tang, S. Leng, and G.-H. Chen. Tomosynthesis via total variation minimization reconstruction and prior image constrained compressed sensing (PICCS) on a C-arm system. In *Proceedings of SPIE.* 2008, vol. 6913, p. 69132D. Orlando, FL.

Noo, F., R. Clackdoyle, and J.D. Pack. A two-step Hilbert transform method for 2D image reconstruction. *Physics in Medicine and Biology* 49, 17: (2004) 3903–3923.

Ogawa, K., M. Nakajima, and S. Yuta. A reconstruction algorithm from truncated projections. *IEEE Transactions on Medical Imaging* 3, 1: (1984) 34–40.

Ohnesorge, B., T. Flohr, C. Becker, A.F. Kopp, U.J. Schoepf, U. Baum, A. Knez, K. Klingenbeck-Regn, and M.F. Reiser. Cardiac imaging by means of electrocardiographically gated multisection spiral CT: Initial experience. *Radiology* 217, 2: (2000a) 564–571.

Ohnesorge, B., T. Flohr, K. Schwarz, J.P. Heiken, and K.T. Bae. Efficient correction for CT image artifacts caused by objects extending outside the scan field of view. *Medical Physics* 27, 1: (2000b) 39–46.

Pack, J.D. and F. Noo. Dynamic computed tomography with known motion field. In *Proceedings of SPIE.* 2004, vol. 5370, pp. 2097–2104. Orlando, FL.

Pan, X., Y. Zou, and D. Xia. Image reconstruction in peripheral and central regions-of-interest and data redundancy. *Medical Physics* 32, 3: (2005) 673–684.

Prummer, M., J. Hornegger, G. Lauritsch, L. Wigstrom, E. Girard-Hughes, and R. Fahrig. Cardiac C-arm CT: A unified framework for motion estimation and dynamic CT. *IEEE Transactions on Medical Imaging* 28, 11: (2009) 1836–1849.

Prummer, M., L. Wigstrom, J. Hornegger, J. Boese, G. Lauritsch, N. Strobel, and R. Fahrig. Cardiac C-arm CT: Efficient motion correction for 4D-FBP. In IEEE *Nuclear Science Symposium Conference Record.* 2006, vol. 4, pp. 2620–2628.

Chapter 14

Qi, Z. and G.-H. Chen. Extraction of tumor motion trajectories using PICCS-4DCBCT: A validation study. *Medical Physics* 38, 10: (2011a) 5530.

Qi, Z. and G.-H. Chen. Performance studies of four-dimensional cone beam computed tomography. *Physics in Medicine and Biology* 56, 20: (2011b) 6709.

Ramirez-Giraldo, J.C., J. Trzasko, S. Leng, L. Yu, A. Manduca, and C.H. McCollough. Nonconvex prior image constrained compressed sensing (NCPICCS): Theory and simulations on perfusion CT. *Medical Physics* 38, 4: (2011) 2157.

Ramirez-Giraldo, J.C., J.D. Trzasko, S. Leng, C.H. McCollough, and A. Manduca. Non-convex prior image constrained compressed sensing (NC-PICCS). In *Proceedings of SPIE*. 2010, vol. 7622, p. 76222C. Orlando, FL.

Ritchie, C.J., C.R. Crawford, J.D. Godwin, KF King, and Y. Kim. Correction of computed tomography motion artifacts using pixel-specific back-projection. *IEEE Transactions on Medical Imaging* 15, 3: (1996) 333–342.

Ritchie, C.J., J.D. Godwin, C.R. Crawford, W. Stanford, H. Anno, and Y. Kim. Minimum scan speeds for suppression of motion artifacts in CT. *Radiology* 185, 1: (1992) 37–42.

Robb, R.A., E.A. Hoffman, L.J. Sinak, L.D. Harris, and E.L. Ritman. High-speed three-dimensional x-ray computed tomography: The dynamic spatial reconstructor. *Proceedings of the IEEE* 71, 3: (1983) 308–319.

Rockmore, A.J. and A. Macovski. A maximum likelihood approach to transmission image reconstruction from projections. *IEEE Transactions on Nuclear Science* 24, 3: (1977) 1929–1935.

Roux, S., L. Desbat, A. Koenig, and P. Grangeat. Exact reconstruction in 2D dynamic CT: Compensation of time-dependent affine deformations. *Physics in Medicine and Biology* 49, 11: (2004) 2169.

Ruchala, K.J., G.H. Olivera, J.M. Kapatoes, P.J. Reckwerdt, and T.R. Mackie. Methods for improving limited field-of-view radiotherapy reconstructions using imperfect a priori images. *Medical Physics* 29, 11: (2002) 2590–605.

Sauer, K. and C. Bouman. A local update strategy for iterative reconstruction from projections. *IEEE Transactions on Signal Processing* 41, 2: (1993) 534–548.

Schafer, S., P.B. Noel, A.M. Walczak, and K.R. Hoffmann. Filtered region of interest cone-beam rotational angiography. *Medical Physics* 37, 2: (2010) 694–703.

Sidky, E.Y., C.M. Kao, and X. Pan. Accurate image reconstruction from few-views and limited-angle data in divergent-beam CT. *Journal of X-Ray Science and Technology* 14, 2: (2006) 119–139.

Sourbelle, K., M. Kachelriess, and W.A. Kalender. Reconstruction from truncated projections in CT using adaptive detruncation. *European Radiology* 15, 5: (2005) 1008–14.

Starman, J., N. Pelc, N. Strobel, and R. Fahrig. Estimating 0th and 1st moments in C-arm CT data for extrapolating truncated projections. In *Proceedings of the SPIE*. 2005, vol. 5747, p. 378.

van Stevendaal, U., J. von Berg, C. Lorenz, and M. Grass. A motion-compensated scheme for helical cone-beam reconstruction in cardiac CT angiography. *Medical Physics* 35, 7: (2008) 3239–3251.

Supanich, M., Y. Tao, B. Nett, K. Pulfer, J. Hsieh, P. Turski, C. Mistretta, H. Rowley, and G.-H. Chen. Radiation dose reduction in time-resolved CT angiography using highly constrained back projection reconstruction. *Physics in Medicine and Biology* 54, 14: (2009) 4575.

Szczykutowicz, T.P. and G.-H. Chen. Dual energy CT using slow kVp switching acquisition and prior image constrained compressed sensing. *Physics in Medicine and Biology* 55, 21: (2010) 6411.

Taguchi, K. and H. Anno. High temporal resolution for multislice helical computed tomography. *Medical physics* 27: (2000) 861.

Taguchi, K. and H. Anno. Motion compensated fan-beam reconstruction for computed tomography using derivative backprojection filtering approach. In *Proceedings of the IXth International Conference on Fully 3D Reconstruction in Radiology and Nuclear Medicine*, Lindau, Germany. 2007, pp. 433–436.

Taguchi, K. and H. Kudo. Motion compensated fan-beam reconstruction for nonrigid transformation. *IEEE Transactions on Medical Imaging* 27, 7: (2008) 907–917.

Taguchi, K., J. Xu, S. Srivastava, B.M. Tsui, J. Cammin, and Q. Tang. Interior region-of-interest reconstruction using a small, nearly piecewise constant subregion. *Medical Physics* 38, 3: (2011) 1307–1312.

Tang, J., J. Hsieh, and G.-H. Chen. Temporal resolution improvement in cardiac CT using PICCS (TRI-PICCS): Performance studies. *Medical Physics* 37, 8: (2010) 4377–4388.

Tang, J., B.E. Nett, and G.-H. Chen. Performance comparison between total variation (TV)-based compressed sensing and statistical iterative reconstruction algorithms. *Physics in Medicine and Biology* 54, 19: (2009) 5781.

Tang, J., P. Thériault Lauzier, and G.-H. Chen. Dose reduction using prior image constrained compressed sensing (DR-PICCS). In *Proceedings of SPIE*. 2011, vol. p. 7961. Orlando, FL.

Ter-Pogossian, M.M. The physical aspects of diagnostic radiology. *The American Journal of the Medical Sciences* 254, 6: (1967) 918.

Thériault Lauzier, P. Quantitative assessment of prior image constrained compressed sensing in the context of cardiac CT imaging. Thesis, Madison, WI: University of Wisconsin–Madison, 2012.

Thériault Lauzier, P., J. Tang, and G.-H. Chen. Time-resolved cardiac interventional cone-beam CT reconstruction from fully truncated projections using the prior image constrained compressed sensing (PICCS) algorithm. *Physics in Medicine and Biology* 57, 9: (2012a) 2461.

Thériault Lauzier, P., J. Tang, M.A. Speidel, and G.-H. Chen. Noise spatial nonuniformity and the impact of statistical image reconstruction in CT myocardial perfusion imaging. *Medical Physics* 39, 7: (2012b) 4079–4092.

Thibault, J.B., K.D. Sauer, C.A. Bouman, and J. Hsieh. A three-dimensional statistical approach to improved image quality for multislice helical CT. *Medical Physics* 34, 11: (2007) 4526.

Whiting, B.R. Signal statistics in x-ray computed tomography. In *Proceedings of SPIE*, eds. L.E. Antonuk, and M.J. Yaffe. 2002, vol. 4682, pp. 53–60. Orlando, FL.

Wiegert, J., M. Bertram, T. Netsch, J. Wulff, J. Weese, and G. Rose. Projection extension for region of interest imaging in cone-beam CT. *Acadademic Radiology* 12, 8: (2005) 1010–1023.

Wintermark, M., W.S. Smith, N.U. Ko, M. Quist, P. Schnyder, and W.P. Dillon. Dynamic perfusion CT: Optimizing the temporal resolution and contrast volume for calculation of perfusion CT parameters in stroke patients. *American Journal of Neuroradiology* 25, 5: (2004) 720–729.

Yang, J., H. Yu, M. Jiang, and G. Wang. High order total variation minimization for interior tomography. *Inverse Problems* 26, 3: (2010) 350131–3501329.

Ye, Y., H. Yu, and G. Wang. Exact interior reconstruction with cone-beam CT. *International Journal of Biomedical Imaging* 2007: (2007a) 10693.

Ye, Y., H. Yu, and G. Wang. Exact interior reconstruction from truncated limited-angle projection data. *International Journal of Biomedical Imaging* 2008: (2008) 427989.

Ye, Y., H. Yu, Y. Wei, and G. Wang. A general local reconstruction approach based on a truncated hilbert transform. *International Journal of Biomedical Imaging* 2007: (2007b) 63634.

Yu, H. and G. Wang. Compressed sensing based interior tomography. *Physics in Medicine and Biology* 54, 9: (2009) 2791–2805.

Yu, H., J. Yang, M. Jiang, and G. Wang. Supplemental analysis on compressed sensing based interior tomography. *Physics in Medicine and Biology* 54, 18: (2009) N425–N432.

Yu, H., Y. Ye, and G. Wang. Interior reconstruction using the truncated Hilbert transform via singular value decomposition. *Journal of X-Ray Science and Technology* 16, 4: (2008) 243–251.

Zamyatin, A.A. and S. Nakanishi. Extension of the reconstruction field of view and truncation correction using sinogram decomposition. *Medical Physics* 34, 5: (2007) 1593–1604.

Zhao, S., K. Yang, and X. Yang. Reconstruction from truncated projections using mixed extrapolations of exponential and quadratic functions. *Journal of X-Ray Science and Technology* 19, 2: (2011) 155–172.

Zhuang, T., S. Leng, B.E. Nett, and G.-H. Chen. Fan-beam and cone-beam image reconstruction via filtering the backprojection image of differentiated projection data. *Physics in Medicine and Biology* 49, 24: (2004) 5489–5503.

Zou, Y. and X. Pan. Exact image reconstruction on PI-lines from minimum data in helical cone-beam CT. *Physics in Medicine and Biology* 49, 6: (2004) 941–959.

15. Assessment of Plaque Features in Atherosclerosis
Quantification of Coronary Calcium

Robert C. Detrano and Sabee Molloi

15.1 Scope of the Problem

Twice each minute an American suffers a coronary event; half are fatal [1]. In a substantial proportion of all persons who eventually suffer clinical manifestations of coronary atherosclerosis, both the initial and final indication of the presence of the disease is sudden coronary death [1]. Myocardial infarction is the single largest killer of Americans, and in survivors of myocardial infarction, the costs, in both human and monetary terms, are tremendous [1,2]. Reducing the devastating societal impact of coronary heart disease is a major global health concern [3–6]; the success of such efforts will depend to a large extent on the accuracy with which we can distinguish those individuals destined to suffer coronary death or infarction from those who will not.

15.2 Theoretical Model behind the Calcium Hypothesis

Coronary calcium and its relationship to coronary atherosclerosis have been studied in vitro (Figure 15.1). The results of these studies form the basis of the coronary calcium hypothesis, the theoretical model that supports coronary calcium screening. This model is dependent on two surrogate indices: (1) Calcium quantity is a

Chapter 15

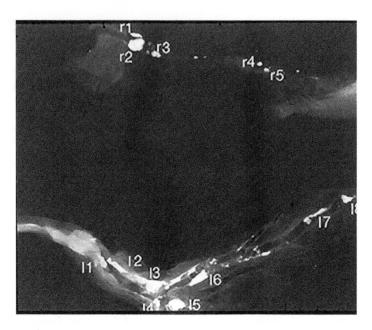

FIGURE 15.1 Radiograph of excised coronary arteries from a patient who died from acute myocardial infarction. The letters r1–r5 are calcifications in the right coronary artery and the numbers l1–l8 are calcification in the left coronary artery.

surrogate for the extent of atherosclerosis and (2) the extent of atherosclerosis is a surrogate for the probability of a coronary event. If both of these surrogate hypotheses are valid, calcium screening should be clinically useful in identifying persons at risk. Aside from this theoretical basis, there is now empirical validation of prognostic accuracy from multiple studies conducted independently.

15.3 Relationship of Coronary Calcium to Coronary Atherosclerosis

Early post mortem investigations established that the presence [7,8] of calcium deposits in the coronary arteries always indicates concomitant atherosclerotic plaque. Mautner and colleagues [9] performed quantitative histo-morphometric analysis on 4298 coronary artery segments from 50 hearts. Cross-sectional area stenosis and the area of calcium deposition were determined for each section. Average percent stenosis and the average percentage of the total calcium in each artery were plotted as a function of distance from the ostium. One striking finding from this study is that the amount of calcium varied both according to the artery examined and also according to the distance from the ostia. Sangiorigi and co-workers [10] analyzed 723 coronary artery sections obtained from all three coronary arteries from 37 human hearts without first decalcifying the arterial samples. After appropriate histologic staining, computerized planimetry was used to obtain the calcium area and the plaque area,

and these latter two variables were then compared. The correlation between the square root of calcific area and the square root of plaque area was $r = 0.52$ ($p < 0.0001$). Despite this significant correlation, examination of the plotted relationship between these two variables indicated wide variability in the amount of calcium for a given plaque area. This was particularly true for smaller values of calcium area, which tended to be associated with a wider range in the associated plaque area. In general, there was usually much more atherosclerosis than calcium, but the amount of plaque varied over a 325-fold range (0.1–32.5 mm^2), while the amount of calcium varied over only a 7.7-fold range (0.13–6.8 mm^2). Calcium areas of zero were associated with plaque areas that ranged from about 0.2 mm^2 to over 4.0 mm^2.

Thus, the relationship between calcium and atherosclerosis is definite and significant but is also highly variable.

15.4 Extent of Coronary Atherosclerosis and the Probability of Coronary Events

Pathologic studies have demonstrated that atherosclerosis begins at a very early age and exhibits a startlingly high prevalence in the adult population [11–13]. Although coronary atherosclerosis increases in severity with advancing age, the rate of increase is not constant: atherosclerosis increases rapidly in the 30–49 year age group, and this rate of increase reaches a maximum in the subsequent decade, and thereafter remains constant [14]. Angiographic studies have also shown that in some plaques the progression of atherosclerosis is very slow [15–17]. Autopsy studies conducted during World War I [18,19], the Korean War [13,20–22], and the Vietnam War [23] showed that many previously healthy young combat victims nevertheless had significant coronary atherosclerosis. In a large study of over 3000 young (ages 15–34) Americans dying of non cardiac causes, the Patho-biological Determinants of Atherosclerosis in Youth (PDAY) study documented the high prevalence of significant coronary atherosclerosis in young people and its association with risk factors for coronary heart disease events [24]. However, there was wide variation in the age at which intermediate lesions develop from fatty streaks, and a variable rate of lesion progression.

The International Atherosclerosis Project performed post mortem analysis of 23,207 sets of coronary arteries from 19 location-race groups in 14 countries who died of coronary heart disease as well as other causes [25].

Significant geographic and ethnic differences in the prevalence and extent of coronary atherosclerosis were found. The extent of atherosclerosis corresponded roughly with the coronary heart disease mortality rates observed at that location. However, notable discordances did occur. For example, in subjects from Bogota, the extent of coronary atherosclerosis was very low, but the death rate from coronary heart disease was intermediate. Conversely, the severity of coronary atherosclerosis in subjects from Manila was high, despite the relatively low coronary heart disease mortality. Although there is a relationship between the extent of coronary atherosclerosis and the probability of coronary heart disease death, there is some degree of variability in this relationship, for reasons that are not understood. From the preceding considerations, it can be appreciated that even if calcium screening correctly identified atherosclerosis with perfect accuracy, it might still fail to reliably identify victims of future events because atherosclerosis is only a fair surrogate for unstable plaque. Therefore, it was necessary to empirically verify that coronary calcium does indeed relate to incident coronary heart disease before it could justifiably be applied clinically.

Radiographic methods are able to detect coronary calcium. They also result in its quantification. In the following sections, we review the methods that have been used to detect and quantify coronary calcium.

15.5 History of Radiographic Coronary Calcium Evaluation

15.5.1 Chest Radiography

Imaging systems for acquiring plain chest x-rays are inexpensive and universally available throughout the world. Chest radiography was investigated during the early part of this century as a potential modality to assess coronary calcium. A summary of this research is indicated in the following text.

There are reports of chest radiographs showing left coronary calcification as early as 1927 [26–29]. In 1978, Souza proposed the examination of the "coronary artery calcification (CAC) triangle" in every adult chest radiograph [30] (Figure 15.2). In 1983, Kelley et al. [31] did an extensive review of the literature on the subject of coronary calcium detected by plain chest radiography. They reported that by using a higher energy

x-ray beam in a film-based system, the sensitivity of the chest radiograph could thus be greatly increased for detecting coronary calcium and almost half of the cases otherwise requiring fluoroscopy could be thus determined.

Despite encouraging initial results, we know of no further research or clinical interest in the detection of coronary calcium in the plain chest radiograph for diagnostic purposes during the latter half of the twentieth century. Chest x-rays, when displayed at clinical conferences, are examined for just about everything except the presence of coronary calcium. Now that we know the clinical importance of coronary calcium in predicting events and in managing risk, it might be worthwhile for us to take another look at Souza's CAC triangle on the chest x-ray.

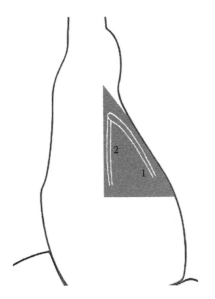

FIGURE 15.2 Souza's CAC triangle. (1) represents the left circumflex coronary artery and (2) the left anterior descending coronary artery.

15.5.2 Fluoroscopy

The fluoroscopic x-ray imaging has been standard equipment in every hospital for decades. The first reports on the use of noninvasive fluoroscopy for detecting coronary calcium appeared in the 1960s [32,33]. A number of reports showed fair to good diagnostic accuracy compared to the coronary angiogram [34–42]. There was a surge of interest in this aspect that culminated in the late 1980s with a meta-analysis [43] and the development of two modifications that increased sensitivity and accuracy. [44,45]. Most of these reports and the meta-analysis [44] concluded that though the sensitivity of fluoroscopy appeared to be low, its specificity was high and its accuracy compared well to most forms of stress testing for predicting coronary stenosis on the coronary arteriogram. One group found a much higher sensitivity and a significantly greater accuracy by applying digital subtraction [44] and taking advantage of the resultant misregistration artifact from the heart's cyclic motion (Figure 15.3). The most striking and encouraging finding was that of Margolis [46] who followed a group of symptomatic adults who had undergone cardiac fluoroscopy and angiography and found that fluoroscopic calcium predicted a fourfold increase in future mortality and that calcium was an independent predictor of mortality even after controlling for extent of angiographic disease.

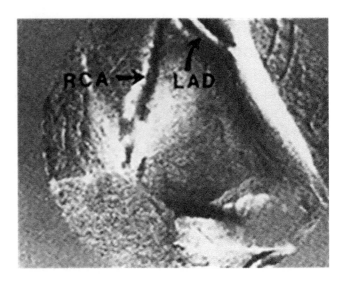

FIGURE 15.3 Digital subtraction images of severe coronary calcification. (Detrano, Digital Subtraction Fluoroscopy JACC; Xu, T. et al., *Med. Phys.*, 33(6), 1612, 2006.)

15.5.3 Dual-Energy Fluoroscopy

Energy subtraction is used as a means of improving object conspicuity [47–49]. The specific tissues are isolated by exploiting the energy dependence of material attenuation coefficients. Energy subtraction has been applied to the detection of calcium in pulmonary nodules [50,51], bone mineral analysis [52], and cardiac imaging [53–60].

A study using dual-energy subtraction chest radiographs with flat-panel detector showed an improved evaluation of CAC [61]. Dual-energy fluoroscopy in conjunction with a flat-panel detector for cardiac imaging requires a flat panel with real-time image acquisition capability along with the ability to switch the x-ray beam energy in real time. The inherent limitation of the traditional temporal subtraction is its susceptibility to misregistration artifacts resulting from cardiac and patient motion. These misregistration artifacts result in signals that cannot be separated from the desired signal for quantitative analysis. A real-time dual-energy subtraction technique [59] addresses the problem of misregistration artifacts associated with temporal subtraction. Furthermore, dual-energy imaging has been used to quantify coronary calcium mass [45].

The main advantage of dual-energy images over standard projection radiographs is the ability to suppress unwanted tissue signals from an image. In the case of chest radiography, the rib shadows are considered as anatomical background noise, which limit lung nodule detection. It is well established that dual energy improves nodule detection in chest radiography [62]. The application of temporal subtraction to cardiac imaging is hampered by motion misregistration artifacts. In contrast, dual-energy imaging is motion insensitive and provides contrast enhancement by eliminating the overlying soft-tissue signals, which improves the detection of calcification.

The technique of dual-energy fluoroscopy in conjunction with a flat panel has previously been implemented for the quantification of CAC [63–65]. However, the limitation of this technique for clinical implementation is its limited sensitivity as compared to the well-established computed tomography (CT) technique as well as the necessary modifications to the x-ray system.

15.5.4 Early Generation Computed Tomography

Computed tomographic (CT) scanners have been used since the 1960s. CT scanners, like fluoroscopic x-ray imaging systems, are expensive. They serve for imaging multiple organs both with and without contrast injections. The first report of the application of computed tomography to diagnose coronary atherosclerosis was that of Reinmuller [66] in 1987 who reported CT to be more sensitive but less specific than fluoroscopy for predicting angiographic stenoses. Masuda [67] and Timmins [68] also found CT to have increased sensitivity for stenoses and Moore [69] found that calcium from CT scans predicted a poorer outcome in symptomatic patients. The increased contrast resolution of computed tomography is the reason for its expected increase in sensitivity. Localization of calcifications is also more accurate and easier with tomography than with projection imaging like fluoroscopy. Despite these advantages and its ubiquitous availability, computed tomography did not meet with better success than fluoroscopy in its acceptance as a clinical diagnostic tool. The reasons for this failure also derive from the fact that CT scanner use was dedicated to proven medical indications that kept them busy around the clock and also to the fact that awareness of the ease with which these instruments could detect coronary calcium had not fully developed.

15.5.5 Electron-Beam Computed Tomography

Electron-Beam Computed Tomography (EBCT) or EBT scanners produce high temporal and spatial

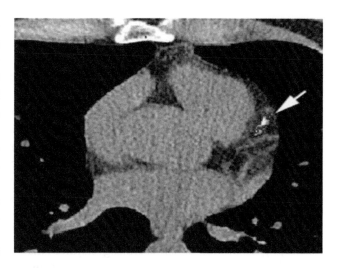

FIGURE 15.4 Electron-beam scan of calcification in the left coronary artery (arrow, outlined in red).

resolution but somewhat noisy images of coronary artery calcifications (Figure 15.4). EBCT scanners were introduced into the American market in the 1980s. In EBCT, the x-ray tube itself is large and stationary, and partially surrounds the imaging circle. The tube does not move but instead, an electron-beam focal point is swept electronically along a tungsten anode in the tube, tracing a large partial circle on its inner surface. The electron energy, when striking the tungsten target, is partially converted into photons. The electron sweep is magnetically aimed as in a cathode ray tube.

The cost of purchase and installation approached two million dollars. Their high temporal resolution made them ideal for imaging the beating heart. Reports on their potential to image coronary calcifications began very early with the papers of Tannenbaum [70] and Agatston [71]. Tannenbaum et al. [70] were the first group of investigators to report the association of EBCT coronary calcium with angiographic stenoses using EBCT. Several other investigators [72–75] reported accuracies for predicting angiographic stenoses that are comparable to those of exercise testing. There was one meta-analysis that showed encouraging results in this regard [75].

Some investigators did studies to determine the value of coronary calcium detected with this technique for predicting future coronary events in asymptomatic adults [76,77]. These studies in asymptomatic subjects received enormous attention and produced widely divergent results and conclusions. In 1996, the American Heart Association, at the persistent suggestion of the proponents of scanning, assembled a

Chapter 15

writing group on the application of EBCT coronary calcium scanning. This group concluded that EBCT had not yet been proven to be useful for screening asymptomatic subjects and that more research was needed [78,79].

Interestingly, this enormously expensive technology was the first technology that found some degree of popularity in the medical community. Of the more than 40 EBCT scanners in the year 2000 in the United States, most were used almost exclusively for coronary artery calcium screening. This technology, not fluoroscopy, not conventional CT, was the subject of not one but three reviews by national consensus panels [78–80]. The proponents of EBCT scanning claimed that its popularity was due to its increased accuracy in predicting prevalent, incident, and angiographic disease. There is only one study that we know of that directly compared EBCT with fluoroscopy for predicting coronary heart disease events. Taylor et al. [81] concluded in their report from that study that coronary fluoroscopy and CT imaging both provide prognostic information on the risk for developing coronary heart disease and that the prognostic accuracy of both techniques is modest, but is slightly greater for CT imaging. In fact, the ROC areas for both radiographic techniques were different but *not that different* (Figure 15.5).

In 2000, the application of CT calcium assessments was confined to screening asymptomatic populations for whom there was only scantly supportive data. This use began long before the publication of any prognostic data and before most of the cross-sectional studies were reported.

15.5.6   Multidetector CT Scanning for Calcium

Multidetector CT scanners produce low noise, high spatial resolution images with some motion artifact (Figure 15.6). These scanners were introduced on the American market in the late 1990s [82]. They were almost immediately applied to scanning for coronary calcium. Qanadli et al. reported on scanning 50 patients using a double detector array scanner [83]. However it was not until Carr et al. [84] and others [85] did comparison studies showing that EBCT and MDCT results were similar that the use of multidetector scanning got an endorsement in an application to coronary calcium in a large U.S. multicenter study, the Multi-Ethnic Study of Atherosclerosis (MESA) [86]. These scanners can be used in spiral [87] and in sequential [88] modes. The MESA study [86], mostly because of radiation dose considerations [89], used the sequential mode that has become standard. The MESA study also used EBCT scanners and used several models and generations of MDCT scanners during its course (4–16 slices). Choice of scanners was made based on availability and expertise of local epidemiological and imaging professionals. There was great concern regarding measurement error due to

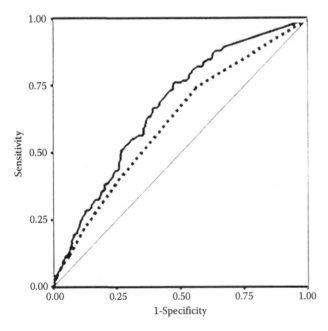

FIGURE 15.5 ROC curves comparing fluoroscopy (dotted) and EBCT (solid) for predicting coronary events. (From Qanadli, S.D. et al., *J. Comput. Assist. Tomogr.*, 25(2), 278, 2000.)

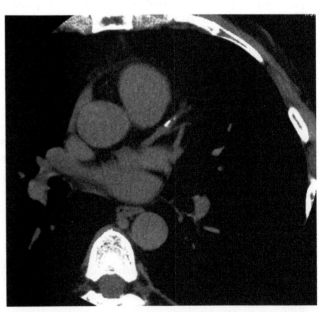

FIGURE 15.6 Multidetector CT scan image of coronary calcification.

scanner differences. When new scanners were added to MESA, both phantom and human inter-scanner reproducibility studies were done. Bone mineral density phantoms were scanned under every MESA participant in order to calibrate images and correct for attenuation differences.

15.6   Standard Method for Evaluating Coronary Risk versus Coronary Calcium

Cardiovascular risk scoring systems give an estimate of the probability, based on demographic and risk factors, that a person will develop cardiovascular disease within a specified amount of time, usually 10–30 years. The standard method in the United States is to use the Framingham risk score. This score for predicting future coronary heart disease events is given by the following equation:

$$p = 1 - \exp(-\exp(u))$$

where $u = (\log(10) - \mu)/\sigma$ length of follow-up = 10 years and $\mu = 15.5303 - 0.9119 \times \log(\text{systolic blood pressure}) - 0.2767 \times \text{smoking} - 0.7181 \times \log(\text{total/high density lipoprotein cholesterol}) - 0.5865 \times \text{electrocardiographic left ventricular hypertrophy} - 1.4792 \times \log(\text{age}) - 0.1759 \times \text{diabetes}$ and $\sigma = \exp(-0.3155 - 0.2784 \times (\mu - 4.4181))$.

Variables smoking, electrocardiographic left ventricular hypertrophy, and diabetes are set to 1 when present and 0 when absent. Systolic blood pressure is measured in mm Hg and age in years.

MESA showed that coronary calcium combined with Framingham risk score was superior in discriminating future events than were coronary calcium or Framingham risk score alone and that Caucasians, African Americans, Chinese Americans, and Hispanics with high scores were much more likely to suffer events than were those with zero or low scores (Figures 15.7 and 15.8). One surprising but heartening finding was that coronary calcium scores or volumes were quite similar and correlated almost perfectly between the different models, modes, generations of scanners, showing that the coronary calcium variable was very robust due to large inter-subject variability compared to inter-scanner variability.

The MESA study showed conclusively that coronary calcium is a significant and strong predictor of coronary events in four American ethnic groups, Caucasians, African Americans, Hispanics, and Chinese Americans [90]. This and subsequent reports [90–95] established strong experimental support for using coronary calcium scanning in asymptomatic subjects of intermediate risk for coronary heart disease. For those with risk higher than 3%, a coronary score over 100 increases the probability of a future event several times (Figure 15.9).

The MESA study publication and another publication [91] resulted in new guidelines which stated that coronary calcium had an accepted niche in preventive medicine that was to increment probabilistic diagnostic information for individuals at intermediate risk for coronary heart disease events. The new guidelines had

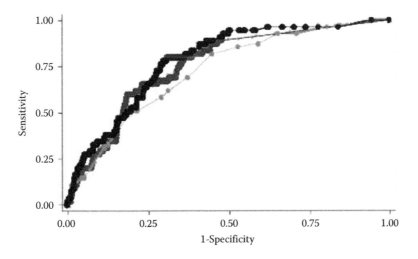

FIGURE 15.7   ROC curves show that coronary calcium score combined with Framingham risk score (black curve) has best discrimination for future coronary heart disease events when compared with coronary calcium score alone (dark gray curve) and Framingham risk alone (light gray curve).

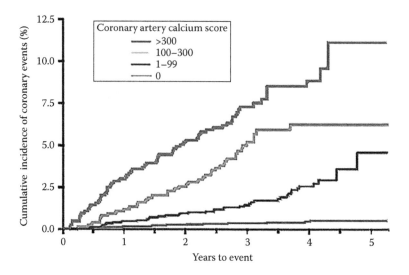

FIGURE 15.8   Cumulative incidence of coronary heart disease events for those with zero, 1–99, 100–300, >300 coronary calcium score in the MESA study.

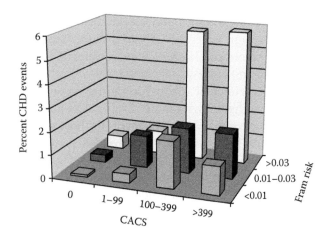

FIGURE 15.9   Change in risk of a coronary event caused by calcium scores in four different ranges.

already been released in 2007 [96] based on the findings of a separate report in *JAMA* [91] but were reinforced by the MESA results [90].

Specifically, the 2007 ACC/AHA guidelines recommended that it was reasonable to use CAC measurement in persons with a 10 year risk of a coronary event (sudden death, myocardial infarction, congestive heart failure [CHF], new angina) of 10%–20% but that it was not indicated to use coronary calcium scanning in those at lower or higher risk of such an event. This guideline was later updated to include those at "low to intermediate risk" meaning those between a 6% and 10% risk of a 10 year coronary heart disease event in addition to those between 10% and 20% risk.

Today, in 2014, there are hundreds of imaging centers using multidetector scanners, produced by a handful of companies, and a few using old, now out of production,

electron-beam scanners, which are actively conducting coronary calcium scanning in the United States. Many go beyond the guidelines and are scanning, not only those at low to intermediate risk (6%–20%) but also those at high (>20%) and low risk (<6%). However, there is an effort to stay within the guidelines.

15.6.1   Measuring Coronary Calcium

The Agatston calcium score is the primary measure of coronary calcium today.

The Agatston score is derived from the work of Agatston and Janowitz [71]. The score is calculated using a weighted value assigned to the highest density of calcification in a given coronary artery. The score is weighted with a coefficient of 1 for 130–199 HU, 2 for 200–299 HU, 3 for 300–399 HU, and 4 for 400 HU and greater. This weighted score is then multiplied by the volume (in cubic millimeters) of the coronary calcification in an image slice. The calcium score of every calcification in each coronary artery for all of the tomographic slices is then summed up to give the total coronary artery calcium score (CAC score).

Other measures have been proposed and tried. One that appeared intuitively promising was the calcium mass score. The mass score is simply a summation of Hounsfield units calibrated to the known mass of calcium phosphate in a phantom scanned with the subject. The units are in milligrams of calcium phosphate. Theoretically, the mass score should have less interscan variability and measurement error.

In reality, the mass score and the Agatston score are very closely correlated over individual patient measurements because the inter-subject variances of both scores are so high. There has not been any report of significant increase in accuracy of either score for predicting angiographic stenoses or coronary heart disease events.

Only one method has been presented that has any promise for improving on the utility of the Agatston coronary calcium score. This is the calcium spread score proposed by Brown et al. [97]. Brown re-scored all of the MESA scans using a method that included in the quantification score a measure of calcium spatial distribution over the lengths of the coronary arteries. She found that by including spatial distribution she could derive a variable, the spread score, which was more powerful in predicting coronary heart disease events than was the calcium score. Calcium that was spread "thin" over the arteries was more significant than calcium that was clumped together (usually near the ostia). This correlated with current thinking regarding the pathophysiology of coronary calcification and holds great promise. Unfortunately, Brown's method did not come into use because it required special software that was never commercialized.

15.6.2   Progression of Calcification

It was not long after imaging scientists started conducting coronary calcium scans that they also started studying changes or progression in coronary

calcification [98–102]. A few interesting facts emerged from this research on coronary calcium progression. First, calcium begets calcium [103]. The progression of calcium is directly and strongly related to the amount of calcium on the initial scan. Second, individuals with no calcium have very low probability of progressing to any detectable calcium within 5 years after the initial scan. This has been called the "warranty period" [98].

Despite an early start and much research, the optimal follow-up of those patients with repeat CAC testing remains unclear. The existing literature has covered two uses of coronary calcium progression in the management of subclinical coronary heart disease. On one hand, calcium progression has been used as a surrogate marker to test the efficacy of inhibition of atherosclerosis with medicines like statins. To date, study results have been mostly disappointing and progression of coronary calcium appears resistant to medications such as statins [102,104,105]. The second use of progression is as a kind of surrogate for the progression of atherosclerosis [106] suggesting that an individual is at even higher risk of an event if calcium progresses rapidly. As explained earlier, the surrogate approach has its downside in that multiple factors are at play in the calcification process itself and in the progression of calcification. More research is needed to determine whether repeat scanning for coronary calcium has a clinical role.

15.7   Conclusions

Coronary calcium presence and amount are reasonable surrogate markers for coronary atherosclerosis presence and amount, which in turn are moderately good predictors of coronary heart disease events. This is the theoretical model upon which the relevant research of the past 30 years has been based. Coronary calcium can be detected and evaluated using radiographic methods that utilize x-ray sources and detectors. The first reports of in vivo coronary calcium detection used standard chest x-rays. This was followed by fluoroscopy and cinefluoroscopy and then by computed tomographic methods. The latter can be divided into three types: standard early generation CT scanning, electron-beam CT scanning, and multidetector CT scanning (MDCT). Most coronary calcium exams today use multidetector scanning. The quantitative scoring systems using all of these technologies correlate reasonably well. The scores for different methods of

analyzing coronary calcium from computed tomographic scans correlate almost perfectly and there is also excellent correlation between scores of the same subjects deriving from different models of scanners. The reason why the technology does not make much difference in the relative scores between individuals is that calcium scores have large inter-individual variance, which makes these calcium scores a very robust variable for which different scanners and methods of analysis can be used with similar results.

Coronary calcium is usually measured as the Agatston calcium score. Other measures including calcium volume and calcium mass have been used but these measures correlate so well with the standard calcium score that they add almost no new information in ranking individuals in terms of amount or severity of coronary calcium deposition. The only calcium measure that has been shown to add significant diagnostic information is the calcium spread score,

Chapter 15

which is not generally used because of lack of availability of the necessary software.

Coronary calcium has been shown to predict coronary heart disease events and is now accepted and recommended for screening those at low to intermediate Framingham risk for a coronary heart disease event (6%–20% risk) but not for other persons. More research is needed to determine the clinical niche for repeat coronary calcium scanning.

Researchers should also consider research as to how less expensive technologies can be applied so as to make coronary calcium assessments more available and affordable to the general population. This is especially relevant because simpler less expensive technologies like fluoroscopy and less advanced and expensive CT scans have been shown to result in estimates of calcium amount that correlate closely to those of EBCT and MDCT scanners.

References

1. American Heart Association. *1999 Heart and Stroke Statistical Update.* American Heart Association, Dallas, TX; 1999.
2. Stone NJ. The clinical and economic significance of atherosclerosis. *Am J Med.* 1996;101(Suppl. 4A):6S–9S.
3. Husten L. Global epidemic of cardiovascular disease predicted. *Lancet.* 1998;352(9139):1530.
4. Murray JL, Lopez AD. Alternative projections of mortality and disability by cause 1990–2020: Global Burden of Disease Study. *Lancet.* 1997;349(9064):1498–1504.
5. Pearson TA, Smith SC Jr, Poole-Wilson P. Cardiovascular specialty societies and the emerging global burden of cardiovascular disease. A call to action. *Circulation.* 1998;97(6):602–604.
6. Reddy KS, Yusuf S. Emerging epidemic of cardiovascular disease in developing countries. *Circulation.* 1998;97(6):596–601.
7. Blankenhorn DH. Coronary arterial calcification: A review. *Am J Med Sci.* 1961;242:1–49.
8. Blankenhorn DH, Stern D. Calcification of the coronary arteries. *Am J Roentgenol.* 1959;81(5):772–777.
9. Mautner GC, Mautner SL, Froehlich J, Feuerstein IM, Proschan MA, Roberts WC, Doppman JL. Coronary artery calcification: Assessment with electron beam CT and histomorphometric correlation. *Radiology.* 1994;192(3):619–623.
10. Sangiorgi G, Rumberger JA, Severson A, Edwards WD, Gregoire J, Fitzpatrick LA, Schwartz RS. Arterial calcification and not lumen stenosis is highly correlated with atherosclerotic plaque burden in humans: A histologic study of 723 coronary artery segments using nondecalcifying methodology. *J Am Coll Cardiol.* 1998;31(1):126–133.
11. Stary HC. The sequence of cell and matrix changes in atherosclerotic lesions of coronary arteries in the first forty years of life. *Eur Heart J.* 1990;11(Suppl. E):3–19.
12. Stary HC. Evolution and progression of atherosclerotic lesions in coronary arteries of children and young adults. *Arteriosclerosis.* 1989;9(Suppl. I):I19–I32.
13. Strong JP. Landmark perspective: Coronary atherosclerosis in soldiers. A clue to the natural history of atherosclerosis in the young. *JAMA.* 1986;256:2863–2866.
14. White NK, Edwards JE, Dry TH. A correlation of the degree of coronary atherosclerosis with age in men. *Circulation.* 1959;19:47–69.
15. Lichtlen PR, Nikutta P, Jost S et al. Anatomical progression of coronary artery disease in humans as seen by prospective, repeated quantitated coronary angiography: Relation to clinical events and risk factors. *Circulation.* 1992;86(3):828–838.
16. Kaski JC, Tousoulis D, Pereira WI, Crea F, Maseri A. Progression of complex coronary artery stenosis in patients with angina pectoris: Its relation to clinical events. *Coron Artery Dis.* 1992;3:305–312.
17. Rafflenbeul W, Nellessen U, Galvao P, Kreft M, Peters S, Lichtlen P. Progression and regression of coronary artery disease as assessed by sequential coronary arteriography. *Z Kardiol.* 1984;73(Suppl. II):II-33–II-40.
18. Mönckeberg JG. Über die Atherosklerose der Kombattanten (nach Obdurtionsbefunden). *Zentralbl Herz Gefässkrankheiten.* 1915;7:7–10.
19. Mönckeberg JG. Anatomische Veränderungen am Kreislaufsystem bei Kreigsteilnehmern. *Zentralbl Herz Gefässkrankheiten.* 1915;7:336–343.
20. Enos WF, Holmes RH, Beyer J. Coronary disease among United States soldiers killed in action in Korea. *JAMA.* 1953;152:1090–1093.
21. Virmani R, Robinowitz M, Geer JC, Breslin PP, Beyer JC, McAllister HA. Coronary artery atherosclerosis revisited in Korean war combat casualties. *Arch Pathol Lab Med.* 1987;111(10):972–976.
22. Enos WF, Beyer JC, Holmes RH. Pathogenesis of coronary disease in American soldiers killed in Korea. *JAMA.* 1955;158(11):912–914.
23. McNamara JJ, Molot MA, Stremple JF, Cutting RT. Coronary artery disease in combat casualties in Vietnam. *JAMA.* 1971;216(7):1185–1187.
24. Wissler RW, Strong JP, and the PDAY Research Group. Risk factors and progression of atherosclerosis in youth. *Am J Pathol.* 1998;153(4):1023–1033.
25. McGill HC Hr (ed.). *The Geographic Pathology of Atherosclerosis.* Williams & Wilkins Co., Baltimore, MD; 1968.
26. Lenk R. X-ray diagnosis of coronary atherosclerosis in vivo. *Fortschr Rontgenstr.* 1927;35:1265.
27. Parade GW, Kuhlman F. Coronary atherosclerosis in vivo. *Klin Wsch.* 1933;12:1247.
28. Snellen HA, Nauta JH. Roentgen diagnosis of coronary heart disease. *Fortschr Rontgenstr.* 1937;56:577.
29. Hobbe JE, Wright HH. Roentgenographic detection of coronary arteriosclerosis. *Am J Roentgenol.* 1950;63:50.
30. Souza AS, Bream PR, Elliott LP. Chest film detection of coronary calcification: The value of the CAC triangle. *Radiology.* 1978;129:7–10.
31. Kelley MJ, Newell JD. Chest radiography and cardiac fluoroscopy in cardiac disease. *Cardiol Clin.* 1983;1(4):575–595.
32. Jorgens J, Blank N, Wilcox WA. The cinefluorographic detection and recording of within the heart: Results of 803 examinations. *Radiology.* 1960;74:550–554.

33. Lieber A, Jorgens J. Cinefluorography of coronary artery calcifications. Correlation with clinical arteriosclerotic heart disease and autopsy findings. *Am J Roentgenol.* 1961;86:1063.

34. Aldrich B, Brensike C, Battaglini A. Coronary calcifications in the detection of coronary artery disease and comparison with electrocardiographic exercise testing. Results from the National Heart Lung, and Blood Institute's Type II Coronary Intervention Study. *Circulation.* 1979;59:1113–1124.

35. Catellier T, Chua B, Youmans B, Waller MD. Calcific deposits in the heart. *Clin Cardiol.* 1990;13(4):287–294.

36. Chen JT. The significance of cardiac calcification. *Appl Radiol.* October 1992;11–19.

37. Detrano R, Froelicher V. Logical approach to screening for coronary artery disease. *Ann Int Med.* 1987;106(6):846–852.

38. Detrano R, Salcedo E, Hobbs J, Yiannikas J. Cardiac cinefluoroscopy as an inexpensive aid in the diagnosis of coronary artery disease. *Am J Cardiol.* 1986;57(13):1041–1046.

39. Detrano R, Simpfendorfer C, Day L, Salcedo E, Rincon M, Kramer, Hobbs J, Shirey J. Comparison of stress digital ventriculography, stress thallium scintigraphy, and digital fluoroscopy in the diagnosis of coronary artery disease in subjects without prior myocardial infarction. *Am J Cardiol.* 1985;56(7):434–440.

40. Froelicher V, Maron L. Exercise testing and ancillary techniques to screen for coronary heart disease. *Prog Cardiovasc Dis.* 1981;24(3):261–274.

41. Uretsky BF, Rifkin RD, Sharma SC, Reddy PS. Value of fluoroscopy in the detection of coronary stenosis: Influence of age, sex, and number of vessels calcified on diagnostic efficacy. *Am Heart J.* 1988;115(2):323–333.

42. Oliver MF. The diagnostic value of detecting coronary calcification. *Circulation.* 1970;42(6):981–982.

43. Gianrossi R, Detrano R, Colombo A et al. Cardiac fluoroscopy for the diagnosis of coronary disease: A meta-analytic review. *Am Heart J.* 1990;120:1179–1188.

44. Detrano R, Markovic D, Simpfendorfer C et al. Digital subtraction fluoroscopy, a new method of detecting coronary calcifications with improved sensitivity for the prediction of coronary disease. *Circulation.* 1985;71(4):725–732.

45. Molloi S, Detrano R, Ersahin A, Roeck W, Morcos C. Quantification of coronary arterial calcium by dual energy digital subtraction fluoroscopy. *Med Phys.* 1991;18(2):295–298.

46. Margolis JR, Chen JT, Kong Y, Peter RH. The diagnostic and prognostic significance of coronary artery calcification. A report of 800 cases. *Radiology.*1980;137(3):609–616.

47. Brody WR, Cassel DM, Sommer FG et al. Dual-energy projection radiography: Initial clinical experience. *AJR Am J Roentgenol.* 1981;137(2):201–205.

48. Riederer SJ, Kruger RA, Mistretta CA. Three-beam K-edge imaging of iodine using differences between fluoroscopic video images: Theoretical considerations. *Med Phys.* 1981;8(4):471–479.

49. Brody WR, Butt G, Hall A, Macovski A. A method for selective tissue and bone visualization using dual energy scanned projection radiography. *Med Phys.* 1981;8(3):353–357.

50. Fraser RG, Hickey NM, Niklason LT et al. Calcification in pulmonary nodules: Detection with dual-energy digital radiography. *Radiology.* 1986;160(3):595–601.

51. Fraser RG, Barnes GT, Hickey N et al. Potential value of digital radiography. Preliminary observations on the use of dual-energy subtraction in the evaluation of pulmonary nodules. *Chest.* 1986;89(4 Suppl.):249S–252S.

52. Kruger RA. Dual-energy electronic scanning-slit fluorography for the determination of vertebral bone mineral content. *Med Phys.* 1987;14(4):562–566.

53. Molloi S, Ersahin A, Tang J, Hicks J, Leung CY. Quantification of volumetric coronary blood flow with dual-energy digital subtraction angiography. *Circulation.* 1996;93(10):1919–1927.

54. Molloi S, Ersahin A, Hicks J, Wallis J. In-vivo validation of video densitometric coronary cross-sectional area measurement using dual-energy digital subtraction angiography. *Int J Card Imaging.* 1995;11(4):223–231.

55. Molloi S, Qian YJ, Ersahin A. Absolute volumetric blood flow measurements using dual-energy digital subtraction angiography. *Med Phys.* 1993;20(1):85–91.

56. Molloi SY, Ersahin A, Roeck WW, Nalcioglu O. Absolute cross-sectional area measurements in quantitative coronary arteriography by dual-energy DSA. *Invest Radiol.* 1991;26(2):119–127.

57. Weber DM, Molloi SY, Folts JD, Peppler WW, Mistretta CA. Geometric quantitative coronary arteriography. A comparison of unsubtracted and dual energy-subtracted images. *Invest Radiol.* 1991;26(7):649–654.

58. Molloi SY, Weber DM, Peppler WW, Folts JD, Mistretta CA. Quantitative dual-energy coronary arteriography. *Invest Radiol.* 1990;25(8):908–914.

59. Molloi SY, Mistretta CA. Quantification techniques for dual-energy cardiac imaging. *Med Phys.* 1989;16(2):209–217.

60. Van Lysel MS. Optimization of beam parameters for dual-energy digital subtraction angiography. *Med Phys.* 1994;21(2):219–226.

61. Gilkeson RC, Novak RD, Sachs P. Digital radiography with dual-energy subtraction: Improved evaluation of cardiac calcification. *AJR Am J Roentgenol.* 2004;183(5):1233–1238.

62. Fischbach F, Freund T, Rottgen R, Engert U, Felix R, Ricke J. Dual-energy chest radiography with a flat-panel digital detector: Revealing calcified chest abnormalities. *AJR Am J Roentgenol.* 2003;181(6):1519–1524.

63. Ducote JL, Xu T, Molloi S. Optimization of a flat-panel based real time dual-energy system for cardiac imaging. *Med Phys.* 2006;33(6):1562–1568.

64. Ducote JL, Xu T, Molloi S. Dual-energy cardiac imaging: An image quality and dose comparison for a flat-panel detector and x-ray image intensifier. *Phys Med Biol.* 2007;52(1):183–196.

65. Xu T, Ducote JL, Wong JT, Molloi S. Feasibility of real time dual-energy imaging based on a flat panel detector for coronary artery calcium quantification. *Med Phys.* 2006;33(6):1612–1622.

66. Reinmuller R, Lipton MJ. Detection of coronary artery calcification by computed tomography. *Dyn Cardiovasc Imaging.* 1987;1:139–145.

67. Masuda Y, Naito S, Aoyagi Y et al. Coronary artery calcification detected by CT: Clinical significance and angiographic correlates. *Angiology.* 1990;41(12):1037–1047.

68. Timmins ME, Pinsk R. The functional significance of calcification of the coronary arteries as detected on CT. *J Thorac Imaging.* 1991;7(1):79–82.

69. Moore E, Greenberg RW, Merrick SH et al. Coronary artery calcification: Significance of incidental detection on CT scans. *Radiology.* 1989;172:711–716.

70. Tanenbaum SR, Kondos GT, Veselik KE et al. Detection of calcific deposits in coronary arteries by ultrafast computed tomography and correlation with angiography. *Am J Cardiol.* 1989;63(2):870–872.

Chapter 15

71. Agatston AS, Janowitz WR, Hildner FJ, Zusmer N, Viamonte M, Detrano RC. Quantification of coronary artery calcium using ultrafast computed tomography. *J Am Coll Cardiol.* 1990;15(4):827–832.

72. Detrano R, Hsiai T, Wang S et al. Prognostic value of coronary calcification and angiographic stenoses in patients undergoing coronary angiography. *JACC.* 1996;27:285–290.

73. Breen JB, Sheedy PF, Schwartz RS et al. Coronary artery calcification detected with ultrafast CT as an indication of coronary artery disease. *Radiology.* 1992;185(2):435–439.

74. Fallavolita JA, Brody JS, Bunnell IL et al. Fast computed tomography detection of coronary artery calcification in the diagnosis of coronary artery disease. *Circulation.* 1994;89(1):285–289.

75. Nallamothu P, Saint S, Bielak LF et al. Electron beam computed tomography in the diagnosis of coronary artery disease: A meta-analysis. Personal communication. *Arch Intern Med.* 2001;161(6):833–838.

76. Detrano RC, Wong ND, Doherty TM, Shavelle R, Tang W, Gintzton LE, Budoff MJ, Narahara KE. Coronary calcium does not accurately predict near-term coronary events in high risk adults. *Circulation.* 1999;99(20):2633–2638.

77. Arad Y, Spadaro LA, Goodman K, Lledo-Perez A, Sherman S, Lerner G, Guerci AD. Predictive value of electron beam computed tomography of the coronary arteries. 19-month follow-up of 1173 asymptomatic subjects. *Circulation.* 1996;93(11):1951–1953.

78. Gibbons RJ, Balady GJ, Beasley JW et al. ACC/AHA guidelines for exercise testing: A report of the American College of Cardiology/American Heart Association Task Force on Practice Guidelines (Committee on Exercise Testing). *J Am Coll Cardiol.* 1997;30(1):260–311.

79. O'Rourke RA, Brundage BH, Froelicher VF et al. American College of Cardiology/American Heart Association expert consensus document on electron-beam computed tomography for the diagnosis and prognosis of coronary artery disease. *J Am Coll Cardiol.* 2000;36(1):326–340.

80. Grundy SM, Balady GJ, Criqui MH et al. Primary prevention of coronary heart disease: Guidance from Framingham: A statement for healthcare professionals from the AHA Task Force on Risk Reduction. *Circulation.* 1998;97(18):1876–1887.

81. Taylor AJ, O'Malley PG, Detrano RC. Comparison of coronary artery computed tomography versus fluoroscopy for the assessment of coronary artery disease prognosis. *Am J Cardiol.* 2001;88(6):675–677.

82. Hu H, He D, Foley D, Fox SH. Four multidetector-row helical CT: Image quality and volume coverage speed. *Radiology.* 2000;215:55–62.

83. Qanadli SD, Mesurollen B, Aegerter P, Joseph T, Oliva VL, Guertin M, Dubourg O, Fauchet M. Volumetric quantification of coronary artery calcifications using dual-slice spiral CT scanner: Improved reproducibility of measurements with 180° linear interpolation algorithm. *J Comput Assist Tomogr.* 2000;25(2):278–286.

84. Carr JJ, Crouse, JR, Goff, DC, Agostino RB, Peterson NP, Burke GL. Evaluation of subsecond gated helical CT for quantification of coronary artery calcium and comparison with electron beam CT. *AJR Am J Roentgenol.* 2000;174(4):915–921.

85. Danielli AL, Wong ND, Friedman JD, Ben-Yosef N, Berman DS. Concordance of coronary artery calcium estimates between MDCT and electron beam tomography. *AJR Am J Roentgenol.* 2005;185(6):1542–1545.

86. Carr JJ, Nelson JC, Wong ND, McNitt-Gray M, Arad Y, Jacobs DR Jr, Sidney S, Bild DE, Williams OD, Detrano R. Measuring calcified coronary plaque with cardiac CT in population based studies: The standardized protocol of the multi-ethnic study of atherslcerosis (MESA) and coronary artery risk development in young adults. *Radiology.* 2005:234:35–43.

87. Knez A, Becker C, Becker A, Leber A, White C, Reiser M, Steinbeck G. Determination of coronary calcium with multi-slice spiral computed tomography: A comparative study with electron-beam. *Int J Cardiovasc Imaging.* 2002;18(4):295–303.

88. Dijkstra H, Greuter MJ, Groen JM et al. Coronary calcium mass scores measured by identical 64-slice MDCT scanners are comparable: A cardiac phantom study. *Int J Cardiovasc Imaging.* 2010;26(1):89–98.

89. Bischoff B, Hein F, Meyer T, Krebs M, Hadamitzky M, Martinoff S, Schömig A, Hausleiter J. Comparison of sequential and helical scanning for radiation dose and image quality: Results of the prospective multicenter study on radiation dose estimates of cardiac CT angiography (PROTECTION) I study. *AJR Am J Roentgenol.* 2010;194(6):1495–1499.

90. Detrano R, Guerci A, Carr JJ et al. Coronary calcium as a predictor of near-term coronary heart disease events in major American ethnic groups: The Multi-Ethnic Study of Atherosclerosis (MESA). *NEJM.* 2008;358(13):1336–1345.

91. Greenland P, LaBree L, Azen S, Doherty T, Detrano R. Coronary artery calcium score combined with Framingham score for risk prediction in asymptomatic individuals. *JAMA.* 2004;291(2):210–215.

92. Folsom AR, Kronmal RA, Detrano RC, O'Leary DH, Bild DE, Budoff MJ, Liu K. Coronary artery calcification compared with carotid intima-media thickness in prediction of cardiovascular disease incidence: The Multi-Ethnic Study of Atherosclerosis (MESA). *Arch Intern Med.* June 23, 2008;168(12):1333–1339.

93. Budoff MJ, Nasir K, McClelland RL, Detrano R, Wong ND, Blumenthal RS, Kondos G, Kronmal RA. Coronary calcium predicts events better than age-sex-race/ethnicity percentiles. *J Am Coll Cardiol.* 2009;53(4):345–352.

94. Elias-Smale SE, Vliegenthart Proença R, Koller MT, Kavousi M, van Rooij FJA, Hunink MG, Oudkerk M, Witteman JCM. Coronary calcium score improves classification of coronary heart disease risk in the elderly. The Rotterdam Study. *J Am Coll Cardiol.* 2010;56(17):1407–1414.

95. Greenland P, Bonow RO, Brundage BH, Budoff BJ, Eisenberg MJ, Scott M, Grundy SM. ACCF/AHA 2007 Clinical expert consensus document on coronary artery calcium scoring by computed tomography in global cardiovascular risk assessment and in evaluation of patients with chest pain. *J Am Coll Cardiol.* 2007;49(3):378–402.

96. Greenland P, Alpert JS, Beller GA, Benjamin EJ, Budoff MJ, Fayad ZA. 2010 ACCF/AHA Guideline for Assessment of Cardiovascular Risk in Asymptomatic Adults: A Report of the American College of Cardiology Foundation/American Heart Association Task Force on Practice Guidelines Writing Committee Members. *Circulation.* 2010;122;e584–e636.

97. Brown ER, Kronmal RA, Bluemke DA, Guerci AD, Carr JJ, Goldin J, Detrano R. Coronary calcium coverage score: Determination, correlates, and predictive accuracy in the Multi-Ethnic Study of Atherosclerosis. *Radiology.* 2008;247(3):669–675.

98. Min JK, Lin FY, Gidseg DS et al. Determinants of coronary calcium conversion among patients with a normal coronary calcium scan: What is the "warranty period" for remaining normal? *J Am Coll Cardiol*. 2010;55(11):1110–1117.

99. Gopal A, Nasir K, Liu ST, Flores FR, Chen L, Budoff MJ. Coronary calcium progression rates with a zero initial score by electron beam tomography. *Int J Cardiol*. 2007;117(2):227–231.

100. Hokanson JE, MacKenzie T, Kinney G et al. Evaluating changes in coronary artery calcium: An analytic method that accounts for interscan variability. *AJR Am J Roentgenol*. 2004;182(5):1327–1332.

101. Kronmal RA, McClelland RL, Detrano R et al. Risk factors for the progression of coronary artery calcification in asymptomatic subjects: Results from the Multi-Ethnic Study of Atherosclerosis (MESA). *Circulation*. 2007;115(21):2722–2730.

102. Raggi P, Callister TQ, Shaw LJ. Progression of coronary artery calcium and risk of first myocardial infarction in patients receiving cholesterol-lowering therapy. *Arterioscler Thromb Vasc Biol*. 2004;24(7):1272–1277.

103. Yoon HC, Emerick AM, Hill JA, Gjertson DW, Goldin JG. Calcium begets calcium: Progression of coronary artery calcification in asymptomatic subjects. *Radiology*. 2002;224(1):236–241.

104. Arad Y, Spadaro LA, Roth M, Newstein D, Guerci AD. Treatment of asymptomatic adults with elevated coronary calcium scores with atorvastatin, vitamin C, and vitamin E: The St. Francis Heart Study randomized clinical trial. *J Am Coll Cardiol*. 2005;46(1):166–172.

105. Burgstahler C, Reimann A, Beck T et al. Influence of a lipid-lowering therapy on calcified and noncalcified coronary plaques monitored by multislice detector computed tomography: Results of the New Age II Pilot Study. *Invest Radiol*. 2007;42(3):189–195.

106. Budoff MJ, Hokanson JE, Nasir K et al. Progression of coronary artery calcium predicts all-cause mortality. *JACC Cardiovasc Imaging*. 2010;3:1229–1236.

Chapter 15

Time-Resolved Imaging IV

16. Motion Control in Cardiovascular Imaging

Michael Grass and Dirk Schäfer

16.1 Introduction

The motion of human organs is one of the major challenges in medical imaging. While an imaging system may deliver excellent images of the static anatomy, results strongly depend on the data acquisition speed when imaging moving organs. The motion occurring during the acquisition of the image yields motion blurring or artifacts. While blurring describes a loss of spatial resolution, image artifacts refer to additional artificial high-frequency structures in the images that obscure the anatomical information. In order to remove these limitations, data acquisition speed was increased and strategies to control motion during the image generation process have been developed. These strategies are based on the measurement of a signal representing the motion occurrence or strength, a signal describing its temporal evolution or even spatially resolved measurements of motion strength and direction. Subsequently, this information is integrated into the data acquisition process or into the image generation and reconstruction process.

In this chapter, different strategies to control motion in cardiovascular imaging are reviewed. The review includes prospective and retrospective gating approaches using physiological signals, motion vector field estimation, and motion-compensated reconstruction. The state of the art in motion control in cardiovascular imaging is presented and related to the current performance of today's most widely used medical imaging modalities.

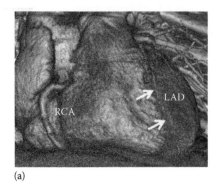

(a)   (b)   (c)

FIGURE 16.1   Retrospectively gated cardiac CT scan of a patient with 82 bpm heart rate. (a) Shows an anterior–posterior volume-rendered view at end systole (40% RR interval) showing a sharp right (RCA) and a blurred left coronary artery (LCA). (b) and (c) Compare a zoomed reconstruction of the LCA reconstructed at 40% and 34% RR interval. (Reprinted from Schmitt, H. et al., Spatially resolved automatic cardiac rest phase determination in coronary computed tomography angiography (CTA), *IFMBE Proceedings of the World Congress on Medical Physics and Biomedical Engineering 2009*, Munich, Germany, vol. 25, issue no. 2, pp. 162–165, 2009.)

16.2   Motion in Cardiovascular Imaging

There are basically two types of motion, which dominate in the arena of cardiovascular imaging. These are the motion of the heart itself and breathing motion. Both are quasiperiodic movements, which differ in their frequency and motion pattern.

Cardiac motion is the motion of the heart muscle triggered by the electrical signal from the sinuatrial node. The electrical excitation pattern leads to a fast contraction of the heart with a subsequent relaxation and resting phase. The length of a cardiac cycle, which may be observed in cardiac imaging, is on the order of 400–1500 ms, resulting in heart rates of 150 beats per minute (bpm) down to 40 bpm. Practical values occurring during scans are in the range of 50–80 bpm. The motion pattern is quasiperiodic with variations of the motion from cycle to cycle. In addition, a heart rate dependency of the motion signal can be observed. With increasing heart rate, the relaxation and resting phase may decrease in its duration; however, the contraction phase remains almost unchanged. Moreover, it can be determined that different parts of the heart have their resting period at different relative time points in the cardiac cycle (see Figure 16.1) (Vembar 2003, Schmitt 2009).

The duration of a breathing cycle and the breathing motion can be controlled within limits.

The duration of a typical breathing cycle is on the order of 3–6 s, resulting in breathing rates of 10–20 cycles per minute. In contrast to cardiac motion, the motion pattern of the breathing motion can be modified, leading, for example, to a deep or a shallow breath or to thoracic or abdominal breathing. The motion pattern itself can show strong variations and is significantly less reproducible than the cardiac motion pattern.

It should be noted that cardiac and breathing motion not only overlap in the temporal domain, but also in the spatial domain. Lung tissue and pulmonary vessels close to the heart may move due to the neighboring cardiac motion (von Berg 2007). At the same time, the overall motion of the diaphragm and the thorax due to breathing may displace and deform the heart. In addition to the pure motion of the heart, secondary effects may occur due to the pulsation of the blood. This leads to periodic changes of the position and size of vascular structures.

In this chapter, we focus on techniques to control cardiac motion in medical imaging. Breathing motion will be discussed for imaging modalities with a scan time exceeding a single breath hold. Other types of motion like bowel-gas motion or muscular patient motion are not covered.

16.3   Measuring the Motion Signal

In order to measure the motion signal of the heart, most frequently an electrocardiogram (ECG) is acquired in parallel to the acquisition of the image data. From the ECG (see Figure 16.2), the R-peaks are determined and serve as reference for the start and end points of each cardiac cycle. A multitude of algorithms are available to detect them in a fully automatic manner (see Köhler 2002 and references therein).

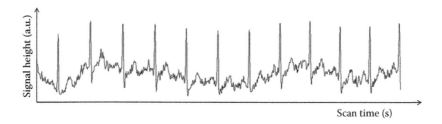

FIGURE 16.2   Measured ECG of a patient. The strong peaks clearly indicate the position of the R-peaks.

Assuming a measurement of the ECG from the beginning of the image data acquisition t_s to the end t_e, a list of N_r R-peaks is measured at the temporal positions T_r during the acquisition time of the imaging system. In order to rescale the cardiac cycles of varying lengths, a warped time axis τ may be introduced, which maps each cardiac cycle to a cycle of equal length (see Figure 16.3 for a schematic representation of the mapping process). In its simplest form, the mapping is represented by a description of each cardiac cycle as a percentage of the RR-interval. More complex, nonlinear mapping functions to account for the nonlinear scaling of the motion pattern with increasing heart rate have been introduced (Vembar 2003) but are not widely used today.

In case that the ECG is only available during the scan time of the imaging system, the first R-peak prior and post the acquisition window is not accessible. A natural strategy to handle this issue is to mirror the first and last cardiac cycle inside the acquisition window across this border.

Finally, a list of R-peaks T_r is available, with $r \in [-1, N_r + 1]$ and $T_{-1} = T_0 - (T_1 - T_0)$ and $T_{N_r+1} = T_{N_r} + (T_{N_r} - T_{N_r-1})$. Phase points P_r for the definition of cardiac motion states during the cardiac cycle may either be defined on the true (P_r^{true}) or warped (P_r^{warp}) temporal axis. In case of prospective cardiac triggering, as it is, for example, used in step and shoot cardiac CT acquisitions (Hsieh 2006) or in cardiac gated x-ray fluoroscopy (Wang 2011), the true temporal axis is used. With, for example, $P_r^{true} = T_r + \Delta T$ and ΔT specified in milliseconds, the start of the image data acquisition can be controlled. It is just based on the last detected R-peak and the length of the prior RR-interval. The end of the current cardiac cycle does not need to be known, though strong intercycle variations of the RR interval will lead to reduced image quality.

The warped time axis is most widely used in combination with retrospectively gated reconstruction methods, for example, not only in computed tomography (CT), magnetic resonance imaging (MRI), or nuclear medicine examinations but also in 3D ultrasound

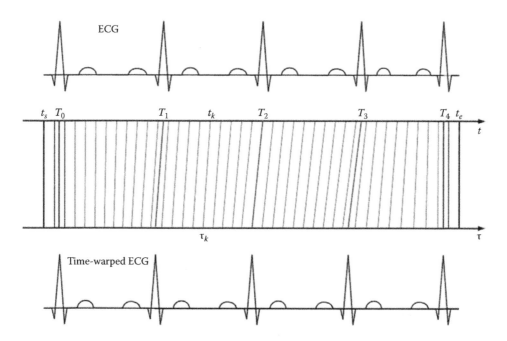

FIGURE 16.3   Schematic representation of an ECG measured on the true time axis (upper graph). Mapping of the true measured R-peaks on a warped time scale and resulting time-warped ECG with equidistant R-peaks (lower graph).

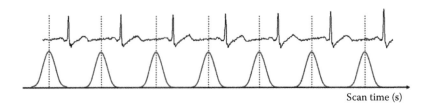

Scan time (s)

FIGURE 16.4 Schematic representation of an ECG and the selected phase points in each cardiac cycle (dashed lines) as well as cardiac gating function, which is nonzero around the phase point and has decreasing weights with increasing distance from the phase point (solid lower lines).

acquisitions. Here, $P_r^{warp} = \tau_r + \Delta\tau$ and $\Delta\tau$ are specified as a percentage of the RR-interval.

Each phase point may be associated with a gating window width ΔW_r and a cardiac weighting function ΔF_r, which is only nonzero inside the gating window. Usually, the phase point defines the center or the start point of the gating window. Most often the cardiac weighting function is used to weigh the measured data. It is a function decreasing with increasing temporal distance from the chosen phase point (see, e.g., Figure 16.4). Various functions have been suggested as cardiac gating functions, and they determine the temporal resolution and the image quality of the resulting gated image (Manzke 2004a, Nielsen 2005).

As an alternative to the ECG waveform data, which reflects the electrical potential propagation and which are subject to noise within a clinical environment, there has been a multitude of attempts to extract motion information from different sources. For example, in rotational x-ray and in CT imaging, the projection data themselves have been evaluated (Kachelriess 2002, Bruder 2003, Blondel 2006). Such techniques showed initially promising results, but did not manifest themselves in the clinical routine yet. A potential reason is the fact that depending on the heart rate and the scanner rotation it is difficult to distinguish the real cardiac signal from other signal variations, produced by, for example, noise, the scanner rotation, or breathing motion.

Similarly, in cardiac MRI, several techniques are known to determine an appropriate trigger delay for selecting stable cardiac phases. For example, patient-specific rest phase models or navigator echoes placed on specific landmarks of the heart can be used to control the acquisition and optimize the image quality (Wang 1996, Stuber 1999, Wang 2001). Within the MR domain, these navigator techniques that represent local image information are a very flexible tool to determine characteristics of both cardiac and respiratory motions.

Phonocardiography and Doppler ultrasound information may be taken as an additional source of information about the patient-specific heart motion (Wang 2002, Greenberg 2003). Merging such techniques with different imaging modalities to obtain scan-specific motion information is challenging due to the increased complexity of the procedure and technical limitations.

16.4 Motion-Correlated Cardiovascular Imaging

A signal representing the cardiac motion like the ECG can be used to correlate the acquisition of image data in a prospective manner. Alternatively, the signal can be used to reorder measured image data retrospectively. It is either the data acquisition speed or the data acquisition control that determines the chosen approach.

An overview on the different imaging modalities and their most prominent approaches to control motion in cardiovascular imaging is given in the following sections.

16.4.1 Computed Tomography

In its early phase, CT has not been a cardiac imaging modality. The first cardiac experiments with slow single-slice imaging have already been attempted in the 1980s (Lackner 1981). Due to the limited temporal and spatial resolution, these experiments did not yield results of broad clinical relevance. In the late 1990s, the availability of CT gantries with a rotation time of a fraction of a second and multi-row detectors led to renewed interest in cardiac CT. Modern scanners offer gantries with rotation times of less than 0.3 s and up to 320 detector rows, thus enabling noninvasive, high-quality cardiac and coronary imaging.

CT imaging of moving objects requires, on the one side, the selection of an acquisition time window with little motion and, on the other side, that for each object point in the field of view sufficient data for image reconstruction have been acquired. The latter so-called completeness criterion in 2D CT means that projection rays covering and angular range of 180° have to be available for every voxel. Data can be acquired with a multitude of axial scans at different patient positions (Köhler 2001, Hsieh 2006) or with continuous gantry rotation and table movement, that is, helical scanning (Kachelriess 2000, Grass 2003). Cardiac imaging can be achieved in the most straightforward manner when acquiring CT datasets with prospective triggering with step and shoot acquisitions (Hsieh 2006). In retrospectively gated helical scanning, the pitch has to be chosen sufficiently low to ensure sufficient projection data for every voxel in the field of view (Grass 2003).

The temporal resolution in cardiac CT scanning is determined by the temporal range of the projection data contributing to the reconstruction of each voxel. For single-cycle reconstructions, which are used most often for step and shoot acquisitions, the temporal resolution is determined by the rotation speed of the scanner and corresponds to the projection acquisition time of 180° plus fan angle. The data for each voxel are collected during the same cardiac cycle. In the case of retrospectively gated multicycle reconstruction, the temporal resolution is dependent on rotation time, heart rate, and pitch for helical data (see Figure 16.5).

As a consequence, the so-called scanner harmonics result from certain combinations of the heart rate, pitch, and rotation time (for details, see Manzke 2003). At these harmonics, the temporal resolution of the image is limited—as in single-cycle reconstructions—to the time required for a 180° acquisition. A comparable pattern for the behavior of the temporal resolution can be observed for retrospectively gated multicycle circular cardiac reconstruction (Stevendaal 2006).

With 4D image data being available from a retrospectively gated helical scan, image-based rest phase determination became feasible (Manzke 2004b, Hoffmann 2006). In this "motion map" concept, the similarity between successive temporal images along a cardiac cycle is calculated. Low similarity represents strong motion, while high similarity corresponds to little motion. Thus, the gating window position can be optimized based on the image data themselves (see Figure 16.6). This approach can be seen as a prestep towards active motion compensation as it is currently the subject of ongoing medical imaging research.

A cardiac CT image acquired with step and shoot acquisition can be found in Figure 16.7. Next to the heart itself, gated acquisition and reconstruction is, for example, applied for aortic imaging (see Figure 16.8).

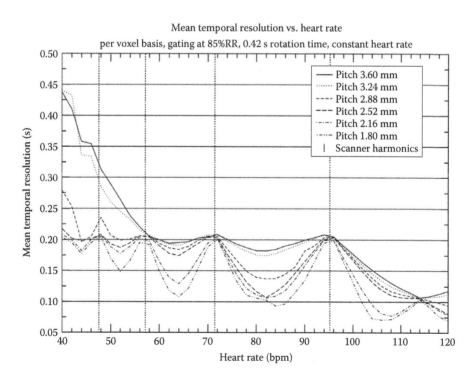

FIGURE 16.5 Exemplary representation of the temporal resolution of a helical CT scanning system with fixed rotation time as a function of the heart rate and the scanner pitch. (Reprinted from Manzke, R. et al., *Med. Phys.*, 30(12), 3072, 2003.)

Chapter 16

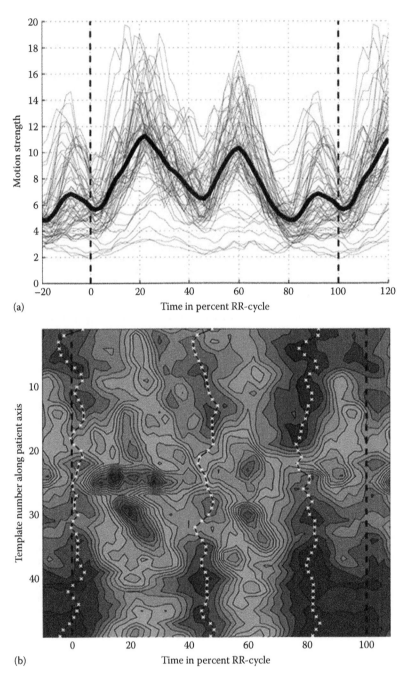

FIGURE 16.6 Motion strength curve and "motion map" sample. (a) Representative mean curve of the motion strength function for all voxels within a single axial plane. Motion strength (corresponding to inverse template similarity) is plotted against the phase point within one cardiac cycle. The curve shows little motion troughs for the end-systolic phase (approx. 46%) and the mid-diastolic phase (approx. 80%). (b) Corresponding "motion map" (dark: little motion; bright: strong motion). The motion strength is plotted depending on the cardiac phase (x-axis) and the patient axis (y-axis). The y-axis spans from the aortic root position ($y = 0$) down to the diaphragmatic surface of the heart ($y = 49$). Phases of little motion become apparent as dark vertical valleys between systolic contraction (0%–40%) and rapid diastolic filling (50%–70%). Atrial contraction is apparent as a hump around 90%. Lines with crosses track the valleys of least motion. Dashed vertical lines mark the beginning and the end of the cardiac cycle (R to R peak). Each horizontal line of (b) is a motion strength curve as illustrated in (a) at successive patient axis positions (or template positions). (Reprinted from Hoffmann, M. et al., *Eur. Radiol.*, 16, 365, 2006.)

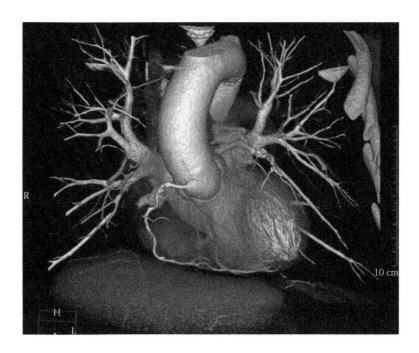

FIGURE 16.7 Volume rendering of a cardiac CT datasets acquired with step and shoot imaging. (Image by courtesy of Philips Healthcare Haifa, Israel.)

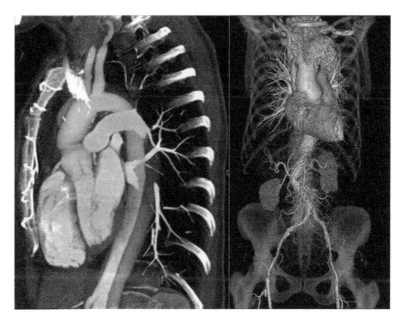

FIGURE 16.8 Left: A volume intensity projection (VIP) of a prospectively gated aorta scan performed at 100 kVp, 140 mAs, length 66 cm; right: A volume-rendered representation of a follow-up routine helical aorta scan performed for the assessment of aortic stent at 80 kVp, 130 mAs, length 63 cm. (Images by courtesy of Philips Healthcare Cleveland, USA.)

16.4.2   Magnetic Resonance Imaging

The first motion pattern that has to be controlled in cardiac MRI imaging is breathing motion. Due to the high flexibility of MRI, imaging free breathing cardiac MRI can be achieved using navigator measurements (Ehmann 1989, Nehrke 1999). An MR-navigator pulse is a rod-shaped excitation pulse, which is positioned at the top of the right hemidiaphragm and determines when the diaphragm is in a reproducible, for example, end-expiration position (see Figure 16.9).

In order to avoid motion artifacts due to cardiac motion, prospective ECG triggering is the most prominent approach being used today in cardiac MRI. The R-peak is used as a navigator signal inside the

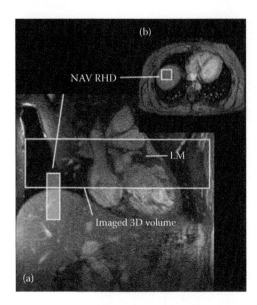

FIGURE 16.9 Planning of the three-dimensional (3D) volume and the navigator (NAV) at the dome of the right hemidiaphragm (RHD). The images show a coronal (a) and a transverse (b) view as acquired during the first scout scan. (Images from Stuber, M. et al., *J. Am. Coll. Cardiol.*, 34(2), 524, 1999b.)

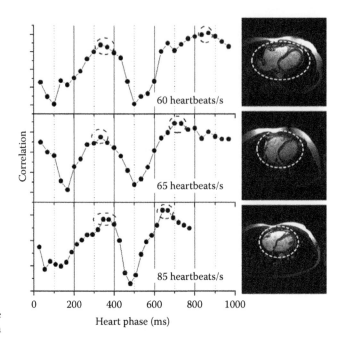

FIGURE 16.10 Determination of the optimal data acquisition window: Automatic determination is performed by calculating the cross-correlation (left) between subsequent cardiac cine images (right) for three different volunteers. The maxima of the curves coincide with the manually determined resting phases. (Image by courtesy of K. Nehrke, Philips Research, Hamburg, Germany)

cardiac cycle. The cardiac phase and gating window are selected according to the measured mean heart rate. In MRI coronary angiography, an acquisition window of 80–150 ms is selected in the diastole, representing the phase of least motion in the cardiac cycle (Botnar 1999). Different empirical formulae to estimate the optimal acquisition window as a function of the heart rate can be found in the literature (Stuber 1999). More accurate estimations can be achieved when using an image-based patient-specific window estimation. It has been suggested to determine the phase of least motion manually or fully automatic from a multiphase cardiac overview acquisition (see Figure 16.10) (Kim 2001a, Nehrke 2003). These approaches correspond to the CT "motion map" techniques mentioned in the preceding section.

Alternatively, cardiac navigators can be used for this purpose (Wang 2001). Widening the acquisition window so that it exceeds the resting phase requires motion correction techniques. For MRI, vessel tracking approaches have been suggested (Saranathan 2001), which enable a slice positioning adapted to the vessel motion during data acquisition. Here, patient adaptive and patient-independent motion models have been used. In order to adapt a cardiac motion model in a patient-specific manner, labeling techniques have been applied. Thus, a selected slice can be tracked throughout the cardiac cycle and is directly related to the measured ECG (Kozerke 1999).

Navigator-based real-time respiratory gating and ECG triggering are technically robust methods, which

enable clinically relevant high-resolution cardiac MR imaging (Kim 2001b) and deliver excellent image quality (see Figure 16.11). Nevertheless, approaches to compensate motion during the image acquisition and reconstruction process are an area of ongoing research. The combination of compressed sensing approaches and motion correction techniques to approach free breathing cardiac MRI is a topic of current research (see Usman 2013 and reference therein).

16.4.3   Ultrasound

Cardiac ultrasound or echocardiography is probably the most widely used imaging tool in cardiology. It is a noninvasive imaging tool, which delivers real-time 2D and 3D information of the heart's anatomy. In general, echocardiography is ideal for the evaluation of cardiac motion due to its acquisition speed. The excellent speed is enhanced by methods to measure motion and thereby functional parameters like Doppler tissue velocity measurements (Yoshida 1961, McDicken 1992) or speckle tracking echocardiography (Leitman 2004, Pirat 2008). Motion parameters for displacement, velocity, and strain can be derived (see, e.g., de Craene 2010). An overview about echocardiography techniques for evaluating the cardiac

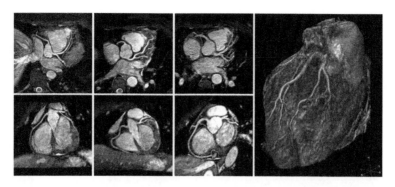

FIGURE 16.11   Free-breathing, whole heart 3D coronary MR angiograms. Different reformattings (top row left: LM/LAD/LCX, bottom row: RCA/LCX) are shown for three selected volunteers (from left to right). A volume-rendered reconstruction of such a dataset shows the LM/LAD branch. Measurements were performed using a 32-channel MRI system with a total scan time of four minutes during free breathing.

motion can be found in Mor-Avi (2011) and references therein.

Due to the fast acquisition speed, techniques to control motion in echo imaging are rarely used. Gated acquisitions are primarily employed when 3D echocardiography examinations are performed. This is due to physical limitations in the measurement of ultrasound signals, which propagate at a speed of approximately 1500 m/s in the human body. For a typical 16 cm scanning depth, an ultrasound pencil beam can be transmitted and received in only 220 µs. For a 2D image with 90 ultrasound lines, about 20 ms is required. The frame

rate for such an image is 50 Hz (Houck 2006). For 2D cardiac imaging at normal heart rates, this frame rate is adequate. For full 3D echocardiography, this is a limitation and various techniques have been applied to overcome this problem. These include electronic slice selection and beam narrowing that basically correspond to a limitation of the field of view. Three-dimensional imaging with sparse acquisition matrices is also applied to overcome this problem (Houck 2006). Retrospective gating and image stitching is required to generate full-resolution 3D images of each single cardiac phase (see Figure 16.12). The combination of

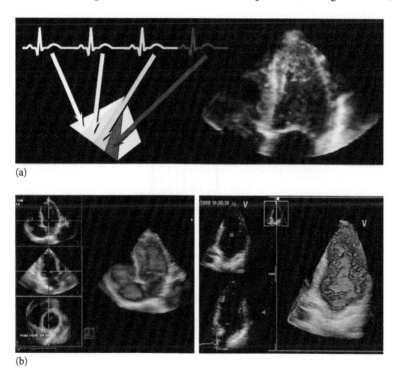

(a)

(b)

FIGURE 16.12   (a) Example of electrocardiographically triggered multiple-beat 3D data acquisition from a transthoracic apical window. Narrow pyramidal volumes from four cardiac cycles (left) are stitched together to form a single volumetric dataset (right). (b) Real-time or live 3D single-beat acquisition of the whole heart (left) and the left ventricle (right) from the transthoracic apical window. (Images from Lang, R. et al., *J. Am. Soc. Echocardiogr.*, 25, 3, 2012.)

Chapter 16

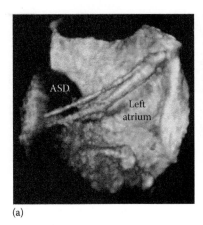
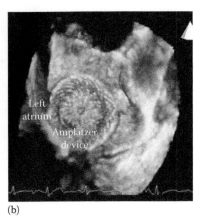

(a)   (b)

FIGURE 16.13 3D TEE acquired during an intracardiac intervention; arterial septal defect (ASD) closure procedure. (a) Monorail catheter passing through an ASD as seen from the left atrial perspective (3D zoom mode acquisition). (b) Closure device properly placed, as seen from the left atrial perspective (3D zoom mode acquisition). (Images from Perk, G. et al., *J. Am. Soc. Echocardiogr.*, 22(8), 865, 2009.)

information acquired at the same cardiac phase in four to seven cardiac cycles delivers full volumetric information. Examples of transthoracic (TTE) (with and without stitching) and transesophageal (TEE) (without stitching) echo images are given in Figures 16.12 and 16.13.

16.4.4   Nuclear Imaging

Using retrospectively ECG-gated image reconstruction is a standard approach in cardiac single photon emission tomography (SPECT) and positron emission tomography (PET) today. Clearly, this is due to the long acquisition time caused by low isotope concentrations. Prospective triggering is not feasible.

Cardiac SPECT is the most widespread cardiac examination in the nuclear imaging domain. The radioactive tracers mostly used for imaging and quantification of myocardial perfusion are ^{99m}Tc and ^{201}Tl. Due to the long acquisition time of up to 20–30 min, both breathing and cardiac motion occur. While breathing motion is mainly neglected, cardiac motion is corrected via retrospective ECG gating (see Figures 16.14 and 16.15 and, e.g., Cullom (1998) and Paul (2004) and references therein).

In cardiac PET, the most frequently used tracers are ^{82}Rb and ^{13}N-ammonia. Again, as in SPECT, retrospective gating is used for motion control during the scan duration of 5–20 min. The gated frames are reordered into 8–16 cardiac gating windows and

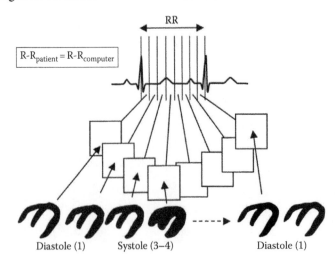

FIGURE 16.14 Principle of ECG-gated SPECT acquisition (eight-frame model). For each angular projection, separate temporal frames are acquired. Perfusion projection images are obtained from summation of the individual frames. (Images from Cullom, S. et al., *J. Nucl. Cardiol.*, 5, 418, 1998.)

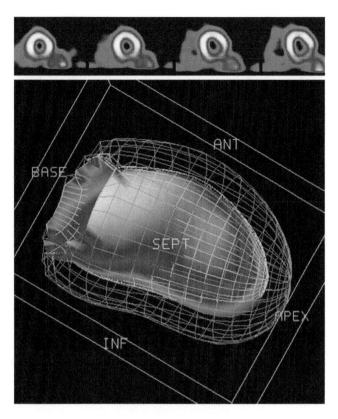

FIGURE 16.15 Slices reconstructed from ECG-gated SPECT measurements for assessment of myocardial perfusion and corresponding 3D functional map. (Images by courtesy of Philips Healthcare, Cleveland, USA.)

reconstructed (see Di Carli 2007). Data acquired during ectopic beats are typically rejected. Approaches for parallel cardiac and respiratory gating have been applied to cardiac PET imaging in order to improve the spatial resolution of the images (Martinez-Moeller 2007).

16.4.5 Interventional X-Ray

Fluoroscopic x-ray projections with a C-arm are acquired during cardiovascular interventions for real-time 2D visualization and guidance of guide wires and catheters using low x-ray dose with frame rates of 15 or 30 Hz. ECG-triggered fluoroscopy acquires only a single 2D projection per cardiac cycle (Wang 2011) for dose saving during long interventions such as electrophysiology ablation procedures.

Vascular structures filled with contrast agent are assessed by angiographic 2D image sequences. Coronary angiograms, for example, are acquired from a fixed C-arm angle for several heart beats to assess coronary stenosis, flow, and motion patterns. These acquisition schemes have the advantage of delivering temporal resolution on the order of the x-ray pulse length in the range of 2–10 ms but may present vascular structures in suboptimal directions with overlaps to other structures or foreshortened.

The disadvantage of the projective geometry in 2D imaging may be overcome by rotating the C-arm gantry around the patient while acquiring the angiograms (Klein 2009). Three-dimensional information can be extracted from these so-called rotational x-ray data using tomographic reconstruction methods (Lauritsch 2006). However, a typical C-arm rotation with 50°/s is much slower compared to CT impeding complete acquisition of projection data in a single resting phase of the heart.

For some clinical applications, the short-term suppression of the cardiac motion, for example, by fast pacing (hyperpacing) or medication with adenosine provides a solution. Hyperpacing is used, in minimally invasive transcatheter aortic valve replacement (TAVR) together with 3D rotational angiography (Lehmkuhl 2013) for 3D imaging of the aortic root. For 3D imaging of the left atrium (LA) and pulmonary veins (PV) during electrophysiological ablation procedures, either hyperpacing or administration of adenosine may be used (Kriastelis 2009).

Chapter 16

For interventions concerning the coronary arteries (percutaneous coronary interventions [PCIs]), retrospective ECG gating may be used combined with rotational angiography. The projection data can be used in two view or multiview modeling (Chen 2009) or 3D reconstruction (Hansis 2008a, 2010). The advantage of 3D coronary angiography becomes apparent in the assessment of complex vascular lesions. The 2D angiogram in Figure 16.16a shows a lesion in the left anterior descending (LAD) artery. The 3D reconstruction (Figure 16.16b) characterizes the nature of this lesion better. ECG-gated cardiac reconstructions of the whole heart have also been tested in combination with intravenous contrast injection and single and multiple rotational C-arm acquisitions (Lauritsch 2006, Schwartz 2011, Mory 2014)

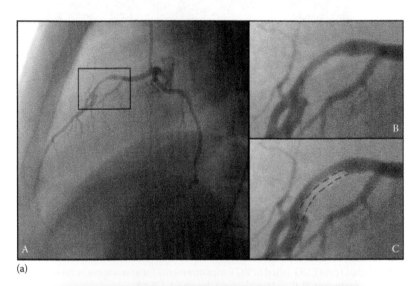

(a)

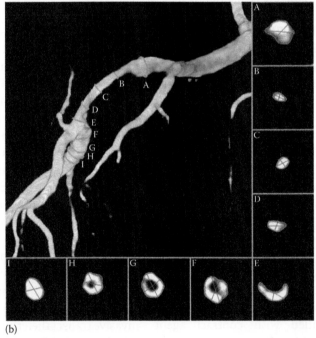

(b)

FIGURE 16.16   (a) 2D angiogram from rotational coronary angiography: (A) Lateral left view of the LCA injection. Lesion location is indicated with a box and enlarged images (B and C). (C) Thrombus is indicated with dashed line. (Images from Schoonenberg, G. et al., *Catheter. Cardiovasc. Interv.*, 74, 97, 2009.) (b) 3D reconstruction from rotational coronary angiography corresponding to Figure 16.16a: (A, I) Normal vessel lumen proximal and distal to the thrombus, respectively. (B–D) The proximal part of the thrombus causes a reduced area in which the contrast is clearly visible. (E) The midsection shows a moon-shaped area. (F–H) More distally a donut shape can be seen where the thrombus is free from the vessel wall. (Images from Schoonenberg, G. et al., *Catheter. Cardiovasc. Interv.*, 74, 97, 2009.)

16.5 Motion-Compensated Cardiovascular Imaging

Active correction of motion in cardiovascular imaging has been and still is a topic of ongoing research in all major imaging modalities. Although there has been significant improvement in the imaging performance of different systems over the years with respect to temporal and spatial resolution, the task to compensate motion inside the imaging process is still a challenge. In order to actively compensate motion during imaging, it is insufficient to measure a signal representing the motion as, for example, an ECG, a navigator, or the difference of subsequent images. It is required to determine the motion vector field representing the true object displacement as a dense motion vector field for the time interval needed to measure the data that are used for the reconstruction of the final image. Methods to calculate motion vector fields from image sequences are described in the next section, while the use of these data in motion-compensated reconstruction is sketched in Section 16.5.2.

16.5.1 Calculation of Motion Vector Fields

In case that a 4D image sequence is available for the estimation of a motion vector field, there are a multitude of different methods to determine the object motion. These methods mainly originate from the image-processing domain, where a deep knowledge on image registration and analysis methods is available.

A motion vector field $\vec{m} = \vec{m}(\vec{x}(p_r^{warp}), p_r^{warp}, p^{warp})$ connects the points of the reconstructed object volume $f(\vec{x}(p_r^{warp}))$ in the reference phase p_r^{warp} with each time point p^{warp} inside the measurement interval ΔW_r contributing to the image $f(\vec{x}(p^{warp}))$, with $\vec{x}(p^{warp}) = \vec{x}(p_r^{warp}) + \vec{m}(\vec{x}(p_r^{warp}), p_r^{warp}, p^{warp})$ (see Figure 16.17).

There are different approaches to estimate motion vector fields depending on the application and the available image data:

- *Local motion estimation using image registration*: Local motion estimation is based on the definition of landmarks or the selection of a region of interest in the image dataset. The landmark or region of interest is selected either manually or automatically in the reference phase. Subsequently, it is registered to the other temporal phases of an image sequence. Due to the local character of the estimated motion pattern, affine transformations are able to describe the motion pattern with sufficient accuracy. Mostly optical flow methods using different similarity measures are applied to estimate the motion for the selected gray value pattern (Schirra 2009, Isola 2010a,b).

- *Global motion estimation using image registration*: Instead of selecting a local structure of interest or landmark, the complete heart or reconstruction volume may be selected for the registration process. Here, elastic registration is required for cardiac motion estimation in order to represent the volumetric cardiac motion with sufficient accuracy (Blondel 2006, Rohkohl 2010, Tang 2012).

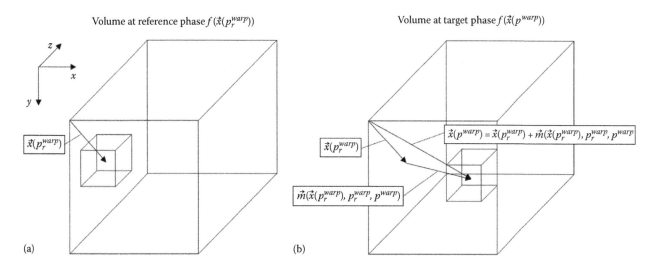

FIGURE 16.17 Schematic representation of the motion vector field estimated in between two time frames of a reconstructed image volume. (a) Image volume at the reference phase. (b) Image volume at the target phase with the shifted voxel position.

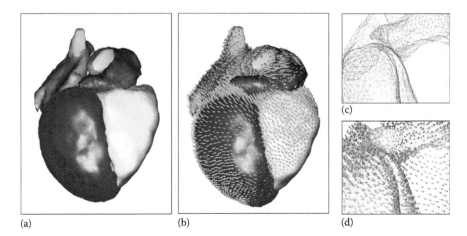

FIGURE 16.18 (a) Adaptive cardiac shape model with different anatomic parts. (b) Estimated motion vector field on the surface of a cardiac shape model after adaption to two different cardiac phases (blue vectors on the surface). (c) Shape model displayed as a grid. (d) Shape model displayed as a grid with the estimated motion vector field connected to the grid triangles (blue vectors).

- *Surface model–based motion estimation*: Adaptive shape models are increasingly used for the segmentation of complex organs from image data acquired with different modalities. These models are trained based on the image features. They can be used for the segmentation of a single image, but also model propagation can be applied when using these models on a 4D image sequence. In the following, the displacement of the triangle nodes resulting from the segmentation of each temporal frame is interpreted as the motion vector field (van Stevendaal 2008, Peters 2009). Since these motion vector fields are only represented on the surface of the model (see Figure 16.18), motion vector field interpolation is required prior to using this motion vector field in motion-compensated reconstruction (Forthmann 2008).

There are a number of challenges in motion estimation for motion-compensated reconstruction. In order to generate a realistic motion vector field for cardiac motion, invertibility of the estimated motion vector field should be enforced (Isola 2010a). Moreover, in case of motion estimation for a complete cardiac cycle, a periodicity assumption may be integrated in the estimation process (Peters 2009). Finally, the finite temporal resolution leads to a temporal blurring of the time frames used for the motion estimation. This limits the quality of the motion estimation process due to blurring and motion artifacts in phases of fast cardiac motion. In order to circumvent this problem, approaches to estimate the motion in akinetic phases of the cardiac cycle have been tested in combination with a temporal interpolation of the motion vector field in phases of fast motion (Schirra 2009).

Alternatively, direct raw-data-based motion estimation can be applied to address the problem of limited quality of the reconstructed temporal frames. These approaches have been applied in, for example, tomographic reconstruction for interventional x-ray imaging (Movassaghi 2006, Hansis 2008b). In general, these approaches can be applied when the target image feature of the motion-compensated reconstruction is also represented in the raw data. However, in x-ray tomography, these methods are limited when 3D objects overlay in the projection data.

16.5.2 Motion–Compensated Image Generation

The active compensation of cardiac motion within the image reconstruction process has the potential to deliver images with significantly increased temporal resolution and image quality. At the same time, additional complexity is added to the image reconstruction process, which needs to be evaluated carefully.

Motion-compensated reconstruction results in a number of different approaches for the different imaging modalities. In ultrasound imaging with its high frame rate, an estimated motion vector field can basically be used to perform an image-based correction of the different partial images from subsequent cardiac cycles. Similar approaches can be used to compensate artifacts in cardiac CT step and shoot images. For the imaging modalities that are based on tomographic reconstruction, the motion vector field needs to be integrated into the inversion process. For x-ray tomography, first approaches for motion compensation inside the reconstruction process have been formulated for the problem of breathing motion (Ritchie 1996). A scheme based on partial scans has been introduced (Grangeat 2002) and an approximate cone-beam-filtered back projection reconstruction has been formulated (Schäfer 2006). In order to perform the motion compensation during the reconstruction, it is required that the filtered back projection algorithm integrates a motion vector field representing the objects motion state for each of the raw projections being processed. This adds a significant amount of complexity to the image reconstruction process. On the other side, the reconstruction results show a significant lower artifact level and sharper anatomic delineation, which may be interpreted as improved temporal resolution (see Figure 16.19 for a motion-compensated CT example).

Exact reconstruction algorithms based on direct inversion are known for various data acquisition geometries in medical tomography. Even in the case of accurately known motion vector fields for each raw data measurement, exact motion-compensated reconstruction methods are only known for a very limited number of acquisition geometries and motion representations (see, e.g., Roux 2004, Katsevich 2010 and references therein for x-ray tomography).

Clearly, the motion vector field can be integrated into motion-compensated iterative reconstruction. In these algorithms, a forward operator is required in addition to the inversion operator. Thus, the motion vector field needs to be integrated in the forward process of the iterative algorithm too. As a consequence, the homogeneous volume representation will be distorted. This is especially true for divergent motion vector fields. Motion vector field–dependent volume representations are required to handle motion compensation in this framework with sufficient accuracy (Isola 2008).

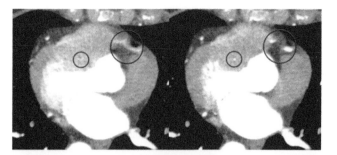

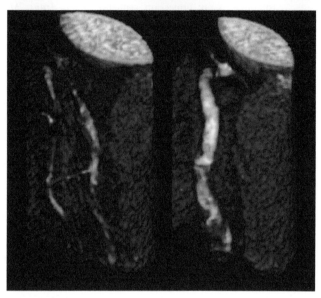

FIGURE 16.19 Gated helical cardiac CT reconstruction at 20% RR (top row, from van Stevendaal 2008) and at 50% RR (bottom row, from Isola 2010a) with a gating window width of 40% cardiac cycle without motion compensation (left column) and with motion compensation (right column).

On the other side, simultaneous reconstruction of the motion vector field and the image reconstruction process can be applied when using iterative reconstruction schemes. First approaches are known today for x-ray and nuclear imaging (Qi 2002, Gravier 2006, Schomberg 2007, Hansis 2009).

16.6 Summary and Conclusion

Motion control and motion compensation are highly desirable features when imaging moving objects like the heart. In order to control motion, prospective and retrospective gating strategies play an important role in cardiac imaging for all major modalities. Cardiac gating is mostly based on ECG data acquired in parallel to the image data acquisition. Both, prospective and retrospective gatings rely on limited intercycle variability of the cardiac motion. Motion control using gating strategies does not reduce the strong requirements imposed on the design of the acquisition system with respect to scan speed when aiming for motion artifact–free cardiovascular imaging.

Motion estimation and compensation is an active area of research in the medical imaging domain. These techniques promise lower artifact levels, lower dose, higher SNR, and improved temporal resolution without increased system requirements. However, the robustness of the approaches for motion estimation and motion-compensated reconstruction require further improvement before these methods can enter clinical practice. The accuracy of current approximate motion-compensated reconstruction methods and their ability to improve the temporal resolution beyond the speed of the acquisition system needs to be quantified. New, potentially exact motion-compensated reconstruction schemes and robust methods for motion estimation, especially for low contrast objects, require further research.

References

(Blondel 2006) C. Blondel, G. Malandain, R. Vaillant, and N. Ayache, Reconstruction of coronary arteries from a single rotational x-ray projection sequence, *IEEE Trans. Med. Imaging*, 25(5), 653–663, 2006.

(Botnar 1999) R. Botnar, M. Stuber, P. Danias, K. Kissinger, and W. Manning, Improved coronary artery definition with T2-weighted, free-breathing, three-dimensional coronary MRA, *Circulation*, 99(24), 3139–3148, 1999.

(Bruder 2003) H. Bruder, E. Maguet, K. Stierstorfer, and T. Flohr, Cardiac spiral imaging in computed tomography without ECG using complementary projections for motion detection, *Proc. SPIE Med. Imaging Conf.*, 5032, 1798–1809, 2003.

(Chen 2009) S. Chen and D. Schäfer, 3D coronary visualization Part 1: Modelling, *Cardiol. Clinics*, 27(3), 433–452, 2009.

(Cullom 1998) S. Cullom, J. Case, and T. Bateman, Electro-cardiographically gated myocardial perfusion SPECT: Technical principles and quality control Considerations, *J. Nucl. Cardiol.*, 5, 418–425, 1998.

(de Craene 2010) M. de Craene, G. Piella, N. Duchateau, E. Silva, A. Doltra, H. Gao, J. D'hooge et al., Temporal diffeomorphic free-form deformation for strain quantification in 3D-US images, *Lect. Notes Comput. Sci.*, 6362, 1–8, 2010.

(Di Carli 2007) M. Di Carli, S. Dorbala, J. Meserve, G. El Fakhri, A. Sitek, and S. Moore, Clinical myocardial perfusion PET/CT, *J. Nucl. Med.*, 48(5), 783–793, 2007.

(Ehmann 1989) R. Ehman and J. Felmlee, Adaptive technique for high-definition MR imaging of moving structures, *Radiology*, 173, 255–263, 1989.

(Forthmann 2008) P. Forthmann, U. van Stevendaal, M. Grass, and T. Köhler, Vector field interpolation for cardiac motion compensated reconstruction, *Proceedings of the IEEE Nuclear Science Symposium Conference Record*, M06-333, Dresden, Germany, 2008.

(Grangeat 2002) P. Grangeat, A. Koenig, T. Rodet, and S. Bonnet, Theoretical framework for a dynamic cone-beam reconstruction algorithm based on a dynamic particle model, *Phys. Med. Biol.*, 47, 2611–2625, 2002.

(Grass 2003) M. Grass, R. Manzke, T. Nielsen, P. Koken, R. Proksa, M. Natanzon, and G Shechter, Helical cardiac cone beam reconstruction using retrospective ECG gating, *Phys. Med. Biol.*, 48, 3069–3084, 2003.

(Gravier 2006) E. Gravier, Y. Yang, M. King, and M. Jin, Fully 4D motion-compensated reconstruction of cardiac SPECT images, *Phys. Med. Biol.*, 51, 4603–4619, 2006.

(Greenberg 2003) N. Greenberg, A. Borowski, and M. Garcia, Optimal cardiac phase selection for multi-slice CT using tissue Doppler ultrasound, *Proceedings of the Fourth International Conference on Cardiac Spiral CT*, Cambridge, MA, p. 13, 2003.

(Hansis 2008a) E. Hansis, D. Schäfer, O. Dössel, and M. Grass, Evaluation of iterative sparse object reconstruction from few projections for 3-D rotational coronary angiography, *IEEE Trans. Med. Imaging*, 27(11), 1548–1555, 2008a.

(Hansis 2008b) E. Hansis, D. Schäfer, O. Dössel, and M Grass, Projection-based motion compensation for gated coronary artery reconstruction from rotational x-ray angiograms, *Phys. Med. Biol.*, 53, 3807–3820, 2008b.

(Hansis 2009) E. Hansis, H. Schomberg, K. Erhard, O. Dössel, and M. Grass, Four-dimensional cardiac reconstruction from rotational x-ray sequences: First results for 4D coronary angiography, In E. Samei and J. Hsieh (eds.) *Medical Imaging 2009: Physics of Medical Imaging*, Bellingham, WA: SPIE, p.72580B, 2009.

(Hansis 2010) E. Hansis, J. Carroll, D. Schäfer, O. Dössel, and M. Grass, High-quality 3-D coronary artery imaging on an interventional C-arm x-ray system, *Med. Phys.*, 37(4), 1601–1609, 2010.

(Hoffmann 2006) M. Hoffmann, J. Lessick, R. Manzke, F. Schmid, E. Gershin, D. Boll, S. Rispler, A. Aschoff, and M. Grass, Automatic determination of minimal cardiac motion phases for computed tomography imaging: Initial experience, *Eur. Radiol.*, 16, 365–373, 2006.

(Houck 2006) R. Houck, J. Cooke, and E. Gill, Live 3D echocardiography: A replacement for traditional 2D echocardiography?, *Am. J. R*, 187(4), 1092–1106, 2006.

(Hsieh 2006) J. Hsieh, J. Londt, M. Vass, J. Li, X. Tang, and D. Okerlund, Step-and-shoot data acquisition and reconstruction for cardiac x-ray computed tomography, *Med. Phys.*, 33(11), 4236–4248, 2006.

(Isola 2008) A. Isola, A. Ziegler, T. Köhler, W. Niessen, and M. Grass, Motion-compensated iterative cone-beam CT image reconstruction with adapted blobs as basis functions, *Phys. Med. Biol.*, 53, 6777–6797, 2008.

(Isola 2010a) A. Isola, M. Grass, and W. Niessen, Fully automatic nonrigid registration-based local motion estimation for motion-corrected iterative cardiac CT reconstruction, *Med. Phys.*, 37(3), 1093–1109, 2010a.

(Isola 2010b) A. Isola, A. Ziegler, D. Schäfer, T. Köhler, W. Niessen, and M. Grass, Motion compensated iterative reconstruction of a region of interest in cardiac cone-beam CT, *Comput. Med. Imaging Graph.*, 34, 149–159, 2010b.

(Isola 2012) A. Isola, C. Metz, M. Schaap, S. Klein, M. Grass, W. Niessen, Cardiac motion-corrected iterative cone-beam CT reconstruction using a semi-automatic minimum cost path-based coronary centerline extraction, *Comput. Med. Imaging Graph.*, 36, 215–226, 2012.

(Kachelriess 2000) M. Kachelriess, S. Ulzheimer, and W. Kalender, ECG correlated imaging of the heart with subsecond multislice spiral CT, *IEEE Trans. Med. Imaging*, 19(9), 888–901, 2000.

(Kachelriess 2002) M. Kachelrieß, D. Sennst, W. Maxlmoser, and W. Kalender, Kymogram detection and kymogram-correlated image reconstruction from subsecond spiral computed tomography scans of the heart, *Med. Phys.*, 29(7), 1489–1503, 2002.

(Katsevich 2010) A. Katsevich, An accurate approximate algorithm for motion compensation in two-dimensional tomography, *Inv. Probl.*, 26, 1–16, 2010.

(Kim 2001a) W. Kim, R. Botnar, M. Stuber, K. Kissinger, and W. Manning, Patient-specific diastolic are required for free breathing right coronary MR vessel wall imaging, *J. Cardiovasc. Magn. Reson.* 50, 2001a.

(Kim 2001b) Y. Kim, P. Danias, M. Stuber, S. Flamm, S. Plein, E. Nagel, S. Langerak et al., Coronary magnetic resonance angiography for the detection of coronary stenoses, *N. Engl. J. Med.*, 345(26), 1863–1869, 2001b.

(Klein 2009) A. Klein and J. Garcia, Rotational coronary angiography, *Cardiol. Clinics*, 27(3), 395–406, 2009.

(Kozerke 1999) S. Kozerke, M. Scheidegger, E. Pedersen, and P. Boesiger, Heart motion adapted cine phase-contrast flow measurements through the aortic valve, *Magn. Reson. Med.*, 42(5), 970–978, 1999.

(Kriastelis 2009) C. Kriatselis, M. Tang, M. Roser, E. Fleck, and H. Gerds-Li, A new approach for contrast-enhanced x-ray imaging of the left atrium and pulmonary veins for atrial fibrillation ablation: Rotational angiography during adenosine-induced asystole, *Europace*, 11, 35–41, 2009.

(Köhler 2001) T. Köhler, R. Proksa, and M. Grass, A fast and efficient method for sequential cone-beam tomography, *Med. Phys.*, 28(11), 2318–2327, 2001.

(Köhler 2002) B.-U. Köhler, C. Hennig, and R. Orglmeister, The principles of software QRS detection, *IEEE Eng. Med. Biol.*, 1(2), 42–57, 2002.

(Lackner 1981) K. Lackner and P. Thurn, Computed tomography of the heart: ECG-gated and continuous scans, *Radiology*, 140, 413–420, 1981.

(Lang 2012) R. Lang, L. Badano, W. Tsang, D. Adams, E. Agricola, T. Buck, F. Faletra et al., MD, EAE/ASE recommendations for image acquisition and display using three-dimensional echocardiography, *J. Am. Soc. Echocardiogr.*, 25, 3–46, 2012.

(Lauritsch 2006) G. Lauritsch, J. Böse, L. Wigström, H. Kemeth, and R. Fahrig, Towards cardiac C-arm computed tomography, *IEEE Trans. Med. Imaging*, 25(7), 922–934, 2006.

(Lehmkuhl 2013) L. Lehmkuhl, K. von Aspern, B. Foldyna, M. Grothoff, S. Nitzsche, J. Kempfert, A. Rastan et al., Comparison of aortic root measurements in patients undergoing transapical aortic valve implantation (TA-AVI) using threedimensional rotational angiography (3D-RA) and multislice computed tomography (MSCT): Differences and variability, *Int. J. Cardiovasc. Imaging*, 29(3), 693–703. Epub 2012, March 18, 2013 (appeared online).

(Leitman 2004) M. Leitman, P. Lysyansky, S. Sidenko, V. Shir, E. Peleg, M. Binenbaum, E. Kaluski, R. Krakover, and Z. Vered, Twodimensional strain—A novel software for real-time quantitative echocardiographic assessment of myocardial function, *J. Am. Soc. Echocardiogr.*, 17, 1021–1029, 2004.

(Manzke 2003) R. Manzke, M. Grass, T. Nielsen, G. Shechter, and D. Hawkes, Adaptive temporal resolution optimization in helical cardiac cone beam CT reconstruction, *Med. Phys.*, 30(12), 3072–3080, 2003.

(Manzke 2004a) R. Manzke, M. Grass, and D. Hawkes, Artifact analysis and reconstruction improvement in helical cardiac cone beam CT, *IEEE Trans. Med. Imaging*, 23(9), 1150–1164, 2004a.

(Manzke 2004b) R. Manzke, T. Köhler, T. Nielsen, D. Hawkes, and M. Grass, Automatic phase determination for retrospectively gated cardiac CT, *Med. Phys.*, 31(12), 3345–3362, 2004b.

(Martinez-Moeller 2007) A. Martinze-Moeller, D. Zikic, R. Botnar, R. Bundschuh, W. Howe, S. Ziegler, N. Navab, M. Schwaiger, and S. Nekolla, Dual cardiac-respiratory gated PET: Implementation and results from a feasibility study, *Eur. J. Nucl. Med. Mol. Imaging*, 34, 1447–1454, 2007.

(McDicken 1992) W. McDicken, G. Sutherland, C. Moran, and L. Gordon, Colour Doppler velocity imaging of the myocardium, *Ultrasound Med. Biol.*, 18, 651–654, 1992.

(Mor-Avi 2011) V. Mor-Avi, R. Lang, L. Badano, M. Belohlavek, N. Cardim, G. Derumeaux, M. Galderisi et al., Current and evolving echocardiographic techniques for the quantitative evaluation of cardiac mechanics: ASE/EAE consensus statement on methodology and indications endorsed by the Japanese Society of Echocardiography, *Eur. J. Echocard.*, 12, 167–205, 2011.

(Mory 2014) C. Mory, V. Auvray, B. Zhang, M. Grass, D. Schäfer, S. Chen, J. Carroll et al., Cardiac C-arm computed tomography using a 3D+time ROI reconstruction method with spatial and temporal registration, *Med. Phys.*, 41(2), 021903, 2014.

(Movassaghi 2006) B. Movassaghi, D. Schäfer, M. Grass, V. Rasche, O. Wink, J. Garcia, J. Chen, J. Messenger, and J. Carroll, 3D reconstruction of coronary stents in vivo based on motion compensated x-ray angiograms, In R. Larsen, M. Nielsen, and J. Sporring (eds.) *Medical Image Computing and Computer-Assisted Intervention—MICCAI Copenhagen*, LNCS 4191, Denmark 2006, pp. 177–184, 2006.

(Nehrke 1999) K. Nehrke, P. BörnertJ. Groen J. SminkJ. Böck, On the performance and accuracy of 2D navigator pulses, *Magn. Reson. Imaging*, 17(8), 1173–1181, 1999.

(Nehrke 2003) K. Nehrke, P. Börnert, and T. Netsch, Automatic selection of cardiac acquisition window using an image-based global cross-correlation of multi heart phase cine scans, *Proc. Int. Soc. Magn. Reson. Med.*, 11, 1623, 2003.

(Nielsen 2005) T. Nielsen, R. Manzke, R. Proksa, and M. Grass, Cardiac cone-beam CT volume reconstruction using ART, *Med. Phys.*, 32(4), 851–860, 2005.

(Paul 2004) A. Paul and H. Nabi, Gated myocardial perfusion SPECT: Basic principles, technical aspects, and clinical applications, *J. Nucl. Med. Tech.*, 32(4), 179–187, 2004.

(Perk 2009) G. Perk, R. Lang, M. Garcia-Fernandez, J. Lodato, L. Sugeng, J. Lopez, B. Knight et al., Use of real time three-dimensional transesophageal echocardiography in intracardiac catheter based interventions, *J. Am. Soc. Echocardiogr.*, 22(8), 865–882, 2009.

(Peters 2009) J. Peters, O. Ecabert, H. Schmitt, M. Grass, and J. Weese, Local cardiac wall motion estimation from retrospectively gated CT images, In N. Ayache, H. Delingette, and M. Sermesant (eds.) *FIMH 2009*, LNCS 5528, Nice, France, pp. 191–200, 2009.

(Pirat 2008) B. Pirat, D. Khoury, C. Hartley, L. Tiller, L. Rao, D. Schulz, S. Nagueh, and W. Zoghbi, A novel feature tracking echocardiographic method for the quantitation of regional myocardial function, *J. Am. Coll. Cardiol.*, 51, 651–659, 2008.

(Qi 2002) J. Qi and R. Huesman, List mode reconstruction for PET with motion compensation: A simulation study, *Proceedings of the IEEE International Symposium on Biomedical Imaging*, Washington DC, USA, pp. 413–416, 2002.

(Ritchie 1996) C. Ritchie, C. Crawford, J. Godwin, K. King, and Y. Kim, Correction of computed tomography motion artifacts using pixel-specific backprojection, *IEEE Trans. Med. Imaging*, 15(3), 333–342, 1996.

Chapter 16

(Rohkohl 2010) C. Rohkohl, G. Lauritsch, L. Biller, M. Prümmer, J. Boese, and J. Hornegger, Interventional 4D motion estimation and reconstruction of cardiac vasculature without motion periodicity assumption, *Med. Image Anal.*, 14, 687–694, 2010.

(Roux 2004) S. Roux, L. Desbat, A. Koenig, and P. Grangeat, Exact reconstruction in 2-D dynamic CT: Compensation of time-dependent affine deformations, *Phys. Med. Biol.*, 49, 2169–2182, 2004.

(Saranathan 2001) M. Saranathan, V. Ho, M. Hood, T. Foo, and C. Hardy, Adaptive vessel tracking: Automated computation of vessel trajectories for improved efficiency in 2D coronary MR angiography, *J. Magn. Reson. Imaging*, 14(4), 368–373, 2001.

(Schirra 2009) C. Schirra, C. Bontus, U. van Stevendaal, O. Dössel, and M. Grass, Improvement of cardiac CT reconstruction using local motion vector fields, *Comput. Med. Imaging Graph.*, 33, 122–130, 2009.

(Schmitt 2009) H. Schmitt, J. Peters, J. Lessick, J. Weese, and M. Grass, Spatially resolved automatic cardiac rest phase determination in coronary computed tomography angiography (CTA), *IFMBE Proceedings of the World Congress on Medical Physics and Biomedical Engineering 2009*, Munich, Germany, vol. 25, issue no. 2, pp. 162–165, 2009.

(Schomberg 2007) H. Schomberg, Time-resolved cardiac cone beam CT, *Proceedings of the Fully 3D Conference*, Lindau, Germany, 2007.

(Schoonenberg 2009) G. Schoonenberg, J. Garcia, and J. Carroll, Left coronary artery thrombus characterized by a fully automatic three-dimensional gated reconstruction, *Catheter. Cardiovasc. Interv.*, 74, 97–100, 2009.

(Schäfer 2006) D. Schäfer, J. Borgert, V. Rasche, and M. Grass, Motion-compensated and gated cone beam filtered backprojection for 3-D rotational x-ray angiography, *IEEE Trans. Med. Imaging*, 25(7), 898–906, 2006.

(Schwartz 2011) J. Schwartz, A. Neubauer, T. Fagan, N. Noordhoek, M. Grass, and J. Carroll, Potential role of three-dimensional rotational angiography and C-arm CT for valvular repair and implantation, *Int. J. Cardiovasc. Imaging*, 27, 1205–1222, 2011.

(Stevendaal 2006) U. van Stevendaal, P. Koken, P. Begemann, R. Koester, G. Adam, and M. Grass, ECG gated continuous circular cone-beam multi-cycle reconstruction for in-stent coronary artery imaging: A phantom study, *Proc. SPIE Med. Imaging Conf.*, 6142, 61420L-1–61420L-10, 2006.

(Stuber 1999) M. Stuber, R. Botnar, P. Danias, K. Kissinger, and W. Manning, Submillimeter three-dimensional coronary MR angiography with real-time navigator correction: Comparison of navigator locations, *Radiology*, 212, 579–587, 1999.

(Stuber 1999a) M. Stuber, R. Botnar, P. Danias, D. Sodickson, K. Kissinger, M. Van Cauteren, J. De Becker, and W. Manning, Double-oblique free-breathing high resolution three-dimensional coronary magnetic resonance angiography, *J. Am. Coll. Cardiol.*, 34(2), 524–531, 1999.

(Tang 2012) Q. Tang, J. Cammin, S. Srivastava, and K. Taguchi, A fully four-dimensional, iterative motion estimation and compensation method for cardiac CT, *Med. Phys.*, 39(7), 4291–4305, 2012.

(Usman 2013) M. Usman, D. Atkinson, F. Odille, C. Kolbisch, G. Vaillant, T. Schaeffter, P. Batchelor, and R. Razavi, Motion corrected compressed sensing for free-breathing dynamic cardiac MRI, *Magn. Reson. Med.*, 70, 504–516, 2013.

(van Stevendaal 2008) U. van Stevendaal, J. von Berg, C. Lorenz, and M. Grass, A motion-compensated scheme for helical cone-beam reconstruction in cardiac CT angiography, *Med. Phys.*, 35(7), 3239–3251, 2008.

(Vembar 2009) M. Vembar, M. Garcia, D. Heuscher, R. Haberl, D. Matthews, G. Böhme, and A. Greenberg, A dynamic approach to identifying desired physiological phases for cardiac imaging using multi-slice spiral CT, *Med. Phys.*, 30(7), 1683–1693, 2003.

(von Berg 2007) J. von Berg, H. Barschdorf, T. Blaffert, S. Kabus, and C. Lorenz, Surface based cardiac and respiratory motion extraction for pulmonary structures from multi–phase CT, *Proceedings of the SPIE Medical Imaging Conference*, San Diego, CA, vol. 6511, pp. 65110Y-1–65110Y-11, 2007.

(Wang 1996) Y. Wang, R. Grimm, J. Felmlee, S. Riederer, and R. Ehman, Algorithms for extracting motion information from navigator echoes, *Magn. Reson. Med.*, 36, 117–123, 1996.

(Wang 2001) Y. Wang, R. Watts, I. Mitchell, T. Nguyen, J. Bezanson, G. Bergman, and M. Prince, Coronary MR angiography: Selection of acquisition window of minimal cardiac motion with electrocardiography-triggered navigator cardiac motion prescanning–initial results, *Radiology*, 218(2), 580–585, 2001.

(Wang 2002) G. Wang, S. Zhao, and D. Heuscher, A knowledge based cone beam x-ray CT algorithm for dynamic volumetric cardiac imaging, *Med. Phys.*, 29(8), 1807–1822, 2002.

(Wang 2011) P. Wang, T. Chen, O. Ecabert, S. Prummer, M. Ostermeier, D. Comaniciu, Image-based device tracking for the co-registration of angiography and intravascular ultrasound images, In G. Fichtinger, A. Martel, and T. Peters (eds.), *14th International Conference on Medical Image Computing and Computer-Assisted Intervention—MICCAI 2011*, Part I, LNCS 6891, Toronto, Ontario, Canada, pp. 161–168, 2011.

(Yoshida 1961) T. Yoshida, M. Mori, Y. Nimura, G. Hikita, G. Taka, K. Nakanishi, and S. Satomura, Analysis of heart motion with ultrasonic Doppler method and its clinical application. *Am. Heart J.*, 61(1), 61–75, 1961.

17. Time-Resolved Neurovascular 2D X-Ray Imaging

Ciprian N. Ionita

17.1 Introduction

This chapter will present methods to estimate physiological flow conditions in the neurovasculature by analyzing the contrast propagation through the arterial network in 2D digital subtraction angiography (DSA) image sequences. The aims are (1) to introduce the basic physics aspects of two component mixture flow and how it can be used to obtain information about flow conditions, (2) to describe established methods to measure flow parameters, and (3) to convey how the flow information is used to describe pathologies associated with stenosis, ischemic stroke, aneurysms, and arteriovenous malformations (AVMs).

The circulatory system is the main supply route of oxygen and nutrients to all the organs in the human body. Various pathologic afflictions of the vascular network could result in a serious injury or malfunction of such organs. This aspect becomes critical in the case of the brain, which is the control center of other organ functions and human activities. Optimal brain functioning is blood flow dependent. The blood is supplied through the carotid and vertebral arteries into an arterial network at the base of the brain known as the circle of Willis and distributed to the brain tissue through an intricate arterial network (Figure 17.1).

Recent technological and clinical advances have tremendously improved the diagnosis, treatment, and recovery of patients suffering from various neurovascular afflictions such as ischemic or hemorrhagic strokes due to stenosis or aneurysms, AVMs,[1] or other pathologies.

DSA[2,3] is the golden investigational tool for neurovascular imaging due to its broad availability and high imaging resolution. In general, the procedure is used to evaluate the structural aspects of neurovascular disease rather than the physiological aspects. In such procedures,

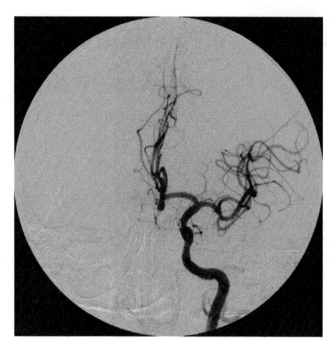

FIGURE 17.1   Typical intracranial DSA snapshot.

a catheter is advanced using x-ray fluoroscopy guidance in one of the main arteries leading to the neuroarterial network: left or right carotid artery or basilar artery.

In the beginning of the sequence, a few images are acquired without any contrast injection and used to create a mask image. All subsequent images in the DSA sequence are logarithmically subtracted as previously described in Section 4.2.3. The iodine contrast bolus, typically 2–10 cm³, is injected using a manual or automatic injector. At the tip of the catheter, the contrast mixes with the blood, creating a bolus that propagates upstream through the arterial network. The subtraction process removes all the bony and tissue background structures, allowing a clear view of the arteries.

In the last two decades, due to the availability of digital data, new methods have been developed for the evaluation of flow aspects related to neurovascular disease. The main way to analyze flow is by monitoring the contrast media flowing in the arteries through a method referred to as video densitometry employing time–density or contrast curves.

The methods used for analysis can be divided into two classes, tracking or computational methods, as will discussed in the next section of this chapter. Tracking methods rely on measuring the bolus passage, while computational methods rely on implementation of physics-based models such as mass conservation and convection.

For both cases, a number of simplifying assumptions are used; most of them will be described in the appropriate chapter section or references.

17.2   Fundamentals of Blood Flow Measurements Using High Rate 2D X-Ray Imaging

17.2.1   Aspects of Blood Flow and Contrast Media Transport

Blood flow and interaction with the surrounding vasculature structures can be described using the classical mechanical physics laws such as: conservation of mass, energy, or momentum, etc. The current limitation is the incomplete knowledge of the initial/boundary conditions to the problem. Information such as viscosity, wall elasticity, blood pressure wave, pulsation, velocity, etc., is known only approximately, resulting in an incomplete description of the problem. However, this does not deter the scientific community from attempting to understand and build models for description of pathology and prediction of future outcomes if treatment is attempted.

The most important parameter of flow is the velocity; hence, most of the approaches presented in this chapter will be focusing on it. The best way to measure flow is to evaluate the dispersion of a tracer, also referred to as contrast media, using a radiographic dynamic imaging modality such as DSA. Evaluation of the contrast flow could reveal details about the hemodynamic conditions in the arterial network such as velocity, flow rate, or regional blood flow.

Contrast propagation in a moving fluid such as blood is done through two physical processes: convection (second term in Equation 17.1) and diffusion (third term in Equation 17.1) so that the rate of change with time of contrast media concentration, C, is governed by the following equation[4]:

$$\frac{\partial C}{\partial t} + \vec{\nabla} \cdot (\vec{v} \cdot C) - D\Delta C = S \qquad (17.1)$$

where

D is the diffusion coefficient
ΔC is the Laplacian of the concentration
v is the fluid velocity
S describes the sources: in our case, the injection point.

In most of the situations, Equation 17.1 cannot be solved exactly and various simplifications are assumed such as: that the injection point is far away from the evaluation area, and the fluid is incompressible, etc.

Manual or automatic contrast injection is a very complex process[5]; it depends on the catheter, applied pressure, contrast flow rate and viscosity, cardiac cycle, and location. However, if the distance between the region of interest (ROI) and the injection point is at least ten vessel diameters, the contrast can be considered fully mixed with the blood.[5] Assuming fluid incompressibility, then Equation 17.1 reduces to

$$\frac{\partial C}{\partial t} = D\Delta C - \vec{v} \cdot \nabla C \tag{17.2}$$

Equation 17.2 states that the variation of the contrast in a certain ROI, far from any sources, depends on only two phenomena: diffusion and convection. Further, the diffusion process, described by the first term on the right-hand side of Equation 17.2, is a very slow process and depends mostly on the thermal molecular motion. These assumptions single out the convection as the dominant process, for propagation of the contrast through the arterial network.

$$\frac{\partial C}{\partial t} \approx -\vec{v} \cdot \nabla C \tag{17.3}$$

Hence, important information about physiological aspects of the cerebral blood flow, especially velocity, and cerebral disease can be inferred by observing the propagation of contrast in time and space.

17.2.2 Time–Density Curve Measurements and Characterization

Time–density curves (TDCs) are derived by selecting a ROI over a certain area, and monitoring the pixel intensity. The ROI can be a line, a point, or a finite area Σ. Each pixel or the average pixel is subtracted from the average background measured in the close proximity of the selected area of interest.

The curves describe the contrast distribution with time for a specific area, and are characterized by an arrival time when there is an increase of the contrast due to flow, by the time of saturation, and by the time for washout.

A typical curve for such a process is shown in Figure 17.2. Based on such temporal distributions, a few parameters are used to describe the bolus transition as described in Table 17.1.

One of the main concerns in the derivation of the TDCs is the dependence of the contrast distribution on the injection profiles. The most used methods to limit the influence of the contrast injection profile on the TDC can be divided in mechanical, mathematical or a combination of both. Mechanical approaches ensure full contrast mixing by injecting far away from the ROI location or designing special catheter tips or injectors. Mathematical methods include deconvolution and normalization models.

17.2.2.1 Velocity Determination Using Bolus Tracking Methods

Tracking methods are used to determine the time it takes for a bolus to travel between two points specified by Two ROI's (Figure 17.3). The contrast is monitored in the two ROIs, twice determined from the measured

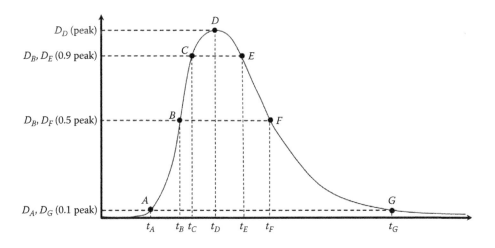

FIGURE 17.2 Landmark points on a TDC; *x*-axis represents the time and *y*-axis represents the contrast density measured in a ROI. The parameters indicated in the figure are used for the parameter definition in Table 17.1.

Chapter 17

Table 17.1 Definition of Parameters Used for Flow Measurements Using Temporal Monitoring of a Bolus in One or Multiple ROIs

Name	Notation	Condition/Formula
Time of peak[1] opacification[1,6–20] (peak arrival time)[9,18,21,22]	TPO	t_D
Time to leading half peak opacification[6,8,13,18,20–28]	TLHP	t_B
Time to trailing half peak opacification[7,21]	TTHP	t_F
Downslope time difference[18,21]	DTD	$\dfrac{\sum_n t_n}{n} \begin{cases} D(t_n) = \gamma D(t_D) \\ 0.5 \le \gamma \le 0.95 \end{cases}$
Time of peak gradient arrival[21,29]	TPGA	$\underset{t_A \le t \le t_B}{\arg\max} \dfrac{\partial D}{\partial t}$
Mean bolus arrival time[7,8,21,22,30–35]	MBAT	$\dfrac{1}{D_D} \displaystyle\int_{t_A}^{t_D} [D_D - D(t)] \cdot dt$
Time of level of area under the curve[20,21]	TLAC	$\dfrac{\gamma \displaystyle\int_{t_A}^{t_G} D(t) \cdot dt}{\displaystyle\int_{t_A}^{t_{ref}} D(t) \cdot dt} = 1; \quad (0 \le \gamma \le 1)$
Time of contrast arrival[11,20,21,24,25,36–39]	TCA	t_A
Isodensity level transit time[21,40](Figure 17.3)	ILTT	$t_j - t_i; \quad (D_i(t_i) = D_j(t_j))$
Mean transit time[7,18,21]	MTT	$\dfrac{\displaystyle\int_{t_A}^{t_G} t \cdot D(t) \cdot dt}{\displaystyle\int_{t_A}^{t_G} D(t) \cdot dt}$

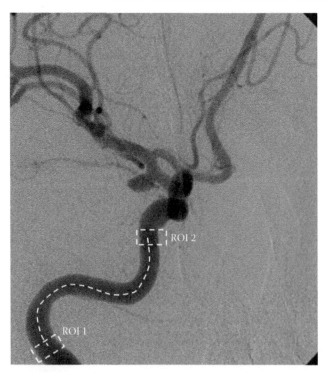

FIGURE 17.3 Selection of ROIs for the determination of velocity in an internal carotid.

TDC. By using various data from the TDCs, the time span needed for the bolus to travel between the two points is calculated and mean blood velocity is found by dividing the distance between the two ROIs by the time interval. Various landmark features used by various authors are summarized in Table 17.1.

The distance between the two ROIs is not easy to estimate from one view. Various views can be obtained using biplane or tomographic systems so that the 3D information can be used to estimate the distance when tortuous vessels are present. TDCs obtained at each ROI are shown in Figure 17.4. As the contrast propagates through the arterial system, it diffuses, making the curve flatter at ROI 2. If the users are interested only in the temporal parameterization of the curves, then it is customary to normalize the curves to the maximum peak value. Differences between bolus arrival times at each ROI denoted by t_i and t_j are taken to estimate the bolus travel time between the two regions.

One easy way to define the arrival time is to record it as the time at which the TDC value increases above the background by 5%–10% or less or it reaches a specified value. The definitions are somehow arbitrary, and

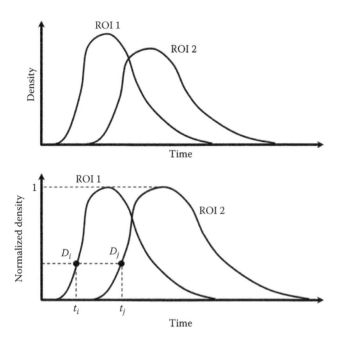

FIGURE 17.4 TDCs measured at different locations (Figure 17.3), top figure shows raw data, bottom figure shows each curve normalized to its peak value.

various authors use different names to refer to the same parameter. To avoid confusion, the most accepted definitions are given in Table 17.1 with some of the parameters indicated in Figure 17.2.

The time to peak opacification (TPO) is measured by taking the time it takes for the TDC to increase from 1% to 95% of the peak value. The time to peak (TTP) is inversely proportional to the velocity of the bolus propagation.

The peak of the TDC is related to the concentration of the iodine contrast, while there is a direct relation with the velocity of the flow, interpretation of its clinical meaning is dependent on the vascular disease.

Mean transit time (MTT) is measured by taking the absolute value difference between time it took the TDC to pass half of the peak value and the second time the TDC value reaches half of the peak. The value is related to the residence time of the contrast in a certain area, but is also inversely related to the blood flow. Its interpretation is also dependent on the vascular disease image.

17.2.2.2 Gamma Variate Fitting Method of TDCs

The TDC of contrast medium following a very short (i.e., delta function) intra-arterial time of injection of contrast medium can be described by a gamma variate function.[30,41] Equation 17.1 can be solved if we considered the 1D case and the source on the right side to be zero. For detailed description of the solution, the readers are advised to see the work of Harpen et al.[30]

The solution is a gamma variate function that can be expressed by

$$D(t) = K(t - t_0)^\alpha e^{-(t - t_0)/\beta} \tag{17.4}$$

where
    K is a constant scale factor
    t_0 is the time of appearance of the contrast medium
    α and β are parameters of the distribution

Using the definition MTT from Table 17.1 and the properties of the gamma variate functions, it can be shown that

$$MTT = t_0 + \beta(\alpha + 1) \tag{17.5}$$

The TTP can be calculated by finding the maximum of the gamma variate function:

$$TTP = \beta\alpha \tag{17.6}$$

17.2.2.3 Indicator-Dilution Methods

Indicator-dilution methods are based on the assumption that the indicator dispersion, in our case the iodine contrast, is related to the blood flow volume element in space with respect to time. This concept is based on the physics transport phenomena laws. Thus, knowledge of the volume or the rate and the diluted concentration at a sampling site over a period of time can be used to calculate flow and blood volume.

The aforementioned statement is known as Hamilton–Stewart principle[42–44] and assumes the following constraints: a system with one inflow and one outflow, all indicator injected must leave the system, distribution of the indicator transit times remains constant in time and is identical with the distributions of the blood flow transit times, and the injection is either sudden or at constant rate over a long period of time.

We start by assuming a sudden injection where m_0 units of iodine is injected initially into the system's entrance, and the measured concentration at exit as a function of time is $c(t)$. Then the amount of contrast leaving the system between t and $t + dt$ is given by

$$dm = c(t)Qdt \tag{17.7}$$

where Q is the blood flow in units of volume/time. Integrating Equation 17.7, we obtain

$$Q = \frac{m_I}{\displaystyle\int_0^\infty c(t)dt} \tag{17.8}$$

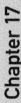

where m_I is the contrast mass injected contained within the measurement space. It is necessary to introduce now a frequency function, $h(t)$, which describes the fraction of injected indicator mass leaving the system per unit time at time, t, defined as

$$h(t) = \frac{Qc(t)}{m_I} \qquad (17.9)$$

If we solve for Q and substitute it in Equation 17.9, we can express $h(t)$ as

$$h(t) = \frac{c(t)}{\int_0^\infty c(t)dt} \qquad (17.10)$$

which states that the frequency (probability distribution) function of the tracer transit time can be obtained by taking the TDC and dividing it by the area under the curve.

Another important curve is the cumulative distribution function, $H(t)$, which is the integral of the distribution function, $h(t)$, or

$$H(t) = \int_0^t h(s)ds \qquad (17.11)$$

(see Figure 17.5). The cumulative function multiplied with the quantity of tracer injected gives the amount that left the ROI. The cumulative function can be further used to calculate the residual distribution function or the total fraction of contrast mass remaining at a given time t as

$$R(t) = 1 - H(t) \qquad (17.12)$$

The residual function multiplied with the amount of contrast injected gives the mass of contrast still present in the system.

Now we assume a constant injection of iodine contrast agent at a constant rate $\dot{m}_I$ in units of mass per time; if mixing at the inflow is complete, the concentration of indicator admitted to the system is $\dot{m}_I/Q$.

Such information can be derived by monitoring the concentration of the contrast in a specific arterial segment, integrating across the center line, and plotting it versus time deriving what is called a contrast wave map.

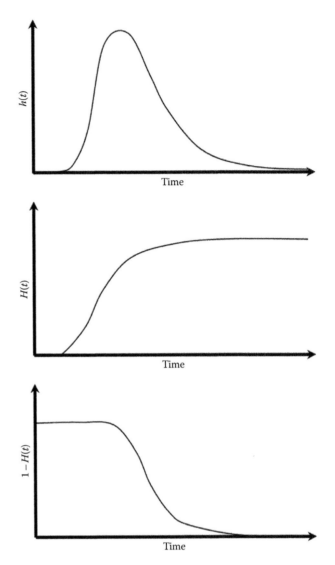

FIGURE 17.5 Frequency distribution curve, distribution function, and residual function.

17.2.2.4 Distance–Density Curves

Distance–density curves, instead of monitoring contrast variation at two locations in the artery, use the vessel centerline for a given distance to monitor the contrast distribution. Two distance–density curves are derived for two times (usually two consecutive frames). As shown in Figure 17.6, the second curve is shifted forward in space and the root mean square (RMS) difference is calculated. The shift that gives the minimum RMS is considered to be the bolus travel distance. Velocity is found by dividing the distance by the time difference between the two frames.

Other methods use a combination of the distance–density curves and TDC to monitor the contrast flow in a vessel by plotting the intensity of each

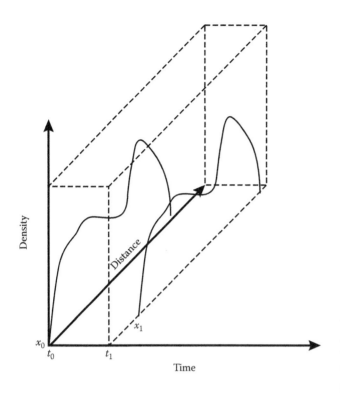

FIGURE 17.6 Distance–density curves for two consecutive frames.

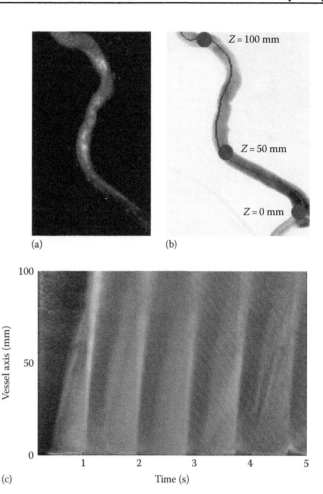

FIGURE 17.7 Contrast intensity map derivation: (a) an inverted DSA snapshot of an artery segment and (b) an outline of the vessel centerline. (c) The pixel intensities along the center line are recorded in each image in the DSA sequence and inserted in the intensity graph. (From Bonnefous, O. et al., *Med. Phys.*, 39, 6264, 2012.)

distance–density curve for an entire run. Thus, one obtains an intensity map of temporal and spatial distribution of the contrast (Figure 17.7).

Thus, according to Equation 17.3, one can find instantaneous velocity at each moment of time by simply calculating the gradients and time derivative from the intensity map.

The primary drawback of this method is that errors in the estimation of velocity can be large if the absolute value of ∇C in Equation 17.3 is small. One can avoid such a problem by maintaining a very large contrast material spatial gradient using very short injections.

17.2.2.5 Optical Flow Methods

Optical flow methods[45,46] use contrast density patterns in an image sequence to calculate the apparent velocity distribution of contrast flow movement. The delay in appearance of the contrast bolus for different point positions downstream in the artery can be related to blood velocities. The main problem associated with this approach is the choice of a suitable parameters from the image (brightness of an image point [pixel] as a function of time) to compute the time delay between the appearance of the contrast bolus at different positions, so as to evaluate the velocity. We start by

assuming that the changes in intensity are very small for small changes in space and time:

$$I(x,y,t) = I(x+\delta x, y+\delta y, t+\delta t) \tag{17.13}$$

For small variation, we can expand right side term in the equation using a Taylor expansion and we can rewrite the expression as:

$$I(x,y,t) = I(x,y,t) + \frac{\partial I}{\partial x}\delta x + \frac{\partial I}{\partial y}\delta y + \frac{\partial I}{\partial t}\delta t + O(x,y,t) \tag{17.14}$$

If we ignore the higher terms, eliminate $I(x,y,t)$, and divide by δt, we have the following:

$$\frac{\partial I}{\partial x}\frac{\delta x}{\delta t} + \frac{\partial I}{\partial y}\frac{\delta y}{\delta t} + \frac{\partial I}{\partial t} = 0 \tag{17.15}$$

Chapter 17

For the limit $\delta t \rightarrow 0$, the relation becomes

$$\frac{\partial I}{\partial x}\frac{dx}{dt}+\frac{\partial I}{\partial y}\frac{dy}{dt}+\frac{\partial I}{\partial t}=0 \qquad (17.16)$$

The goal is to calculate velocity as

$$\omega=(u,v)=\left(\frac{dx}{dt};\frac{dy}{dt}\right) \qquad (17.17)$$

Using Equation 17.17, one can rewrite Equation 17.16 as

$$-\frac{\partial I}{\partial t}=\left(\vec{\nabla}\cdot I(x,y)\right)\cdot\omega \qquad (17.18)$$

Equation 17.18 says that if we measure the variation of intensity in two adjacent frames and the spatial gray level difference, represented by the gradient term, we should be able to estimate the velocity components. However, for a 2D problem, the equation does not specify the velocity vector components completely. We can estimate only the velocity component in the direction of the intensity gradient, $((\partial I/\partial x),(\partial I/\partial y))$

$$\omega_{\vec{k}}=\frac{(\partial I/\partial t)}{\sqrt{(\partial I/\partial x)^2+(\partial I/\partial y)^2}} \qquad (17.19)$$

where $\vec{k}$ represents the direction of the intensity gradient. In case of flow through vasculature where the vessel is parallel to the imager, velocity distribution can be considered uniform across the vessel lumen and intensity variation occurs only along the vessel centerline. Using this approximation, Equation 17.19 can be used directly for flow estimation. However, when the flow is complex, as in the case of aneurysms, Equation 17.19 cannot be solved without additional constraints. One assumption is that the flow changes are smooth and a smoothness condition can be applied to Equation 17.19. Horn and Schunck[45] used the square of the magnitude of the gradient as a smoothness measure. Using their method, the problem reduces to minimizing an error defined as

$$\varepsilon=\iint\left(\frac{\partial I}{\partial x}u+\frac{\partial I}{\partial y}v+\frac{\partial I}{\partial t}\right)^2$$

$$+\lambda\left[\left(\frac{\partial u}{\partial x}\right)^2+\left(\frac{\partial u}{\partial y}\right)^2+\left(\frac{\partial u}{\partial x}\right)^2+\left(\frac{\partial u}{\partial y}\right)^2\right]dxdy \quad (17.20)$$

where λ are Lagrange multipliers. Using calculus of variation Equation 17.20 leads to the following differential equations:

$$\begin{cases}\left(\dfrac{\partial I}{\partial x}\right)^2 u+\dfrac{\partial I}{\partial x}\dfrac{\partial I}{\partial y}v=\lambda^2\Delta u-\dfrac{\partial I}{\partial x}\dfrac{\partial I}{\partial t}\\[4mm]\dfrac{\partial I}{\partial x}\dfrac{\partial I}{\partial y}u+\left(\dfrac{\partial I}{\partial y}\right)^2 v=\lambda^2\Delta v-\dfrac{\partial I}{\partial y}\dfrac{\partial I}{\partial t}\end{cases} \qquad (17.21)$$

The Laplacian is estimated by subtracting the value at a point from a weighted average of the values at neighboring points (for detailed derivation of the solution, the readers should consult the work of Horn and Schunck[45]):

$$\Delta u\approx\bar{u}-u \qquad (17.22)$$

Using this approximation, Equation 17.21 can be rewritten as

$$\begin{cases}\left[\lambda^2+\left(\dfrac{\partial I}{\partial x}\right)^2\right]u+\dfrac{\partial I}{\partial x}\dfrac{\partial I}{\partial y}v=\lambda^2\bar{u}-\dfrac{\partial I}{\partial x}\dfrac{\partial I}{\partial t}\\[4mm]\dfrac{\partial I}{\partial x}\dfrac{\partial I}{\partial y}u+\left[\lambda^2+\left(\dfrac{\partial I}{\partial y}\right)^2\right]v=\lambda^2\bar{v}-\dfrac{\partial I}{\partial y}\dfrac{\partial I}{\partial t}\end{cases} \qquad (17.23)$$

Solving for u and v, we find that

$$\begin{cases}\left[\lambda^2+\left(\dfrac{\partial I}{\partial x}\right)^2+\left(\dfrac{\partial I}{\partial y}\right)^2\right]\\[4mm]\quad u=\left[\lambda^2+\left(\dfrac{\partial I}{\partial y}\right)^2\right]\bar{u}-\dfrac{\partial I}{\partial x}\dfrac{\partial I}{\partial y}\bar{v}-\dfrac{\partial I}{\partial x}\dfrac{\partial I}{\partial t}\\[4mm]\left[\lambda^2+\left(\dfrac{\partial I}{\partial x}\right)^2+\left(\dfrac{\partial I}{\partial y}\right)^2\right]\\[4mm]\quad v=-\dfrac{\partial I}{\partial x}\dfrac{\partial I}{\partial y}\bar{u}+\left[\lambda^2+\left(\dfrac{\partial I}{\partial x}\right)^2\right]\bar{v}-\dfrac{\partial I}{\partial y}\dfrac{\partial I}{\partial t}\end{cases} \quad (17.24)$$

These equations can be rearranged in an alternate form:

$$\begin{cases}\left[\lambda^2+\left(\dfrac{\partial I}{\partial x}\right)^2+\left(\dfrac{\partial I}{\partial y}\right)^2\right](u-\bar{u})=-\dfrac{\partial I}{\partial x}\left[\dfrac{\partial I}{\partial x}\bar{u}+\dfrac{\partial I}{\partial y}\bar{v}+\dfrac{\partial I}{\partial t}\right]\\[4mm]\left[\lambda^2+\left(\dfrac{\partial I}{\partial x}\right)^2+\left(\dfrac{\partial I}{\partial y}\right)^2\right](v-\bar{v})=-\dfrac{\partial I}{\partial y}\left[\dfrac{\partial I}{\partial x}\bar{u}+\dfrac{\partial I}{\partial y}\bar{v}+\dfrac{\partial I}{\partial t}\right]\end{cases}$$

$$(17.25)$$

The solution can be computed using iterative methods such as Gauss–Seidel:

$$\begin{cases} u^{n+1} = \overline{u}^n - \dfrac{(\partial I/\partial x)\left[(\partial I/\partial x)\overline{u}^n + (\partial I/\partial y)\overline{v}^n + (\partial I/\partial t)\right]}{\left[\lambda^2 + (\partial I/\partial x)^2 + (\partial I/\partial y)^2\right]} \\[4ex] v^{n+1} = \overline{v}^n - \dfrac{(\partial I/\partial y)\left[(\partial I/\partial x)\overline{u}^n + (\partial I/\partial y)\overline{v}^n + (\partial I/\partial t)\right]}{\left[\lambda^2 + (\partial I/\partial x)^2 + (\partial I/\partial y)^2\right]} \end{cases}$$

$$(17.26)$$

A typical optical flow showing velocity magnitude is shown in Figure 17.8. The images were acquired in a patent with an arterial venous malformation. The analysis shows flow changes before (Figure 17.8a and c) and after AVM embolization (Figure 17.8b and d). Further, the authors split the DSA sequence into two parts: one corresponding to the arterial phase (Figure 17.8a and b) and the second to the venous phase (Figure 17.8c and d). The start of the venous phase was established by setting a threshold in an ROI placed over the venous superior sagittal sinus.

17.2.2.6 Parametric Imaging

Parametric imaging, also referred to as colored coded DSA, is done by analyzing the contrast behavior at each pixel in the image. Since the early development of DSA,

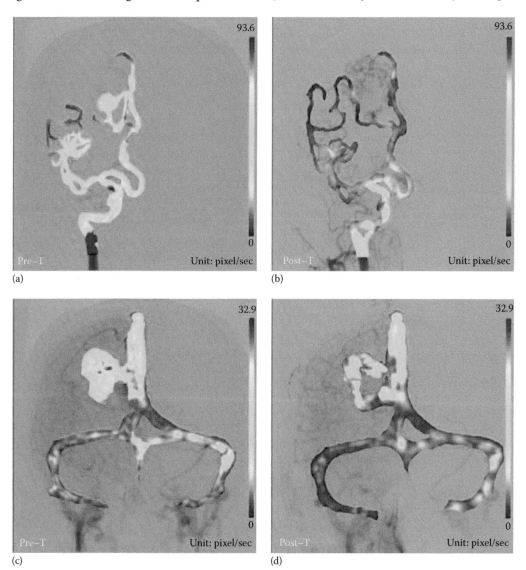

FIGURE 17.8 Example of optical flow[47]: The velocity of the blood flow is calculated using Equation 17.26. For each value at each location, a color is assigned using a linear mapping displayed in each segment. Figures show the results pre- and post-treatment/embolization of an arteriovenous malformation (AVM). The flow has been divided in two phases arterial and venous. (a) and (b) show arterial-phase results before and after treatment while (b) and (d) show venous-phase results.

Chapter 17

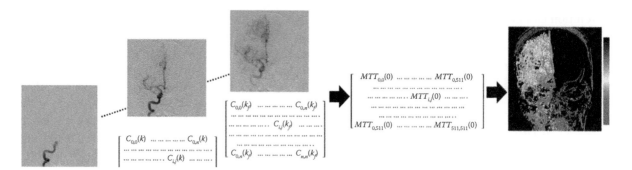

FIGURE 17.9 The colored DSA or parametric imaging is exemplified. Each image is recorded as a matrix, TDC is derived at each matrix element by reading the values in each frame. Using approaches outlined earlier, TDCs are analyzed and parameterized (e.g., MTT, TTP, etc.). Finally using an intensity mapping convention, the parameters are displayed as an intensity distribution of the parameters.

there were many attempts to incorporate the temporal information of the contrast propagation. Such temporal information, as shown in the previous sections, could be used to create a blood flow parametric description over the entire imaged vasculature. Despite these attempts, this imaging processing method did not become a clinical reality. The implementation delays were due to various constraints and limitations such as computational speed or contrast flow dependence on injection technique, or limited knowledge of flow parameters in the patient physiology. The computational aspect meanwhile has been addressed using preformat hardware and advanced programming techniques. As far as the last two aspects are concerned, there is a significant effort to understand and solve these challenges.

A typical algorithm is exemplified in Figure 17.9. The value of individual pixels is tracked in each frame and a TDC is built. Each curve is analyzed, and curve parameters are calculated at each pixel. Various matrices containing parameters such as TTP, MTT, arrival time, etc. can be calculated and later displayed using a colored mapping convention.

To speed up the process, binning as well as intensity thresholds can be used. In the first case, binning is performed on the actual DSA images. For the second case, the user can define a contrast threshold for which TDC measurement and parameter calculation is entirely avoided. In general, the threshold can be a few percent above the quantum mottle of the background.

This method is still in its infancy, and its potential remains to be proven. For the time being, the benefit of such methods consists in the additional information that can be obtained from an already existing DSA. In particular, the method has great potential in the cases when one can compare pathology with a healthy region in the same patient (e.g., left carotid injection vs. right carotid injection) or immediately after treatment in order to evaluate the treatment effect.

17.3 Use of Time–Density Curves for Neurovascular Disease Characterization

17.3.1 Carotid and Intracranial Stenosis

A stenosis is an endoluminal atherosclerotic deposition, which causes an irregular narrowing of the vessel (Figure 17.10). Such deposition can induce severe alteration of the normal hemodynamics or even occlusion. In neurovascular cases, if an atherosclerotic lesion occurs prior to the circle of Willis in one of the arteries, a dilation of the other arteries is observed in order to maintain adequate brain tissue perfusion. The second effect is elevated flow rates across the lesion accompanied by turbulence due to the constriction of the vessel. Such flow conditions can trigger a clotting mechanism; thrombus that forms in this way can travel to more distal locations where it can cause

ischemic stroke. Stenoses have various grades, ranging from mild (1%–39% arterial lumen occlusion) to critical (80%–99% arterial lumen occlusion). The first effect of such malformations is the reduction of blood flow. Figure 17.10 (top rows) shows perfusion of the brain before and after balloon angioplasty for a middle cerebral artery stenosis. Qualitative analysis of the angiograms indicates increased perfusion after the procedure.

For estimation of stenosed vessels, DSA is one of the most used methods for morphology estimation. If the lesion is outside the cranial cavity, Doppler ultrasound is the most used technique for hemodynamics estimation. Where Doppler ultrasound is not readily available or an intracranial location presents a challenge in achieving

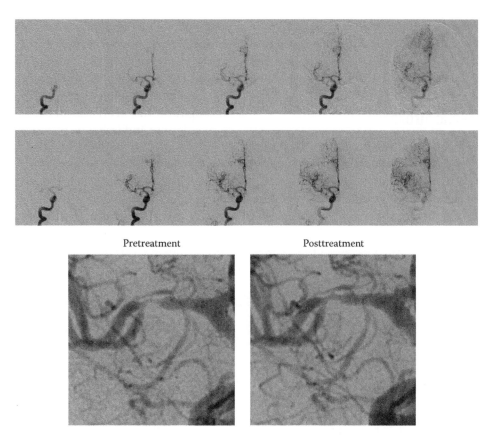

Pretreatment   Posttreatment

FIGURE 17.10 Patient showing with a severe stenosis (85%) in the middle communicating artery. After balloon angioplasty, the stenosis was alleviated to 70%. Frontal angiograms show an increased tissue perfusion after intervention.

such measurements, high frame sequence DSA can be used for flow estimation in these conditions.[9]

The most commonly used method for blood flow velocity estimation is bolus tracking. Most of the arteries in the brain are very tortuous and estimation of the vessel length between two ROIs can be challenging; however, stereoscopic reconstruction or 3D-DSA is often used to estimate the artery length.

Another reason for the hemodynamic evaluation using DSA in a patient presenting with stenosis is to evaluate the compensation of blood supply by the collateral circulation. In general, those patients with a poor blood flow compensation could be at high risk for infarction. For this kind of analysis, parametric imaging can be a very useful tool. Figure 17.11 shows such an analysis for the case depicted in Figure 17.10.

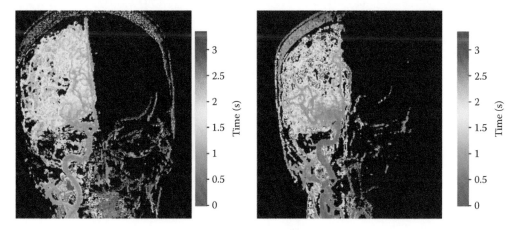

FIGURE 17.11 Parametric imaging of bolus arrival time for pre- and post-intracranial balloon angioplasty in a patient suffering from severe stenosis (Figure 17.10). The parametric imaging in the right shows faster contrast arrival after procedure, indicating restored flow in the middle cerebral region.

Chapter 17

Before angioplasty, the parametric imaging in the right side shows a delayed flow. After angioplasty, the arteries past the middle cerebral artery show a shorter arrival time that is equivalent with faster flow.

17.3.2 Ischemic Stroke

Ischemia describes a condition in which blood supply decreases to a level where temporary or permanent loss of neurological function may occur. Ischemia can be global or local depending on the affected brain area. Unlike an infarct, which is an area of coagulation necrosis following an ischemic event, ischemia is reversible and has various levels, from mild to severe. A severe ischemic event is defined by cerebral blood flow values below 10 mL/100 g/min.

Most of the patients with ischemic events typically go through a perfusion CT scan to establish the extent of the ischemic core and ischemic penumbra. If endovascular thrombectomy is needed, the patient will be taken to an angiographic suite. To monitor progress, angiographic sequences will be acquired frequently to evaluate the progress of the procedure. Typical DSA sequences for an ischemic event are shown in Figure 17.12. In this case,

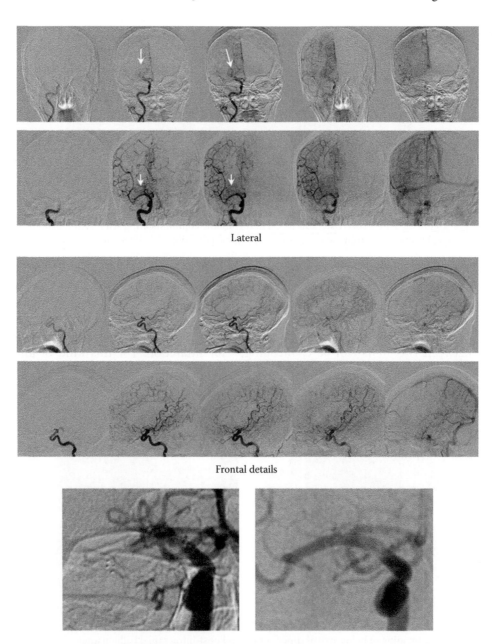

Lateral

Frontal details

FIGURE 17.12 Ischemic stroke after occlusion of an intracranial stent. Lateral and frontal runs show deficit in perfusion of the left side of the brain, after clot removal circulation is restored.

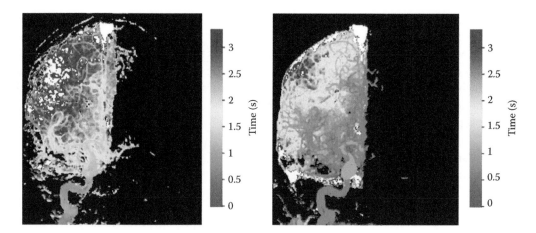

FIGURE 17.13 Parametric imaging of bolus arrival time for pre- and postischemic stroke treatment (Figure 17.11). The parametric imaging in the left shows delayed flow mostly associated with the posterior circulation. The right image shows faster contrast arrival after clot removal and restored flow in the middle cerebral region.

the patient suffered a stent occlusion after an angioplasty procedure. The stent is indicated by the white arrows in the frontal view. Details of the occluded region are shown in the frontal detail in the bottom row of Figure 17.12. Next the patient underwent a clot retrieval procedure, the DSA sequence following the procedure shows re-established flow in the region past the stent.

The main aspects evaluated in the DSA runs in Figure 17.12 are structural problems such as intraluminal thrombus or vessel occlusion, or flow-related defects such as in a partially occluded artery, collateral filling of the affected tissue, or bare areas (no contrast observed, see Figure 17.12). Parametric imaging again can be used to add more information for the interventionalist performing the procedure. In Figure 17.13, we show maps for the bolus arrival time before and after the treatment of the case shown in Figure 17.12.

Before the treatment parametric imaging reveals delayed inflow and outflow. In the case of the affected tissue, slow, delayed flow can occur also through collateral supply. After the treatment, the affected region fills nearly instantaneously indicating re-established flow past the occlusion site. All these aspects can be characterized semiquantitatively using TDC curves with ROIs over the affected areas. In addition, other parameters such as MTT and area under the curve can be used to parameterize the contrast flow in these situations.

17.3.3 Intracranial Aneurysms

Aneurysms are ballooning of the arterial walls occurring in areas subject to high hemodynamic stresses such as arterial bifurcations and tortuous vessels. Many of such malformations are asymptomatic until they rupture, resulting in subarachnoid hemorrhage (SAH). Intracranial aneurysm (IA) SAH mortality occurs at a younger age, producing a large burden of premature mortality, which can be compared with ischemic stroke.[48,49] The median age of death from SAH is 59 years compared with 73 years for intra-cerebral hemorrhage and 81 years for ischemic stroke.[48]

The most used method to assess in vivo aneurysmal hemodynamics modification following treatment remains digital subtracted angiography (DSA) analysis.[50–58] The flow of iodine contrast, intra-arterially injected, is monitored before and after treatment. Treatment outcome could be predicted based on the observed differences. Contrast TDCs derived from videodensitometry measurements of the aneurysmal contrast filling are used to assess the flow modifications.

Analyses of the DSA and TDCs focus on qualitative and quantitative evaluation of the contrast flow. The qualitative analysis focuses on the presence of a jet-like appearance impinging on the aneurysmal wall.

The quantitative analysis focuses on the determination of TDC parameters and relates them to the risk of rupture or treatment success. Unlike the velocity measurements for a stenosed artery, in the case of aneurysms, it is rather difficult to estimate blood velocity. This is due to the presence of special flow patterns such as recirculation zones with various degrees of coupling with the main flow, injection conditions, and the fact that depending on the malformation geometry, only a certain part of the contrast enters the aneurysms dome. Consequently, various authors have proposed different ways to reduce this dependence using various normalization constants. One method[6] is to normalize density curves to the bolus injection derived from the DSA runs. The new normalized time–density curves (NTDCs) are later used to derive a set of parameters described earlier (MTT, TTP, wash-out time (WOT), etc.).

The calculations and the quantities used to derive the NTDCs for the aneurysms are indicated in Figure 17.14.

$$NTDC(t) = \frac{\iint_{ROI}(Bckg - PI_{ROI})}{N} \qquad (17.27)$$

where
  PI_{ROI} represents the individual pixel intensity
  Bckg is the average background value, and the double integral indicates summation of all the pixels contained in the ROI area

The numerator represents the total amount of contrast material entering the aneurysm ROI and is obtained by integrating over the entire ROI for each frame.

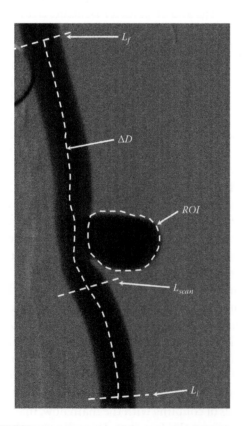

FIGURE 17.14   Determination of the normalized TDC in a canine aneurysm model.

The normalization constant N is equal to the bolus velocity multiplied with the total mass of material passing through a vessel cross section and integrated over the entire run. Using Figure 17.14 and considering the cross section along L_{scan}, N can be written as

$$N = V_{bolus} \int_0^T \left[\int_{L_{scan}} (Bckg - PI_{line})dL \right] dT \qquad (17.28)$$

where
  PI_{line} represents individual pixels in L_{scan}
  Bckg is the average background value
  V_{bolus} is the velocity of the bolus, which is obtained by dividing the vessel path length ΔD, between locations L_i and L_f, by the transit time, $T_{transit}$, of the bolus between the two points indicated in Figure 17.14 using a bolus peak tracking method

Even with such normalization, the curve parameters are very difficult to interpret in relation with aneurysm rupture or treatment outcome. TDC parameters are related to the physiological flow; however, the relation between blood flow conditions and aneurysm evolution, rupture, or treatment outcome is not fully understood. One thing is sure: reducing the flow-induced stresses on the aneurysm dome reduces the risk of rupture. This suggest that post-/pre-treatment NTDC relative parameters could be used to evaluate flow reduction and ultimately the success of aneurysms treatment with flow diverters.[51,54,56] As shown in previous work, such relative parameters tend to correlate very well with the treatment outcome evaluated ex vivo for animal models using histology.

17.3.4   Arteriovenous Malformations

AVMs (Figure 17.15) are complex knot-like structures (nidus) made of abnormal arteries and veins linked by one or more fistulas.[22,59,60] They are believed to have a congenital origin and mostly develop between the third and seventh week of gestation. The vessels in the nidus are weak and enlarged, and they serve as direct connection between the high pressurized arterial network and the low-pressure venous network. The high pressure difference results in a large pressure gradient that in time fatigues the vessels, causing rupture and hemorrhage. AVMs play a role in a smaller number of strokes. They affect mostly healthy young adults who most commonly present with headache, seizure, or

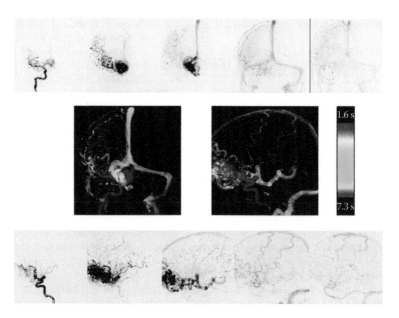

FIGURE 17.15 AP and lateral DSA images from a right internal carotid angiogram from a patient with a large frontal AVM (top and bottom rows). AP and lateral color-coded images from these acquisitions (middle row). The complex circulation of this AVM is clearly depicted on the color-coded composite images. (From Strother, C.M. et al., *Am. J. Neuroradiol.*, 31, 919, 2010.)

intracranial hemorrhage (ICH). Methods of treatment include microsurgery, radiosurgery, and minimally invasive endovascular image-guided intervention to embolize the AVM.

Rapid sequence DSA is one of the most used methods to assess such malformations. Besides the morphology, the DSA could offer information about the hemodynamic properties of the AVM. One parameter of interest is how fast is the transition of the blood from the arterial network to the venous one. A fast transition time of the contrast is usually associated with a high risk of rupture.

Temporal analysis can be used to assess the treatment effectiveness. After treatment, the interventionalist would like to observe delayed contrast arrival to the venous phase. Such delay means slower flow, hence less mechanical stress on the venous structure. Increase in the delayed time can be assessed very well using TDC or parametric imaging. Increased bolus arrival time, TTP, or MTT is associated with delayed flow.

17.4   Conclusions

This chapter introduces the reader to the DSA-based methods, currently used to evaluate flow in the neurovascular application as well as some methods to account for injection variability. Time-resolved neurovascular imaging has been a secondary topic so far, due to its semiquantitative nature. The measurements depend on the injection technique and could lead to inaccurate conclusions. Also patient variability is a hurdle in establishing what range of values are significant and enable clinical judgments to be made. We can account for such obstacles by establishing a ground truth. Such a reference could be a healthy hemisphere versus an affected one, or before and after treatment measurements.

While there are various challenges in interpreting the TDC results, such methods can relay important aspects about the flow condition in neurovascular disease and should be used for immediate posttreatment evaluation in the clinical suites.

References

1. Norris JS, Valiante TA, Wallace MC, Willinsky RA, Montanera WJ, terBrugge KG, Tymianski M. A simple relationship between radiological arteriovenous malformation hemodynamics and clinical presentation: A prospective, blinded analysis of 31 cases. *Journal of Neurosurgery.* 1999;90:673–679.

2. Butler P. Digital subtraction angiography (dsa): A neurosurgical perspective. *British Journal of Neurosurgery.* 1987;1:323–333.

3. Mistretta CA, Crummy AB, Strother CM. Digital angiography: A perspective. *Radiology.* 1981;139:273–276.

4. Landau LD, Lifshits EM. *Fluid Mechanics*. Oxford, U.K.: Pergamon Press; 1987.

5. Lieber BB, Sadasivan C, Hao Q, Seong J, Cesar L. The mixability of angiographic contrast with arterial blood. *Medical Physics*. 2009;36:5064–5078.

6. Yerushalmi S, Itzchak Y. Angiographic methods for blood flow measurements. *Medical Progress through Technology*. 1976;4:107–115.

7. Tenjin H, Asakura F, Nakahara Y, Matsumoto K, Matsuo T, Urano F, Ueda S. Evaluation of intraaneurysmal blood velocity by time-density curve analysis and digital subtraction angiography. *American Journal of Neuroradiology*. 1998;19:1303–1307.

8. Forbes G, Gray JE, Felmlee JP. Phantom testing of peripheral artery. Absolute blood flow measurement with digital arteriography. *Investigative Radiology*. 1985;20:186–192.

9. Kwan ESK, Hall A, Enzmann DR. Quantitative-analysis of intracranial circulation using rapid-sequence dsa. *American Journal of Roentgenology*. 1986;146:1239–1245.

10. Kedem D, Kedem D, Smith CW, Dean RH, Brill AB. Velocity distribution and blood flow measurements using videodensitometric methods. *Investigative Radiology*. 1978;13:46–56.

11. Kruger RA, Bateman W, Liu PY, Nelson JA. Blood flow determination using recursive processing: A digital radiographic method. *Radiology*. 1983;149:293–298.

12. Hohne KH, Bohm M, Erbe W, Nicolae GC, Pfeiffer G, Sonne B. Computer angiography: A new tool for x-ray functional diagnostics. *Medical Progress through Technology*. 1978;6:23–28.

13. Slutsky RA, Carey PH, Bhargava V, Higgins CB. A comparison of peak-to-peak pulmonary transit time determined by digital intravenous angiography with standard dye-dilution techniques in anesthetized dogs. *Investigative Radiology*. 1982;17:362–366.

14. Smedby O, Hogman N, Nilsson S, Erikson U. Flow disturbances in early femoral atherosclerosis—An in vivo study with digitized cineangiography. *Journal of Biomechanics*. 1993;26:1105–1115.

15. Ratib O, Chappuis F, Rutishauser W. Digital angiographic technique for the quantitative assessment of myocardial perfusion. *Annales de Radiologie*. 1985;28:193–197.

16. Nissen SE, Elion JL, Booth DC, Evans J, DeMaria AN. Value and limitations of computer analysis of digital subtraction angiography in the assessment of coronary flow reserve. *Circulation*. 1986;73:562–571.

17. Heintzen PH, Bürsch JH. *Progress in Digital Angiocardiography*. Dordrecht, the Netherlands: Kluwer Academic; 1988.

18. Fencil LE, Doi K, Chua KG, Hoffman KR. Measurement of absolute flow-rate in vessels using a stereoscopic dsa system. *Physics in Medicine and Biology*. 1989;34:659–671.

19. Smedby O, Fuchs L, Tillmark N. Separated flow demonstrated by digitized cineangiography compared with LDV. *Journal of Biomechanical Engineering*. 1991;113:336–341.

20. Mygind M, Engell L, Mygind T. Flow measurements with digital subtraction densitometry in a steady flow experimental model. *Acta Radiologica*. 1995;36:402–409.

21. Shpilfoygel SD, Close RA, Valentino DJ, Duckwiler GR. X-ray videodensitometric methods for blood flow and velocity measurement: A critical review of literature. *Medical Physics*. 2000;27:2008–2023.

22. Todaka T, Hamada J, Kai Y, Morioka M, Ushio Y. Analysis of mean transit time of contrast medium in ruptured and unruptured arteriovenous malformations: A digital subtraction angiographic study. *Stroke; A Journal of Cerebral Circulation*. 2003;34:2410–2414.

23. Bursch JH. Use of digitized functional angiography to evaluate arterial blood-flow. *CardioVascular and Interventional Radiology*. 1983;6:303–310.

24. Silverman NR, Rosen L. Arterial blood flow measurement: Assessment of velocity estimation methods. *Investigative Radiology*. 1977;12:319–324.

25. Swanson DK, Myerowitz PD, Vanlysel MS, Peppler WW, Fields BL, Watson KM, Oconnor J. Arterial blood-flow measurement using digital subtraction angiography (DSA). *Proceedings of the International Society of Photo-Optical Instrumentation Engineers*. 1984;486:122–128.

26. Riediger G, Gravinghoff LM, Hohne KH, Keck EW. Digital cine angiographic evaluation of pulmonary blood flow velocity in ventricular septal defect. *Cardiovascualar and Interventional Radiology*. 1988;11:1–4.

27. Heintzen PH. Review on the research and some aspects upon the modern development of densitometry, particularly roentgen-video-computer techniques. *Annales de Radiologie*. 1978;21:343–348.

28. Cusma JT, Toggart EJ, Folts JD, Peppler WW, Hangiandreou NJ, Lee CS, Mistretta CA. Digital subtraction angiographic imaging of coronary flow reserve. *Circulation*. 1987;75:461–472.

29. Simon R, Herrmann G, Amende I. Comparison of three different principles in the assessment of coronary flow reserve from digital angiograms. *International Journal of Cardiac Imaging*. 1990;5:203–212.

30. Harpen MD, Lecklitner ML. Derivation of gamma variate indicator dilution function from simple convective dispersion model of blood flow. *Medical Physics*. 1984;11:690–692.

31. Proenca J, Muehlsteff J, Aubert X, Carvalho P. Is pulse transit time a good indicator of blood pressure changes during short physical exercise in a young population? *Proceedings of the Annual International Conference of the IEEE Engineering in Medicine and Biology Society*. 2010;2010:598–601.

32. Zhou X, Strobel D, Haensler J, Bernatik T. Hepatic transit time: Indicator of the therapeutic response to radiofrequency ablation of liver tumours. *The British Journal of Radiology*. 2005;78:433–436.

33. Heuck F. *Radiological Functional Analysis of the Vascular System: Contrast Media, Methods, Results*. Berlin, Germany: Springer-Verlag; 1983.

34. Imamura T, Tsuburaya K, Yamadori A. Peak time difference of time-density curve in contrast media transit as an indicator of asymmetric cerebral perfusion. *Journal of the Neurological Sciences*. 1994;126:197–201.

35. Hudson M, Ninan T, Russell G. Evaluation of mean transit time in children as an indicator of airways obstruction. *Respiratory Medicine*. 1992;86:301–304.

36. Geddes LA, Voelz MH, Babbs CF, Bourland JD, Tacker WA. Pulse transit time as an indicator of arterial blood pressure. *Psychophysiology*. 1981;18:71–74.

37. Lin YC. A programmable calculator program for rapid logarithmic extrapolation, and calculation of mean transit time from an indicator-dilution curve. *Computer Programs in Biomedicine*. 1979;9:135–140.

38. Rothe CF, Johns BL, Bennett TD. Vascular capacitance of dog intestine using mean transit time of indicator. *The American Journal of Physiology*. 1978;234:H7–H13.

39. Lagerlof HO, Ekelund K, Johansson C. Studies of gastrointestinal interactions. I. A mathematical analysis of jejunal indicator concentrations used to calculate jejunal flow and mean transit time. *Scandinavian Journal of Gastroenterology*. 1972;7:379–389.

40. von Spreckelsen M, Wolschendorf K. A method to determine the instantaneous velocity of pulsatile blood flow from rapid serial angiographies. *IEEE Transactions on Bio-Medical Engineering*. 1985;32:380–385.

41. Thompson HK, Jr., Starmer CF, Whalen RE, McIntosh HD. Indicator transit time considered as a gamma variate. *Circulation Research*. 1964;14:502–515.

42. Stewart GN. Researches on the circulation time in organs and on the influences which affect it: Parts i–iii. *The Journal of Physiology*. 1893;15:1–89.

43. Kinsman JM, Moore JW, Hamilton WF. Studies on the circulation: I. Injection method: Physical and mathematical considerations. *American Journal of Physiology—Legacy Content*. 1929;89:322–330.

44. Hamilton WF, Moore JW, Kinsman JM, Spurling RG. Studies on the circulation: IV. Further analysis of the injection method, and of changes in hemodynamics under physiological and pathological conditions. *American Journal of Physiology—Legacy Content*. 1932;99:534–551.

45. Horn BKP, Schunck BG. Determining optical flow. *Artificial Intelligence*. 1981;17:185–203.

46. Sonka M, Hlavac V, Boyle R. *Image Processing, Analysis, and Machine Vision*. Toronto, Ontario, Canada: Thompson Learning; 2008.

47. Huang TC, Wu TH, Lin CJ, Mok GS, Guo WY. Peritherapeutic quantitative flow analysis of arteriovenous malformation on digital subtraction angiography. *Journal of Vascular Surgery*. 2012;56:812–815.

48. Johnston SC, Selvin S, Gress DR. The burden, trends, and demographics of mortality from subarachnoid hemorrhage. *Neurology*. 1998;50:1413–1418.

49. Harrigan MR, Deveikis JP. *Handbook of Cerebrovascular Disease and Neurointerventional Technique*. Dordecht, the Netherlands: Humana Press; 2009.

50. Ionita CN, Suri H, Nataranjian S, Siddiqui A, Levy E, Hopkins NL, Bednarek DR, Rudin S. Angiographic imaging evaluation of patient-specific bifurcation-aneurysm phantom treatment with pre-shaped, self-expanding, flow-diverting stents: Feasibility study. *Proceedings of the Society of Photo-Optical Instrumentation Engineers*. 2011;7965:79651H79651–79651H79659.

51. Ionita CN, Natarajan SK, Wang W, Hopkins LN, Levy EI, Siddiqui AH, Bednarek DR, Rudin S. Evaluation of a second-generation self-expanding variable-porosity flow diverter in a rabbit elastase aneurysm model. *American Journal of Neuroradiology*. 2011;32:1399–1407.

52. Ionita CN, Wang W, Bednarek DR, Rudin S. Assessment of contrast flow modification in aneurysms treated with closed-cell self-deploying asymmetric vascular stents (savs). *Proceedings of the Society of Photo-Optical Instrumentation Engineers*. 2010;7626:76260I.

53. Rangwala HS, Ionita CN, Rudin S, Baier RE. Partially polyurethane-covered stent for cerebral aneurysm treatment. *Journal of Biomedical Materials Research. Part B, Applied Biomaterials*. 2009;89:415–429.

54. Ionita CN, Paciorek AM, Dohatcu A, Hoffmann KR, Bednarek DR, Kolega J, Levy EI, Hopkins LN, Rudin S, Mocco JD. The asymmetric vascular stent: Efficacy in a rabbit aneurysm model. *Stroke; A Journal of Cerebral Circulation*. 2009;40:959–965.

55. Ionita CN, Dohatcu A, Sinelnikov A, Sherman J, Keleshis C, Paciorek AM, Hoffmann KR, Bednarek DR, Rudin S. Angiographic analysis of animal model aneurysms treated with novel polyurethane asymmetric vascular stent (P-AVS): Feasibility study. *Proceedings of the Society of Photo-Optical Instrumentation Engineers*. 2009;7262:72621H1–72621H10.

56. Ionita CN, Paciorek AM, Hoffmann KR, Bednarek DR, Yamamoto J, Kolega J, Levy EI, Hopkins LN, Rudin S, Mocco J. Asymmetric vascular stent: Feasibility study of a new low-porosity patch-containing stent. *Stroke; A Journal of Cerebral Circulation*. 2008;39:2105–2113.

57. Dohatcu A, Ionita CN, Paciorek A, Bednarek DR, Hoffmann KR, Rudin S. Endovascular image-guided treatment of in-vivo model aneurysms with asymmetric vascular stents (AVS): Evaluation with time-density curve angiographic analysis and histology. *Proceedings of the Society of Photo-Optical Instrumentation Engineers*. 2008;6916:6916OP.

58. Hoi Y, Ionita CN, Tranquebar RV, Hoffmann KR, Woodward SH, Taulbee DB, Meng H, Rudin S. Flow modification in canine intracranial aneurysm model by an asymmetric stent: Studies using digital subtraction angiography (DSA) and image-based computational fluid dynamics (CFD) analyses. *Proceedings of the Society of Photo-Optical Instrumentation Engineers*. 2006;6143:61430J.

59. Wilkins RH. Natural history of intracranial vascular malformations: A review. *Neurosurgery*. 1985;16:421–430.

60. Arteriovenous malformations of the brain in adults. *The New England Journal of Medicine*. 1999;340:1812–1818.

61. Strother CM, Bender F, Deuerling-Zheng Y, Royalty K, Pulfer KA, Baumgart J, Zellerhoff M, Aagaard-Kienitz B, Niemann DB, Lindstrom ML. Parametric color coding of digital subtraction angiography. *American Journal of Neuroradiology*. 2010;31:919–924.

62. Bonnefous O, Pereira VM, Ouared R, Brina O, Aerts H, Hermans R, van Nijnatten F, Stawiaski J, Ruijters D. Quantification of arterial flow using digital subtraction angiography. *Medical Physics*. 2012;39:6264.

Chapter 17

18. 3D Perfusion Imaging in Cardiac and Neurologic Applications

Arundhuti Ganguly

18.1 Emergence of Perfusion Imaging

18.1.1 Introduction

Perfusion imaging provides functional information of a tissue or organ of interest by imaging the uptake of contrast agent in the targeted anatomy. The contrast-enhancing agent serves as a surrogate for blood [1]. While vascular imaging mostly involves visualizing the larger blood vessels, perfusion imaging evaluates blood flow in the capillaries and/or microvessels [2]. Exchange of oxygen and nutrient and removal of cellular waste

occur at this level [3]. Thus, tissue perfusion is directly related to tissue health and the changes in tissue perfusion in affected regions can be indicative of whether the damage is reversible or not. It can serve as a predictor of therapeutic outcomes. In some areas such as in the brain and heart, perfusion imaging is quite often part of the standard of care for diseases such as stroke and coronary artery disease (CAD). In other parts of the anatomy such as the liver and kidneys, it is an area of growing clinical interest and research. In addition, when correctly calibrated for cardiac output, perfusion measurements can directly relate to the level of tumor angiogenesis [4]. In this chapter, the fundamental principles, methods, and applications of perfusion imaging will be described for 3D modalities. For two-dimensional (2D) modalities, the reader is referred to Chapter 21.

18.1.2 Perfusion 3D Imaging Modalities

18.1.2.1 CT

Perfusion imaging with contrast enhancement was first introduced by Axel in 1980 [5,6] shortly after the emergence of CT. It was based on the earlier work by Meier and Zierler [7,8]. Improvements in image acquisition speeds resulting from faster gantry rotation and multislice acquisitions have made clinical perfusion CT feasible. This involves superior temporal sampling (0.33–0.5 s) and volume coverage beyond a single CT slice (256 × 0.5 mm to up to 320 × 0.5 mm).

The noninvasive determination of tissue perfusion depends on the accurate measurement of contrast agent concentration as a function of time. In dynamic CT, the actual concentration of iodinated contrast medium in a given voxel is linearly related to the Hounsfield unit (HU) or CT number in the reconstructed images. Thus, CT number changes over time can be directly mapped to the contrast agent concentration versus time. The ability to accurately quantify perfusion parameters such as the flow, volume, and transit times has allowed CT perfusion to become a widely accepted diagnostic tool particularly in the case of acute stroke.

18.1.2.2 C-Arm CT

A C-arm system typically consists of an x-ray tube and a flat panel–based detector on the opposing ends of a C-shaped support arm (Figure 18.1). Both radiographic and fluoroscopic imaging are available on such systems and are most commonly used for guidance during minimally invasive endovascular surgeries. These procedures involve real-time visualization of contrast agent flow in blood vessels and the localization of treatments and endovascular devices. While the common imaging modes used for guidance are 2D projection images, the modern C-arm systems are enabled and increasingly being used for 3D volume imaging as well. Further modification of image acquisition protocols and reconstruction algorithms has enabled the establishment of the feasibility of C-arm CT-based perfusion imaging of the brain and liver. The biggest advantage of cone-beam CT (CBCT) perfusion

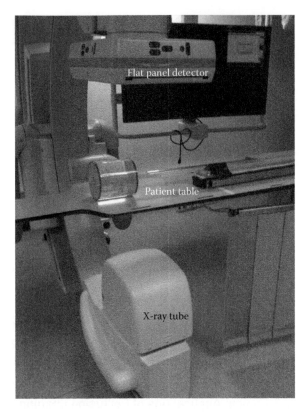

FIGURE 18.1 C-arm imaging system capable of performing CBCT (Siemens Axiom Artis dTA and syngo DynaCT™, Siemens AG, Healthcare Sector, Forchheim, Germany). In the earlier orientation, the C-arm supports a flat panel detector on the top and an x-ray tube at the other (bottom) end and rotates about the axis along the patient's table.

imaging is the large area coverage without table toggle (currently up to 15 cm), higher-resolution acquisition with isotropic voxels (e.g., currently acquisition in 4 × 4 binned mode with 0.496 × 0.496 mm² pixels and reconstruction with 0.5 × 0.5 × 0.5 mm³ available on Artis dTA System and 0.48 × 0.48 mm² or 0.6 × 0.6 mm² for 3 s fast scan on the Artis Zeego System, both from Siemens Healthcare, Forcheim, Germany), and availability of intraprocedural imaging [9].

In the case of acute cerebral ischemia (blockage) where intravenous (IV) delivery of tissue Plasminogen Activator (tPA) or clot-busting drugs is not suitable, patients are treated with selective endovascular therapies. These include localized delivery of tPA and/or mechanical endarectomy. Such endovascular procedures are guided using real-time imaging, with the most prevalent method being fluoroscopy and radiography using C-arm-based x-ray systems. The feasibility for the use of flat panel detector-based C-arm CBCT systems for perfusion imaging was first demonstrated for cerebral blood volume (CBV) measurement by Ahmed et al. in 2010 [10]. The ability to obtain serial

C-arm CT measurements of CBV before and during an intervention could add value in determining the effectiveness of revascularization and also in determining the end point when the risk of injury from treatment exceeds further possible benefit. Complete acquisition of cerebral flow and perfusion has been more recently demonstrated by Ganguly et al. [9], making intraprocedural perfusion imaging an attractive tool.

18.1.2.3   MR

The earliest work in the use of MR for perfusion imaging was reported almost 25 years ago by Villringer et al. [11]. Using a mouse model, the images of gadolinium-based chelates for enhancing T1 were obtained as a function of time. These images were acquired in one dimension to accommodate the imaging limitations of early MRI. Perfusion imaging using MR is also called dynamic susceptibility-contrast (DSC) MRI and can be performed with or without injected contrast agents [3]. Spin labeling of the blood is commonly used for noncontrast enhanced scans that are less popular because of increased susceptibility to motion, increased artifacts, and low signal in case of slow flow [12]. More commonly, gadolinium-based tracers such as Gd-DTPA-enhanced imaging is used that relies on local changes in magnetic susceptibility. Popular imaging sequences are either based on spin echo (SE) or gradient echo (GRE) methods. While SE sequences are less influenced by the flow in neighboring large vessels, the GRE sequences have the advantage of higher signal-to-noise ratio (SNR) in comparison to SE sequences.

18.1.2.4   PET and SPECT

Both positron emission tomography (PET) and single photon emission computer tomography (SPECT) are particularly popular for cardiac perfusion imaging. Both methods use radioactive drugs that specifically collect in the organ of interest. In case of SPECT, the radioisotope emits gamma rays that are collected by a gamma camera. It is the most common diagnostic and therapeutic decision-making tool particularly in cardiac atherosclerotic disease (CAD) patients [13].

In the case of PET, the radionuclide emits a positron in the tissue of interest. The positron travels in the tissue until it combines with an electron and ejects a pair of gamma ray photons that are detected by the imaging system. Cardiac PET imaging is an area of rapid growth. This has been facilitated by the wider availability of PET tracers such as rubidium ^{82}Rb, detector hardware and software advancement, improved display, and better diagnostic and prognostic accuracy [13]. Since PET involves positron annihilation radiation emitted at 511 keV (compared to 140 keV for ^{99m}Tc), it suffers less attenuation. Reported image quality of PET can be as much as 70% better than SPECT in the same patient [14,15]. Further enhancement can be achieved in combined PET/CT systems similar to SPECT/CT systems that provide anatomic overlay with additional attenuation correction. PET is particularly advantageous in obese patients [16] and, in cases, where SPECT was equivocal [13].

18.2   Theoretical Background: Development of Models for Calculating Tissue Perfusion

Perfusion imaging and its quantification is based on the indicator dilution theory [2,17]. It was explained in the context of physiological monitoring by Zierler [7] detailing the associated assumptions, validity, and the effects of specific violations. An indicator is essentially an additive to a fluid that allows measurement of the fluid properties. The ideal case in which the measurements accurately represent the fluid behavior includes single points for inflow and outflow with no recirculation. The indicator is introduced as a sharp bolus or at a constant rate, and it is required that the indicator mixes uniformly with the fluid. The flow and volume during the period of observation should remain constant. Finally, the system should be stationary: the local and global transit times must have

the same distribution. In real systems, not all conditions are met exactly, and hence, the theory can be modified to accommodate certain modification of these conditions.

Various methods have been proposed for calculating perfusion parameters. Some common assumptions are made for the various physiological models used for this purpose. They can be summarized as follows:

1. The physiological behavior of blood flow in the tissue is represented by the behavior of tracers. The most common methods described refer to the behavior of nondiffusible tracers. In case of CT, these are iodine based, and in case of MR, they are usually gadolinium-based contrast agents.

Chapter 18

2. The simple models assume a fixed volume of tissue with a single inlet through which the blood flows in and a single outlet for the blood to exit.

3. The tissue is perfused by the flow of blood in capillaries with varying transit times that can be represented by a probability distribution.

4. Some models do not allow for recirculation of blood within the volume and are suitable for handling cases such as brain perfusion when the blood–brain barrier (BBB) is still intact. When recirculation is allowed, the resultant perfusion parameters are suitable for assessing perfusion in tumors and in the heart.

18.2.1 Indicator Dilution Theory (Moments Method)

The concept of indicator or a substance introduced into blood circulation that is used to measure hemodynamic properties of blood was introduced by Hering in 1829 [18]. The indicator dilution method in the context of perfusion imaging was developed and described by Meier [8], Zierler [7], and Antman [19] for calculating the blood flow per unit of total tissue volume. The specific case of nondiffusible iodinated tracers used in CT imaging in normal brain tissue (intact BBB) was reviewed by Axel [17]. The basic concept involves plotting the time density curve (TDC) or time attenuation curve (TAC). This represents the variation of the contrast density (mg/mL) in each voxel as a function of time (s or min). The area under the tissue TDC is directly related to the tissue blood volume. This value is normalized by the area under the arterial TDC of the feeding artery to remove the dependence on the contrast injection and flow characteristics of this blood vessel.

Some very simplistic assumptions are made in this model:

1. The flow is stationary with one inlet and one outlet, and all contrast material eventually leaves the volume, that is, there is no recirculation.

2. The tracer is "suddenly" injected at the inlet, meaning the injected bolus has a sharp leading edge or is a step-function, when introduced at time $t = 0$.

3. The tracer is well mixed with the blood, and the amount is small enough to not disturb the flow that is represented by the contrast agent flow.

Three basic equations can be used to describe the behavior mathematically [5]:

$$F = \frac{m}{\int_0^\infty c(t)dt} \tag{18.1}$$

$$t^* = \frac{\int_0^\infty tc(t)dt}{\int_0^\infty c(t)dt} \tag{18.2}$$

$$V = Ft^* \tag{18.3}$$

The constant flow denoted by F (mL/s) is given by the known amount of injected indicator m (g) divided by the total contrast measured at the output integrated over time. The instantaneous contrast concentration is denoted by $c(t)$ (g/mL) and is measured from the time the contrast agent appears to when it disappears in the volume of interest (VOI). The quantity t^* is the mean transit time (MTT) with the total volume of contrast agent that flows through in the given time, V (mL). Equation 18.1 is also referred to as the *Stewart–Hamilton equation*.

The formulation described by Axel was further developed by Gobbel [20] for refining the measurement parameters particularly in the context of cerebral perfusion. It included practical considerations such as allowing for the arterial and venous TDC measurements from a large artery and vein instead of the one immediately feeding or exiting the tissue region of interest (ROI). The assumption made here is that the area under the curve (AUC) should be the same for all arteries visible in the CT scan slice. It avoids deconvolution methods that can be difficult to implement. In addition, the calculations also accounted for the fact that the blood volume to be considered consists of iodinated contrast, blood plasma, and hematocrit (primarily red blood cells). The fact that the fraction of hematocrit H_{CT} (or red blood cells) in the periphery and in the tissue is different was accounted for in order to obtain accurate perfusion numbers. A "center of gravity"-based approach was hence formulated by Gobbel [20], resulting in the following key equation for the regional cerebral blood flow (rCBF):

$$\text{rCBF} = \frac{V_B}{\langle t_f \rangle} \tag{18.4}$$

here

V_B represents the fractional vascular blood volume

the quantity t_f is the center of gravity of the transit time distribution and is called an MTT

Both these quantities are not directly measurable from CT images. Also the flow of iodine that is imaged in CT scans relates to the flow of the blood plasma only. Hence, the fraction of blood volume that results from the hematocrits (blood solids like the red and white blood cells, platelets, etc.) needs to be corrected for. Another correction needs to be included for the fact that the bolus arrives at a peripheral artery at a different time than it does to the tissue ROI and is not a sharp delta function. Hence, by substituting quantities that are more easily measurable in a contrast-enhanced CT scan, and correcting for the nonideal factors, the following expression is obtained for the regional CBF that accounts for the hematocrit fraction and the plasma volume in the tissue ROI:

$$rCBF = n \left[\frac{AUC_{ROI}/AUC_{PA}}{\langle t_{ROI} \rangle - \langle t_{PA} \rangle} \right] \tag{18.5}$$

where

$n = ((1 - Hct_{artery})(1 + k^2))/(2(1 - rHct_{artery})(1 - q))$

$r = Hct_{tissue}/Hct_{artery}$

$k = \sqrt{\langle t_f^2 \rangle - \langle t_f \rangle^2}/\langle t_f \rangle$

q is a correction factor

Specifically q is the fraction of $\langle t_{ROI} \rangle - \langle t_{PA} \rangle$ that results only from the difference in the time of bolus arrival in the artery used for measurement, and the ROI and $q(\langle t_{ROI} \rangle - \langle t_{PA} \rangle)$ can be treated as a delay. The remaining fractional difference $(1 - q)(\langle t_{ROI} \rangle - \langle t_{PA} \rangle)$ is the actual difference in the mean of the time distribution in the ROI and the peripheral artery. Here PA represents the peripheral artery used for measuring the arterial input function (AIF). The ratio of the AUCs relates to the fraction of the blood volume that comes in via the peripheral artery and actually goes only to the ROI and hence is related to fractional blood volume V_B. The term k is related to the standard deviation of the TDC from a curve fitted to the data. The TDC curves are fitted to a gamma variate function to evaluate this quantity and is shown in Equation 18.7. It is assumed that the value of k is constant across the tissue. The ratio r has a typical value of 0.7 in adults and 0.85 in infants. The proportionality term n contains all the correction factors

discussed earlier. Typically CBF is expressed in mL/min/100 g of tissue. In the earlier derivation following Gobbel's method, the units for rCBF are in units of inverse time. To express it in the more historically conventional units of mL/min/100 g of tissue, the value would need to be divided by the density of tissue (which is assumed to be 1.05 g/mL for brain) and the volume of the VOI or equivalently the mass of tissue in the VOI. This normalization is explained by Guibert et al. [21] as follows: "But for all those methods, a current convention states that part of the methodological variability can be rescaled into a volume normalization (generally expressed in 100 mL volume or in 100 g tissue weight)."

The mean or expectation value involves calculating the first moment of the distribution, which is also the centroid and is given by

$$\langle t \rangle = \frac{\int_0^\infty tc(t)dt}{\int_0^\infty c(t)dt} \tag{18.6}$$

For using this method, the common practice is to model the TDC as a gamma variate function of the form:

$$c(t) = \frac{A(t - t_0)^a}{(a \cdot b \cdot e^{-1})^a} e^{-(t - t_0)/b} u(t - t_0) \tag{18.7}$$

where

A is the peak enhancement

a and b are shape parameters

$u(t - t_0)$ is a step function starting at time t_0

Variable a has a typical value of 3 and the variable b is 1.5 or 1 for IV and intra-arterial (IA) injections, respectively, when c(t) represents the AIF. Use of the gamma variate function allows the integration of the integrals to infinity. Usually the average enhancement within the tissue VOI is more than an order of magnitude lower than the enhancements of the feeding artery and the draining vein.

18.2.2 Linear Systems Approach (Deconvolution Method)

The approach used in the moments method formulations requires the use of the gamma variate fits to the actual TDC. This can be a cause for errors as has been shown by Wen et al. [22]. Another approach that is fairly immune to such errors uses a linear systems formalism that involves deconvolution techniques. More recent perfusion literatures report on the use of this technique because of its advantages.

A linear system is an idealized system that can be treated as a black box with an input and an output. It obeys the rules of (a) superposition: if input x_1 gives output y_1 and input x_2 gives output y_2, then $x_1 + x_2$ gives $y_1 + y_2$ and (b) scaling: if input x results in output y, then input ax will result in output ay. A linear time invariant (LTI) system has the additional property that the system response to a signal at a given time is the same at a different time except for a time delay. In the simplest case, the input is a delta function, which then travels through the system or equivalently is deconvolved with the system response prior to output.

output = input $\otimes$ *response*, which can also be written in the integral form as

$$output = \int_{-\infty}^{+\infty} input(\xi)h(t-\xi)d\xi \tag{18.8}$$

The physiological equivalence of Equation 18.8 involves a sharp instantaneous bolus of imaging contrast agent injected at the entrance of the blood vessel and observation of changes at the output that reflect the health of the perfused tissue. The response of the tissue in the form of the contrast enhancement to such an input function is the impulse response function (IRF). Typically, the quantity to be determined is the impulse residue function $r(t)$ describing the amount of contrast agent remaining in the system at any given time. Please refer to Chapter 17 for a detailed explanation of residue function. Mathematically, it is written as

$$r(t) = \begin{cases} 1 - \int_0^t h(\tau)d\tau, & t \geq 0 \\ 0, & t < 0 \end{cases} \tag{18.9}$$

The function $h(\tau)$ describes the probability density function of tissue transit times. The MTT parameter can be described from this distribution by

$$MTT = \frac{\int_0^\infty \tau h(\tau)d\tau}{\int_0^\infty h(\tau)d\tau} \tag{18.10}$$

which is the same as Equation 18.6 in the moments method.

For a physiological system from Equation 18.8, we get

$$c_v(t) = \int_{-\infty}^\infty c_a h(t-\xi)d\xi \tag{18.11}$$

where $c_v(t)$ and $c_a(t)$ are the contrast concentration (g/mL) over time measured in the vein and artery associated with the tissue of interest. Using the principle of conservation of mass of the contrast agent within the VOI at any given time,

$$m_{voi}(t) = m_a(t) - m_v(t) \tag{18.12}$$

where $m_{voi}(t)$, $m_a(t)$, and $m_v(t)$ are the mass of contrast in the VOI, in the artery, and in the vein associated with the VOI, respectively. Also

$$m_{voi}(t) = F \int_0^t c_{voi}(\tau)d\tau = F \int_0^t \left[c_a(\tau) - c_v \right] d\tau \tag{18.13}$$

where F is the flow in mL/s. CBF or the blood flow normalized by the mass of the VOI (V_{voi}) comprising the mass of the parenchyma and the interstitium is

$$CBF = \frac{F}{V_{voi}\rho_{voi}} \tag{18.14}$$

Here is ρ_{voi} the density of tissue in the VOI. Substituting Equation 18.11 in Equation 18.13 and using the residue function definition in Equation 18.9, one arrives at the main equation for perfusion using this method. The intermediate steps involve (a) using the integral form of the convolution operation as shown in Equation 18.8, (b) reordering of the terms in the integral, and then (c) converting it back to a compact equation involving a convolution. For a complete derivation, refer to the article by Fieselmann et al. [2]:

$$c_{voi}(t) = CBF \cdot \rho_{voi} \cdot \left\{ c_a(t) \otimes r(t) \right\} \tag{18.15}$$

or

$$c_{voi}(t) = c_a(t) \otimes k(t) \quad \text{with } k(t) = CBF \cdot \rho_{voi} \cdot r(t) \tag{18.16}$$

The contrast enhancement over time in the VOI ($c_{voi}(t)$) and the artery ($c_a(t)$) can be determined from the dynamic CT images acquired over time. The term $k(t)$ (units of 1/s) is calculated from Equation 18.16 using deconvolution of the contrast concentrations equation. This value of $k(t)$ is then used to obtain the value of the CBF. Since at $t = 0$, $r(t)$ has a value of 1, $k(t)$ reaches its maximum value. Therefore, the volume flow can be calculated using

$$CBF = \frac{1}{\rho_{voi}} \cdot \max[k(t)] \qquad (18.17)$$

CBV in the VOI is derived from the CBF using the central volume theorem [2,23] that states that within the tissue VOI, the blood volume is related to the blood flow rate by the mean time it takes for the blood to traverse through the capillary bed by

$$CBV = MTT \cdot CBF \qquad (18.18)$$

Various approaches have been used for solving the convolution Equation 18.16. The main difficulty with this approach is the fact that Equation 18.16 is an ill-posed problem. Hence, any noise in the measured TDC and AIF results in significant errors in the residue function and therefore the perfusion. The three main approaches to finding a solution include (a) model based, (b) algebraic, and (c) Fourier transform based. Further details on these methods and relevant references can be found in Reference 2. Briefly, the model-based approach assumes a certain shape for the residue function. A common form reported in the literature includes a decaying exponential function. The algebraic method involves the following steps:

1. Discretization of the integral in Equation 18.13 into a summation that then allows switching over to matrix algebra
2. Solving of the matrix equation using approaches such as singular value decomposition (SVD) techniques
3. Regularization using a filter to obtain a stable and physiologically viable solution

The Fourier transform method is mathematically similar to the SVD approach with the useful property that upon undergoing Fourier transform, a convolution becomes a multiplication operation. The regularization step is an equivalent filtering step with the use of a common Weiner filter.

18.2.3 Compartmental Analysis (Slope Method)

The main advantage of the slope method is the need for a shorter time series dataset for evaluating the perfusion parameters. It involves calculating the maximum slope of the time enhancement curve and measuring the peak enhancement of the AIF. It is obtained by differentiating Equation 18.16. The relation needs to hold at all times, t, and hence needs to hold at the time point when the slope of the TDC is maximum. The primary assumption is that there is no venous outflow (and hence recirculation), thereby requiring a sharp bolus and quick data acquisition before venous outflow starts. Hence, in Equations 18.15 and 18.16, setting $c_v(t)$ to 0 followed by the differentiation of the time integral of $c_a(t)$ reduces simply to $c_a(t)$. The primary equation used to explain the measurement is given by

$$\left[\frac{dc_{voi}(t)}{dt}\right]_{max} = \rho_{voi} \cdot CBF \cdot [c_a(t)]_{max} \qquad (18.19)$$

Here $dc_{voi}(t)/dt$ is obtained by differentiating the tissue TDC to find the maximum slope. The method primarily requires four images: a baseline, peak arterial enhancement, and two sequential tissue enhancement images with the biggest change in intensity. While the method is fast, it is very sensitive to noise in any of the four images since it involves a differentiation operation. The slope method is preferred where speed is of paramount importance.

18.3 Clinical Applications of Perfusion Imaging

In addition to CBF (or more generally perfusion blood flow [PBF]), CBV (or perfusion blood volume [PBV]), and MTT described earlier, there are several other quantitative parameters that can be calculated and provide additional diagnostic information [24]. Figure 18.2 describes some of these parameters graphically. They include:

1. *Tmax*: Time-to-maximum of the residue function (*Tmax*) is a parameter that has also received increasing interest in acute stroke studies. It primarily reflects a combination of the delay in bolus arrival and it dispersion. It relates to macrovascular features and to collateral flow.

Chapter 18

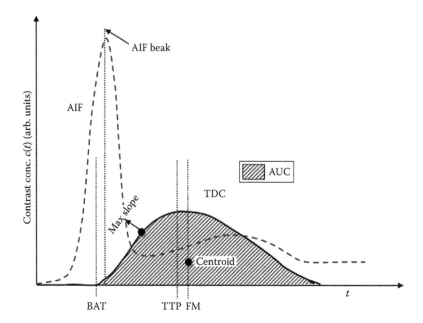

FIGURE 18.2 The AIF and the tissue TDC shown as a function of time. The tissue TDC can be as much as an order of magnitude lower than the AIF. The AIF shows a secondary bump during recirculation. The different perfusion parameters including BAT, TTP, and FM are shown on the time axis. The gray shaded region is the AUC for the tissue TDC with the centroid location indicted by the black dot. For calculating perfusion using the slope method, the steep region on the rising edge of the TDC is used.

2. Time-to-peak (TTP): This is the time it take from the start of scanning ($t = 0$) to the time it takes for the signal in the tissue to reach the peak contrast concentration value.

3. Bolus arrival time (BAT): There is a finite time difference between when the bolus peaks in the artery to when it peaks in the tissue of interest downstream. This relates to the microvascular distribution in the tissue and hence to the state of the tissue health. The time at which the bolus arrives in the tissue is the BAT.

4. Maximum-peak concentration (MPC or c_{max}): This is the maximum value of the tissue TDC.

5. Full-width at half-maximum (FWHM): This is the width of the TDC at half the peak value.

6. First moment (FM) of the tissue concentration: The first moment is calculated from the centroid of the TDC. This centroid when projected onto the time axis gives the FM in units of time.

Given this background, a typical perfusion study involves the following steps [3]:

Step 1. Selecting an appropriate image acquisition protocol.

Step 2. Selecting an appropriate contrast agent injection protocol.

Step 3. Assessing the need of motion correction as a preprocessing step.

Step 4. Estimating the time-dependent contrast agent concentration.

Step 5. Measuring the AIF.

Step 6. Performing the deconvolution analysis to remove the AIF contribution.

Step 7. Quantifying the hemodynamic parameters of interest.

Step 8. Scaling the measurements and displaying, if values in absolute units are required.

18.3.1 Cerebral Perfusion Imaging

18.3.1.1 CT Perfusion

Perfusion CT and where possible perfusion MR imaging is fast becoming part of the standard of care for the diagnosis and treatment of stroke. Ischemic stroke, which constitutes 85% of all stroke cases mainly, benefits from perfusion imaging [25]. This additional information helps to identify the extent of tissue damage and delineates the regions that can be salvaged by thrombolysis. Large clinical trials have established that administration of thrombolytic drugs within 3–6 h of ictus or the onset of stroke is feasible and can result in successful re-perfusion [25,26]. Typically, it involves IV administration of the imaging contrast agent followed by a rapid sequence of scans for about a minute. Initially, PCT imaging was restricted to a single slice that passes through the basal ganglia to allow for a clear visualization of carotid artery branches since they are most commonly affected. With table toggling techniques and with the advent of multidetector row CT, larger areas of the brain can be covered including full brain coverage in some of the most recent scanners.

Brain perfusion parameters have been very well quantified, and the typical values for healthy and damaged tissue are well known. In animal studies, the combined assessment of CT CBV and CBF from CT measurements was found to have sensitivity of 90.6% and specificity of 93.3% when compared with histological measurements for discerning ischemic and oligemic tissues [27] (i.e., irreversible versus reversibly damaged tissue). Normal cerebral perfusion or CBF rate is around 50–60 mL/min/100 g of tissue. A value below 20 mL/min/100 g indicates affected tissue metabolism though irreversible tissue damage is further dependent on the time lapse since the advent of the event and the absolute perfusion value. Other than the flow values, CBV and MTT need to be considered. These parameters in addition to the conventional CT imaging and CT angiograms can provide a fairly complete assessment of the condition. Figure 18.3 shows the perfusion CT parameter maps (CBV, CBF, MTT, and *Tmax*) from multiple slices in a stroke patient.

A typical decision process would include an evaluation of the MTT map as the first step. A normal MTT and perfusion value for a stroke patient would indicate a transient ischemic attack (TIA). Normal to reduced MTT with normal perfusion would indicate a tumor. If the MTT is normal, then assessment of lacunae (internal spaces in the brain) and evaluation of alternate etiology is required. For abnormal MTT, the next step involves looking at the CBV map followed by the CBF map. This is essential since depending on the situation the body very rapidly enters an autoregulatory mode to compensate for blood flow problems. If both CBF and CBV values are high, then it indicates a hyperperfusion or a hemorrhagic event. A small decrease in cerebral perfusion results in vasodilation that allows the CBV and MTT to increase and restore the CBF value. With further decrease, despite the increase in CBV and MTT, the CBF value is unable to return to normal levels. The vasodilation in this case is unable to compensate. Finally, when the CBF value is too low, the tissue metabolism ceases, and in due course, the tissue

is irreversibly damaged. The autoregulatory process at this point totally fails. However, if both CBF and CBV have low values, then this indicates an infarcted region. By comparing the CBF and CBV maps and studying their mismatch, the regions that still have intact autoregulation can be identified and, depending on the actual values and time lapse, can be treated either by system or by localized tPA [26]. Table 18.1 shows an example of the decision-making process based on perfusion information.

18.3.1.2 MR Perfusion

Perfusion MRI or DSC-MRI most commonly uses T_2^*-weighted imaging sequence such as GRE and faster gradient echo planar imaging (EPI) sequences that use T_2^*-weighted imaging. Guidelines for selection of imaging parameters for brain imaging have been provided by an international consortium [28]. The recommendations for the various parameters are listed in Table 18.2.

A single AIF is selected for all the slices and is assumed to have a behavior that is representative of the entire volume. It is measured from some vessel close to the middle cerebral artery (MCA). Most commonly, the deconvolution technique is used for determining perfusion. Most commonly, the CBF, CBV, MTT, and *Tmax* are evaluated and the corresponding maps are generated. An issue with MR perfusion data is the difficulty in quantifying the results (other than timing values) in absolute units. This is because of the unknowns present in the conversion of absolute amount of contrast agent concentration in any given voxel that is a complex function of several factors. It is possible to provide subject-specific scaling that involves other scan sequences or a PET-based lookup table.

In addition to perfusion, MR imaging is capable of providing tissue diffusion parameter, which essentially maps out the interaction of water molecules within tissue fibers and across membranes. Mismatch between tissue diffusion and perfusion regions has been very successfully related to defining the ischemic penumbra. These regions have normal perfusion but abnormal diffusion profiles and indicate affected tissue that is still viable and can be salvaged with rapid therapy. Data from several large ongoing and completed clinical trials involving acute stroke are being used to define the optimal threshold for outlining the regions that comprise the penumbra. Figure 18.4 shows perfusion MR images acquired in a stroke case that was evaluated for endovascular therapy.

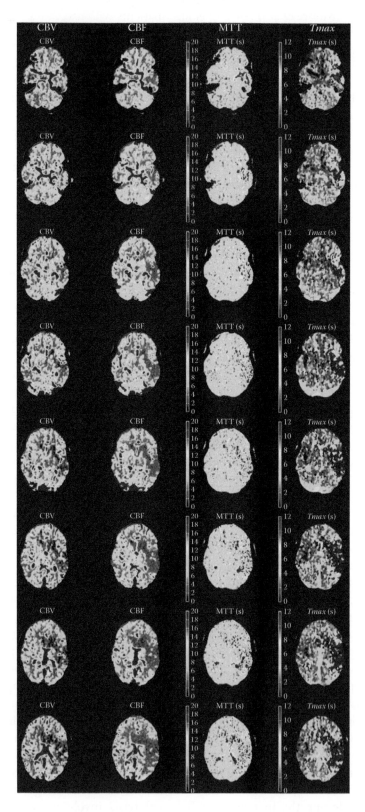

FIGURE 18.3 Perfusion CT maps showing CBV, CBF, MTT, and *Tmax* in a 70-year-old female stroke patient. Imaging was performed with the following imaging parameters: 80 kVp 200 mA, exposure of 500 ms, with a 233 mm field of view. The image acquisition and reconstruction parameters were: pixel spacing 0.455 mm × 0.455 mm (512² matrix) resampled to 256² matrix, pixel size 0.911 × 0.911 mm, and slice thickness 4.78 mm. A 40 mL bolus of iodinated contrast agent was used and images were acquired every 2.05 s for 30 s with 0.5 s gantry rotation on an Ingenuity CT System (Philips Healthcare, Best, the Netherlands). (Image courtesy of M. Straka, Stanford University, Stanford, CA.)

Table 18.1 Decision-Making Process in Stroke Diagnosis Based on CBF and CBV Maps

CBF	CBV High	CBV Normal	CBV Low (<67%)
High	Hyperperfusion	Hemorrhagic stroke	Hemorrhagic stroke
Normal	Compensatory autoregulation	Normal	Hemorrhagic stroke
Low (<67%)	Ischemia	Ischemia	Infarction

Table 18.2 List of Scanning Parameters for MR Imaging in Stroke

Scan Parameter	1.5T	3T
Echo time (TE) ms	35–45	25–30
Flip angle (θ)	60°–90°	60°
Repetition time (TR) s	≤1.5–2	
In-plane resolution (mm)	1.8–2	
Slice thickness (mm)	5	
Imaging contrast agent	Gadolinium-based single dose	
Contrast agent volume	0.1–0.2 mmoL/kg body weight	
Contrast injection rate	4–6 mL/s using MR-compatible power injector	
Injection protocol	Use 18–20 gauge needle and use 20–40 mL saline flush	

18.3.1.3   MR CT Correlation

Most cerebral perfusion measurements from contrast-enhanced CT and from DSC-MRI have shown significant correlation. In a study reported by Eastwood et al. [29], statistically significant correlation was obtained between the CBF (τ = 0.60, P = 0.003) and MTT (τ = 0.65, P = 0.001) values measured using CT and MR. However, the CBV values had very low correlation. This could possibly be attributed to the fact that his study involved CT scans with a small thickness of tissue coverage ranging between 5 and 10 mm. This extent of coverage is quite common though with increased use of multidetector row CT, whole brain coverage is achievable in CT scans. However, another report [30] on the difference in results comparing small thickness coverage (2 slices of 10 mm each) with 256 slices finds similar discrepancy where the CBF and MTT values when compared between the healthy and the affected hemisphere show large differences as expected but the CBV value is very similar in both regions. However, large area CT was able to find lesions that were missed in the two slice series (3/50 missed lesions detected).

18.3.1.4   C-Arm CBCT

The first attempts to demonstrate perfusion imaging in the brain with a C-arm CBCT system allowed for measurement of the CBV alone [10,31]. Instead of the normal IV bolus injection of the contrast agent, the injection protocol involved a steady-state contrast infusion to the tissue during imaging. This allowed sufficient images (275 projections) to be acquired for CBCT during a 10 s rotation of the C-arm. Pre- and postcontrast volumes were obtained and subtracted. The AIF is obtained from the angiographic vessel tree map. A CBV map was then calculated, and the results were compared with perfusion CT. The results comparing the number of areas of reduced CBV detected on C-arm CT (83%–87%) were comparable to PCT (70%–75%). Diffusion-weighted MR imaging was used as the ground truth. However, CBF data were not measurable using this method.

The measurement of the CBF using a C-arm CBCT system was first reported by Ganguly et al. [9]. Five adult female swine were used for this study following institutional animal care and use committees (IACUC) approval. To overcome the limitation of the C-arm system in terms of the slow rotation speed of 4.3 s (plus 1.25 s turn-around time) compared to conventional clinical scanners (0.33–0.5 s), multiple scans were acquired at different delays with respect to the start of contrast agent injection (each scan contains six bidirectional sweeps of the arm). This allowed better temporal sampling but required special reconstruction software to reorder and interpolate between projections [32,33]. In short, the reconstruction algorithm involves dividing the projections obtained from each scan set (consisting of six bidirectional sweeps) into blocks. Reconstruction of the blocks and interpolation of data between time blocks were used to obtain the complete reconstruction at each 1 s time interval. Optimal results were obtained with an IA injection of 50% dilute iodine contrast agent (350 mg I/mL) injected at 6 mL/s. The optimal dataset in terms of the fewest number of scans with the best correlation to PCT was found to be one containing two scan sets, with the scan starting 4.6 and 0.9 s before the injection of the contrast agent. As described earlier, the data from these two different delays between the start of the contrast agent injection

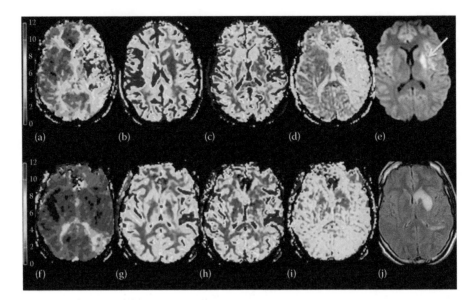

FIGURE 18.4 13-year-old boy with a sudden onset of right hemiparesis and aphasia, NIHSS 13; (a) MR perfusion imaging demonstrates a large region of tissue at risk with *Tmax* > 6 s (74 cc). The perfusion deficit is less apparent on (b) rCBF, (c) rCBV, and (d) MTT maps. (e) The initial diffusion lesion (yellow arrow) was small (8 cc), indicating that the patient is a good candidate for endovascular therapy. The patient was taken to the neurointerventional suite, where the left M1 vessel was opened. (f–i) are maps of the corresponding parameters as in a–e above postintervention. Perfusion imaging shows improvement in hemodynamics in the left hemisphere 18 h after the procedure. (j) The region of infarcted tissue did not grow significantly, as seen on FLAIR imaging at 5 days.

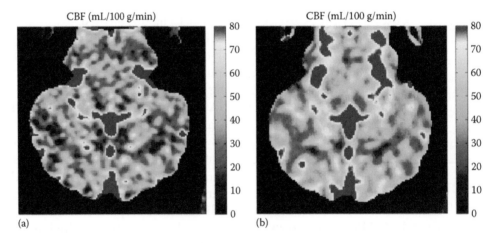

FIGURE 18.5 CBF maps obtained from an swine study (a) using clinical CT (Somatom Sensation 64, Siemens) (b) using C-arm CBCT system (Axiom Artis dTA and syngo DynaCT; Siemens, Erlangen, Germany) and modified acquisition and reconstruction protocol. All images were acquired using an IA injection of iodinated contrast injected at 3 mL/s with a total volume of 50 mL. Imaging was performed at 80 kVp on the CT scanner and at 81 kVp on the C-arm system.

and the start of the scan allow finer temporal sampling than a single scan set with a single injection. Figure 18.5 shows the CBF maps obtained in the same image slice in one animal in this study using a clinical CT scanner and a C-arm CBCT system. The interpolated temporal sampling in CBCT results in the differences in the absolute CBF values between CBCT and CT, giving an overall smoothing effect in the former modality. The difference in the CBF is more pronounced near the larger blood vessels. However, correlation between the CBF values measured using the C-arm CBCT system compared to PCT when averaged for the five animals was high (correlation coefficient of 0.88 and mean difference in CBF value of 8.78 ± 7.68 mL/100 g/min) or 20.11 ± 2.20% difference.

Faster C-arm systems, such as the Artis Zeego (Siemens Healthcare, Forcheim, Germany), allow about 360° rotation in 6 s and a faster one (~3 s) for short scans. This would further alleviate the need for multiple scans.

18.3.2 Cardiac Perfusion Imaging

Perfusion imaging in cardiac disease is most often related to evaluating effects of atherosclerosis (blockage) of the major epicardial arteries. Very commonly, this involves diagnosis related to CAD. Other common conditions include hypertrophic cardiomyopathy or wall thickening and myocardial wall motion abnormalities. While angiography is the gold standard for evaluating the patency of the coronary artery and its branches, it is invasive. CT and MR angiography are noninvasive alternatives but suffer from artifacts including those related to motion and calcification. Myocardial perfusion imaging (MPI) is an important addition that assesses the microvascular blood flow within the myocardium or heart wall and directly relates to ischemic conditions. Myocardial perfusion changes between rest and exercise or stress conditions due to autoregulation. Imaging under conditions of rest and controlled stress provides more complete information on the extent of ischemia. The current clinical standard for imaging is myocardial perfusion single photon emission computed tomography (SPECT). Where available, PET imaging is another reliable alternative and is further enhanced by combining it with CT (PET-CT). MPI with CT and MR have been areas of active research and have been clinically adopted. Other methods include echocardiography and ultrasound. Described in the following are the important features of the first four imaging modalities with respect to MPI.

18.3.2.1 SPECT

In SPECT MPI, 3D tomographic images of the distribution of a radioactive pharmaceutical in the myocardium are obtained under stress and rest conditions. It is an established clinical standard, and standardized guidelines exist for image interpretation [34]. Diagnosis is made by comparing the two sets of images with the stress image being obtained first. Since the radiopharmaceutical redistribution is slow, typically the second imaging set is obtained 1–4 h apart. SPECT-based MPI has been shown to have a sensitivity of 85%–88%, specificity of 72%–74%, and accuracy of 83%–84% [35,36] and is comparable with (or better than) other noninvasive tests for ischemic heart disease. More recently, SPECT is being combined with CT on the same system, allowing colocated scans that are attenuation corrected. This allows combined anatomical information from CT and the functional information from SPECT MPI. Improved risk stratification is observed with the

combined system [37]. However, SPECT suffers from long image acquisition time and corresponding susceptibility to motion, low resolution, and radiation dose. With recent development in image acquisition hardware and reconstruction software, these timings have reduced from 15 to 20 min to as low as 2 min [38]. It is well recognized that SPECT techniques frequently underestimate the degree of ischemia and therefore the presence of multivessel CAD [13].

In SPECT MPI, radioactive thallium ^{201}Tl, technetium ^{99m}Tc-sestamibi, and ^{99m}Tc-tetrofosmin are three routinely used tracers [39]. Since ^{99m}Tc has a shorter half-life (6 h) compared to ^{201}Tl (73 h), a higher dose can be administered, resulting in better signal statistics and improved suitability for rapid gated imaging. ^{201}Tl emits two gamma rays of 136 and 157 keV. The gamma from the daughter ^{201}Hg resulting from the decay of ^{201}Tl has an energy of about 71 keV. The abundance of the higher-energy gamma (136 and 157 keV) is only about 10% that of the gamma from ^{201}Hg, and hence, these lower energy gammas are used for imaging. This further reduces the counts, and hence, image quality of ^{201}Tl compared to ^{99}Tc. However, with the development of multidetector cameras that have high count efficiencies, gated acquisition with ^{201}Tl is now feasible [40].

In stress-induced MPI, adenosine is administered intravenously at a standard infusion rate of 140 µg/kg/min for 5 min. The radiopharmaceutical such as ^{99m}Tc with an activity rate of 25 mCi (925 MBq) is injected 3 min after the adenosine infusion. Gated SPECT images are obtained within 60 min of injection [39]. The rest images using the same dose of ^{99m}Tc are obtained the following day, typically, only if the stress images show abnormality.

The detector systems in SPECT use a variety of designs for the geometry, collimators, and detector elements [38]. The number of individual cameras varies from two to three heads or a large arc of detectors. Some use thallium-doped cesium iodide (CsI:Tl) coupled to silicon photodiodes, cadmium zinc telluride (CZT), and more commonly sodium iodide (NaI:Tl) with photomultiplier tubes. The collimators' holes are usually parallel and are made to match the pixelation of the detector elements. More recently, pinhole-based collimators have been developed that remove the need for electromechanical movement of the collimator and/or detector [41]. Image reconstruction typically consists of filtered back projection techniques though more recently faster algorithms such as iterative reconstruction that include the imaging physics and geometry models are being used [42].

Chapter 18

18.3.2.2 PET

A typical MPI protocol using PET requires rapid imaging following stress induction since the half-life of ^{82}Rb is only 73 s. After an initial scout image, a set of transmission and emission scans are obtained. This is followed by pharmacologically induced stress using either adenosine, dipyridamole, or dobutamine. After 3 min, another set of transmission and emission scans are obtained. The total time each for the set of pre- and poststress scans is 7 min each compared to the typical timing of about 15 min each in case of SPECT. In the case of PET/CT, the CT scan is obtained immediately following the initial scout image. Typical radiation dosage with ^{82}Rb is 15 mCi for the scout and 55 mCi each for the rest and stress scans. Total protocol time for a PET MPI is between 35 and 40 min. Overall PET using ^{82}Rb technology provides greater myocardial count density in a much shorter acquisition time and dose than ^{99}Tc SPECT [16].

18.3.2.3 MR

Myocardial perfusion using imaging was introduced in 1990 by Atkinson et al. using first-pass gadolinium (Gd-DTPA) contrast-enhanced MR (CMR) imaging. With improvements in MR hardware, imaging sequence design, improved contrast agent, and image analysis methods, CMR has been shown to compare well with SPECT and PET for myocardial imaging. It has the added advantage of being noninvasive, nonionizing, and have high soft tissue contrast. In fact, a comparison of MR perfusion in the diagnosis of CAD has been shown it to be superior to SPECT and echocardiographic measurements [43]. Recently, a large clinical trial (MR-IMPACT II) produced similar conclusion with CMR showing higher sensitivity (0.67 vs. 0.59) for CAD than SPECT though the specificity turned out to be lower (0.61 vs. 0.72) [44]. MR images have the added advantage of high contrast-to-noise ratio (CNR) between the infarcted region and the normal myocardium, particularly when compared to CT.

MR MPI involves rapid imaging of relevant myocardial segments with adequate spatial resolution and enhanced T1 contrast [45]. Initially, inversion recovery (IR)-based sequence was used. Saturation recovery (SR)-based sequences have gained popularity because of better image contrast and availability of larger number of image slices from interleaved scanning. Fast sequences such as rapid GRE [46] and more recently, advanced acceleration techniques such as k-t sensitivity encoding (SENSE) [47], compressed sensing [48], and highly constrained back projection [49] allow increased spatial coverage and spatial resolution.

Recently reported 3D techniques such as stack-of-spiral acquisition [50] enable entire myocardial coverage with an in-plane resolution of 2.4 mm. The clinical protocol used in the MR-IMPACTII trials includes CMR after stress and after rest. The stress is induced with a 3 min IV infusion of adenosine at 0.14 mg/min/kg. This was followed by an IV injection of Gd-DTPA contrast agent using a dose of 0.075 mmol/kg as per FDA recommendations with a 25 mL saline flush. Cardiac gated images of slices across the short axis and located along the length of the left ventricle (LV) are acquired during bolus arrival. Fast GRE sequences with an echo-planar component were used, resulting in a slice thickness of 8–10 mm and a spatial resolution of 2–3 × 2–3 mm^2.

18.3.2.4 CT

Myocardial perfusion CT is an area of fairly recent development and has been greatly facilitated by the advent of multidetector row CT (MDCT) and fast gantry rotations [51]. A complete cardiac CT dataset including CT angiography of the coronary and its branches, stress perfusion, rest perfusion, and delayed enhancement can provide diagnostic accuracy that is comparable to SPECT and at similar radiation doses [52–54]. The combined dataset provides structural information about the stenosis in addition to the physiological extent of damage. A recent study reported on the combined sensitivity, specificity, positive predictive value, and negative predictive value of CT perfusion for identifying segments with perfusion defects. These values, respectively, on a per-vessel basis were [1.00, 0.757, 0.541, 1.00] for CT perfusion alone and [0.90, 0.814, 0.581, and 0.966] for CT perfusion combined with CTA when compared against stress and rest SPECT data (1.00, 0.815, 0.676, and 1.00) for the same patient population [54].

A typical cardiac CT protocol [51,53,54] starts with rest CTA and CTP during infusion of 50–75 mL of iodinated contrast agent injected at 5mL/s followed by a 50 mL saline chaser. Next adenosine is infused intravenously (0.14 mg/kg/min) under constant ECG monitoring. At 3–4 min following the stress induction, a PCT scan is obtained with similar volumes of contrast agent as at rest. At 5–10 min after stress PCT, another scan is obtained to study the delayed enhancement of the tissue. Typically imaging parameters are: kV 100–120, mA of 320–550, slice thickness 0.5–2 mm, and scan length 12–16 cm. Data are typically obtained with prospective gating at 60%–75% of the R–R phase of the cardiac cycle. For perfusion calculations, the AIF is obtained in the descending aorta.

18.3.3   Other Perfusion Imaging Applications

In addition to cerebral and cardiac applications, perfusion measurements are also useful in evaluating the extent of tumors and their growth. It is gaining acceptance as an oncological tool in assessing tumors in organs including the liver, lung, and kidneys. Typically, tumors are associated with neovasculature, resulting from angiogenesis. The increased vascular development in most cases is an indicator of poor prognosis [4,55,56]. It results in increased blood flow and an associated measurable increase in perfusion [57,58]. Since CT imaging is currently the primary imaging modality for oncology, addition of a perfusion sequence in a patient workup is fairly practical.

In lung imaging, perfusion CT has been shown to be able to allow distinction between benign and malignant nodules in cases that failed to be classified using conventional CT [59–61]. The progression of tumors following therapy has been shown to correlate well with perfusion measurements [62–66].

Renal perfusion CT research started early with the introduction of electron beam CT scanners. It has been demonstrated to be useful for measuring perfusion in hypertension, renal stenosis and obstruction, and drug-induced toxicity [58]. Renal perfusion CT is a dose-intensive protocol and hence is not used extensively. Recent research developments in image reconstruction techniques have demonstrated feasibility of renal perfusion CT in an animal model with 10-fold dose reduction [67].

Perfusion CT imaging in hepatocellular carcinoma (HCC) has been shown to be able to demonstrate significant increase in the arterial perfusion in the tumor compared to the liver parenchyma [68]. Further changes in perfusion parameters over time following transarterial chemoembolization (TACE) were quantitatively correlated with therapeutic outcome [69].

Recently, the feasibility of perfusion imaging has been demonstrated using C-arm CBCT system [70] and shown in Figure 18.6. In particular, a method for

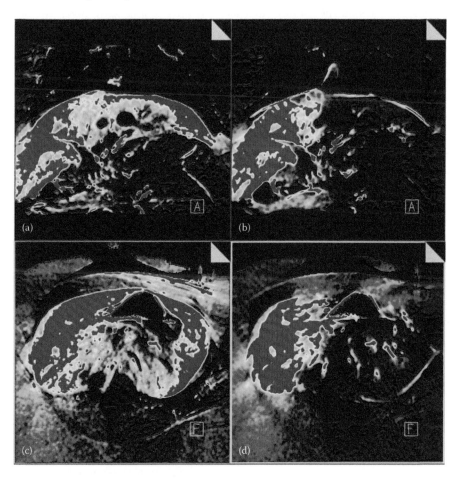

FIGURE 18.6   C-arm CBCT images showing PBV in swine liver. (a and b) show the coronal view and (c and d) show the axial view for the PBV data. (a) and (c) show the prepartial and (b) and (d) show the postpartial embolization of the left lobe of the liver as described in Section 18.3.3.

quantitative measurement of the hepatic blood volume obtained using such a system was found to correlate well with clinical CT results. For this study, six swine underwent partial TACE of the left lobe of the liver. Embolization was achieved by injecting 1.5 mL of 150–300 µm diameter microspheres into the left hepatic artery through a 2.3 Fr catheter. An IA injection of 50% dilute iodinated contrast (350 mg/mL, 3 mL/s for 14 s) was used for enhancement. As described in Section 18.3.1.4, optimized parameters for multisweep imaging (2-sweep 8 s-rotation) with 6 s delay between the start of the injection and image acquisition was used for obtaining pre- and postembolization images. The contrast-enhanced images were obtained during steady state, and the nonenhanced images were subtracted to obtain the difference image. A specially developed software prototype previously tested in humans [71] and canine [10] CBV measurements was used for image reconstruction and PBV calculation from the difference data. Clinical CT images were obtained post embolization using a 52.2 s scan with a 6s delay prior to injection of 9 mL of 100% 350 mg I/mL at 3mL/s intra-arterially for 3 s. A significant ($P = 0.005$) reduction in the PBV value was seen following embolization. The left to right lobe PBV ratios in C-arm CT and helical CT showed reasonable agreement (concordance correlation coefficient = 0.72; $r = 0.84$). These results are an encouraging indicator of the feasibility of using intraprocedural PCT in the treatment of HCC.

18.4 Summary and Conclusion

Perfusion imaging is currently the standard of care in the management of some diseases such as in stroke and is an important diagnostic tool in the complete assessment of myocardial pathology. Various modalities such as CT, MR, PET, and SPECT have been used for perfusion imaging, with each modality having its own strength and weakness and associated suitability for imaging a given anatomy. The temporal variation of imaging contrast agents over the tissue or organ of interest is measured and, with the use of various models, described here, and quantified for blood flow and volume measurements. Diagnosis is based on comparison with the known range for normal values and when possible by comparing with symmetric anatomy. Given the additional information provided by perfusion imaging and the increasing availability of tools for acquisition and evaluation, perfusion imaging should see wider clinical applications. With perfusion imaging on C-arm CBCT systems becoming feasible, intraprocedural perfusion imaging should see wider use in the interventional suite.

References

1. P. Dawson, Functional imaging in CT, *Eur J Radiol* 60 (2006), 331–340.
2. A. Fieselmann, M. Kowarschik, A. Ganguly, J. Hornegger, R. Fahrig, Deconvolution-based CT and MR brain perfusion measurement: Theoretical model revisited and practical implementation details, *Int J Biomed Imaging* 2011 (2011), 467–563.
3. F. Calamante, Perfusion magnetic resonance imaging quantification in the brain, in: E. Badoer (ed.), *Visualization Techniques*, Humana Press, 2012, pp. 283–312.
4. K.A. Miles, Functional CT imaging in oncology, *Eur Radiol* 13 Suppl 5 (2003), M134–M138.
5. L. Axel, Cerebral blood flow determination by rapid-sequence computed tomography: Theoretical analysis, *Radiology* 137 (1980), 679–686.
6. L. Axel, Tissue mean transit time from dynamic computed tomography by a simple deconvolution technique, *Invest Radiol* 18 (1983), 94–99.
7. K.L. Zierler, Theoretical basis of indicator-dilution methods for measuring flow and volume, *Circulation Research* 10 (1962), 15.
8. P. Meier, K.L. Zierler, On the theory of the indicator-dilution method for measurement of blood flow and volume, *J Appl Physiol* 6 (1954), 731–744.
9. A. Ganguly, A. Fieselmann, M. Marks, J. Rosenberg, J. Boese, Y. Deuerling-Zheng, M. Straka, G. Zaharchuk, R. Bammer, R. Fahrig, Cerebral CT perfusion using an interventional C-arm imaging system: Cerebral blood flow measurements, *AJNR—Am J Neuroradiol* 32 (2011), 1525–1531.
10. A.S. Ahmed, M. Zellerhoff, C.M. Strother, K.A. Pulfer, T. Redel, Y. Deuerling-Zheng, K. Royalty, D. Consigny, D.B. Niemann, C-arm CT measurement of cerebral blood volume: An experimental study in canines, *AJNR—Am J Neuroradiol* 30 (2009), 917–922.
11. A. Villringer, B.R. Rosen, J.W. Belliveau, J.L. Ackerman, R.B. Lauffer, R.B. Buxton, Y.S. Chao, V.J. Wedeen, T.J. Brady, Dynamic imaging with lanthanide chelates in normal brain: Contrast due to magnetic susceptibility effects, *Magn Reson Med* 6 (1988), 164–174.
12. A.R.A.M. Reto, Imaging viable brain tissue with CT scan during acute stroke, *Cerebrovasc Dis* 17 Suppl 3 (2004), 28–34.
13. G.V. Heller, D. Calnon, S. Dorbala, Recent advances in cardiac PET and PET/CT myocardial perfusion imaging, *J Nucl Cardiol* 16 (2009), 962–969.
14. K. Yoshinaga, B.J. Chow, K. Williams, L. Chen, R.A. deKemp, L. Garrard, A. Lok-Tin Szeto, M. Aung, R.A. Davies, T.D. Ruddy, R.S. Beanlands, What is the prognostic value of

myocardial perfusion imaging using rubidium-82 positron emission tomography?, *J Am Coll Cardiol* 48 (2006), 1029–1039.

15. B.A. Mc Ardle, T.F. Dowsley, R.A. deKemp, G.A. Wells, R.S. Beanlands, Does Rubidium-82 PET have superior accuracy to SPECT perfusion imaging for the diagnosis of obstructive coronary disease? A systematic review and meta-analysis, *J Am Coll Cardiol* 60 (2012), 1828–1837.

16. T.M. Bateman, G.V. Heller, A.I. McGhie, J.D. Friedman, J.A. Case, J.R. Bryngelson, G.K. Hertenstein, K.L. Moutray, K. Reid, S.J. Cullom, Diagnostic accuracy of rest/stress ECG-gated Rb-82 myocardial perfusion PET: Comparison with ECG-gated Tc-99m sestamibi SPECT, *J Nucl Cardiol* 13 (2006), 24–33.

17. L. Axel, Cerebral perfusion CT techniques, *Radiology* 233 (2004), 935; author reply 935.

18. E. Hering, Versuche, die Schnelligkeit des Blutlaufs und der Absonderung zu Bestimmen, *Z Physiol* 3 (1829), 85.

19. S. Antman, Foundations of indicator-dilution theory, in: D.A. Bloomfield (ed.), *Dye Curves: The Theory and Practice of Indicator Dilution*, HM+M Medical & Scientific, Aylesbury, U.K., 1974, pp. 21–40.

20. G.T. Gobbel, C.E. Cann, J.R. Fike, Measurement of regional cerebral blood flow using ultrafast computed tomography. Theoretical aspects, *Stroke* 22 (1991), 768–771.

21. R. Guibert, C. Fonta, F. Esteve, F. Plouraboue, On the normalization of cerebral blood flow, *J Cereb Blood Flow Metab* 33 (2013), 669–672.

22. Z. Wen, K.K. Vigen, W. Shin, T.J. Carroll, S.B. Fain, Optimization of arterial input function selection for cerebral perfusion imaging with quantitative blood vloume correction, *Proc Intl Soc Mag Reson Med* 15 ISMRM (2007).

23. T.G. Stewart, A clinical lecture on a case of perverted localisation of sensation or allachaesthesia: Delivered in the University of Edinburgh, *Br Med J* 1 (1894), 1–4.

24. A. Fieselmann, *Interventional Perfusion Imaging Using C-Arm Computed Tomography: Algorithms and Clinical Evaluation*, Department of Computer Science, Pattern Recognition Lab, Friedrich-Alexander University of Erlangen-Nuremberg, Erlangen, Germany, 2011, p. 148.

25. A. Furlan, R. Higashida, L. Wechsler, M. Gent, H. Rowley, C. Kase, M. Pessin et al., Intra-arterial prourokinase for acute ischemic stroke. The PROACT II study: A randomized controlled trial. Prolyse in Acute Cerebral Thromboembolism, *JAMA* 282 (1999), 2003–2011.

26. W. Hacke, M. Kaste, C. Fieschi, R. von Kummer, A. Davalos, D. Meier, V. Larrue et al., Randomised double-blind placebo-controlled trial of thrombolytic therapy with intravenous alteplase in acute ischaemic stroke (ECASS II). Second European-Australasian Acute Stroke Study Investigators, *Lancet* 352 (1998), 1245–1251.

27. B.D. Murphy, X. Chen, T.Y. Lee, Serial changes in CT cerebral blood volume and flow after 4 hours of middle cerebral occlusion in an animal model of embolic cerebral ischemia, *AJNR—Am J Neuroradiol* 28 (2007), 743–749.

28. M. Wintermark, G.W. Albers, A.V. Alexandrov, J.R. Alger, R. Bammer, J.C. Baron, S. Davis et al., Acute stroke imaging research roadmap, *Stroke* 39 (2008), 1621–1628.

29. J.D. Eastwood, M.H. Lev, M. Wintermark, C. Fitzek, D.P. Barboriak, D.M. Delong, T.Y. Lee, T. Azhari, M. Herzau, V.R. Chilukuri, J.M. Provenzale, Correlation of early dynamic CT perfusion imaging with whole-brain MR diffusion and perfusion imaging in acute hemispheric stroke, *AJNR—Am J Neuroradiol* 24 (2003), 1869–1875.

30. F. Dorn, D. Muenzel, R. Meier, H. Poppert, E.J. Rummeny, A. Huber, Brain perfusion CT for acute stroke using a 256-slice CT: Improvement of diagnostic information by large volume coverage, *Eur Radiol* 21 (2011), 1803–1810.

31. T. Bley, C.M. Strother, K. Pulfer, K. Royalty, M. Zellerhoff, Y. Deuerling-Zheng, F. Bender, D. Consigny, R. Yasuda, D. Niemann, C-arm CT measurement of cerebral blood volume in ischemic stroke: An experimental study in canines, *AJNR—Am J Neuroradiol* 31 (2010), 536–540.

32. A. Fieselmann, F. Dennerlein, Y. Deuerling-Zheng, J. Boese, R. Fahrig, J. Hornegger, A model for filtered backprojection reconstruction artifacts due to time-varying attenuation values in perfusion C-arm CT, *Phys Med Biol* 56 (2011), 3701–3717.

33. A. Fieselmann, A. Ganguly, Y. Deuerling-Zheng, M. Zellerhoff, C. Rohkohl, J. Boese, J. Hornegger, R. Fahrig, Interventional 4-D C-arm CT perfusion imaging using interleaved scanning and partial reconstruction interpolation, *IEEE Trans Med Imag* 31 (2012), 892–906.

34. C. Members, F.J. Klocke, M.G. Baird, B.H. Lorell, T.M. Bateman, J.V. Messer, D.S. Berman et al., ACC/AHA/ASNC guidelines for the clinical use of cardiac radionuclide imaging—Executive summary, *Circulation* 108 (2003), 1404–1418.

35. A. Elhendy, J.J. Bax, D. Poldermans, Dobutamine stress myocardial perfusion imaging in coronary artery disease, *J Nucl Med* 43 (2002), 1634–1646.

36. M.L. Geleijnse, A. Elhendy, P.M. Fioretti, J.R. Roelandt, Dobutamine stress myocardial perfusion imaging, *J Am Coll Cardiol* 36 (2000), 2017–2027.

37. A.P. Pazhenkottil, J.-R. Ghadri, R.N. Nkoulou, M. Wolfrum, R.R. Buechel, S.M. Küest, L. Husmann, B.A. Herzog, O. Gaemperli, P.A. Kaufmann, Improved outcome prediction by SPECT myocardial perfusion imaging after CT attenuation correction, *J Nucl Med* 52 (2011), 196–200.

38. P. Slomka, J. Patton, D. Berman, G. Germano, Advances in technical aspects of myocardial perfusion SPECT imaging, *J Nucl Cardiol* 16 (2009), 255–276.

39. A.K. Paul, H.A. Nabi, Gated myocardial perfusion SPECT: Basic principles, technical aspects, and clinical applications, *J Nucl Med Technol* 32 (2004), 179–187.

40. C. Maunoury, C.C. Chen, K.B. Chua, C.J. Thompson, Quantification of left ventricular function with thallium-201 and technetium-99m-sestamibi myocardial gated SPECT, *J Nucl Med Off Publ, Soc Nucl Med* 38 (1997), 958–961.

41. R.J. Jaszczak, J. Li, H. Wang, M.R. Zalutsky, R.E. Coleman, Pinhole collimation for ultra-high-resolution, small-field-of-view SPECT, *Phys Med Biol* 39 (1994), 425–437.

42. E.V. Garcia, T.L. Faber, F.P. Esteves, Cardiac dedicated ultrafast SPECT cameras: New designs and clinical implications, *J Nucl Med* 52 (2011), 8.

43. M.C. de Jong, T.S. Genders, R.J. van Geuns, A. Moelker, M.G. Hunink, Diagnostic performance of stress myocardial perfusion imaging for coronary artery disease: A systematic review and meta-analysis, *Eur Radiol* 22 (2012), 1881–1895.

44. J. Schwitter, C.M. Wacker, N. Wilke, N. Al-Saadi, E. Sauer, K. Huettle, S.O. Schonberg et al., MR-IMPACT II: Magnetic resonance imaging for myocardial perfusion assessment in coronary artery disease trial: Perfusion-cardiac magnetic resonance vs. single-photon emission computed tomography for the detection of coronary artery disease: A comparative multicentre, multivendor trial, *Eur Heart J* 34 (2013), 7.

Chapter 18

45. B.L. Gerber, S.V. Raman, K. Nayak, F.H. Epstein, P. Ferreira, L. Axel, D.L. Kraitchman, Myocardial first-pass perfusion cardiovascular magnetic resonance: History, theory, and current state of the art, *J Cardiovasc Magn Reson* 10 (2008), 18.

46. B. Hargreaves, Rapid gradient-echo imaging, *J Magn Reson Imag* 36(6) (2012), 1300–1313.

47. S. Plein, S. Ryf, J. Schwitter, A. Radjenovic, P. Boesiger, S. Kozerke, Dynamic contrast-enhanced myocardial perfusion MRI accelerated with k-t sense, *Magn Reson Med* 58 (2007), 777–785.

48. R. Otazo, D. Kim, L. Axel, D.K. Sodickson, Combination of compressed sensing and parallel imaging for highly accelerated first-pass cardiac perfusion MRI, *Magn Reson Med* 64 (2010), 767–776.

49. L. Ge, A. Kino, M. Griswold, C. Mistretta, J.C. Carr, D. Li, Myocardial perfusion MRI with sliding-window conjugate-gradient HYPR, *Magn Reson Med* 62 (2009), 835–839.

50. T. Shin, K.S. Nayak, J.M. Santos, D.G. Nishimura, B.S. Hu, M.V. McConnell, Three-dimensional first-pass myocardial perfusion MRI using a stack-of-spirals acquisition, *Magn Reson Med* 69(3) (2012), 839–844.

51. R.T. George, A. Arbab-Zadeh, J.M. Miller, A.L. Vavere, F.M. Bengel, A.C. Lardo, J.A.C. Lima, Computed tomography myocardial perfusion imaging with 320-row detector computed tomography accurately detects myocardial Ischemia in patients with obstructive coronary artery disease/clinical perspective, *Circ Cardiovasc Imag* 5 (2012), 333–340.

52. B.S. Ko, J.D. Cameron, T. DeFrance, S.K. Seneviratne, CT stress myocardial perfusion imaging using Multidetector CT—A review, *J Cardiovasc Comp Tomogr* 5 (2011), 345–356.

53. R. Blankstein, L.D. Shturman, I.S. Rogers, J.A. Rocha-Filho, D.R. Okada, A. Sarwar, A.V. Soni et al., Adenosine-induced stress myocardial perfusion imaging using dual-source cardiac computed tomography, *J Am Coll Cardiol* 54 (2009), 1072–1084.

54. Y. Wang, L. Qin, X. Shi, Y. Zeng, H. Jing, U.J. Schoepf, Z. Jin, Adenosine-stress dynamic myocardial perfusion imaging with second-generation dual-source CT: Comparison with conventional catheter coronary angiography and SPECT nuclear myocardial perfusion imaging, *Am J Roentgenol* 198 (2012), 521–529.

55. M.K. Brawer, R.E. Deering, M. Brown, S.D. Preston, S.A. Bigler, Predictors of pathologic stage in prostatic carcinoma. The role of neovascularity, *Cancer* 73 (1994), 678–687.

56. G. Fontanini, M. Lucchi, S. Vignati, A. Mussi, F. Ciardiello, M. De Laurentiis, S. De Placido, F. Basolo, C.A. Angeletti, G. Bevilacqua, Angiogenesis as a prognostic indicator of survival in non-small-cell lung carcinoma: A prospective study, *J Natl Cancer Inst* 89 (1997), 881–886.

57. K.A. Miles, C. Charnsangavej, F.T. Lee, E.K. Fishman, K. Horton, T.Y. Lee, Application of CT in the investigation of angiogenesis in oncology, *Acad Radiol* 7 (2000), 840–850.

58. K.A. Miles, M.R. Griffiths, Perfusion CT: A worthwhile enhancement?, *Br J Radiol* 76 (2003), 220–231.

59. C.I. Henschke, D. Yankelevitz, J. Westcott, S.D. Davis, H. Fleishon, W.B. Gefter, T.C. McLoud et al., Work-up of the solitary pulmonary nodule. American College of Radiology. ACR Appropriateness Criteria, *Radiology* 215 Suppl. (2000), 607–609.

60. S.J. Swensen, Functional CT: Lung nodule evaluation, *Radiographics* 20 (2000), 1178–1181.

61. S.J. Swensen, R.W. Viggiano, D.E. Midthun, N.L. Muller, A. Sherrick, K. Yamashita, D.P. Naidich, E.F. Patz, T.E. Hartman, J.R. Muhm, A.L. Weaver, Lung nodule enhancement at CT: Multicenter study, *Radiology* 214 (2000), 73–80.

62. F. Fraioli, M. Anzidei, F. Zaccagna, M.L. Mennini, G. Serra, B. Gori, F. Longo, C. Catalano, R. Passariello, Whole-tumor perfusion CT in patients with advanced lung adenocarcinoma treated with conventional and antiangiogenetic chemotherapy: Initial experience, *Radiology* 259 (2011), 574–582.

63. J. Wang, N. Wu, M.D. Cham, Y. Song, Tumor response in patients with advanced non-small cell lung cancer: Perfusion CT evaluation of chemotherapy and radiation therapy, *AJR—Am J Roentgenol* 193 (2009), 1090–1096.

64. Q.S. Ng, V. Goh, H. Fichte, E. Klotz, P. Fernie, M.I. Saunders, P.J. Hoskin, A.R. Padhani, Lung cancer perfusion at multidetector row CT: Reproducibility of whole tumor quantitative measurements, *Radiology* 239 (2006), 547–553.

65. Q.S. Ng, V. Goh, E. Klotz, H. Fichte, M.I. Saunders, P.J. Hoskin, A.R. Padhani, Quantitative assessment of lung cancer perfusion using MDCT: Does measurement reproducibility improve with greater tumor volume coverage?, *AJR—Am J Roentgenol* 187 (2006), 1079–1084.

66. M.S. Park, E. Klotz, M.J. Kim, S.Y. Song, S.W. Park, S.W. Cha, J.S. Lim, J. Seong, J.B. Chung, K.W. Kim, Perfusion CT: Noninvasive surrogate marker for stratification of pancreatic cancer response to concurrent chemo- and radiation therapy, *Radiology* 250 (2009), 110–117.

67. X. Liu, A.N. Primak, J.D. Krier, L. Yu, L.O. Lerman, C.H. McCollough, Renal perfusion and hemodynamics: Accurate in vivo determination at CT with a 10-fold decrease in radiation dose and HYPR noise reduction, *Radiology* 253 (2009), 98–105.

68. D.V. Sahani, N.S. Holalkere, P.R. Mueller, A.X. Zhu, Advanced hepatocellular carcinoma: CT perfusion of liver and tumor tissue—Initial experience, *Radiology* 243 (2007), 736–743.

69. L. Yang, X.M. Zhang, B.X. Tan, M. Liu, G.L. Dong, Z.H. Zhai, Computed tomographic perfusion imaging for the therapeutic response of chemoembolization for hepatocellular carcinoma, *J Comput Assist Tomogr* 36 (2012), 226–230.

70. A. Ganguly, N. Kothary, K. Blum, T. Moore, Y. Deuerling-Zheng, A. Fieselmann, R. Fahrig, Measurement of hepatic perfusion blood volume using C-arm CT, *SIR, J Vasc Interv Radiol* 23(3) (2012), S97.

71. T. Struffert, Y. Deuerling-Zheng, S. Kloska, T. Engelhorn, C.M. Strother, W.A. Kalender, M. Kohrmann, S. Schwab, A. Doerfler, Flat detector CT in the evaluation of brain parenchyma, intracranial vasculature, and cerebral blood volume: A pilot study in patients with acute symptoms of cerebral ischemia, *AJNR—Am J Neuroradiol* 31 (2010), 1462–1469.

19. Flow Imaging with MRI

M.J. Negahdar, Mo Kadbi, and Amir A. Amini

Chapter 19

19.1 Introduction

Magnetic resonance imaging (MRI) is a powerful non-invasive technique for diagnosis, evaluation, and monitoring of disease and is based on the magnetization properties of atomic nuclei. MRI is based on nuclear magnetic resonance (NMR) phenomena [1]. How to induce, detect, and quantify the NMR signal from bulk materials was first proposed by Bloch and Purcell in 1946. Subsequently, in the 1970s, the idea of using NMR for imaging living organisms was proposed by Lauterbur who called it zeugmatography. Subsequent improvements in image acquisition and signal processing were carried out by Peter Mansfield and others, leading to more clear images with anatomical detail and more diverse applications. Lauterbur and Mansfield were awarded the 2003 Nobel Prize in medicine and physiology for the invention of MRI [2–4].

This chapter is a short review of MR imaging of flow and hemodynamic in cardiovascular and neurovascular arteries. In Sections 19.1 and 19.2, a brief overview of MR imaging is provided. Section 19.3 gives an overview of blood composition, cardiac phases, and significance

of flow waveform pattern in the arterial tree. Section 19.4 gives a synopsis of flow regimes that may be encountered in the human circulatory system. Section 19.5 provides an overview about flow imaging methods in general (flow angiography and flow quantification). Sections 19.6 and 19.7 explain physical and mathematical underpinnings of flow measurement techniques in MRI. In particular, we will focus on and discuss the major MR velocimetry techniques based on phase-contrast (PC) MRI as well as their advantages and disadvantages. We will also explain new 4D PC MRI techniques typically referred to as "4D flow," which provides for time-efficient 3D cine acquisition of flow velocities. Section 19.8 will briefly explain Fourier velocity encoding (FVE) techniques. Sections 19.9 and 19.10 discuss artifacts and errors typically encountered in PC MRI and Section 19.11 summarizes the chapter and proposes future research directions in the MR velocity measurement area. Due to space limitations, authors assume some familiarity with MR concepts. For more detailed information about the physics of MRI, the reader is referred to Reference 1.

19.2 MR Physics

The basis for NMR is the microscopic magnetic character of the atomic nucleus. A nucleus that contains an odd number of protons possesses a magnetic moment called μ, which describes the strength and direction of a microscopic magnetic field surrounding the nucleus. In the presence of a strong static external magnetic field, such as that produced inside an imaging magnet,

a small excess fraction of nuclei align their magnetic moment with the magnetic field (in the case of ^{1}H nucleus that consists of a single proton, there are two energy states—parallel and aligned or antiparallel to the external magnetic field), producing a macroscopic and measurable bulk magnetic moment (Figure 19.1) [5].

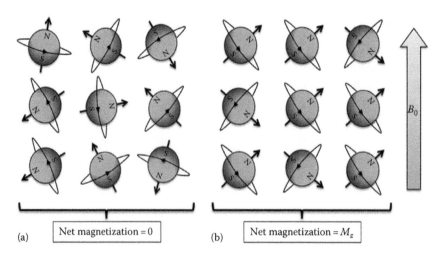

(a) Net magnetization = 0   (b) Net magnetization = M_z

FIGURE 19.1 Proton alignments (a) without and (b) with existence of external static magnetic field. In the absence of an external magnetic field, the magnetic moments are completely random, resulting in a summed zero net magnetization. In the presence of a large external magnetic field, the protons become either parallel or antiparallel to the field, though the number that is in the parallel state is slightly larger. The resulting sum becomes nonzero, is detectable, and is the basis for the NMR signal.

In addition, the interaction between the magnetic moment of the nucleus and the external magnetic field causes each nucleus to precess about the external static magnetic field. Each nucleus precesses at a characteristic (resonant) frequency that is proportional to the strength of the external field. For example, a hydrogen nucleus precesses at a frequency of 42.57 MHz/T, while a carbon-13 nucleus precesses at 10.71 MHz/T. Because of the high abundance of hydrogen (^{1}H) nuclei in tissue compared with other atomic nuclei, the NMR signal from ^{1}H is used in virtually all clinical MR scans.

The longitudinal magnetization (the fraction of protons aligned with the external magnetic field) can be tilted to transverse plane by application of a radiofrequency (RF) pulse at the exact resonant frequency, which perturbs the individual magnetic moments. It takes some time (on the order of few milliseconds to about a second) for nuclei to return to their equilibrium—this is called the relaxation time. During relaxation, the precession of the bulk magnetic moment in the transverse plane leads to a current to be detected in an RF coil when placed next to the subject—the resulting current is the basis for image formation. By applying RF pulses at different frequencies while turning on the so-called magnetic field gradients and changing their timings, MRI is capable of generating images with a variety of different contrasts and can be made sensitive to various physiological processes including proton density, $T1$ and $T2$ relaxation, diffusion, and motion, among others. Image formation in MRI involves spatial encoding; we need to spatially resolve the MR signal to specific tissue locations.

In MRI, in addition to the large static magnetic field generated by the main MR system coil, which is on the order of 1.5–3 T, low-intensity transient magnetic field gradients are present that are generated by the gradient coils—these transient magnetic fields are on the order of 40–80 mT/m and result in a range of resonance frequencies that can be traced back to individual tissue points. Specifically, to create a 2D image of a tissue slab, phase-encoding and frequency-encoding gradients (orthogonal to one another) are applied on the imaging slice before signal detection. This results in the protons in a small voxel, to resonate at a unique frequency and to have a unique phase, permitting tracing back of the NMR signal to a specific tissue location.

19.3   Physiology of Blood Flow in Arteries

Blood is comprised of various components such as protein, red blood cell, white blood cell, lipoproteins, and ions by which nutrients and wastes are transported. The red blood cell is one of the main ingredients of blood and comprises approximately 40% of blood [6]. Red blood cells affect the viscosity of blood due to them being semisolid particles and lead to four time higher viscosity compared to water. Viscosity defines the tendency of a fluid to resist flow. Viscosity of blood in low flow rates is higher because red blood cells stick together and aggregate in the middle of the vessel. The heart pumps the blood throughout the vessel in a cyclic manner. This cyclic pumping is divided into systolic and diastolic phases [7]. In the systolic phase, the blood is pumped out of the heart and transfers to organs. In this phase, blood flow and pressure are higher in arteries. In the diastolic phase, the heart is in its resting phase and does not pump blood. Blood flow and pressure drop in arteries in the diastolic filling phase when arterial and venous blood return to the heart. In general, blood flow is unsteady and pulsatile during the cardiac cycle and varies in arteries and veins with respect to the cardiac phase. The flow waveform is affected by atherosclerotic disease and can be an indicator of vascular function. Figure 19.2 exhibits the blood velocity and pressure in a dog's arterial system. The blood leaves the heart in a large artery called the aorta. The pulsatile blood pressure in aorta is high in systole and becomes small but does not become zero during diastole. On the other hand, blood flow can be zero and even negative in diastole in some arteries such as the external carotid artery and the femoral artery [6]. These arteries have high downstream resistance during rest and the flow is on and off with each cardiac cycle. The internal carotid artery and the renal artery, on the other hand, can have higher flows during diastole with low downstream resistance.

19.4   Flow Regimes

The phase of the MR signal is sensitive to motion and can be used to measure velocity and flow in the phase-contrast MRI (PC-MRI) technique. In this section, we will review the basic fluid mechanics of blood flow before embarking on a description of PC-MRI and other MRI techniques, which are commonly utilized for imaging and visualization of flow in the vasculature.

Chapter 19

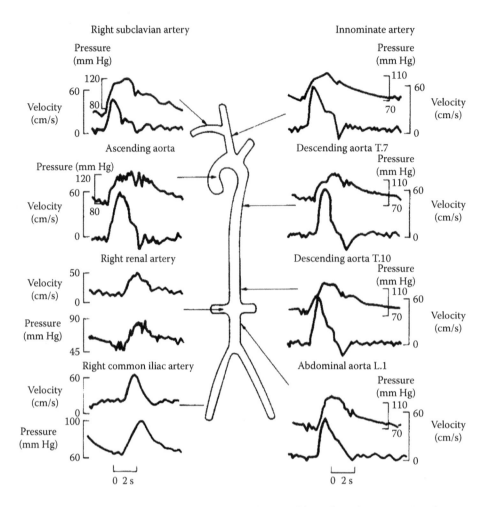

FIGURE 19.2 Pressure and velocity pulse waveforms in the aorta and arterial branches of a dog. (Taken from Ku, D.N., *Annu. Rev. Fluid Mech.*, 29, 399, 1997.)

Laminar flow: Flowing blood is faced with frictional forces from the vessel wall and neighboring fluid elements in its way through the vessel, so blood flows with various velocities across the diameter of the vessel and this gives rise to the shape of the flow profile and pattern in different regions of healthy and diseased vessels. The best known flow regime in healthy human subjects where there is a constant vessel thickness is the laminar flow regime. In laminar flow, the velocity profile looks parabolic where the velocities at the center of the vessel are higher than the surrounding ones with zero velocity at the vessel wall. Figure 19.3 shows the laminar flow distribution in a normal vessel.

Turbulent and disturbed flow: In the case of high velocity through a stenosis or vessel branches like the carotid and iliac bifurcations, there is a different story; here we have complex flow patterns with flow

vortices and turbulent flow distal to the stenosis. A flow vortex is usually caused by rapid deceleration of the jet after a stenosis, or at the peripheral area of arterial bifurcations. In a vortex, we have slowly swirling flows close to the vessel wall after the stenosis. Figure 19.4 demonstrates flow vortices distal to a stenosis. Flow vortices and turbulent flow can cause MR signal loss (also referred to as signal dephasing) because of the protons (spins) moving in

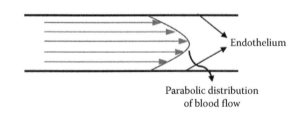

FIGURE 19.3 Laminar flow distribution in healthy constant vessel thickness.

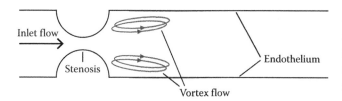

FIGURE 19.4 Vortex formation in the case of stenosis.

random directions. Fluid's Reynolds number (Re) is defined as:

$$Re = \frac{\rho DV}{\eta} \qquad (19.1)$$

where

ρ is the fluid density
D is the vessel diameter
V is the flow velocity
η is the fluid viscosity

19.5 Flow Imaging

Flow visualization and quantification in vivo is very helpful for diagnosis and monitoring of many vascular and cerebrospinal fluid flow diseases. Doppler ultrasound is currently the most widely adopted method for cardiovascular flow imaging. But the existence of air, bone, or surgical scar is a significant barrier to accurate evaluation [14,15]—for example, due to the presence of the skull, Doppler has found limited applications to neurovascular flow imaging.

MRI is a unique imaging modality with superb anatomic imaging capability with excellent soft-tissue contrast that also has the capability to image hemodynamic parameters such as velocity and flow. In MR flow imaging, scan time is intimately related to both spatial and temporal resolution [16] with short imaging times achievable by reducing temporal and spatial resolution. However, low temporal resolution leads breath hold of peak velocities, while the total imaging time is usually limited to the length of a breath-hold. On the other hand, poor spatial resolution can lead to data inconsistency, partial volume effects, and phase dispersion, which can degrade the image quality and accuracy [17,18]. Fourier velocity encoding (FVE) is more accurate than PC-MRI in quantification of high-speed and complex flows where a range of flow velocities may be present in an imaged voxel [19,20]. This idea eliminates partial volume artifact; however, it leads to a considerably longer scan time.

In general, MR flow imaging methods can be categorized into two major groups: flow angiography and flow quantification. In flow angiography, we are seeking to

Reynolds numbers (Re) less than 1000 imply laminar flow, whereas Reynolds numbers (Re) more than 2000 imply the possibility of turbulent flow [8,9]; Re above 1000 and less than 2000 result in transitional flows.

Pulsatile flow: In this case we have a complex time-varying flow waveform. The waveform depends on many known and unknown parameters in the human body including vessel wall compliance and stiffness. The temporal changes in flow can lead to spatial artifacts in image reconstruction and spatial position. Pulsatile flow is more significant in arteries than veins. Peak systolic velocity in normal subjects generally decreases with distance from the heart. Rapid changes in velocity are troublesome in velocity quantification with MR. Many fast imaging methods have been proposed to overcome artifacts due to pulsatile flow [10–13].

visualize the flowing blood (or other fluids such as the cerebrospinal fluid or urine). The essential goal of magnetic resonance angiography (MRA) is to enhance the flowing blood in blood vessels relative to the stationary tissues in order to provide physiological information and to evaluate them for pathologies such as stenosis, aneurysm, dissection, and coarctation. Results provide valuable information about blood supplied to organs. Methods such as black blood imaging and bright blood imaging, which use the so-called time-of-flight (TOF) effect, have been developed to achieve this goal [20]. Also, using intravenous contrast agents shortens T1 of blood, resulting in blood having higher signal than surrounding tissues [21].

Figure 19.5 demonstrates carotid anatomy in a healthy volunteer visualized with TOF MRA. Time-resolved MRA methods have also been developed that provide valuable diagnostic information about vascular anatomy and function. These techniques generally offer less spatial resolution and temporal resolution than CT angiography; however, new improvements could accelerate their speed significantly while preserving signal-to-noise ratio (SNR) [21]. Use of parallel imaging and use of highly undersampled 3D radial acquisitions can accelerate conventional MRA methods significantly [22,23].

Quantitative flow measurement methods provide us with a numerical tool to evaluate the amount of flow. Clearly, comparison of measured flow in a diseased vessel with the expected normal flow can be helpful in the

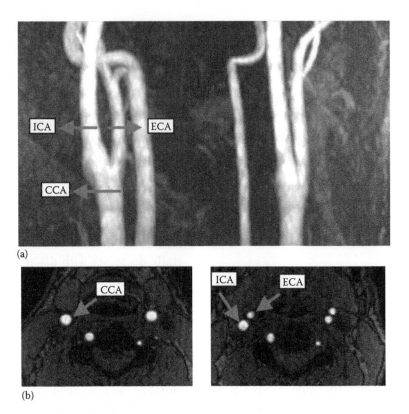

(a)

(b)

FIGURE 19.5 TOF MRA of carotid arteries in a healthy volunteer. (a) Maximum intensity projection from an MRA study. (b) Axial magnitude images before (left) and after (right) carotid bifurcation.

diagnosis of patients and can help in understanding and monitoring of the disease process. Quantitative flow methods are essentially based on the accumulated phase

of moving spins against stationary ones. Two major phase-based methods are PC-MRI and FVE techniques [24]. These methods are explained in the following sections.

19.6 Velocity and Flow Quantification Techniques

MR flow quantification methods are highly versatile and can in principle determine all three components of velocities within an image voxel. To achieve this goal, new elements in the gradient waveforms of the imaging pulse sequence are introduced. These elements make the phase of the MR signal for each pixel dependent on velocity.

19.6.1 Bipolar Gradients

Bipolar gradients are combinations of two identical gradient pulses with opposite polarity that are added to the gradient waveform for a particular direction (i.e., *x*, *y*, or *z*) where velocity measurement is required [24]. Bipolar gradients can be applied along the phase-encoding, frequency-encoding, or slice-select direction subsequent to the excitation. They can encode velocity of moving spins in the phase of the acquired signal. Figure 19.6 shows a general bipolar gradient. Based on the theory of MR physics, a spin at position *r*

will accumulate a phase shift that is equal to the time integral of the precessional frequency:

$$\phi(r,t) = \gamma \int_0^t G(\tau) . r(\tau) d\tau, \qquad (19.2)$$

where

$G(\tau)$ is the magnetic experienced gradient experienced by a spin

γ is a constant (the gyromagnetic ratio)

$r(\tau)$ is the location of the spin

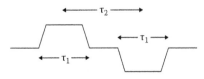

FIGURE 19.6 Trapezoid bipolar velocity-encoding gradients. The amplitude of the positive lobe of the bipolar gradient is G and the amplitude of the negative lobe of the bipolar gradient is -G.

When bipolar gradients are applied, the phase shift accumulated after the first gradient pulse will be canceled by the second pulse in case of static tissues. However, if the location of the spin changes between the two gradient pulses, the phase shifts no longer cancel each other. The residual phase shift depends on the distance that the spin has moved between the two gradient pulses and is proportional to the spin's mean velocity.

If a spin is at initial position x_0 when the first pulse of the bipolar gradient is applied and it is assumed that the position of the spin is constant during the first pulse, the accumulated phase shift according to Equation 19.2 becomes

$$\phi_1 = \gamma G x_0 \tau_1. \tag{19.3}$$

If the spin moves with constant velocity v, its position when the second pulse of the bipolar gradient is applied is given by

$$x_2(t) = x_0 + v.\tau_2. \tag{19.4}$$

Then the accumulated phase shift by the second pulse according to Equation 19.2 is

$$\phi_2 = -\gamma G (x_0 + v.\tau_2) \tau_1 \tag{19.5}$$

and the total accumulated phase shift is obtained by summation of (19.3) and (19.5):

$$\phi = \phi_1 + \phi_2 = -\gamma G v \tau_1 \tau_2. \tag{19.6}$$

Equation 19.6 shows that a bipolar gradient waveform results in a phase shift proportional to the gradient amplitude, the spin velocity, the pulse duration, and the separation time between two lobes of bipolar gradient. Figure 19.7 shows a bipolar gradient and its effect on static spins at 3 different locations and a spin with constant velocity (it is assumed there is no gap between the gradient lobes).

19.6.2   Gradient Moment Analysis

One can get an expression for motion-induced phase shifts of any order using gradient moment analysis [26]. Using an nth order Taylor series expansion, the position of a spin at time t, $x(t)$, can be approximated as

$$x(t) \cong x_0 + \frac{\partial x(t)}{\partial t} \cdot t + \frac{\partial^2 x(t)}{2!^* \partial t^2} \cdot t^2 + \cdots + \frac{\partial^n x(t)}{n!^* \partial t^n} \cdot t^n. \tag{19.7}$$

By plugging in Equation 19.7 into Equation 19.2 (assuming r(t) = x(t)), the phase shift at time T is obtained

$$\phi(T) \cong \gamma \sum_{i=0}^{n} \frac{\partial^n x(T)}{n!^* \partial t^n} \cdot M_n(T), \tag{19.8}$$

where $M_n(T)$ is the gradient moment of order n for the gradient pulse $G(t)$ and is defined by

$$M_n(T) = \int_0^T G(t) \cdot t^n \, dt. \tag{19.9}$$

Nulling a particular order of gradient moment by adding extra lobes to the pulses will make an MR pulse sequence insensitive to that order of motion. For example, in MR pulse sequences that are insensitive to constant velocity, $M_1 = 0$. With increasing order of moment nulling, the pulse sequence duration also increases, and the longer the pulse sequence, the more the artifacts and the less the temporal resolution. Therefore, there is a trade-off between pulse sequence duration and moment nulling order.

19.6.3   Velocity-Encoding (V_{enc}) Value

According to the gradient moment analysis, a zeroth-order gradient moment and a nonzero first-order

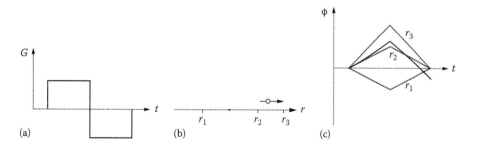

FIGURE 19.7   (a) Bipolar gradient lobe, (b) spins with different position (r), and (c) effect of bipolar gradient on the cumulative phase of three static spins at different locations, r_1, r_2, r_3, and a moving spin, moving from position r_2 to position r_3. (Taken from Pelc, N.J. et al., *Magn. Reson. Q.*, 10, 125, 1994.)

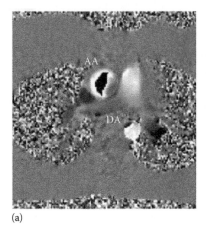
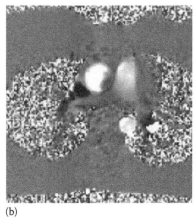

(a)   (b)

FIGURE 19.8 Velocity measurement in the ascending aorta: (a) aliasing due to small V_{enc} and (b) correct velocity image after increasing V_{enc}. (Taken from Gatehouse, P. et al., *Eur. Radiol.*, 15, 2172, 2005.)

gradient moment are needed for velocity encoding in PC imaging. Note that we are assuming that acceleration and higher-order motion terms are zero during the data acquisition. The phase shift due to motion in that case is given by

$$\phi = \gamma v M_1, \tag{19.10a}$$

$$\Delta\phi = v \left(\gamma \Delta M_1\right). \tag{19.10b}$$

Hence, velocity measurements are obtained from the phase values of the acquired images. Phase values that are greater than π radians cannot be discriminated from their modulo 2π counterparts. For example, a phase shift of $3\pi/2$ is indistinguishable from a phase shift of $-(\pi/2)$ radians. So the velocity corresponding to a phase shift of π radians defines the upper limit and $-\pi$ the lower limit on the range of velocities that can be accurately measured. This upper limit is referred

to as velocity-encoding value or V_{enc}, for short. From Equation 19.10b, V_{enc} is defined as

$$V_{enc} = \frac{\pi}{\gamma \Delta M_1} \frac{\pi}{\gamma \Delta M_1}. \tag{19.11}$$

Any velocity value outside of the range $[-V_{enc}, +V_{enc}]$ will be aliased and assigned to a smaller value. Although many methods have been proposed to unwrap the aliased phase values [27], choosing V_{enc} larger than the maximum expected velocity values avoids velocity aliasing and is preferred in PC imaging. However, V_{enc} cannot be made arbitrarily large because of SNR considerations [28,29]—a large V_{enc} results in more velocity noise [30]. Conversely, a small V_{enc} tends to increase ghosting in velocity images.

Therefore, choosing the proper value for the V_{enc} requires some consideration. Figure 19.8 shows the effect of aliasing due to choosing a V_{enc} value smaller than the peak velocity when imaging the ascending aorta. The problem is fixed by increasing the value for V_{enc}.

19.7 Phase-Contrast MRI

The basis of the PC technique, first proposed by Hahn [31] in 1960, is that spins moving in the presence of a magnetic field gradient accumulate a different phase from stationary ones. The first application of this method was developed by Moran [24] and subsequently applied in human cases by Van Dijk [32]. The drawback of the PC technique is that the phase can be affected by many undesirable factors like magnetic field inhomogeneity, pulse sequence tuning, acceleration, partial volume artifact, and eddy current (these artifacts are explained in detail in Sections 19.9 and 19.10). Numerous papers have been published on how

to correct and compensate for the undesirable factors noted earlier. To remove signal from static tissue and constant noise, it is suggested to use two different sequences with identical zero moments and different first moment and subtract them. Figure 19.9 shows a reference and a velocity-encoded scan and the result of phase map subtraction, yielding the velocity map. Stationary tissues have midgray intensities (zero velocity), while positive velocities have bright intensities (e.g., the ascending aorta in Figure 19.9), and negative velocities have darker intensities (descending aorta in Figure 19.9). As shown in Figure 19.9, phase errors are

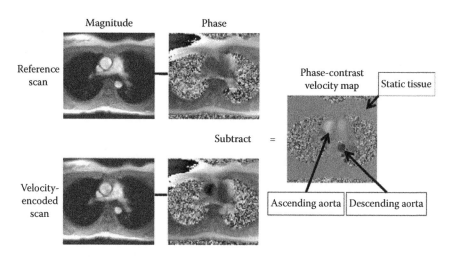

Magnitude Phase

Reference scan

Velocity-encoded scan

Subtract =

Phase-contrast velocity map Static tissue

Ascending aorta Descending aorta

FIGURE 19.9 In PC-MRI, the magnitude and phase images are reconstructed from the reference scan (top row) and also from the velocity-encoded scan (bottom row). The corresponding phase images are subtracted resulting in the PC velocity image. (Taken from Gatehouse, P. et al., *Eur. Radiol.*, 15, 2172, 2005.)

present even in static tissues within the reference and velocity-encoding scans that are removed by the subtraction step in the final velocity map.

19.7.1 PC-MRI Techniques

In this section, we consider different data acquisitions methods to cover *k*-space in PC-MRI and compare them to in vivo experiment. There are three major *k*-space coverage methods: (1) conventional Cartesian sequence, (2) radial sequence, and (3) spiral sequence.

19.7.1.1 Cartesian Trajectory

This method uses conventional Cartesian coordinates to cover the *k*-space (Figure 19.10). Implementation of this method is straightforward; it is sufficient to add a bipolar gradient in the desired flow measurement direction to the regular imaging sequence. However, this method can lead to long echo times that can result in signal loss. Additionally, relatively long repetition times decrease

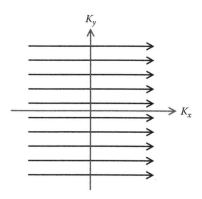

FIGURE 19.10 *k*-Space trajectory for Cartesian acquisition.

the temporal resolution in the case of CINE imaging. Motion artifacts are another disadvantage of Cartesian scans, leading to ghosting of the moving object in the phase-encoded direction. Figure 19.11 demonstrates the 2D PC pulse sequence with Cartesian readout.

19.7.1.2 Radial Trajectory

Radial PC has also been developed but has primarily been used for vessel visualization and angiography, though subsequently, it was extended for quantitative flow measurement. Barger [33] introduced phase contrast with interleaved projections (PIPR), an interleaved undersampled projection technique for contrast-enhanced PC angiography. PIPR uses radial trajectory as shown in Figure 19.12 to fill the *k*-space. The motivation behind PIPR is that with PC acquisitions, the degree of undersampling can be even larger than contrast-enhanced MRI that is typically used for vessel visualization, since in the case of PC-MRI, background tissue will be subtracted and will not contribute to artifacts. In PIPR, since every velocity encoding (V_x, V_y, and V_z) applies to a different projection angle, reducing the number of velocity encoding from six to four, as is typical with Cartesian readout, is no longer possible in PIPR.

Relative to Cartesian PC, radial acquisition can significantly reduce ghosting artifacts. However, because of the reduced number of acquisition angles, smearing and streaking artifacts will be visible. PIPR suffers from low SNR in the case of high-resolution acquisitions and suffers from undersampling artifacts in the case of low-resolution acquisitions. Barger et al. also showed that PIPR can achieve good results in constant flow measurements. Generally, with the radial

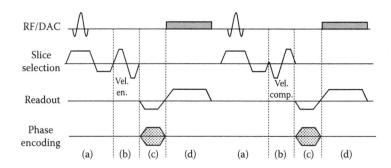

FIGURE 19.11 Cartesian PC-MRI pulse sequence. It consists of (a) slice-selective RF excitation pulse, (b) bipolar velocity-encoding gradient, (c) phase-encoding gradient, and (d) Cartesian readout.

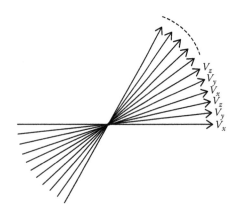

FIGURE 19.12 *k*-Space trajectory for PIPR radial acquisition (see text).

acquisition it is possible to obtain a higher spatial resolution per unit time than with Cartesian PC.

In blood flow imaging, often, where there is disturbed flow or a flow jet, standard techniques fail due to intravoxel dephasing and fluid mixing, leading to flow artifacts and signal dropouts. Additionally, in standard PC-MRI, it is often assumed (see Equations 19.8 through 19.10) that during echo time, no acceleration or higher orders of motion exist. There are two approaches to compensate for the effect of higher-order motion terms. The first approach uses complicated velocity-encoding gradients to have zero-, second-, and third-order moments. This approach leads to longer TE and TR that is not desirable. The second approach reduces TE in order to diminish the effect of higher-order motion terms on the phase. This can be done through the use of non-Cartesian radial sampling of the free induction decay (FID) signal, through a combination of slice-selective and flow-encoding gradients. Ultrashort TE (UTE) flow imaging was developed based on radial trajectory [34,35], which can significantly reduce the TE. Prior to its application to flow imaging, UTE MRI was developed for imaging tissues where there is a deficit of protons and/or in situations where $T2$ is small (such as

bone, cartilage, and the lungs) [36]. The readout trajectory in UTE is based on radial traversal of evenly spaced *k*-space lines starting from the center of the *k*-space and ending on the surface of a sphere with radius K_{max} determined by the spatial resolution in three spatial directions. A notable difference between radial UTE and a standard radial trajectory is that readout starts at the beginning of gradient ramp-up, requiring nonlinear data samplings. Center-out *k*-space radial trajectories help to reduce the effect of intravoxel dephasing and related phase artifacts due to inherent minimization of first moment of readout gradient by oversampling the center of *k*-space [34]. However, center-out *k*-space lines are sensitive to phase errors due to several parameters such as delays on physical gradients, eddy current, and B_0 field inhomogeneity.

Figure 19.13 shows an UTE PC sequence diagram [37] where the slice excitation gradient is combined with flow-encoding/compensation gradient. Sampling the FID is started at the rising slope of the readout gradient. The minimum TE for $V_{enc} = 500$ cm/s is 0.76 ms.

19.7.1.3 Spiral Trajectory

The implementation of PC-MRI with parallel lines in Cartesian trajectory suffers from several issues including

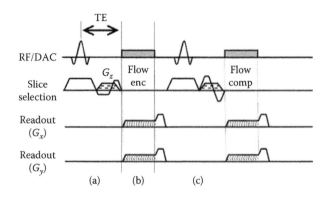

FIGURE 19.13 Radial UTE PC pulse sequence: (a) slice-selective excitation pulse and bipolar velocity-encoding gradient, (b) radial readouts, and (c) spoiler and refocusing gradients.

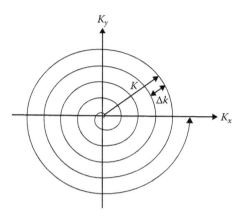

FIGURE 19.14 *k*-Space trajectory of a single spiral interleaf showing sample spacing, Δk, in the radial direction.

error due to acceleration, partial volume artifact, and off-resonance (these will be explained in Sections 19.9 and 19.10 in detail). Essentially the basic problem with the Cartesian sequence is its long acquisition time that can lead to many other artifacts. Additionally due to the increase in acquisition times in breath-held scans, more severe respiratory motion artifacts in the abdomen and thorax areas may be encountered. The long acquisition time also decreases the temporal resolution, which is undesirable and prevents real-time imaging. Moreover, sampling the center of *k*-space on every interleaf reduces artifacts from pulsatile flow [38].

The aforementioned considerations provide incentive for having a new strategy for covering *k*-space; instead of using horizontal readouts in *k*-space, a spiral trajectory may be used for filling the *k*-space. In this method, the repetition time T_R and scan time decrease significantly.

Figure 19.14 shows the *k*-space trajectory of a single spiral interleaf. The shape of the gradient waveforms in spiral MRI is different from the gradient waveform in Cartesian MRI [39]; the following equations show the expressions for the spiral gradient waveforms:

$$K_x = k(t)\cos(\theta(t)), \tag{19.12}$$

$$K_y = k(t)\sin(\theta(t)), \tag{19.13}$$

$$k(t) = A\,\theta(t), \tag{19.14}$$

where A is a constant and is determined by the Nyquist criterion. If Δk is the radial distance in *k*-space advanced by one rotation, D is the size of the Field of View (FOV), to satisfy the Nyquist criterion for uniform density spiral trajectories, the following condition needs to be satisfied.

$$\Delta k \le \frac{1}{D} \tag{19.15}$$

Pike [40] suggested a rapid, interleaved, spiral *k*-space acquisition. The advantage of this method is the capability for single breath-hold imaging, significantly decreasing time-related artifacts such as respiratory ghosting and acceleration-related errors, among others. In interleaved spiral PC instead of using one long spiral arm with *N* rotations, *N* short spiral interleaves with one rotation can be used (Figure 19.15). With this approach, the total readout time required to cover the entire *k*-space stays the same but density of data sampling increases at the origin, leading to higher accuracy and SNR in reconstruction. Figure 19.15 demonstrates conventional single shot spiral and interleaved spiral trajectories for covering identical regions in *k*-space and the pulse sequence for spiral acquisition.

Although the Nyquist rate is satisfied in both cases, with interleaved spiral, a higher data density is achieved at the origin of *k*-space. Additionally, with interleaved acquisitions (with *M* interleaves), the total readout time is separated into *M* shorter readout times, resulting in less dephasing in the outer part of *k*-space and off-resonance artifacts [41].

19.7.2 Comparison of Cartesian, UTE Radial, and Spiral Phase–Contrast MRI

In this section, three major PC-MRI methods (Cartesian, UTE radial, and spiral) are investigated in order to compare their physiologic accuracy and reproducibility when imaging the common, internal, and external Carotid arteries [42]. All imaging experiments were performed on a Philips Achieva 1.5 T scanner (Philips Healthcare, Best, NL) using a combined 16-element neurovascular coil capable of imaging carotid vessels from the aortic arch to the circle of Willis. Three normal male volunteers (29–30-year-old males) were scanned, each with the three imaging techniques described in Section 19.7.1. For all cases, the scan volume included 10 successive slices without gap, covering common, internal, and external carotid arteries including the carotid bifurcation. The imaging volume covered the axial volume starting at 1.5 cm proximal to the bifurcation to 3.5 cm distal to the bifurcation. The remaining sequence parameters were as follows: $FOV = 160 \times 160$ mm, 1.5 × 1.5 mm in-plane resolution, 5 mm slice thickness, flip angle = 12°, matrix size = 108 × 108, number of signal averages = 1, $TE = 2.9$ ms for Cartesian, $TE = 2.6$ ms for spiral, and $TE = 1.08$ ms for UTE acquisitions. To avoid velocity aliasing, a V_{enc} of 100 cm/s was used in all experiments. For the spiral acquisitions, we used 36 interleaves with 4 ms readout length each to cover the entire *k*-space. In order to reduce total imaging time, for

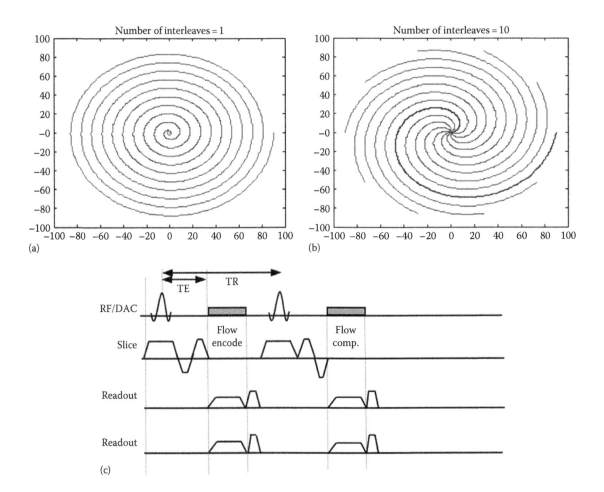

FIGURE 19.15 Demonstration of (a) conventional single-shot spiral acquisition with 10 rotations and (b) interleaved spiral acquisition with 10 interleaves. (c) 2D Pulse sequence of spiral acquisition with two back-to-back acquisitions: flow-encoded and flow-compensated acquisitions.

radial UTE acquisitions, we undersampled the k-space with 75% angle density. The resulting total scan times for the 10 slices were 3.81, 2.18, and 5.65 min for each of the Cartesian, spiral, and radial acquisitions, respectively.

In order to compare the accuracy between Cartesian as reference with the two other trajectories, the following normalized root mean square error (RMSE) is used:

$$RMSE = \sqrt{\frac{1}{T}\sum\left(\frac{Q_{ref}(t)-Q_x(t)}{Q_{ref}(t)}\right)^2}, \qquad (19.16)$$

where

   $Q_{ref}(t)$ is flow at time t as measured by the Cartesian sequence

   $Q_x(t)$ is flow as measured by radial UTE or spiral sequences

   T is the total number of time points where measurements are available (i.e., number of images collected during the cardiac cycle)

To investigate noise variation in time and space, for each acquisition method, we also defined $RMSE_{Total}$. In principle, summation of flow waveforms in the internal and external carotid arteries should equal the flow waveform in the common carotid artery. However, due to noise, a discrepancy will exist when measurements are made with MRI. This discrepancy is captured by the following equations:

$$Error(t) = Q_{CCA}(t)-(Q_{ICA}(t)+Q_{ECA}(t)), \qquad (19.17)$$

$$RMSE_{Total} = \sqrt{\frac{1}{T}\sum\left(\frac{Error(t)}{Q_{CCA}(t)}\right)^2}, \qquad (19.18)$$

where

   Q_{CCA} is the flow waveform in the common carotid artery

   Q_{ICA} is flow waveform in the internal carotid artery

   Q_{ECA} is the flow waveform in external carotid artery

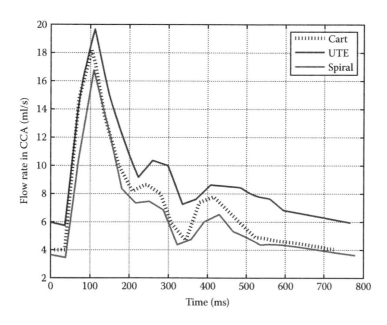

FIGURE 19.16   Flow waveform in the right common carotid artery for Cartesian, spiral, and radial UTE over the cardiac cycle.

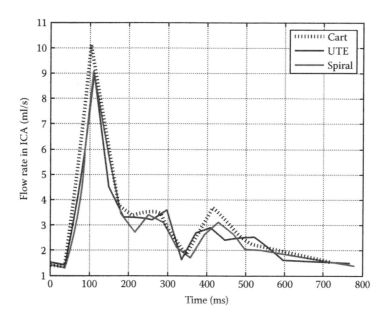

FIGURE 19.17   Flow waveform in the right internal carotid artery for Cartesian, spiral, and radial UTE over the cardiac cycle.

Figures 19.16 through 19.18 show flow waveform in CCA, ICA, and ECA for all three trajectories. In comparison to the Cartesian acquisition, UTE slightly overestimated flow and spiral slightly underestimated flow in the CCA and ECA. The mean flow waveforms in the right ICA were very close. Table 19.1 shows RMSE (Equation 19.16) for different trajectories, with Cartesian once again used as a reference including $RMSE_{Total}$ error for each trajectory.

Figure 19.19 shows the error signal as defined in Equation 19.17 for each acquisition method during the cardiac cycle, showing radial UTE, spiral, and Cartesian trajectories all having similar error variations through time and space. It should be mentioned that since each vessel has a different distance from the heart, the peak flow rates for each of the CCA, ICA, and ECA do not occur at precisely the same time. This accounts for some of the error in the error signals displayed in Figure 19.19.

Also to assess variations of measured flow rate across slices during the cardiac cycle, standard deviations for each flow measurement over axial locations

Chapter 19

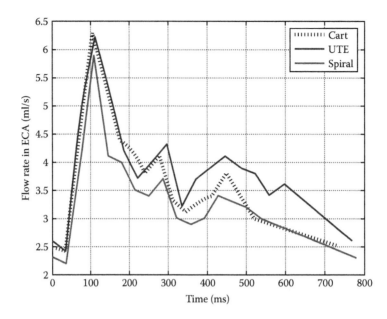

FIGURE 19.18 Flow waveform in the right external carotid artery for Cartesian, spiral, and radial UTE over the cardiac cycle.

Table 19.1 Normalized RMSE for Each of the Trajectories for Each Vessel (Equation 19.7) as well as the RMSETotal (Equation 19.18) for Each of the Trajectories

	Cartesian (Reference)	Spiral	Radial UTE
RMSE in CCA (%)	—	10	17
RMSE in ICA (%)	—	7.6	5.9
RMSE in ECA (%)	—	9.6	7.2
RMSE$_{Total}$ (%)	18	26	23
Echo time (ms)	2.9	2.6	1.08
Scan time (s)	228	130	339

and for each time point in the ICA and the ECA are calculated. The first observation was that compared to the ECA, the standard deviation in the ICA was less. A second observation was that the maximum standard deviation occurred in the systolic phase of the cardiac cycle—though as noted earlier, some of this variability is attributable to delay in pulse wave propagation. The maximum standard deviations in the ECA were found to be 2.2, 1.7, and 1.3 mL/s for Cartesian, spiral, and radial UTE acquisitions, respectively, which shows that the radial UTE has less flow variation compared to Cartesian and spiral trajectories. Also maximum standard deviations were 0.8, 0.9, and 1.0 mL/s in the ICA, for each of the Cartesian, spiral, and radial UTE acquisitions,

respectively (Figures 19.20). Based on this figure, it may be concluded that on average, the radial UTE sequence results in reduced standard deviations for the flow measurements.

As may be gathered, each *k*-space coverage method has its advantages and disadvantages. Based on the specific application, the user may choose to adopt a different method. In comparison to the Cartesian acquisition, spiral acquisitions can significantly drop the total imaging time, beneficial in breath-hold imaging [12,13]. On the other hand, UTE can significantly reduce the echo time, resulting in reduction of flow-related artifacts in the presence of complex flow or distal to stenosis [34,37].

Many other velocity acquisition and *k*-space coverage techniques exist that have not been discussed in this chapter. Luk Pat et al. [11] proposed one-shot velocity imaging by bow tie k-space coverage. Johnson et al. [43] proposed dual-echo vastly undersampled isotropic projection reconstruction (VIPR), an improved version of VIPR (see Section 19.7.3.2) in which two echo signals are obtained during each repetition time. The common disadvantage of non-Cartesian trajectories comes from the basic definition of the discrete Fourier transform that requires Cartesian gridded data as input. As a result, all non-Cartesian acquired data first need to be regridded as Cartesian data prior to inverse Fourier transformation, resulting in interpolation errors.

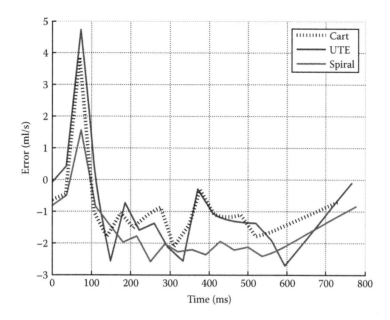

FIGURE 19.19   Total error signal (Equation 10.17) for each trajectory as a function of time.

19.7.3   4D PC-MRI

Traditional PC-MRI is performed using methods that encode velocity in one direction (often through plane) in a 2D imaging slice. Time-resolved 2D-CINE-PC-MRI is an acquisition gated to the ECG, which makes it possible to observe temporal variation of velocities (and as a result, flow) at a single 2D slice location within a breath hold [13,25]. A 4D flow imaging is a 3D gated acquisition and offers the ability to encode all three velocity directions in a 3D spatial volume through time. Results permit time-resolved high spatial resolution imaging of complex flow patterns in a 3D imaging volume. 4D PC-MRI can be adopted to visualize global and local blood flow in various vascular regions [44,45,74,75]. Recent studies show the potential of 4D PC-MRI for detection and visualization of complex flow patterns associated with vascular diseases such as an aneurysm or stenosis in a variety of vasculatures such as the aorta, carotid arteries, and intracranial arteries [46–51,74,75]. As 4D flow requires acquisitions with three velocity-encoding gradient directions in three spatial dimensions through time, data acquisition requires much longer scan times, no longer possible in a single breath hold. Therefore, several strategies have been suggested to overcome breathing artifacts in the thic and abdominal regions by using navigator echoes at the lung–liver interface. These methods that provide respiratory gating permit imaging during free breathing, providing significant benefit in clinical applications [45].

19.7.3.1   4D Cartesian PC-MRI

As opposed to 2D acquisition that was explained in Section 19.7.1, in 4D acquisition, a sequence of temporally gated 3D k-space acquisitions is performed where for the third dimension, an additional phase encoding is required (therefore, 3D acquisitions require readout along x, phase encoding along y, and a third phase encoding along z).

To acquire 4D PC data, further, each line of 3D k-space should be acquired twice (once with flow encoding and once with flow compensation), resulting in temporal resolution of $6TR$. Assuming N_y and N_z are the number of k-space lines in Y and Z directions, the total scan time in that case will be $N_y{}^*N_z$ heartbeats, which can be long in case of a large imaging volume. To speed up the imaging time, the segmented k-space acquisition method is adopted, which acquires N_{seg} k-space lines of all required phase-encoding steps in each cardiac phase, reducing the total imaging time to $N_y N_z/N_{seg}$ heartbeats. This reduction of the total imaging time is, however, at the cost of decreased temporal resolution, becoming $6TR^*N_{seg}$. As an example, consider $N_y = 128$, $N_z = 32$, $TR = 7$ ms, and $N_{seg} = 2$. These parameters will lead to a total imaging time of $N_y{}^*N_z/N_{seg} = 128^*32/2 = 2048$ heartbeats and result in a temporal resolution of 84 ms, leaving only a few acquired

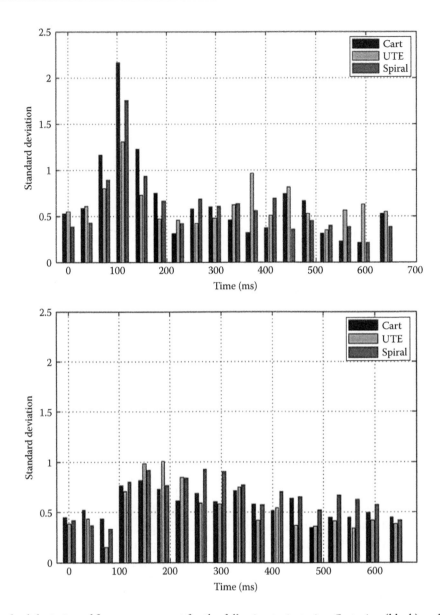

FIGURE 19.20 Standard deviation of flow measurement for the following trajectories: Cartesian (black), radial UTE (light gray), and spiral (dark gray), respectively. Top panel displays the standard deviation in the ECA and the bottom panel displays the standard deviation in the ICA.

cardiac phases, not suitable for the purpose of flow calculation, especially through the rapidly changing systolic phase of the cardiac cycle.

In order to remedy the poor temporal resolution, a four-point acquisition technique was proposed [52]. As described in Section 19.7.1, velocity acquisition for a specific direction requires velocity encoding in two separate scans: a flow-encoded and a flow-compensated scan. Therefore, to acquire three components of velocities, this leads to six separate scans, causing long scan times or poor temporal resolution with segmented *k*-space acquisitions. The four-point method instead requires a flow-compensated and three velocity-encoded

acquisitions (one for each of the *x*, *y*, and *z* velocities). With this approach, to arrive at velocities in a specific direction, the same flow-compensated acquisition is subtracted from each of the velocity-encoded acquisitions, thus improving the temporal resolution to $4TR$. Additionally, at the cost of worsening the temporal resolution, the segmented *k*-space approach may be adopted, which reduces the total imaging time while resulting in a temporal resolution of $4*TR*N_{seg}$. Returning to the example previously given, the four-point approach once again results in 2048 heartbeats for the imaging time but improves the temporal resolution from 84 to 56 ms. However, if N_{seg} is increased to 3, it will result in a

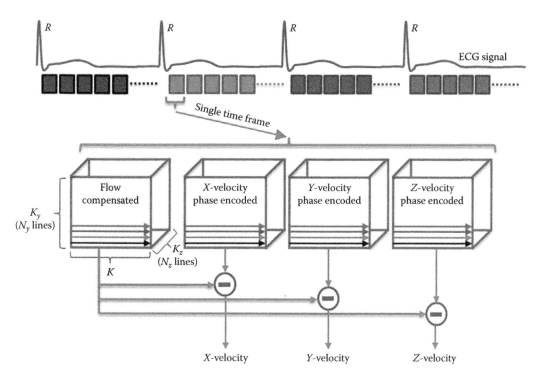

FIGURE 19.21 Time diagram of 4D Cartesian PC-MRI. In each phase, four successive acquisitions are collected. The whole 3D k-space is filled in N_yN_z/N_{seg} heartbeats.

temporal resolution of 84 ms while improving the total imaging time to 1365 heartbeats. In summary, the four-point method improves the temporal resolution or the total scan times by a factor of 4/6.

Figure 19.21 illustrates 4D k-space data collection with the four-point acquisition method with $N_{seg} = 1$. In addition to simple four-point method, other balanced four-point methods like Hadamard and five-point balanced acquisition method have been proposed, leading to SNR improvement of velocity data [53,54].

19.7.3.2 4D Radial PC-MRI

The main concern in 4D Cartesian PC-MRI is the long scan time in the case of large 3D volumes and low spatial resolution. A second concern is the relatively long echo time leading to intravoxel dephasing and signal loss specifically in the presence of disturbed and turbulent flows. The scan time for radial PC-MRI acquisition is longer than conventional Cartesian PC-MRI. This is due to the required pi^*N k-space lines to cover the whole k-space for an N^*N image. Undersampled radial acquisition can reduce scan time with the cost of less SNR and appearance of streak artifacts. Some parts of streak artifacts due to undersampled radial acquisition are removed by subtraction of flow-encoded and flow-compensated images. The remaining portion is tolerable due to high contrast and sparse signal distribution from the vessels [55].

A 4D PC VIPR was developed, which provides high spatial resolution and temporal resolution in large imaging volumes [56,57]. It is shown that radial sampling can reduce the severity of motion artifacts in comparison to Cartesian acquisitions [43]. A second option is the combination of 4D PC-MRI and UTE [75], which can offer 4D acquisition in shorter TE with improvement in flow quantification in disturbed and turbulent blood flows [58,75].

Figures 19.23 and 19.24 illustrate the flow path lines acquired using the 4D UTE PC [75] and 4D Cartesian sequence in a Gaussian-shaped, rigid, stenotic flow phantom machined from transparent acrylic with 87% area narrowing at the throat. Figure 19.22 shows the stenotic phantom setup in a closed-loop flow system. A CardioFlow 5000 programmable pump (Shelley Medical Imaging Technologies, London, Ontario, Canada) was

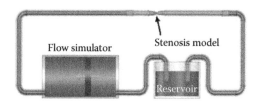

FIGURE 19.22 Schematic of the experimental setup showing the test section with 87% area stenosis that was machined in-house from clear acrylic plastic.

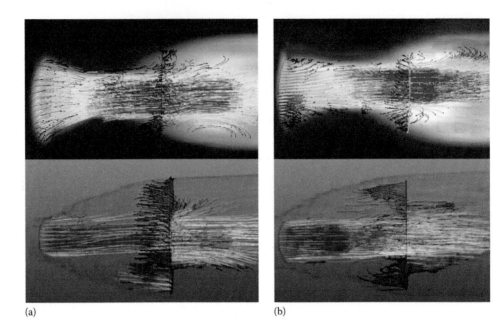

(a)   (b)

FIGURE 19.23 Flow path lines in sagittal view at the pulsatile flow rate with Q_{max} = 50 mL/s (top row) and 150 mL/s (bottom row) using (a) conventional 4D flow MRI and (b) 4D UTE flow MRI. The flow path lines are in good agreement for the case of Q_{max} = 50 mL/s. A slight difference is visible between flow path lines at Q_{max} = 150 mL/s due to moderately high flow rates and turbulence distal to the center of stenosis.

used for generating flows. Results were displayed with GTFlow software (Gyrotools, Zurich, Switzerland) with both 4D Cartesian and 4D UTE PC-MRI.

Figure 19.23 demonstrates the flow path lines in a sagittal view of the phantom for the peak systolic flow of Q_{max} = 50 mL/s (top row) and Q_{max} = 150 mL/s (bottom row) pulsatile flow rates using conventional 4D flow MRI (1) and 4D UTE flow MRI (2). Flow path lines for Q_{max} = 50 mL/s show good agreement between conventional 4D flow MRI and 4D UTE flow MRI. The correlation between the two results may be assessed based on the color map as well as the flow path lines. At Q_{max} = 150 mL/s, a slight difference between the flow path lines obtained from the two sequences can be observed in locations distal to the stenosis where turbulence appears at this moderately high flow rate. Nevertheless, the correlation between the color map and flow path lines using the two sequences can be appreciated.

Figure 19.24 demonstrates the flow path lines proximal to the stenosis, at the throat, and distal to the stenosis (top row) using conventional 4D flow MRI (1) and 4D UTE flow MRI (2) at pulsatile flow with Q_{max} = 300 mL/s pulsatile flow rate at peak systolic time. Due to intravoxel dephasing caused by high flow jet and turbulence distal to narrowing, the flow path lines from the conventional 4D flow seem inaccurate. The bottom row shows an enlarged view of phantom distal to the throat of the stenosis.

Figure 19.25 illustrates the flow path lines acquired using both 4D Cartesian and 4D UTE PC-MRI sequences in the carotid artery of a normal volunteer during the peak systolic phase—reslts were displayed and analyzed with GTFlow software. Although for the case of normal volunteer, the results are comparable, the benefit of 4D UTE PC becomes apparent in the case of stenotic flows and in the presence of flow jets, which cause intravoxel dephasing and inaccuracy in velocity quantification.

19.7.3.3 4D Spiral PC-MRI

Spiral trajectory results in shorter echo time and reduces the total scan time [12,13]. In [59,60,74], a 4D spiral PC-MRI technique was designed to reduce the total scan time compared to conventional Cartesian. 4D spiral PC-MRI has the added advantage of reducing the phase error and signal loss by providing shorter echo time. This benefit comes due to removal of the phase-encoding gradient and the rephasing part of the readout gradient.

In a separate study [61,74], this approach was improved upon by reducing the *TE* and was compared with standard 4D flow imaging. The scan parameters for two sequences were *TE/TR* = 4.4/7.7 ms (for Cartesian trajectory), *TE/TR* = 2.1/9.3 ms (for spiral trajectory), *FOV* = 160*160*50 mm, V_{enc} = 150 cm/s in all three flow directions, flip angle = 10, spatial

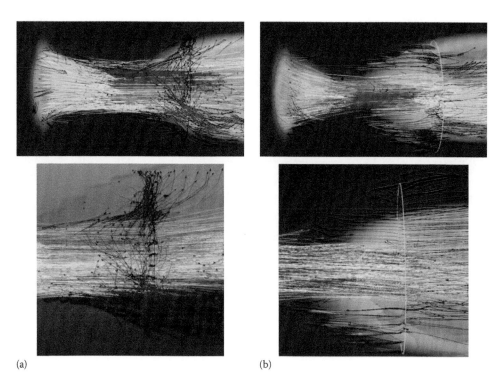

(a)   (b)

FIGURE 19.24   Flow path lines at peak systolic time for the pulsatile flow with Q_{max} = 300 mL/s using (a) conventional 4D flow MRI and (b) 4D UTE flow MRI. The top row shows a sagittal view of the phantom including regions proximal and distal to the stenosis. The bottom row shows an enlarged view of distal regions. The error and inaccuracy in flow path lines in the conventional sequence are evident.

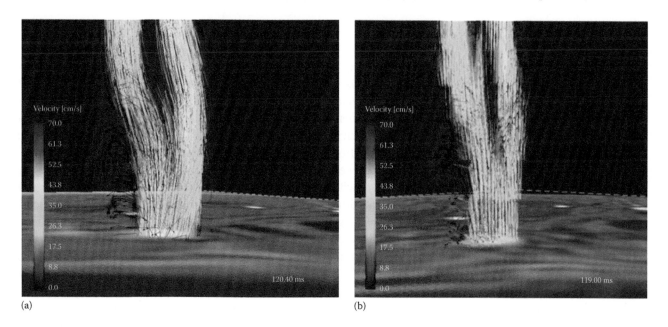

(a)   (b)

FIGURE 19.25   Flow path lines in a normal carotid artery during the peak systolic phase of the cardiac cycle using (a) conventional Cartesian technique and (b) 4D UTE PC-MRI.

resolution = 1.5*1.5*5.0 mm, and 12 cine frames in each cardiac cycle. All other imaging parameters were identical. Figure 19.26 shows blood flow path lines for a volunteer in the systolic cardiac phase of the cardiac cycle in the right carotid artery with both Cartesian and spiral 4D flow techniques. There is a relatively good agreement between both techniques and reducing scan time to half is a significant achievement; however, spiral shows slight erroneous flow path lines in the ICA possibly due to off-resonance artifacts.

(a) (b)

FIGURE 19.26 Flow path lines during a systolic phase of the cardiac cycle in the right carotid artery (magnified at carotid bifurcation) acquired using (a) conventional 4D Cartesian PC-MRI and (b) 4D spiral PC-MRI.

In summary, we can conclude that the main advantage of the 4D spiral acquisition technique is the shorter scan time in comparison to the conventional method. On the other hand, the 4D UTE technique has the advantage of very short echo time, which is beneficial in the quantification of stenotic flows.

19.8 Fourier Velocity Encoding

FVE involves an additional Fourier encoding along a velocity dimension [10,24]. The velocity variable is v, and the velocity frequency variable is $k_v = \gamma/\pi\, M_1$ s/cm, where γ is the gyromagnetic ratio and M_1 is first moment of the velocity-encoding gradient. Several velocity-encoding levels are used to obtain desirable resolution for flow encoding, and different velocity-encoding levels are typically achieved by acquiring velocity images with different bipolar gradient amplitudes. The procedure essentially phase encodes along the k_v axis. An image plane acquired with a particular value of k_v is denoted by $S_{k_x,k_y}(k_v)$. For a specific sample (k_x, k_y), this represents one sample from the Fourier transform of the velocity distribution of all spins in all voxels. The voxel velocity distribution is denoted $s_{x,y}(v)$ and is obtained by 1D inverse Fourier transformation along k_v. Velocity range (maximum velocity) is determined by the value of the increment between successive bipolar velocity-encoding gradient amplitudes, and the velocity resolution is determined by the amplitude of the largest bipolar velocity-encoding gradient. Improving velocity-encoding resolution leads to larger bipolar gradients and decreases temporal resolution [10]. Placing the bipolar gradient along the z-axis will encode through-plane velocities. Placing the bipolar gradient along x or y will encode in-plane velocities. Oblique flow can be encoded using a combination of bipolar gradients along the x, y, and z axes. Carvalho et al. [10] demonstrate the application of spiral FVE acquisitions technique in the aortic valve and the carotid artery and compared the results with ultrasound. Results showed that the peak and the time velocity waveforms in both the aortic valve and carotid artery were in good agreement with ultrasound (Figure 19.27).

19.9 Artifacts

There are many artifacts in MRI, and many solutions have been suggested in order to avoid or reduce them. In this section, we consider artifacts relevant to flow and velocity measurement techniques.

19.9.1 Partial Saturation

This artifact is more severe in sequences that have a high temporal resolution (short TR) and a wide variety

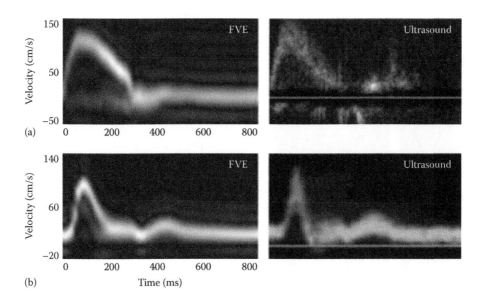

FIGURE 19.27 Comparison of FVE and ultrasound waveform in a healthy volunteer in the (a) aortic valve and (b) carotid artery. (Taken from Carvalho, J.L.A. and Nayak, K.S., *Magn. Reson. Med.*, 57, 639, 2007.)

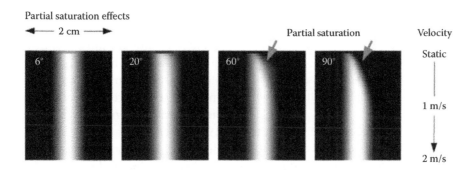

FIGURE 19.28 Bloch simulation of the partial saturation effects particularly affecting high flip angles and low velocities. Slice thickness is 2 cm; flip angles are 6°, 20°, 60°, and 90°; and velocity varies between 0 and 2 m/s in the vertical direction. (Taken from Nayak, K.S. et al., *Magn. Reson. Med.*, 50, 366, 2003.)

of velocity distribution in the imaging slice. This artifact occurs because of partial saturation of spins moving slowly perpendicular to the slice orientation. That is, presaturated spins in the previous TR neither have enough time to relax nor enough time to leave the slice. Nayak et al. [12] performed Bloch simulations to find out the effective steady-state slice profile for different velocities and flip angles. The results in Figure 19.28 are images of effective slice profile (horizontal axis) for velocity between 0 and 2 m/s (vertical axis), for different flip angles. Each horizontal line in each image contains a magnitude slice profile for that particular flip angle and through-plane velocity (black = low signal, white = high signal)

It was concluded that, with flip angles higher than 20°, partial saturation artifacts are more considerable. Also, it was shown that spins moving slower than 50 cm/s can be troublesome. The best way to avoid

partial saturation artifact is to use small flip angles. This is especially the case if slow-moving spins are present within the region of interest.

19.9.2 In-Plane Flow Artifact

Another important artifact related to high-velocity flows is in-plane flow during readout and acquisition. In-plane flow artifacts lead to spatial dispersion of flow in the transverse plane. Velocity point spread function (PSF) is a useful tool for illustrating this artifact. The PSF describes the response of an imaging system to a point source or point object. A more general term for the PSF is a system's impulse response [62]. To obtain velocity PSF, Bloch equations are simulated. Simulations for different in-plane flow showing in-plane velocities less than 2 m/s exhibit small artifacts, and for velocities higher than 2 m/s, artifacts are more

Velocity point spread functions

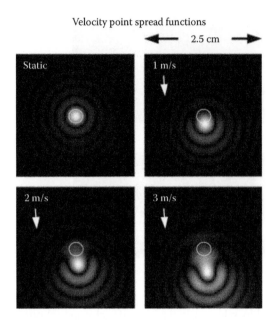

FIGURE 19.29 Bloch equation simulation to depict velocity PSF in the case of in-plane flow during readout. The white empty circle is a special resolution element. As shown in the bottom row, for in-plane velocities higher than 2 m/s, spatial displacement accrues. (Taken from Nayak, K.S. et al., *Magn. Reson. Med.*, 50, 366, 2003.)

severe. Nayak et al. [12] performed this simulation to find out the dependency of the PSF on different velocities (Figure 19.29).

PSF distortion causes blurring in the direction of flow, displacement artifacts, loss of spatial resolution, and partial volume effect. Using high TR and short TE methods can reduce this artifact.

19.9.3 Partial Volume Artifact

Probably partial volume artifact is the most critical in affecting the accuracy of PC velocity measurement. Assume an imaging voxel volume contains 50% static spins and 50% spins moving at a constant velocity. Suppose the phase for moving spins is ϕ_m and for static spins is zero. The measured signal for this voxel is then a summation of two phase vectors from static and moving spins. The phase of measured signal ϕ will have a value somewhere between ϕ_m and zero (Figure 19.30).

19.10 Errors

There are several factors that can cause imperfection in MRI flow quantification accuracy (phase-based methods). The major sources of inaccuracy in velocity-encoded images are eddy currents, Maxwell terms,

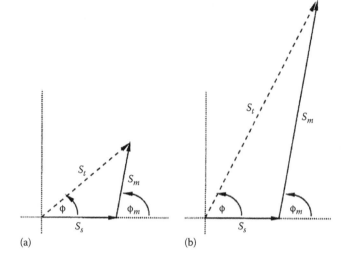

FIGURE 19.30 The total signal S_t in voxel is the vector sum of the signals from static spins S_S and moving spins S_M. (a) Without flow-related enhancement and (b) with flow-related enhancement. (Taken from Pelc, N.J. et al., *Magn. Reson. Q.*, 10, 125, 1994.)

Correction methods for this artifact depend on our goal from measurement. If we want to measure only velocity of moving spins, the desired measurement is $\phi = \phi_m$, obtained by suppressing signal intensity from static tissue. Many approaches have been proposed in order to suppress or reduce the signal from static tissues and to enhance flow-related signal (Figure 19.30b). In the case of volume flow calculation, it is desirable that moving and static spins contribute based on their fraction in the voxel volume. For the example given earlier, the resultant phase should be $\phi = \phi_m/2$ (Figure 19.30a). When multiplied by the total area of imaged vessel in the slice, this value will result in the correct flow [25].

Typically, use of flow enhancement methods results in the moving spins to have a more significant contribution than their actual fraction to the final signal. Figure 19.30b demonstrated this situation in which moving spins have three times the intensity of static spins. This enhancement would lead to overestimation of flow measurement. Statistical approaches have also been adopted for modeling and estimation of partial volume errors [63,64].

gradient field distortions, B_0 inhomogeneity, and flow acceleration effects [65–68].

Eddy currents: In PC-MRI, two consecutive bipolar velocity-encoding gradients cause different eddy currents to be induced in RF coils—this is due to magnetic field gradient switching, leading to phase errors in each phase image. Since the errors are unrelated, subtraction of two phase images does not remove errors due to eddy currents. Many proposed eddy current-related phase error correction methods are based on subtraction of phase images of static tissues from phase images of moving flows [69].

Maxwell terms: In MRI, to encode spatial information over the volume of interest, linear magnetic field gradients are employed. However, from Maxwell's equation, the magnetic field must have zero divergence and negligible curl—this results in higher orders of dependency of the linear magnetic field to location. The resulting nonlinear terms are typically referred to as Maxwell terms, causing phase errors.

These phase errors are usually corrected during image reconstruction; however, knowledge of the gradient waveforms is required [66,70].

Acceleration: In Section 19.6.2, we assumed constant velocity without acceleration during imaging. This is clearly an oversimplification and in many instances is not satisfied in cardiovascular imaging. By considering nonzero acceleration (and assuming zero higher-order terms),

$$\hat{v} = v + a\,.\bar{t}, \tag{19.19}$$

where

  a is the acceleration
  v is correct velocity
  $\hat{v}$ is the measured value

The parameter $\bar{t}$ depends on flow-encoding gradients, but essentially is the time interval between the application of the excitation pulse and the flow-encoding gradient [25].

As a result, the most effective way to reduce acceleration effect is to reduce TE, TR, and the time interval between excitation pulse and flow-encoding gradient; this goal is satisfied by using stronger gradients and short TE/TR sequences. Many techniques have been proposed for measuring the effect of acceleration on phase images—many techniques have also been proposed for compensating for acceleration [71].

19.11 Conclusion

MRI is a noninvasive modality for accurate and clear anatomical imaging with superb soft-tissue contrast. The intrinsic sensitivity of the MRI signal to spin motion provides a unique opportunity to acquire flow- and velocity-related information through the use of PC sequences.

A disadvantage of phase-based methods (such as PC and FVE) is the sensitivity of the phase image to many factors that can cause artifacts and errors in the measurements, as outlined in Sections 19.9 and 19.10 of the chapter. Many approaches have been proposed for estimating and correcting them; however, there are still errors and artifacts that have not yet been fully explained. In volumetric imaging, a long scan time is another factor that has impeded clinical applications of PC techniques. However, new 4D PC-MRI methods based on fast *k*-space sampling strategies such as spiral and radial *k*-space trajectories have been proposed, which to some extent alleviate the scan time problem [55,74]. It should be mentioned that alternative approaches to fast imaging include compressive sampling [72] and parallel imaging [73], which may be incorporated into PC acquisitions but that were not explored in this chapter. Also a combination of short echo like UTE and 4D acquisition methods has reduced dephasing and signal loss issues in the case of stenoses and aneurysms in vascular diseases [75].

References

1. Haacke, E. M., R. W. Brown, M. R. Thompson, R. Venkatesan. 1999. *Magnetic Resonance Imaging: Physical Principles and Sequence Design*. New York: Wiley-Liss.
2. Lauterbur, P. C. 1980. Progress in n.m.r. zeugmatography imaging. *Philos Trans R Soc Lond B Biol Sci* 289:483–487.
3. Damadian, R. 1980. Field focusing n.m.r. (FONAR) and the formation of chemical images in man. *Philos Trans R Soc Lond B Biol Sci* 289:489–500.
4. Lai, C. M. and P. C. Lauterbur. 1981. True three-dimensional image reconstruction by nuclear magnetic resonance zeugmatography. *Phys Med Biol* 26:851–856.
5. Webb, A. G. 2003. *Introduction to Biomedical Imaging*. Hoboken, NJ: IEEE Press Series on Biomedical Engineering.
6. Ku, D. N. 1997. Blood flow in arteries. *Annu. Rev. Fluid Mech.* 29:399–434.

7. John, E. H. and A. C. Guyton. 2011. *Textbook of Medical Physiology*. Philadelphia, PA: Saunders/Elsevier.

8. Allison, J., B. S. Chirs Wright, and M. S. M. P. Tom Lavin. 2009. *MRI Physics Course*. Augusta, GA: Department of Radiology, Medical College of Georgia.

9. Farzan, G., D. Xiaoyan, C. Alain De, D. Yvan, K. Martin, and G. Robert. 1998. Low Reynolds number turbulence modeling of blood flow in arterial stenoses. *Biorheology* 35:281–294.

10. Carvalho, J. L. A. and K. S. Nayak. 2007. Rapid quantitation of cardiovascular flow using slice-selective fourier velocity encoding with spiral readouts. *Magn Reson Med* 57:639–646.

11. Luk Pat, G. T., J. M. Pauly, B. S. Hu, and D. G. Nishimura. 1998. One-shot spatially resolved velocity imaging. *Magn Reson Med* 40:603–613.

12. Nayak, K. S., B. S. Hu, and D. G. Nishimura. 2003. Rapid quantitation of high-speed flow jets. *Magn Reson Med* 50:366–372.

13. Negahdar, M. J., M. Kadbi, M. Kotys, M. Alshaher, S. Fischer, and A. A. Amini. 2011. Rapid flow quantification in iliac arteries with spiral phase-contrast MRI. In *2011 Annual International Conference of the IEEE Engineering in Medicine and Biology Society, EMBC*, Boston, Massachusetts. pp. 2804–2808.

14. Winkler, A. J. and J. Wu. 1995. Correction of intrinsic spectral broadening errors in doppler peak velocity measurements made with phased sector and linear array transducers. *Ultrasound Med Biol* 21:1029–1035.

15. Hoskins, P. R. 1996. Accuracy of maximum velocity estimates made using Doppler ultrasound systems. *Br J Radiol* 69:172–177.

16. O'Donnell, M. 1985. NMR blood flow imaging using multiecho, phase contrast sequences. *Med Phys* 12:59–64.

17. Clarke, G. D., W. G. Hundley, R. W. McColl, R. Eckels, D. S. C. Chaney, H.-F. Li, and R. M. Peshock. 1996. Velocity-encoded, phase-difference cine MRI measurements of coronary artery flow: Dependence of flow accuracy on the number of cine frames. *J Magn Reson Imaging* 6:733–742.

18. Tang, C., D. D. Blatter, and D. L. Parker. 1993. Accuracy of phase-contrast flow measurements in the presence of partial-volume effects. *J Magn Reson Imaging* 3:377–385.

19. Mohiaddin, R. H., P. D. Gatehouse, M. Henien, and D. N. Firmin. 1997. Cine MR fourier velocimetry of blood flow through cardiac valves: Comparison with doppler echocardiography. *J Magn Reson Imaging* 7:657–663.

20. Nishimura, D. G. 2010. *Principles of Magnetic Resonance Imaging*. Stanfoed, CA: Stanford University.

21. Grist, T. M., C. A. Mistretta, C. M. Strother, and P. A. Turski. 2012. Time-resolved angiography: Past, present, and future. *J Magn Reson Imaging* 36:1273–1286.

22. Haider, C. R., J. F. Glockner, A. W. Stanson, and S. J. Riederer. 2009. Peripheral vasculature: High-temporal- and high-spatial-resolution three-dimensional contrast-enhanced MR angiography. *Radiology* 253:831–843.

23. Trzasko, J. D. 2010. Nonconvex compressive sensing with parallel imaging for highly accelerated 4D CE-MRA. In *International Society for Magnetic Resonance in Medicine (ISMRM)*, Stockholm, Sweden.

24. Moran, P. R. 1982. A flow velocity zeugmatographic interlace for NMR imaging in humans. *Magn Reson Imaging* 1:197–203.

25. Pelc, N. J., F. G. Sommer, K. C. Li, T. J. Brosnan, R. J. Herfkens, and D. R. Enzmann. 1994. Quantitative magnetic resonance flow imaging. *Magn Reson Q* 10:125–147.

26. Simonetti, O. P., R. E. Wendt, 3rd, and J. L. Duerk. 1991. Significance of the point of expansion in interpretation of gradient moments and motion sensitivity. *J Magn Reson Imaging* 1:569–577.

27. Xiang, Q. S. 1995. Temporal phase unwrapping for CINE velocity imaging. *J Magn Reson Imaging* 5:529–534.

28. Andersen, A. H. and J. E. Kirsch. 1996. Analysis of noise in phase contrast MR imaging. *Med Phys* 23:857–869.

29. Gatehouse, P., J. Keegan, L. Crowe, S. Masood, R. Mohiaddin, K.-F. Kreitner, and D. Firmin. 2005. Applications of phase-contrast flow and velocity imaging in cardiovascular MRI. *Eur Radiol* 15:2172–2184.

30. Negahdar, M. J., J. Cha, M. Shakeri, M. Kendrick, M. Alshaher, I. Khodarahimi, M. K. Sharp, A. Yancey, J. Heidenreich, and A. Amini. 2012. Effect of Venc on accuracy of velocity profiles in multi-slice phase-contrast MR imaging of stenotic flow. In *2012 Ninth IEEE International Symposium on Biomedical Imaging (ISBI)*, Barcelona, Spain. pp. 820–823.

31. Hahn, E. L. 1960. Detection of sea-water motion by nuclear precession. *J Geophys Res* 65:776–777.

32. Van Dijk, P. 1984. Direct cardiac NMR imaging of heart wall and blood flow velocity. *J Comput Assist Tomogr* 8:429–436.

33. Barger, A. V., D. C. Peters, W. F. Block, K. K. Vigen, F. R. Korosec, T. M. Grist, and C. A. Mistretta. 2000. Phase-contrast with interleaved undersampled projections. *Magn Reson Med* 43:503–509.

34. Tyler, D. J., M. D. Robson, R. M. Henkelman, I. R. Young, and G. M. Bydder. 2007. Magnetic resonance imaging with ultrashort TE (UTE) PULSE sequences: Technical considerations. *J Magn Reson Imaging* 25:279–289.

35. Kadbi, M., H. Wang, M. Traughber, M. Alshaher, A. Yancey, J. Heidenreich, and A. A. Amini. 2012. 3D Cine Ultra-short TE (UTE) phase contrast imaging in carotid artery: Comparison with conventional technique. In *International Society for Magnetic Resonance in Medicine (ISMRM)*, Melbourne, Victoria, Australia.

36. Gatehouse, P. D. and G. M. Bydder. 2003. Magnetic resonance imaging of short T2 components in tissue. *Clin Radiol* 58:1–19.

37. Kadbi, M., H. Wang, M. J. Negahdar, L. Warner, M. Traughber, P. Martin, and A. A. Amini. 2012. A novel phase-corrected 3D cine ultra-short te (UTE) phase-contrast MRI technique. In *2012 Annual International Conference of the IEEE Engineering in Medicine and Biology Society (EMBC)*, Chicago, IL. pp. 77–81.

38. Nishimura, D. G., P. Irarrazaval, and C. H. Meyer. 1995. A velocity k-space analysis of flow effects in echo-planar and spiral imaging. *Magn Reson Med* 33:549–556.

39. King, K. F., T. K. Foo, and C. R. Crawford. 1995. Optimized gradient waveforms for spiral scanning. *Magn Reson Med* 34:156–160.

40. Pike, G. B., C. H. Meyer, T. J. Brosnan, and N. J. Pelc. 1994. Magnetic resonance velocity imaging using a fast spiral phase contrast sequence. *Magn Reson Med* 32:476–483.

41. Carvalho, J. L. A., J.-F. Nielsen, and K. S. Nayak. 2010. Feasibility of in vivo measurement of carotid wall shear rate using spiral fourier velocity encoded MRI. *Magn Reson Med* 63:1537–1547.

42. Negahdar, M. J., M. Kadbi, J. Heidenreich, A. Yancey, and A. A. Amini. 2013. Comparison of Cartesian, UTE radial, and spiral phase-contrast MRI in measurement of blood flow in extracranial carotid arteries: Normal subjects. In *Proceedings of SPIE on Medical Imaging*, Orlando, Florida.

43. Johnson, K. M., D. P. Lum, P. A. Turski, W. F. Block, C. A. Mistretta, and O. Wieben. 2008. Improved 3D phase contrast MRI with off-resonance corrected dual echo VIPR. *Magn Reson Med* 60:1329–1336.

44. Wigström, L., L. Sjöqvist, and B. Wranne. 1996. Temporally resolved 3D phase-contrast imaging. *Magn Reson Med* 36:800–803.

45. Markl, M., A. Harloff, T. A. Bley, M. Zaitsev, B. Jung, E. Weigang, M. Langer, J. Hennig, and A. Frydrychowicz. 2007. Time-resolved 3D MR velocity mapping at 3T: Improved navigator-gated assessment of vascular anatomy and blood flow. *J Magn Reson Imaging* 25:824–831.

46. Frydrychowicz, A., J. T. Winterer, M. Zaitsev, B. Jung, J. Hennig, M. Langer, and M. Markl. 2007. Visualization of iliac and proximal femoral artery hemodynamics using time-resolved 3D phase contrast MRI at 3T. *J Magn Reson Imaging* 25:1085–1092.

47. Wetzel, S., S. Meckel, A. Frydrychowicz, L. Bonati, E. W. Radue, K. Scheffler, J. Hennig, and M. Markl. 2007. In vivo assessment and visualization of intracranial arterial hemodynamics with flow-sensitized 4D MR imaging at 3T. *Am J Neuroradiol* 28:433–438.

48. Harloff, A., F. Albrecht, J. Spreer, A. F. Stalder, J. Bock, A. Frydrychowicz, J. Schöllhorn et al. 2009. 3D blood flow characteristics in the carotid artery bifurcation assessed by flow-sensitive 4D MRI at 3T. *Magn Reson Med* 61:65–74.

49. Uribe, S., P. Beerbaum, T. S. Sørensen, A. Rasmusson, R. Razavi, and T. Schaeffter. 2009. Four-dimensional (4D) flow of the whole heart and great vessels using real-time respiratory self-gating. *Magn Reson Med* 62:984–992.

50. Hope, M. D., T. A. Hope, A. K. Meadows, K. G. Ordovas, T. H. Urbania, M. T. Alley, and C. B. Higgins. 2010. Bicuspid aortic valve: Four-dimensional MR evaluation of ascending aortic systolic flow patterns. *Radiology* 255:53–61.

51. Stankovic, Z., A. Frydrychowicz, Z. Csatari, E. Panther, P. Deibert, W. Euringer, W. Kreisel et al. 2010. MR-based visualization and quantification of three-dimensional flow characteristics in the portal venous system. *J Magn Reson Imaging* 32:466–475.

52. Pelc, N. J., M. A. Bernstein, A. Shimakawa, and G. H. Glover. 1991. Encoding strategies for three-direction phase-contrast MR imaging of flow. *J Magn Reson Imaging* 1:405–413.

53. Jung, B. and M. Markl. 2012. Phase-contrast MRI and flow quantification. In *Magnetic Resonance Angiography*. J. C. Carr, and T. J. Carroll, eds. New York: Springer, pp. 51–64.

54. Johnson, K. M. and M. Markl. 2010. Improved SNR in phase contrast velocimetry with five-point balanced flow encoding. *Magn Reson Med* 63:349–355.

55. Markl, M., A. Frydrychowicz, S. Kozerke, M. Hope, and O. Wieben. 2012. 4D flow MRI. *J Magn Reson Imaging* 36:1015–1036.

56. Gu, T., F. R. Korosec, W. F. Block, S. B. Fain, Q. Turk, D. Lum, Y. Zhou, T. M. Grist, V. Haughton, and C. A. Mistretta. 2005. PC VIPR: A high-speed 3D phase-contrast method for flow quantification and high-resolution angiography. *Am J Neuroradiol* 26:743–749.

57. Jing, L., M. J. Redmond, E. K. Brodsky, A. L. Alexander, L. Aiming, F. J. Thornton, M. J. Schulte, T. M. Grist, J. G. Pipe, and W. F. Block. 2006. Generation and visualization of four-dimensional MR angiography data using an undersampled 3-D projection trajectory. *IEEE Trans Med Imaging* 25:148–157.

58. Kadbi, M., M. Traughber, P. Martin, and A. A. Amini. 2013. 4D UTE flow: A novel 4D ultra-short TE phase-contrast MRI technique for assessment of flow and hemodynamics. In *International Society for Magnetic Resonance in Medicine (ISMRM)*, Salt Lake City, Utah.

59. Sigfridsson, A., S. Petersson, C. J. Carlhall, and T. Ebbers. 2012. Four-dimensional flow MRI using spiral acquisition. *Magn Reson Med* 68:1065–1073.

60. Janiczek, R. L., C. H. Meyer, S. T. Acton, B. R. Blackman, and F. H. Epstein. 2009. 4D spiral phase-contrast MRI of wall shear stress in the mouse aorta. In *International Society for Magnetic Resonance in Medicine (ISMRM)*, Honolulu, Hawaii.

61. Kadbi, M., MJ. Negahdar, M. Traughber, P. Martin, and A. A. Amini. 2013. A fast reduced TE 4D spiral PC MRI sequence for assessment of flow and hemodynamics. In *International Society for Magnetic Resonance in Medicine (ISMRM)*, Salt Lake City, Utah.

62. Gatehouse, P. D. and D. N. Firmin. 1999. Flow distortion and signal loss in spiral imaging. *Magn Reson Med* 41:1023–1031.

63. González Ballester, M. Á., A. P. Zisserman, and M. Brady. 2002. Estimation of the partial volume effect in MRI. *Med Image Anal* 6:389–405.

64. van Osch, M. J. P., E.-j. P. A. Vonken, C. J. G. Bakker, and M. A. Viergever. 2001. Correcting partial volume artifacts of the arterial input function in quantitative cerebral perfusion MRI. *Magn Reson Med* 45:477–485.

65. Markl, M. 2005. *Velocity Encoding and Flow Imaging*. Freiburg, Germany: University of Hospital Freiburg, Deptartment of Diagnostic Radiology, Medical Physics.

66. Walker, P. G., G. B. Cranney, M. B. Scheidegger, G. Waseleski, G. M. Pohost, and A. P. Yoganathan. 1993. Semiautomated method for noise reduction and background phase error correction in MR phase velocity data. *J Magn Reson Imaging* 3:521–530.

67. Miller, T., A. Landes, and A. Moran. 2009. Improved accuracy in flow mapping of congenital heart disease using stationary phantom technique. *J Cardiovasc Magn Reson* 11:52.

68. Peeters, J. M., C. Bos, and C. J. G. Bakker. 2005. Analysis and correction of gradient nonlinearity and B0 inhomogeneity related scaling errors in two-dimensional phase contrast flow measurements. *Magn Reson Med* 53:126–133.

69. K. M. Johnson, D. L., O. Wieben. 2009. Eddy current corrections for phase contrast MRI using gradient calibration. In *International Society for Magnetic Resonance in Medicine (ISMRM)* Honolulu, Hawaii.

70. Bernstein, M. A., X. J. Zhou, J. A. Polzin, K. F. King, A. Ganin, N. J. Pelc, and G. H. Glover. 2006. Concomitant gradient terms in phase contrast MR: Analysis and correction. *Magn Reson Med* 39:300–308.

71. Thunberg, P., L. Wigström, B. Wranne, J. Engvall, and M. Karlsson. 2000. Correction for acceleration-induced displacement artifacts in phase contrast imaging. *Magn Reson Med* 43:734–738.

72. Lustig, M., D. Donoho, and J. M. Pauly. 2007. Sparse MRI: The application of compressed sensing for rapid MR imaging. *Magn Reson Med* 58:1182–1195.

Chapter 19

73. Niendorf, T. and D. K. Sodickson. 2006. Parallel imaging in cardiovascular MRI: Methods and applications. *NMR Biomed* 19:325–341.

74. Negahdar M, M. Kadbi M, Kendrick M, Stoddard M, Amini A. (in press). 4D Spiral imaging of flows in stenotic phantoms and subjects with aortic stenosis. *Magnetic Resonce in Medicine*. doi 10.1002/mrm.25636.

75. Kadbi M, Negahdar M, Traughber M, Martin P, Stoddard M, Amini A. 2015. 4D UTE flow: A phase-contrast MRI technique for assessment and visualization of stenotic flows. *Magnetic Resonance in Medicine* 73(3):939–950.

20. Physics and Engineering Principles of Fluid Dynamics

Giovanni Puppini and Gabriele Meliadò

20.1 Introduction to Computational Fluid Dynamics (CFD): What Is CFD and Why Use It?

Computational fluid dynamics (CFD) is a mechanical engineering field that deals with analyzing fluid flow, heat transfer, and associated phenomena, using computer-based simulation. CFD is a widely adopted methodology for solving complex problems in many modern engineering fields. The merit of CFD lies in the development of new and improved devices and system designs, and existing equipments are optimized through computational simulations, resulting in an enhanced efficiency and lower operating costs. However, in the biomedical field, CFD is still emerging. The main reason why CFD in the biomedical field has lagged behind is the tremendous complexity of human body fluid behavior. Recently, CFD biomedical research is relatively easy and possible, because high-performance hardware and software are easily available with advances in computer science. All CFD processes contain three main components to provide useful information, namely, preprocessing, solving mathematical equations, and postprocessing. Initial accurate geometric modeling and boundary conditions are essential to achieve adequate results.

Gas and liquid flows are governed by partial differential equations (PDEs), which represent conservation laws for the mass, momentum, and energy. CFD is the art of replacing such PDE systems by a set of algebraic equations that can be solved using digital computers.

As reported in Anderson (1995), historically, the early development of CFD in the 1960s and 1970s was driven by the need of the aerospace community. However, modern CFD cuts across all disciplines where the flow of a fluid is important.

The role of CFD in engineering predictions has become so strong that today it can be viewed as a new "third dimension" in fluid dynamics, the other two dimensions being the classical cases of pure experiment and pure theory. From 1687, with the publication of Isaac Newton's Principia, to the mid-1960s, advancements in fluid mechanics were made with the synergistic combination of pioneering experiments and basic theoretical analyses—analyses that almost always required the use of simplified models of the flow to obtain closed-form solutions of the governing equations. These closed-form solutions have the distinct advantage of immediately identifying some of the fundamental parameters of a given problem, and explicitly demonstrating how the answers to the

Chapter 20

problems are influenced by variations in the parameters. They frequently have the disadvantage of not including all the requisite physics of the flow. Into this picture stepped CFD in the mid-1960s. Thanks to its ability to manage the flow equations in the "exact" form, CFD has soon become a popular instrument in engineering analyses. Today, CFD supports and complements both pure experiment and pure theory, and it is the opinion of several authors that, from now on, it always will. CFD is not a passing fad; rather, with the advent of the high-speed digital computer, CFD will remain a third dimension in fluid dynamics, of equal stature and importance to experiment and theory. It has taken a permanent place in all aspects of fluid dynamics, from basic research to engineering design.

It is important to emphasize that the use of CFD in preliminary design can affect the basic research: a given numerical solution to a basic experimental problem, like the flow behavior in presence of a backward facing step, can be exploited as a tool to realize computational experiments with the purpose of studying the fundamental characteristics of the flow itself. These numerical experiments are directly analogous to actual laboratory experiments.

Of course, inherent in the earlier discussion is the assumption that CFD results are accurate as well as cost-effective; otherwise, any assumption of the role of CFD would be foolish. The results of CFD are only as valid as the physical models incorporated in the governing equations and boundary conditions, and therefore are subject to error, especially for turbulent flows. Truncation errors associated with the particular algorithm used to obtain a numerical solution, as well as round-off errors, both combine to compromise the accuracy of CFD results.

What can CFD not do? The fundamental answer to this question is that it cannot reproduce physics that is not properly included in the formulation of the problem. The most important example is turbulence. Most CFD solutions of turbulent flows now contain turbulence models that are just approximations of the real physics and that depend on empirical data for various constants that go into the turbulence models. Therefore, all CFD solutions of turbulent flows are subject to inaccuracy, even though some calculations for some situations are reasonable.

Again, emphasis is made that CFD solutions are secondary to the degree of physics that goes into their formulation. CFD is used in biomedical applications with different purposes: modeling the flow of blood in the heart and in the vessels, modeling the flow of blood in the cardiac assist devices, modeling the flow of the solutions contained in drug delivery devices. The use of CFD reduces the need for testing on human beings and animals until the last stage of the process (Abdulnaser 2009).

Recently, medical researchers have used simulation tools to assist in predicting the behavior of circulatory blood flow inside the human body. Computational simulations provide invaluable information that is extremely difficult to obtain experimentally and is one of the many CFD sample applications in the biomedical area in which blood flow through an abnormal artery can be predicted. CFD analysis is increasingly performed to study fluid phenomena inside the human vascular system Medical simulations of circulatory function offer many benefits. They can lower the chances of post-operative complications, assist in developing better surgical procedures, and deliver a good understanding of biological processes, as well as more efficient and less destructive medical equipment such as blood pumps. Furthermore, medical applications using CFD have expanded not only into the clinical situations of disease, but also into healthy life support fields, such as sport medicine and rehabilitation medicine.

20.2   Basic Principles of CFD

The dynamics of fluids is governed by three principles of classical physics:

- Conservation of mass (continuity equation)
- Newton's second law: F = ma (the momentum equation)
- Conservation of energy

From these laws, PDFs, which represent the relationship between the flow variables and their evolution in space and time, are derived and simplified under specific circumstances.

The conservation laws are formulated under the continuum hypothesis, that is, the fluid is a continuum medium. It is useful to remember that for a liquid flow, the continuum hypothesis is always satisfied in practice.

20.2.1 Equations of Fluid Mechanics

Definitions, symbols, variables, and mathematical notations used in this chapter are now presented.

First a discussion of finite control volume and infinitesimal control volume follows. Consider a general flow field as represented by the streamlines in Figure 20.1. Let us imagine an infinitesimally (from the point of view of differential calculus) small fluid element in the flow, with a differential volume, dV, however large in comparison with the mean spacing between molecules so that it can be viewed as a continuous medium.

The fluid element may be fixed in space (Eulerian point of view) with the fluid moving through it, as shown at the top of Figure 20.1. Alternatively, it may be moving along a streamline with a vector velocity V equal to the flow velocity at each point (Lagrangian point of view).

PDEs obtained from the fluid element fixed in space are named the *conservative* form of the equations; the ones obtained from the moving fluid element are called the *nonconservative* form of the equations.

Instead of looking at the whole flow field at once, the fundamental physical principles are applied to just the fluid element itself.

The vectors are represented with capital letters with an arrow on top of the symbol.

p and T are the static pressure and temperature, respectively.

$$\rho = \rho(x, y, z, t)$$

is a scalar field, which represents the density of the fluid element.

$$\vec{V} = u\vec{i} + v\vec{j} + w\vec{k}$$

is the vector velocity field in Cartesian space where the x, y, and z components of velocity are given, respectively, by

$$u = u(x, y, z, t)$$

$$v = v(x, y, z, t)$$

$$w = w(x, y, z, t)$$

Stress is termed as the force per unit area on a surface. We distinguish between the normal stress (τ_{zz} in Figure 20.2) when the force and the normal surface vector are in the same (or opposite) direction, and shear stress when the force and the normal surface vector are perpendicular with each other (τ_{zx} and τ_{zy} in Figure 20.2).

τ_{ij} denotes the element of the stress tensor with the usual convention, that is, τ_{ij} is the stress exerted on a

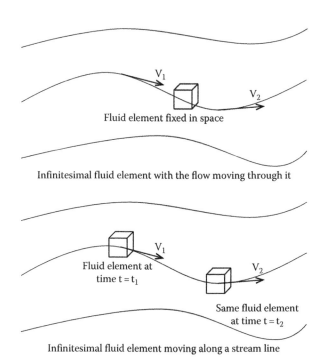

FIGURE 20.1 Flow field representations.

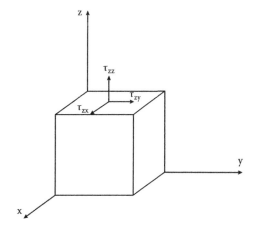

FIGURE 20.2 Normal and shear stresses.

plane perpendicular to the i-axis along the j-direction. In Figure 20.2, as examples, the three elements (normal and shear stresses) acting on a cubic volume element on a plan perpendicular to the z axis are shown.

Since, in general, both the force and the normal surface vector can be decomposed in three orthogonal directions (x, y, and z in a Cartesian coordinate system), a complete stress should be represented by a tensor with nine elements:

$$\tau = \begin{pmatrix} \tau_{xx} & \tau_{xy} & \tau_{xz} \\ \tau_{yx} & \tau_{yy} & \tau_{yz} \\ \tau_{zx} & \tau_{zy} & \tau_{zz} \end{pmatrix} \tag{20.1}$$

The elements in a row are the components acting on a plane, whereas the ones in a column are the components acting in one direction.

e and q̇ are the internal energy per unit mass and the rate of volumetric heat addition per unit mass, respectively; k is the thermal conductivity;

$$\nabla \equiv \vec{i}\frac{\partial}{\partial x} + \vec{j}\frac{\partial}{\partial y} + \vec{k}\frac{\partial}{\partial z}$$

is the vector operator in Cartesian's coordinates;

$$\frac{D}{Dt} \equiv \frac{\partial}{\partial t} + u\frac{\partial}{\partial x} + v\frac{\partial}{\partial y} + w\frac{\partial}{\partial z} \equiv \frac{\partial}{\partial t} + (\vec{V} \cdot \nabla)$$

is the substantial derivative, which is physically the time rate of change following a moving fluid element; ∂/∂t is called the local derivative, which is physically the time rate of change at a fixed point;

$$\vec{V} \cdot \nabla$$

is the convective derivative (some authors prefer the term advective), which is physically the time rate of change due to the movement of the fluid element from one location to another in the flow field where the flow properties are spatially different.

$$\nabla \cdot \vec{V}$$

represents the time rate of change of the volume of a moving fluid element per unit volume.

Starting from the equations of classical physics (continuity equation, the momentum equation, and conservation of energy), using the notations introduced

earlier, it is possible to derive the equations governing the fluids' motion (for a step-by-step demonstration, see Anderson 1995, Blazek 2001, and Wendt 2009); we will present them in both the nonconservative form and the conservative one.

Nonconservative form
Continuity equation

$$\frac{D\rho}{Dt} + \rho\nabla \cdot \vec{V} = 0 \tag{20.2}$$

The following are the momentum equations or Navier–Stokes equations:

$$\rho\frac{Du}{Dt} = -\frac{\partial p}{\partial x} + \frac{\partial \tau_{xx}}{\partial x} + \frac{\partial \tau_{yx}}{\partial y} + \frac{\partial \tau_{zx}}{\partial z} + \rho f_x$$

$$\rho\frac{Dv}{Dt} = -\frac{\partial p}{\partial y} + \frac{\partial \tau_{xy}}{\partial x} + \frac{\partial \tau_{yy}}{\partial y} + \frac{\partial \tau_{zy}}{\partial z} + \rho f_y \tag{20.3}$$

$$\rho\frac{Dw}{Dt} = -\frac{\partial p}{\partial z} + \frac{\partial \tau_{xz}}{\partial x} + \frac{\partial \tau_{yz}}{\partial y} + \frac{\partial \tau_{zz}}{\partial z} + \rho f_z$$

Conservation of energy

$$\rho\frac{D}{Dt}\left(e + \frac{V^2}{2}\right)$$

$$= \rho\dot{q} + \frac{\partial}{\partial x}\left(k\frac{\partial T}{\partial x}\right) + \frac{\partial}{\partial y}\left(k\frac{\partial T}{\partial y}\right) + \frac{\partial}{\partial z}\left(k\frac{\partial T}{\partial z}\right)$$

$$- \frac{\partial(up)}{\partial x} - \frac{\partial(vp)}{\partial y} - \frac{\partial(wp)}{\partial z} + \frac{\partial(u\tau_{xx})}{\partial x} + \frac{\partial(u\tau_{yx})}{\partial y}$$

$$+ \frac{\partial(u\tau_{zx})}{\partial z} + \frac{\partial(v\tau_{xy})}{\partial x} + \frac{\partial(v\tau_{yy})}{\partial y} + \frac{\partial(v\tau_{zy})}{\partial z}$$

$$+ \frac{\partial(w\tau_{xz})}{\partial x} + \frac{\partial(w\tau_{yz})}{\partial y} + \frac{\partial(w\tau_{zz})}{\partial z} + \rho\vec{f} \cdot \vec{V} \tag{20.4}$$

Conservative form
Continuity equation

$$\frac{\partial\rho}{\partial t} + \nabla \cdot (\rho\vec{V}) = 0 \tag{20.5}$$

The following are the momentum equations or Navier–Stokes equations:

$$\frac{\partial(\rho u)}{\partial t}+\nabla\cdot(\rho u\vec{V})=-\frac{\partial p}{\partial x}+\frac{\partial\tau_{xx}}{\partial x}+\frac{\partial\tau_{yx}}{\partial y}+\frac{\partial\tau_{zx}}{\partial z}+\rho f_x$$

$$\frac{\partial(\rho v)}{\partial t}+\nabla\cdot(\rho v\vec{V})=-\frac{\partial p}{\partial y}+\frac{\partial\tau_{xy}}{\partial x}+\frac{\partial\tau_{yy}}{\partial y}+\frac{\partial\tau_{zy}}{\partial z}+\rho f_y$$

$$\frac{\partial(\rho w)}{\partial t}+\nabla\cdot(\rho w\vec{V})=-\frac{\partial p}{\partial z}+\frac{\partial\tau_{xz}}{\partial x}+\frac{\partial\tau_{yz}}{\partial y}+\frac{\partial\tau_{zz}}{\partial z}+\rho f_z$$

$$(20.6)$$

Conservation of energy

$$\frac{\partial}{\partial t}\left[\rho\left(e+\frac{V^2}{2}\right)\right]+\nabla\cdot\left[\rho\left(e+\frac{V^2}{2}\right)\vec{V}\right]$$

$$=\rho\dot{q}+\frac{\partial}{\partial x}\left(k\frac{\partial T}{\partial x}\right)+\frac{\partial}{\partial y}\left(k\frac{\partial T}{\partial y}\right)+\frac{\partial}{\partial z}\left(k\frac{\partial T}{\partial z}\right)-\frac{\partial(up)}{\partial x}$$

$$-\frac{\partial(vp)}{\partial y}-\frac{\partial(wp)}{\partial z}+\frac{\partial(u\tau_{xx})}{\partial x}+\frac{\partial(u\tau_{yx})}{\partial y}+\frac{\partial(u\tau_{zx})}{\partial z}$$

$$+\frac{\partial(v\tau_{xy})}{\partial x}+\frac{\partial(v\tau_{yy})}{\partial y}+\frac{\partial(v\tau_{zy})}{\partial z}+\frac{\partial(w\tau_{xz})}{\partial x}+\frac{\partial(w\tau_{yz})}{\partial y}$$

$$+\frac{\partial(w\tau_{zz})}{\partial z}+\rho\vec{f}\cdot\vec{V}$$

$$(20.7)$$

Note that for the momentum and energy equations, the nonconservative forms differ from the conservative forms only for the left-hand side. The right-hand sides of the equations in the two forms are the same.

Moreover, we can see that all the previous equations in the conservative form have a divergence term on the left-hand side involving the divergence of the flux of some physical quantity, such as

mass flux—$\rho\vec{V}$

flux of x, y, and z-component of momentum—$\rho u\vec{V}$, $\rho v\vec{V}$, and $\rho w\vec{V}$, respectively

flux of internal and total energy—$\rho e\vec{V}$ and $\rho(e+V^2/2)\vec{V}$, respectively

For this reason, the conservative form of the equations is sometimes called the divergence form.

We can write the conservative form of the whole governing equations—continuity, momentum, and energy—in the same generic form, given by

$$\frac{\partial U}{\partial t}+\frac{\partial\Phi x}{\partial x}+\frac{\partial\Phi y}{\partial y}+\frac{\partial\Phi z}{\partial z}=J \qquad (20.8)$$

where U, Φx, Φy, Φz, and J are column vectors, given by

$$U=\begin{Bmatrix}\rho\\\rho u\\\rho v\\\rho w\\\rho(e+V^2/2)\end{Bmatrix}$$

$$\Phi x=\begin{Bmatrix}\rho u\\\rho u^2+p-\tau_{xx}\\\rho vu-\tau_{xy}\\\rho wu-\tau_{xz}\\\rho(e+V^2/2)u+pu-k\dfrac{\partial T}{\partial x}-u\tau_{xx}-v\tau_{xy}-w\tau_{xz}\end{Bmatrix}$$

$$\Phi y=\begin{Bmatrix}\rho v\\\rho uv-\tau_{yx}\\\rho v^2+p-\tau_{yy}\\\rho wv-\tau_{yz}\\\rho(e+V^2/2)v+pv-k\dfrac{\partial T}{\partial y}-u\tau_{yx}-v\tau_{yy}-w\tau_{yz}\end{Bmatrix}$$

$$\Phi z=\begin{Bmatrix}\rho w\\\rho uw-\tau_{zx}\\\rho vw-\tau_{zy}\\\rho w^2+p-\tau_{zz}\\\rho(e+V^2/2)w+pw-k\dfrac{\partial T}{\partial z}-u\tau_{zx}-v\tau_{zy}-w\tau_{zz}\end{Bmatrix}$$

$$J=\begin{Bmatrix}0\\\rho f_x\\\rho f_y\\\rho f_z\\\rho(uf_x+vf_y+wf_z)+p\dot{q}\end{Bmatrix}$$

Chapter 20

The column vectors Φx, Φy, and Φz are named the flux vectors (or flux terms), and J represents a "source term" (which is zero if body forces are negligible). For an unsteady problem, U is called the solution vector because the elements in U (ρ, ρu, ρv, etc.) are the dependent variables that are usually solved numerically in steps of time. Please note that, in this formalism, the elements of U are obtained computationally, that is, numbers are obtained for the products ρu, ρv, ρw, and ρ(e + V²/2) rather than for the primitive variables u, v, w, and e by themselves. Hence, in a computational solution of an unsteady flow problem using Equation 20.8, the dependent variables are treated as ρ, ρu, ρv, ρw, and ρ(e + V²/2). Of course, once numbers are known for these dependent variables (which includes ρ by itself), obtaining the primitive variables is trivial.

Equations for inviscid flow: inviscid flow is a flow where the dissipative, transport phenomena of viscosity, mass diffusion, and thermal conductivity are neglected. The governing equations for an unsteady, 3D, compressible inviscid flow are obtained by dropping the viscous terms in the earlier equations. So Equation 20.8 remains the same, except that the elements of the flow vectors are simplified and become

$$\Phi x = \begin{Bmatrix} \rho u \\ \rho u^2 + p \\ \rho vu \\ \rho wu \\ \rho(e + V^2/2)u + pu \end{Bmatrix} \quad \Phi y = \begin{Bmatrix} \rho v \\ \rho uv \\ \rho v^2 + p \\ \rho wv \\ \rho(e + V^2/2)v + pv \end{Bmatrix}$$

$$\Phi z = \begin{Bmatrix} \rho w \\ \rho uw \\ \rho vw \\ \rho w^2 + p \\ \rho(e + V^2/2)w + pw \end{Bmatrix}$$

For an unsteady inviscid flow, the solution is the vector U again, and the dependent variables for which numbers are directly obtained are, as already seen, ρ, ρu, ρv, ρw, and ρ(e + V²/2). For a steady inviscid flow (∂U/∂t = 0), the numerical solution often takes the form of "marching" techniques; for example, if the solution is being obtained by marching in the x-direction, then Equation 20.8 can be written as

$$\frac{\partial \Phi x}{\partial x} = J - \frac{\partial \Phi y}{\partial y} - \frac{\partial \Phi z}{\partial z} \qquad (20.9)$$

Here, Φx is the "solution" vector, and the dependent variables for which numbers are obtained are ρu, (ρu² + p), ρuv, ρuw, and [ρu(e + V²/2) + pu]. Of course, it is still possible to obtain the primitive variables from the dependent variables mentioned earlier even if the algebra is more complex than in the previous case (Kutler 1975). The governing equations, when written in the form of Equation 20.8, have no flow variables outside the single x, y, z, and t derivatives. Actually, the terms in Equation 20.8 have everything contained inside these derivatives. The flow equations in the form of Equation 20.8 are said to be in a strong conservative form. In contrast, examine the form of Equation 20.6. These equations have a number of x, y, and z derivatives explicitly appearing on the right-hand side and are called the weak conservative form of the equations.

Summarizing what we have seen so far, the equations governing the motion of a fluid (henceforth the governing equations) are most simply described as a statement of Newton's second law of motion as it applies to the movement of a mass of fluid (whether that be air, water, or a more exotic fluid): mass x acceleration = force acting on a body.

For a fluid, the "mass" is the mass of the fluid body; the "acceleration" is the acceleration of a particular fluid particle; and the "forces acting on the body" are the total forces acting on our fluid.

Newton's second law produces a system of differential equations relating rates of change of fluid velocity to the forces acting on the fluid. We require other two physical constraints to be applied on our fluid, which can be most simply stated as follows: mass is conserved; that is, fluid neither appears nor disappears from our system and energy is conserved. The solution can be broken into two parts. The first focuses on the existence of solutions to the equations. The second focuses on whether these solutions are bounded (remain finite).

A closed analytical solution of the governing equations has not been found yet (it is one of the six millennium problems remaining unsolved), and we must resort to solving the equations on a computer using numerical approximations such as finite difference method, finite element method, and finite volume method (Hirsch 1988, Wesseling 2001).

20.3 CFD and Clinical Practice

Using CFD, medical researchers can gain knowledge of how body fluids and system components are expected to perform, to make the required improvements for biofluid physiology studies, and to develop medical devices. CFD offers chances for simulation before a real commitment is undertaken to execute any medical design alteration and may provide the correct direction to develop medical interventions.

The technique of "image-guided modeling" involves a combination of imaging and computational modeling and offers a method to obtain additional information not easily obtained from imaging. Computational modeling of blood flow, referred to as FD, from which 3D blood velocities may be estimated, allows measurement of 3D phenomena, such as visualization of flow patterns in disease, measurements related to the spatial distribution of velocity and turbulence, and estimation of wall shear stress (WSS).

Actually the most important source for the blood flow information is magnetic resonance imaging (MRI) with 4D phase-contrast (PC) imaging, that is, time-resolved 3D PC imaging with velocity sensitivity in all three spatial directions, in some respects, a "one-stop-shop" MRI technique.

Computational modeling of the mechanical behavior of the arteries is achieved through solid modeling, from which 2D or 3D tissue stresses may be estimated.

The measurement of wall stresses (WSS and tissue stress) is of particular importance, as they are linked to disease initiation and development and to the rupture of atherosclerotic plaque and aneurysm. Wall stresses measured using image-guided modeling have the potential for improved diagnosis and selection of patients for interventional procedures.

Since both morphologic and dynamic information are simultaneously available, these 4D PC techniques are especially interesting for cardiovascular diseases, neurovascular imaging, and organ hemodynamics where comws are present and systemic and localized changes interact. Furthermore, the opportunity to simultaneously obtain morphological and functional results in a reasonable scan time without the restraints of ionization radiation is particularly interesting in pediatric imaging and congenital heart disease in particular (Frydrychowicz et al. 2011).

20.3.1 Pulse Wave Velocity

The term pulse wave velocity (PWV) characterizes the velocity by which the pulse wave propagates through the vessel or system of interest. Although PWV usually receives little attention during clinical diagnosis, it is a well-established parameter that is frequently used in research studies of vascular pathology. The PWV has been shown to correlate well with vascular stiffness and risk for coronary artery disease and stroke as shown in the Rotterdam Study and by others (Laurent et al. 2003, Mattace-Raso et al. 2006). Therefore, it could be valuable in preemptive diagnostic settings or in the follow-up of therapy, for example, in lipid-lowering attempts in diabetic children with already increased aortic wall stiffness (Urbina et al. 2010). In addition, the PWV is a better marker in hypercholesterolemic children than the intima-media thickness (Riggio et al. 2010).

Markl et al. have shown that the global aortic PWV can be estimated from 4D PC-MRI datasets. They were able to show the expected PWV changes between young volunteers (4.39 ± 0.32 m/s) and older subjects with cardiovascular disease (7.03 ± 0.24 m/s; p < 0.001) despite a suboptimal temporal resolution. They also analyzed the different quantitative approaches including two arrival time methods and cross-correlation (Markl et al. 2010).

With improved spatiotemporal resolution, the analysis of regional PWV changes could lead to an interesting application in clinical diagnostics. Especially locations prone to stenosing arteriosclerosis such as the aortic arch, the carotid arteries, and the renal arteries qualify for this assessment. The analysis of PWV could also be transferred to other vascular territories such as the pulmonary arteries in pulmonary artery hypertension or altered pulmonary flow physiology. Thereby, entirely new information could lead to the detection of interesting additional biomarkers for the severity of disease (Lakoma et al. 2010).

20.3.2 Wall Shear Stress and Oscillatory Shear Index

The term "stress" refers to the force per unit area. There are two main forces applied to the arterial wall by blood. Blood pressure is a force that is directed perpendicular to the wall and is responsible for the cyclical distension of the vessel wall. As the blood pressure changes during the cardiac cycle, the vessel wall extends and then distends. The second force is called the "WSS." This is the force acting on the inner lining of the artery wall, the endothelium, and is a frictional force resulting from the viscous drag of blood on the wall. The WSS changes through the cardiac cycle in a cyclical manner. Typically, WSS has values of 0–20 Pa in a healthy artery, compared with the 10,000 Pa of blood pressure. The unit "Pa" stands for "Pascal."

WSS acts locally by inducing the secretion of nitric oxide (NO) and prostacyclin resulting in vessel dilatation, hindrance of platelet activation, and attenuation of smooth muscle cell proliferation. It also directly interacts with endothelial gene expression and maintains an atheroprotective phenotype in high and elevated WSS levels. In conditions where WSS is low and the directional changes of the WSS, expressed as oscillatory shear index (OSI), are high, atherosclerotic lesions develop (Malek et al. 1999).

The stretching of the artery during the cardiac cycle induces stress or tension within the artery wall. In healthy people, the increased stress helps the arterial wall to return to its "resting" position in the same way that a spring will return to its resting position once the stretching force is released. The stress field produced within the arterial wall will be referred to here as "tissue stress." Like blood pressure, tissue stress is a large force in comparison with WSS. In general, for materials, both biological and nonbiological, when the tissue stress exceeds the tissue strength, the tissue will break. It is the high tissue stress caused by blood pressure that is responsible for the eventual rupture of both atherosclerotic plaque and aneurysms, not WSS, which is a relatively tiny force incapable of producing the stresses required for rupture.

The absence of noninvasive methods to measure WSS and OSI in humans is what has largely prevented this research in the past from moving forward. Only a few studies using 2D PC-MRI have aimed at analyzing these forces in vivo (Frayne and Rutt 1995, Tsuji et al. 2002). In principle, the data provided by PC imaging contain most information necessary for WSS computations. The vessel radius can be derived from the images of magnitude acquisition, whereas the blood flow information from the images of phase acquisition. Only the blood flow viscosity has to be accepted as a constant with relative narrow inter- and intraindividual changes.

Four-dimensional PC-MRI may catalyze novel in vivo WSS research and transfer of technology to preemptive medicine. Limited by its spatial resolution and the assumption that flow at the border to the vessel wall is known, 4D PC-MRI allows one to make consistent WSS and OSI estimations (Stalder et al. 2008). In comparison to CFD that relies on numerous assumptions, 4D PC-MRI is mainly restricted in its estimations by a spatial resolution between 1.4 and 2.4 mm. While PC-MRI currently underestimates the WSS from CFD by an order of magnitude, it can provide such data in vivo and within a diagnostic time frame suitable for most patients and disease-related questions.

The calculation of WSS and OSI could also provide vital information in nonatherosclerotic diseases. Conditions such as aneurysm formation, whether degenerative nature, associated with connective tissue disorders, or following surgery such as in coarctation, have been speculated to be at least associated with if not triggered by hemodynamics. So far, CFD is the reliable approach to derive vessel wall parameters in such conditions. However, initial 4D PC-MRI results in coarctation and coarctation repair have been presented and are promising for future follow-up results (Frydrychowicz et al. 2008a,b, 2009).

20.3.3   Extracranial Carotid Disease

The carotid arteries play a key role in the timely development of atherosclerosis, the principal cause of heart attack, stroke, and gangrene of the extremities.

Atherosclerosis is a disease of large arteries in which gradual thickening of the vessel wall occurs, and localized plaques arise. However, despite the importance of global risk factors, atheroma is essentially a focal disease. A large number of human autopsy and in vivo studies have shown that atheromatous lesions occur preferentially in regions having low shear stress, for example, along the outer wall of bifurcations and along the inner wall of bends (Caro et al. 1969, Glagov et al. 1988). Recent studies have also suggested relationships between WSS and intimal thickening in the affected vessels (Ku et al. 1985, Moore et al. 1994).

The results of this study show that there is a strong secondary, spiral flow at the carotid bifurcation, especially in the internal carotid artery (ICA) due to nonplanar features of the in vivo geometry.

The values for instantaneous WSS indicate that WSS patterns on the anterior walls are different from those on the posterior walls and that WSS may vary significantly in different regions of the same vessel. The asymmetry in WSS patterns in the common carotid artery (CCA) suggests the important influence of inlet velocity profiles on predicted WSS distributions. In the carotid sinus, WSS is permanently low throughout the cardiac cycle, supporting the hypothesis that low WSS may be a localizing factor for atherosclerosis, as the carotid sinus has been found to be prone to atherosclerotic plaques.

In atherosclerosis, initiation of disease has long been recognized as occurring at regions of low WSS (Ku et al. 1985). Early studies, on flow models and casts, noted the presence of low WSS in the bulb region of the carotid bifurcation (Moore et al. 1994, Jou et al. 1996, Vantyen et al. 1994) and an inverse relationship between intima–media thickness (IMT) and WSS (Siegel et al. 1995, 1997). In vivo studies also demonstrated that IMT increased in regions where the WSS is low (Steinman et al. 1996, 1997). Reviews of the role of WSS in disease development have emphasized the importance of low WSS in disease initiation and positive arterial remodeling whereby artery diameter increases with lumen preservation (Steinman and Rutt 1998, Weston et al. 1998). Once the plaque has grown to the point at which the lumen is diminished, the high WSS in the upstream region is then thought to play a key role. It was hypothesized that high WSS stimulates macrophage activity, leading to thinning of the fibrous cap, which is then at the risk of rupture through high tissue stress (Steinman and Rutt 1998).

20.3.4 Thoracic Aorta

Aneurysm of the proximal ascending aorta represents a life-threatening disease; if our understanding of blood flow patterns in and around the aneurysm were more complete, then one theoretically could create risk prediction models that could predict the likelihood of serious adverse events, including fatal aortic rupture and/or acute type A aortic dissection.

Several studies have demonstrated initial detailed information about local changes in hemodynamic parameters in patients with different aortic diseases as revealed in flow-sensitive 4D MRI investigations. These included patients with aortic aneurysms (Hope et al. 2007, Markl et al. 2004), aortic dissections, and partial thrombosis (Frydrychowicz et al. 2006), patients after aortic root replacement (Markl et al. 2005), after ascending–descending aortic bypass surgery (Frydrychowicz et al. 2007b) and with a persistent ductus arteriosus (Frydrychowicz et al. 2007a). The flow-sensitive 4D MRI data reveal a marked variety of pathological local blood flow patterns within the ascending aortic aneurysms that have been investigated. There are clear differences in pathological local

flow patterns, not just when compared to the aortic flow situation in healthy volunteers, but among the patients themselves as well.

Hope et al. (2007) recently reported on 4D MRI investigations in patients suffering from ascending aortic aneurysms. Their report describes how anatomical changes in ascending aortic aneurysms can skew normal flow patterns, changing physiologically observed retrograde and helical flow patterns. The vortices in patients with ascending aortic aneurysm were significantly larger in diameter (48.5%–53.1% of the lumen diameter in volunteers as opposed to 77.1%–99% of the same inpatients), and they also lasted longer.

Weigang et al. (2008) identified five different flow patterns in dilated ascending aortas. The most common pattern (54% of included patients) included initial systolic flow along the right anterior aortic wall and infolding before entering the transverse arch, in turn creating a large area of retrograde flow in the left coronary sinus area. This pattern appears as one prominent vortex in the ascending aorta.

Artificial aneurysm models indicate that hemodynamics and wall mechanics play an important role in aortic disease development (Vorp et al. 1998). In particular, the kinetic energy occurring during turbulence was found to have a significant effect on pressure distribution along aneurysmatic vessel walls, considered an important factor in aneurysm growth and rupture (Schafers et al. 2007). Several studies have included attempts to define mathematical models to predict aneurysm growth and rupture and different computer models for estimating vessel wall forces (Berguer et al. 2006). It was shown that wall shear–stress is one of the most important factors contributing to aneurysm rupture (Deplano et al. 2007). Changes in local mechanical forces can lead to acceleration and aggravation of pathological processes including atherosclerotic plaque development, loss of elastic fibers within the vessel wall (Prado and Rossi 2006, Prado et al. 2006), and, finally, wall dilatation and dissection.

Presumably, the individual aneurysm's morphology thus plays a role in generating different levels of mechanical forces on the vessel wall, thereby completing some kind of vicious cycle in aneurysm development and progression. In this context, others have demonstrated that asymmetric aneurysms are exposed to higher stress levels than those with more fusiform morphology (Scotti et al. 2005).

It remains unclear as to what extent and how exactly pathological flow effectively initiates or aggravates aneurysm development.

Chapter 20

Recently, some studies have investigated the in vivo flow patterns in the normal aortic root and sinuses of Valsalva by using MRI and to contrast them with flow patterns associated with valve-sparing aortic surgery with straight Dacron grafts (Kvitting et al. 2004).

The objective of the study of Markl et al. (2005) was to demonstrate that 3D MR velocity mapping is a reliable technique for the qualitative and quantitative assessment and visualization of blood flow in the ascending aorta and that this technique can portray flow features such as vortex formation in the sinuses of Valsalva or helical aortic flow in volunteers and patients who have undergone valve-sparing aortic root replacement with the T. David reimplantation method.

The observations of Markl et al (2005) demonstrate that time-resolved 3D velocity mapping is a useful tool for the analysis and visualization of blood flow characteristics in patients with valve-sparing aortic root replacement.

Bicuspid aortic valve (BAV) disease is the most common form of congenital heart disease, affecting 0.5%–2% of the population (Guntheroth 2008). It includes different morphological phenotypes and predisposes to aortic valvar pathology (stenosis, regurgitation, or both) and aortic aneurysms at different levels, even in children and young adults, irrespective of the severity of valvar dysfunction (Gurvitz et al. 2004). The pathogenesis of aortic dilatation in the presence of BAV disease is still controversial. Histopathologic changes such as cystic medial necrosis of the proximal aortic wall causing abnormal aortic distensibility and stiffness were identified in patients with BAV disease, not different from Marfan patients (Bonderman et al. 1999). A genetic basis accounting for both valve and wall defects was thus postulated (BAV syndrome).

Another possible explanation for aortic aneurysms in BAV patients is a pathophysiological phenomenon due to increased wall stress caused by abnormal blood flow in the aortic root through a stenotic BAV. Nonetheless, aortic dilation is noted also in patients with a functionally normal or regurgitant valve. One hypothesis is that the abnormal opening of the BAV, even if not stenotic or only mildly stenotic, may cause increased hemodynamic wall stress leading to aneurysm formation (Bauer et al. 2006).

Recently reported applications indicated altered WSS in the presence of BAV-related flow abnormalities (Poullis et al. 2008). This study showed that BAV disease was associated with altered aortic flow patterns (nested flow, helical flow, or retrograde flow) and elevated or asymmetrically expressed WSS along the circumference of the aorta wall.

The type B aortic dissection medical management can result in progressive dilation of the false lumen and poor long-term outcome. Recent studies using models of aortic dissection have suggested flow characteristics, such as stroke volume, velocity, and helicity, are related to aortic expansion. The aim of the study of Clough et al. (2012) was to assess whether 4D phase-contrast magnetic resonance imaging (4D PC-MRI) can accurately visualize and quantify flow characteristics in patients with aortic dissection and whether these features are related to the rate of aortic expansion. Computer simulations have shown that areas of the false lumen close to entry tears are associated with turbulent flow and high WSS, and these regions are prone to aneurysm formation. This study demonstrates the potential of 4D PC-MRI as a technique to identify patients with higher rates of aortic expansion.

Midulla et al. (2012) presented an imaging method combining MRI and CFD to obtain a patient-specific hemodynamic analysis of patients treated with thoracic endovascular aortic repair (TEVAR), to detect characteristics of locally altered flow, particularly of velocity and WSS, related to vascular lesions and stent configuration, and to improve the ability to identify endograft failure and to assist future endograft design, treatment planning, and eventually in identifying new patient-specific prognostic factors.

20.3.5 Abdominal Aortic Aneurysm

Abdominal aortic aneurysm (AAA) disease is characterized by a breakdown of the constituents of the arterial wall and, in most cases, the formation of thrombus within the lumen. Unlike atherosclerosis, aneurysm disease is caused by a breakdown of the middle and outer layers of the vessel wall. The expansion of the aneurysm cavity causes the formation of vortices and turbulent flow, leading to irregular WSS patterns (Frydrychowicz et al. 2008a). As WSS has a regulatory effect on inflammation, it will affect the initiation and evolution of AAA disease and, potentially, the formation of thrombus (Frydrychowicz et al. 2008b). There has been little research to quantify the effects of WSS on AAA growth in vivo, but an animal study demonstrated greater increase in diameter when WSS was low (Frydrychowicz et al. 2009). A model of aneurysm growth was proposed, which concentrated on the role of tissue stress, ignoring the role of WSS (Salonen and Salonen 1990).

These studies demonstrate that wall stresses are closely associated with the disease process, and therefore, their measurement may be useful in providing diagnostic indices directly related to the disease process.

Frauenfelder et al. (2006) have evaluated quantitatively and qualitatively the hemodynamic changes in AAAs after stent-graft placement based on multidetector CT angiography (MDCT-A) datasets using the possibilities of CFD. After stenting, the simulation shows a reduction of wall pressure and WSS and a more equal flow through both external iliac arteries after stenting. The postimplantation flow pattern is characterized by a reduction of turbulences. New areas of high pressure and shear stress appear at the stent bifurcation and docking area. CFD is a versatile and noninvasive tool to demonstrate changes of flow rate and flow pattern caused by stent-graft implantation. The desired effect and possible complications of a stent-graft implantation can be visualized. They also concluded that CFD improves our understanding of the local structural and fluid dynamic conditions for abdominal aortic stent placement.

References

Abdulnaser S. 2009. Computational fluid dynamics. Bookboon www.bookboon.com (accessed October 17, 2012).

Anderson John D Jr. 1995. *Computational Fluid Dynamics: The Basics with Applications*. McGraw-Hill Science, Singapore.

Bauer M, Siniawski H, Pasic M et al. 2006. Different hemodynamic stress of the ascending aorta wall in patients with bicuspid and tricuspid aortic valve. *J Card Surg* 21: 218–220.

Berguer R, Bull JL, Khanafer K. 2006. Refinements in mathematical models to predict aneurysm growth and rupture. *Ann N Y Acad Sci* 1085: 110–116.

Blazek J. 2001. *Computational Fluid Dynamics: Principles and Applications*. Elsevier Science Ltd, Amsterdam.

Bonderman D, Gharehbaghi-Schnell E, Wollenek G et al. 1999. Mechanisms underlying aortic dilatation in congenital aortic valve malformation. *Circulation* 99: 2138–2143.

Caro CG, Fitz-Gerald JM, Schroter RC. 1969. Arterial wall shear and distribution of early atheroma in man. *Nature* 223: 1159–1161.

Clough RE, Waltham M, Giese D et al. 2012. A new imaging method for assessment of aortic dissection using four-dimensional phase contrast magnetic resonance imaging. *J Vasc Surg* 55: 914–923.

Deplano V, Knapp Y, Bertrand E et al. 2007. Flow behavior in an asymmetric compliant experimental model for abdominal aortic aneurysm. *J Biomech* 40 (11): 2406–2413.

Frauenfelder T, Lotfey M, Boehm T et al. 2006. Computational fluid dynamics: Hemodynamic changes in abdominal aortic aneurysm after stent-graft implantation. *Cardiovasc Intervent Radiol* 29: 613–623.

Frayne R, Rutt BK. 1995. Measurement of fluid-shear rate by Fourier-encoded velocity imaging. *Magn Reson Med* 34 (3), 378–387.

Frydrychowicz A, Arnold R, Hirtler D et al. 2008a. Multidirectional flow analysis by cardiovascular magnetic resonance in aneurysm development following repair of aortic coarctation. *J Cardiovasc Magn Reson* 10 (1), 30.

Frydrychowicz A, Berger A, Russe MF et al. 2008b. Time-resolved magnetic resonance angiography and flow-sensitive 4-dimensional magnetic resonance imaging at 3 Tesla for blood flow and wall shear stress analysis. *J Thorac Cardiovasc Surg* 136 (2): 400–407.

Frydrychowicz A, Bley TA, Dittrich S et al. 2007a. Visualization of vascular hemodynamics in a a case of a large patent ductus arteriosus using flow sensitive 3D CMR at 3T. *J Cardiovasc Magn Reson* 9 (3), 585–587.

Frydrychowicz A, Francois CJ, Turski PA. 2011. Four-dimensional phase contrast magnetic resonance angiography: Potential clinical applications. *Eur J Radiol* 80: 24–35.

Frydrychowicz A, Hirtler D, Arnold R et al. 2009. Analysis of aortic hemodynamics after treatment for coarctation using flow-sensitive 4D MRI at 3T. *Proc Intl Soc Magn Reson Med* 17: 321.

Frydrychowicz A, Schlensak C, Stalder A et al. 2007b. Ascending-descending aortic bypass surgery in aortic arch coarctation: Four-dimensional magnetic resonance flow analysis. *J Thorac Cardiovasc Surg* 133 (1): 260–262.

Frydrychowicz A, Weigang E, Harloff A et al. 2006. Images in cardiovascular medicine. Time-resolved 3-dimensional magnetic resonance velocity mapping at 3T reveals drastic changes in flow patterns in a partially thrombosed aortic arch. *Circulation* 113 (11): e460–e461.

Glagov S, Zarins C, Giddens DP, Ku DN. 1988. Haemodynamics and atherosclerosis—Insights and perspectives gained from studies of human arteries. *Arch Pathol Lab Med* 112: 1018–1031.

Guntheroth WG. 2008. A critical review of the American College of Cardiology/American Heart Association practice guidelines on bicuspid aortic valve with dilated ascending aorta. *Am J Cardiol* 102: 107–110.

Gurvitz M, Chang RK, Drant S, Allada V. 2004. Frequency of aortic root dilation in children with a bicuspid aortic valve. *Am J Cardiol* 94: 1337–1340.

Hirsch, C. 1988. *Numerical Computation of Internal and External Flows*, Vols. 1 and 2. John Wiley & Sons, New York.

Hope TA, Markl M, Wigstrom L et al. 2007. Comparison of flow patterns in ascending aortic aneurysms and volunteers using four-dimensional magnetic resonance velocity mapping. *J Magn Reson Imaging* 26 (6): 1471–1479.

Jou LD, Vantyen R, Berger A, Saloner D. 1996. Calculation of the magnetization distribution for fluid-flow in curved vessels. *Magn Reson Med* 35: 577–584.

Ku DN, Giddens DP, Zarins CK, Glagov S. 1985. Pulsatile flow and atherosclerosis in the human carotid bifurcation—Positive correlation between plaque location and oscillating shear stress. *Atherosclerosis* 5: 293–302.

Kutler, P. 1975. Computation of three-dimensional, inviscid supersonic flows. In *Progress in Numerical Fluid Dynamics*, ed. H.J. Wirz, pp. 293–374. Springer-Verlag, Berlin, Germany.

Chapter 20

Kvitting JP, Ebbers T, Wigstrom L et al. 2004. Flow patterns in the aortic root and the aorta studied with time-resolved, 3-dimensional, phase-contrast magnetic resonance imaging: Implications for aortic valve-sparing surgery. *J Thorac Cardiovasc Surg* 127 (6): 1602–1607.

Lakoma A, Tuite D, Sheehan J, Weale P, Carr JC. 2010. Measurement of pulmonary circulation parameters using time-resolved MR angiography in patients after Ross procedure. *Am J Roentgenol* 194 (4): 912–919.

Laurent S, Katsahian S, Fassot C et al. 2003. Aortic stiffness is an independent predictor of fatal stroke in essential hypertension. *Stroke* 34 (5): 1203–1206.

Malek AM, Alper SL, Izumo S. 1999. Hemodynamic shear stress and its role in atherosclerosis. *JAMA* 282 (21): 2035–2042.

Markl M, Draney MT, Hope MD et al. 2004. Time-resolved 3-dimensional velocity mapping in the thoracic aorta: Visualization of 3-directional blood flow patterns in healthy volunteers and patients. *J Comput Assist Tomogr* 28 (4): 459–468.

Markl M, Draney MT, Miller DC et al. 2005. Time-resolved three-dimensional magnetic resonance velocity mapping of aortic flow in healthy volunteers and patients after valve-sparing aortic root replacement. *J Thorac Cardiovasc Surg* 130: 456–463.

Markl M, Wallis W, Brendecke S, Simon J, Frydrychowicz A, Harloff A. 2010. Estimation of global aortic pulse wave velocity by flow-sensitive 4D MRI. *Magn Reson Med* 63 (6): 1575–1582.

Mattace-Raso FU, van der Cammen TJ, Hofman A et al. 2006. Arterial stiffness and risk of coronary heart disease and stroke: The Rotterdam study. *Circulation* 113 (5): 657–663.

Midulla M, Moreno R, Baali A et al. 2012. Haemodynamic imaging of thoracic stent-grafts by computational fluid dynamics (CFD): Presentation of a patient-specific method combining magnetic resonance imaging and numerical simulations. *Eur Radiol* 22: 2094–2102.

Moore JE, Xu C, Glagov S et al. 1994. Fluid wall shear stress measurements in a model of the human abdominal aortic oscillatory behavior and relationship to atherosclerosis. *Atherosclerosis* 110: 225–240.

Poullis MP, Warwick R, Oo A, Poole RJ. 2008. Ascending aortic curvature as an independent risk factor for type A dissection, and ascending aortic aneurysm formation: A mathematical model. *Eur J Cardiothorac Surg* 33: 995–1001.

Prado CM, Ramos SG, Alves-Filho JC et al. 2006. Turbulent flow/low wall shear stress and stretch differentially affect aorta remodeling in rats. *J Hypertens* 24 (3): 503–515.

Prado CM, Rossi MA. 2006. Circumferential wall tension due to hypertension plays a pivotal role in aorta remodeling. *Int J Exp Pathol* 87 (6): 425–436.

Riggio S, Mandraffino G, Sardo MA et al. 2010. Pulse wave velocity and augmentation index, but not intima-media thickness, are early indicators of vascular damage in hypercholesterolemic children. *Eur J Clin Invest* 40 (3): 250–257.

Salonen R, Salonen JT. 1990. Progression of carotid atherosclerosis and its determinants: A population-based ultrasonography study. *Atherosclerosis* 81: 33–40.

Schafers HJ, Aicher D, Langer F et al. 2007. Preservation of the bicuspid aortic valve. *Ann Thorac Surg* 83 (2): S740–S745.

Scotti CM, Shkolnik AD, Muluk SC et al. 2005. Fluid-structure interaction in abdominal aortic aneurysm: Effects of asymmetry and wall thickness. *Biomed Eng Online* 4: 64.

Siegel JM, Oshinski JN, Pettigrew RI, Ku DN. 1995. Comparison of phantom and computer-simulated MR-images of flow in a convergent geometry—Implications for improved 2-dimensional MR-angiography. *J Magn Reson Imaging* 5: 677–683.

Siegel JM, Oshinski JN, Pettigrew RI, Ku DN. 1997. Computational simulation of turbulent signal loss in 2D time-of-flight magnetic resonance angiograms. *Magn Resonan Med* 37: 609–614.

Stalder AF, Russe MF, Frydrychowicz A, Bock J, Hennig J, Markl M. 2008. Quantitative 2D and 3D phase contrast MRI: Optimized analysis of blood flow and vessel wall parameters. *Magn Reson Med* 60 (5): 1218–1231.

Steinman DA, Ethier CR, Rutt BK. 1997. Combined analysis of spatial and velocity displacement artifacts in phase contrast measurements of complex flows. *J Magn Reson Imaging* 7: 339–346.

Steinman DA, Frayne R, Zhang XD, Rutt BK, Ethier CR. 1996. MR measurement and numerical-simulation of steady flow in an end-to side anastomosis model. *J Biomech* 29: 537–542.

Steinman DA, Rutt BK. 1998. On the nature and reduction of plaque-mimicking flow artifacts in black blood MRI of the carotid bifurcation. *Magn Reson Med* 39: 635–641.

Tsuji T, Suzuki J, Shimamoto R et al. 2002. Vector analysis of the wall shear rate at the human aortoiliac bifurcation using cine MR velocity mapping. *Am J Roentgenol* 178 (4): 995–999.

Urbina EM, Wadwa RP, Davis C et al. 2010. Prevalence of increased arterial stiffness in children with type 1 diabetes mellitus differs by measurement site and sex: The SEARCH for Diabetes in Youth Study. *J Pediatr* 156 (5): 731–737.

Vantyen R, Saloner D, Jou LD, Berger S. 1994. MR imaging of flow through tortuous vessels—A numerical simulation. *Magn Reson Med* 31: 184–195.

Vorp DA, Trachtenberg JD, Webster MW et al. 1998. Arterial hemodynamics and wall mechanics. *Semin Vasc Surg* 11 (3): 169–180.

Weigang E, Karl FA, Beyersdorf F et al. 2008. Flow sensitive four-dimensional magnetic resonance imaging: Flow patterns in ascending aortic aneurysms. *Eur. J Cardiothor Surg* 34: 11–16.

Wendt JF (ed.). 2009. *Computational Fluid Dynamics*, 3rd edn. Springer-Verlag, Berlin, Germany.

Wesseling P. 2001. *Principles of Computational Fluid Dynamics*. Springer series in computational mathematics, No. 29; Springer-Verlag, Berlin, Germany.

Weston SJ, Wood NB, Tabor G, Gosman AD, Firmin D. 1998. Combined MRI and CFD analysis of fully developed steady and pulsatile laminar flow through a bend. *J Magn Reson Imaging* 8: 1158–1171.

21. Computational Fluid Dynamics
Current Techniques and Future Perspectives

Juan R. Cebral

21.1 Introduction

Computational fluid dynamics (CFD) is increasingly being used to investigate the role of hemodynamics in a variety of vascular diseases and to evaluate and design medical devices and treatments. When combined with medical imaging, computational models can be made patient specific and therefore can be used in clinical studies. The last decade has seen a sharp increase in the use of image-based CFD to investigate clinical questions. The primary reason is that these models allow the connection of patient-specific hemodynamic conditions

Chapter 21

to clinical events or observations, thus allowing researchers to formulate new hypotheses and test them through computational experiments.

Cerebral aneurysms is one of the fields where the use of image-based CFD is growing very fast. The reason is perhaps a reflection of the pressing need in this field to understand the mechanisms underlying the formation, growth, and eventual rupture or stabilization of cerebral aneurysms. This understanding is the key to identify conditions that predispose aneurysms for rupture and thus enable an objective assessment of their natural rupture risk, as well as to design new devices and therapies that directly attack the factors involved in their evolution. Image-based CFD provides information about the in vivo hemodynamics environment in aneurysms on a patient-specific basis. This is important because hemodynamics is thought to play a fundamental role in the mechanisms governing aneurysm development and evolution (Sforza et al., 2009). The line of thought is that biological processes in the aneurysm wall in response to abnormal hemodynamic

loads (in particular, wall shear stress [WSS]) tend to degenerate the wall structure, resulting in a weakening of the wall and subsequent enlargement of the aneurysm until it either stabilizes or ruptures (when wall stress exceeds wall strength). However, the exact hemodynamic conditions and their link to the mechanobiological responses at the wall that drive aneurysm initiation and progression are still poorly understood. Image-based CFD is a promising approach to study these and other clinically relevant questions.

The purpose of this chapter is to provide an overview of the methods, assumptions, approximations, modeling choices, and decisions faced by researchers using image-based CFD, with the goal of guiding and helping them make the most appropriate and convenient decisions when selecting imaging modalities, mathematical models, numerical approaches, and modeling software. Although we focus our discussion around cerebral aneurysms, the issues considered are encountered in most applications of CFD for vascular flows.

21.2 Anatomy Modeling

The first choice the modeler faces when constructing blood flow models is whether to use idealized or realistic vascular geometries.

Idealized geometrical models are constructed by combining simple 3D shapes such as cylinders and spheres. This is typically done by employing CAD software to draw the geometry in 3D, or by fusing triangulated data of the simpler shapes (Cebral et al., 2001, Cebral and Löhner, 2005a). The main characteristic of this type of models is their geometrical simplicity and symmetry (see Figure 21.1). Models with idealized geometries are useful to perform controlled parametric studies of the effects of different geometrical characteristics (e.g., arterial curvature, tapering, bifurcation angles, etc.) separately. For many years, computational studies have been performed on idealized geometries of cerebral aneurysms or approximations of a specific patient geometry (Liepsch et al., 1987, Steiger et al., 1987, Liou and Liou, 1999). These experiments have greatly influenced thinking about the mechanisms of aneurysm development as researchers have attempted to make generalized statements about aneurysmal hemodynamics from these simplified models. While studies based on idealized geometries have characterized the complexity of intra-aneurysmal hemodynamic patterns, they have had significant limitations in connecting the hemodynamic variables studied to

clinical events. In other words, results based on idealized geometries are difficult to extrapolate to an individual patient's anatomy or used in population studies. In addition, it has been shown that idealized models can be very sensitive to small geometrical deviations that break their high degree of symmetry (Ford et al., 2008).

On the other hand, the geometry of human cerebral arteries and aneurysms can be quite complex and variable among subjects (see Figure 21.2). Typical arteries follow curved, nonplanar, and sometimes tortuous paths; have nonuniform or noncircular cross sections; exhibit some degree of tapering; and bifurcate in a nonsymmetrical manner. Likewise, aneurysms can be roughly spherical or elongated along a single or multiple axes, they can have smooth or irregular surfaces, they can be single lobulated or harbor secondary lobulations or blebs (also called daughter aneurysms), and can have small arteries branching off their necks or dome. The geometric complexity of arteries and aneurysms has a large impact on the distributions of hemodynamic variables of interest such as WSS, making them quite different from the ones obtained with idealized models. For this reason, many researchers have focused on realistic or patient-specific vascular geometries. Although current imaging modalities are limited for accurate in vivo quantification of cerebral

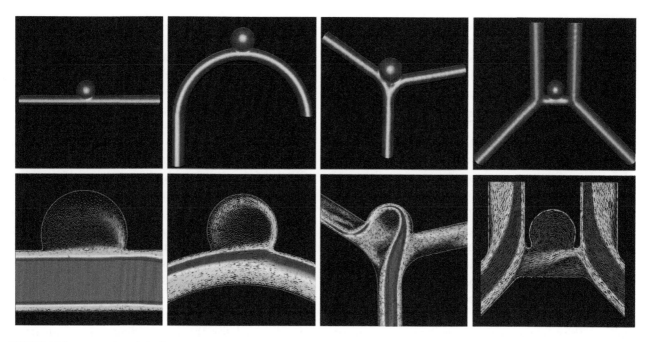

FIGURE 21.1   Examples of idealized models of cerebral aneurysms (top row) and corresponding velocity distributions on symmetry plane (bottom row). From left to right: aneurysm in straight vessel, aneurysm in curved vessel, aneurysm in arterial bifurcation, and aneurysm at the anterior communicating artery complex.

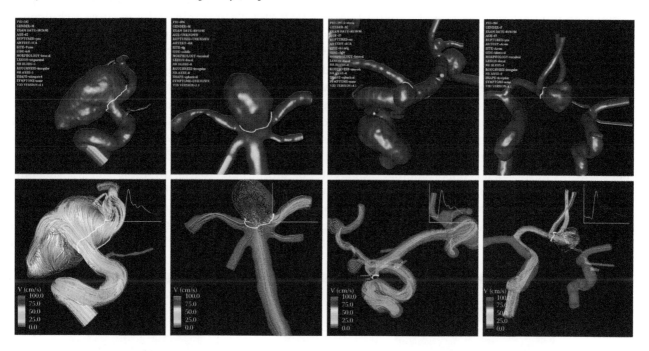

FIGURE 21.2   Examples of patient-specific models of cerebral aneurysms (top row) and corresponding flow patterns (bottom row). From left to right: lateral aneurysm in the internal carotid artery; bifurcation aneurysm at the tip of the basilar artery; multiple aneurysms in the internal carotid artery, middle cerebral artery and anterior communicating artery; and aneurysm in the anterior communicating artery complex.

blood flow patterns, pressure gradients, and WSS distributions, the geometrical shape of the blood vessels and aneurysms can be accurately reconstructed from anatomical images. This has allowed the construction of subject-specific computational models of cerebral aneurysms from medical images (Butty et al., 2002, Jou et al., 2003, Steinman et al., 2003, Cebral et al., 2004, Hassan et al., 2004, Shojima et al., 2004, Cebral et al., 2005), and has led to the development of the field of image-based CFD (Taylor et al., 1998, Taylor and

Steinman, 2010). This approach allows the connection of hemodynamic characteristics and clinical events or observations such as aneurysm growth or rupture, and thus enables statistical studies aiming at identifying hemodynamic conditions associated with rupture risk and understanding the corresponding mechanisms underlying the development and evolution of cerebral aneurysms (Cebral and Putman, 2008).

21.3   Vascular Imaging

The construction of patient-specific vascular models from medical images is not trivial and will in general depend on the imaging modality used to depict the blood vessels (see Figure 21.3).

The gold standard for imaging and quantifying blood vessel geometries is x-ray digital subtraction angiography (DSA), which is commonly used to diagnose and treat cerebral aneurysms. Most modern x-ray angiography systems are capable of acquiring a sequence of images from different viewpoints as the x-ray emitter and detector mounted on a C-arm rotate around the patient. The sequence of projection images is then combined to create a 3D volume image composed of isotropic voxels called a 3D rotational angiography (3DRA). X-ray angiography is an invasive technique, because it requires an intra-arterial injection of a contrast agent while imaging the blood vessels with x-rays. The risks for the patient include exposure to ionizing radiation (x-rays), arterial puncture, and catheterization, and injection of contrast material (which could adversely affect patients with kidney problems or allergies to the contrast agent). However, this modality provides the highest resolution and contrast between blood vessels and surrounding tissue that is digitally subtracted. For modeling purposes, it is important that the parent artery and aneurysm be completely filled with contrast during the acquisition of the entire sequence of images. This means that the contrast injection needs to be long enough. Good results are obtained with injections of

4 cc/s for a total of 6 s while acquiring 120 images spanning an 180° rotation. Besides possible incomplete filling of the vessels, other potential problems associated with this imaging modality include low contrast doses because of patient intolerance; noise/reflections from metallic implants such as from previous treatments with clips, coils, or stents; and presence of intra-aneurysmal thrombus, which will appear as a dark bubble inside the aneurysm (this modality only images the injected contrast).

The other two imaging modalities that are commonly used for the initial diagnosis of aneurysms or for their follow-up after treatment or conservative observation are computed tomography angiography (CTA) and magnetic resonance angiography (MRA). CTA is a minimally invasive technique that employs x-rays and intravenous contrast injection. Contrast-enhanced MRA uses intravenous contrast and nonionizing radiation, while time-of-flight MRA is noninvasive (no contrast injection) but suffers from artifacts related to signal loss in regions of low flow velocities or disturbed flows. Both CTA and MRA techniques provide 3D images that typically have lower resolution than 3DRA images and are composed of anisotropic voxels. In these images, anatomical structures other than the blood vessels are also visible, that is, not subtracted. This can be an advantage if one is interested in studying the interaction of the aneurysm with other structures of the perianeurysmal

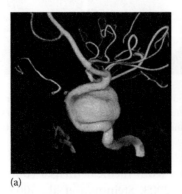 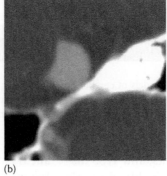 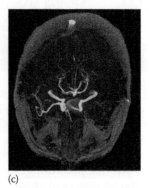

(a)   (b)   (c)

FIGURE 21.3   (a) 3DRA image of a giant aneurysm showing incomplete filling of the proximal parent artery, (b) slice of a CTA image of an aneurysm in contact with bone, (c) time-of-flight MRA of a large intracranial aneurysm showing signal loss within the aneurysm sac due to slow and disturbed flow.

environment such as bones or visualization of intra-aneurysmal thrombi. However, the presence of bright structures such as bones or calcifications complicates vessel reconstruction, especially if these structures are in direct contact with the vessel wall. Additionally, nonisotropic voxels with large aspect ratios may result in inaccurate vessel reconstructions where the vessel runs perpendicularly to the elongated voxels, since in these regions, the vessel cross section will contain only a few voxels.

21.4   Arterial Network Modeling

The second choice the modeler faces is what portion of the arterial network to include in the models.

The vascular architecture of the brain is very particular, because it needs to maintain a constant blood flow, since neural tissues are extremely sensitive to ischemia. For instance, if the blood flow to the brain is stopped instantaneously in humans, unconsciousness occurs in six or seven seconds and certain portions of the cerebral cortex are irreversibly damaged after about four minutes (de Groot, 1991). Consequently, the vascular system of the brain functions to ensure a fairly constant perfusion through a complex system of autoregulation and collateral circulation. The brain vasculature is anatomically different from that of other organs, in which the main arterial branches first penetrate to a more or less central position in the organ, and then supply the blood via terminal branches that extend from the center to the periphery (Martinez, 1982). In contrast, the oxygenated blood is transported to the brain by the shortest route, through vessels without major ramifications: the carotid and vertebral systems. These major arteries are joined to one another at the base of the brain in the form of an arterial polygon called the circle of Willis (Gray, 1901) (see Figure 21.4). From this basal origin, the cerebral arteries are distributed over the brain surface, from which they ramify penetrating perpendicularly to the brain tissue. Each major artery supplies a certain territory. Normally, there is little exchange of blood between the main arteries through the slender communicating vessels. The arterial circle provides alternative routes when one of the major arteries leading to it becomes occluded.

Cerebral aneurysms are most frequently located near arterial bifurcations in the circle of Willis (Stehbens, 1972, Foutrakis et al., 1999, Weir, 2002). Therefore, it is important to include in the CFD models all the arterial branches that will have a significant effect on the intra-aneurysmal hemodynamics. This is particularly important for branches located proximally (upstream) of the aneurysm, since they could influence the way the blood flows into the aneurysm. For instance, the ophthalmic artery is very likely to influence the hemodynamics of periophthalmic or superior hypophyseal aneurysms, the posterior communicating artery (Pcom) is likely to influence aneurysms located near the Pcom and anterior choroidal

(a)

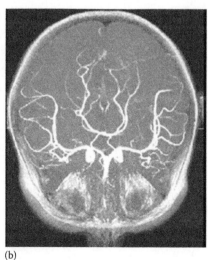

(b)

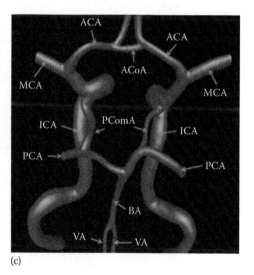

(c)

FIGURE 21.4   (a) MRA of the neck showing the carotid and vertebral arteries from their origin in the aortic arch to the circle of Willis, (b) MRA of the cerebral arteries at the level of the circle of Willis, (c) reconstruction of the circle of Willis and main arteries (ICA, internal carotid artery; MCA, middle cerebral artery; ACA, anterior cerebral artery; PCA, posterior cerebral artery; VA, vertebral artery; BA, basilar artery; ACoA, anterior communicating artery; PComA, posterior communicating artery).

Chapter 21

arteries and at the internal carotid artery (ICA) terminus or bifurcation, and so on. Similarly, aneurysms located at the anterior communicating artery (Acom) could in principle receive blood from the A1 segments of the left and right anterior cerebral arteries (ACA), and therefore should include these two feeding vessels (unless of course one of them is inexistent or hypoplastic as it frequently happens). Building models of bilateral Acom aneurysms from 3DRA can be complicated, because these images are typically acquired during contrast injection into one of the two internal carotid arteries, and therefore, they do not properly depict the contralateral feeding artery. A possible solution would be to acquire 3DRA images in such a way that both ICAs are filled with contrast, for instance, by injecting in the aortic arch, or by simultaneously injecting on both ICAs with two catheters. Another alternative is to fuse 3DRA images

acquired separately by injection into the left and right ICAs, or to build independent vascular models from these images and then fuse these models (see Figure 21.5). This approach requires alignment of either the 3DRA images or the vascular models reconstructed from each of them (Castro et al., 2006b). This alignment has to be done carefully, because most image-to-image or surface-to-surface registration algorithms will fail, since typically, there is little overlap between the left and right portions of the vasculature depicted on both images.

In addition to deciding what arterial branches to include in the vascular model, the modeler has to select the location where each branch will be cut or truncated and boundary conditions will be applied. Typically at the model inlet(s), flow boundary conditions are applied by prescribing a simplified velocity profile (e.g., uniform velocity profile, Womersley

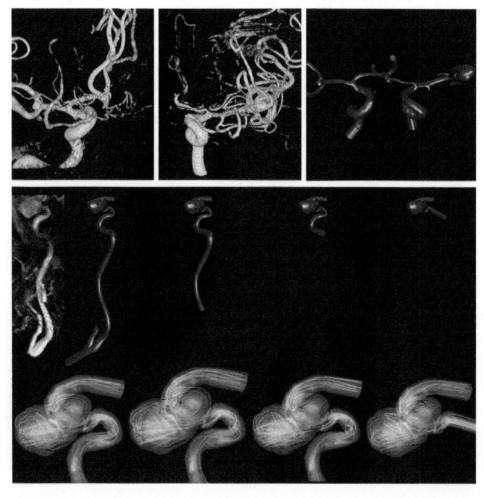

FIGURE 21.5 Top: example of an aneurysm in the anterior communicating artery imaged with bilateral 3DRA by selective injection in the left and right internal carotid arteries and corresponding computational model. Bottom: example of the effects of parent artery truncation at different distances from the aneurysm neck on intra-aneurysmal blood flow pattern.

profile, etc.—see section on physiologic modeling). Most profiles consist only of axial velocity components, that is, they neglect secondary or rotational velocities. However, the swirling of blood flow in the parent artery can significantly affect the manner and the location where the blood enters the aneurysm, which in turn determines the overall intra-aneurysmal flow structure and the corresponding WSS distribution (see Figure 21.5). Therefore, it is important to capture the correct swirling of flow upstream of the aneurysm. This requires including enough of the proximal parent artery in order to allow for the development of secondary flows as blood travels along the curved segments of the parent artery before entering the aneurysm (Castro et al., 2006a). Exactly how much of the parent artery needs to be considered has not been quantitatively established, and it is likely that it will depend on the geometry and curving characteristics of the parent artery. Sensitivity studies (Castro et al., 2006a) suggest that a few turns of a tortuous parent artery would be sufficient to develop swirling flows and that if the parent artery is fairly straight upstream of the aneurysm, it may be adequate to truncate it closer to the aneurysm, since in this case the velocity profile would be closer to fully developed. Major arterial bifurcations or junctions located proximally and close to the aneurysm need also to be considered in order to properly represent the skewed velocity profiles produced by bifurcations or the mixing of flow produced by the confluence of flow streams at arterial junctions. For example, it may be important to include the ICA bifurcation for adequate modeling of the hemodynamics in middle cerebral artery (MCA) bifurcation aneurysms. Similarly, it may be necessary to include the junction of the left and right vertebral arteries (VA) in models of aneurysms located along the basilar artery (BA) trunk or tip. Finally, the truncation of afferent vessels (downstream of the aneurysm) is likely to have less influence on the intra-aneurysmal flow structure. However, major bifurcations near the aneurysm may affect the way the blood flows out of the aneurysm and into each of the bifurcating branches.

21.5 Modeling the Vascular Geometry

The construction of patient-specific anatomical models from medical images is far from trivial due to the limited image resolution (and possible anisotropy), the presence of noise in these images, and possibly the contact of vessel walls with other bright structures. Accurate computational modeling requires the following: (1) a water-tight surface representation of the vascular structures, that is, no holes, gaps, or intersecting triangles; (2) the anatomical model must have the correct topology of the vessel network, that is, no self-intersecting or disconnected vessels; and (3) the geometrical model must accurately represent the shape of the blood vessels and aneurysms, that is, the surface of the geometrical model must coincide with the blood vessel boundaries. Several strategies have been used to construct patient-specific vascular models from anatomical images (Beutel and Sonka, 2000) (see Figure 21.6). These approaches can be classified into three main categories: (1) intensity-based vessel segmentation techniques, (2) geometry-based or deformable model techniques, and (3) partial differential equation (PDE)-based techniques or level set methods. In the first approach, the image intensity levels are used to identify the vascular structures using methods such as thresholding, region growing, clustering, etc. In the second approach, an initial surface triangulation representing the vascular structures of interest is constructed, for instance, by extracting a gray level isosurface. This surface is then allowed to deform under internal elastic forces and external forces computed from the image intensity and its gradients. In contrast to the other approaches, deformable models have the capacity of preserving the topology of the original surface if desired, that is, no leaking into undesired branches or other anatomical structures. This could be an advantage in problematic cases with vessel walls in contact with bone structures, for example. In the third approach, the vessel surface is implicitly represented as the zero-level isosurface of a signed distance map, which is obtained by solving a partial differential equation in the spatial domain of the anatomical image. The various methods in this last category differ in the PDE used and the method employed to find its numerical solution. For a more in-depth review of vessel reconstruction techniques, the reader is referred to Kirbas and Quek (2004). Many of these techniques have been implemented and are available in public domain software such as itksnap (http://www.itksnap.org) or vmtk (http://www.vmtk.org) as well as commercial packages like Mimics (http://www.materialise.com/mimics) or Amira (http://www.amira.com). An important aspect that needs to be considered when selecting the appropriate method or software to

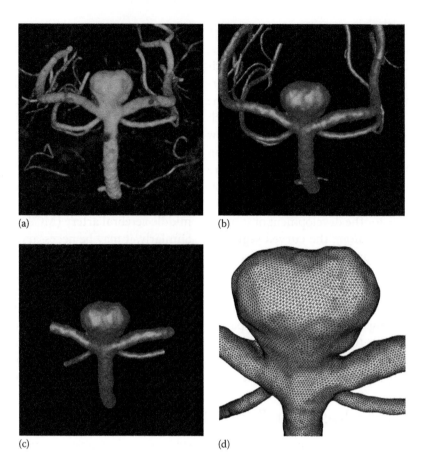

FIGURE 21.6 Example of model construction from 3DRA image. (a) volume-rendered 3DRA image, (b) segmented vasculature, (c) final vascular model, and (d) finite element mesh.

build vascular models is that the reconstructed models should be independent of threshold values or other parameters used by the reconstruction algorithms.

The numerical solution of the governing equations (see below) requires the generation of a computational grid in order to discretize the flow fields and differential operators. Grid generation imposes a number of requirements on the vascular model used to specify the computational domain: for example, the surface triangulation should contain no holes, no overlapping or intersecting triangles, no topological defects (e.g., edges shared by more than two triangles), and should be fairly smooth. Therefore, before proceeding to grid generation, the vascular model is typically further processed to satisfy these requirements. In particular, the surface is smoothed to remove high-frequency noise (Taubin, 1995); the triangulation is optimized using edge collapses and diagonal swaps to improve the quality of the triangles (Cebral and Löhner, 2001); holes are closed or triangulated; intersecting triangles, duplicated points and triangles, and topological defects are repaired (Cebral et al., 2005); and finally, the surface

model is trimmed or cut at the desired locations where boundary conditions will be applied, typically perpendicularly to the vessel axis. These operations can be applied using, for example, freeware software packages such as ReMesh (http://remesh.sourceforge.net). Next, a computational grid that fills the volume inside the vascular model is generated. This can be done with either freely available software such as Netgen (http://sourceforge.net/projects/netgen-mesher) or commercially available grid generators that typically are provided together with commercial flow solvers such as Fluent (http://www.ansys.com) or CFX (http://www.ansys.com). Some grid generators work only with computational domains defined via analytical surfaces such as NURBS, not triangulations. In such a case, the triangulated vascular model needs to be converted to a collection of analytical surface patches. This is typically done using CAD packages. Other grid generators operate directly on a triangulated surface model and therefore do not require this step. These are some of the considerations that one must take into account when selecting the grid generation software.

21.6   Blood Flow Modeling

The next choice faced by the modeler is the selection of the physical and mathematical fluid models to represent the dynamics of the blood flow, and the numerical approach or scheme to solve the corresponding equations.

21.6.1   Microscopic Models

Microscopic hemodynamic models consider blood as an ensemble of interacting discrete objects or particles. Discrete particle methods have been used to model blood flows in capillaries (Dzwinel et al., 2003), and particle transport and aggregation processes in blood flows (Federspiel and Popel, 1986, Boryczko et al., 2004). Essentially these methods solve the Newton's equations for a system of interacting discrete particles representing the different components of blood. Different interparticle force terms are included in order to model the interactions between each set of particles and between the particles and the vessel walls. These methods require the use of a large number of particles, and thus are commonly limited to small vessels.

Lattice Boltzmann methods (LBMs) can also be considered microscopic or mesoscopic methods. These techniques are based on the Boltzmann equation, which describes the evolution of a particle probability density function (Chen and Doolen, 1998):

$$\frac{\partial f}{\partial t} + v_i \frac{\partial f}{\partial x_i} + F_i \frac{\partial f}{\partial v_i} = \Omega \tag{21.1}$$

where
$f(x, v, t)$ is the particle probability density
v is the velocity
F is the external force
Ω is a collision integral

The LBM consists in discretizing this equation by limiting space to a lattice and the velocity space to a discrete set of velocity directions (Succi et al., 1990). Macroscopic variables such as the density and momentum of the fluid flow are then obtained from the moments of the probability density function:

$$\rho = \sum_i f_i, \quad \rho\mathbf{u} = \sum_i f_i \mathbf{v}_i \tag{21.2}$$

where the summation is taken over the lattice points. LBMs have been used to model blood flows in cerebral aneurysms (Hirabayashi et al., 2003, Beronov and Durst, 2005), to study the non-Newtonian properties of blood (Sun and Munn, 2005, Dupin et al., 2006), and to study cell wall interactions (Sun et al., 2003). It has been argued that the main advantages of these techniques are that they are naturally suited for incorporating microscopic effects and interactions, and that the resulting algorithms are fast and easy to parallelize. These methods also have several limitations. For instance, LBM does not guarantee a divergence-free velocity field for incompressible flows. In a sense, these methods are similar to the pseudocompressibility methods used to solve the incompressible Navier–Stokes equations.

21.6.2   Macroscopic Models

Most hemodynamic models are based on the continuum hypothesis and consider blood as an incompressible fluid (Fung, 1984, Mazumdar, 1992). Under these assumptions, the most general governing equations are the unsteady 3D Navier–Stokes equations for an incompressible fluid (Kundu and Cohen, 2004):

$$\rho\left(\frac{\partial \mathbf{u}}{\partial t} + \mathbf{u} \cdot \nabla \mathbf{u}\right) = -\nabla p + \nabla \cdot \tau + \mathbf{F} \tag{21.3}$$

$$\nabla \cdot \mathbf{u} = 0 \tag{21.4}$$

where
$\mathbf{u}$ is the velocity field
ρ is the density
p is the pressure
τ is the deviatoric stress tensor
$\mathbf{F}$ represents any externally applied body forces such as gravity

For Newtonian fluids (see next section), the deviatoric stress tensor can be written as

$$\tau_{ij} = 2\mu\varepsilon_{ij} \tag{21.5}$$

where μ is the viscosity, and the strain rate tensor is defined as

$$\varepsilon_{ij} = \frac{1}{2}\left(\frac{\partial u_i}{\partial x_j} + \frac{\partial u_j}{\partial x_i}\right) \tag{21.6}$$

Chapter 21

These equations have been used in numerous computational hemodynamics models in a wide variety of applications (Steinman and Taylor, 2005). Two-dimensional versions of these equations have also been used in studies of cerebral blood flows (Ferrandez et al., 2000, Ferrandez et al., 2002). However, these 2D models commonly neglect secondary 3D flow structures generated by the arterial curvature, and therefore the results are difficult to extrapolate to the in vivo conditions or to generalize to entire patient populations or to extrapolate to a patient's individual anatomy.

21.6.3 Numerical Schemes

The next set of choices faced by the modeler has to do with the selection of the computational approach for the numerical solution of the governing equations. This involves selecting the type of computational grid, the spatial and temporal discretization schemes, and the space and time resolutions.

Since patient-specific vascular models are characterized by complex geometrical shapes, unstructured grids have been the method of choice in most studies. Besides providing great flexibility to deal with complex geometries, these grids allow nonuniform distributions of element sizes, thus permitting locally increasing the spatial resolution in smaller vessels, high curvature regions, vessel narrowings, etc. Additionally, since these grids are body conforming, that is, element faces are aligned with the domain boundary, the calculation of quantities such as WSS that require proper representation of surface normals and velocity gradients at the wall do not present any difficulty. Furthermore, they allow the development of higher-order schemes to recover boundary gradients and thus reduce numerical errors associated with computing WSS vectors (Löhner et al., 2007). This meshing approach also allows the use of grids with increased resolution in layers next to the vessel wall (i.e., boundary layer grids) in order to properly represent boundary layers and accurately compute pressure drops and flow separation regions in flows with high Reynolds numbers. Different discretization approaches have been used in combination with unstructured grids, including finite element methods, finite volume methods, and spectral element methods. These techniques produce similar results (Radaelli et al., 2008) provided they use meshes with equivalent resolutions (similar number of degrees of freedom), but their computational cost may be different, typically favoring the low-order elements.

The second strategy is to use structured grids and finite difference methods. Finite difference schemes are much faster than finite element methods (and easier to parallelize), because they avoid all the indirect addressing and the associated disordered memory access imposed by the element connectivity of unstructured grids. However, generating structured grids that conform to the patient-specific vascular geometries is quite challenging and not automatic. For this reason, structured grids have been used for hemodynamics calculation in combination with immersed boundary methods (Usera and Mendina, 2012). The basic idea is to create a Cartesian grid covering the bounding box of the computational domain, identify nodes that lie inside or outside the vascular domain, and then impose boundary conditions at the intersections between the vessel boundary and the mesh lines. Since vascular geometries are tubular in nature, most of the volume covered by the Cartesian grid may lie outside the region of interest. Thus, researchers have developed narrow band and multiblock techniques to avoid the unnecessary computations outside the physical domain, thus increasing the performance of these codes at the expense of increasing the complexity of finite difference schemes (Usera and Mendina, 2012). The main limitations with this kind of grids are that they do not allow local adaptation and that they are not body conforming. Local adaptation allows increasing the mesh resolution where it is required (e.g., small vessels) instead of having to reduce the grid spacing globally everywhere. Using nonbody conforming grids introduces two main difficulties: the computation of WSS is less accurate than with body conforming grids, and increasing the mesh resolution close to the vessel wall to resolve boundary layers is not straightforward.

When selecting the numerical approach, the modeler will have to weigh the relative performance of the different algorithms and codes, the ability to deal with complex geometries and representation of WSSs of each grid type, as well as the throughput required by the particular study (i.e., is it necessary to create only a few or many models? Is it necessary to perform only a few CFD simulations or many?). This decision will also be influenced by the availability of the approaches discussed before in open source software such as OpenFoam (http://www.openfoam.com), or commercial codes such as Fluent or CFX (http://www.ansys.com), or in in-house developed codes.

21.7 Modeling the Blood Rheology

Next the modeler has to select a model of the rheological behavior of blood in order to approximate the deformation and stresses of material elements of blood.

21.7.1 Newtonian Models

The simplest rheological model for blood is a Newtonian fluid, which assumes that the fluid is homogeneous and isotropic, and that the stress tensor and is a linear function of the strain rate tensor (Fung, 1984). Under these assumptions, the viscosity can be regarded as constant $\mu = \mu_0$ for all shear strain rates, and typically, the values used are $\mu_0 = 0.035 - 0.04$ dyne s/cm. However, blood is mainly composed of red and white blood cells and platelets suspended in the plasma (an aqueous medium). As a result, the behavior of blood deviates from that of a Newtonian fluid at low shear rates where a markedly increase of the apparent viscosity is observed, or in small vessels, where the apparent viscosity is smaller than in larger vessels (Mazumdar, 1992). Thus, it has been argued that Newtonian models constitute a reasonable approximation for the behavior of blood in large vessels (Perktold and Rappitsch, 1995, Zhao et al., 2000, Steinman, 2004, Stuhne and Steinman, 2004, Taylor and Draney, 2004). However, for cerebral aneurysms, this approximation may not be appropriate, since aneurysms may exhibit regions of slow flow and low shear strain rate where non-Newtonian effects may become important (Basombrio et al., 2000).

21.7.2 Non-Newtonian Models

The simplest and most commonly used non-Newtonian models for blood are time-independent fluid models in which the shear stress is a nonlinear function of the shear strain rate, independent from shearing time and the previous history of the fluid (Mazumdar, 1992). One of the most popular non-Newtonian models for blood flows is the Casson model (Casson, 1959). The relationship between the shear stress and strain rate is given by

$$\sqrt{\tau} = \sqrt{\tau_0} + \sqrt{\mu_0 \dot{\gamma}} \tag{21.7}$$

where
τ_0 is the yield stress
μ_0 is the asymptotic Newtonian viscosity

Typical values used for these constants that fit empirical data for blood are $\tau_0 = 0.04$ dyne/cm^2 and $\mu_0 = 0.04$ dyne s/cm. The values of these constants depend on the hematocrit, which is the volume fraction of red blood cells. The strain rate is computed from the second invariant of the strain rate tensor (Perktold et al., 1989), which for incompressible fluids takes the form

$$\dot{\gamma} = 2\sqrt{\varepsilon_{ij}\varepsilon_{ij}} \tag{21.8}$$

The apparent viscosity can be written as

$$\mu = \left(\sqrt{\frac{\tau_0}{\dot{\gamma}}} + \sqrt{\mu_0}\right)^2 \tag{21.9}$$

Since the apparent viscosity grows indefinitely as the strain rate is reduced, alternative expressions have been used for numerical simulations. One such formula is (Papanastasiou, 1987)

$$\mu = \left(\sqrt{\frac{\tau_0(1 - e^{-m\dot{\gamma}})}{\dot{\gamma}}} + \sqrt{\mu_0}\right)^2 \tag{21.10}$$

with values of $m > 100$ reported to produce satisfactory results (Neofitou and Drikakis, 2001). Other non-Newtonian models have been used to approximate the rheological behavior of blood (Neofitou, 2004), including power-law fluids (Walburn and Schneck, 1976), the Carreau model (Gijsen et al., 1999), and the Quemada model (Quemada, 1978). Interestingly, researchers who investigated the effects of Newtonian and non-Newtonian models have concluded that non-Newtonian effects are insignificant (Perktold et al., 1989, Lee and Steinman, 2007), while others have concluded that non-Newtonian effects are important and must be taken into consideration (Neofitou and Tsangaris, 2006). Further research is needed to clarify this issue, in particular with regard to aneurysmal flows characterized by slow and disturbed flow patterns. However, since the incorporation of non-Newtonian models is quite simple and the CPU requirements of non-Newtonian computations are slightly increased (about 10%–20% more expensive than Newtonian calculations), one could argue in favor of always using non-Newtonian models.

Chapter 21

21.8 Modeling the Physiologic Conditions

The next important decision that the modeler needs to make is the selection of appropriate physiologic flow conditions and the method to impose these conditions in the computational models.

21.8.1 Blood Flow Imaging

Patient-specific blood flow conditions can be obtained from phase-contrast magnetic resonance (PC-MR) or Doppler ultrasound (DUS) measurements (see Figure 21.7).

PC-MR techniques can provide reliable estimations of the volumetric flow rate. These images typically consist of a time sequence of 2D images acquired perpendicularly to the vessel. At each time or phase, the pixel values are proportional to the tissue (blood)

velocity normal to the imaging plane. Imaging the other velocity components (in plane) is possible but requires longer scan times. Image acquisition is synchronized with the cardiac cycle (i.e., cardiac gated), and images acquired over several cardiac cycles are averaged. Measurements of blood flow velocity based on PC-MR are mainly affected by (1) spatial resolution that provides an average velocity in each imaging pixel; (2) temporal resolution that can smooth out high-frequency velocity fluctuations; (3) specification of the encoding velocity value (Venc), which determines the velocity resolution, and if chosen below the maximum flow velocity results in velocity aliasing; and (4) scan time that depends on the number of velocity components being imaged and affects the signal-to-noise ratio or image quality. Quantification

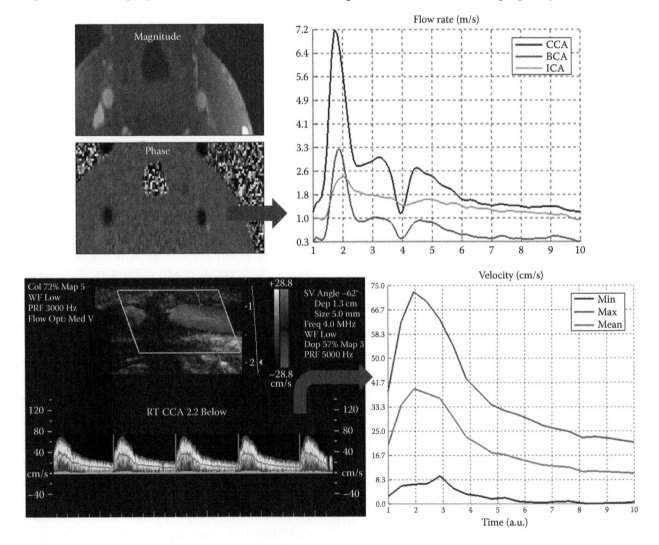

FIGURE 21.7 Examples of measurement of blood flow rates in human carotid arteries (CCA, common; ECA, external; ICA, internal) using phase-contrast MR (top) and Doppler ultrasound (bottom) in the CCA.

of the volumetric flow rate from 2D PC-MR measurements requires integration of the velocity profile over the cross section of the vessel. This in turn requires masking the vessel lumen to define the integration region. This can be accomplished by manual delineation or segmentation of the vessel (Cebral et al., 2002), or using automatic algorithms that perform vessel segmentation based on cross-correlations of velocity waveforms at each pixel (Alperin and Lee, 2003). Since PC-MR averages velocities over the pixel region and over temporal intervals, it can provide reasonable estimations of the flow rates (integral over many pixels) but perhaps not so accurate measurements of peak velocities. In larger vessels that contain several pixels over the cross section, reasonable measurements of the velocity profiles can also be obtained. In addition, recent advances in the PC-MR field have enabled the measurement of 3D unsteady flow fields (4D PC-MR) with reasonable accuracy (Cebral et al., 2009). These techniques suffer from the same limitations listed earlier.

Doppler ultrasound techniques can provide reliable estimations of the peak flow velocity. Typically, the operator places a small window near the center of the vessel and time-dependent velocities are sampled within this window with high temporal resolution over a few cardiac cycles. The peak or maximum velocity over this window can be quantified by reconstructing the envelope velocity curve. Waveform reconstructions can be obtained by manually tracing the peak velocity on the DUS images or by image segmentation algorithms. Quantification of blood flow rates is more problematic, because it requires measurement of the mean velocity across the vessel lumen and multiplication by the cross-sectional area. Mean velocity estimations can be obtained by placing the measurement window on the vessel walls, thus sampling all velocities over the cross section, and computing the mean velocity at each instant of time. Alternatively, one could assume a velocity profile (e.g., parabolic) and estimate the mean velocity from the peak velocity measurements. The vessel diameter and area can be obtained from the same DUS image or from a 3D image, but in this latter case, the exact location of the DUS measurements is not known. Additionally, DUS is limited to exposed extracranial vessels such as the common carotid artery. While trans-cranial Doppler (TCD) ultrasound can yield velocity waveforms in intracranial vessels, the exact location of the measurements is largely unknown. These limitations and approximations introduce errors that will affect the

quantification of blood flow rates. Nevertheless, DUS measurements can provide rough estimations of the flow division among the different branches of the arterial tree, and provide accurate measurements of peak velocity waveforms and their variability from cycle to cycle.

21.8.2 Flow Pulsatility

Steady flow conditions have been used assuming that they represent the average hemodynamics over many cardiac cycles (Bluestein et al., 1996). The greatest advantage of steady flow calculations is their speed. Many specialized numerical techniques have been developed to accelerate the convergence of steady flow calculations. The result is that steady computations can be carried out in about 5% of the time required for pulsatile flows. On the other hand, their main drawback is that they do not provide information about the stability of the flow patterns, the range of variability of hemodynamic variables during the cardiac cycle, their temporal gradients, or measures of quantities such as the oscillatory shear index that have been proposed as possibly important factors in the evolution of cerebral aneurysms (Valencia et al., 2006).

Pulsatile flow conditions can be classified into three main categories: (1) idealized, (2) average, and (3) personalized. Idealized conditions correspond to analytically created flow waveforms, for instance, sinusoidal pulsations (Chatziprodromou et al., 2003). Average flow conditions correspond to population averaged or "typical" flow conditions derived from measurements performed on a (normal) subject population (Ford et al., 2005). When patient-specific flow conditions are not available, the idealized or typical flow waveforms can be scaled with the inflow vessel area to estimate the mean and peak flow using empiric flow-area curves derived from PC-MR measurements in normal subjects (Ford et al., 2005, Cebral et al., 2008). Personalized flow conditions correspond to patient-specific measurement of blood flows (Cebral et al., 2003, Jou et al., 2003). While this last option would be the optimal choice, flow measurements are not typically performed as a part of the routine clinical examinations and therefore, this information is not commonly available. Furthermore, it is known that the normal physiologic flow conditions change during the day due to moment-to-moment variations in the heart rate, cardiac output, and peripheral vascular resistances. These variations are associated with disease states, seasonal variations, emotional status, activity, etc. So, no single measurement could be expected to represent the physiologic condition for any individual. Rather, a range of flow conditions would probably give a better description of the in vivo hemodynamics of a given individual. Thus, performing a series of CFD calculations under a range of physiologic flow conditions can provide a range of expected values of hemodynamic variables for each individual.

21.8.3 Inflow Conditions

Typically, velocity boundary conditions are prescribed at the model inlet(s). In the majority of the models, this is achieved by imposing an idealized velocity profile at the inlet boundary corresponding to the prescribed flow rate. The most commonly used velocity profiles are parabolic profiles for steady flows (Jou et al., 2003), and profiles based on the Womersley solution for pulsatile flows (Womersley, 1955, Taylor et al., 1998). These profiles correspond to the analytic solution of the fully developed flow of an incompressible Newtonian fluid in a straight rigid tube of circular cross section. Uniform or flat velocity profiles (also called plug flow) have also been used for both steady and pulsatile calculations (Oshima et al., 2001). While these options are easy to implement and use, they all neglect secondary flows at the inlet boundaries. This can have an important effect on the aneurysm hemodynamics when the inflow vessels are truncated too close to the region of interest as discussed earlier. Presumably, if the inflow vessels are sufficiently long, secondary flows will develop as blood travels along the curves and bends of the arteries (Castro et al., 2006a). One possibility to overcome this problem would be to measure the three velocity components in the vessel cross section where boundary conditions will be applied. The measured velocities would then be interpolated to the inlet mesh points. These measurements could be performed with PC-MR techniques. However, the main limitations of this approach are the limited resolution of these images, which could be insufficient to accurately represent the velocity profile in intracranial vessels, and the long scanning times required.

21.8.4 Outflow Conditions

At the model outlets, three types of boundary conditions can be applied: (1) prescribed outflow rates, (2) traction-free boundary conditions, and (3) boundary conditions based on coupling to reduced models. The first possibility is to impose flow rates in all but one outflow boundaries and let the solution algorithm determine the flow through the remaining outlet, assuming incompressible flows. With this approach, velocity profiles similar to the ones prescribed at the inlets are used, which dissipate secondary flows. The second possibility is to prescribe traction-free boundary conditions with the same pressure level at all the outlets. With this approach, the flow division among the different arterial branches is determined by the geometry of the model. However, it is known that the flow divisions are actually determined by the relative resistances of the distal vascular beds. Modeling 3D blood flows in the entire vascular trees up to the capillaries is currently impossible due to the large number of arterial branches in each tree and the large number of scales involved. For this reason, investigators have tried to develop boundary conditions based on coupling the 3D models to reduced blood flow models of the distal vascular trees (Formaggia et al., 1999). Estimations of the distal vascular resistances have been obtained from fractal tree models (Olufsen, 1999, Cebral et al., 2000, Olufsen et al., 2000), tree models based on optimization techniques (Schreiner et al., 2000), and trees constructed from morphometric information (Spilker et al., 2007). Additionally, this approach offers the possibility of incorporating autoregulation models, for instance, by changing the resistances of the vascular trees based on the flow rate or pressure in each tree (Ferrandez et al., 2002).

21.8.5 Multiscale Circulation Models

A further possibility that has been proposed is to connect the 3D flow models to a multiscale model of the entire circulation (Formaggia et al., 1999). These models are typically based on reduced 1D and/or 0D models obtained by integration of the 3D Navier–Stokes equations in cylindrical vascular districts.

One-dimensional models of blood flow assume that vessels are composed of cylindrical compliant segments. The governing equations can be derived by integrating the 3D Navier–Stokes equations over the vessel cross section (Formaggia et al., 1999, Sherwin et al., 2003a). The continuity equation becomes

$$\frac{\partial A}{\partial t} + \frac{\partial Q}{\partial x} = 0 \qquad (21.11)$$

and the momentum equation becomes

$$\frac{\partial Q}{\partial t} + \frac{\partial}{\partial x}\left(\alpha \frac{Q^2}{A}\right) + \frac{A}{\rho}\frac{\partial P}{\partial x} + F_r = F \qquad (21.12)$$

where

Q is the flow rate across the vessel cross section

A is the area of the cross section

x is the direction along the vessel axis

P is the average pressure over the cross section

F_r is the frictional force per unit length, which is usually assumed to be proportional to the mean velocity and in the direction opposite to the flow ($F_r = -k \, Q/A$, with k a constant)

The parameter α is related to the velocity profile u by

$$\alpha = \frac{\int u^2 dA}{Q^2/A} \qquad (21.13)$$

Most 1D models assume a particular velocity profile in order to derive the value of α. For instance, assuming a uniform or flat velocity profile (also referred to as "plug flow"), this parameter becomes $\alpha = 1$. The system of equations is closed by a pressure–area relationship. Assuming a thin, homogeneous elastic wall, this relationship takes the form (Sherwin et al., 2003b, Alastruey et al., 2006)

$$P = P_0 + \beta \frac{(\sqrt{A} - \sqrt{A_0})}{\sqrt{A_0}} \qquad (21.14)$$

where the "0" subscript indicates the undeformed reference state. The parameter β is given by

$$\beta = \frac{\sqrt{\pi} h E}{(1 - \upsilon^2)} \qquad (21.15)$$

where

h is the wall thickness

E is the Young's modulus

υ is the Poisson ratio

The parameter β is related to the pulse wave propagation speed c by

$$c^2 = \frac{\beta \sqrt{A}}{2\rho A_0} \qquad (21.16)$$

At arterial bifurcations, continuity of Q and P is assumed. One-dimensional models have recently gained increased interest and are being used for estimations of blood flow and pressure distributions in a variety of arterial networks including aortas and bypass grafts (Wan et al., 2002), peripheral arteries (Sherwin et al., 2003b), and the circle of Willis (Moore et al., 2005, Alastruey et al., 2006). Numerical methods of varying degrees of sophistication are used to solve these systems of equations. In particular, models of the circle of Willis are used to study the flow distribution for different normal variations of the circle of Willis and under pathological vessel occlusions. Early models of the circle of Willis used similar formulations and simple finite difference methods (Kufahl and Clark, 1985, Clark et al., 1989). However, an interesting property of some of these earlier models is that they used a polynomial approximation for the velocity profile, and additional equations were derived in order to find the coefficient of the polynomials. This allowed these models to provide an estimation of the WSS distribution in the arterial network (Kufahl and Clark, 1985).

Compartment or 0D models (also called lumped parameter models) are obtained by linearizing the 1D models and integrating over a length L (Formaggia et al., 1999, Cassot et al., 2000):

$$C\frac{dP}{dt} + Q_2 - Q_2 = 0 \qquad (21.17)$$

$$L\frac{dQ}{dt} + RQ + P_2 - P_1 = 0 \qquad (21.18)$$

with

$$C = \frac{3\pi R_0^3 l}{2Eh}, \quad L = \frac{\rho l}{\pi R_0^2}, \quad R = \frac{8\mu l}{\pi R_0^4} \qquad (21.19)$$

where R_0 is the undeformed radius of the artery. These equations are analogous to the equations describing an electric circuit composed of resistors (R), inductances (L), and capacitances (C). At arterial bifurcations, continuity of Q and P is applied, which corresponds to the Kirchoff law for electric networks. While 1D models are effective for the calculation of average mass flux and pressure distributions along arterial branches, 0D models are useful to describe the average flow and pressure of entire vascular trees.

Chapter 21

21.9 Modeling the Wall Compliance

Finally, the modeler needs to decide whether to incorporate the effects of moving or compliant vascular walls. This decision will depend on the particular application and purpose of the computational models. For instance, if wall biomechanic quantities such as cyclic straining of the wall are required, the modeler does not have any choice but to include wall compliance effects. However, if one is only interested in characterizing the hemodynamic environment within arteries and aneurysms, it may be acceptable to neglect wall deformations.

21.9.1 Fluid–Structure Interaction Models

Vessel wall compliance can be incorporated into the computational models via fluid-structure interaction (FSI) techniques. This approach requires (1) a biomechanics (solid) model of the arterial walls, (2) determination of the distribution of material properties of the vessel walls, (3) determination of the pulsatile intra-arterial pressure wave, (4) a model of contacts and support provided by the extravascular environment, and (5) a fluid–solid coupling algorithm.

Biomechanics models of varying degrees of sophistication have been developed over the years. They include membrane and solid walls, with material models ranging from simple elastic vessels to complex models considering the main constituents of the vessel walls (Humphrey, 1995, Humphrey and Canham, 2000, Gasser et al., 2006).

Different material models require the specification of a number of material properties, which are extremely difficult or impossible to measure in vivo with current imaging technology. Even the simplest elastic models require the specification of the distribution of vessel wall thickness and elastic modulus. This is a challenging problem especially for cerebral arteries and aneurysms because of their small size and location inside the cranium. For this reason, many researchers assume simplistic or idealized distributions of wall thickness and stiffness. However, this distribution can have a large influence on the resulting wall stresses and strains.

FSI models also require the prescription of the pulsatile intra-arterial pressure that drives the deformation of the vascular wall. These pressures could potentially be measured with catheter-mounted pressure probes during angiographic imaging. Although these probes are likely to provide good estimates of the mean arterial pressure, they may not be reliable for the accurate quantification of the pressure waveforms. Furthermore, these measurements are not a part of the routine clinical protocols. So, the modeler would have to prescribe a typical pressure waveform, making sure that it is compatible with the flow waveform used to impose inflow conditions; otherwise, spurious wave reflections could be created.

Fluid-structure coupling algorithms can be classified into two main groups: loose or strong coupling. Loose coupling algorithms combine two separate computer codes, one for the fluid and one for the solid, which work concurrently to solve the coupled problem (Löhner et al., 1995). Information is transferred from one code to the other during the solution process. Both explicit and implicit loose coupling algorithms have been developed (Cebral and Löhner, 2005b). In the strong coupling approach, a single code must be written to solve the coupled problem (Figueroa et al., 2006). The main advantage of loose coupling algorithms is that they allow different modeling approaches, different solution strategies, different discretizations, different timesteps, etc., for the fluid and the solid.

The extravascular environment is also important, because during the cardiac cycle, the vessel walls, and especially cerebral aneurysms, can come in contact with other anatomical structures such as bone or dura matter (Ruiz et al., 2006). These extravascular structures can provide extra support to the arterial walls. New techniques that account for the support provided by extravascular environment have recently been developed (Moireau et al., 2012). In the case of cerebral aneurysms, it has been shown that contact with smooth structures can be protective, while contact with sharp structures can cause a local increase of the wall stresses and an associated damaging effect (Seshaiyer and Humphrey, 2001).

21.9.2 Moving Wall Models

Arterial wall motion can be measured in vivo by using dynamic imaging techniques. Recent studies have used high-frame-rate angiography (Dempere-Marco et al., 2006) and cardiac-gated multislice CTA (Ishida et al., 2005) to image the wall motion of cerebral aneurysms. Quantifications of the wall excursion during the cardiac cycle were obtained by selecting a series of landmark points on the vessel boundary and tracking them with nonrigid image registration algorithms

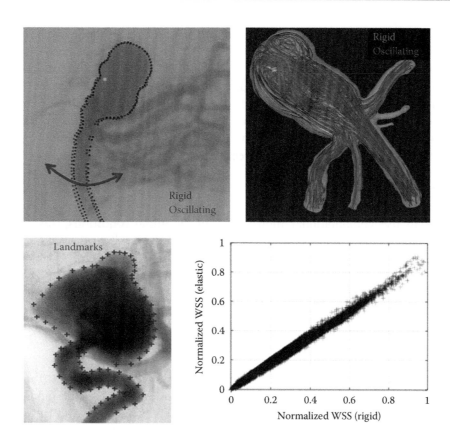

FIGURE 21.8   Top: example of aneurysm with oscillatory motion of the parent artery quantified by tracking landmark points on dynamic DSA images (left) and comparison of flow patterns obtained with rigid (blue streamlines) and moving wall models (red streamlines) showing very similar flow structures. Bottom: example of quantification of aneurysm pulsation from dynamic DSA images by tracking landmark points (left) and comparison of normalized WSS in models with rigid (x-axis) and pulsating walls (y-axis) showing small overestimation of WSS by rigid wall model.

in subsequent frames (Dempere-Marco et al., 2006). Estimations of the wall motion obtained from dynamic images can be used directly to prescribe wall deformation in computational models (see Figure 21.8). The main advantage of this approach is that it does not require any approximation or assumption about the wall biomechanics or the intra-arterial pressure waveform, or the extravascular environment. In addition, the computational expense of these models is only slightly higher than rigid walled models. Therefore, this is an attractive method to study the effects of wall compliance on the hemodynamic patterns and to relate hemodynamic variables such as WSS to abnormal wall motions. However, this strategy will not provide biomechanic quantities such as wall stresses or strains.

21.9.3   Rigid Wall Models

Rigid walled hemodynamics models have been extensively used for a wide variety of applications. The main reason is that these models are much simpler than compliant wall models, they do not require

specification of material properties or pressure waveforms, and are computationally cheaper. Studies based on idealized vascular geometries have shown little differences between rigid and compliant walls (Perktold and Rappitsch, 1995). More recent studies based on realistic geometries of cerebral aneurysms indicate that although rigid models tend to overestimate the WSS compared to compliant models, the qualitative appearance of the WSS distribution and flow patterns is not significantly different (Dempere-Marco et al., 2006). An example is presented in Figure 21.8 (bottom row). The left panel shows landmark points placed on 2D dynamic DSA images and used to quantify the wall motion during the cardiac cycle. The right panel shows the normalized WSS (normalized to the maximum WSS) obtained with a rigid model (x-axis) and with an elastic or pulsating wall model (y-axis). Each point corresponds to a grid point, and the deviation from the identified 45° line indicates an overestimation of WSS by the rigid model with respect to the pulsating model. Similarly, another study (Sforza et al., 2010) suggested that the

Chapter 21

translational motion of the parent artery estimated from dynamic DSA images did not have a substantial effect on the intra-aneurysmal hemodynamics or the WSS distribution. An example is presented in Figure 21.8 (top row). This figure shows landmarks used to track the wall motion on 2D dynamic DSA images (red dots), the contour of the rigid model (blue dots), the center of rotation of the parent artery (green dot), and the oscillatory rotational motion direction (red arrow). The right panel shows flow streamlines obtained with the rigid model (blue lines) and with the moving walls model (red lines), indicating that both models produce consistent flow patterns. Thus, these studies suggest that rigid wall models provide a reasonable approximation of the aneurysm blood flow dynamics.

21.10   Verification and Validation

Before proceeding to use computational models to investigate a particular clinical or biomedical problem, it is necessary to conduct verification and validation studies.

Verification is defined as "the process of determining that a model implementation accurately represents the developer's conceptual description of the model and the solution to the model" (AIAA, 1998). In other words, verification checks that the numerical model and software produce correct solutions to the governing equations of the mathematical model, or simply "solving the equations right." Two aspects of verification have been proposed: (1) verification of a simulation code, which aims at error evaluation or finding and removing errors and bugs in the code; and (2) verification of a calculation, which aims at error estimation or determining the accuracy of a calculation. Thus, verification assesses the errors associated with round-off, convergence, discretization, programming (bugs), and usage (inputs).

Verification of a code typically involves (1) testing individual routines, functions, and modules to verify that they produce the desired outputs; (2) comparing numerical solutions against analytic solutions; (3) comparing to other simulation codes; and (4) comparing against known benchmark solutions.

Verification of a calculation includes (1) checking the convergence of the numerical algorithm by monitoring the residuals of the equations and the numerical values of the quantities of interest; (2) checking mesh convergence or independence, which involves solving the equations with increasing mesh resolution to quantify the spatial discretization error and the order of convergence; (3) checking temporal convergence, which involves examining the sensitivity of the solution to the magnitude of the time step and estimating the temporal discretization error and order of convergence; (4) checking cardiac cycle convergence to verify that the numerical solution is independent of the initial conditions, this is typically achieved by running several cardiac cycles until the solutions of two subsequent cycles are equivalent; (5) checking the consistency of the solution, typically by monitoring global quantities such as mass conservation; and (6) checking that the results are physiologically reasonable by verifying that the values of computed quantities (e.g., pressure drops, etc.) are within expected physiological ranges.

Validation is defined as "the process of determining the degree to which a model is an accurate representation of the real world from the perspective of the intended uses of the model" (AIAA, 1998). In other words, validation determines if the conceptual model as implemented conforms to the real world and agrees with real-world observations, or simply if we are "solving the right equations." The general strategy is to identify and quantify errors by comparing simulation results to experimental data and through sensitivity and parametric studies.

Validation against experimental data is challenging, because many of the quantities of interest are extremely difficult or impossible to measure directly in vivo in human subjects. Thus, numerical models are typically compared against experimental data obtained from in vitro or animal models that allow the measurement of relevant quantities. Experimental in vitro models can be based on idealized or anatomically realistic geometries, and validation proceeds by systematically comparing numerical results to a set of increasingly complex cases. Examples are presented in Figure 21.9 (top row). The top-left panel shows comparisons of CFD (black line) and experimental (red diamonds) velocity profiles at four locations (a, b, c, d) in a 90° bend with square cross section, showing excellent agreement between the numerical and experimental data. The top-right panel shows an in vitro silicon model of a cerebral aneurysm constructed from a CTA image and flow

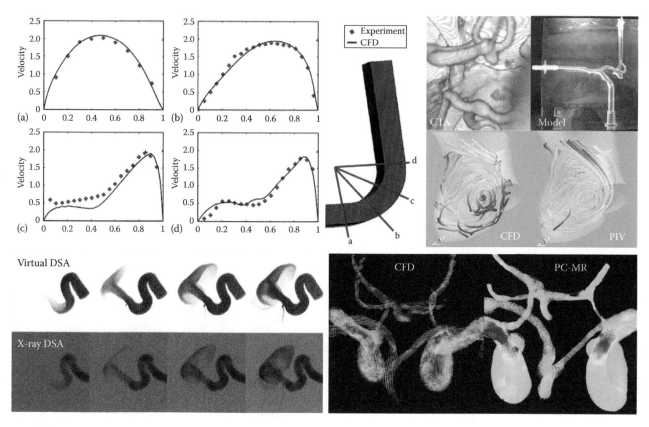

FIGURE 21.9 Examples of validation studies. Top left: validation of CFD against experimental velocity measurements at four selected locations (a–d) along a 90° bend pipe with square cross section. Top right: validation of CFD against PIV in a silicon model of an intracranial aneurysm constructed from CTA images. Bottom left: comparison of CFD-based "virtual angiogram" against in vivo x-ray angiography images showing consistent flow structures. Bottom right: comparison of image-based CFD model and in vivo 4D phase-contrast MR measurements around the circle of Willis of a normal subject.

patterns obtained with CFD and particle image velocimetry (PIV), demonstrating the consistency of flow structures obtained with both techniques. One must keep in mind that experimental data contain errors which should be quantified and documented, and that accuracy requirements are dependent on the particular application and the specific quantities of interest. Furthermore, while comparing numerical models to measurements performed on experimental in vitro or animal models, it is very important to note that these models are imperfect representations of the actual system that one is trying to model, for example, the in vivo hemodynamics in humans with cerebral aneurysms. Experimental or animal models are also based on a set of assumptions and approximations that need to be evaluated, verified, and validated. Ideally, computational models should be validated with in vivo data from human subjects. This is possible but limited. Imaging techniques that allow quantification of blood flow velocities such as phase-contrast MR and Doppler ultrasound are

limited by spatial and temporal resolutions, and circumscribed to limited regions (e.g., individual slices for 2D PC-MR, or small probe window for DUS, etc.). Other imaging techniques can also provide data that can be used for indirect validation: for example, "virtual angiograms" (simulations of the transport of a contrast agent) can be compared to in vivo DSA data. Examples are presented in Figure 21.9 (bottom row). The bottom-left row shows flow structured in a cerebral aneurysm obtained with a CFD-based virtual angiogram and the corresponding in vivo x-ray angiogram. Similar flow structures are seen with both techniques. The bottom-right panel shows blood velocities obtained with an image-based CFD model and with 4D phase-contrast MR in the circle of Willis of a normal subject. These images show comparable velocity patterns and magnitudes.

Finally, as with any experimental technique, it is important to quantify the errors of computational models and provide error bars along with the values of the relevant quantities. This can be achieved by

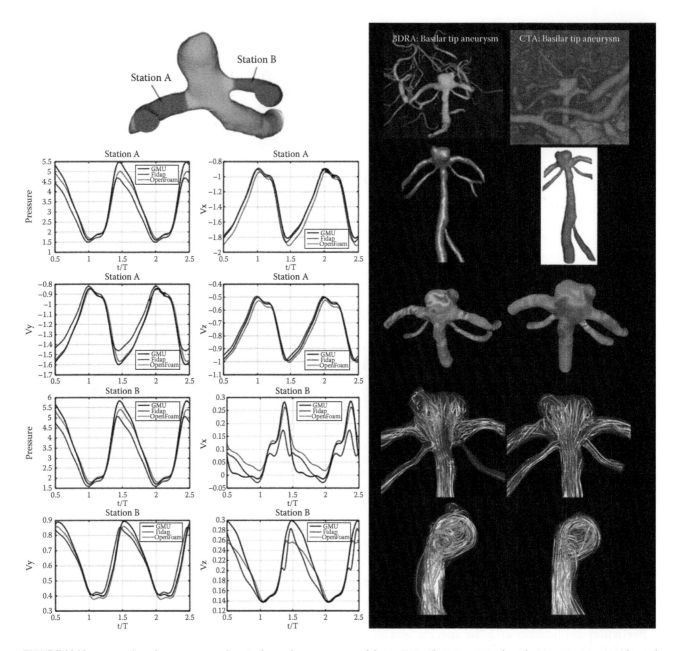

FIGURE 21.10 Examples of sensitivity studies. Left panel: comparison of three CFD solvers, a research code, an open source code, and a commercial code, used to compute the flow field in a cerebral aneurysm. Time-dependent pressure and velocity curves obtained with the different software are shown to be in good agreement at two particular locations. Right panel: comparison of the WSS distribution and flow pattern obtained with two models created from 3DRA (left column) and CTA (right column) images of a cerebral aneurysm. Similar WSS distributions and flow structures are obtained, but absolute values of WSS do not exactly match.

performing sensitivity and variability or parametric studies.

Sensitivity studies assess errors that are not due to lack of knowledge and that should be minimized such as physical approximation errors, geometrical errors, and model parameter errors. Sensitivity studies include quantification of errors due to the different choices made during the model selection process, including (1) use of different imaging modalities; (2) use of different segmentation techniques;

(3) inclusion/exclusion of side arterial branches; (4) truncation of vascular models at different locations; (5) prescription of different boundary conditions (e.g., velocity profiles or outflow boundaries); (6) use of steady instead of pulsatile flow conditions; and vii) use of rigid walled models instead of compliant walls. Examples are presented in Figure 21.10. The left panel shows a comparison of three CFD solvers (a research code, a public domain code, and a commercial code) used to compute the hemodynamics

in a cerebral aneurysm. Time-dependent pressures and velocity components are shown at two locations for all solvers, demonstrating the consistency of the solutions obtained with the different software. The right panel shows a comparison of the WSS and flow patterns obtained with two CFD models created from 3DRA (left column) and CTA (right column) images of a cerebral aneurysm. Consistent flow patterns and WSS distributions are obtained from both imaging modalities; however, the absolute magnitude of the WSS is not exactly the same as vessels tend to be a little bit larger in CTA than in 3DRA.

Variability or parametric studies assess errors that are due to lack of knowledge. This uncertainty characterization typically involves inputs to the computational models. Usually, patient-specific flow conditions are not available and "typical" conditions are used to impose physiologic conditions. Therefore, it is important to know the variability of the results when these conditions change. This is especially important, since physiologic conditions are not constant, they may change with the subject's activity, posture, emotional state, etc. Parametric studies that cover the physiologic range of flow conditions, including mean flow rates, flow waveforms, and heart rates, can thus provide us with "error" bars associated to normal variations of the relevant variables.

21.11   Knowledge Discovery

The combination of clinical information, imaging data, and image-based CFD models enables clinical studies of the role of hemodynamics in the mechanisms underlying vascular diseases and the identification of hemodynamic conditions associated with a higher risk of disease development and progression (see Figure 21.11). This is possible, because this approach allows the formulation of hypotheses that can be tested through computational experimentation, and thus generating new knowledge. Testing hypotheses using computational (or other kind of) experiments in general requires performance of many experiments followed by statistical data analysis. This approach requires (1) a well-formulated falsifiable hypothesis that can be unambiguously tested with the available data; (2) data reduction methods to extract the necessary information, variables, etc. from the simulation data; (3) a sample (number of patient-specific CFD models) large enough to achieve statistically significant results; and (4) statistical methods for hypothesis testing, comparison of mean values and proportions, regression and correlation, multivariate logistic regression, survival analysis, etc. Thus, the essential characteristics required by image-based CFD methods and tools include efficient reconstruction of vascular models from medical images, fast numerical solutions of the governing equations for high throughput, and reliable quantification of the relevant variables. These are considerations that should be taken into account when selecting the mathematical and numerical models as well as the software and hardware to run the numerical simulations.

21.12   Future Perspectives

Currently, image-based CFD models are able to provide a quantitative description of the in vivo blood flow patterns and distribution of hemodynamic loads on the vascular walls. However, this is just one factor of much more complex problems such as the mechanisms causing the development and progression of cerebral aneurysms. In order to advance our understanding of such processes, it is necessary to study the connections or interactions between the hemodynamic environment and loads and the biological processes taking place both in the blood and the vascular wall. Computational models offer an attractive research approach, since they can be used to study the effect of different factors separately and in a controlled manner. Therefore, one can envision the integration and coupling of models of the blood flow, blood rheology, thrombosis, wall mechanobiology and biomechanics, and endothelial cell dynamics. This would require the development of computational strategies for coupling models of diverse scales and based on different representations (e.g., particle-based models or PDE-based models). These advances and developments would have to be driven by empirical data such as mechanical testing, microscopy imaging, and histological characterization of tissue samples as well as population- and patient-specific clinical information and imaging data.

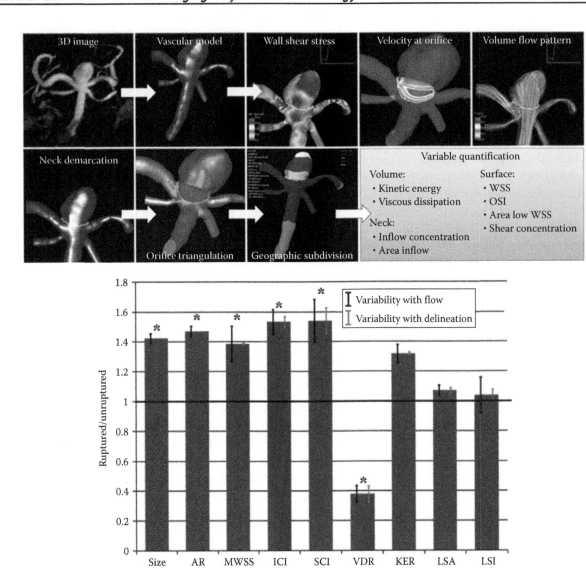

FIGURE 21.11 Example of the use of image-based CFD models to investigate clinical and biomedical problems (Cebral et al., 2011, Mut et al., 2011). This top panel shows the steps in the construction of a patient-specific model of a cerebral aneurysm and the method for quantification of different hemodynamic variables. The bottom panel shows results of a statistical comparison of hemodynamic variables between ruptured and unruptured aneurysms. The bars represent the ratio of the mean values of the ruptured over the unruptured groups. Error bars represent variability of these ratios with respect to physiologic flow conditions and manual delineation of the aneurysm neck.

References

AIAA 1998. AIAA Guide for the Verification and Validation of Computational Fluid Dynamics Simulations. AIAA G-077, American Institute of Aeronautics and Astronautics, 1998. http://dl.acm.org/citation.cfm?id=521779.

Alastruey, J., Parker, K. H., Peiro, J., Byrd, S. M., and Sherwin, S.J. 2007. Modelling the circle of Willis to assess the effects of anatomical variations and occlusions on cerebral flows. *Journal of Biomechanics*, 40(8), 1794–1805.

Alperin, N. and Lee, S. H. 2003. PUBS: Pulsatility-based segmentation of lumens conducting non-steady flow. *Magnetic Resonance in Medicine*, 49, 934–944.

Basombrio, F., Dari, E., Buscaglia, G. C., and Feijoo, R. A. 2000. Numerical experiments in complex haemodynamics flows. non-newtonian effects. *XI ENIEF*, November 24–29, 2000, Bariloche, Argentina.

Beronov, K. N. and Durst, F. 2005. Numericla simulations of pulsating flow in intracranial blood vessels with aneurysms using Lattice Boltzmann methods. *Zeitschrift für Medizinische Physik*, 15, 257–264.

Beutel, J. and Sonka, M. 2000. *Handbook of Medical Imaging, Volume 2: Medical Image Processing and Analysis.* Bellingham, WA: SPIE Press.

Bluestein, D., Niu, L., Schoephoerster, R. T., and Dewanjee, M. K. 1996. Steady flow in an aneurysm model: Correlation between fluid dynamics and blood platelet deposition. *Journal of Biomechanical Engineering*, 118, 280–286.

Boryczko, K., Dzwinel, W., and Yuen, D. A. 2004. Modeling fibrin aggregation in blood flow with discrete particles. *Computer Methods and Programs in Biomedicine*, 73, 181–191.

Butty, V. D., Gudjonsson, K., Buchel, P., Makhijani, V. B., Ventikos, Y., and Poulikakos, D. 2002. Residence times and basins of attraction for a realistic right internal carotid artery with two aneurysms. *Biorheology*, 39, 387–393.

Casson, M. 1959. *Rheology of Dispersive Systems*. New York: Pergamon Press.

Cassot, F., Zagzoule, M., and Marc-Vergnes, J. P. 2000. Hemodynamic role of the circle of Willis in stenosis of internal carotid arteries. An analytical solution to a linear model. *Journal of Biomechanics*, 33, 395–405.

Castro, M. A., Putman, C. M., and Cebral, J. R. 2006a. Computational fluid dynamics modeling of intracranial aneurysms: Effects of parent artery segmentation on intra-aneurysmal hemodynamcis. *AJNR American Journal of Neuroradiology*, 27, 1703–1709.

Castro, M. A., Putman, C. M., and Cebral, J. R. 2006b. Patient-specific computational modeling of cerebral aneurysms with multiple avenues of flow from 3D rotational angiography Images. *Academic Radiology*, 13, 811–821.

Cebral, J. R., Castro, M. A., Appanaboyina, S., Putman, C. M., Millan, D., and Frangi, A. F. 2005. Efficient pipeline for image-based patient-specific analysis of cerebral aneurysm hemodynamics: Technique and sensitivity. *IEEE Transactions in Medical Imaging*, 24, 457–467.

Cebral, J. R., Castro, M. A., Putman, C. M., and Alperin, N. 2008. Flow-area relationship in internal carotid and vertebral arteries. *Physiological Measurement*, 29, 585–594.

Cebral, J. R., Castro, M. A., Soto, O., Löhner, R., and Alperin, N. 2003. Blood flow models of the circle of Willis from magnetic resonance data. *Journal of Engineering Mathematics*, 47, 369–386.

Cebral, J. R., Hernandez, M., Frangi, A. F., Putman, C. M., Pergolizzi, R., and Burgess, J. E. Subject-specific modeling of intracranial aneurysms. *SPIE Medical Imaging*, February 14–19, 2004, San Diego, CA, pp. 319–327.

Cebral, J. R. and Löhner, R. 2001. From medical images to anatomically accurate finite element grids. *International Journal for Numerical Methods in Engineering*, 51, 985–1008.

Cebral, J. R. and Löhner, R. 2005a. Efficient simulation of blood flow past complex endovascular devices using an adaptive embedding technique. *IEEE Transactions in Medical Imaging*, 24, 468–477.

Cebral, J. R. and Löhner, R. 2005b. On the Loose Coupling of Implicit Time-Marching Codes. *AIAA 43rd Aerospace Science Conference and Exhibit*, January 10–13, 2005, Reno, NV.

Cebral, J. R., Löhner, R., and Burgess, J. E. 2000. Computer simulation of cerebral artery clipping: Relevance to aneurysm neuro-surgery planning. *ECCOMAS*, September 11–14, 2000, Barcelona, Spain.

Cebral, J. R., Löhner, R., Choyke, P. L., and Yim, P. J. 2001. Merging of intersecting triangulations for finite element modeling. *Journal of Biomechanics*, 34, 815–819.

Cebral, J. R., Mut, F., Weir, J., and Putman, C. M. 2011. Quantitative characterization of the hemodynamic environment in ruptured and unruptured brain aneurysms. *AJNR American Journal of Neuroradiology*, 32, 145–151.

Cebral, J. R. and Putman, C. M. 2008. Relating cerebral aneurysm hemodynamics and clinical events. In: P. Yim (ed.) *Vascular Hemodynamics: Journal of Vascular Hemodynamics: Bioengineering and Clinical Perspectives.* John Wiley, Hoboken, New Jersey, 2008.

Cebral, J. R., Putman, C. M., Alley, M. T., Hope, T. A., Bammer, R., and Calamante, F. 2009. Hemodynamics in normal cerebral arteries: Qualitative comparison of 4D phase-contrast magnetic resonance and image-based computational fluid dynamics. *Journal of Engineering Mathematics*, 64, 367–378.

Cebral, J. R., Yim, P. J., Löhner, R., Soto, O., and Choyke, P. L. 2002. Blood flow modeling in carotid arteries using computational fluid dynamics and magnetic resonance imaging. *Academic Radiology*, 9, 1286–1299.

Chatziprodromou, I., Butty, V. D., Makhijani, V. B., Poulikakos, D., and Ventikos, Y. 2003. Pulsatile blood flow in anatomically accurate vessels with multiple aneurysms: A medical intervention planning application of computational haemodynamics. *Flow, Turbulence and Combustion*, 71, 333–346.

Chen, S. and Doolen, G. D. 1998. Lattice Boltzmann methods for fluid flows. *Annual Review of fluid Mechanics*, 30, 329–364.

Clark, M. E., Kufahl, R. H., and Zimmerman, F. J. 1989. Natural and surgically imposed anastomoses of the circle of Willis. *Neurological Research*, 11, 217–230.

de Groot, J. 1991. *Correlative Neuroanatomy*. East Norwalk, CT: Appleton & Lange.

Dempere-Marco, L., Oubel, E., Castro, M. A., Putman, C. M., Frangi, A. F., and Cebral, J. R. 2006. Estimation of wall motion in intracranial aneurysms and its effects on hemodynamic patterns. *Lecture Notes in Computer Science*, 4191, 438–445.

Dupin, M. M., Halliday, I., and Care, C. M. 2006. A multi-component lattice Boltzmann scheme: Towards the mesoscale simulation of blood flow. *Medical Engineering and Physics*, 28, 13–18.

Dzwinel, W., Boryczko, K., and Yuen, D. A. 2003. A discrete-particle model of blood dynamics in capillary vessels. *Journal of Colloid and Interface Science*, 258, 163–173.

Federspiel, W. J. and Popel, A. S. 1986. A theoretical analysis of the effect of the particle nature of blood on oxygen release in capillaries. *Microvascular Research*, 32, 164–189.

Ferrandez, A., David, T., Bamford, J., Scott, J., and Guthrie, A. 2000. Computational models of blood flow in the circle of Willis. *Computer Methods in Biomechanics and Biomedical Engineering*, 4, 1–26.

Ferrandez, A., David, T., and Brown, M. D. 2002. Numerical models of auto-regulation and blood flow in the cerebral circulation. *Computer Methods in Biomechanics and Biomedical Engineering*, 5, 7–19.

Figueroa, C. A., Vignon-Clementel, I. E., Jansen, K. E., Hughes, T. J. R., and Taylor, C. A. 2006. A coupled momentum method for modeling blood flow in three-dimensional deformable arteries. *Computer Methods in Applied Mechanics and Engineering*, 195, 5685–5706.

Ford, M. D., Alperin, N., Lee, S. H., Holdsworth, D. W., and Steinman, D. A. 2005. Characterization of volumetric flow rate waveforms in the normal internal carotid and vertebral arteries. *Physiological Measurement*, 26, 477–488.

Ford, M. D., Lee, S. W., Lownie, S. P., Holdsworth, D. W., and Steinman, D. A. 2008. On the effect of parent-aneurysm angle on flow patterns in basilar tip aneurysms: Towards a surrogate geometric marker of intra-aneurismal hemodynamics. *Journal of Biomechanics*, 41, 241–248.

Chapter 21

Formaggia, L., Nobile, F., Quarteroni, A., and Veneziani, A. 1999. Multiscale modeling of the circulatory system: A preliminary analysis. *Computing and Visualization in Science*, 2, 75–83.

Foutrakis, G. N., Yonas, H., and Sclabassi, R. J. 1999. Saccular aneurysm formation in curved and bifurcation arteries. *AJNR American Journal of Neuroradiology*, 20, 1309–1317.

Fung, Y. C. 1984. *Biodynamics: Circulation*. New York: Springer-Verlag.

Gasser, T. C., Ogden, R. W., and Holzapfel, G. A. 2006. Hyperelastic modelling of arterial layers with distributed collagen fibre orientations. *Journal of Royal Society Interface*, 3, 15–35.

Gijsen, F. J. H., van de Vosse, F. N., and Janssen, J. D. 1999. The influence of the non-Newtonian properties of blood on the flow in large arteries: Steady flow in a carotid bifurcation model. *Journal of Biomechanics*, 32, 601–608.

Gray, H. 1901. *Anatomy, Descriptive and Surgical*. Philadelphia, PA: Running Press.

Hassan, T., Ezura, M., Timofeev, E. V., Tominaga, T., Saito, T., Takahashi, A., Takayama, K., and Yoshimoto, T. 2004. Computational simulation of therapeutic parent artery occlusion to treat giant vertebrobasilar aneurysm. *AJNR American Journal of Neuroradiology*, 25, 63–68.

Hirabayashi, M., Ohta, M., and Rufenacht, D. A. 2003. Characterization of flow reduction properties in an aneurysm due to a stent. *Physical Review E: Statistical, Nonlinear, and Soft Matter Physics*, 68, 0219918.

Humphrey, J. D. 1995. Mechanis of the arterial wall: Review and directions. *Critical Reviews in Biomedical Engineering*, 23, 1–162.

Humphrey, J. D. and Canham, P. B. 2000. Structure, mechanical properties, and mechanics of intracranial saccular aneurysms. *Journal of Elasticity*, 61, 49–81.

Ishida, F., Ogawa, H., Simizu, T., Kojima, T., and Taki, W. 2005. Visualizing the dynamics of cerebral aneurysms with four-dimensional computed tomography angiography. *Neurosurgery*, 57, 460–471.

Jou, L. D., Quick, C. M., Young, W. L., Lawton, M. T., Higashida, R., Martin, A., and Saloner, D. 2003. Computational approach to quantifying hemodynamic forces in giant cerebral aneurysms. *AJNR American Journal of Neuroradiology*, 24, 1804–1810.

Kirbas, C. and Quek, F. 2004. A review of vessel extraction techniques and algorithms. *ACM Computing Surveys*, 36, 81–121.

Kufahl, R. H. and Clark, M. E. 1985. A circle of Willis simulation using distensible vessels and pulsatile flow. *Journal of Biomechanical Engineering*, 107, 112–122.

Kundu, P. K. and Cohen, I. M. 2004. *Fluid Mechanics*. New York, Academic Press (Elsevier).

Lee, S. W. and Steinman, D. A. 2007. On the relative importance of rheology for image-based CFD models of the carotid bifurcation. *Journal of Biomechanical Engineering*, 129(2), 273–278.

Liepsch, D. W., Steiger, H. J., Poll, A., and Reulen, H. J. 1987. Hemodynamic stress in lateral saccular aneurysms. *Biorheology*, 24, 689–710.

Liou, T. M. and Liou, S. N. 1999. A review of in vitro studies of hemodynamic characteristics in terminal and lateral aneurysm models. *National Scientific Council Republic of China Part B*, 23(4), 133–148.

Löhner, R., Appanaboyina, S., and Cebral, J. R. 2007. Parabolic recovery of boundary gradients. *Communications in Numerical Methods in Engineering*, 24, 1611–1615.

Löhner, R., Yang, C., Cebral, J. R., Baum, J. D., Luo, H., Pelessone, D., and Charman, C. 1995. Fluid-structure interaction using a loose coupling algorithm and adaptive unstructured meshes. In: Hafez, M. and Oshima, K. (eds.) *Copmutational Fluid Dynamics Review*. John Wiley & Sons, Chichester, U.K.

Martinez, P. F. A. 1982. *Neuroanatomy, Development and Structure of the Central Nervous System*. Philadelphia, PA: Saunders.

Mazumdar, J. 1992. *Biofluid Mechanics*. Singapore: World Scientific.

Moireau, P., Xiao, N., Astorino, M., Figueroa, C. A., Chapelle, D., Taylor, C. A., and Gerbeau, J. F. 2012. External tissue support and fluid-structure simulation in blood flows. *Biomechanics and Modeling in Mechanobiology*, 11, 1–18.

Moore, S. M., Moorhead, K. T., Chase, J. G., David, T., and Fink, J. 2005. One-dimensional and three-dimensional models of cerebrovascular flows. *Journal of Biomechanical Engineering*, 127, 440–449.

Mut, F., Löhner, R., Chien, A., Tateshima, S., Viñuela, F., Putman, C. M., and Cebral, J. R. 2011. Computational hemodynamics framework for the analysis of cerebral aneurysms. *International Journal for Numerical Methods in Biomedical Engineering*, 27, 822–839.

Neofitou, P. 2004. Comparison of blood rheological models for physiological flow simulation. *Biorheology*, 41, 693–714.

Neofitou, P. and Drikakis, D. 2001. Non-Newtonian modeling effects on stenotic channel flows. In: E. Oñate, G. B. A. B. S. (ed.) *ECCOMAS CFD*, September 4–7, 2001, Swansea, U.K.

Neofitou, P. and Tsangaris, S. 2006. Flow effects of blood constitutive equations in 3D models of vascular anomalies. *International Journal for Numerical Methods in Fluids*, 51, 489–510.

Olufsen, M. S. 1999. Structured tree outflow condition for blood flow in larger systemic arteries. *American Journal of Physiology—Heart and Circulatory Physiology*, 276, 257–268.

Olufsen, M. S., Peskin, C. S., Kim, W. Y., Pedersen, E. M., Nadim, A., and Larsen, J. 2000. Numerical simulation and experimental validation of blood flow in arteries with structured-tree outflow conditions. *Annals of Biomedical Engineering*, 28, 1281–1299.

Oshima, M., Torii, R., Kobayashi, T., Taniguchi, N., and Takagi, K. 2001. Finite element simulation of blood flow in the cerebral artery. *Computer Methods in Applied Mechanics and Engineering*, 191, 661–671.

Papanastasiou, T. C. 1987. Flow of materials with Yield. *Journal of Rheology*, 31, 385–404.

Perktold, K., Peter, R., and Resch, M. 1989. Pulsatile non-Newtonian blood flow simulation through a bifurcation with an aneurysm. *Biorheology*, 26, 1011.

Perktold, K. and Rappitsch, G. 1995. Computer simulation of arterial blood flow. Vessel viseases under the aspect of local hemodynamics. *Biological Flows, Plenum Press*, 83–114.

Quemada, D. 1978. Rheology of concentrated disperse systems III. General features of the proposed non-Newtonian model. Comparison with experimental data. *Rheologica Acta*, 17, 643–653.

Radaelli, A., Ausburger, L., Cebral, J. R., Ohta, M., Rufenacht, D. A., Balossino, R., Benndorf, G. et al. 2008. Reproducibility of haemodynamical simulations in a subject-specific stented aneurysm model—A report on the virtual intracranial stenting challenge 2007. *Journal of Biomechanics*, 41, 2069–2081.

Ruiz, D., Yilmaz, H., Dehashti, A. R., Alimenti, A., de Tribolet, N., and Rufenacht, D. A. 2006. The perianeurysmal environment: Influence on saccular aneurysm shape and rupture. *AJNR American Journal of Neuroradiology*, 27, 504–512.

Schreiner, W., Karch, R., Neumann, F., and Neumann, M. 2000. Constrained constructive optimization of arterial tree models. In: Brown, J.H. and West, G.B. (eds.) *Scaling in Biology*. Oxford, U.K.: Oxford University Press.

Seshaiyer, P. and Humphrey, J. D. 2001. On the potentially protective role of contact constraints on saccular aneurysms. *Journal of Biomechanics*, 34, 607–612.

Sforza, D., Löhner, R., Putman, C. M., and Cebral, J. R. 2010. Hemodynamic analysis of intracranial aneurysms with moving parent arteries: Basilar tip aneurysms. *International Journal for Numerical Methods in Biomedical. Engineering*, 26, 1219–1227.

Sforza, D., Putman, C. M., and Cebral, J. R. 2009. Hemodynamics of cerebral aneurysms. *The Annual Review of Fluid Mechanics*, 41, 91–107.

Sherwin, S. J., Formaggia, L., Peiro, J., and Franke, V. 2003a. Computational modelling of 1D blood flow with variable mechanical properties and its application to the simulation of wave propagation in the human arterial system. *International Journal for Numerical Methods in Fluids*, 43, 673–700.

Sherwin, S. J., Franke, V., Peiro, J., and Parker, K. 2003b. One-dimensional modelling of a vascular network in space-time variables. *Journal of Engineering Mathematics*, 47, 217–250.

Shojima, M., Oshima, M., Takagi, K., Torii, R., Hayakawa, M., Katada, K., Morita, A., and Kirino, T. 2004. Magnitude and role of wall shear stress on cerebral aneurysm: Computational fluid dynamic study of 20 middle cerebral artery aneurysms. *Stroke*, 35, 2500–2505.

Spilker, R. L., Feinstein, J. A., Parker, D. W., Reddy, V. M., and Taylor, C. A. 2007. Morphometry-based impedance boundary conditions for patient-specific modeling of blood flow in pulmonary arteries. *Annals of Biomedical Engineering*, 35(4), 546–559.

Stehbens, W. E. 1972. Intracranial aneurysms. *Pathology of the Cerebral Blood Vessels*. St. Louis, MO: CV Mosby.

Steiger, H., Poll, A., Liepsch, D., and Reulen, H. J. 1987. Haemodynamic stress in lateral saccular aneurysms: An experimental study. *Acta Neurchiurgicar (Weinheim)*, 86, 98–105.

Steinman, D. A. 2004. Image-based CFD modeling in realistic arterial geometries. *Annals of Biomedical Engineering*, 30, 483–497.

Steinman, D. A., Milner, J. S., Norley, C. J., Lownie, S. P., and Holdworth, D. W. 2003. Image-based computational simulation of flow dynamics in a giant intracranial aneurysm. *AJNR American Journal of Neuroradiology*, 24, 559–566.

Steinman, D. A. and Taylor, C. A. 2005. Flow imaging and computing: Large artery hemodynamics. *Annals of Biomedical Engineering*, 33, 1704–1709.

Stuhne, G. R. and Steinman, D. A. 2004. Finite element modeling of the hemodynamics of stented aneurysms. *Journal of Biomechanical Engineering*, 126, 382–387.

Succi, S., Benzi, R., and Massaioli, F. 1990. A review of lattice Boltzmann methods. In: Doolen, G. D., Frisch, U., Hasslacher, B., and Wolfram, S. (eds.) *Lattice Gas Methods for Partial Differential Equations*. Addison-Wesley, Santa Fe Institute, New Mexico.

Sun, C., Migliorini, C., and Munn, L. L. 2003. Red blood cells initiate leukocyte rolling in postcapillary expansions: A lattice Boltzmann analysis. *Biophysical Journal*, 85, 208–222.

Sun, C. and Munn, L. L. 2005. Particulate nature of blood determines macroscopic rheology: A 2-D lattice Boltzmann analysis. *Biophysical Journal*, 88, 1635–1645.

Taubin, G. A signal processing approach to fair surface design. *Proceedingsof 22nd Annual Conference on Computer Graphics and Interactive Techniques (SIGGRAPH 1995)*, August 6–11, 1995, Los Angeles, CA, pp. 351–358.

Taylor, C. A. and Draney, M. T. 2004. Experimental and Computational Methods in Cardiovascular fluid Mechanics. *Annual Review of Fluid Mechanics*, 36, 197–231.

Taylor, C. A., Hughes, T. J. R., and Zarins, C. K. 1998. Finite element modeling of blood flow in arteries. *Computer Methods in Applied Mechanics and Engineering*, 158, 155–196.

Taylor, C. A. and Steinman, D. A. 2010. Image-based modeling of blood flow and vessel wall dynamics: Applications, methods and future directions. *Annals of Biomedical Engineering*, 38, 1188–1203.

Usera, G. and Mendina, M. 2012. CFD challenge: Solutions using open source flow solver caffa3d.MBRi with immersed boundary condition. *ASME Summer Bioengineering Conference*. Fajardo, Puerto Rico.

Valencia, A., Guzman, A. M., Finol, E. A., and Amon, C. H. 2006. Blood flow dynamics in saccular aneurysm models of the basilar artery. *Journal of Biomechanical Engineering*, 128, 516–526.

Walburn, F. J. and Schneck, D. J. 1976. A constitutive equation for whole human blood. *Biorheology*, 13, 201–210.

Wan, J., Steele, B., Spicer, S. A., Strohband, S., Feijoo, R. A., Hughes, T. J. R., and Taylor, C. A. 2002. A one-dimensional finite element method for simulation-based medical planning for cardiovascular disease. *Computer Methods in Biomechanics and Biomedical Engineering*, 5, 195–206.

Weir, B. 2002. Unruptured intracranial aneurysms: A review. *Journal of Neurosurgery*, 96, 3–42.

Womersley, J. R. 1955. Method for the calculation of velocity, rate of flow and viscous drag in arteries when the pressure gradient is known. *Journal of Physiology*, 127, 553–563.

Zhao, S. Z., Xu, X. Y., Hughes, A. D., Thom, S. A., Stanton, A. V., Ariff, B., and Long, Q. 2000. Blood flow and vessel mechanics in a physiologically realistic model of a human carotid arterial bifurcation. *Journal of Biomechanics*, 33, 975–984.

Chapter 21

Image-Guided
Therapeutic Procedures

V

22. Vascular Imaging for Image-Guided Interventions

Andrew Kuhls-Gilcrist

22.1 Introduction: Overview of Vascular Imaging for Image-Guided Interventions

Vascular imaging has advanced considerably from the advent of diagnostic x-ray angiography in 1927 with the first angiogram performed by Egas Moniz by injecting radiopaque contrast into the carotid artery to visualize the intracranial internal carotid artery. Moniz and other pioneers were able to visualize and diagnose various disease states of the vasculature to be subsequently cured surgically, resulting in improved outcomes and ushering in a continued era of vascular imaging (Hoeffner et al. 2012). Manual exchange of single-plate angiograms provided the ability to capture a snapshot or two in time, but little more. Image intensifiers with automatic film cassette changers and roll film provided the ability to capture a sequence of images over time, hence enabling visualization of blood flow, movement of anatomical structures, and the navigation and deployment of devices. Use of real-time imaging technology was an important precursor leading up to the first image-guided interventional procedure performed by Charles Dotter in 1964 (Figure 22.1) using a dilating catheter to heal a stenosis in the femoral artery (Payne 2001).

Fast-forward 50 years and the technological advancements in vascular imaging are staggering. Today, image-guided interventional procedures often leverage a wide variety of complementary imaging technologies from preplanning and follow-up CT and MR angiography to real-time fluoroscopy and ultrasound during the course of an intervention. Reflecting the increasing number of

Chapter 22

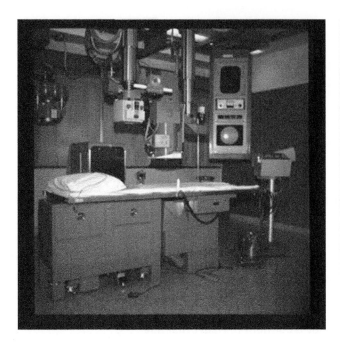

FIGURE 22.1 The 1960 state-of-the-art interventional radiology laboratory at the University of Oregon Medical School used by Charles Dotter. Cineangiography and the ability to capture moving images were important precursors to the first interventional image-guided procedure.

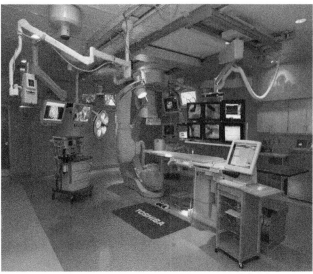

FIGURE 22.2 Today's state-of-the-art interventional system. The importance of imaging technology is evident with 15 monitors each displaying different video feeds.

informational feeds, today's state-of-the-art interventional systems often include a multitude of digital displays, large and small, located throughout the exam and control rooms (Figure 22.2). The fusion of imaging modalities, physiology, and other complementary information promises to improve guidance and enable increasingly complex interventions to be performed with successful outcome. Common image-guided interventional procedures and their associated imaging technologies and applications are described in the next subsection. Brief overviews of the different clinical ailments are also presented to provide a frame of context.

22.2 Common Clinical Procedures and Treatments

The cardiovascular system, comprising the heart, blood vessels, and blood, is responsible for transporting oxygen, nutrients, and hormones throughout the entire body; protecting the body from cellular debris and pathogens and in sealing wounds; and regulating stable temperature and pH. Disease of the cardiovascular system is the leading cause of death and a major cause of disability worldwide (Figure 22.3) and

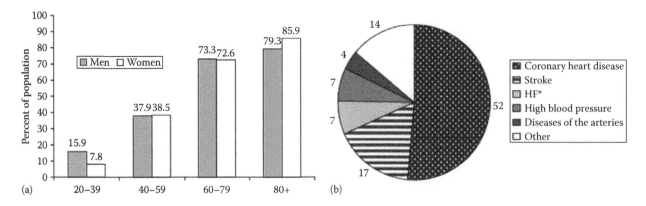

FIGURE 22.3 (a) Prevalence of cardiovascular disease in adults ≥20 years of age by age and sex. (From Lloyd-Jones, D. et al., *Circulation*, 119, e21, 2009.) (b) Percentage breakdown of deaths due to cardiovascular disease in the United States. Heart failure is abbreviated HF and is not a true underlying cause.

is exacerbated by an aging population and suboptimal control of risk factors, including obesity and smoking (Mensah and Brown 2007). The two most common forms of cardiovascular disease (CVD) impact the brain and the heart, and a more detailed description of both stroke and heart disease is provided in the subsequent subsections.

22.2.1 Stroke

A stroke results from disturbances in the blood supply to the brain and leads to rapid deterioration of brain function. A stroke can cause permanent neurological damage, complications, and death. Stroke is a leading cause of serious, long-term disability in the United States and Europe and is the second leading cause of death worldwide, accounting for over 5.7 million deaths annually (Adamson et al. 2004, Lloyd-Jones et al. 2009). This amounts to one death every 6 s on average. It is estimated that in the United Sates alone, some 795,000 people experience a new or recurrent stroke annually with associated costs of over $68.9 billion (Lloyd-Jones et al. 2009).

The two main types of stroke are ischemic and hemorrhagic, accounting for 87% and 13% of all strokes, respectively (Lloyd-Jones et al. 2009). An ischemic stroke results from a decrease in the blood supply to part of the brain. Over time, plaque—comprised of fat, cholesterol, and other substances—builds up on the wall of arteries, inhibiting blood flow. If blood flow is stopped for longer than a few seconds, the brain cannot receive sufficient blood and oxygen and brain cells can die, resulting in irreversible damage. Ischemic strokes are typically caused by clogged arteries. There are two types of clots: a cerebral thrombus, which is the obstruction of a blood vessel by a blood clot forming locally, and a cerebral embolism, which is a clot forming in another area of the vasculature, such as the heart, which is then carried through the blood stream to the brain.

Hemorrhagic stroke results from bleeding in the brain. It can occur when blood vessels in the brain weaken and burst. High blood pressure can stress the artery walls until they break. Cerebral aneurysms (a localized, blood-filled balloon-like bulge of a blood vessel) and arteriovenous malformations (an abnormal network of connections between veins and arteries) are the typical mechanisms that weaken blood vessels and lead to hemorrhagic stroke. Bleeding irritates the brain tissue, causes swelling, and increases pressure on the brain, which can all cause rapid damage to brain cells. Further clinical description can be found in Chapter 2.

22.2.1.1 Treatment of Stroke

The treatment of cerebral stroke requires the restoration of proper blood circulation. The mantra "time is brain" emphasizes the importance of a sooner rather than later treatment approach to minimize damage to the brain and improve outcomes. If an ischemic stroke is diagnosed at an early stage, treatment with medications such as aspirin may be given to minimize clot enlargement. For more advanced cases, a pharmacological approach may be used for "clot busting," in which a drug is administered to dissolve the clot and unblock the artery. However, recent studies indicate that the mortality rate may actually be higher for those receiving certain types of drugs versus a placebo (Dubinsky and Lai 2006). Another minimally invasive approach involves the removal of the clot directly. This is done by inserting a catheter into the femoral artery, which is then navigated to the cerebral vasculature under fluoroscopic image guidance. A clot removal device (such as a corkscrew-like device) is then deployed to clear the vessel. More recently, angioplasty and stenting have become viable treatment options of ischemic stroke. Stents are tube-like devices that are crimped and deployed through catheters. The stent is navigated through the vasculature to the stenosis site. It is then expanded, using the inflation of a balloon upon which the stent was crimped or using self-expanding shape memory alloys, such that the expansion force pushes the vessel open. In a recent systematic review, the rate of technical success for intracranial stenting was more than 90% (Derdeyn and Chimowitz 2007).

The treatment of hemorrhagic stroke has historically involved the surgical clipping of aneurysms, an invasive procedure that involves opening of the skull, exposing the aneurysm, and closing the base of the aneurysm with a clip. With the advancement of improved endovascular image-guided intervention (EIGI) techniques, alternative *minimally invasive* approaches are becoming increasingly successful. EIGI may offer reduced mortality, morbidity, and discomfort to the patient (Higashida et al. 2007). The most common EIGI treatment involves filling the aneurysm with an embolic material, such as detachable coils—delivered through a catheter under fluoroscopic guidance into the aneurysm. This alters the hemodynamic flow into the aneurysm and can eliminate the aneurysm.

With minimally invasive EIGI procedures rapidly replacing many invasive surgical procedures,

Chapter 22

the rudimentary devices currently being used for the treatment of stroke will continue to advance (Rudin et al. 2008). Devices are already becoming more advanced and are moving toward patient-specific designs. The deployment of these devices will require improved imaging capabilities. One such example is the asymmetric vascular stent (AVS), which is being developed at the Toshiba Stroke and Vascular Research Center at the University at Buffalo for the treatment of neurovascular aneurysms (Rudin et al. 2004, 2008). The AVS is constructed by adding an asymmetric, low-porous region to an otherwise standard vascular stent. This low-porosity region is designed to cover the aneurysm neck to reduce and occlude blood flow into the aneurysm, while saving nearby perforators (small "feeder" vessels). Accurate placement of such devices places an increasing burden on imaging technologies (Ionita et al. 2005).

22.2.1.2 Vascular Imaging for Interventional Treatment of Stroke

Minimally invasive treatments for stroke require real-time image guidance to provide immediate visual feedback to the interventionalist. Imaging systems must provide visualization of complex cerebral vasculature and enable accurate deployment of treatment options. X-ray-based interventional fluoroscopy is the most widely and heavily used imaging modality, as it provides superior image quality and both spatial and temporal resolution. Biplane angiographic C-arm systems are utilized to provide both an AP and lateral projection view

simultaneously, improving visualization of the complex 3D vasculature, thereby contributing to safer, quicker, and more effective procedures (Johnson et al. 2001). Midsized flat panel detectors (e.g., 12 × 12 in.) are optimum in that the vasculature of the entire brain can be imaged within the detector field of view (FOV) but are not so large as to obstruct patient access or to sacrifice inherent spatial resolution, which is typically better than 2.5 line pairs per mm. Detectors providing higher resolution in a targeted region of interest (ROI) are being investigated for use in conjunction with a larger FOV detector, to be deployed during critical portions of the interventional procedure where demands for higher–spatial resolution may be of benefit (Kuhls-Gilcrist et al. 2006). For example, during the coiling of an aneurysm, the ability to visualize individual coils entering the aneurysm may lead to improved patient care (Kahn 2012). Chapter 13 provides a detailed description of special detectors for digital angiography, microangiography, and micro-ROI-CBCT.

Due to the limited motion of the anatomical structures within the human head, digital subtraction angiography (DSA) is an effective tool in improving the visualization of the vasculature by removing background boney structures that could otherwise interfere with the specific imaging task. A "map" of the vasculature can be overlaid onto the live fluoroscopic image to provide a 2D roadmap that assists in the navigation of catheters and the placement of devices. Three-dimensional rotational angiography (3D-RA) and DSA (3D-DSA) (see Figure 22.4) can be used similarly to provide a 3D roadmap overlay.

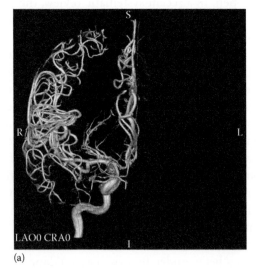

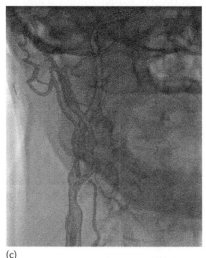

(a) (b) (c)

FIGURE 22.4 (a) Example of a 3D digital subtraction angiography (3D-DSA) image, (b) a 3D-DSA image with improved stent imaging during treatment of an aneurysm, and (c) a 3D roadmap overlaid with live fluoroscopy to improve catheter and device navigation. State-of-the art systems automatically adjust and align the position of the 3D structure with C-arm and table movements, reducing the need to acquire additional angiograms and thereby potentially reducing iodinated contrast and radiation exposure usage.

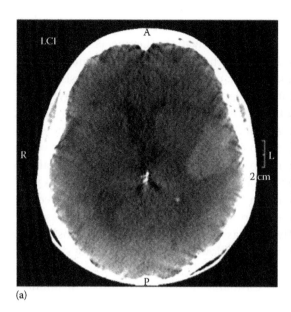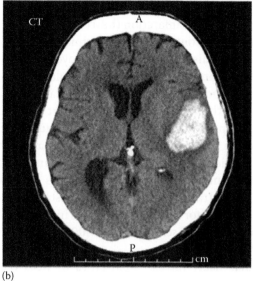

(a)   (b)

FIGURE 22.5 (a) Low-contrast "CT-like" image obtained on an interventional angiographic C-arm compared with (b) a postprocedure CT scan demonstrating the capability of providing improved diagnostic information during the course of an intervention. (Image courtesy of the Gates Vascular Institute, Buffalo, NY.)

Moving to 3D offers the added benefits of not only capturing the true 3D nature of the vasculature but also enables autotracking with both C-arm and table movements such that a single 3D image may be used throughout the course of an intervention, provided patient motion is limited. Continued improvements to the 3D imaging capabilities of interventional C-arm systems are resulting in improved capabilities to view device details, such as apposition of a stent to the vessel wall, and to differentiate soft tissue and to provide "CT-like" low-contrast imaging that enables the detection of bleeding and differentiation between gray and white matter. The clinician can then more immediately diagnose the accuracy of stent placement or possible bleeding during the course of an intervention, rather than having to complete the procedure and wait for a follow-up CT exam, with further iterations until success is achieved. See Figure 22.5 for C-arm cone-beam CT (CBCT) compared with standard CT.

Advancements in neurovascular imaging for image-guided interventions will continue to progress. For example, the use of computational fluid dynamics (CFD) for the assessment of aneurysmal flow has the potential to provide rupture risk predictors to better aid in treatment and decision making and move further toward the deployment of patient-specific treatment options, both of which are becoming increasingly important in the consideration of today's health care environment, moving toward the better management of escalating costs and a greater focus

on treatment outcomes. Chapters 20 and 21 describe CFD principles and applications.

22.2.2   Heart Disease

Heart disease can result from both disturbances in the blood flow related to plaque buildup in the walls of the coronary arteries, an irregular heartbeat caused by abnormal sequences of electrical impulses, heart valve problems, and congenital defects. Further clinical descriptions can be found in Chapter 3 of this volume. Heart disease is the leading cause of death worldwide, accounting for over 7.2 million deaths annually. This amounts to 13.7% of all deaths. An estimated 16.8 million Americans have coronary heart disease with 5.7 million Americans experiencing heart failure annually. The estimated annual economic cost of heart disease in the United States alone is over $300 billion (Lloyd-Jones et al. 2009).

CVD, which includes stroke, results in approximately 7.1 million hospital discharges annually. Coronary heart disease, also referred to as coronary artery disease (CAD), accounts for roughly 25% of all CVD hospital discharges (Lloyd-Jones et al. 2009). CAD is a narrowing of the small blood vessels that supply blood and oxygen to the heart as a result of the buildup of plaque. Without an adequate blood supply, the heart becomes starved of oxygen and other nutrients, which can lead to various types of angina or chest pain. Complete blockage of blood flow to a part of the

Chapter 22

heart can result in damage or death to the heart muscle causing myocardial infarction (heart attack).

Arrhythmias contribute to approximately 12% of all CVD hospital discharges annually (Lloyd-Jones et al. 2009). Arrhythmias result from an irregular heartbeat caused by abnormal sequences of electrical impulses. Severe arrhythmia may result in the heart not being able to pump a sufficient blood supply throughout the body, which can damage the brain, heart, and other organs. Atrial fibrillation, the most common cardiac arrhythmia, causes the heart muscles to quiver instead of contract, during which blood stasis creates an environment for clot formation. This independently increases the risk of ischemic stroke 400%–500% (Wolf et al. 1991).

Valvular heart disease contributes to 1.3% of all the CVD hospital discharges annually (Lloyd-Jones et al. 2009). Valvular heart disease is characterized by damage or a defect in one of the four heart valves: the aortic, mitral, pulmonary, and tricuspid. Incomplete closure of a valve results in regurgitation or backflow. Incomplete opening of a valve due to stenosis results in a restriction in the blood flowing through the valve. Improperly formed valves can cause regurgitation or stenosis. Advanced heart valve disease can cause heart failure, stroke, blood clots, or sudden cardiac arrest. The two most commonly affected valves are the aortic and mitral valves.

Congenital cardiovascular defects account for roughly 1% of all the CVD hospital discharges annually. Congenital heart defects are the most common type of birth defect, impacting approximately 36,000 newborns annually in the United States (Lloyd-Jones et al. 2009). There are many types of defects that alter the normal flow of blood through the heart. In addition to congenital heart valve disease, defects can involve the interior walls and the arteries and veins associated with the heart. Symptoms frequently present early in life and can range from shortness of breath to underdevelopment of limbs and muscles to death.

22.2.2.1 Treatment of Heart Disease

The treatment of heart disease requires the restoration of proper blood circulation and/or the restoration of proper heart function. If stable CAD is diagnosed at an early stage, treatment with life style modifications or medications such as aspirin may slow the progression of disease or remedy symptoms (Henderson et al. 2003, Stergiopoulos and Brown 2012). For more advanced cases, percutaneous coronary intervention (PCI) is used to treat the stenosis, which may include balloon angioplasty or stent implantation, and to promote revascularization (Levine et al. 2011, Bruyne et al. 2012). The balloon or stent is guided through the femoral or radial artery to the stenosis using fluoroscopic guidance, which are then deployed to expand the vasculature. Still a matter of active debate, improvements in device and imaging technologies are leading to increasing acceptance of PCI as a viable alternative to surgical coronary artery bypass graft (CABG) treatment (Teirstein and Price 2012). For example, advances in drug-eluting stents designed to reduce the risk of thrombosis; virtual histology intravascular ultrasound for assessment of the plaque constituents; and fractional flow reserve assessment of stenosis; may all help to reduce the risk of potential adverse events. Recent estimates suggest the number of annual PCI procedures being performed annually is approximately three times greater than the number of CABG procedures at approximately half the cost (Lloyd-Jones et al. 2009).

The treatment of arrhythmia most often involves medical therapy to prevent heart attack and stroke. In more severe/symptomatic cases, treatments involving atrial fibrillation, pacing, and ablation therapy or eletrophysiology (EP), have been shown to improve a broad range of clinical outcomes and have the added benefits of absolute heart rate control, diminished drug burden, improved ventricular function, and freedom from drug side effects (Wood 2000). Pacing involves the use of medical devices such as an implantable cardioverter defibrillator or pacemaker that monitor and correct for irregularities in electrical impulses. Ablation of dysfunctional tissue to correct the electrical activities of the heart is becoming an increasingly accepted treatment option. Catheter ablation involves the navigation of a series of catheters through blood vessels of the arm, groin, and/or neck into the chambers of the heart under fluoroscopic x-ray guidance. The tip of an ablation catheter administers radiofrequency (RF) energy or a refrigerant to damage the tissue that interferes with the normal distribution of the heart's electrical impulses.

Treatment of valvular heart disease may involve administration of antithrombotic medications. Heart valve surgery to repair or replace deficient valves is a well-established treatment option for more progressed disease states. Less invasive alternatives are becoming increasingly common. Balloon valvuloplasty may be performed in an attempt to enlarge a narrowed heart valve and improve blood flow by dilating a balloon-tipped catheter placed across the stenotic valve under fluoroscopic x-ray guidance (Ben-Dor et al. 2013). With the advent of transcatheter valves, hybrid procedures such as transcatheter aortic valve replacement are becoming more commonplace, which involve the replacement of the native valve with an artificial valve that is placed on a stent-like scaffolding that can be crimped onto a catheter tip and navigated through the vasculature. The artificial valve is positioned inside the patient's native valve using fluoroscopic x-ray imaging and expanded, pushing the faulty valve aside. Transcatheter treatments are also being developed to address regurgitation. For example, clip-like devices that attach to the native valve leaflets are being used to treat mitral regurgitation and improve proper blood circulation.

Congenital cardiovascular defects may be treated surgically or with cardiac catheterization procedures to widen arteries, repair or replace heart valves, or to close holes in the heart. Treatment can be complicated as the patient and subsequent anatomy are still growing when the disease is detected in an early state. As such, several treatment procedures may be required into adulthood. Septal defects, the most common form of congenital heart defect, can be cured surgically through the closure of the defect (Lloyd-Jones et al. 2009). Transcatheter closure can be accomplished using occluding devices that are placed across the septal defect to effectively close the opening.

With minimally invasive EIGI procedures rapidly replacing many invasive surgical procedures, the devices and technologies currently being used for treatment of heart disease will continue to advance. Devices are already becoming more advanced, easier to deliver, and more patient-specific with better anatomical size and design configurations being considered. The deployment of these devices will continue to leverage and require improved imaging capabilities.

22.2.2.2 Vascular Imaging for Interventional Treatment of Heart Disease

Minimally invasive treatments for heart disease require real-time image guidance to provide immediate visual feedback to the interventionalist. Imaging systems must provide visualization not only of the vasculature, but also soft tissue structures of the heart (i.e., low contrast), to enable accurate deployment of treatment options. X-ray-based interventional fluoroscopy is the most widely used imaging modality, as it provides superior image quality and both spatial and temporal resolution. Single-plane angiographic C-arm systems are most commonly utilized in adult cardiac catheterization laboratories, while biplane systems are often employed in dedicate pediatric laboratories and

are considered essential for providing a 3D perspective of the treatment of a variety of congenital defects (Feltes et al. 2011). Small- to mid-sized flat panel detectors (e.g., 8×8 to 12×12 in.^{2}) are optimum in that the entire heart can be imaged within the detector FOV but are not so large as to be obtrusive and steep C-arm gantry angulation can be met. Spatial resolution on the order of 2.5 line pairs per mm has historically been sufficient for the treatment of adult vascular heart disease. Neonates and smaller pediatric patients may benefit from higher–spatial resolution than standard detectors due to the small size of the developing anatomy, where the application of a high-resolution ROI detector may be warranted, as discussed further in Chapter 13.

Due to the rapid motion of the anatomical structures within the human torso, there has historically been a greater demand on temporal resolution of the imaging system. However, more and more clinicians are adapting to the use of lower frame rates (15 pulses per second or less) in order to better manage the radiation exposure to patients and staff (Sawdy et al. 2011). Angiography is most commonly used for diagnostic assessment of disease severity and in estimating the magnitude of a stenosis, as illustrated in Figure 22.6. Despite the cardiac motion, 3D-RA that can typically take several seconds of rotational acquisition time is being increasingly used in cardiac applications (Berman et al. 2012). Controlled respirations and interruption of the normal heart rhythm with rapid ventricular pacing or administration

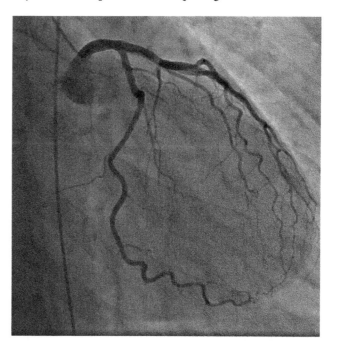

FIGURE 22.6 A coronary angiogram used for diagnostic assessment of disease severity.

of adenosine create a relatively stationary environment long enough to acquire the 3D image. 3D imaging in the interventional lab provides the ability to diagnose for both planning at the beginning of a procedure as well as posttreatment assessment such that adjustments can be made before completion of the intervention.

Successful treatment of nonvascular heart disease often requires visualization of the structures of the heart, which typically cannot be imaged with fluoroscopic x-ray equipment. CT, MR, and ultrasound do a much better job imaging the structures of the heart due to their increased ability to image soft, low attenuating tissues with superior low-contrast resolution. Multimodality fusion with CT or MR images overlaid and registered onto the live fluoroscopy image can be useful for guidance during interventional treatments involving structures of the heart, such as RF ablation of regions within the heart that produce abnormal electrical activity, or in proper placement of a transcatheter aortic valve, as illustrated in Figure 22.7. An important limitation of 3D-RA, CT, and MR fusion is that the overlay images are static. Ultrasound can be used to provide real-time imaging of soft tissues, such as the valve leaflets, which is of particular importance in the treatment of the mitral valve. Intravascular imaging tools are being increasingly utilized in the cardiac catheterization laboratory, for example, in the use of virtual histology intravascular ultrasound to assess atherosclerotic plaque composition, to aid in the

decision-making process of whether a stenosis warrants placement of a stent (Eshtehardi et al. 2011).

Mapping catheters are used to map the electrical activity of the heart and to identify the origins of the arrhythmia. This imaging is essential for identifying the abnormal tissue contributing to an arrhythmia for successful ablation treatment. The fusion of this electrical map overlaid onto the live fluoroscopy image results in improved guidance and placement of the ablation catheter tip.

Innovation in imaging technology continues to progress. Clever consideration of both improvement to image quality and better management of exposure to ionizing radiation continues to push the boundaries of image guidance capabilities. For example, tight collimation is known to benefit image quality by improving contrast resolution through the reduction in the amount of x-ray scatter impingent upon the image receptor. However, this comes at the cost of limiting the visualization of the peripheral anatomy, catheters, and other devices. Spot fluoroscopy, the merging of live fluoroscopy within a collimated region overlaid onto a previously acquired full FOV image, combines the benefits of having both a large field and a collimated field within a single image. ROI fluoroscopy using an x-ray beam attenuator with higher attenuation in the periphery maintains high image quality at the center of the image where the intervention is occurring, while significantly reducing dose in the periphery of the imaging field (Vasan et al. 2012).

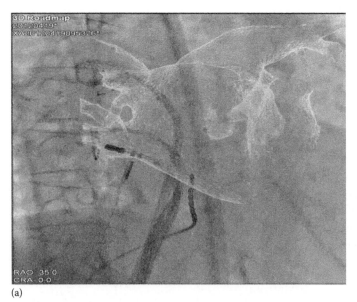
(a)

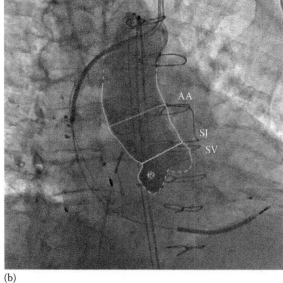
(b)

FIGURE 22.7 (a) Three-dimensional multimodality CT fusion roadmap used during an electrophysiology radiofrequency ablation procedure. The ablation and mapping catheters can be visualized under fluoroscopic guidance, and the 3D overlay provides context relative to the structures of the heart. (b) An aortography x-ray image with an overlay of important anatomical landmarks identified and measurements required for successful transcatheter aortic valve repair treatment.

22.3 Quality Assurance of X-Ray Imaging Equipment in the Interventional Suite

Accurate diagnosis and early detection of vascular disease requires well-performing imaging systems capable of detecting plaque buildup, blood clots, narrowing of vessel lumen, aneurysm, and other anatomical morphologies. Improved EIGIs for minimally invasive treatment of such diseases, including the ability to visualize fine anatomical detail and for accurate placement of treatment devices, also place an ever-increasing demand on imaging technologies. The workhorse imaging technology common to most vascular image-guided interventions continues to be x-ray fluoroscopy. High-quality images must be obtained with minimal x-ray irradiation to the patient to avoid adverse biological effects, thereby also requiring high sensitivity. It is therefore the role of the qualified medical physicist to ascertain imaging system performance, its adequacy for a given imaging task, and the preservation over the lifetime of clinical use.

Modern interventional x-ray equipment is designed to provide images of the highest clinical utility at the lowest possible radiation exposures under the extraordinarily diverse set of imaging conditions experienced from tiny neonates to severely obese adults, head to toe, fingertip to fingertip imaging. As a result, modern interventional x-ray equipment is extraordinarily complex. Automatic exposure control (AEC) and image quality control logic circuits dynamically adjust spectral shaping filters, radiologic technique parameters, and choice of x-ray filaments during the course of image acquisition (Lin et al. 2012). Detectors are configured with multiple modes at the hardware level that adapt to the particular imaging environment from low-dose fluoroscopy to high-dose DSA. Highly nonlinear and adaptive image processing is leveraged to improve the clinical presentation of images. The impacts of such complexities should be well understood and taken into consideration during the course of a quality assurance program.

An idealized quality assurance program is one that

1. Establishes a benchmark or baseline from which subsequent measurements can be compared. Ideally this occurs during the acceptance of the equipment.
2. Is easily reproducible. This facilitates more frequent testing that is less susceptible to measurement error.
3. Requires minimal test devices and objects. This lessens the burden of acquiring, maintaining, and transporting quality assurance tools and creates a greater opportunity for QA program consistency on a larger scale with outside institutions/facilities.
4. Provides comprehensive quantification of overall system performance, encompassing each major component of the imaging chain from the x-ray source to the image presented to the operator. If a system's performance has been found to deviate from the benchmark, the manufacturer's service engineer or facility biomedical engineer can be contacted to further investigate. Alternatively, more focused testing can be accomplished to isolate which component contributes to lost performance (e.g., x-ray source, detector, etc.).
5. Assures the quality of the imaging equipment is maintained.

The National Electrical Manufacturers Association Standard Publication XR 27-2012, x-ray Equipment for Interventional Procedures User Quality Control Mode, defines a set of minimum requirements designed to more easily facilitate quality control at the facility level and creates scaffolding for a robust QA program (NEMA 2012). Implementation of this standard is underway with a commitment from industry to incorporate into new equipment design configurations over the next few years. Quality control user interfaces described in this standard have been defined to provide controlled access for x-ray dose-related constancy testing, access to and export of both "for processing" and "for presentation" imaging data, calibration inputs for fields in the Digital Imaging and Communications in Medicine (DICOM) radiation dose structured report (RDSR), electronic documentation of dose-related parameters in exam protocols, and direct access to RDSRs. This level of equipment access provides the necessary controls to move toward an idealized QA program.

22.3.1 Quality Testing of Individual Components within the Imaging Chain

Modern imaging systems require a chorus of well-performing components within the imaging chain to maximize the quality of the images per unit radiation.

At present, few guidelines exist within the United States describing quality assurance and constancy testing of medical x-ray equipment used for interventional vascular imaging, although efforts are underway to outline a set of recommended tests for acceptance testing and quality control. The charge of AAPM Task Group 150 is to outline a set of tests to be used in the acceptance testing and quality control of digital radiographic imaging systems. Some European countries have been more proactive in developing standards to provide a minimum set of testing: for example, Germany has the DIN 6868-4 Standard describing constancy testing of medical x-ray equipment for fluoroscopy (DIN 2007). Ultimately, flexibility remains in customization of a quality assurance program that most appropriately meets the needs of an individual facility in the context of any local regulatory requirements and recommendations from the equipment manufacturer. Brief descriptions of common quality testing approaches for the major components of the imaging chain are described in the following. A more detailed and comprehensive description can be found elsewhere (Beutel et al. 2000, Shepard et al. 2002).

22.3.1.1 X-Ray Source: Focal Spot, X-Ray Output, Beam Quality, AEC

X-ray tubes tend to totally fail all at once, for example, via arching, and do not tend to degrade significantly over time (Stears et al. 1986, Erdélyi et al. 2009). As such, routine monitoring beyond acceptance testing may be unnecessary. The nonfinite focal spot size of the radiation source contributes to geometric unsharpness, influencing image quality and resolution. Measurement of the 2D focal spot size can be accomplished by observing the regions of blurring in a star or bar resolution test pattern or alternatively imaging of a pinhole or slit test object (NEMA 1992). X-ray output is commonly assessed by measuring linearity of the exposure versus milliamperes for a particular kilovolt and beam quality. Increasing surface roughness of the anode may result in a slight loss in intensity over time, decreasing the patient entrance skin dose over time (Mehranian et al. 2010). Modern fixed fluoroscopic interventional equipment have air kerma and dose area product (DAP) displays, where the air kerma is determined at the patient entrance reference point (PERP)—located 15 cm toward the x-ray source from the isocenter. FDA regulations require a minimum accuracy of ±35%, which can readily be verified by measuring the air kerma at the PERP and comparing with the displayed value (Lin et al. 2013). Interventional

systems are now being designed to include calibration factor inputs that can be stored along with detailed dose information in the DICOM RDSR to enable local adjustments to modify accuracy or to incorporate additional considerations toward estimation of patient dose. For example, including attenuation and scatter effects of the patient support, which are otherwise not included in the air kerma displays. Beam quality is often interpreted through half value layer measurements to ensure minimum requirements are met. AECs dynamically adjust the x-ray exposure conditions to maintain a sufficient level of exposure to the image receptor in order to maintain adequate image quality. Testing of the AEC most commonly involves a compliance check with maximum regulatory limits for fluoroscopic air kerma rates (AKRs). Additional testing may include the periodic measurement of the AKR or entrance skin exposure for a prescribed phantom(s) (Winston et al. 2003). AKR information is now being provided in the operations manual for the different modes of operation using a 20 cm thick polymethylmethacrylate phantom as prescribed in international standards (IEC 2010).

22.3.1.2 Detector: Artifacts, Uniformity, Signal, Noise, Resolution

The detection of detector artifacts and overall uniformity of detector response can be assessed using flat field images acquired under controlled imaging conditions, with known beam quality and exposure conditions. Artifacts in the form of dead or slow-responding pixels and other nonuniformities may be detected upon visual inspection of the flat field image. More subtle deviations can be quantified using a simple image analysis program, such as ImageJ (Rasband 2012), to quantify the behavior of the global image or regions within the image. For example, reviewing the histogram of the overall image enables the detection of outlying pixel performance. Regional analysis in the form of line profiles for signal trending or mean signal and standard deviation measurements within ROIs provide information on signal and noise uniformities (Floyd et al. 2001). Accurate gain and offset corrections at the detector level are often sufficient to correct for and overcome the minor imperfections of well-performing flat panel detectors.

Signal, noise, and signal-to-noise performance at the detector level are often quantified using the modulation transfer function (MTF), noise power spectrum (NPS), and detective quantum efficiency (DQE), which are metrics typically used by design engineers who work on maximizing the capabilities of detector technologies (Kuhls-Gilcrist 2009). The DQE describes the ability of the imaging detector to preserve the signal-to-noise ratio relative to an idealized detector. As the DQE incorporates both MTF and NPS, this is the metric of choice in the scientific community for assessment of imaging detector performance. Further, only the DQE can reliably describe the spatial resolution of an x-ray imaging detector in the presence of noise, parallax, and blurring (Moy 2000). In theory, the DQE can be used to compare absolute detector performance, when measured in carefully controlled laboratory environments. An international standard describes the methodology and conditions required for measurement of the DQE (IEC 2003). However, determination of both the MTF and NPS requires significant processing of the images acquired under the conditions defined in this standard, for which no standard exists to date defining such processing. Inconsistencies in various approaches impede the ability to compare absolute quantification unless identical processing steps were utilized. For example, determination of the MTF from an edge test device requires several image-processing steps to get to the desired end point: first converting the edge image to an edge spread function, next to a line spread function (LSF), and finally to the MTF (Samei et al. 1998). Different results may be obtained with utilization of different calculation methodologies (Neitzel et al. 2004). Inherent inaccuracies may also exist, for example, due to truncation of the LSF tails or from incorrect normalization (Friedman and Cunningham 2008). Additional limitations existing in the clinical environment, including the inability to get the proper imaging geometry (defined as a source-to-image distance of at least 1.5 m), push the DQE into a more qualitative metric that can be valuable in ensuring constancy of detector performance.

22.3.1.3 System: Protocols, Image Processing, Overall Performance

Exam protocol selection buttons (EPSBs), designed to optimize the image acquisition and display for the wide variety of clinical applications encountered by a single imaging system, have evolved to control a full set of programmed technical factors and control algorithms. Different protocols may be configured to modify or influence the behavior of the target detector exposure level, frame rates, beam filtration, detector settings (e.g., binning mode), and other parameters that impact the overall behavior of the imaging system. Several hundred exam protocol selections may be configured on a particular system, so testing of individual protocols is impractical. Recent efforts have focused at the system design level to provide output of electronic documentation of system configuration and technical factors including all dose-related parameters invoked by each EPSB (NEMA 2012). This capability will greatly facilitate local review and audit. Increasingly complex and often highly nonlinear image processing is often implemented to increase image quality through the suppression of noise, increase in contrast, reduction in halation, enhancement of edges, and increased device and catheter visualization. The characteristics of such image processing are often adjusted to meet individual clinician preference and may be incorporated into the EPSBs. Testing of "for presentation" images that ultimately include this processing becomes an object-specific endeavor, as different objects may be impacted differently. Overall system performance from the creation of the x-ray to the displayed image is often assessed using one's favorite test tool, for example, as shown in Figure 22.8. Fluoroscopic test tools often include a variety of test objects imbedded into a single phantom to facilitate more routine testing. Step-wedges, contrast objects, bar patterns, and uniform regions can be effective in providing both a qualitative and/or quantitative assessment of overall system performance in terms of contrast resolution, line-pair resolution, uniformity, radiation field size, and artifact detection. Comparison to measurements obtained during the acceptance testing of the equipment enables confirmation and assurance that the equipment continues to perform optimally.

22.3.2 Quality Assurance: A New Approach

Modern interventional x-ray imaging systems are complex. Modern hospital environments are also complex and often employ multiple interventional x-ray imaging systems in addition to a plethora of other x-ray and non-x-ray imaging equipment. To be most effective, routine quality assurance needs to be comprehensive, reproducible, and practicable. The procurement, maintenance, and implementation of a wide variety of test objects add a layer of complexity that must be managed appropriately. Improvements in understanding of the physical processes contributing to image formation in combination with improvements in the information and data being output from the imaging systems are unlocking the potential to move toward a more idealized quality assurance program (Cunningham 2000, Zhao et al. 2004, Mackenzie and Honey 2007, Hajdok et al. 2008). The following subsections describe

Chapter 22

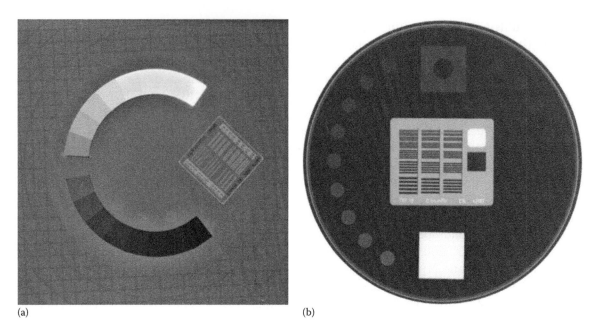

(a)   (b)

FIGURE 22.8   Example of common test objects utilized for image quality assessment of fluoroscopic systems. The (a) PTW NORMI®
RAD/FLU includes a step wedge with embedded contrast detail objects, bar pattern, and other test objects/regions. The (b) Leeds TOR
18FG similarly includes contrast detail objects, bar pattern, and other test objects/regions.

a new approach for quality assurance that is believed
to be easily implemented, highly accurate, and does
not require precisely machined test objects or other
specialized measurement devices. Ultimately, performance of the imaging equipment is judged by the
clinician, who relies on excellent and consistent image
quality in the context of radiation dose appropriateness. With the aid of the qualified medical physicist,
this can be best assured.

22.3.2.1   Instrumentation Noise Equivalent Exposure

The performance of high-sensitivity x-ray imagers
may be limited by additive electronic instrumentation
noise rather than by quantum noise when operated
at the low exposure rates used in fluoroscopic procedures. Instrumentation noise is often specified by vendors in units of electrons and medical physicists often
use the normalized noise power spectrum in units of
mm², both of which are generally not as intuitive or
as informative as would be a direct radiological representation in terms of exposure. Previous attempts to
gauge the quantum-noise-limited performance range,
most common of which examines the constancy of the
product of the NNPS and exposure (or similarly, the
constancy of DQE) with exposure to verify instrumentation noise-free operation over a particular range of
exposures, fall short of providing a direct quantitative
measure (Dobbins et al. 2006). The instrumentation

noise equivalent exposure (INEE) addresses this
need by providing a *direct* quantitative measure of
the quantum-noise-limited exposure range of x-ray
detectors by providing the exposure at which the additive instrumentation noise equals the quantum noise
(Yadava et al. 2008). Hence, the INEE moves toward
a clinically relevant representation of noise in terms
of detector exposure. The INEE can be directly determined through experimental measurements of image
noise (output signal variance or NPS for inclusion
of spatial frequency effects) as a function of incident
exposure, and does not require any specialized test
phantom/device outside of added spectral filtration.

Through the acquisition of flat field "for processing"
images obtained at known detector exposure levels,
ideally spanning the dynamic range of the image
receptor, the INEE can be readily quantified using the
following equation:

$$N^2 = k(E + \text{INEE})$$

where N^2 is the measured total noise or digital output
signal variance whose square is proportional to the signal, which is in turn proportional by a constant, k, to the
detector exposure, E, combined with the *INEE* (Rudin
et al. 2006). Further simplification can be achieved by
reducing the measurement to the collection of two sets
of images: dark field ($E = 0$) and flat field images at a single known exposure, $E_{\text{flat field}}$ (Kuhls-Gilcrist et al. 2010)

where the flat field noise includes both quantum noise and dark field or instrumentation noise:

$$INEE = \left(\frac{E_{\text{flat field}}}{N_{\text{flat field}}^2 - N_{\text{dark field}}^2} \right) N_{\text{dark field}}^2$$

At this simplest level, the frequency-independent INEE can be determined, representing the exposure at which the additive instrumentation noise, which is constant irrespective of exposure, equals the quantum noise, which scales with exposure. Detector exposures below this value would be considered to be instrumentation noise limited and would suffer additional degradation in image quality due to the electronic noise of the imager. For routine quality assurance, verification of the constancy of the frequency-independent INEE is likely sufficient to ensure the low-dose performance of the detector remains constant. Inclusion of spatial frequency effects using the NPS in place of the signal variance is described elsewhere and may be considered in more comprehensive assessments (Kuhls-Gilcrist 2009).

The INEE metric addresses the need for a direct, quantitative measure of the quantum-noise-limited exposure range of x-ray detectors by providing the threshold exposure at which the detector instrumentation noise exceeds the quantum noise. The INEE also provides insight into the overall detector performance in terms of the behavior of the DQE as a function of exposure, as described in further detail in the next subsection. When the radiation exposure reaching the detector approaches the INEE, the DQE will degrade by approximately a factor of two, resulting in a proportional loss in image quality (Kuhls-Gilcrist 2009). Thus, assessment of the INEE in conjunction with the DQE provides a complete assessment of detector performance. Confirming the constancy of the INEE assures that the quality of the detector in terms of electronic/instrumentation noise performance is maintained, hence confirming that the inherent low-dose object-independent imaging quality is similarly maintained. The next subsection describes how the DQE can be obtained using the same dataset acquired for INEE determination, consisting of dark field and flat field images (Kuhls-Gilcrist et al. 2011).

22.3.2.2 Detective Quantum Efficiency from Flat Field Images Using the Noise Response Approach

DQE is a powerful tool for assessment of imaging detector performance, as discussed in Section 22.3.1.2, as it incorporates both signal and noise performance

and provides insights into the dose efficiency of a given system. The DQE is a ratio of the output to input signal-to-noise ratio squared and is determined using the following equation:

$$DQE(f) = \frac{\text{Signal-to-Noise Ratio}_{\text{Out}}^2}{\text{Signal-to-Noise Ratio}_{\text{In}}^2} = \frac{\text{MTF}^2(f)\,\Phi}{\text{NPS}(f)}$$

where Φ is the x-ray quanta per unit area at the detector input, which is proportional to the detector exposure. The absolute magnitude of the DQE depends upon a variety of factors, including how the images used to measure the MTF and the NPS are acquired, processed, and analyzed. If a consistent approach is taken, the DQE can be effective in verifying if the performance is maintained within a given system (Cunningham 2008). Any deviations in measured DQE would be reflective of variations in performance of not only the detector, but also inclusive of the x-ray source and other x-ray system components, as described further in the following texts.

With knowledge of the DQE and the INEE, the detector is completely characterized and can be inferred as a function of detector entrance exposure. However, these metrics are not typically measured as part of a routine quality assurance program. This is in part due to the fact that previous methods for MTF (and hence DQE) measurement required precision test objects (slits, edges, etc.) that needed to be precisely aligned relative to the detector input (Fujita et al. 1992, Samei et al. 1998). Such methods have several inherent drawbacks (Kuhls-Gilcrist et al. 2010): they require a precision test object that may not be readily available; the test object needs to be precisely aligned relative to the detector input, which can be time consuming and a source of error; use of a slit or edge to infer detector resolution response itself is potentially prone to errors; and standard methods require considerable understanding of the physical processes involved with the measurement to ensure accuracy and are prone to "user" error. Such measurements have been further impeded by the fact that systems do not routinely enable an option to disable the highly nonlinear image processing used to improve the perceived image quality. Historically, it has not been practical to expect the medical physicist to perform these measurements routinely in the field. As a result, QA programs have often relied on less objective testing, such as examining limiting spatial or contrast resolution using phantoms similar to those shown in Figure 22.8. However, the paradigm may be shifting with two recent developments. Systems will increasingly offer quality control modes that provide access to and exportation for processing images, which include no image processing outside of detector level corrections (NEMA 2012), the first prerequisite for DQE measurement. Secondly, the advent of new approaches for measurement of the MTF (hence the DQE) that eliminate the need for and challenges associated with precision test objects by using the noise response of the detector promise to greatly simplify determination of the DQE while also reducing potential variability (Kuhls-Gilcrist et al. 2010).

Noise response, that is, the NPS, inherently includes the resolution response, that is, the MTF. Measured NPS is the sum of quantum and additive noise contributions and is a function of spatial frequency f and detector exposure E (Mackenzie and Honey 2007, Kuhls-Gilcrist et al. 2010):

$$\text{NPS}_{\text{Measured}}\left(f,E\right) = \text{NPS}_{\text{Quantum}}\left(f,E\right) + \text{NPS}_{\text{Additive}}\left(f\right)$$

which is similar to the equation using standard deviation for determination of the INEE. Flat field corrections sufficiently minimize structured noise, which would otherwise scale proportionally with the squared exposure; hence, this term is omitted. The additive electronic noise is a constant irrespective of exposure and can be determined from dark field images. Quantum noise contains contributions from primary quantum noise resulting from variations in the number of incident x-ray quanta, the Poisson excess noise resulting from variations in energy absorption in the detection layer, and secondary quantum noise due to variations in the number of detected secondary quanta, and can be written generically as

$$\text{NPS}_{\text{Quantum}}\left(f,E\right) = k_1 \frac{\text{MTF}^2\left(f\right)}{A_S\left(f\right)} + k_2$$

$$= \text{NPS}_{\text{flat field}}\left(f,E\right) - \text{NPS}_{\text{dark field}}\left(f\right)$$

where

k_1 and k_2 are system-dependent constants

$A_S(f)$ represents the frequency-dependent Swank factor, which is determined from knowledge of the absorption layer material and thickness (Hajdok et al. 2006, 2008, Kuhls-Gilcrist 2009)

The MTF can then be determined from a simple fitting technique using either a Gaussian mixture model or complementary error function to describe the MTF, using the same dataset acquired for determination of the INEE with several dark field and flat field images at a single exposure. Results of the fit can be used to directly obtain the MTF or to first determine k_1 and k_2 for subsequent calculation of the MTF (Kuhls-Gilcrist et al. 2010):

$$\text{MTF}(f)$$
$$= \sqrt{\frac{\text{NPS}_{\text{flat field}}\left(f,E\right) - \text{NPS}_{\text{dark field}}\left(f\right) - k_2}{k_1} A_S\left(f\right)}$$

Calculation of the DQE is then straightforward. In this manner, it is plausible to determine both INEE and DQE without using specialized test objects, thereby alleviating the associated burdens and potential sources of variability.

Ideally, systems primarily operate at exposure levels where quantum noise dominates and the additive noise contribution is relatively negligible, that is, in a quantum-noise-limited state. This ensures the most effective utilization of x-rays for a given system. At the INEE, the exposure where quantum noise and additive electronic instrumentation noise are equal, the DQE reduces to the following equation:

$$\text{DQE}(f,E) = \frac{\text{MTF}^2\left(f\right)\Phi}{\text{NPS}_{\text{Quantum}}\left(f,E\right) + \text{NPS}_{\text{Additive}}\left(f\right)}$$

$$= \frac{1}{2}\frac{\text{MTF}^2\left(f\right)\Phi}{\text{NPS}_{\text{Quantum}}\left(f,\text{INEE}\right)} \text{ at } E = \text{INEE}$$

which more clearly demonstrates that the INEE also represents the exposure at which the additive electronic noise contribution reduces the maximum measured DQE under quantum-limited conditions by a factor of two.

As described in Section 22.3.1.1, x-ray output is commonly assessed by measuring linearity of the exposure versus milliamperes for a particular kilovolt and beam quality. These data can be leveraged to determine Φ based on the technique utilized, reducing or eliminating the need to measure exposure in every instance. Any degradation in the measured INEE or DQE beyond a facilities audit level would be indicative that further more detailed testing is warranted and could be a result of changes in performance of the x-ray tube, detector, antiscatter grid, or other components described in Section 22.3.1.

22.3.2.3 Monitoring of Exposure Output via DICOM Radiation Dose Structured Report Data

Radiation exposure usage will be monitored increasingly from a clinical perspective with the advent of DICOM RDSRs consisting of both high-level total procedural dose data (e.g., cumulative air kerma or the total dose at the reference point) in addition to detailed dose and system information provided for each irradiation event. Dose analysis and reporting is further facilitated via the Integrating the Healthcare Enterprise Radiation Exposure Monitoring profile (IHE 2011). Changes in behavior in both system and clinician performance may be observable in examination of aggregate patient dose information as a function of time. Monitoring and analyzing the average procedural dose provides the ability to examine upward or downward trends that could be indicative of degradation or improvement in performance (Sawdy et al. 2011, Fetterly et al. 2012). A transition from machine dose metrics, such as air kerma and dose area product, to more meaningful patient dose metrics, such as peak skin dose, further refines the ability for quality audit as these metrics are inclusive of additional details such as backscatter, source-to-skin distance, and beam angulations (Bednarek et al. 2011). Chapter 24 describes dosimetric techniques in greater detail.

The establishment of reference levels in a variety of interventional imaging procedures is already taking place (Miller et al. 2009, 2012). With the advent of a national dose index registry specific to interventional image-guided procedures, capabilities to benchmark performance in the administration of radiation exposure will lend an opportunity to improve the state of practice and to perhaps more readily maintain or improve the quality of the imaging equipment and how it is operated.

22.4 Summary and Conclusion

Vascular disease, accounting for over 21% of all deaths worldwide, continues to be a major epidemic. Significant progress and rapid advancement of technologies are reducing morbidity and improving quality of life. New and exciting areas of and methodologies for healing, such as renal denervation for the treatment of hypertension (Esler et al. 2012), will continue to be discovered, while existing approaches steadily progress. Vascular imaging, as the cornerstone for both diagnosis and treatment, plays a critical role in any intervention and its quality assurance is equally as vital. A continued trend toward merging of complementary imaging modalities and informatics promises to enable increasingly complex and precise interventions. The future is clear.

References

Adamson, J., Beswick, A., and Ebrahim, S. 2004. Is stroke the most common cause of disability? *J. Stroke Cerebrovasc. Dis.* 13:171–177.

Bednarek, D.R., Barbarits, J., Rana, V.K., Nagaraja, S.P., Josan, M.S., and Rudin, S. 2011. Verification of the performance accuracy of a real-time skin-dose tracking system for interventional fluoroscopic procedures. *Proc. SPIE* 796127:1–8.

Ben-Dor, I., Maluenda, G., Dvir, D. et al. 2013. Balloon aortic valvuloplasty for severe aortic stenosis as a bridge to transcatheter/surgical aortic valve replacement. *Catheter. Cardiovasc. Interv.* 82:632–637.

Berman, D.P., Khan, D.M, Gutierrez, Y., and Zahn, E.M. 2012. The use of three-dimensional rotational angiography to assess the pulmonary circulation following cavo-pulmonary connection in patients with single ventricle. *Catheter. Cardiovasc. Interv.* 80:922–930.

Beutel, J., Kundel, H.L., and Van Metter, R.L. 2000. *Handbook of Medical Imaging*, Volume 1—Physics and Psychophysics. Bellingham, WA: SPIE.

Cunningham, I.A. 2000. Applied linear-systems theory. In *Handbook of Medical Imaging: Physics and Psychophysics*, Vol. 1, eds. J. Beutel, H.L. Kundel, and R.L. Van Metter, pp. 82–156. Bellingham, WA: SPIE.

Cunningham, I.A. 2008. Use of the detective quantum efficiency in a quality assurance program. *Proc. SPIE* 69133I:1–4.

De Bruyne, B., Pijls, N.H., Kalesan, B. et al. 2012. Fractional flow reserve-guided PCI versus medical therapy in stable coronary disease. *N. Engl. J. Med.* 367:991–1001.

Derdeyn, C. and Chimowitz, M. 2007. Angioplasty and stenting for atherosclerotic intracranial stenosis: Rationale for a randomized clinical trial. *Neuroimaging Clin. N. Am.* 17:355–363.

DIN 6868-4. 2007. Image quality assurance in diagnostic X-ray departments—Part 4: Constancy testing of medical X-ray equipment for fluoroscopy. German National Standard.

Dobbins, J.T., Samei, E., Ranger, N.T., and Chen, Y. 2006. Intercomparison of methods for image quality characterization. II. Noise power spectrum. *Med. Phys.* 33:1466–1475.

Dubinsky, R. and Lai, S. 2006. Mortality of stroke patients treated with thrombolysis: Analysis of a nationwide inpatient sample. *Neurology* 66:1742–1744.

Erdélyi, M., Lajkó, M., Kákonyi, R., and Szabó, G. 2009. Measurement of the x-ray tube anode's surface profile and its effects on the x-ray spectra. *Med. Phys.* 36:587–593.

Eshtehardi, P., Luke, J., McDaniel, M.C., and Samady, H. 2011. Intravascular imaging tools in the cardiac catheterization laboratory: Comprehensive assessment of anatomy and physiology. *J. Cardiovasc. Transl. Res.* 4:393–403.

Esler, M.D., Krum, H., Schlaich, M. et al. 2012. Renal sympathetic denervation for treatment of drug-resistant hypertension: One-year results from the Symplicity HTN-2 randomized, controlled trial. *Circulation* 126:2976–2982.

Feltes, T.F., Bacha, E., Beekman, R.H. et al. 2011. Indications for cardiac catheterization and intervention in pediatric cardiac disease: A scientific statement from the American Heart Association. *Circulation* 123:2607–2652.

Fetterly, K.A., Mathew, V., Lennon, R., Bell, M.R., Holmes, D.R., and Rihal, C.S. 2012. Radiation dose reduction in the invasive cardiovascular laboratory: Implementing a culture and philosophy of radiation safety. *JACC Cardiovasc. Interv.* 5:866–873.

Floyd, C.D., Warp, R.J., Dobbins, J.T. et al. 2001. Imaging characteristics of an amorphous silicon flat-panel detector for digital chest radiography. *Radiology* 218:683–688.

Friedman, S.N. and Cunningham, I.A. 2008. Normalization of the modulation transfer function: The open-field approach. *Med. Phys.* 35:4443–4449.

Fujita, H., Tsai, D.Y., Itoh, T. et al. 1992. A simple method for determining the modulation transfer function in digital radiography. *IEEE Trans. Med. Imaging* 11:34–39

Greer, P.B. and Van Doorn, T. Evaluation of an algorithm for the assessment of the MTF using and edge method. *Med. Phys.* 27:2048–2059.

Hajdok, G., Battista, J.J., and Cunningham, I.A. 2008. Fundamental x-ray interaction limits in diagnostic imaging detectors: Frequency-dependent Swank noise. *Med. Phys.* 35:3194–3204.

Hajdok, G., Yao, J., Battista, J.J., and Cunningham, I.A. 2006. Signal and noise transfer properties of photoelectric interactions in diagnostic x-ray imaging detectors. *Med. Phys.* 33:3601–3620.

Henderson, R.A., Pocock, S.J., Clayton, T.C. et al. 2003. Seven-year outcome in the RITA-2 trial: Coronary angioplasty versus medical therapy. *J. Am. Coll. Cardiol.* 42:1161–1170.

Higashida, R., Lahue, B., Torbey, M., Hopkins, L.N., Leip, E., and Hanley, D. 2007. Treatment of unruptured intracranial aneurysms: A nationwide assessment of effectiveness. *Am. J. Neuroradiol.* 28:146–151.

Hoeffner, E.G., Mukerji, S.K., Srinivasan, A., and Quint, D.J. 2012. Neuroradiology Back to the Future: Brain Imaging. *AJNR Am. J. Neuroradiol.* 33(1):5–11.

IEC 60601-2-43 Ed. 2.0:2010. 2010. Medical electrical equipment—Part 2-43: Particular requirements for the basic safety and essential performance of X-ray equipment for interventional procedures.

IEC 62220-1 Ed. 1.0:2003. 2003. Medical electrical equipment—Part 1: Determination of the detective quantum efficiency.

IHE Radiology Technical Framework, Volume 1 Revision 11.0 2012. Integrating the Healthcare Enterprise (IHE). http://www.ihe.net.

Ionita, C.N., Rudin, S., Hoffmann, K.R., and Bednarek, D.R. 2005. Microangiographic image guided localization of a new asymmetric stent for treatment of cerebral aneurysms. *Proc. SPIE* 5744:354–365.

Johnson, D., Coley, S.J., Kyrion, J., and Taylor, W.J. 2001. Comparing the performance of mono- and biplane fluoroscopy systems in diagnostic and interventional neuroangiography using the dose-area product. *Neuroradiology* 43:728–734.

Kahn, P., Yashar, P., Ionita, C.N., Jain, A., Rudin, S., Levy, E.I., and Siddiqui, A.H. 2012. Endovascular coil embolization of a very small ruptured aneurysm using a novel micro-angiographic technique: Technical note. Accepted for publication in Journal of NeuroInterventional Surgery (JNIS).

Kuhls-Gilcrist, A., Bednarek, D.R., and Rudin, S. 2009. Component analysis of a new solid state X-ray image intensifier (SSXII) using photon transfer and instrumentation noise equivalent exposure (INEE) measurements. *Proc. SPIE* 725817:1–10.

Kuhls-Gilcrist, A.T. 2009. Frequency dependent instrumentation noise equivalent exposure (INEE). In Development and evaluation of a new radiographic and fluoroscopic imager based on electron-multiplying CCDS: The solid state x-ray image intensifier. PhD dissertation. Ann Arbor, MI: Proquest LLC.

Kuhls-Gilcrist, A.T., Bednarek D., and Rudin S. 2010. A new simple, accurate, and quantitative approach for routine quality assurance in digital radiography. *Med. Phys.* 37:3359.

Kuhls-Gilcrist, A.T., Jain, A., Bednarek, D.R., Hoffmann, K.R., and Rudin, S. 2010. Accurate MTF measurement in digital radiography using noise response. *Med. Phys.* 37:724–735.

Kuhls-Gilcrist, A.T., Jain, A., Bednarek, D.R., and Rudin, S. 2011. Measuring the presampled MTF from a reduced number of flat-field images using the Noise Response (NR) method. *Proc SPIE* 79614G:1–6.

Kuhls-Gilcrist, A.T., Patel, V., Ionita, C. et al. 2006. New microangiography system development providing improved small vessel imaging, increased contrast to noise ratios, and multiview 3D reconstructions. *Proc. SPIE* 61423M:1–9.

Kuhls-Gilcrist, A.T, Yadava, G., Patel, V., Jain, A., Bednarek, D.R., and Rudin S. 2008. The solid-state x-ray image intensifier (SSXII): An EMCCD-based x-ray detector. *Proc. SPIE* 69130K:1–10.

Levine, G.N., Bates, E.R., Blankenship, J.C. et al. 2011. 2011 ACCF/AHA/SCAI Guideline for Percutaneous Coronary Intervention: A report of the America College of Cardiology Foundation/American Heart Association Task Force on Practice Guidelines and the Society for Cardiovascular Angiography and Interventions. *Circulation* 124:e574–e651

Lin, P., Rauch, P., Balter S. et al. 2012. Functionality and operation of fluoroscopic automatic brightness control/automatic dose rate control logic in modern cardiovascular and interventional angiography systems. A Report of AAPM Task Group 125 Radiography/Fluoroscopy Subcommittee, Imaging Physics Committee, Science Council, One Physics Ellipse, College Park, Washington, DC.

Lin, P., Schueler, B.A., Balter, S. et al. 2013. DRAFT Report. Accuracy and calibration of integrated radiation output indicators in radiology—A Report of AAPM Task Group 190, Radiography and Fluoroscopy Sub-committee, Imaging Physics Committee, Science Council, One Physics Ellipse, College Park, Washington, DC.

Lloyd-Jones, D., Adams, R., Carnethon, M. et al. 2009. Heart disease and stroke statistics—2009 Update: A report from the American Heart Association Statistics Committee and Stroke Statistics Subcommittee. *Circulation* 119: e21–e181.

Mackenzie, A. and Honey, I.D. 2007. Characterization of noise sources for two generation of computed radiography systems using powder and crystalline photostimulable phosphors. *Med. Phys.* 34:3345–3357.

Mehranian, A., Ay, M.R., Riyahi Alam, N., and Zaidi, H. 2010. Quantifying the effect of anode surface roughness on diagnostic x-ray spectra using Monte Carlo simulation. *Med. Phys.* 37:742–752.

Mensah, G.A. and D.W. Brown. 2007. An overview of cardiovascular disease burden in the United States. *Health Aff.* 26(1):38–48.

Miller, D.L., Hilohi, C.M., and Spelic, D.C. 2012. Patient radiation doses in interventional cardiology in the U.S.: Advisory data sets and possible initial values for U.S. reference levels. *Med. Phys.* 39:6276–6285.

Miller, D.L., Kwon, D., and Bonavia, G.H. 2009. Reference levels for patient radiation doses in interventional radiology: Proposed initial values for U.S. practice. *Radiology* 253:753–764.

Moy, J. 2000. Signal-to-noise ratio and spatial resolution in x-ray electronic imagers: Is the MTF a relevant parameter? *Med. Phys.* 27:86–93.

Neitzel, U., Gunther-Kohfahl, S., Borasi, G., and Samei, E. 2004. Determination of the detective quantum efficiency of a digital x-ray detector: Comparison of three evaluations using a common image data set. *Med. Phys.* 31:2205–2211.

NEMA XR 5-1992. 1992. *Measurement of Dimensions and Properties of Focal Spots of Diagnostic X-Ray Tubes*. Rosslyn, VA: National Electrical Manufacturers Association.

NEMA XR 27-2012. 2012. *X-Ray Equipment for Interventional Procedures User Quality Control Mode*. Rosslyn, VA: National Electrical Manufacturers Association.

Payne, M.M. 2001. Charles Theodore Dotter—Father of Intervention. *Tex. Heart Inst. J.* 28:28–38.

Rasband, W.S. 2012. ImageJ, U.S. National Institutes of Health. Bathesda, MD. http://imagej.nih.gov/ij/.

Rowlands, J.A. and Taylor, K.W. 1983. Absorption and noise in cesium iodide x-ray image intensifiers. *Med. Phys.* 10:786–795.

Rudin, S., Bednarek, D.R., and Hoffmann, K.R. 2008. Endovascular image-guided interventions EIGIs. *Med. Phys.* 35:301–309.

Rudin, S., Kuhls, A.T., Yadava, G.K. et al. 2006. New light-amplifier-based detector designs for high spatial resolution and high sensitivity CBCT mammography and fluoroscopy. *Proc. SPIE* 61421R:1–11.

Rudin, S., Wang, Z., Kyprianou, I. et al. 2004. Measurement of flow modification in phantom aneurysm model: Comparison of coils and a longitudinally and axially asymmetric stent—Initial findings. *Radiology* 231:272–276.

Samei, E., Flynn, M.J., and Reimann, A. 1998. A method for measuring the presampled MTF of digital radiographic systems using an edge test device. *Med. Phys.* 25:102–113.

Sawdy, J.M., Kempton, T.M., Oishove, V. et al. 2011. Use of a dose-dependent follow-up protocol and mechanisms to reduce patients and staff radiation exposure in congenital and structural interventions. *Catheter. Cardiovasc. Interv.* 78:136–142.

Shepard, S.J., Lin, P., Boone, J.M. et al. 2002. AAPM Report No. 74—Quality control in diagnostic radiology. Report of Task Group #12, Diagnostic X-Ray Imaging Committee.

Siewerdsen, J.H., Antonuk, L.E., El-Mohri, Y. et al. 1997. Empirical and theoretical investigation of the noise performance of indirect detection, active matrix flat-panel imagers (AMFPIs) for diagnostic radiology. *Med. Phys.* 24:71–89.

Stears, J.G., Felmlee, J.P., and Gray, J.E. 1986. Half-value-layer increase owing to tungsten buildup in the x-ray tube: Fact or fiction. *Radiology* 160:837–838.

Stergiopoulos, K. and Brown, D.L. 2012. Initial coronary stent implantation with medical therapy vs medical therapy alone for stable coronary artery disease. *Arch. Intern. Med.* 172:312–319.

Teirstein, P.S. and Price, M.J. 2012. Left main percutaneous coronary intervention. *J. Am. Coll. Cardiol.* 60:1605–1613.

Vasan, S.N., Panse, A., Jain, A. et al. 2012. Dose reduction technique using a combination of a region of interest (ROI) material x-ray attenuator and spatially different temporal filtering for fluoroscopic interventions. *Proc. SPIE* 831357.

Vedantham, S., Karellas, A., and Suryanarayanan, S. 2004. Solid-state fluoroscopic imager for high-resolution angiography: Parallel-cascaded linear systems analysis. *Med. Phys.* 31:1258–1268.

Winston, J.P, Best, K., Plusquiellic, L., Thoma, P., and Gilley, D.B. 2003. Conference of Radiation Control Program—Patient exposure and dose guide. Publication E-03-2.

Wolf, P.A., Abbott, R.D., and Kannel, W.B. 1991. Atrial fibrillation as an independent risk factor for stroke: The Framingham Study. *Stroke* 22:983–988.

Wood, M.A., Brown-Mahoney, C., Kay, G.N., and Ellenbogen, K.A. 2000. Clinical outcomes after ablation and pacing therapy for atrial fibrillation. *Circulation* 101:1138–1144.

Yadava, G.K, Kuhls-Gilcrist, A.T., Rudin S., Patel, V.K., Hoffmann, K.R., and Bednarek, D.R. 2008. A practical exposure-equivalent metric for instrumentation noise in x-ray imaging systems. *Phys. Med. Biol.* 53:5107–5121.

Zhao, W., Ji, W.G., Debrie, A., and Rowlands, J.A. 2003. Imaging performance of amorphous selenium based flat-panel detectors for digital mammography: Characterization of a small area prototype detector. *Med. Phys.* 30:254–263.

Zhao, W., Ristic, G., and Rowlands, J.A. 2004. X-ray imaging performance of structured cesium iodide scintillators. *Med. Phys.* 31:2594–2605.

Chapter 22

23. Angiography for Radiosurgery and Radiation Therapy

Carlo Cavedon and Joseph Stancanello

23.1 Introduction

Stereotactic radiosurgery (SRS) is an image-based treatment modality characterized by high spatial accuracy and the delivery of high doses of ionizing radiation within a well-defined, usually small intracranial target.[1] In its conventional implementation, the treatment is generally performed in one single fraction or—in more recent applications—in a hypofractionated regime.[2,3] The technique dates back to the 1950s when Lars Leksell—a prominent Swedish neurosurgeon—first proposed to focus collimated x-rays (gamma rays in later developments) onto a small volume to act as a noninvasive alternative to open-skull neurosurgery; this pioneering idea led to the introduction of the gamma knife in 1968.[4] Several different approaches followed, including cyclotron-based irradiation[5] and—above all—linear accelerator–based SRS,[6,7] which paved the way to the worldwide diffusion of the technique.

It was soon clear that the method could be applied to the treatment of vascular disorders of the brain such as arteriovenous malformations (AVMs), especially if deep-sited and difficult or impossible to treat surgically.[8–12] Even more than neoplastic lesions, however, AVMs need to be treated with high doses of ionizing radiation within a volume characterized by sharply defined boundaries, which prompts the need for accurate, precise, and robust

image-based targeting of the lesion. As a development of the (surgical) stereotactic approach, SRS is founded on the translation of image features onto a Cartesian or pseudo-Cartesian reference frame for accurate localization, aided by stereotactic frames rigidly connected to the patient's skull. The concept is quite straightforward as long as the correspondence between single points in space is concerned, but might be tricky when extensive volumes shall be dealt with, especially if image guidance is provided by projective images only. Anyway, the stereotactic technique applied to projective angiography formed the basis for the angiographic guidance of SRS of AVMs.[8,9]

Several improvements have been developed since then and continue to be proposed nowadays. One of the most important was probably the translation from 2D approaches to 2D–3D techniques and then to full 3D implementations, which also allowed frameless methods to be developed. This chapter deals with the methods used to guide SRS and radiotherapy by means of angiographic images; this will be done by analyzing both conventional frame-based approaches and modern frameless techniques. The chapter also includes considerations on quality assurance and a tentative projection into future developments of the current standard of practice.

23.2 Frame-Based Angiographic Localization

Treatment of AVMs by means of SRS requires accurate localization of the target to irradiate, in order to achieve the desired effect with the delivery of high doses while sparing nearby tissues and critical structures. This has been possible since the clinical introduction of AVM radiosurgery through the use of angiographic stereotactic localizers.[13]

Stereotactic frames connect rigidly to the patient's skull, thus providing a reliable reference frame for the stereotactic coordinates. Imaging localizers are typically connected to the stereotactic frame to provide a spatial correspondence between anatomic details seen on the images and the stereotactically defined space (Figure 23.1).

Localizers have been developed to target details visible on projective images since the beginning of stereotaxy, not only for SRS but also for surgical applications, for example, stereotactic biopsy. With the availability of CT scanners, however, stereotactic localizers began to be used to provide a correspondence between details visible on tomographic axial slices and coordinates of a stereotactic space. A typical example of CT stereotactic localizer is the Brown–Roberts–Wells system[14] (Figure 23.2).

Transverse coordinates (i.e., 2D coordinates in the axial plane) are directly derived from in-plane distances—once calibration has been performed—while the position of the slice on the longitudinal axis (SUP-INF direction) is given by the distance between radio-opaque details of objects mounted on the stereotactic frame and extending longitudinally to cover the treatable volume (Figure 23.3).

In the case of angiography, however, 3D techniques have become available much later compared to the clinical introduction of CT scanners, and the stereotactic localization of vascular targets has relied on projective images for much longer than, for example, for tumors. In addition, an essential information in the management of AVMs is the distinction between arterial and venous phases, which requires a temporal resolution of the order of no more than a few tens of milliseconds, readily available in 2D-projective angiography but achievable with 3D techniques only with specialized applications[15] (see also Sections 23.3 and 23.5). For these reasons, 2D techniques are still in widespread use for the stereotactic localization of brain AVMs.

A stereotactic angiographic localizer is pictured in Figure 23.1, and an example of anterior-posterior (AP) and latero-lateral (LL) projections is reported in Figure 23.4. Due to the typical magnification of details closer to the radiation source compared to objects closer to the detector plane, the localizing box seen in Figure 23.1 is characteristically distorted on both projections of Figure 23.4. In addition, no information on the position of the target along the line connecting the source to the detector is obviously available from one projection alone. A point in the 3D space is traceable only combining two (possibly orthogonal) or more projections, while the shape of extensive objects is difficult or impossible to reconstruct from two planar images. As a result, projective angiographic targeting of AVMs—though essential because of the valuable timing information associated with the distinction between arterial and venous phases—is limited by the difficult geometric reconstruction of irregularly

(a)   (b)

FIGURE 23.1 (a) Localizer box mounted on a stereotactic frame. The black crosses ("×" and "+") identify fiducial points of reference visible on projections. (b) Localizer box and stereotactic frame disconnected to each other.

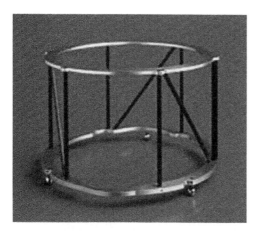

FIGURE 23.2 The Brown–Roberts–Wells stereotactic CT localizer. Notice the three N-shaped sets of rods.

shaped targets. This limitation was less important in the early years of SRS, when spherically shaped dose distributions were often used, but constitutes a severe restriction in modern techniques that exploit the full capability of current radiotherapy and radiosurgery systems to obtain irregularly shaped dose distributions, highly conformal to the volumes to treat.

Two strategies are possible to overcome the limitations of projective (sometimes called 2D–3D)

angiographic localizers: (1) using true 3D information, such as data provided by 3D rotational angiography (3DRA)[16] or CT angiography (CTA),[17] however losing much of the timing information inherent to angiographic datasets, and (2) combining 2D–3D data to tomographic datasets through image registration, at the expense of a lower spatial accuracy compared to direct frame-based angiographic techniques. The development of full 3D angiographic techniques with a sufficient time resolution to be used alone in SRS is still under investigation.[18]

3DRA is a technique generally performed without a stereotactic frame, while CTA and MR angiography (MRA) can be performed with the aid of a stereotactic frame fixed to the patient's skull. However, since the information obtained by means of these techniques is generally acquired in frameless examinations and used in SRS treatment planning through image registration, these modalities will be treated in some detail in Section 23.3.

SRS treatment planning is usually performed on CT volumes because of the high spatial accuracy and freedom from distortion of CT datasets, and because Hounsfield units—the quantity used to describe the

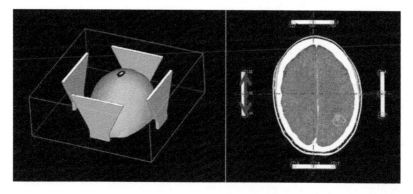

FIGURE 23.3 On the Fischer stereotactic frame as in many frame types, the longitudinal coordinate (SUP-INF direction) is provided by the distance between radiopaque details on the four localizers (right panel). A similar concept is used in N-shaped localizers (e.g., Figure 23.2).

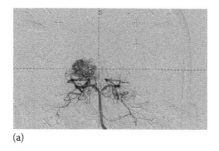

(a)

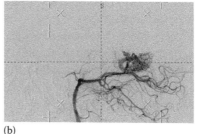

(b)

FIGURE 23.4 (a) AP and (b) LL angiographic views obtained with the stereotactic localizer box pictured in Figure 23.1. The face marked with four "×" markers is closer to the x-ray source than the face with the five "+" markers in the AP view; the opposite is seen on the LL view.

content of CT voxels—are strictly related to attenuation coefficients and hence adequate to be used in dose calculation algorithms and models. In selected applications, treatment planning is more often performed on MR volumes than on CT datasets. This happens, for example, in brain radiosurgery with the gamma knife, when MR geometrical distortion is known to be within acceptable values and information on the distribution of attenuation coefficients is not needed for dose calculation, to take advantage of the superior capacity of MRI to highlight tissue differences. Anyway, in both cases (primary CT or MR data), angiographic localization must be reported to tomographic volumes by means of image registration techniques. The class of algorithms that allow the registration of projective images to tomographic modalities to be performed is usually referred to as 2D–3D image registration.[19–21] 2D–3D image registration can be based on the mutual mechanical coherence of stereotactic localizers, on features segmented from the images or on the mutual information between different modalities.[18] In the latter two cases, the application of 2D–3D methods to angiographic datasets requires unsubtracted images to be used in order to provide the system with sufficient common details between angiography and CT; however, these methods belong mostly to the use of frameless radiosurgery systems. Therefore, they will be treated in more detail in the next section.

In conclusion, frame-based SRS of AVMs or other vascular disorders of the brain often requires angiographic images to be coregistered to CT and/or MR datasets, in order to help targeting the disease both from the geometrical and functional standpoints (e.g., considering also the distinction between arterial and venous phases). The use of a stereotactic frame helps in improving the overall accuracy of the procedure. However, full 3D information is generally not available from angiographic projections; this restriction is increasingly considered a severe limitation as even more conformal dose distributions can be obtained with modern treatment technology.

23.3 Frameless Angiographic Localization

The development of medical imaging technology, including robust and accurate image registration techniques, has led to the introduction of frameless radiosurgery and to stereotactic radiotherapy of anatomical regions other than the brain.[22] The avoidance of a stereotactic frame rigidly connected to the patient's skull has obvious implications on patient's comfort, but probably its most important advantage is the possibility to perform staged (i.e., fractionated) treatments, which allows clinicians to widen the spectrum of indications in terms of maximum treated volumes and disease type. In particular, stereotactic approaches are increasingly being proposed for extracranial applications including lung, liver, and prostate tumors. Even the treatment of vascular malformations has been possible outside the brain, thanks to frameless techniques, for example, in the case of spinal AVMs.[23]

Frameless radiosurgery can be performed by means of conventional linear accelerators or by dedicated systems. Examples of dedicated, frameless radiosurgery systems are the CyberKnife (Accuray Inc., CA) and the Novalis (Brainlab AG, Germany), pictured in Figures 23.5 and 23.6.

Both systems exploit 2D–3D registration algorithms to perform target localization through the comparison of projective images acquired at the time of treatment to CT datasets used for treatment planning.[24,25] In this scenario, however, projective angiography in the form used in frame-based radiosurgery is not readily available, because of the lack of a common reference frame. 3DRA has become an essential source of information for the treatment of vascular disorders by means of frameless radiosurgery.[12,26] 3DRA datasets for treatment

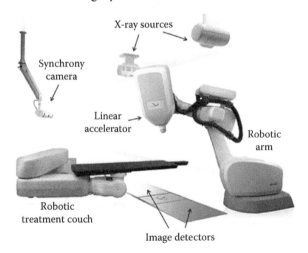

FIGURE 23.5 The CyberKnife robotic radiosurgery system (Accuray Inc., CA). Notice the stereoscopic x-ray system with ceiling-mounted sources and in-floor imagers. Projective images obtained by this system are coregistered to pretreatment planning CT to provide 2D–3D image guidance.

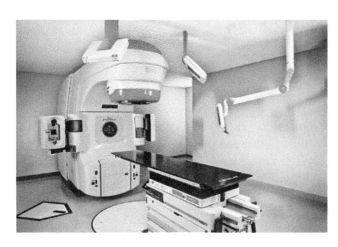

FIGURE 23.6 The Novalis radiosurgery system (Brainlab AG, Germany). The source-to-detector geometry is inverse compared to the CyberKnife system (ceiling-mounted imagers and in-floor x-ray sources).

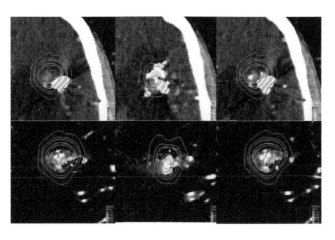

FIGURE 23.7 Example of coregistration between a CT and a 3D rotational angiography (3DRA) dataset, shown on axial slices with isodose lines from a radiosurgery treatment plan. The target is a previously embolized arteriovenous malformation (embolization material seen on CT—subtracted on 3DRA).

planning are usually integrated to other pretreatment imaging modalities by means of image registration. An example of integration between 3DRA and CT is reported in Figure 23.7.

3DRA is a tomographic modality obtained by means of a cone-beam CT technique; therefore, image formation is based on multiple projections acquired through x-ray transmission from several angles, much like conventional fan-beam CT.[27] The similar nature of the tomographic data obtained by the two techniques facilitates image registration, but special attention is needed because the voxel content of CT and 3DRA is not necessarily correspondent (not even proportional), and 3DRA images are

typically processed to enhance the vasculature at the expense of skeletal details—in a more complex and articulate way compared to subtractive angiography. For this reason, registration between 3DRA and CT should always be considered a multimodality problem and addressed by registration algorithms optimized for multimodality image registration, such as normalized mutual information.[23,28] Intramodality registration techniques such as cross-correlation-based algorithms might work properly under optimal assumptions but fail to provide sufficient accuracy in general; therefore, they should be avoided or used very carefully after sufficient validation. In addition, not all vendors allow 3DRA tomographic datasets to be extracted from the angiographic system for post-processing. This aspect should be considered when a decision is made on the acquisition of an angiographic system if SRS of vascular malformations is performed in the medical center.

Unfortunately, 3DRA cannot provide high time resolution, and therefore, it is not always adequate for the distinction between arterial and venous phases in the characterization of a vascular lesion.[15,29] Attempts are being made to solve this problem using different strategies[30] and difficulties are not yet overcome; however, the prospected technological development in the next few years might include readily available imaging protocols capable of including timing information in full 3D angiographic datasets.

An alternative to the use of 3DRA is CTA.[31] Modern CT scanners offer an imaging speed that enables the accurate characterization of vascular malformations, but difficulties remain such as poorer selectivity compared to angiographic imaging and the need to meet patient's safety standards that is not always easy to achieve in a general-purpose CT room.[32] Similar considerations hold for MRA, for which the time resolution is even poorer and patient's safety requirements prevent some specialized angiographic procedures to be performed. Nonetheless, due to the "true" tridimensional nature of CT/MR datasets and the relative easiness of registration to other CT volumes used for planning, CTA and MRA remain important sources of information for treatment planning of vascular malformations of the brain and of extracranial sites. In addition, ever-faster applications are being proposed in recent years, possibly leading to the development of extremely high temporal resolution methods that will eventually take the place of 2D and 3D angiography in SRS and radiotherapy planning.[33]

Chapter 23

23.4 Quality Assurance in Angiography-Guided Radiosurgery and Radiotherapy

Accuracy and precision are essential factors in treatment modalities that deliver high doses in a limited number of treatment fractions. Radiosurgery is a well-established modality for which quality assurance guidelines generally exist.[22,34–36] However, the attention to vascular imaging for treatment planning is rather weak in QA protocols adopted by scientific societies and/or published on the literature, possibly due to the limited number of centers that perform radiosurgery of vascular malformations compared to SRS in general.

Despite the scarcity of guidelines, people involved in SRS or radiotherapy who use angiographic imaging for treatment planning should set up their own quality assurance program. In particular, care should be taken to test for spatial accuracy and distortion, as angiography is generally used for diagnostic and interventional procedures that are not especially sensitive to small deviations from the optimal geometric behavior, being used in an autoconsistent imaging environment. On the other hand, use of angiographic datasets for radiosurgery and radiotherapy treatment planning requires their superposition/registration to independent, 3D-tomographic datasets characterized by high spatial accuracy like CT volumes. The possible spatial incoherence between the two modalities might cause significant mismatch and eventually patient harm if not properly accounted for.

Aspects for which more guidance on quality assurance would be needed in angiography-guided radiosurgery and radiotherapy include

- Spatial accuracy in general, including image distortion over the whole field of view
- Visibility of critical details in the presence of high-density embolization materials, and their impact on dose calculation (CTA used for treatment planning)[37]
- Suitability of the angiographic datasets for coregistration with CT and/or MR used for treatment planning
- Stability over time of parameters and quantities used to define the target volume based on image details (especially in case of automatic segmentation)

Hopefully, working groups will be organized by scientific societies to produce documents that set the quality standards in this field, especially in view of the expected increase of multimodality imaging use in radiosurgery and radiotherapy treatment planning.

23.5 Future Prospect

23.5.1 MR Vascular Imaging

Ultrahigh field MR scanners are still considered investigational devices. As of now, about 40 7T scanners for humans are installed worldwide and several applications are being investigated to evaluate the benefit-to-cost ratio of 7T over lower field scanners, such as 1.5T and 3T.

MRA is an area of active investigation at 7T, in particular aiming at identifying microvessels with high spatial resolution (about 0.1 mm) in a clinically acceptable time frame.[38] Diagnosis of small aneurysms[39] and AVMs[40] are likely to benefit from such a spatial resolution, in addition to visualization of structures historically difficult to image by means of MRA such as renal[41] or lower extremity arteries,[42] especially in noncontrast-enhanced acquisitions.

Because of their improved signal-to-noise and contrast-to-noise ratios, one of the advantages of ultrahigh field scanners is the capability to detect small tissue differences and the possibility to avoid contrast agent administration in selected cases. MRA techniques usually employed at 7T are time-of-flight (TOF) (Figure 23.8) or phase contrast (PC). Both methods have proven to add significant value over lower field acquisitions in small vessel detection using noncontrast-enhanced techniques.[40]

The future role of 7T MRA in radiosurgery and radiation therapy might be at each step of the theragnostic chain:

- Accurate identification of vessel structures difficult to visualize at a lower field strength, such as small AVMs with the related feeding and draining vessels
- Identification of microvessels around tumor to assess the stage/aggressiveness of the tumor itself[43]
- Visualization of microvessels around tumor to indicate regions of angiogenesis to be considered in treatment planning phase, for example, increasing the CTV or boosting the dose in a given region

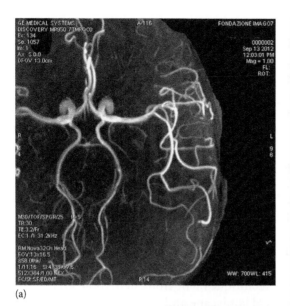

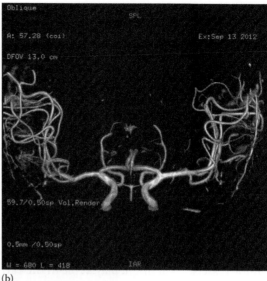

(a)   (b)

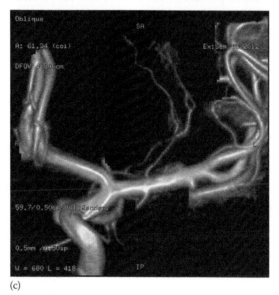

(c)

FIGURE 23.8 (a) Example of time-of-flight (TOF) 7T MR angiography. Axial maximum intensity projection of the circle of Willis with a depiction of distal branch of the main intracranial arteries. (b) 3D volume reconstruction of TOF acquisition of carotid intracranial branches. (c) Magnification of the left carotid siphon bifurcation. Note ventricular striate arteries originating from M1 segment of the middle cerebral artery. (Courtesy of Mirco Cosottini and Michela Tosetti, IMAGO7 Pisa, Italy.)

- Visualization of microvessels as a predictor of treatment success/failure in posttreatment follow-up, in the scenario of a personalized (radio)therapy approach

Ongoing investigations worldwide will clarify the potential role of 7T MR scanner, especially in view of the cost of such equipment versus the actual clinical value, beyond the research arena where 7T is currently placed.

Recently proposed approaches aim at describing the nature of vessels in terms of macro- versus microvasculature. MRI techniques based on spin echo (SE) and gradient echo (GRE) in a single, fast sequence, for example, a specifically designed echo planar imaging, offer the advantage of visualizing vessel size imaging (VSI)[44] or calculating vessel size index.[45] The rationale behind such an approach is that GRE, capable to calculate T2*-weighted images or maps, is sensitive to a broad range of "macro" vessel diameter (hiding microvasculature) while SE, capable to calculate T2-weighted images or maps, exhibits a peak sensitivity to microvasculature (arterioles, capillaries, and venules).

Chapter 23

VSI holds the potential to improve all the steps of the theragnostic chain, as already highlighted for 7T MR scanner, such as staging, treatment plan, and follow-up. In particular, VSI can also be used when a new treatment modality is proposed, for example, radiotherapy by means of microbeams (MRT). This technique is based on highly collimated, quasiparallel arrays of x-ray microbeams of 50–600 kV, produced by third-generation synchrotron sources.[46] MRT uses arrays of narrow (approximately 25–100 μm wide) microplanar beams separated by wider (100–400 μm center-to-center) microplanar spaces. The height of these microbeams typically varies from 1 to 100 mm. Results suggest that VSI can be more accurate than other parameters to optimize the MRT with the aim to affect tumor vessels.[47] VSI might play a role in the early detection of responders versus nonresponders before results become visible at a morphological level. VSI is likely also to benefit from ultrahigh field MR scanner, a subject of ongoing investigation.

To tackle the drawback of poor time resolution of MRA compared to x-ray-based angiography such as biplanar or 3DRA, a few methods have been recently introduced. Among them, a highly undersampled 3D radial acquisition called PC with vastly undersampled isotropic projection reconstruction (PC-VIPR) has been introduced and is able to significantly speed up acquisition time (up to a factor equal to 30 when using contrast agent with respect to Cartesian 3D PC MRA) while decreasing streak and pulsatile flow artifacts.[48] Improved techniques based on dual echo approaches have been proposed to limit the sensitiveness to blurring and artifacts from off-resonance and trajectory errors of PC-VIPR: this improves image quality in AVM visualization while preserving short acquisition time,[49] leading to time-resolved 4D PC-VIPR-based MRA.[50] The possibility of calculating parameters such as flow, velocity, and stress condition of AVMs opens the way to have new tools for the selection of patients to be treated (e.g., by radiosurgery) and for the posttreatment follow-up, which can be significantly long due to the obliteration time after treatment by radiosurgery.[51] In parallel, Cartesian acquisition methods to speed up acquisition based also on undersampling have been proposed like the Cartesian acquisition with projection reconstruction-like (CAPR) based on 2D parallel imaging, with improved methods up to a 40-fold acquisition speedup.[52] CAPR-based MRA

has also proven their usefulness in distinguishing feeding and draining vessels.[53] Further acceleration is achievable using highly constrained back projection reconstruction,[54] up to a factor of 1000.[33]

Cartesian and radial acquisitions have both advantages and disadvantages.[55] Radial acquisition is less prone to motion artifacts, while Cartesian acquisitions do not require a correction for eddy currents or smaller datasets for image storage/processing. Ideally, we would get feedback comparing the methods. In general, square field-of-view (FOV) and arrhythmia patients benefit more from PC-VIPR (e.g., whole head, liver) while tailored FOV and regular heart-rate patients will benefit more from CAPR (e.g., heart, aorta). It is likely both techniques will find their room in the future clinical landscape and will definitively contribute also in the identification/follow-up of AVM treatment, for example, by radiosurgery.

Recently introduced MRI techniques aim at creating sequences with very short echo time (TE). These can be classified as ultrashort TE,[56,57] in the range of 8–30 μs, and as zero TE (ZTE),[58,59] in the range of 3 μs. Initially thought for the visualization of very short T2 structures,[60] such as cortical bone,[61] ligaments,[62] and lungs,[63] they have found several applications, in particular for silent MRI.[59] Acquisition can be based on conical (hence also the name of 3D cones) or on radial trajectories in k-space.

Adding a preparation pulse tagging the arterial blood to ZTE sequence, it is possible to obtain an MRA based on ZTE pulse, which, in addition to be silent, offers also the advantage of accurately visualizing regions affected by flow dephasing in state-of-the-art MRA techniques like TOF MRA.[64] Also, on the base of the nature of the ZTE acquisition scheme, images are likely to be less affected by motion artifacts. In addition, due to the low flip angle,[59] specific absorption rate is likely not to be a limiting factor in repeated acquisitions to produce time-resolved MRA. This is likely to benefit the accurate visualization of aneurysms and AVMs.

23.5.2 Other Prospected Developments

As stated earlier, many efforts are being made to improve the time resolution of 3D angiographic modalities. This will probably be one of the most immediate developments regarding the use of angiography in radiosurgery and radiotherapy.

Other more fundamental fields where significant advancement is expected are the translation of stereotactic approaches to sites affected by organ motion[65,66] and the personalized treatment of neoplastic disease to include specific dose prescriptions to volumes of increased neoangiogenesis, possibly guided by vascular imaging techniques.[67]

Angiography in the presence of organ motion is well developed in the case of coronary examinations, where the problem of heart motion is addressed by means of multiple strategies.[68–70] For radiosurgery and radiotherapy applications, on the other hand, the type of organ motion that shall be addressed is usually respiratory motion.[71] Several techniques exist to account for breathing movements in various medical imaging modalities,[72–74] even if limited experiences are reported for angiographic imaging. Radiosurgery of vascular malformations generally involves sites that are not significantly affected by respiratory motion, such as the brain or the spine (even though spinal treatments—especially if performed in the prone position—are actually subject to inaccuracies caused by displacements induced by breathing movements). On the other hand, as radiosurgical concepts extend to treatment protocols involving anatomical sites usually treated with conventional techniques, there is an increasing need for angiographic modalities capable of motion compensation. It is likely that such methods will be available in the near future, but more research is needed in order to bring them to the clinical routine.

Novel imaging techniques such as imaging of neoangiogenesis through MR- or PET-specialized applications are emerging and/or developing in recent years.[75,76] Neoangiogenesis is thought to be an indicator of tumor proliferation and might indicate a sensitive volume to treat with a higher dose of radiation in treatment protocols that adapt the dose distribution to the actual tumor architecture, in a patient-personalized vision of radiotherapy that is often called "dose painting by numbers," a strategy that exploits modern imaging modalities to provide a voxel-by-voxel guidance and build an optimized dose distribution.[77] In this context, it is likely that imaging of the vasculature assumes an increasingly important role in radiotherapy treatment planning, in analogy to surgical procedures where information on the vascular structure of a lesion is often needed for presurgical planning. Up-to-date, vascular information in radiotherapy treatment planning is seldom acquired, and experiences are limited to specialized applications such as integration of functional MR based on the blood oxygen level–dependent effect in treatment planning, usually involving radiosurgery more than conventional radiotherapy.[78] However, significant development might be seen in the future, driven above all by emerging high-dose, hypofractionated treatment protocols that exploit the high spatial accuracy of modern radiotherapy systems.

Vascular imaging is also emerging as an essential tool for the assessment of treatment response in SRS and radiotherapy.[79] Tumor hypoxia is considered an indicator of treatment response as well as an important factor to consider in treatment planning.[80] In this field, geometric accuracy requirements might be less stringent than in treatment planning, but high accuracy is anyway needed to allow for quantitative analysis and in case of replanning, for example, to address partial response or recurrence.[81] The field of quantitative analysis in medical imaging is increasingly being considered as one of the most important supports to treatment planning, and a field where medical physics is going to contribute significantly in the near future.[82]

Finally, vascular imaging is an increasingly important tool in treatment modalities other than radiosurgery, including stereotactic neurosurgery (e.g., image-guided biopsy[83]) and emerging treatment modalities such as focused ultrasound[84] or targeted drug delivery.[85] Image-guided neurosurgery has seen a rapid development since the introduction of neuronavigation systems.[86] Inclusion of vascular information in such systems is not trivial, in analogy to the integration between angiography and CT or MR in radiosurgery treatment planning. However, the vasculature itself is increasingly being used as a way to reach deeply seated targets; this includes futuristic nanorobotic techniques that might be able to deliver drugs or miniaturized surgical agents to selected locations with limited damage to healthy tissues and greatly reduced side effects.[87,88]

Within this prospected scenario, vascular imaging for radiosurgery, radiotherapy, and image-guided intervention in general is going to be a field of rapid development and intense research in the near future.

References

1. Rahman M, Murad GJ, Bova F, Friedman WA, Mocco J, Stereotactic radiosurgery and the linear accelerator: Accelerating electrons in neurosurgery, *Neurosurg Focus.* September 2009;27(3):E13.

2. Chang SD, Martin DP, Lee E, Adler JR Jr, Stereotactic radiosurgery and hypofractionated stereotactic radiotherapy for residual or recurrent cranial base and cervical chordomas, *Neurosurg Focus.* March 15, 2001;10(3):E5.

3. Bugoci DM, Girvigian MR, Chen JC, Miller MM, Rahimian J, Photon-based fractionated stereotactic radiotherapy for postoperative treatment of skull base chordomas, *Am J Clin Oncol.* August 2013;36(4):404–410.

4. Monaco EA, Grandhi R, Niranjan A, Lunsford LD, The past, present and future of Gamma Knife radiosurgery for brain tumors: The Pittsburgh experience, *Expert Rev Neurother.* April 2012;12(4):437–445.

5. Amin-Hanjani S, Ogilvy CS, Candia GJ, Lyons S, Chapman PH, Stereotactic radiosurgery for cavernous malformations: Kjellberg's experience with proton beam therapy in 98 cases at the Harvard Cyclotron, *Neurosurgery.* June 1998;42(6):1229–1236.

6. Colombo F, Benedetti A, Pozza F, Zanardo A, Avanzo RC, Chierego G, Marchetti C, Stereotactic radiosurgery utilizing a linear accelerator, *Appl Neurophysiol.* 1985;48(1–6):133–145.

7. Betti OO, Galmarini D, Derechinsky V, Radiosurgery with a linear accelerator. Methodological aspects, *Stereotact Funct Neurosurg.* 1991;57(1–2):87–98.

8. Betti OO, Munari C, Rosler R, Stereotactic radiosurgery with the linear accelerator: Treatment of arteriovenous malformations, *Neurosurgery.* March 1989;24(3):311–321.

9. Colombo F, Benedetti A, Pozza F, Marchetti C, Chierego G, Linear accelerator radiosurgery of cerebral arteriovenous malformations, *Neurosurgery.* June 1989;24(6):833–840.

10. Colombo F, Cavedon C, Casentini L, Francescon P, Causin F, Pinna V, Early results of CyberKnife radiosurgery for arteriovenous malformations, *J Neurosurg.* October 2009;111(4):807–819.

11. Rubin BA, Brunswick A, Riina H, Kondziolka D, Advances in radiosurgery for arteriovenous malformations of the brain, *Neurosurgery.* February 2014;74 Suppl 1:S50–S59.

12. See AP, Raza S, Tamargo RJ, Lim M, Stereotactic radiosurgery of cranial arteriovenous malformations and dural arteriovenous fistulas, *Neurosurg Clin N Am.* January 2012;23(1):133–146.

13. Vandermeulen D, Suetens P, Gybels J, Oosterlinck A, Angiographic localizer ring for the BRW stereotactic system, *Acta Neurochir Suppl* (Wien). 1987;39:15–17.

14. Brown RA, Nelson JA, Invention of the N-localizer for stereotactic neurosurgery and its use in the Brown-Roberts-Wells stereotactic frame, *Neurosurgery.* June 2012;70(2 Suppl Operative):173–176.

15. Gupta R, Mehndiratta A, Mitha AP, Grasruck M, Leidecker C, Ogilvy C, Brady TJ, Temporal resolution of dynamic angiography using flat panel volume CT: In vivo evaluation of time-dependent vascular pathologies, *AJNR Am J Neuroradiol.* October 2011;32(9):1688–1696.

16. Cavedon C, Bova F, Hendee WR, Three-dimensional rotational angiography (3DRA) adds substantial information to radiosurgery treatment planning of AVM'S compared to angio-CT and angio-MR, *Med Phys.* 2004;31:2181.

17. Chen W, Xing W, Peng Y, He Z, Wang C, Wang Q, Cerebral aneurysms: Accuracy of 320-detector row nonsubtracted and subtracted volumetric CT angiography for diagnosis, *Radiology.* December 2013;269(3):841–849.

18. Conti A, Pontoriero A, Faragò G, Midili F, Siragusa C, Granata F, Pitrone A, De Renzis C, Longo M, Tomasello F, Integration of three-dimensional rotational angiography in radiosurgical treatment planning of cerebral arteriovenous malformations, *Int J Radiat Oncol Biol Phys.* November 1, 2011;81(3):e29–e37.

19. Penney GP, Weese J, Little JA, Desmedt P, Hill DL, Hawkes DJ, A comparison of similarity measures for use in 2-D-3-D medical image registration, *IEEE Trans Med Imaging.* August 1998;17(4):586–595.

20. Varnavas A, Carrell T, Penney G, Increasing the automation of a 2D-3D registration system, *IEEE Trans Med Imaging.* February 2013;32(2):387–399.

21. Markelj P, Tomaževič D, Likar B, Pernuš F, A review of 3D/2D registration methods for image-guided interventions, *Med Image Anal.* April 2012;16(3):642–661.

22. Benedict SH, Yenice KM, Followill D, Galvin JM, Hinson W, Kavanagh B, Keall P et al., Stereotactic body radiation therapy: The report of AAPM Task Group 101, *Med Phys.* August 2010;37(8):4078–4101.

23. Sinclair J, Chang SD, Gibbs IC, Adler JR Jr, Multisession CyberKnife radiosurgery for intramedullary spinal cord arteriovenous malformations, *Neurosurgery.* June 2006;58(6):1081–1089.

24. Adler JR Jr, Chang SD, Murphy MJ, Doty J, Geis P, Hancock SL, The Cyberknife: A frameless robotic system for radiosurgery, *Stereotact Funct Neurosurg.* 1997;69(1–4 Pt 2):124–128.

25. De Salles AA, Pedroso AG, Medin P, Agazaryan N, Solberg T, Cabatan-Awang C, Espinosa DM, Ford J, Selch MT, Spinal lesions treated with Novalis shaped beam intensity-modulated radiosurgery and stereotactic radiotherapy, *J Neurosurg.* November 2004;101 Suppl 3:435–440.

26. Stancanello J, Cavedon C, Francescon P, Cerveri P, Ferrigno G, Colombo F, Perini S, Development and validation of a CT-3D rotational angiography registration method for AVM radiosurgery, *Med Phys.* 2004;31:1363.

27. Chang HH, Duckwiler GR, Valentine DJ, Chu WC, Computer-assisted extraction of intracranial aneurysms on 3D rotational angiograms for computational fluid dynamics modeling, *Med Phys.* December 2009;36(12):5612–5621.

28. Kin T, Nakatomi H, Shojima M, Tanaka M, Ino K, Mori H, Kunimatsu A, Oyama H, Saito N, A new strategic neurosurgical planning tool for brainstem cavernous malformations using interactive computer graphics with multimodal fusion images, *J Neurosurg.* July 2012;117(1):78–88.

29. Lauritsch G, Boese J, Wigström L, Kemeth H, Fahrig R, Towards cardiac C-arm computed tomography, *IEEE Trans Med Imaging.* July 2006;25(7):922–934.

30. Jandt U, Schäfer D, Grass M, Rasche V, Automatic generation of time resolved motion vector fields of coronary arteries and 4D surface extraction using rotational x-ray angiography, *Phys Med Biol.* January 7, 2009;54(1):45–64.

31. Babin D, Vansteenkiste E, Pizurica A, Philips W, Segmentation of brain blood vessels using projections in 3-D CT angiography images, *Conf Proc IEEE Eng Med Biol Soc.* 2011;2011:8475–8478.

32. Gross BA, Du R, Diagnosis and treatment of vascular malformations of the brain, *Curr Treat Options Neurol.* January 2014;16(1):279.

33. Grist TM, Mistretta CA, Strother CM, Turski PA, Time-resolved angiography: Past, present, and future, *J Magn Reson Imaging.* December 2012;36(6):1273–1286.

34. Schell MC, Bova FJ, Larson DA, Leavitt DD, Lutz WR, Podgorsak EB, Wu A, Stereotactic radiosurgery—Report of AAPM Task Group 42, AAPM-AIP 1995.

35. Langen KM, Papanikolaou N, Balog J, Crilly R, Followill D, Goddu SM, Grant W III, Olivera G, Ramsey CR, Shi C, AAPM Task Group 148, QA for helical tomotherapy: Report of the AAPM Task Group 148, *Med Phys.* September 2010;37(9):4817–4853.

36. Dieterich S, Cavedon C, Chuang CF, Cohen AB, Garrett JA, Lee CL, Lowenstein JR et al., Report of AAPM TG 135: Quality assurance for robotic radiosurgery, *Med Phys.* June 2011;38(6):2914–2936.

37. Roberts DA, Balter JM, Chaudhary N, Gemmete JJ, Pandey AS, Dosimetric measurements of Onyx embolization material for stereotactic radiosurgery, *Med Phys.* November 2012;39(11):6672–6681.

38. von Morze C, Purcell DD, Banerjee S, Xu D, Mukherjee P, Kelley DA, Majumdar S, Vigneron DB, High-resolution intracranial MRA at 7T using autocalibrating parallel imaging: Initial experience in vascular disease patients, *Magn Reson Imaging.* December 2008;26(10):1329–1333.

39. von Morze C, Xu D, Purcell DD, Hess CP, Mukherjee P, Saloner D, Kelley DA, Vigneron DB, Intracranial time-of-flight MR angiography at 7T with comparison to 3T, *J Magn Reson Imaging.* October 2007;26(4):900–904.

40. Stamm AC, Wright CL, Knopp MV, Schmalbrock P, Heverhagen JT, Phase contrast and time-of-flight magnetic resonance angiography of the intracerebral arteries at 1.5, 3 and 7 T, *Magn Reson Imaging.* May 2013;31(4):545–549.

41. Umutlu L, Maderwald S, Kraff O, Kinner S, Schaefer LC, Wrede K, Antoch G, et al., New look at renal vasculature: 7 tesla nonenhanced T1-weighted FLASH imaging, *J Magn Reson Imaging.* September 2012;36(3):714–721.

42. Johst S, Orzada S, Fischer A, Schäfer LC, Nassenstein K, Umutlu L, Lauenstein TC, Ladd ME, Maderwald S, Sequence comparison for non-enhanced MRA of the lower extremity arteries at 7 tesla, *PLoS One.* January 16, 2014;9(1):e86274.

43. Cuenod CA, Fournier L, Balvay D, Guinebretière JM, Tumor angiogenesis: Pathophysiology and implications for contrast-enhanced MRI and CT assessment, *Abdom Imaging.* March–April 2006;31(2):188–193.

44. Kording F, Weidensteiner C, Zwick S, Osterberg N, Weyerbrock A, Staszewski O, von Elverfeldt D, Reichardt W, Simultaneous assessment of vessel size index, relative blood volume, and vessel permeability in a mouse brain tumor model using a combined spin echo gradient echo echo-planar imaging sequence and viable tumor analysis, *J Magn Reson Imaging.* December 2014;40(6):1310–1318.

45. Lemasson B, Valable S, Farion R, Krainik A, Rémy C, Barbier EL, In vivo imaging of vessel diameter, size, and density: A comparative study between MRI and histology, *Magn Reson Med.* January 2013;69(1):18–26.

46. Bräuer-Krisch E, Serduc R, Siegbahn EA, Le Duc G, Prezado Y, Bravin A, Blattmann H, Laissue JA, Effects of pulsed, spatially fractionated, microscopic synchrotron X-ray beams on normal and tumoral brain tissue, *Mutat Res.* April–June 2010;704(1–3):160–166.

47. Serduc R, Christen T, Laissue J, Farion R, Bouchet A, Sanden BV, Segebarth C et al., Brain tumor vessel response to synchrotron microbeam radiation therapy: A short-term in vivo study, *Phys Med Biol.* July 7, 2008;53(13):3609–3622.

48. Gu T, Korosec FR, Block WF, Fain SB, Turk Q, Lum D, Zhou Y, Grist TM, Haughton V, Mistretta CA, PC VIPR: A high-speed 3D phase-contrast method for flow quantification and high-resolution angiography, *AJNR Am J Neuroradiol.* April 2005;26(4):743–749.

49. Johnson KM, Lum DP, Turski PA, Block WF, Mistretta CA, Wieben O, Improved 3D phase contrast MRI with off-resonance corrected dual echo VIPR, *Magn Reson Med.* December 2008;60(6):1329–1336.

50. Nett EJ, Johnson KM, Frydrychowicz A, Del Rio AM, Schrauben E, Francois CJ, Wieben O, Four-dimensional phase contrast MRI with accelerated dual velocity encoding, *J Magn Reson Imaging.* June 2012;35(6):1462–1471.

51. Chang W, Loecher MW, Wu Y, Niemann DB, Ciske B, Aagaard-Kienitz B, Kecskemeti S et al., Hemodynamic changes in patients with arteriovenous malformations assessed using high-resolution 3D radial phase-contrast MR angiography, *AJNR Am J Neuroradiol.* September 2012;33(8):1565–1572.

52. Haider CR, Borisch EA, Glockner JF, Mostardi PM, Rossman PJ, Young PM, Riederer SJ, Max CAPR: High-resolution 3D contrast-enhanced MR angiography with acquisition times under 5 seconds, *Magn Reson Med.* October 2010;64(4):1171–1181.

53. Mostardi PM, Young PM, McKusick MA, Riederer SJ, High temporal and spatial resolution imaging of peripheral vascular malformations, *J Magn Reson Imaging.* October 2012;36(4):933–942.

54. Huang Y, Wright GA, Time-resolved MR angiography with limited projections, *Magn Reson Med.* August 2007;58(2):316–325.

55. Jahnke C, Paetsch I, Schnackenburg B, Gebker R, Köhler U, Bornstedt A, Fleck E, Nagel E, Comparison of radial and Cartesian imaging techniques for MR coronary angiography, *J Cardiovasc Magn Reson.* 2004;6(4):865–875.

56. Robson MD, Gatehouse PD, Bydder M, Bydder GM, Magnetic resonance: An introduction to ultrashort TE (UTE) imaging, *J Comput Assist Tomogr.* November–December 2003;27(6):825–846.

57. Du J, Bydder M, Takahashi AM, Carl M, Chung CB, Bydder GM, Short T2 contrast with three-dimensional ultrashort echo time imaging, *Magn Reson Imaging.* May 2011;29(4):470–482.

58. Weiger M1, Pruessmann KP, Hennel F, MRI with zero echo time: Hard versus sweep pulse excitation, *Magn Reson Med.* August 2011;66(2):379–389.

59. Weiger M, Brunner DO, Dietrich BE, Müller CF, Pruessmann KP, ZTE imaging in humans, *Magn Reson Med.* August 2013;70(2):328–332.

60. Gatehouse PD, Bydder GM, Magnetic resonance imaging of short T2 components in tissue, *Clin Radiol.* January 2003;58(1):1–19.

61. Reichert IL, Robson MD, Gatehouse PD, He T, Chappell KE, Holmes J, Girgis S, Bydder GM, Magnetic resonance imaging of cortical bone with ultrashort TE pulse sequences, *Magn Reson Imaging.* June 2005;23(5):611–618.

62. Robson MD, Gatehouse PD, Bydder M, Bydder GM, Magnetic resonance: An introduction to ultrashort TE (UTE) imaging, *J Comput Assist Tomogr.* November–December 2003;27(6):825–846.

Chapter 23

63. Lederlin M, Crémillieux Y, Three-dimensional assessment of lung tissue density using a clinical ultrashort echo time at 3 tesla: A feasibility study in healthy subjects, *J Magn Reson Imaging*. October 2014;40(4):839–847.

64. Wilcock DJ, Jaspan T, Worthington BS, Problems and pitfalls of 3-D TOF magnetic resonance angiography of the intracranial circulation, *Clin Radiol*. August 1995;50(8):526–532.

65. Worm ES, Høyer M, Fledelius W, Poulsen PR, Three-dimensional, time-resolved, intrafraction motion monitoring throughout stereotactic liver radiation therapy on a conventional linear accelerator, *Int J Radiat Oncol Biol Phys*. May 1, 2013;86(1):190–197.

66. Chin E, Loewen SK, Nichol A, Otto K, 4D VMAT, gated VMAT, and 3D VMAT for stereotactic body radiation therapy in lung, *Phys Med Biol*. February 21, 2013;58(4):749–770.

67. Meijer TW, Kaanders JH, Span PN, Bussink J, Targeting hypoxia, HIF-1, and tumor glucose metabolism to improve radiotherapy efficacy, *Clin Cancer Res*. October 15, 2012;18(20):5585–5594.

68. Baka N, Metz CT, Schultz C, Neefjes L, van Geuns RJ, Lelieveldt BP, Niessen WJ, van Walsum T, de Bruijne M, Statistical coronary motion models for 2D+t/3D registration of X-ray coronary angiography and CTA, *Med Image Anal*. 2013;17:698–709.

69. Uribe S, Beerbaum P, Sørensen TS, Rasmusson A, Razavi R, Schaeffter T, Four-dimensional (4D) flow of the whole heart and great vessels using real-time respiratory self-gating, *Magn Reson Med*. 2009;62:984–992.

70. Lai P, Larson AC, Park J, Carr JC, Li D, Respiratory self-gated four-dimensional coronary MR angiography: A feasibility study, *Magn Reson Med*. 2008;59:1378–1385.

71. Keall P, Mageras GS, Balter JM, Emery RS, Forster KM, Jiang SB, Kapatoes, JM et al., The management of respiratory motion in radiation oncology report of AAPM TG 76, *Med. Phys*.2006;33:3874–3900.

72. Lee SM, Lee HJ, Kim JI, Kang MJ, Goo JM, Park CM, Im JG, Adaptive 4D volume perfusion CT of lung cancer: Effects of computerized motion correction and the range of volume coverage on measurement reproducibility, *AJR Am J Roentgenol*. June 2013;200(6):W603–W609.

73. Nehmeh SA, Erdi YE, Pan T, Pevsner A, Rosenzweig KE, Yorke E, Mageras, GS et al., Four-dimensional (4D) PET/CT imaging of the thorax, *Med Phys*. 2004;31:3179–3186.

74. von Siebenthal M, Székely G, Gamper U, Boesiger P, Lomax A, Cattin PH, 4D MR imaging of respiratory organ motion and its variability, *Phys. Med. Biol*. 2007;52:1547–1564.

75. Faye N, Clément O, Balvay D, Fitoussi V, Pidial L, Sandoval F, Autret G et al., Multiparametric optical and MR imaging demonstrate inhibition of tumor angiogenesis natural history by mural cell therapy, *Magn Reson Med*. September 2014;72(3):841–849.

76. Oxboel J, Brandt-Larsen M, Schjoeth-Eskesen C, Myschetzky R, El-Ali HH, Madsen J, Kjaer A., Comparison of two new angiogenesis PET tracers ^{68}Ga-NODAGA-E[c(RGDyK)]$_2$ and ^{64}Cu-NODAGA-E[c(RGDyK)]$_2$; in vivo imaging studies in human xenograft tumors, *Nucl Med Biol*. March 2014;41(3):259–267.

77. Bentzen SM, Theragnostic imaging for radiation oncology: Dose-painting by numbers, *Lancet Oncol*. February 2005;6(2):112–117.

78. Stancanello J, Cavedon C, Francescon P, Causin F, Avanzo M, Colombo F, Cerveri P, Ferrigno G, Uggeri F, BOLD FMRI integration into radiosurgery treatment planning of cerebral vascular malformations, *Med Phys*. 2007;34:1176.

79. Guan LM, Qi XX, Xia B, Li ZH, Zhao Y, Xu K, Early changes measured by CT perfusion imaging in tumor microcirculation following radiosurgery in rat C6 brain gliomas, *J Neurosurg*. June 2011;114(6):1672–1680.

80. Thorwarth D, Mönnich D, Zips D, Methodological aspects on hypoxia PET acquisition and image processing, *Q J Nucl Med Mol Imaging*. September 2013;57(3):235–243.

81. Leinders SM, Breedveld S, Méndez Romero A, Schaart D, Seppenwoolde Y, Heijmen BJ, Adaptive liver stereotactic body radiation therapy: Automated daily plan reoptimization prevents dose delivery degradation caused by anatomy deformations, *Int J Radiat Oncol Biol Phys*. December 1, 2013;87(5):1016–1021.

82. Giger ML, Chan HP, Boone J, History and status of CAD and quantitative image analysis: The role of Medical Physics and AAPM, *Med Phys*. December 2008;35(12):5799–5820.

83. Gempt J, Buchmann N, Ryang YM, Krieg S, Kreutzer J, Meyer B, Ringel F, Frameless image-guided stereotaxy with real-time visual feedback for brain biopsy, *Acta Neurochir* (Wien). September 2012;154(9):1663–1667.

84. Martin E, Jeanmonod D, Morel A, Zadicario E, Werner B: High-intensity focused ultrasound for noninvasive functional neurosurgery, *Ann Neurol*. 2009;66:858–861.

85. Aryal M, Arvanitis CD, Alexander PM, McDannold N, Ultrasound-mediated blood-brain barrier disruption for targeted drug delivery in the central nervous system, *Adv Drug Deliv Rev*. June 2014;72:94–109.

86. Ishii M, Gallia GL, Application of technology for minimally invasive neurosurgery, *Neurosurg Clin N Am*. October 2010;21(4):585–594.

87. Zamorano L, Li Q, Jain S, Kaur G, Robotics in neurosurgery: State of the art and future technological challenges, *Int J Med Robotics Comput Assist Surg*. 2004;1:7–22. Doi:10.1002/rcs.2

88. Cavalcanti A, Shirinzadeh B, Fukuda T, Ikeda S, Nanorobot for brain aneurysm, *Int J Robotics Res*. 2009;28:558–570.

Dosimetry and Radiation Protection

24. Dosimetric Techniques in Cardiovascular and Neurovascular Imaging

Daniel R. Bednarek

This chapter provides a survey of methods for determining the radiation dose to the patient, including both measurement and computational techniques. Different considerations are needed when evaluating risk to the patient depending on the type of effect (i.e., whether deterministic or stochastic) and different dosimetric quantities may be appropriate.

24.1 Quantities and Units

In order to quantify the amount of radiation in an x-ray beam and the relation to biological effects and risk, it is necessary to define the quantities and their units of measure.

24.1.1 Flux and Fluence

For a beam of x-rays, the number of photons passing per unit time is called the flux (e.g., photons per second) and the number passing per unit area is called the fluence (e.g., photons per square meter) [1]. The flux density and the fluence rate are equivalent quantities representing the number of particles passing through a unit area per unit time (e.g., photons per square meter per second). These quantities can be converted to the energy flux (e.g., keV/s) or energy fluence (e.g., keV/m^2) by multiplying the photon flux or fluence, respectively, by the energy of each photon.

For a polychromatic x-ray beam as would be produced by an x-ray tube, both quantities can be represented as a spectrum or distribution as a function of photon energy where the fluence or the flux is sorted into energy bins. Figure 24.1 shows the photon number fluence spectra for x-ray beams produced at 80 kVp with 2.5 mm Al and with 0.2 mm Cu of added filtration.

24.1.2 Exposure

One of the earliest units to be used to quantify an x-ray beam was the roentgen (R), so-named in honor of Wilhelm Conrad Roentgen, who discovered x-rays in 1895. The roentgen is a unit of exposure and this quantity arose because of the ability to measure the ionization of air by radiation. X-rays cause electrons to be liberated from air molecules, resulting in the creation of negative and positive ions. Exposure is the quantity of charge released in this manner per mass of air in the measurement volume. The ICRU defines exposure to be "the quotient of dQ by dm where dQ is the absolute

value of the total charge of the ions of one sign produced in air when all the electrons liberated by photons in a volume element of air having a mass dm are completely stopped in air" [3]. If 1 electrostatic unit of charge is released in 1 cm^3 of air at standard temperature and pressure (STP: 0°C and 760 mm Hg), then we can say the exposure is equal to 1 R. One roentgen is equal to 2.58×10^4 C of charge produced per kg of air. In 1960, the SI committee recommended the quantity of exposure be represented in the basic units of C/kg rather than the traditional unit of the roentgen [4].

24.1.3 Kerma

Kerma is an acronym for kinetic energy released in matter and is defined as the total initial kinetic energy of all charged particles resulting from interactions of indirectly ionizing radiation per unit mass of the absorber. It has units of the gray (Gy) where one Gy equals one J/kg [6]. An exposure of 1 R corresponds to an air kerma of 8.73 mGy since air requires an average of 33.85 eV to be absorbed to produce one ion pair of charge (1.6×10^{-19} C) or 33.85 J/C of charge. This conversion provides what has been called the collision air kerma since exposure does not include that part of kerma that is carried off by bremsstrahlung radiation. This distinction is important only at high energies, and the conversion is correct for diagnostic energy x-rays. Multiplying the value of exposure expressed in C/kg by 33.85 J/C will provide the collision air kerma in Gy.

24.1.4 Absorbed Dose

Absorbed dose is the amount of energy absorbed from ionizing radiation per mass of matter and has units of the gray (Gy), which equals 1 J of energy absorbed in 1 kg of matter. The traditional unit for absorbed dose is the rad that equals one hundred ergs of energy absorbed in 1 g of matter. One gray equals 100 rads. For air, it takes 33.85 eV of energy to produce one ion pair, so that 1 R of exposure is equivalent to the absorption of 0.873 rads in air. This conversion factor from roentgens to rads is referred to as the f-factor. The corresponding roentgen to mGy conversion factor would be 8.73 mGy/R. f-factors can be defined for any material to allow the conversion from exposure to dose in that material and are proportional to the ratio of mass energy absorption coefficients of the material to that of air. For example, the f-factor for soft tissue for photons of 50 keV is 0.92 rad/R or 9.2 mGy/R.

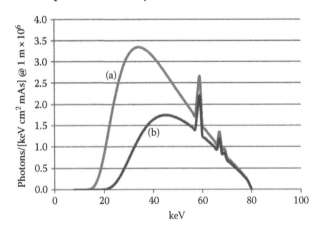

FIGURE 24.1 Photon fluence spectrum for 80 kVp beam filtered by (a) 2.5 mm Al and (b) 0.2 mm Cu. (These spectra were produced by the program SpekCalc [2].)

The absorbed dose is equal to the kerma when the amount of the kinetic energy of secondary charged particles that is converted to bremsstrahlung radiation is small and there is electronic equilibrium; this is the case for diagnostic energy x-rays.

24.1.5 Equivalent Dose

Different types of ionizing radiation (x-rays, protons, alpha particles, neutrons, etc.) have a different rate of energy transfer as their secondary particles are slowed down along their path, and thus, they have a different effectiveness in producing biological damage. A radiation weighting factor (w_r) is assigned to each type of radiation so that an equivalent dose (H) can be calculated as the product of the absorbed dose (D) and the weighting factor.

$$H = Dw_r \tag{24.1}$$

Equivalent dose is expressed in sieverts (Sv) in SI units or rem in the traditional system, where 1 Sv is equal to 100 rem. Equal equivalent dose from different forms of ionizing radiation should have equal risk of biological effect. This quantity is primarily used in radiation protection and is a way to estimate the relative biological hazard to individuals exposed to different types of radiation. The weighting factor is 1 for x-rays (and electrons, gamma-rays, and positrons) so that other types of radiation have their risk expressed relative to that of x-rays. For example, neutrons of 1 MeV have a weighting factor of 20 so that 1 Gy of absorbed dose from neutrons of this energy would have an equivalent dose of 20 Sv, while the same 1 Gy of absorbed dose from x-rays would have an equivalent dose of only 1 Sv.

24.1.6 Organ Dose

Organ dose is the absorbed dose calculated for a particular organ in the body and is usually represented as the average energy absorbed per unit mass of that organ. Organ dose is composed of both a primary beam component and a component of scattered radiation from other tissues and organs. For external beam exposure, the organ dose is dependent on the depth of the organ and its size since the primary x-ray beam is exponentially attenuated as it passes through the body. In many cases, organ dose is calculated using Monte Carlo software [5]. Estimates can also be obtained using values of tissue air ratios (TARs) that are published values of the dose at a depth in water to the dose at the same point measured in air [6]. These values are dependent on x-ray beam energy and field size.

24.1.7 Effective Dose

Tissue weighting factors were established so that the different radiosensitivities for stochastic effects of different organs and tissues in the body could be accounted for in estimating the risk to an individual. The effective dose (E) is the sum of the products of the equivalent dose to each organ or tissue irradiated (H_T) and their corresponding tissue weighting factor (w_T).

$$E = \sum w_T \times H_T \tag{24.2}$$

The sum of the weighting factors for all organs in the body is equal to 1.0, so the effective dose is the equivalent dose that if absorbed by all organs equally (total body exposure) would result in the same stochastic risk as the actual equivalent dose received by the individual organs. Effective dose was formulated to estimate the relative risk to individuals from nonuniform exposure. Table 24.1 gives a listing of the effective dose tissue weighting factors from ICRP Report 60 [7] and Report 103 [8]. These factors only consider stochastic effects such as cancer and hereditary genetic effects. The risk for deterministic effects such as skin erythema or cataract induction is not included in effective dose.

Many investigators have used this concept to compare the risk between different diagnostic radiological procedures. Mettler et al. provide a comparative summary of effective dose for a number of standard diagnostic procedures including radiographic, computed tomography, and radioisotope studies [9]. Table 24.2 provides a comparison between several interventional and CT examinations. Although useful for comparing the relative risk between procedures, it is not recommended that effective dose be used to estimate the risk for individual patients since the weighting factors were developed for a reference population.

PCXMC is a software program for calculating organ and effective dose using Monte Carlo techniques [10]. It allows the calculation of effective dose for rectangular x-ray beams for the operator-selected projections based on the hermaphrodite models whose size is adjustable by weight and height. Figure 24.2 shows a typical setup screen in the program and the type of output obtained. The calculation can be based on an input value of incident air kerma, dose-area product, entrance exposure, exposure-area product or mAs.

Table 24.1 Organ Tissue Weighting Factors from ICRP Reports Used for the Calculation of the Effective Dose

Organ or Tissue	Tissue Weighting Factor w_T (ICRP 60)[a]	Tissue Weighting Factor w_T (ICRP 103)
Active bone marrow	0.12	0.12
Breasts	0.05	0.12
Colon	0.12	0.12
Lungs	0.12	0.12
Stomach	0.12	0.12
Gonads	0.20	0.08
Liver	0.05	0.04
Esophagus	0.05	0.04
Thyroid	0.05	0.04
Urinary bladder	0.05	0.04
Brain	R	0.01
Bone surface	0.01	0.01
Salivary glands	—	0.01
Skin	0.01	0.01
Remainder	0.05	0.12
Total	**1.0**	**1.0**

[a] The remainder includes adrenals, brain, upper large intestine, small intestine, kidney, muscle, pancreas, spleen, thymus, and uterus.

Weighting factors labeled as "R" belong to the "remainder tissues" of ICRP Publication 60. The remainder weighting factor is applied to the mass-averaged dose in the remainder organs and tissues. However, if any of these organs receives a dose that is higher than the dose to any of the 12 organs for which a weighting factor is specified, a weighting factor of 0.025 is applied to that tissue or organ and 0.025 is applied to the mass-averaged dose in the other remainder organs and tissues.

Another program (Kramer [11]) is available that provides effective dose for models, which are more anthropometric but is limited to a selected number of defined projections [11]. Others have calculated conversion factors to obtain effective dose for interventional procedures [12].

24.1.8 Skin Dose

Skin dose is the absorbed dose to the skin and is usually given for a particular point on the body. It is composed of two components: the dose due to the external primary beam that is absorbed by the skin and the dose from the radiation backscattered to the skin from the patient. (For calculation of effective dose, the average dose to the skin over the entire body is used for the skin dose value.)

Table 24.2 Adult Effective Doses for Various Examinations

Examination	Average Effective Dose (mSv)	Values Reported in the Literature (mSv)
Head and/or neck angiography	5	8–19.6
Coronary angiography (diagnostic)	7	2.0–15.8
Coronary percutaneous transluminal angioplasty, stent placement, or radiofrequency ablation	15	6.9–57
Head CT	2	0.9–4.0
CT coronary angiography	16	5.0–32

Source: Adapted from Mettler, F.A. et al., *Radiology*, 248(1), 254.

The skin dose for the beam entering the patient is the product of the entrance air kerma times the backscatter factor times the f-factor or conversion factor from air kerma to skin dose. The skin on the exit surface of the patient receives much less dose than that at the entrance surface because of the attenuation of the primary beam by the intervening tissue and inverse square reduction of the beam intensity with distance from the focal spot; for a cardiac procedure, the exit skin dose is only about 2% of that at the entrance surface and, in most cases, is neglected when estimating deterministic risk.

24.1.8.1 Backscatter Factor

The amount of radiation backscattered from the patient depends on a number of factors including beam area, thickness and composition of tissue, and radiation beam quality. The backscatter factor is the ratio of the total dose at the entrance surface including radiation backscattered from the object to the dose of the primary beam entering the object. Figure 24.3 gives backscatter factors for water as a function of field area for various half value layers (HVLs) as measured using thermoluminescent dosimeters (TLDs) by Harrison [13].

Although many calculations of skin dose use the backscatter factor measured at the center of the beam, backscatter is not uniform within the x-ray beam but is highest in the center of the field and falls off at the edge as the scatter going outward is not compensated by backscatter coming inward. Likewise, backscatter

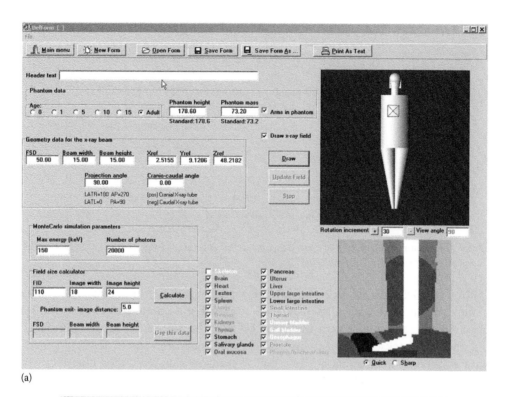

(a)

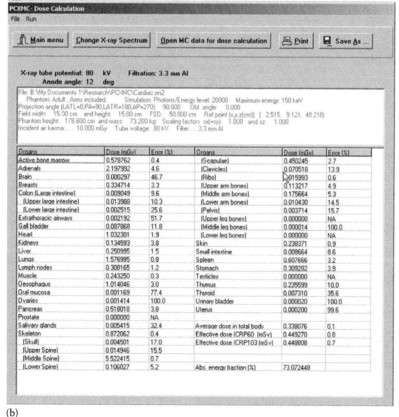

(b)

FIGURE 24.2 (a) Example of the setup page of PCXMC showing the conditions for a PA cardiac x-ray projection. Image in lower right shows organs included in the projection excluding the skeleton. (b) Display of results of Monte Carlo simulation of the cardiac projection in 2a, giving the organ doses and the effective dose values for a 10 mGy incident air kerma.

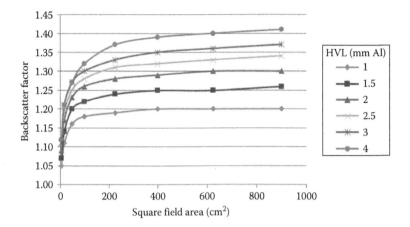

FIGURE 24.3 Backscatter factors as a function of x-ray field area at the entrance for square fields and different beam HVLs. (Adapted from Harrison, R.M., *Phys. Med. Biol.*, 27(12), 1465, 1982.)

continues to fall off and exposes the skin outside the beam. For example, Rana et al. have shown in measurements with a 30 cm square by 20 cm thick solid water phantom that the backscatter drops by about 10% at the beam edge compared to that at the center; outside the primary beam, backscatter was measured to be over 20% of the primary and drops to about 3% at 6 cm from the beam edge [14].

24.1.9 Peak Skin Dose

The peak skin dose (PSD) is the maximum value of the accumulated dose at any point of the patient's skin and is related to the risk of deterministic skin effects. For interventional procedures, PSD is usually located in areas of the skin over which the x-ray beam was stationary for extended periods or in areas of beam overlap from different projections.

24.1.10 Dose Area Product and Kerma Area Product

The dose area product (DAP) and the kerma area product (KAP) are quantities that have been used to designate the total radiation in the beam. The DAP is calculated as the product of the cross-sectional area of an x-ray beam in a plane perpendicular to the beam central axis and the average dose over that area. A number of other terms such as exposure area product (EAP) and roentgen area product (RAP) have also been used in the past to represent corresponding products.

Most current x-ray machines use the quantity KAP (mGy cm²) to monitor the radiation output of the machine during an exposure. If the x-ray field is uniform in intensity, the KAP is just the product of the beam cross-sectional area and the kerma at the beam axis. For a nonuniform beam, the kerma is integrated over the area of the beam to provide an average value. The measurement of KAP is generally done using a transmission ionization chamber whose sensitive area is larger than the beam dimensions. When a transmission ionization chamber is used to measure the KAP, integration over the field is performed automatically and the KAP is the product of beam area and average kerma in the beam. Most transmission chambers on interventional units are mounted inside the collimator housing after the collimator blades, so the KAP value is integrated over the projected beam area. If compensation filters are used before the transmission chamber, the KAP value will be reduced by the attenuation of the filter proportionally with the area of the beam attenuated. If compensation filters are mounted on the outside of the collimator housing after the transmission chamber, the kerma reduction afforded by the filter will not be reflected in the KAP reading.

Although the KAP or DAP may be measured with a transmission chamber at the collimator, it also provides the value at the patient entrance as long as there are no attenuating objects in the beam path. In fact, these products are independent of distance from the focal spot since the beam cross-sectional area increases with the square of the distance from the focal spot, while the radiation fluence decreases as the square of that distance; this just means that the amount of radiation in the beam does not change. However, when the beam passes through an attenuating material before the patient such as the patient table, the KAP or DAP value entering the patient will be reduced [15]. Since KAP and DAP readouts are usually calibrated for the beam in air without attenuation outside of the x-ray tube, the patient dose would be overestimated if the beam were to pass through the table before reaching the patient.

If the entire beam is intercepted by the patient, the KAP is proportional to the total radiation absorbed by the patient or the integral dose if the radiation transmitted through and scattered out of the patient is negligible; with this assumption, KAP has been considered an indicator of stochastic risk where an average tissue sensitivity is assumed and without considering the tissue sensitivity of the actual organs exposed [16].

24.1.11   Reference Point Air Kerma

Reference point air kerma ($K_{a, r}$) was a concept introduced by the IEC in 2000 (60601-2-43) [17] and is defined as the total air kerma accumulated during a procedure at a particular point in the beam called the interventional reference point (IRP). In the 2010 revision of this standard, this point was renamed the patient entrance reference point (PERP) [18]. For an isocentric C-arm fluoroscopic system, the IRP is located in the center of the beam at 15 cm from the isocenter in a direction toward the x-ray tube. This point is meant to approximate the entrance skin location for an average adult patient for a procedure during which the feature of interest is located at the isocenter.

The FDA defines a similar quantity called the cumulative air kerma (CAK). For fluoroscopic systems manufactured after June 10, 2006, the FDA requires that the CAK and the air kerma rate (AKR) (when the frame rate is greater than six images per second) be displayed. Both quantities are values determined at the reference point defined earlier. However, for non-C-arm fluoroscopes, the reference point is the location used for measuring compliance for maximum AKR limits, usually 30 cm from the image receptor input surface or 1 cm above the tabletop for undertable units. Alternatively, the FDA allows the reference point to be "specified by the manufacturer to represent a location of the intersection of the x-ray beam with the patient's skin" [19]. The CAK must be displayed within 5 s following termination of an exposure or continuously displayed and updated at a minimum every 5 s. The AKR must be continuously displayed and updated at least once each second. The accuracy of the RAK display required by IEC standard 60601–2–43 was 50% [17] and 35% in the revision [18], while the accuracy of the CAK display required by FDA's CFR 1020.30 is 35% [19].

Both of these quantities have limitations in their ability to provide a measure of the PSD. Since they provide a measure at a fixed reference point, they do not necessarily provide the air kerma (rate) at the entrance to the skin because of variations in the actual source to skin distance (SSD) used. Further, the value represents the kerma in air and thus would need a conversion factor to obtain the dose to skin; the f-factor for air is 8.76 mGy/R while that for soft tissue is 9.2 at 50 keV (and varies with energy), so that the conversion factor from air kerma to skin dose at this energy would be 1.05. Also, these quantities do not include the effect of backscatter from the patient, which can result in a 30%–40% increase in skin dose. Finally, these quantities do not include the attenuation and scatter produced by intervening structure such as the table and table pad, which can result in a 10%–15% net reduction in dose [20,21].

24.1.12   Maximum AKR Compliance Limits

In the United States, there is no limit per se on the entrance dose or dose rate to the patient during a medical fluoroscopic procedure. The use of the imaging system is considered to be a medical decision governed by the needs of the physician to provide health care to the patient, and he or she is not restricted in the amount of radiation used. However, in the nonrecording mode of fluoroscopy, which is generally used for patient and device positioning, there are limits on the maximum AKR at a compliance location. For normal fluoroscopy, the output cannot exceed 88 mGy/min (10 R/min), while in high-level control (HLC) fluoroscopy, the output limit is 176 mGy/min. The compliance point is at the center of the beam 30 cm from the input surface of the image receptor for C-arm units, 1 cm above the table for undertable systems, 15 cm tube side of the table center for lateral tubes, and 30 cm above the tabletop for overtable tubes [19, § 1020.32(d)(3)]. For a C-arm fluoroscope with an SID of less than 45 cm such as the Hologic Fluoroscan Insight for extremity imaging, the compliance point is at the minimum SSD. Whereas in HLC mode, an audible tone must be produced that alerts the operator that a higher level of exposure rate is allowed.

24.1.13   FDA Restrictions

Other restrictions also apply by FDA regulation to encourage limitation of patient dose. An audible signal should be produced at the end of each 5 min interval of fluoroscopy time, and this signal needs to be reset by the operator. The minimum SSD shall be no less than 38 cm for fixed installation fluoroscopes and no less than 30 cm on mobile units, and these restrictions are usually enforced by construction of the tube enclosure. A removable spacer cone can be provided that allows the minimum SSD to be reduced to no less than 20 cm when necessary for specific surgical procedures [19]. The purpose of this SSD restriction is to limit the maximum exposure to the entrance skin surface since exposure increases rapidly by the inverse square law as the distance from the focal spot is reduced.

24.2 Detectors and Measurement

24.2.1 Ionization Chambers

An ionization chamber detector consists of an air-filled volume with positive and negative electrodes to collect charge produced by the ionization of the air due to the passage of ionizing radiation. Sufficient voltage must be applied across the electrodes to collect all the charged particles produced before recombination of the positive and negative ions occurs. If not, a lower reading will be obtained and the determination of exposure will be in error. To determine if saturation exists, readings can be taken at two different voltages. If the readings do not differ, then saturation exists; if they differ, then saturation may not exist for the lower voltage value. This is not normally a problem at the intensities measured for most chambers when used as intended. However, a chamber with a large volume and large interelectrode spacing might have too low of an electric field strength to collect all the charge at higher intensities if placed in the beam close to the x-ray source.

Parallel-plate ionization chambers consist of planar electrodes with the sensitive volume between them. Examples of chambers commonly used to measure entrance air kerma for diagnostic energy beams are the PTW SDF 6 cc chamber and the Keithley model 35050 15 cc chamber shown in Figure 24.4. These chambers have a central plate that is used as the collector with outer plates on either side creating an electrical potential difference.

Thimble-type ionization chambers consist of a cap-shaped outer shell, the inner surface of which is conducting and used as an electrode, and an inner, rod-shaped central electrode.

Pencil chambers are thimble chambers with a long pencil-like thimble. Such chambers are typically used for CT dose measurements (computed tomography

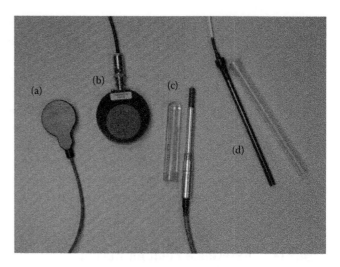

FIGURE 24.4 Ionization chambers: (a) PTW SFD 6 cc parallel-plate chamber, (b) Keithley Model 35050 15 cc parallel-plate chamber, (c) Nuclear Enterprises Farmer type 0.6 cc thimble chamber with buildup cap, (d) Keithley CTDI pencil chamber shown with Lucite cap for insertion in phantom.

dose index—CTDI) and are placed with their axis parallel to the rotational axis of the scanner; they typically have a 10 cm long sensitive volume and only the central section of the chamber is exposed during axial scans with a fan beam. Farmer-type ionization chambers have a very small sensitive volume (0.6 cc) and thus need a high radiation exposure to obtain a reasonable reading. They are normally used in radiation therapy calibration applications. However, they have been used in diagnostic energy x-ray beams where high spatial resolution is needed for a dose measurement such as in determining the point dose in multidetector and cone-beam CT where it is impractical to use the pencil chamber or in beam profile mapping [22]. Different thimble materials are used for different chambers and will give different sensitivities since most of the electrons depositing energy in the air are produced through photoelectric or Compton interactions in the wall material and thus care must be taken in applying the correct calibration factor for the chamber and energy used.

Transmission chambers (Figure 24.5) are parallel-plate ionization chambers with a sensitive area large enough to intercept the entire x-ray beam. It can be used to monitor the output of an x-ray machine if the beam area does not change. If the beam area changes, then the reading gives a value proportional to the product of beam area and average exposure over the field. These chambers are uniform in construction over the sensitive area so that they do not project shadows in the beam and thus are not seen on the image;

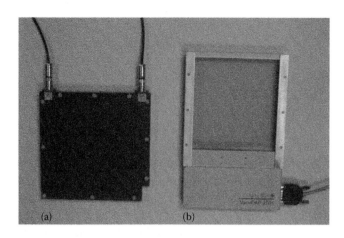

FIGURE 24.5 Transmission ionization chambers: (a) custom transmission dual chamber with carbon conductor surfaces that measures dose area product as well as dose over a 2.5 cm diameter central volume [24]; (b) optically transparent Nuclear Associates VacuDAP chamber for dose area product measurement on radiographic x-ray tubes with a light localizer.

they do provide some hardening of the beam and the spectrum is shifted after passing through the chamber. Some chambers are placed after the light localizer on the x-ray tube and are made optically transparent by using indium tin oxide (ITO) for the conductive coating over clear PMMA or polycarbonate plates. The use of ITO results in greater energy dependence of response of the chamber compared to plates coated with a graphite-based conductor [23]. Transmission chambers can be made to measure the charge collected over the entire beam for DAP determination and simultaneously over a selected fixed area, which can be at the center of the field to measure the dose (rate) [24].

Normally, the electric field lines defining the sensitive volume are shaped by using a control electrode placed between the collector and the high voltage. This electrode is typically at the same electrical potential as the collector and intercepts any leakage charge in the insulation separating it from the high-voltage electrode. When an external electrometer is used to measure the charge collected, a triax cable may be used to connect it to the chamber. In this cable, the central conductor carries the signal from the collector while the outer conductor sheath is at the high-voltage potential and the middle sheath conductor acts as a guard and is at the same potential as the central conductor. (High voltage as used here is with reference to the collector. In many cases, the collector and control electrodes are at high voltage with respect to ground and the "high-voltage" electrodes are at ground potential.)

The charge detected by the chamber is usually measured with an electrometer. The electrometer is essentially a very high input impedance operational amplifier; placing resistors across the operational amplifier allows it to be used in the current reading mode and placing capacitors across the amplifier allows the integration of charge and total exposure to be measured.

The center of the chamber in the beam direction is usually taken as the point of measurement although that point may be shifted due to field (inverse square) or attenuation (beam hardening) variation of the absorption profile as the beam passes through the chamber. The reading of the chamber is the integral over the sensitive volume and thus is an average of the field over the volume. If only part of the chamber is exposed, the reading will be an average of the x-ray intensity in the beam and the intensity outside the beam weighted by the respective chamber volumes.

Condenser chambers have been used in the past for beam calibration (Victoreen R-Meter) and are used in direct reading pocket chambers. These chambers have a built-in capacitor that is charged before exposure. The amount of discharge is read following an exposure and can be calibrated to give a reading of exposure in roentgens or C/kg.

Integral dosimeters are available that have an ionization chamber built into a unit containing the measurement electronics (e.g., Rad-Check Plus, Fluke Biomedical, Cleveland, OH). These dosimeters can be used to measure the output of a machine when fixed settings are used but should not be used when automatic exposure control is used since the dosimeter attenuation will influence the x-ray output and thus the dosimeter reading. Ionization chambers are also used in survey meters to monitor radiation levels in areas outside the x-ray beam. They have a relatively large sensitive volume to provide very high sensitivity and can read down to 10 mR/h. They also can read at much higher levels in higher radiation fields (limited by the electric field applied) and are not paralyzable because of dead time losses like a Geiger counter would be.

Most ionization chamber detectors are open to the atmosphere (vented), and thus, the mass of air contained in the sensitive volume is dependent on the ambient temperature and pressure. In many cases, the chambers are calibrated at 1 atmosphere and 22°C so that measurements obtained under normal indoor conditions without correction would only result in an error of a few percent. When operating at extremes of pressure or temperature, the following factor can be used as a correction to the exposure reading:

$$\frac{273.15+T}{273.15+T_{c}} \times \frac{P_{c}}{P} \qquad (24.3)$$

where

T_{c} is the calibration temperature
T is the operating temperature in °C
P_{c} is the calibration pressure
P is the operating pressure

Dose calibrators used in nuclear medicine use a sealed ionization chamber, and thus, no temperature and pressure correction is needed. They use pressurized argon gas to increase the detection sensitivity.

24.2.2 TLDs

TLDs consist of a phosphor that can store x-ray energy and re-emit that energy in the form of light when heated. The amount of light emitted is proportional to the amount of energy or dose absorbed by the phosphor. The process involves the raising of electrons in the phosphor from the valence band into energy traps through the absorption of energy from recoil and photoelectrons when x-rays interact. The traps for various phosphors have various energy levels above the valence band. As the phosphor is heated, sufficient energy may be given to the electron to raise it out of its trap with the subsequent release of visible light energy. A curve showing the light output as a function of temperature is called a glow curve. Each peak in the curve corresponds to the thermal energy necessary to raise the electron from the characteristic trap with the emission of a photon as the electron drops to its ground

Table 24.3 TLD Phosphor Characteristics

	Effective Z	Density (g/cc)	Main Glow Peak (°C)	Energy Response (30 keV/ Co-60)
LiF	8.2	2.64	195	1.25
Li₂B₄O₇:Mn	7.4	2.3	200	0.9
CaF₂:Mn	16.3	3.18	260	13
CaF₂:nat	16.3	3.18	260	13

Source: From Cameron, J.R. et al., *Thermoluminescent Dosimetry,* The University of Wisconsin Press, Madison, WI, 1968.

state. Electrons can drop to their ground state over time without heating so that the peaks exhibit fading and the time of readout after exposure can affect the results. Normally, the area under a stable peak of the curve is used as a measure of energy absorbed for dosimetry purposes.

Several different phosphor materials exhibit thermoluminescence including calcium fluoride, lithium fluoride, calcium sulfate, lithium borate, calcium borate, potassium bromide, and feldspar. Table 24.3 gives a listing of several TLD phosphors and their characteristics.

TLDs are not used as absolute dosimeters but must be calibrated by exposing them to a known x-ray beam that may have been calibrated using a calibration lab traceable ionization chamber. It is important that the spectrum of the calibration beam be nearly the same as that of the unknown field or that the energy response of the TLD does not vary between the spectra of the two beams. LiF–Mg,Ti is fairly tissue equivalent and has a response for measuring tissue dose, which is relatively energy independent [26].

The readout process generally does not release all of the energy of trapped electrons, and TLDs are annealed before exposure by heating them in an oven. Careful dosimetry requires that the same sequence be followed for the unknown exposure as for the calibration process. That is, the same annealing sequence is performed, and the same time period transpires between the end of annealing and exposure and the end of exposure and readout.

TLDs usually take the form of small chips or rods of mm dimension. Their small size makes them ideal for measuring dose in spatially varying fields. For example, a series of TLD rods could be placed in a line parallel to the axis of a CT gantry to obtain a profile of the dose across the beam. Likewise, TLD chips could be placed on the patient during radiographic or fluoroscopic procedures to determine the entrance skin dose; their small size generally makes them transparent enough to be used in the beam without interfering with the x-ray image. Disadvantages of using TLDs for monitoring skin dose during fluoroscopic procedures include the fact that the readout must be made after the procedure, so monitoring cannot be done in real time and it is not known a priori what point on the skin will receive the maximum or PSD so it may not be known where to place the individual dosimeters. Some investigators have placed TLDs in an array on the patient's back to determine the PSD [27].

TLDs are also used in personal dosimeters for monitoring personnel and radiation areas for radiation exposure.

24.2.3 PSP

Phostimulable phosphor (PSP) or optically stimulable luminescence (OSL) technology is similar to that of TLDs except the trapped electrons are released using visible light. Upon exposure to x-rays, electrons are raised into higher energy traps. Subsequent exposure of the phosphor to a red laser light releases the electrons from their traps and they fall back to their valence band, emitting blue light in the process. This is a similar process to computed radiography imaging plates that typically use barium strontium bromide. Aluminum oxide (Al_2O_3) phosphors have been used for personal dosimeters provided by a commercial vendor (Luxel, Landauer, Inc., Glenwood, IL). The readout process is essentially nondestructive in that most of the electrons remain in their traps and the phosphor can undergo several readouts so that it provides a "permanent" record [28].

24.2.4 MOSFET Solid–State Diode Radiation Sensors

Metal-oxide semiconductor field-effect transistor (MOSFET) detectors are solid-state silicon-based radiation detectors that utilize a p-n junction. The VeriDose solid-state diode detectors use a low-noise coaxial cable to connect to an electrometer for readout. MOSFET dosimeters have been reported to overrespond at diagnostic energies by a factor of about 3–4 compared to the response to high-energy photons, so that they should be calibrated at the energy at which they are intended to be used. MOSFET dosimeters have an advantage over TLDs in that they can be read out during the procedure, but the visibility of the detectors and their leads may be objectionable [29,30]. There are cable-free detectors available, so the dose can be read out remotely [31].

24.2.5 X–Ray Film

X-ray film can be used as a dosimeter since the darkening of the film following exposure and processing increases monotonically with dose. A plot of the optical density produced versus the logarithm of exposure is known as the characteristic or H&D curve of the film. Film is composed of silver halide crystals fixed in a gelatin binder and has an energy dependence of its response that is dependent on the absorption characteristics of these elements. Film exposed in the diagnostic energy range is quite energy dependent because of the photoelectric interaction of the x-rays with the silver ($Z = 45$) in the emulsion. Processing conditions greatly affect the film response and include the chemistry used for development, the temperature of the developer, and the time the film remains in the developer. All of these factors must be carefully controlled to obtain reproducible results.

Optical density is defined as the logarithm of the film opacity or logarithm of the ratio of the incident light to the light transmitted through the film. Optical density at individual points on a film can be measured with a densitometer. Film digitizers can be used to determine the density distribution over a film and, if properly calibrated, the absorbed dose distribution. Direct exposure of film is not affected by reciprocity law failure so that the same density is obtained for a given product of intensity and exposure time independent of the individual values of these parameters. Film exposed to light such as from intensifying screens exhibits reciprocity law failure such that the density is typically less for the same exposure made over a long time.

A number of investigators have used film for patient dosimetry in interventional procedures [32–36]. An advantage of film is that it can provide a distribution of dose at the skin and can help locate the region of PSD and, when placed under the patient, it is not visible on the image and does not interfere with the procedure. However, the skin dose for interventional procedures is unpredictable and may cover a wide dynamic range between different patients; the film may reach its maximum density for exposures less than 2 Gy, which is just the threshold for temporary erythema and may not be able to provide details of exposure above this level. Furthermore, film must be processed and does not provide immediate feedback of the dose distribution. Most hospitals no longer have darkrooms with automatic film processors and the film may have to be manually processed.

24.2.6 Radiochromic Film

Radiochromic film has been used for measuring dose distributions in radiation therapy for many years [37] and has more recently been specifically developed for measuring skin dose for interventional procedures [38]. The technology is based on solid-state polymerization

Chapter 24

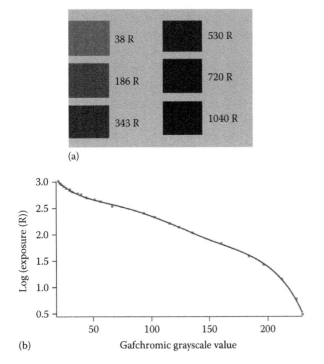

(a)

(b)

Log (exposure (R))

Gafchromic grayscale value

FIGURE 24.6 (a) Sections of Gafchromic XR-RV3 film that were exposed to different exposure levels to generate a calibration curve. (b) A calibration curve generated to determine the exposure to the Gafchromic XR-RV3 film for a given gray scale value measured for the red channel of the scanner output. (Taken from Rana, V. et al., *Med. Phys.*, 38(6), 3702, 2011.)

FIGURE 24.7 Gafchromic XR-RV3 film that was wrapped around an Alderson adult anthropomorphic torso phantom and exposed during a simulated cardiac procedure showing the skin dose distribution. (Taken from Rana, V. et al., *Med. Phys.*, 38(6), 3702, 2011.)

such that the film increases in optical density with increasing radiation exposure. The film is available in both transmission and reflection density forms and offers high spatial resolution measurement of dose with acceptable accuracy for clinical measurements. The film does not require development and darkens immediately with exposure to x-rays, but is not sensitive to visible light and so can be handled without using light-tight packaging. The film is thin and flexible and can be easily placed under or around the patient for clinical measurements. The GafChromic XR-RV3 film (International Specialty Corporation, Wayne, NJ) has been developed specifically for use in interventional fluoroscopic

procedures. It is a reflection density film with an amber base, available in 35 × 43 cm sheets. McCabe et al have presented the results of a calibration evaluation of this film using a flatbed scanner that showed a dependence of response on film orientation relative to the source and on the kVp of the beam [39]. The manufacturer indicates energy independence for XR-RV2 film from 30 keV to 30 MeV. A low-resolution scan (~72 dpi) is recommended to reduce the noise in the readout. It is also recommended that the measurement of the density be done 24 h after exposure since there is some initial postexposure reduction in density. Figure 24.6a shows the density change of samples of the film exposed to different levels of exposure at 80 kVp, and Figure 24.6b shows a calibration curve using the gray levels from the red channel obtained by scanning the samples with a Canon MX310 scanner (75 dpi, color mode). Figure 24.7 illustrates the appearance of the film following a simulated cardiac procedure on an anthropomorphic phantom.

24.3 Skin Dose Mapping

As interventional procedures have become longer and more complicated, the radiation dose to the patient has increased to the point where deterministic effects have become a concern. The primary deterministic effects of concern have been to the patient's skin since the radiation beam has the highest intensity on the patient at its entry point. The determination of the PSD has been a main dosimetric goal in interventional procedures so

that it may be determined if the threshold dose for the skin effect has been reached. Table 24.4 provides a listing of some of these effects and their threshold as well as the latent period before the effect manifests itself [41].

PSD is not a quantity that is generally measured since the beam location and intensity varies during a dynamic procedure and it is not known beforehand where this peak dose might be located. Thus, unless an array of

Table 24.4 Deterministic Effects of X-Ray Exposure to the Skin

Effect	Single Dose Threshold (cGy)	Onset
Early transient erythema	200	2–24 h
Temporary epilation	300	3 weeks
Main erythema	600	10 days
Permanent epilation	700	3 weeks
Dermal atrophy (phase 1)	1000	>12 weeks
Dermal atrophy (phase 2)	1000	>1 year
Telangiectasia	1000	>1 year
Dermal necrosis	>1200?	>1 year
Dry desquamation	1400	4 weeks
Moist desquamation	1800	4 weeks
Ischemic dermal necrosis	1800	>10 weeks
Secondary ulceration	2400	>6 weeks
Skin cancer	Not known	>5 years

Source: Adapted from Wagner, L.K., *Biomed. Imaging Interv. J.*, 3(2), e22, 2007.

closely spaced detectors is placed on the patient covering the area of beam entry, the peak value may not be recorded if the detector is not placed in the correct location. Both x-ray film and radiochromic film are available in large sheets and have high resolution and have been used for determining PSD. However, both film types are read out after the procedure and do not provide immediate feedback to the physician. Such methods are not automatic and are relatively cumbersome to employ on a routine basis for clinical procedures and are thus not routinely used. A number of surrogates have been used for PSD estimation or, at least, as a flag to indicate the potential for having reached the deterministic threshold. These have included fluoroscopic exposure time, KAP or DAP, and reference point air kerma. All of these have been shown to be unreliable indicators. Figure 24.8a through c show the lack of correlation observed between these surrogates and PSD for cardiac procedures; use of a single conversion factor from the slope of these curves would introduce a large uncertainty in PSD determination.

Guidelines have been provided by the Society of Interventional Radiologists recommending follow-up of patients who have received a significant dose where the significant dose level is taken to be 3000 mGy PSD. If PSD is not available, then surrogate recommended levels are 5000 mGy reference point air kerma, 500 Gy·cm² KAP (assuming a field size at the entrance to the patient of 100 cm²), or 60 min of fluoroscopy time

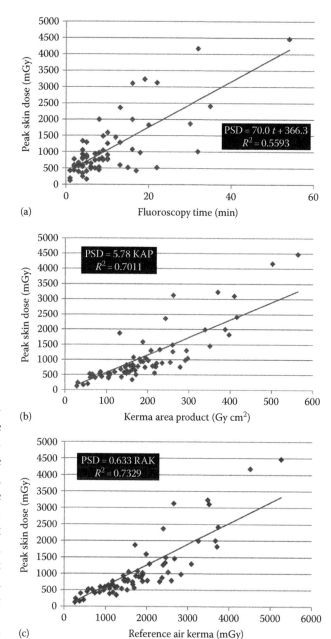

(a)

(b)

(c)

FIGURE 24.8 PSD for cardiac catheterization procedures versus the recorded (a) fluoroscopy time, (b) kerma area product, and (c) reference air kerma. R^2 is the square of Pearson's product-moment correlation coefficient for the least-squares linear regression of the data shown. (From Bednarek, D.R. et al., Peak skin dose for cardiac catheterization procedures determined with an automatic dose tracking system, in *Scientific Program and 97th Scientific Assembly and Annual Meeting of RSNA*, Meeting Program, Chicago, IL, November 27–December 2, 2011, p. 196, Scientific presentation. SSK15-02. Online program abstract: http://rsna2011.rsna.org/search/search.cfm?action=add&filter=Subspecialty&value=301102130.)

in the absence of other data [43]. The Joint Commission has designated 15000 mGy PSD to be a sentinel event that should be investigated and a report voluntarily submitted [44].

To provide a more automatic system for determining PSD, a number of calculation-based methods have been developed.

24.3.1 PEMNET

The PEMNET system (Clinical Microsystems, Arlington, VA) was one of the earliest to be marketed to estimate skin dose. The system is electronically connected to the x-ray system to extract information about the technique parameters used during the procedure, and calibration data are used to calculate the dose to the skin using a microprocessor. Ultrasonic proximity sensors and information on the table height are used to obtain the SSD, so the dose rate during fluoroscopy and the skin dose can be calculated. However, the system does not track the location on the skin where the dose is absorbed, so only the cumulative skin dose and not the PSD is determined [45,46].

24.3.2 CareGraph® System

The CareGraph system was developed by Siemens Medical Systems (Iselin, NJ) to provide a graphic representation of skin dose distribution and was available for several years as an option for their interventional fluoroscopic systems. A prototype was first described by den Boer in 2001 [47]. The system used a graphic representation of the patient based on the patient's weight and height and information about the system to determine the location of the beam on the patient and the readout from a DAP meter to determine the dose rate and cumulative dose. The display is recalculated every 500 ms and provides a numerical display of the PSD as the "Max HotSpot" and the area of the skin receiving 95% of the PSD. The skin dose is also displayed as a color-coded mapping on the graphic of the patient, which is shown with the skin unwrapped and flattened on a 2D representation [48].

24.3.3 Mapping Software

A number of efforts are underway to develop skin dose mapping software using the information provided in the DICOM radiation dose structured report (SDR) [49,50]. The purpose of this software has been to determine the distribution of dose given to the skin

during interventional fluoroscopic procedures so that the area of the skin at risk for deterministic effects can be identified. At this time, the RSDR only provides the dose for individual exposure events; the cumulative dose over the period when the fluoroscopy switch is depressed until it is released is recorded so that any changes in position or technique during that interval are not tracked. If movement of the beam or change in technique parameters occurs during an exposure activation, then the distribution displayed may not represent the true distribution.

24.3.4 Real-Time Skin Dose Tracking

A skin dose tracking system (DTS) has been developed, so the skin dose distribution could be determined in real time, recording and displaying each exposure as it is made. In this way, the dose information is provided to the interventionalist during the procedure, so an informed decision can be made about the risk versus benefit trade-offs of his or her actions. The system was initially developed using analog-to-digital converters to provide the position feedback sensor information of the imaging system as input to a PC to determine the intersection of the beam with the patient; a DAP transmission chamber was used for dose information and analysis of the clinical images was used to determine the collimator position [51].

Subsequently, the hardware requirements of the system were able to be simplified with the introduction of a controller area network (CAN) digital bus by the imaging system manufacturer that provided not just the positional information but also exposure technique parameter information. Tapping into this digital bus provided all the needed information to determine the intersection of the beam with the patient table and, with calibration, the radiation output of the x-ray tube. Selection of a patient graphic model and positioning of this model to match the shape of the patient and the position of the patient relative to the table allow calculation of the dose to each point on the patient's skin.

In this system, the calculation of patient skin dose uses a calibration file that contains the exposure output of the interventional unit measured at the reference point 15 cm tube side of isocenter. The exposure is measured with an ionization chamber placed under 20 cm of solid water material to provide backscatter and above the table for exposures with the x-ray tube under the table so that the attenuation and scatter from the table is included. Measurements are made for each beam filter available, for fluoroscopy and digital fluorography (cine), and for every 10th kVp value. The exposure per mAs is fit to a fourth-order polynomial as a function of kVp, and the values are calculated at every kVp from 50 to 125 to populate the output data calibration files.

The data in the calibration file is then used to calculate the dose to the skin using the exposure parameters that are provided by the CAN bus for each x-ray pulse. The entrance skin exposure at the reference point (ESE_{ref}) is first calculated as

$$ESE_{ref}(R) = C\left(\frac{R}{mAs}\right) \times \frac{PW(ms)}{1000} \times I(mA) \qquad (24.4)$$

where

C is the calibration factor from the data file for the imaging mode, beam filter, and kVp used
PW is the pulse width in ms
I is the x-ray tube current

The absorbed dose to each point on the patient graphic is then calculated using

$$Dose(mGy) = ESE_{ref} \times f \times \left(\frac{dr}{d}\right)^2 \qquad (24.5)$$

where

f is the roentgen to dose conversion factor for soft tissue (9.2 mGy/R)
The squared term gives the distance correction for each point on the patient graphic that is in the beam where dr is the distance from the focal spot to the calibration reference point and d is the distance from the focal spot to the individual point on the graphic surface

This method showed excellent agreement with values of skin dose measured on an anthropomorphic phantom with ionization chambers and Gafchromic film [52,53].

The patient graphic is composed of surface elements defined by vertices. Each vertex point contains the cumulative value of dose for the entire procedure, and the surface elements are then color-coded and shaded to the values of the vertices. An example of the display is shown in Figure 24.9, where, in this case, red corresponds to a skin dose of 1000 mGy. Bar graphs are also provided that give the peak value of cumulative dose at the current location of the x-ray beam, so the interventionist can easily see if any point on the patient's skin in the beam has reached a threshold for deterministic effects. The second bar graph gives the real-time value of the maximum skin dose rate in the current beam, which is calculated from the maximum dose per pulse

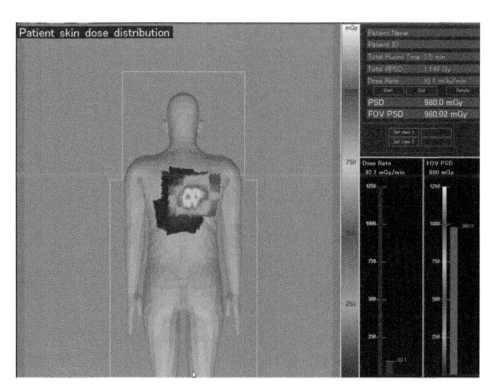

FIGURE 24.9 Example of graphic display from the DTS software showing skin dose distribution as color-coded mapping.

Chapter 24

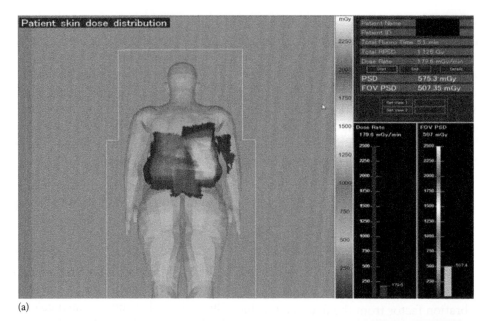

(a)

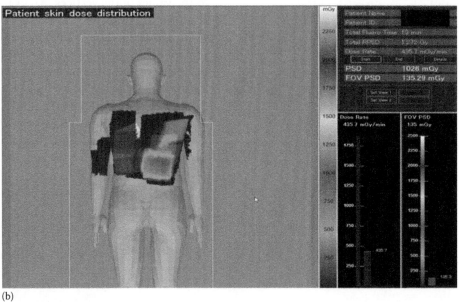

(b)

FIGURE 24.10   (a) Display of the DTS showing the skin dose distribution at the end of a diagnostic coronary angiogram: 69-year-old female, 5′8″, 310 lbs; right ulnar artery access. The fluoroscopy time was 5.1 min and the digital angiography time was 54 s at 15 f/s, resulting in a total KAP of 16,702 cGy cm^2 and a PSD of 575 mGy. (b) Display of the DTS showing the skin dose distribution at the end of a percutaneous coronary intervention: 58-year-old male, 5′6″, 150 lbs; emergency STEMI, femoral artery access. The fluoroscopy time was 13 min and the digital angiography time was 105 s at 15 f/s, resulting in a total KAP of 20,108 cGy cm^2 and a PSD of 1,026 mGy.

times the pulse rate; this bar provides immediate feedback to the interventionist about the dose-rate consequence of his actions, whether it is changing the SSD or changing the pulse rate or angling the beam. Likewise, he will immediately see the dose-rate penalty in changing from fluoroscopy to fluorographic (cine) mode.

Figures 24.10a and b show examples of the final dose distribution at the end of clinical cardiac procedures [54]. The color-coded dose distribution recorded for a diagnostic procedure is seen in Figure 24.10a and that for a percutaneous coronary intervention involving a stent placement is seen in Figure 24.10b. Typically, the distribution for the diagnostic procedure is more uniformly spread over the back, while the intervention gives higher skin dose and it is more concentrated over a small area since the beam is not moved much while the angioplasty or stent placement is occurring. In these displays as well as in most clinical application of the software, red is set to correspond to 2000 mGy since this dose corresponds to the threshold for temporary erythema.

References

1. Johns, H.E. and Cunningham, J.R. *The Physics of Radiology*, 4th ed. Charles C Thomas, Springfield, IL, 1983.
2. Poludniowski, G., Landry, G., DeBlois, F., Evans, P.M., and Verhaegen, F. SpekCalc: A program to calculate photon spectra from tungsten anode x-ray tubes. *Phys. Med. Biol.* 54, N433–N438, 2009.
3. ICRU Report 19. *Radiation Quantities and Units*. ICRU Publications, Washington, DC, 1971.
4. Measurement Techniques *11th General Conference on Weights and Measures*, 3(11): 909–912. The International System of Units, Washington, DC, 1960.
5. Rosenstein, M. *Organ Doses in Diagnostic Radiology*. FDA 76–830, Bureau of Radiological Health, Rockville, MD, 1976.
6. Meriçl, N., Bor, D., Büget, N., and Özkırlı, M. The use of Monte Carlo Technique for the determination of tissue -air ratios (TAR) in diagnostic energy range. *Phys. Med.* XIV, 1, January–March 1998.
7. ICPRP 60. *1990 Recommendations of the International Commission on Radiological Protection*, Vol. 21, pp. 1–3. Pergamon/ICRP Publication 60, Oxford, U.K., 1991.
8. ICPRP 103. *The 2007 Recommendations of the International Commission on Radiological Protection*, Vol. 37, pp. 2–4. ICRP Publication 103, Oxford, U.K., 2007.
9. Mettler, F.A. Jr, Huda, W., Yoshizumi, T.T., and Mahesh, M. Effective doses in radiology and diagnostic nuclear medicine: A catalog. *Radiology* 248(1), 254–263, July 2008.
10. Tapiovaara, M., Lakkisto, M., and Servomaa, A. PCXMC: A PC-based Monte Carlo program for calculating patient doses in medical x-ray examinations. Report STUK-A139, Finnish Center for Radiation and Nuclear Safety, Helsinki, Finland, 1997.
11. Kramer, R., Khoury, H.J., and Vieira, J.W. CALDose_X—A software tool for the assessment of organ and tissue absorbed doses, effective dose and cancer risks in diagnostic radiology. Available from http://www.caldose.org/CaldoseEng1.aspx (accessed April 1, 2015), 2008.
12. Compagnone, G., Giampalma, E., Domenichelli, S., Renzulli, M., and Golfieri, R. Calculation of conversion factors for effective dose for various interventional radiology procedures. *Med. Phys.* 39, 2491, 2012.
13. Harrison, R.M. Backscatter factors for diagnostic radiology (1–4 mm Al HVL). *Phys. Med. Biol.* 27(12), 1465–1474, 1982.
14. Rana, V.K., Gill, K., Rudin, S., and Bednarek, D.R. Significance of including field non-uniformities such as the heel effect and beam scatter in the determination of the skin dose distribution during interventional fluoroscopic procedures. In *SPIE: Proceedings from Medical Imaging 2012: Physics of Medical Imaging*, San Diego, CA, Vol. 8313–8359, 2012, paper 83131N:1–9. NIHMSID 391950.
15. Bednarek, D.R. and Rudin, S. Comparison of two dose-area-product ionization chambers with different conductive surface coating for over-table and under-table tube configurations. *Health Phys.* 78, 316–321, March 2000.
16. Alm Carlsson, G., Carlsson, C.A., and Persliden, J. Energy imparted to the patient in diagnostic radiology: Calculation of conversion factors for determining the energy imparted from measurements of the air collision kerma integrated over beam area. *Phys. Med. Biol.* 29(11), 1329–1341, 1984.
17. International Electrotechnical Commission. Report 60601 medical electrical equipment–part 2–43: Particular requirements for the safety of x-ray equipment for interventional following repeated coronary procedures. IEC, Geneva, Switzerland, 2000.
18. International Electrotechnical Commission. *Report 60601 Medical Electrical Equipment–Part 2–43: Particular Requirements for the Safety of X-ray Equipment for Interventional Procedures*, 2nd ed. IEC, Geneva, Switzerland, 2010.
19. FDA 1020.32(k)(4)(ii) "Fluoroscopic Equipment", Federal Register/Vol. 70, No. 111/Friday, June 10, 2005/Rules and Regulations. http://www.accessdata.fda.gov/scripts/cdrh/cfdocs/cfcfr/CFRSearch.cfm?fr=1020.32 (accessed April 1, 2015).
20. Rana, V.K., Rudin, S., Bednarek, D.R. Updates in the real-time Dose Tracking System (DTS) to improve the accuracy in calculating the radiation dose to the patient's skin during fluoroscopic procedures. In *Proceedings from Medical Imaging 2013: Physics of Medical Imaging*, Orlando, FL, SPIE Vol. 8668–143, 2013.
21. Geiser, W.R., Huda, W., and Gkanatsios, N.A. Effect of patient support pads on image quality and dose in fluoroscopy. *Med. Phys.* 24, 377–382, 1997.
22. Rana, V.K., Gill, K., Rudin, S., and Bednarek, D.R. Significance of including field non-uniformities such as the heel effect and beam scatter in the determination of the skin dose distribution during interventional fluoroscopic procedures. In *Proceedings from Medical Imaging 2012: Physics of Medical Imaging*, San Diego, CA, SPIE Vol. 8313–59, 2012, paper 83131N:1–9. NIHMSID 391950.
23. Bednarek, D.R. and Rudin, S. Comparison of two dose-area-product ionization chambers with different conductive surface coating for over-table and under-table tube configurations. *Health Phys.* 78, 316–321, March 2000.
24. Bednarek, D.R. and Rudin, S. A double transmission ionization chamber for dose and dose-area-product monitoring (WIP-D-03). *Med. Phys.* 25(8), A1585, August 1998.
25. Cameron, J.R., Suntharalingam, N., and Kenney, G.N. *Thermoluminescent Dosimetry*. The University of Wisconsin Press, Madison, WI, 1968.
26. Nunn, A.A., Davis, S.D., Micka, J.A., and DeWerd, L.A. LiF:Mg, Ti TLD response as a function of photon energy for moderately filtered x-ray spectra in the range of 20–250 kVp relative to 60Co. *Med. Phys.* 35, 1859–1869, 2008.
27. Working, K., Hood, R., Slowey, T., Slowey, T., and Pagel, J. A survey of radiation skin dose received by patients undergoing cardiac catheterization. In *40th AAPM*, San Antonio, CA, 1998; *Med. Phys.* 25(7), A166, TH-C4–01, 1998.
28. Reft, C.S. The energy dependence and dose response of a commercial optically stimulated luminescent detector for kilovoltage photon, megavoltage photon, and electron, proton, and carbon beams. *Med. Phys.* 36, 1690–1700, 2009.
29. Peet, D.J. and Pryor, M.D. Evaluation of a MOSFET radiation sensor for the measurement of entrance surface dose in diagnostic radiology. *Br. J. Radiol.* 72, 562–568, 1999.
30. Jones, A.K., Pazik, F.D., Hintenlang, D.E., and Bolch, W.E. MOSFET dosimeter depth-dose measurements in heterogeneous tissue-equivalent phantoms at diagnostic x-ray energies. *Med. Phys.* 32(10), 3209–3213, October 2005.

Chapter 24

31. Falco, M.D., D'Andrea, M., Strigari, L., D'Alessio, D., Quagliani, F., Santoni, R., and Bosco, A.L. Characterization of a cable-free system based on p-type MOSFET detectors for "in vivo" entrance skin dose measurements in interventional radiology. *Med. Phys.* 39, 4866, 2012.

32. Vañó, E., Gonzalez, L., Ten, J.I., Fernandez, J.M., Guibelalde, E., and Macaya, C. Skin dose and dose-area product values for interventional radiology procedures. *Br. J. Radiol.* 74, 48–55, 2001.

33. Vañó, E., Goicolea, J., Galvan, C., Gonzalez, L., Meiggs, L., Ten, J.I., and Macaya, C. Skin radiation injuries in patients angioplasty procedures. *Br. J. Radiol.* 74, 1023–1031, 2001.

34. Vañó, E., Guibelalde, E., Fernández, J.M., and González, L. Patient dosimetry in interventional radiology using slow films. *Br. J. Radiol.* 70, 195–200, 1997.

35. Geise, R.A. and Ansel, H.J. Radiotherapy verification film for estimating cumulative entrance skin exposure for fluoroscopic examinations. *Health Phys.* 59, 295–298, 1990.

36. Geise, R.A. and O'Dea, T.J. Radiation dose in interventional fluoroscopic procedures. *Appl. Radiat. Isot.* 50, 173–184, 1999.

37. Niroomand-Rad, A., Blackwell, C.R., Coursey, B.M., Gall, K.P., Galvin, J.M., McLaughlin, W.L., Meigooni, A.S., Nath, R., Rodgers, J.E., and Soares, C.G. Radiochromic film dosimetry: Recommendations of AAPM Radiation Therapy Committee Task Group 55. *Med. Phys.* 25, 2093–2115, 1998.

38. Giles, E.R. and Murphy, P.H. Measuring skin dose with radiochromic dosimetry film in the cardiac catheterization laboratory. *Health Phys.* 82(6), 875–880, 2002.

39. McCabe, B.P., Speidel, M.A., Pike, T.L., and Van Lysel, M.S. Calibration of GafChromic XR-RV3 radiochromic film for skin dose measurement using standardized x-ray spectra and a commercial flatbed scanner. *Med. Phys.* 38, 1919, 2011.

40. Rana, V., Bednarek, D.R., Josan, M., and Rudin, S. Comparison of skin-dose distributions calculated by a real-time dose-tracking system with that measured by Gafchromic film for a fluoroscopic C-arm unit. *Med. Phys.* 38(6), 3702, June 2011.

41. Wagner, L.K. Radiation injury is a potentially serious complication to fluoroscopically-guided complex interventions. *Biomed. Imaging Interv. J.* 3(2), e22, 2007.

42. Bednarek, D.R., Rana, V.K., Rudin, S. Peak skin dose for cardiac catheterization procedures determined with an automatic does-tracking system. In *Scientific Program and 97th Scientific Assembly and Annual Meeting of RSNA*, Chicago, IL, p. 196, November 27–December 2, 2011, Meeting Program. Scientific presentation. SSK15–02. Online program abstract: http://rsna2011.rsna.org/search/search.cfm?action = add&filter = Subspecialty&value = 301102130

43. Stecker, M.S. Guidelines for patient radiation dose management. *J. Vasc. Interv. Radiol.* 20, S263–S273, 2009.

44. The Joint Commission. Radiation overdose as a reviewable sentinel event. http://www.jointcommission.org/assets/1/18/Radiation_Overdose.pdf. Accessed March 22, 2013.

45. Gkanatsios, N.A., Huda, W., Peters, K.R., Freeman, J.A. Evaluation of an on-line patient exposure meter in neuroradiology. *Radiology* 203, 837–842, 1997.

46. Cusma, J.T., Bell, M.R., Wondrow, M.A., Taubel, J.P., and Holmes, D.R., Jr. Real-time measurement of radiation exposure to patients during diagnostic coronary angiography and percutaneous interventional procedures. *J. Am. Coll. Cardiol.* 33, 427–435, 1999.

47. den Boer, A., de Feijter, P.J., Serruys, P.W., and Roelandt, J.R. Real-time quantification and display of skin radiation during coronary angiography. *Circulation* 104, 1779–1784, 2001.

48. Ozeroglu, M.A. Verification of Caregraph' peak skin dose data using radiochromic film. Master of Science in Public Health thesis, Department of Preventive Medicine and Biometrics. Uniformed Services University of the Health Sciences, Bethesda, Maryland. http://www.openthesis.org/documents/Verification-peak-skin-dose-data-158245.html (accessed April 1, 2015).

49. Johnson, P.B., Borrego, D., Balter, S., Johnson, K., Siragusa, D., and Bolch, W.E. Skin dose mapping for fluoroscopically guided interventions. *Med. Phys.* 38(10), 5490–5499, October 2011.

50. Khodadadegan, Y., Zhang, M., Pavlicek, W., Paden, R.G., Chong, B., Schueler, B.A., Fetterly, K.A., Langer, S.G., and Wu, T. Automatic monitoring of localized skin dose with fluoroscopic and interventional procedures. *J. Digit Imaging* 24(4), 626–639, August 2011.

51. Chugh, K., Dinu, P., Bednarek, D.R., Wobschall, D., Rudin, S., Hoffmann, K.R., Peterson, R., and Zeng, M. A computergraphic display for real-time operator feedback during interventional x-ray procedures. In *Proceedings from Medical Imaging 2004: Physics of Medical Imaging*, San Diego, CA, SPIE Vol. 5367, pp. 464–447, 2004.

52. Bednarek, D.R., Barbarits, J., Rana, V.K., Nagaraja, S.P., Josan, M.S., and Rudin, S. Verification of the performance accuracy of a real-time skin-dose tracking system for interventional fluoroscopic procedures. In *Proceedings from Medical Imaging 2011: Physics of Medical Imaging*, Orlando, FL, SPIE Vol. 7961–7978, 2011, paper 796127:1–8. NIHMSID 303276.

53. Rana, V., Bednarek, D.R., Josan, M., Rudin, S. Comparison of skin-dose distributions calculated by a real-time dose-tracking system with that measured by gafchromic film for a fluoroscopic c-arm unit. *Med. Phys.* 38(6), 3702, June 2011, SU-F-BRA-09.

54. Dashkoff, N., Bednarek, D.R., Rana, V.K., and Rudin, S. Implementation of a real-time skin dose tracking system in a cardiac catheterization laboratory. Poster presentation. *J. Am. Coll. Cardiol.* 60(17_S), Abstract TCT-317, October 23, 2012. http://content.onlinejacc.org/article.aspx?articleid=1383587

25. Patient Dose Control in Fluoroscopically Guided Interventions

Daniel R. Bednarek

Fluoroscopy and fluorography are real-time, high-frame-rate imaging modalities used for interventional procedures in which the information may be used to aid in the diagnosis or treatment of a patient pathology. Fluoroscopy involves a lower dose rate and the images are typically not saved for future review, while fluorography involves a higher dose rate and acquisition of the images. Many fluoroscopically guided interventional (FGI) procedures involve visualization of the vasculature, and iodine injection provides vessel lumen contrast, in which case the technique is referred to as angiography. There are many factors that influence the dose received by the patient in performing such a procedure.

This chapter reviews those factors that determine the radiation dose to the patient, ways by which this dose can be reduced, and the compromises that may need to be made in the image information. Very often, the quality of the information obtained in an image has an inverse relation to patient dose such that more dose is required to obtain an improvement in the image quality. It must be kept in mind that, if the information is sufficient to perform the procedure at a given dose, then any additional dose is unwarranted and wasted. Therefore, one should operate on the principle that the minimum dose needed to make a diagnosis or to perform a task should be the goal. However, this is a difficult balance to achieve and typically higher dose is used than absolutely needed since patient care should not be compromised.

25.1 Collimation

The region of interventional activity (ROIA) can be much smaller than the available field of view (FOV) of the image intensifier or flat panel detector (FPD) being used. If it is not necessary to visualize features outside of the treatment area during the intervention, the field size can be collimated significantly and live zoom could be used to fill the display monitor with the collimated image. The size of the x-ray beam determines the FOV and also, to a large extent, determines the integral dose received by the patient. The collimators are used to adjust the field size and usually consist of two pairs of lead shutters, which move in or out on opposing sides from the center of the FOV. The minimum field size necessary to perform the task should be used since it results in reduced radiation exposure to the patient, less scatter to personnel, and improved image quality.

Collimation does not reduce the dose rate to the tissue within the beam from primary radiation, but some reduction may be realized if the technique factors are unchanged because of the reduction of scattered radiation to internal organs or of backscatter to the skin. The effect of field size on backscatter can be substantial and is demonstrated in Figure 24.3. Reducing the field size will reduce the volume of tissue or the organs exposed to radiation and thus will reduce the effective dose, resulting in a reduction of stochastic risk. It has been shown that in neuroimaging, the effective dose can be reduced between 1 and 2 orders of magnitude when using 3.5 cm FOV of the microangiographic fluoroscope (MAF) compared to a full 20 cm FPD FOV [1]. Approximately proportional changes in effective dose can be expected for other FOV sizes. Collimation can also facilitate a reduction of the tissue subject to deterministic risk as discussed later in the section on dose spreading.

When collimating, care must be taken that the beam area is not smaller than the area of the automatic exposure rate control (AERC) sensor, since the x-ray tube output will be increased to compensate for the decreased signal if the sensor is blocked. Even if the sensor is not blocked, the AERC will likely increase the tube output to compensate for decreased scattered radiation resulting from the smaller FOV. This will result in increased dose rate to the patient in the beam area, while the integral dose is decreased over the volume exposed. Reduced scattered radiation reaching the image receptor by collimation will improve the contrast-to-noise ratio (CNR) or allow the dose rate to be reduced at equivalent CNR [2]. An AERC system that maintains constant CNR rather than constant signal would not increase the dose with collimation and would be optimal for balancing risk vs. benefit. If the beam size is reduced sufficiently, the scattered radiation may be reduced enough to warrant removal of the grid, resulting in even further dose savings [3]. This is common practice when doing pediatric fluoroscopic procedures. If the scatter is low because of a small field size or a small volume of tissue exposed, use of a grid may degrade image quality by causing further hardening of the beam without significant improvement of CNR, while absorbing primary radiation. Some procedures may not require the highest image quality and the grid may be removed even in the presence

of scatter and this should always be considered as an option for dose reduction [4,5].

Collimation when using the normal nonmagnification mode of the image receptor will result in a smaller dose rate than when using the same reduced FOV in a magnification (mag) mode. Fluoroscopic imaging systems are typically set up to increase exposure rate with increasing mag mode to reduce the perception of noise by maintaining the x-ray quanta per unit display area. If the input exposure to the receptor were set to increase inversely with the field area as the mag mode is changed to keep the signal per pixel area constant (as on many early image intensifier systems), the integral dose rate (or kerma-area-product (KAP) rate) would be essentially independent of mag mode if the kVp did not change; if the kVp is increased by the AERC to achieve the increased signal for increased mag mode, the integral dose to the patient will be reduced because of the increased penetration of the beam. For FPDs, there is no need to increase the exposure rate with increasing mag mode since the detector element size is constant except when pixels are binned for large field sizes to reduce the image matrix size. As a general rule for both types of detector, imaging systems are set up so that the patient entrance air kerma rate increases, while the integral dose decreases with increasing mag mode (e.g., changing from 20 to 10 cm mode with corresponding change in collimation). Most manufacturers increase air kerma rate by less than the inverse of the beam area when increasing the magnification modes.

Reducing the beam size by collimation does not improve the intrinsic spatial resolution of the image even if live zoom is used since the limiting resolution is determined by the detector element size and the FPD or image intensifier element size is not changed by collimation. However, live zoom with collimation may allow better utilization of the display matrix and the human visual system, resulting in increased perception of smaller features. When the field size is reduced by increasing the magnification mode with an image intensifier, intrinsic spatial resolution will increase proportionally with FOV size since the same number of pixels sample a smaller field and the effective pixel size decreases; changing magnification mode with an FPD does not change the intrinsic resolution since the detector element (del) pitch is unchanged unless pixel binning is used with larger FOVs. For large FOV magnification modes, the unbinned matrix size might exceed the data transfer rate capability of the FPD so that the matrix size can be reduced by combining or binning the signal from adjacent elements. In this case, the resolution is reduced by the number of pixels binned in a row or column; 2 × 2 binning is common for larger FOV mag modes, and this reduces the intrinsic resolution by a factor of 2.

25.1.1   ROI Fluoroscopy

Reduction of the FOV by collimation may not be feasible when the region peripheral to the ROIA needs to be visualized for monitoring purposes. This peripheral region may not need the same high-quality image as in the ROIA, and the dose rate in the peripheral areas could be reduced without negatively impacting the conduct of the procedure in many cases. In region-of-interest (ROI) fluoroscopy, a semitransparent beam attenuator with a (central) hole for the ROI is used to reduce the dose to tissue in the periphery typically by about 80% [6]. For a filter with an ROI opening, which is half of the total FOV area and a transmission of 20%, the integral dose is reduced to 60% of that without the attenuator. Vasan et al. have shown from phantom as well as animal studies for neurointerventional procedures that acceptable image quality could be obtained in the periphery with a dose-rate reduction of 5–6 times by the attenuator, resulting in an integral dose reduction as high as 86% [7]. With a special mechanism, this ROI opening can be made to be positioned off-center and to follow the ROIA.

With ROI fluoroscopy, the beam intensity in the peripheral region is reduced relative to that in the ROI and the brightness of the image in the periphery will be reduced. If this brightness reduction is unacceptable, there are a number of ways that the peripheral brightness can be equalized to that in the ROI [8–11]. On angiographic systems, one of the simplest ways to do this is to include the attenuation filter in the mask for digital subtraction angiography (DSA) or road mapping modes. In that case, the attenuation of the filter will be automatically subtracted along with the structure of the patient. If the attenuator alone is to be subtracted, then a mask can be taken of the attenuator alone without the patient and that mask can be subtracted. Even after brightness equalization, the image quality will appear degraded in the periphery because of the increase in quantum mottle due to a reduced dose. This noise increase in the periphery may be acceptable since it is not the ROIA. If desired, the increased quantum mottle can be reduced by using either spatial or temporal filtering; however, this would result in a reduction of either spatial or temporal resolution that may be unacceptable in the ROI if such processing were done uniformly

throughout the field. If that were the case, spatially different processing can be done in real time using a graphics processing unit (GPU) so that any spatial or temporal resolution loss is only in the periphery where it might be acceptable [7,12]. The feasibility of doing this has been shown on a research system, but this capability is not currently available on commercial imaging systems.

ROI attenuators can be made of copper in which case a 0.7 mm thickness should be sufficient to provide about 80% reduction of dose at 70 kVp [13]. Cu filters will likely result in beam hardening that can cause a decrease in subject contrast in the periphery. Many of the attenuators originally used for ROI fluoroscopy consisted of about 4–5 layers of Kodak Lanex Regular intensifying screens. These screens contain gadolinium that has a K-edge of 50 keV so that the beam-hardening effect is reduced. In fact the use of a K-edge filter can increase subject contrast and negate some of the effect of the increase in quantum mottle on contrast signal-to-noise ratio while still providing substantial dose reduction in the periphery [14,15]. Although the net effect of any attenuator is to reduce image quality in the periphery, this reduction should not have a negative impact on the procedure since the information in the periphery is used for reference or orientation while the ROIA is the main focus and high image quality is maintained there.

25.1.2   Spot Fluoroscopy

Spot fluoroscopy is an imaging mode available on Toshiba Medical Systems C-arm fluoroscopes that combines collimation with a form of ROI fluoroscopy [16]. With this technique, real-time fluoroscopy can be performed in a collimated ROI that can be located at any point within the FOV of the image receptor (even off-center), while the remainder of the FOV is filled in with a previously acquired still frame. This allows full dose sparing in the peripheral FOV (except for scattered radiation from the ROI), while maintaining a reference image outside the collimated ROI. However, any real-time activity occurring outside the fluoro "spot" is not seen. The reference image can be updated when needed but with additional radiation exposure. Just as with collimation, this method reduces integral dose to the patient, improves CNR by reducing scattered radiation to the image receptor in the ROI, and reduces scatter dose to the staff.

25.1.3   Dose Spreading

The peak skin dose can be a concern in long interventional procedures because of the increased risk of deterministic effects such as erythema, epilation, dermatitis, desquamation, and necrosis [17]. Since these effects have a threshold below which they are unlikely to occur, their risk can be reduced by limiting the dose to individual points on the skin. This can be done even for long procedures by moving the point on the skin where the beam enters the patient. This is referred to as dose spreading. Figure 25.1 shows that if the ROIA is kept at the C-arm gantry isocenter, the projection angle can be changed so that a different region of the skin is exposed while the ROIA remains in the FOV. The smaller the beam size, the less is the change in projection angle needed to avoid overlap as demonstrated in Figure 25.1. Both collimation and ROI fluoroscopy can thus facilitate dose spreading. As seen in Figure 25.1e, the peripheral region of reduced exposure for

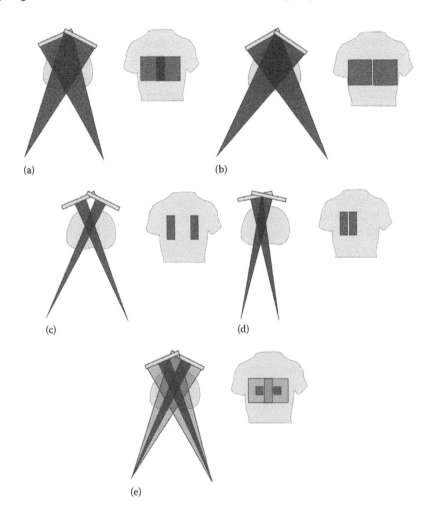

(a)

(b)

(c)

(d)

(e)

FIGURE 25.1 In this series of illustrations, the left drawing in each group shows the profiles of two x-ray beams passing through the chest region of a patient, and the right drawing shows the projection of those beams on the patient's skin at the entrance surface. Note: These figures do not illustrate the effects of beam attenuation in the patient or inverse square reduction of intensity with distance from the focal spot. (a) Two uncollimated beams are shown with a change in RAO/LAO angulation. The region of overlap between the beams has approximately the sum of the skin dose for each of the individual beams. (b) If the RAO/LAO angles for the two uncollimated beams are increased sufficiently, the region of overlap can be prevented. In this case, the dose is "spread" over the skin so that the peak skin dose is reduced. To prevent overlap of the beams, larger angular separation is required for larger uncollimated beams. (c) Shown are two beams with the same RAO/LAO angulation as the beams in (a) but collimated in the transverse direction. The region of overlap on the skin is eliminated, while the ROIA at the rotational isocenter remains in the FOV. Compared to the uncollimated beams, the integral dose can be considerably reduced using collimation. (d) With collimation, the angular separation of the two beams needed to prevent overlap can be much less than for the uncollimated beams. (e) With ROI fluoroscopy, the peripheral region is reduced in x-ray intensity using a beam attenuator; the central region with higher exposure is the ROIA. If the beams are moved about the isocenter by a small angle, overlap of the central ROI beams on the skin can be prevented. If the peripheral areas of the beams overlap as seen here, the total skin dose in that region will be much less than if the full intensity uncollimated beams were to overlap since the peripheral dose rate is typically 20% that of the central ROI. In this case, the integral dose is also considerably reduced compared to using the full uncollimated beam.

two projections can overlap in ROI fluoroscopy, while still reducing the peak skin dose. Generally, lower dose will be given to the patient if dose spreading is achieved by changing the projection angle in the RAO/LAO direction rather than in the CRA/CAU direction since the beam path through the patient is typically longer if the central ray is not perpendicular to the skin surface.

This is true particularly for procedures involving exposure of the chest and abdomen region. As a rule, dose spreading can be achieved without having beam overlap on the entrance surface for a smaller change in projection angle if the ROIA is nearer to the exit surface of the patient (e.g., using posterior-anterior (PA) projections when the ROIA is closer to the anterior surface).

25.2   Beam Spectrum and Filters

The x-ray spectrum plays a major role in determining how much dose the patient absorbs for a given signal at the image receptor. Generally, any factor that increases the average energy of the spectrum (or beam half-value layer) increases the beam penetration and reduces the patient dose for the same signal at the image receptor. Increasing the kVp adds higher-energy photons to the spectrum. Use of a waveform with less ripple for the same kVp results in a higher average energy beam. Use of a grid-controlled tube removes the low-energy tails of the x-ray pulse, resulting in a higher energy beam. Except for K-edge filters, increasing the supplemental filtration added to the beam shifts the energy to higher values since lower-energy photons are preferentially absorbed by the filter. Figure 24.1 shows the effect on the spectrum of using a Cu filter in place of an Al filter; Cu generally provides greater beam hardening and lower patient dose than Al. The goal of adding a supplemental filter is to remove those x-ray photons that have a higher probability of being absorbed in the patient than of reaching the image receptor to contribute to the signal. Photons absorbed in the filter that would have reached the image

receptor must be replaced by additional photons from the x-ray tube, and thus, tube loading is increased. This may limit the ability to perform extended procedures unless a high heat-load capacity tube is used, which would allow increased mAs. Such tubes include the MRC tube of Philips and the Stratton tube produced by Siemens.

Typical added filtration available on interventional C-arm units can include 1.8 mm Al or 0.1, 0.2, 0.3, 0.6, and 0.9 mm Cu. The heaviest filter of 0.9 mm Cu is typically used for dose reduction in pediatric imaging. Most imaging systems have a control logic that automatically selects the filter based on dose reduction, image quality, and tube loading considerations [18]. Tube loading is also affected by the focal spot size selected, and more loading might require a larger focal spot with more geometric unsharpness. Typical focal spot sizes for angiographic systems include 0.3, 0.6, and 1.0 mm. Above a certain value, added filtration reduces image contrast and becomes counterproductive. The value of k-edge filters in retaining contrast is limited to a narrow kVp range or narrow range of patient thicknesses [19,20].

25.3   Compensation (Wedge) Filters

Compensation filters are typically available to be moved into the FOV to reduce the intensity of x-rays over patient image areas that have low attenuation to reduce the glare into regions of higher attenuation. These filters are not meant to cover the entire FOV and can be selectively positioned by the operator. For example, in cardiac imaging, they can be positioned over the lungs to reduce the bright area in that region and equalize the range of intensities in the image while concentrating on the heart. Use of such filters reduces the dose to the patient under the filter while still allowing visualization of structures under the filter

similar to an ROI attenuator. This is preferable to software equalization of the brightness range since the compensation filter generally reduces dose over regions where dose is not needed, reduces the scattered radiation reaching the image receptor to improve image quality, and prevents signal saturation of the image receptor by the unattenuated beam. Software equalization results in no dose reduction. Since wedge filters are placed before the transmission ionization chamber in the beam, their use generally causes the reference point air kerma values to be underestimated when obtained from the measured kerma area product.

25.4   Image Processing

Image processing can allow a reduction of dose by reducing the visualization of quantum mottle in the

image. Noise reduction processing typically involves averaging either over pixels or over frames. Spatial averaging over pixels tends to reduce the spatial resolution and in its most basic form involves the global convolution of the image matrix with an averaging kernel (e.g., 3 × 3 kernel filled with a value of 1/9 for each element). The larger the kernel, the greater the noise reduction and the greater the reduction of spatial resolution. This is referred to as low pass filtering, since the low spatial frequency content is retained. Adaptive spatial filtering can be used to apply less filtering over regions of high spatial frequency content and more filtering over regions of low spatial frequency content, thereby minimizing the spatial resolution loss.

Temporal filtering by averaging individual pixels of successive image frames in a series can also be used to reduce the statistically varying signal fluctuations [21]. Simple averaging of four frames would be the equivalent of increasing the dose by a factor of 4 and would reduce the quantum mottle by a factor of 2. More commonly, recursive filtering is used whereby a weighted sum of the current frame (Fc) with previous frames (Fp) is displayed as described by the following:

$$D = k\, Fc + (1 - k)\, Fp \qquad (25.1)$$

where
  D is the displayed image
  k is a value from 0 to 1

After a large number of frames, the displayed image noise variance (V_D) approaches a value

$$V_D = \frac{k}{(2-k)V_F} \qquad (25.2)$$

where V_F is the variance of a single frame [22]. Recursive temporal filtering adds a fraction of previous image frames to the current frame so that the effective number of photons used to create the currently viewed image is increased and the variance is reduced. The smaller the value of k, the greater is the weight of previous frames used and the less is the noise. Since information from previous frames is included in the displayed image, moving structures may be blurred by image lag and, the smaller the value of k, the greater is the potential for lag.

Image lag introduced by temporal filtering is most noticeable for fast-moving features. This might interfere with visualization of dynamic interventional activity such as the control of catheter tips and placement of aneurysm coils, and because of the rapid motion of the heart, temporal filtering has not been used in cardiac procedures. However, adaptive temporal filtering can be used to reduce image lag by adjusting the filter weight depending on the correlation of pixels in successive frames to determine regions containing moving structure [23]. A combination of adaptive spatial and temporal filtering can be used to minimize image degradation by the filtering process [24]. Aufrichtig and Wilson estimated that their spatiotemporal filter is able to reduce noise enough to provide an estimated dose reduction of over 80% [25].

25.5 Frame Rate Reduction

Systems with pulsed fluoroscopy allow the image frame capture rate to be selected. Typical frame rates for fluoroscopy are 30, 20, 15, 7.5, and 3 per second. Digital acquisition (DA) and DSA include additional frame acquisition rates of 2, 1, or 0.5 frames per second (fps). Higher frame rates are needed when features important for the intervention are moving rapidly such as the coronary arteries in cardiac catheterization procedures. It is typical to use frame rates of 30, 20, and 15 fps for DA in cardiac procedures, and this is often called "cine" mode in reference to high frame rate cinematography film imaging commonly used in "legacy systems." Although the frame rate may be changed, the display refresh rate remains unchanged in digital systems so that there is no appearance of flicker.

In pulsed fluoroscopy, the kVp, mA, pulse width (ms), or frame rate can be changed. Increasing any of these will increase the output of the x-ray tube and the dose rate to the patient. The quantum mottle per frame will depend on the number of x-rays absorbed by the image receptor for that frame and thus on the kVp and mAs per pulse. If these factors remain constant, the dose rate will be proportional to the frame rate. Since the human visual system integrates the light input to the eye over a time period of about 200 ms, the observed quantum mottle is determined by the total quanta absorbed over 200 ms [26]. The real-time displayed image will thus appear less noisy than a single frame in the sequence, and the real-time image will have better observed contrast resolution. Effectively at 30 fps, the eye will integrate six frames and the noise will be reduced by a factor of √6 compared to a single frame. At 15 fps, the eye will integrate only three frames and the noise will be reduced by $\sqrt{3}$. If the dose per frame were not changed, then the images viewed at 15 fps would appear noisier than those at 30 fps. Aufrichtig et al. derived a relationship between frame rate and the detector input air kerma (IAK) rate necessary to obtain constant perception of quantum mottle [27]. He gives an inverse relation between IAK per frame and the square root of the frame rate so that the IAK at 15 fps should be 1.4 times that at 30 fps to provide equal perception of noise. This would result in a dose rate reduction of a factor of 1.4 rather than 2 times when switching from 30 to 15 fps. This relation does not apply to frame rates less than 5 per second since the interval between frames is more than 200 ms and no frame integration occurs in the human visual system.

Some systems also allow operation with continuous fluoroscopy at 30 fps. In this case, there is no pulsing of the x-ray exposure and the frame capture time is 33 ms. It should be noted that using pulsed fluoroscopy at 30 fps rather than continuous fluoroscopy at 30 fps does not necessarily reduce the dose rate. In most cases, the mA is adjusted to provide the same mAs per frame so that the dose rate and contrast resolution are nearly the same. However, pulsed frame rates below 30 per second normally have a lower dose rate than continuous fluoroscopy at 30 fps. An advantage of pulsed fluoroscopy over continuous fluoroscopy at the same frame rate is that the shorter pulse width (e.g., 5 vs. 33 ms) results in less blur due to the motion of the patient or interventional device.

25.6   Fluoroscopy vs. DA vs. DSA

There are considerable differences in the dose rate per frame between fluoroscopy, DA, and DSA. Compared to standard fluoroscopy that has an IAK to the receptor in the range of 2 microGy per frame, the receptor IAK for DA is typically about 50 times higher, while DSA can be up to 100 times higher. Cineradiography that usually is obtained at frame rates over 15 per second has a per-frame IAK about 10 times that of standard fluoroscopy [28].

If fluoroscopy provides sufficient image quality, then the operator has the option to save the frames of a fluoroscopic run that are not normally saved. This would result in a substantial reduction in patient dose compared to capturing the same number of DA frames. High-level control fluoroscopy should also be avoided unless needed for acceptable image quality with large patients.

This mode allows the air kerma rate to go up to 176 mGy/min compared to a maximum of 88 mGy/min for standard fluoroscopy at the FDA compliance point 30 cm from the image receptor; this mode has a separate switch for activation and should be able to be recognized by an audible beeping tone. However, saving a HLC fluoroscopy sequence would still result in a sizeable dose reduction compared to saving the same number of DA frames and with better image quality than standard fluoroscopy.

When possible, dose can be reduced substantially by using last image hold (LIH) for fluoroscopy or by viewing a saved fluoroscopy or DA run rather than taking additional imaging sequences. Likewise virtual collimation should be used to adjust the field size without viewing the real-time x-ray image whenever possible so the patient does not have to be exposed.

25.7   Exposure Geometry

The distance of the patient from the x-ray tube will have a large effect on dose since the dose rate varies as the inverse square of the distance from the focal spot. To minimize dose, the patient should be positioned as far as possible from the x-ray tube and as near as possible to the image receptor. Moving the image receptor away from the patient and farther from the x-ray tube will cause the AERC to increase the output of the tube to maintain the image signal and thus cause an increase in patient dose. Moving both the image receptor and the patient farther from the x-ray tube will usually result in a reduction of patient dose depending on the parameters changed by the AERC. If moving the receptor farther from the tube results in the AERC increasing the

kVp, the dose to the patient will be decreased due to the increased x-ray transmission through the patient. On some imaging systems, the demand for increased output from the tube causes the added beam filtration to be reduced, which softens the x-ray spectrum, resulting in increased absorption and increased dose to the patient. Because of these factors, it is not always clear to the operator what the actual patient dose rate is. The air kerma rate and cumulative air kerma indicated on the display are at the interventional reference point, which is not necessarily at the patient's entrance skin and does not account for changes in source to skin distance (SSD); changes in dose rate due to changes in SSD are not indicated by the value displayed.

Chapter 25

The larger the patient, the higher the AERC system will drive the output of the x-ray tube and the higher the patient dose will be. The operator should be aware that this can result in a substantial increase in skin dose rate and the threshold for deterministic effect can be reached more quickly. Likewise, using large cranial–caudal angles can increase the path length of the beam through the patient, requiring much more radiation output and much higher skin dose rate.

25.8 Determining Peak Skin Dose

For those facilities accredited by The Joint Commission, a root-cause analysis must be performed if the peak skin dose to the patient exceeds 15 gray over a 6–12-month period [29]. This PSD value is very difficult to determine using just the cumulative air kerma values since the beam is not stationary during a procedure and the $K_{a,r}$ is determined at a reference point, which does not necessarily coincide with the entrance skin location. To get a better estimate of the actual peak skin dose, the DA projections can be obtained from the saved images and the dose to regions of the skin can be calculated. The fluoroscopy projection information is not available from the images since fluoroscopic images are typically not saved and fluoroscopic dose would need to be assigned to the DA projections. Much more information can be obtained from the DICOM Radiation Dose Structured Report (RDSR), which contains air kerma information for each radiation exposure event, including fluoroscopy [30]. Unfortunately, the event currently includes all radiation produced for an exposure sequence and the beam may have been moved during the event. Other ways to obtain the peak skin dose value are described in the earlier section on skin dose mapping and an excellent review of radiation dose management is provided in NCRP Report No. 168 [31].

References

1. K. Gill, C. Ionita, D.R. Bednarek, and S. Rudin. Effective-dose rate comparison between the micro-angiographic fluoroscope (MAF) and the x-ray image intensifier (XII) used during neuro-endovascular device deployment procedures. Abstract SU-E-I-191: *Med. Phys.* 38, 3440 (2011).
2. U. Neitzel. Grids or air gaps for scatter reduction in digital radiography: A model calculation. *Med. Phys.* 19 (2), 475 (1992).
3. P. Sprawls. Scattered radiation and contrast. The physical principles of medical imaging. http://www.sprawls.org/ppmi2/SCATRAD/ (Accessed April 2014).
4. K. Cortis, R. Miraglia, L. Maruzzelli, R. Gerasia, C. Tafaro, and A. Luca. Removal of the antiscatter grid during routine biliary interventional procedures performed in a flat-panel interventional suite: Preliminary data on image quality and patient radiation exposure. *Cardiovasc. Interv. Radiol.* 37(4), 1078–1082. (December 2013). doi:10.1007/s00270-013-0814-9.
5. S. Rudin and D.R. Bednarek. Minimizing radiation dose to patient and staff during fluoroscopic, nasoenteral-tube insertions. *Brit. J. Radiol.* 65 (770), 162–166 (1992).
6. S. Rudin and D.R. Bednarek. Region of interest fluoroscopy. *Med. Phys.* 19 (5), 1183–1189 (1992).
7. S.N. Swetadri Vasan, L. Pope, C.N. Ionita, A.H. Titus, D.R. Bednarek, and S. Rudin. Dose reduction in fluoroscopic interventions using a combination of a region of interest (ROI) x-ray attenuator and spatially different, temporally variable temporal filtering. *Proc. SPIE*, 8668, 86 683Y–86 683Y–8 (2013). [Online]. Available: http://dx.doi.org/10.1117/12.2006277 (Accessed April 1, 2015).
8. S. Rudin, D.R. Bednarek, L.R. Guterman, A. Wakhloo, L.N. Hopkins, L. Fletcher, and P. Massoumzadeh. Implementation of region of interest fluoroscopy using the road mapping mode of a real-time digital radiographic unit. *RadioGraphics* 15 (6), 1465–1470 (November 1995).
9. S. Rudin, L.R. Guterman, W. Granger, D.R. Bednarek, and L.N. Hopkins. Neuro-interventional radiologic application of region of interest (ROI) imaging techniques. *Radiology* 199, 870–873 (1996).
10. L.M. Fletcher, S. Rudin, and D.R. Bednarek. Method for image equalization of ROI fluoroscopic images using mask localization, selection, and subtraction. *Computerized Med. Imag. Graphics,* 20 (2), 89–103 (1996).
11. S. Rudin, D.R. Bednarek, and C.Y. Yang. Real-time equalization of region of interest fluoroscopic images using binary masks. *Med. Phys.* 26 (7), 1359–1364 (July 1999).
12. S.N. Vasan, P. Sharma, C.N. Ionita, A.H. Titus, A.N. Cartwright, D.R. Bednarek, and S. Rudin. Spatially different, real-time temporal filtering and dose reduction for dynamic image guidance during neurovascular interventions. *Conf. Proc. IEEE Eng. Med. Biol. Soc.*, 2011, 6192–6195 (2011).
13. S.S. Nagesh, A. Jain, C. Ionita, A. Titus, D. Bednarek, and S. Rudin. Sud- 134–03: Design considerations for a dose-reducing region of interest (ROI) attenuator built in the collimator assembly of a fluoroscopic interventional c-arm. *Med. Phys.* 40 (6), 112 (2013).
14. P. Massoumzadeh, S. Rudin, and D.R. Bednarek. Filter material selection for region of interest radiologic imaging. *Med. Phys.* 25 (2), 161–171 (February 1998).

15. S. Rudin and D.R. Bednarek. Comparison of filter material and design for use in ROI angiography. *SPIE Phys. Med. Imag.* 1896, 354–364 (1993).

16. T. Takahashi and T. Kurihara. Toshiba medical systems. Infinix-I spot fluoroscopy. Diagnostic x-ray system. US Patent 7,116,752 B2. http://www.medical.toshiba.com/downloads/vl-dose-cs-spot-fluoro (Accessed April 14, 2014).

17. L.K. Wagner, P.J. Eifel, and R.A. Geise. Potential biological effects following high x-ray dose interventional procedures. *J. Vasc. Interv. Radiol.* 5 (1), 71–84 (1994).

18. P. Lin. Technical advances of interventional fluoroscopy and flat panel image receptor. *Health Phys.* 95 (5), 650–657 (November 2008).

19. M. Sandborg, C.A. Carlsson, and G.A. Carlsson. Shaping x-ray spectra with filters in x-ray diagnostics. *Med. Biol. Eng. Comput.* 32 (4), 384–390 (July 1994).

20. R.M. Gagne, P.W. Quinn, and R.J. Jennings. Comparison of beam-hardening and K-edge filters for imaging barium and iodine during fluoroscopy. *Med. Phys.* 21 (1), 107–121 (January 1994).

21. C.L. Chan, A.K. Katsaggelos, and A.V. Sahakian. Image sequence filtering in quantum limited noise with applications to low-dose fluoroscopy. *IEEE Trans. Med. Imag.*, 12 (3), 610 (September 1993).

22. B.H. Hasagawa. *The Physics of Medical X-Ray Imaging*, 2nd edn. Medical Physics Publishers, Madison, WI, 1991, pp. 284–287.

23. E. Dubois and S. Sabri. Noise reduction in image sequences using motion-compensated temporal filtering. *IEEE Trans. Commun.* 32, 826–821 (July 1984).

24. A.S. Wang. Spatial and temporal filtering mechanism for digital motion video signals. Patent No: US 6,281,942 B.1, August 28, 2001.

25. R. Aufrichtig and D.L. Wilson. X-ray fluoroscopy spatio-temporal filtering with object detection. *IEEE Trans. Med. Imag.* 14 (4), 733–746 (December 1995) ISSN:0278-0062.

26. F. Schaeffel. Processing of information in the human visual system. In: *Handbook of Machine Vision*, A. Hornberg (ed.). Wiley-VCH, Weinheim, Germany, 2006, pp. 1–34.

27. R. Aufrichtig, P. Xue, C.W. Thomas, G.C. Gilmore, and D.L. Wilson. Perceptual comparison of pulsed and continuous fluoroscopy. *Med. Phys.* 21, 245–256 (1994).

28. M. Mahesh. Fluoroscopy: Patient radiation exposure issues. *RadioGraphics* 21, 1033–1045 (2001).

29. The Joint Commission. Radiation overdose as a reviewable sentinel event. http://www.jointcommission.org/assets/1/18/Radiation_Overdose.pdf. (Accessed March 22, 2013).

30. National Electrical Manufacturer's Association, Digital imaging and communications in medicine (DICOM) supplement 94: Diagnostic x-ray radiation dose reporting (dose SR). NEMA, Rosslyn, VA, 2005. Available at ftp://medical.nema.org/medical/dicom/final/sup94_ft.pdf (Accessed April 2014).

31. NCRP Report No. 168, Radiation dose management for fluoroscopically-guided interventional medical procedures (July 21, 2010). National Council on Radiation Protection and Measurements, Bethesda, MD.

Chapter 25

26. Radiation Protection of Staff and Patients in Cardiovascular and Neurovascular Imaging

Renato Padovani, Madan Rehani, Eliseo Vano, and Carlo Cavedon

Chapter 26

26.1 Radiation Protection of Staff

26.1.1 Introduction

Interventional procedures are increasing in number and complexity [1]. The benefits of interventional radiology to patients are extensive and beyond dispute, but many of these procedures also have the potential to produce occupational doses to interventional radiologists high enough to cause concern [2,3].

The radiation dose received by interventional radiologists and cardiologists can vary by more than one order of magnitude for the same type of procedure and for similar patient dose. Recently, there has been particular concern regarding occupational dose to the lens of the eye in operators [4]. New data from exposed human populations have suggested that lens opacities (leading to cataracts in some) occur at doses far lower than those previously believed to cause cataracts [5–8]. Additionally, it appears that the latency period for radiation cataract formation is inversely related to radiation dose [5]. The International Commission on Radiological Protection (ICRP) has published a statement that for the lens of the eye, the threshold for tissue reactions is now considered to be 0.5 Gy and recommends a new occupational equivalent dose limit for the lens of the eye of 20 mSv in a year, averaged over defined periods of 5 years, with no single year exceeding 50 mSv [9].

Procedures may involve high radiation dose rates in the interventional laboratory [10,11]. The magnitude and distribution of radiation scattered from the irradiated patients' body is affected by many factors, including patient size, C-arm angulation, equipment settings, and use of shields between patient and operator.

Occupational radiation protection is necessary during CT-guided procedures as well, including CT fluoroscopy. CT fluoroscopy differs from conventional fluoroscopy in both equipment and technique. Radiation protection concerns for CT fluoroscopy differ somewhat from those related to conventional angiography and fluoroscopy, as the beam is highly collimated: the major concern is to avoid excessive radiation dose to the interventional radiologist's hands [11,12].

Occupational radiation protection requires appropriate education and training of the operators and the availability of appropriate protective tools and devices, as well as of adequate equipment. The measures must also comply with local and national regulations.

Protection measures are necessary for all individuals who work in the interventional fluoroscopy suite. This includes not only interventional radiologists and cardiologists, radiographers, and nurses, who spend a substantial amount of time in a radiation environment, but also individuals such as anesthesiologists and ecographists who may be in a radiation environment only occasionally. All of these individuals may be considered radiation workers, depending on their level of exposure and on national regulations. All workers require appropriate radiation monitoring, as well as protection tools. They must also receive education and training appropriate to their jobs and sometimes specific certification issued by the competent authorities [13].

26.1.2 Dose Quantities and Units for Occupational Exposure

International organizations have published recommendations on the quantities and units that should be used in occupational dosimetry [14,15]. National regulations or guidelines provide specific requirements for personal dosimetry in interventional practice [16]. Dose limits to workers are expressed in terms of effective dose (E) for whole body exposure and equivalent dose in an organ or tissue (H_T) for exposure of a part of the body. The SI unit for both quantities is the sievert (Sv).

Equivalent dose and effective dose cannot be measured directly but must be calculated from other, more simply measurable quantities using personal dosimeters. Equivalent dose is the mean absorbed dose in a tissue or organ, T, multiplied by a radiation weighting factor, w_R. For diagnostic x-rays, $w_R = 1$, so the absorbed dose and the equivalent dose are numerically equal. Effective dose is the weighted sum of the equivalent doses (multiplied by a tissue weighting factor) in all specified tissues and organs of the body. These tissue weighting factors, w_T, are highest for red bone marrow, breast, colon, lung, and stomach and lowest for cortical bone, salivary glands, brain, and skin [16].

The operational dose quantity for personal dosimetry is the personal dose equivalent (Hp). Personal dose equivalent Hp(d) is defined as the dose equivalent in soft tissue, at an appropriate depth, d, below a specified point on the body. For calibration purposes, the definition is extended to include a calibration phantom [17]. Personal dosimeters provide two dose values, Hp(0.07) and Hp(10). These represent the dose equivalent in soft tissue at 0.07 and 10 mm below the surface of the body at the location of the dosimeter [17]. Hp(0.07) from the collar dosimeter worn over protective garments (apron, thyroid shield) provides a reasonable estimate of the dose delivered to the surface of the unshielded skin and to the lens of the eye. Consultation with a qualified medical physicist is recommended if dosimeters calibrated in terms of Hp(0.07) are used to estimate dose to the lens of the eye: calibration in terms of Hp(3) (i.e., at a depth of 3 mm) might be preferable, if available. Hp(10) from the dosimeter worn on the anterior chest under protective garments is assumed to be a good estimate of the operator's effective dose for workers in catheterization laboratories, which is considered an adequate indicator of the health detriment from radiation exposure. A single under-apron dosimeter does not provide any information about the eye dose and the dose to the unprotected parts of the body.

When two personal dosimeters are worn, it is possible to do better estimations of effective dose for workers. The formula used to estimate E from dosimeter data may be specified by national regulations or by local hospital policy. In the United States, when a protective apron is worn during diagnostic and interventional medical procedures using fluoroscopy, the National Council on Radiation Protection and Measurements (NCRP) recommends combining the readings from both body and collar dosimeters to estimate effective dose:

$$E \text{ (estimate)} = 0.5 \, H_W + 0.025 \, H_N$$

where

  H_N is the reading from the dosimeter at the neck, outside the protective apron

  H_W is the reading from the dosimeter at the waist or on the chest, under the protective apron [15]

Other algorithms exist to estimate E from the use of a single dosimeter worn over the protective apron or from the use of double dosimetry [18]. All formulas used to estimate E from dosimeter readings are based on certain assumptions about the wearer's radiation protective garments and position of operator with respect to the x-ray beam. For safety reasons, most of the commonly used formulas overestimate the individual's actual effective dose [15,18]. The formula given earlier is unlikely to underestimate E by more than a few percent or overestimate it by more than 100% [15].

Personal dosimeters in the interventional laboratory are exposed to a radiation field composed of both x-rays that irradiate the dosimeter directly, mainly scattered by the patient, and x-rays scattered back from the wearer's body. Accuracy and precision are affected by factors that influence the amount of radiation reaching the dosimeter from these two sources compared to the calibration conditions. The NCRP has published a full report on dosimetric uncertainty [19].

But, most important, inaccurate dosimetry results arise from mistakes or omissions in the chain of events of the monitoring program. These include wearing the dosimeter inappropriately or in the wrong location on the body and leaving the dosimeter in a radiation environment. It is not uncommon that individuals forget to wear or ignore wearing their dosimeter. These actions result in an incorrect personal dose evaluation and occupational risk assessment.

26.1.3 Occupational Dosimetry in the Interventional Laboratory

26.1.3.1 Dosimeter Use

Radiation workers are monitored to determine their level of exposure. In order to allow an adequate time for identification of procedures leading to high personal dose and implementation of work habit changes, monthly monitor replacement is recommended (mandatory in some countries) for operators conducting interventional procedures.

Several international and national organizations have published recommendations on occupational dosimetry that are applicable to workers in interventional laboratories [20–22]. The relatively high occupational exposures in interventional radiology require the use of robust monitoring arrangements for staff. The ICRP recommends that interventional radiology departments develop a policy that staff wears two dosimeters, one under the apron and one at collar level above the lead apron [21]. When high exposure of hands is likely, hand doses may also be monitored by an additional dosimeter [23].

26.1.3.2 Dose Limits

Dose limits for occupational exposures are expressed in equivalent doses for tissue reactions in specific tissues, and as effective dose for stochastic effects.

The occupational dose limits recommended by the ICRP have been adopted by most of the countries in the world. The limits are slightly different in the European Union and the United States. In the European Union, the limit for effective dose is 20 mSv/year, averaged over defined periods of 5 years. The effective dose may not exceed 50 mSv in any 1 year. In the United States, individual state governments set occupational dose limits, but in most cases, the recommendations developed by the NCRP are used: 50 mSv in any 1 year and a lifetime limit of 10 mSv multiplied by the individual's age in years [24].

For pregnant women, the ICRP recommends that the standard of protection for the conceptus should be broadly comparable to that provided for members of the general public [16]. After a worker has declared her pregnancy, her working conditions should ensure that the additional dose to the embryo–fetus does not exceed 1 mSv during the remainder of the pregnancy. In the United States, the NCRP recommends a 0.5 mSv equivalent dose monthly limit for the embryo–fetus (excluding medical and natural background radiation) once the pregnancy is declared [24].

The dose received by the lens of the eye can be estimated by placing a dosimeter near the tissue of interest. The "collar" badge is commonly used to estimate eye dose in interventional laboratories. This method is usually acceptable if the x-ray tube is mounted below the patient.

The evidence that regulatory eye dose limit was too high [4,7,8] made ICRP in 2011 to recommend a new annual limit of 20 mSv [9]. The new limit is currently being implemented in the national regulations, and the process will be completed in a few years, but in the context of the optimization of exposure, it must be implemented in a shorter time. To achieve it, interventional suites with a heavy workload will be required to adopt additional protective measures and devices and a more accurate eye dosimetry.

The annual limit for the hands and feet is 500 mSv. It is not possible to accurately estimate an operator's hand dose using a body or wrist dosimeter because of the proximity of the hands to the x-ray beam. When high doses are expected, for example, in procedures where hands have to be near or inside the x-ray beam, a ring dosimeter could be used to estimate hand dose [20], but sterilization and hygiene issues might interfere with this requirement. Feet doses are usually not monitored. When a curtain protective screen is properly used, the feet dose will be far below the annual dose limit.

In some models, effective dose is thought to be proportional to the risk of radiation-induced cancer. The ICRP occupational limits and limits for the general public are stated in terms of effective dose. National regulations require that a worker receive a radiation dose no greater than the dose limit. Again, because there might be no dose threshold for stochastic effects, the regulation is asking to implement the ALARA (as low as reasonably achievable) principle trying to maintain exposures as low as possible. Well-conducted practice (optimized) is showing that a busy interventionalist who takes appropriate radiation safety precautions is unlikely to have an effective dose exceeding 10 mSv/year and is more likely to have values of 2–4 mSv/year [24–27]. These values are well below the present dose limits.

The risk to specific organs such as the fingers or the lens of the eye is related to the physical dose delivered to these tissues.

26.1.3.3 Evaluation of Personal Dosimetry Data

The information in a personal dose record will vary depending on the number, type, and location of personal dosimeters used. This record will contain information on the effective dose, assessed from the readings of 1 or 2 dosimeters worn on the chest or abdomen under and/or over the lead apron, and may contain information on the equivalent dose to the lens of the eye from the dosimeter worn at the eye level or at collar level over the apron or thyroid collar, and the equivalent dose to the hand from a ring or bracelet dosimeter.

Copies of these dose reports should be sent to each department and individual after each reading and annual values communicated annually.

The facility's radiation safety section should review the personal dose records of individual workers regularly. This review ensures that dose limits are not exceeded. It also evaluates whether the dose received is at the level expected for that worker's particular duties. Typical staff dose readings for different types of procedures have been published in the literature. Depending on the type of procedure and the technique used, the operator dose, per procedure, ranges from 3 to 450 µSv at the neck over protective garments, from <0.1 to 32 µSv at the waist or chest under protective garments, and from 48 to 1280 µSv at the hand [11,27–37]. Most of the published data are stated in terms of dose per procedure, and most of the data are for physicians rather than assistants, nurses, technologists, or other staff. Translating these data into monthly or annual worker doses is difficult. As noted earlier, the effective dose for an interventional radiologist is typically 2–4 mSv/year [24–27].

26.1.3.4 Investigation of High Occupational Dose

The World Health Organization (WHO) recommends investigation when monthly exposure reaches 0.5 mSv for effective dose, 5 mSv for dose to the lens of the eye, or 15 mSv to the hands or extremities [20]. With the new recommended limit for the lens of the eye, the investigation level should be reduced from 5 to 0.5–1 mSv. The medical physicist (or equivalent expert) in charge of radiation safety should contact the worker directly to determine the cause of the unusual dose and to make suggestions about how to keep the worker's dose as low as reasonably achievable (ALARA).

Dosimeter readings for staff in interventional laboratories can be expected to be higher than for most other hospital staff. Investigation of a high personal dose value should start from a check of the validity of the dosimeter reading. Potential sources of invalid dosimeter readings include wearing dosimeters in the wrong location (e.g., under- or overprotective garment), wearing of a different staff's dosimeter, and dosimeter storage in a location where it is exposed to radiation. If an invalid reading is suspected, the reading for the individual's next monitoring period should be reviewed to ensure the problem has been corrected.

If the dosimeters have been stored and worn correctly, the worker will be asked if there was a change in work habits that could explain the increase in radiation exposure. Sometimes, a temporary cause is found. If this is the case, dose levels should return to usual levels during the next monitoring period, when workload returns to normal, equipment settings are corrected, or there is an additional experience with a new procedure or technique.

If the cause is not thought to be temporary, or if no cause can be identified, the individual's working habits should be observed during a series of representative procedures. The observer could be a medical or health physicist or a physician colleague with knowledge of radiation protection principles and the operation of the specific imaging equipment being used. The observer should pay close attention to equipment settings (particularly those that affect patient dose quantities), the staff's proximity to the patient, and the use of equipment-mounted shields and personal protective devices and personal dosimeters.

Real-time dosimeters might be used in cases where frequent feedback of radiation dose levels is important.

With adequate cooperation and attention to dose reduction principles, forced limitation of workload to ensure compliance with dose limit is generally not needed.

26.1.4   Radiation Protection Tools

The greatest source of radiation exposure to the operator and other staff is scatter from the patient. Generally, controlling patient dose also reduces scatter level and thus operator and staff dose. However, chronic radiation exposure in the workplace mandates the use of protective tools in order to reduce occupational radiation dose to an acceptable and optimized level. The purpose of radiation protection tools is to improve operator and staff safety without impeding the procedure or jeopardizing the patient's safety.

26.1.4.1 Shielding

There are three types of shielding: architectural shielding, equipment-mounted shields, and personal protective devices. Architectural shielding is built into the walls of the procedure room. This type of shielding is not discussed further here. In addition, mobile and stationary shields that rest on the floor are available and are useful for providing additional shielding for both operators and staff. They are particularly well suited for use by nurses and anesthesia personnel. Equipment-mounted shielding includes protective drapes suspended from the table. Table-suspended drapes hang from the side of the patient table, between the undertable x-ray tube and the operator. They should always be employed, as they have been shown to substantially reduce operator dose and feet dose. Unfortunately, they sometimes cannot be used if the x-ray gantry (C-arm) is in a steep oblique or lateral position.

Ceiling-suspended shields, or mounted screens should always be used in interventional suite. Properly placed shields, between scattering patients' body and operator, have been shown to dramatically reduce operator eye dose [37,38]. Now with the evidence that the threshold dose for cataract formation can be reached within several years for a moderately busy practitioner, suspended shields or protective glasses should be used by anyone performing interventional procedures on a regular basis [4]. Lens injuries have been reported in both operators and staff when systems that lack ceiling-suspended shields are used for complex interventional procedures or when shields are used irregularly [39].

Disposable, protective patient drapes are now available. These contain metallic elements (bismuth or tungsten–antimony) and are placed on the patient after the operative site has been prepared and draped [40,41]. They have been shown to reduce operator dose substantially, with reported reductions of 12-fold for the eyes, 26-fold for the thyroid, and 29-fold for the hands [41]. While their use adds some cost to the procedure, disposable protective drapes should be considered for complex procedures and procedures where the operator's hands must be near the radiation field (e.g., grafts, biliary interventions) [41].

26.1.4.2 Personal Protective Devices

Personal protective devices include aprons, thyroid shields, eyewear, and gloves. Protective aprons with thyroid shield are the principal radiation protection tool for interventional staff. They should be employed at all times. The vest/skirt configuration is preferred by many operators in order to reduce the risk of musculoskeletal/back injury [42]. These are typically 0.25 mm lead equivalent, so that when worn they provide 0.5 mm lead equivalent anteriorly. Operators and staff who work in the interventional laboratory on a regular basis should be provided with properly fitted aprons, both to reduce ergonomic hazards and to provide optimal radiation protection [43]. Aprons should be inspected fluoroscopically on an annual basis to detect deterioration and defects in the protective material [44]. The advent of the zero-gravity protective shield should also be an option—and may be recommended as a tool to solve problems of strain on back.

Because of the new ICRP recommendation for the eye exposure limit of 20 mSv/year, operators are strongly advised to use eye protection at all times [3,16]. Leaded eyeglasses with large lenses and protective side shields provide more protection than eyeglasses without these features. They help to minimize scatter, which approaches the operator from the side and scatter from the operator's own head. The principal disadvantage of leaded eyeglasses is their weight and discomfort.

In general, the operator's hands should be kept out of the primary radiation beam. Leaded gloves may seem useful for radiation protection on those rare occasions when the operator's hands must be in the primary radiation beam, but they are not. Because of scatter created within the glove, the increased dose when any shielding is placed in the primary beam, and the false sense of security that they provide, protective gloves can result in increased radiation dose to the hand when it is in the primary beam [45]. They are not recommended in this situation. The best way to protect the operator's hands is to keep them out of the radiation field. Leaded gloves may be of benefit if the operator's hands will be near, but not in, the primary radiation beam.

26.1.4.3 Effectiveness of Shielding

The shielding material for protective aprons has evolved from heavy, lead-impregnated vinyl or rubber with a shielding equivalent of 0.5 mm of lead, to lighter, composite (lead plus other high atomic-numbered elements) or entirely lead-free materials. These lighter materials have largely replaced the all-lead aprons of the past, and typically are designed to provide 0.5 mm lead equivalent protection anteriorly [46]. Transmission of 70–100 kVp x-rays through 0.5 mm lead is approximately 0.5%–5% [23,47]. Leaded glasses reduce the dose to the operator's eye from frontal exposure by a factor of approximately 8–10 [38,47]. When side exposure is included (the typical situation in clinical practice), the protection factor is decreased to 2–3 [48]. Design and individual fit of the eyewear has a significant impact on protection. Combining various types of shielding (i.e., table-suspended drapes, ceiling-suspended screens, aprons, leaded glasses, mobile shields, and disposable drapes) results in dramatic dose reduction for the operator [38].

26.1.5 Management Responsibilities

Management should provide an appropriate level of resources to ensure that radiation dose is adequately controlled. Resources include, but are not limited to, shielding, radiation-monitoring instruments, and protective clothing. Quality assurance is an essential component of any monitoring program. Occupational doses should be analyzed by each department; high doses and outliers should be investigated.

Standardized methods for acceptance testing and periodic quality control of protective aprons are also essential [44,46].

Adequate and relevant training programs should be provided for all levels of staff within the organization, including management, to develop a commitment to radiological protection.

26.2 Radiation Protection of Patients

26.2.1 Introduction

X-ray fluoroscopy and x-ray fluorography are imaging techniques that allow invasive cardiovascular and neurovascular procedures to be performed. Patients derive great diagnostic and therapeutic benefits from these procedures, but the use of ionizing radiation represents an associated hazard that must be justified by the procedure's benefits and the exposures optimized.

In recent years, the frequency, capability, and complexity of invasive procedures are increasing. As procedures have become increasingly complex, they may employ greater fluoroscopy time (FT) and a large number of fluorographic, rotational, and DSA images leading to the potential for high patient radiation exposure, possibly causing even skin damage. The new methods for the detailed measurement and display of the distribution of skin dose for interventional procedure were discussed in Chapter 24 of this book.

Technology developments in radiologic equipment have improved image quality while reducing x-ray dose rates, and IEC international standards ask manufacturers to include technological tools in their equipment to help dose reduction and dose management [49]. However, the need for optimization of radiation protection of patients is evident from the results of several surveys [50–56].

Relevant international standards and recommendations have been considered and referenced [49,57–60]. In particular, Publication 85 of ICRP is drawing attention to reported skin injuries and to the means to reduce such events. In fact, skin injuries have steadily increased, spanning the whole spectrum of possible injuries from erythema to ulcers requiring major plastic surgery. Many of these injuries are avoidable, and this is particularly important as the severest lesions can lead to permanent disability and chronic intractable pain. Some skin dose reduction strategies were discussed in this book in Chapter 25. The ICRP underlines that most injuries occur because interventionists are not aware of the radiation doses that are delivered to the skin [57].

26.2.2 Equipment-Related Factors

26.2.2.1 X-Ray Beam Parameters

To produce an optimally exposed image, the x-ray beam energy must be appropriately adjusted for the patient's x-ray attenuation, by varying a number of beam parameters. Optimal x-ray imaging parameters appropriately balance the requirements for contrast (necessary to detect the object), sharpness (necessary to characterize it), and patient dose.

Increasing the kVp of an x-ray beam decreases its absorption, enabling the penetration of dense body parts, and reduces patient exposure, in particular the entrance skin dose, by reducing the fraction of the beam absorbed by the patient. However, the difference in the relative absorption of different tissues decreases when kVp increases. This reduces image contrast. Therefore, optimal x-ray imaging requires a compromise to produce the best balance of penetration power, image contrast, and patient dose.

The increase of tube current (mA) applied to the x-ray tube, with fixed kVp, increases the number of x-ray photons produced with the same penetrating power. This strategy maintains image contrast at the cost of greater patient dose and x-ray tube loading. High tube loading requires high-power generators usually not available on mobile C-arm units.

The gain from this strategy is an image with less noise, greater contrast, at the cost of greater patient dose. Another potential downside of this strategy is that the increased loading may require a larger x-ray tube focal spot, which will reduce image sharpness.

In modern angiographic units, the exposure parameters are set automatically by software installed in the system. The acquisition programs are user-configurable, and it is important that physicians, medical physicists, and radiographers understand the operation and the patient exposure implications of choosing among the different selectable programs.

26.2.2.2 X-Ray Beam Filtration

Because low-energy x-rays have very limited penetration, they are almost completely absorbed in the patient's body without contribution to image formation. Thus, it is important to remove low-energy photons with filtration in the x-ray beam. Aluminum shows a high probability of photoelectric absorption at low x-ray energies, and an Al disk at the exit window of the x-ray tube preferentially removes low-energy x-rays, thus "hardening the beam" and, at the same time, reducing the dose absorbed by the patient. Increasing beam hardness increases the fraction of the beam's photons that successfully penetrate the patient and that

Chapter 26

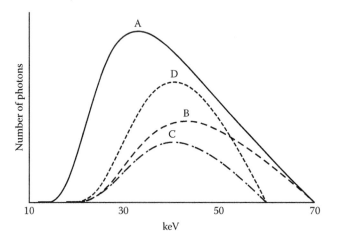

FIGURE 26.1 X-ray beam spectra obtained with different added filtration for x-rays generated at 70 (curve A and B) and 60 kVp (curve C and D).

contribute to the image (Figure 26.1). This means that less radiation must enter the patient in order to produce a given exit dose. All new angiographic systems use combinations of high-power x-ray tubes and high-voltage generators equipped with copper filters up to 0.9 mm thickness, but also aluminum or other elements, automatically inserted according to adjustment curves to produce even greater beam hardening.

26.2.2.3 Pulsed Fluoroscopy

Pulsed fluoroscopy provides brief x-ray pulses (of the order of some milliseconds) to generate images that, digitally processed, are continuously presented. The pulse rate is operator selectable and ranges from 30 frames/s (60 for pediatric applications) down to 1 frame/s. As the pulse rate decreases, the patient exposure rate decreases (at the cost of increasing motion effect in the presentation). Pulse rate reduction can produce images of acceptable quality as a function of the purpose of the examination. The dose reduction of pulsed fluoroscopy is not set to be linear with the decrease of the pulse rate, and it should be assessed by a medical physicist. The x-ray pulse rate determines the image sequence's temporal resolution. More rapid pulse rates provide greater temporal resolution and are useful in imaging of rapidly moving structures, albeit at the price of a greater x-ray exposure. When the pulse rate is less than the video frame rate, the video chain presents the frame acquired during the pulse repeatedly until the next pulse is delivered.

There is a correlation between dose and frame rates. The neuropsychology of vision provides a degree of integration that decreases perceived visual image noise and blurring from object movement. Digital image processing permits the integration of contiguous images using

a process called "recursive filtering" that reduces the impression of noise at the expense of blurring objects that are moving. For example, in cardiovascular imaging, 15 frames/s is often an optimal compromise frame rate. Rates below 15 frames may be adequate for tasks with less organ motion.

26.2.2.4 Digital Image Subtraction

A digitally subtracted image is obtained by subtracting one image from another. This electronically removes information that is identical in two images. The resulting image is a display of the difference between the two images. In angiography, the first image (mask) is obtained before the injection of contrast material and then subsequent images are acquired during the angiographic run. The subtracted image (digital subtraction angiography [DSA]) contains the difference between the two acquired images and emphasizes the structures opacified by the contrast media. The subtraction process accentuates image noise, and for this reason, it is necessary to use a substantially higher dose per frame (as much as 20 times). The increased dose may be partially attenuated by using a lower frame rate.

26.2.2.5 X-Ray Exposure Modulation (Automatic Exposure Control)

The attenuation of the x-ray beam in the patient's body varies with tissue density and other factors such as the projection angle and the distance between the x-ray tube and the image receptor. In the image chain, feedback circuits measure the brightness at the image intensifier output or the digital output from the flat panel detector. This feedback signal is used to modulate the output of the generator in response to changes in patient density and position to maintain predefined dose at the image detector. This system is usually called an automatic exposure control system (AEC). A modern system includes several trajectories that modulate kVp, anode current (mA), and pulse width (ms) as a function of the equivalent patient thickness and the imaging mode selected (fluoroscopy, image acquisition, DSA, pediatric examination, etc.).

As a general rule, attenuation increases as overall tissue thickness increases. This means that the patient entrance skin dose increases substantially with patient thickness and when very angulated projections are employed.

26.2.2.6 Electronic Magnification

The dose rate often increases as the degree of electronic magnification of the image increases. In general, both image intensifier and flat panel detector dose rates are programmed to increase somewhat as the size of the field-of-view decreases. A dose increase is usually required to limit the image noise increase that would otherwise be observed at high degrees of magnification.

For this reason, the operator should avoid electronic magnification if not strictly required by the imaging task.

26.2.2.7 Last Image Hold

This feature presents the last acquired fluoroscopic image on the video monitor, thus providing the possibility to study the image without continuing the exposure.

26.2.2.8 C-Arm Position Memory

Current systems are able to move the C-arm on command to preselected positions, thus enabling the operator to avoid manual movement of the beam.

26.2.3 Patient-Related Factors

Patient-related factors that are not under the control of the operator are mainly the size of the patient, determining the thickness of the body the x-ray beam has to cross to reach the imaging detector, the anatomy and the severity of the pathology, and determining the complexity of the procedure.

Though in principle entrance skin dose rates can increase by a factor of 10 in fluoroscopy mode when patient thickness increases from 16 to 28 cm [60], a study demonstrated a smaller variability related to patient thickness in clinical practice [55]. Anatomy- and pathology-related factors are the most important determinants of the total dose of a procedure. For example, in cardiac procedures, the type of occlusion, the tortuosity of the vessel to cross, and the number of vessels to treat in an angioplasty are all elements that can account for long FTs and high patient doses [52,61–63].

26.2.4 Procedural-Related Factors

26.2.4.1 Scattered Radiation, Organ Dose, and Beam Overlap

Scattered radiation is produced when the x-ray beam interacts with the patient. Scattered x-rays reach the image detector, increasing image noise and reducing image contrast. It is also the principal source of exposure to the patient's body outside the x-ray beam and to the staff. The amount of scatter increases with increasing size and intensity of the x-ray beam and with increasing patient thickness.

Organ dose is higher for the organs inside the primary x-ray beam, while scattered radiation is the mechanism that releases dose outside the beam. The absorbed dose decreases with the distance from the edge of the beam. When beams with different angulations are used, attention must be paid to avoid beam overlapping on the patient's skin. Beam overlapping can be the cause of high skin doses. Probability of beam overlapping is reduced with proper collimation of the field size.

26.2.4.2 Position of the X-Ray Source and Image Receptor

The x-ray system should be positioned so that the distance from the patient to the image detector is minimized and the distance from the patient to the x-ray tube is maximized. It is usually clinically desirable to position the patient's region of interest near the isocenter of the imaging system. Given this constraint, the distance between the x-ray tube and the patient should be practicably maximized; some system designs permit the independent control of this distance.

26.2.4.3 Beam Size

Beam collimation restricts the size of the beam. This is an important tool to limit exposures. The collimator should always be adjusted so that only the region of interest is exposed, thus sparing the surrounding tissue from direct irradiation. The collimation, reducing the amount of scattered x-ray, has also a positive influence on the image quality.

26.2.4.4 Procedure Protocol

Procedure protocols include all technical and clinical factors (CFs) adopted to conduct an interventional procedure. Together with the previously described

Chapter 26

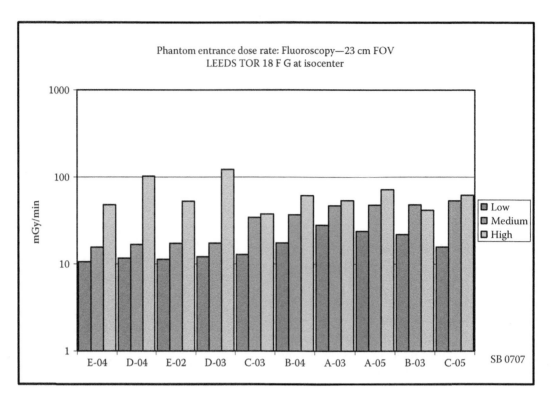

FIGURE 26.2   Dose rates in fluoroscopic modes (low, medium, and high image quality) in a sample of interventional cardiology installations (logarithmic scale). Results are for 20 cm PMMA phantom with Leeds TOR 18FG at isocenter. (From an international survey Prieto, C. et al., *Br. J. Radiol.*, 79(945), 730, 2006.)

factors, other important procedure determinants of patient dose under the control of the operator are the total FT, the number of cine or DSA runs, the total number of acquired images, the equipment protocol used in determining image quality levels, and pulse rates in fluoroscopy and in image acquisition. Examples of dose rate variability due to equipment performance, equipment setup, and different fluoroscopy mode are reported in Figure 26.2. Table 26.1 summarizes the factors affecting patient dose in interventional procedures, grouped as factors related to equipment, patient, and physician.

26.2.5   Patient Dose Assessment

Patients undergoing invasive procedures do not receive uniformly distributed whole-body radiation. The distribution of skin dose was discussed in Chapter 24 of this book. The dose is mostly concentrated in a confined area of the body. The effect of the patient's exposure is related to the dose received by each directly or indirectly exposed structure.

Proper patient dose monitoring is required in interventional procedures and, in particular, in long

and complex cases. Specific dose indicators have been developed to inform the operator in real time during the procedure on the doses received by the patient.

Two dose quantities—the dose at the interventional reference point (IRP) and the dose (or air kerma) area product (DAP or KAP)—are useful for characterizing patient exposure. Currently available angiographic systems determine real-time estimates of the instantaneous and cumulated values for these dose quantities, displayed in the treatment room to inform the operator. The unit's indication of these cumulated values provides valid indicators of a patient's dose and consequent risk for radiation-induced effects.

26.2.5.1 Kerma at the IRP and Trigger Levels

The IRP is located on the x-ray beam axis 15 cm from the isocenter on the side of the x-ray tube side. For an interventional procedure with an adult standard patient and not heavily angulated projections, this location approximates the location of the skin at the beam entrance point when the region of interest is located at the isocenter. Thus, the air kerma at this point, expressed in mGy, is an

Table 26.1 Factors Affecting Dose in Interventional Procedures

Equipment design and settings
Movement capabilities of C-arm, x-ray source, image intensifier
X-ray photon energy spectra
Beam filtration
Fluoroscopy pulse rate and acquisition frame rate
Fluoroscopy and acquisition input dose rates
Automatic dose-rate control including beam energy management options
Field-of-view size
Collimator position
Software image filters
Patient factors
Patient body weight and habitus
Physician procedure conduct
Positioning of image intensifier and x-ray source relative to the patient
Beam orientation and movement
Detector field-of-view size
Collimation
Acquisition and fluoroscopic technique factors on some units
Fluoroscopy pulse rate
Acquisition frame rate
Use of variable beam filtration
Total fluoroscopy time
Total acquisition time
Preventative maintenance and calibration
Quality control

indicator of skin dose. This quantity is not an exact measure of skin dose; the estimate provides a measure by which an assessment can be done regarding the risk of injury to the patient's skin. Trigger or alert levels can be provided in terms of this quantity to inform the operator that skin dose could have reached a dose comparable with a threshold for deterministic injuries.

26.2.5.2 Kerma Area or Dose Area Product

The KAP is the air kerma multiplied by the x-ray beam cross-sectional area at the point of measurement. It is expressed in $Gy \cdot cm^2$, $\mu Gy \cdot m^2$, $dGy \cdot cm^2$, or $mGy \cdot cm^2$, according to the different manufacturers. The cumulated KAP for a procedure is a surrogate measurement for the total amount of x-ray energy delivered to the patient. Consequently, it is a measure of the patient's risk of a stochastic effect.

Contrary to the measurement of dose at the IRP, the value of the KAP of an unattenuated x-ray beam does not depend on the distance from the x-ray source. This is because the dose decreases with distance from the x-ray source, while the beam area increases, both with a square law. The KAP can be measured by means of a transmission ionization chamber placed in the x-ray tube assembly or usually in modern systems, computed by the system.

KAP values should be corrected for the attenuation of the patient table. Attenuation is a complex function of the beam quality (kVp and total filtration) and beam angle, but usually, only one attenuation factor representing the mean attenuation factor for typical beams used in clinical practice is used as an acceptable approximation. With the advent of detailed information on single exposures available in the Radiation Dose Structured Report (RDSR), more attenuation factors can be adopted for a better accuracy of KAP to patient entrance estimation. Real-time dose tracking systems such as that described in Chapter 24 of this book will improve the assessment of patient entrance skin dose distribution.

Chapter 26

26.2.6 Benchmarking Data

Patient dose measurements should be regarded as part of a comprehensive quality assurance and audit program. Patient dose survey results can be analyzed to identify centers, which have average patient doses higher than expected, for instance, in the upper quartile of a patient dose distribution. An audit may then determine the underlying cause of these high doses. High dose values could be due to suboptimal equipment, equipment used in an inappropriate manner, or unavoidable patient-related factors. Concentrating dose reduction efforts on the upper quartile of the patient dose distribution is an efficient method for optimizing a practice.

Air kerma area product (KAP) and dose levels at the IRP are influenced by many variables, not all of which are under the operator's control. Nonetheless, assessment of these parameters provides a measure of a physician's radiation management performance.

Factors not under the operator's control include patient size and disease complexity. However, other variables, such as x-ray system position, collimator position, and appropriateness of beam-on time, are affected by the operator's attention to radiation safety practices.

Thus, although the relationship of KAP to patient injury is indirect, monitoring KAP is a valuable part of overall quality assurance monitoring. The KAP tracking for all procedures provides a measure of appropriateness of patient radiation protection practices.

26.2.6.1 Diagnostic Reference Level

The concept of diagnostic reference level (DRL) refers to "common examinations" done on large numbers of patients in a relatively standardized manner and its use is recommended by the ICRP [58, 59]. Extending this concept to fluoroscopically guided interventions raises several problems. In addition to technical variables (patient size, equipment performance, and operational technique), procedures are often nonstandard for clinical reasons. In fact, the complexity of a procedure is affected by factors related to the patient's anatomy and to the severity of the treated pathology. An index related to CFs affecting an individual procedure could reflect the complexity of the procedure. Appropriate scaling of DRL provides an additional tool for optimization processes in a facility.

Some experiences are available, demonstrating the feasibility to introduce DRLs in interventional cardiology and radiology procedures [50,63–65]. As an example, Figure 26.3 reports the KAP distribution for coronary angiography (CA) and coronary angioplasty (percutaneous cardiac intervention [PCI]) procedures

from a large multinational survey. The 75% percentile value of the KAP distributions has been used to assess international DRLs: 50 $Gy \cdot cm^2$ for CA and 120 $Gy \cdot cm^2$ for PCI [62]. Table 26.2 reports a comparison of DRLs assessed for cardiac procedures, assessed during the period 2003–2009, where it is possible to note that values are very similar and that most of the authors have assessed DRLs, not only in terms of KAP but also using multiple parameters including the dose at IRP, the FT, and the number of acquired images.

Table 26.3 reports DRLs for some neurovascular procedures where less experience in assessing DRLs can be appreciated.

As an example of use of DRLs for benchmarking, European experiences suggest that CA procedures with mean DAP values in excess of 50 $Gy \cdot cm^2$ should be evaluated for appropriateness as part of a quality assurance review. In the mentioned examples, DRLs have been expressed also in terms of FT, number of acquired images, and dose at IRP, providing a set of data useful to benchmarking specific parameters of the practice in a center.

26.2.6.2 Complexity Factors of an Interventional Procedure

For PCI, it has been demonstrated that DRLs can be assessed, taking into account the complexity of the procedure, where the complexity factors are clinical determinants of the total FT of the procedure [65–68].

The index assessed in the reported studies related to aspects of cardiac anatomy and pathology that influence the complexity of an interventional procedure and therefore might influence patient exposure. The study was designed to evaluate the relationships of CFs and anatomical factors (AFs) as a function of FT (FT), number of acquired images, and KAP. The intent was to develop a complexity index capable of predicting the patient's exposure.

The derived complexity index has several potential applications:

- As a means of expressing the average complexity of the mix of procedures performed by an individual physician or center
- As a means of normalizing dosimetric data to account for the complexity of procedures performed by an individual physician or center
- As a means of relating a center's performance to national or local guidance levels for the purpose of quality assurance and the optimization of clinical practice

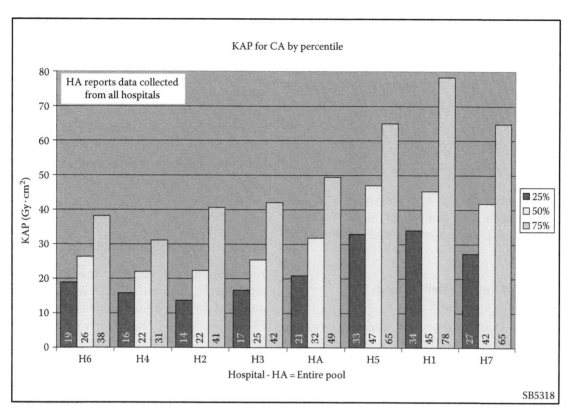

FIGURE 26.3 Kerma area product percentiles for CA. The 25th, 50th, and 75th percentiles in eight hospitals are shown. (From IAEA international survey Vano, E. et al., Radiat. Prot. Dosimetry, 129(1–3), 39, 2008.)

Table 26.2 DRLs Assessed for Cardiac Interventional Procedures

Procedure	Author	KAP (Gy·cm²)	Dose at IRP (mGy)	FT (min)	Number of Images
CA	Neofotistu 2003 [2]	57		6	1270
	Padovani 2008 [17]	45	650	6.5	700
	Balter 2009 [18]	50		9	1000
	Aroua 2007 [19]	80		7	1400
PCIs	Neofotistu 2003 [2]	94		16	1355
	Padovani 2008 [17]	85	1500	15	1000
	Balter 2009 [19]	125		22	1700
	Prieto 2006 [18]	69		23.1	1536
	Aroua 2007 [19]	110		10	1500

Table 26.3 DRLs in Terms of KAP Proposed for Neurovascular Procedures

Procedure	Aroua [19]	Brambilla [22]	Miller [21]	Vano [62]
Cerebral angiography	125	198		120
Carotid angiography				120
Cerebral embolization	440	338	400	

Chapter 26

Table 26.4 DRLs for DAP and FT of PCIs Assessed for Three Levels of Complexity of the Procedures

	Complexity of Procedure		
	Simple	Medium	Complex
DAP (Gy·cm²)	53.8	76.4	126.8
FT (min)	12	20	27

Source: Peterzol, A. et al., *Radiat. Prot. Dosimetry*, 117(1–3), 54, 2005.

From the complexity index, Peterzol et al. introduced three groups of complexity for PCI defined by three sets of DRL values, for both KAP and FT (Table 26.4) [66].

26.2.6.3 Dose at the Interventional Reference Point

The dose at IRP displayed by the angiographic system can be used to estimate the peak skin dose (PSD) with a margin of error that may be as much as a factor of two or greater although the dose-tracking system described in Chapter 24 of this book will improve this. Thus, the measure must be appropriately interpreted. For example, a calculated IRP air kerma of 2 Gy is very unlikely to correspond to an actual skin dose of 6 Gy (the approximate threshold dose for delayed skin erythema). Conversely, if the calculated IRP dose is 4 Gy, it is more likely that the erythema threshold may have been crossed. Thus, patients who receive an IRP dose greater than 4 Gy

should be advised that they might develop a skin erythema, along with instructions on what to do in the event that one is observed.

Modern angiographic equipment are providing an estimation of the dose to the skin delivered by each "irradiation event," assuming a standard geometry for the patient body. In the future as shown in Chapter 24, with a better simulation and tracking of the patient body with respect to the x-ray source, it is possible to have a more accurate estimation of the skin dose distribution and the derived PSD.

26.2.7   Avoiding Skin Injuries

Interventional procedures in radiology and cardiology often involve high radiation doses to the skin, and the potential for skin injury was discussed in 1994 by the FDA (Figure 26.4). The incidence of radiation injuries is small compared with the number of procedures performed, but a serious injury can be debilitating, requiring a prolonged course of intense care that sometimes lasts for years.

Interventionalists are often unaware of the magnitude of the radiation dose to the skin. Many are not aware that such injuries can even occur with modern equipment. Consequently, they, and other physicians, frequently do not recognize the injury as being related to the procedure. For this reason, an underreporting of the number of injuries from interventional work is suspected. To prevent skin injuries, optimized protocols

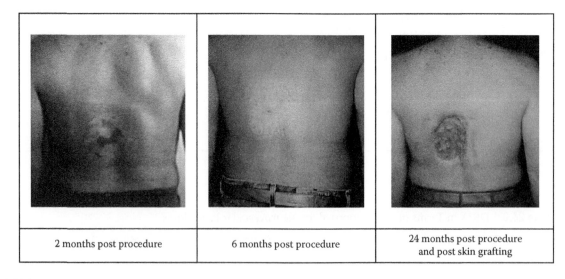

| 2 months post procedure | 6 months post procedure | 24 months post procedure and post skin grafting |

FIGURE 26.4 Time course of a deterministic radiation injury following high-dose coronary angioplasty. (Courtesy of U.S. FDA, Center for Devices and Radiological Health.)

are required, but it is also necessary to consider multiple procedures and it is recommended to implement a periodic skin dosimetry monitoring, the adoption of trigger levels, and follow-up procedures.

26.2.7.1 Considering Multiple Procedures

The risk of skin injuries of a given total radiation dose is clearly reduced by dose fractionation. On the contrary, the linear no-threshold model of stochastic injury indicates that the stochastic radiation risk (the risk of developing a radiation-induced cancer) depends on the total dose accumulated by a patient during his or her lifetime. Thus, splitting a procedure offers no practical protection against radiogenic malignancy.

When multiple procedures are performed on a patient, it is necessary to collect information on all procedures and plan the new one in order to prevent a skin injury.

26.2.7.2 Skin Dose Monitoring

The dose at IRP can be only an indicator of PSD. The assessment of skin dose can be accurately performed with dose-tracking systems (see Chapter 24) or portal films, in particular radiochromic dosimeters. These have the advantage that the readout is directly related to the radiation that enters locally on the skin, it includes backscatter, and it is independent of beam reorientation.

26.2.7.3 Trigger or Alert Levels

Dose thresholds can be adopted to identify procedures where erythema dose threshold can be reached and to alert the operator to modify, when possible, the procedure conduction. These dose levels, usually called "trigger" or "alert" levels, allow also physicians to identify patients to submit to a follow-up program for skin examination and should be assessed for procedure types where IRP doses of several Gy can be reached.

Trigger levels can be assessed by measuring the PSD (PSD is defined as the highest dose produced at any point on the patient's skin) in a sample of procedures by means of dose-tracking systems (see Chapter 24) or large area detectors, for example, radiochromic films. This provides a map of the irradiated area, shows multiple beam orientations employed during the procedure, and enables the estimation of PSD by measurement.

26.2.8 Training and Credentialing of the Staff and Physicians

The physician who performs invasive procedures is responsible for conducting the procedure and for effectively balancing the importance of the procedure with the need to minimize the patient's radiation dose.

To meet this responsibility, the physician must understand the patient's characteristics that determine the risk associated to the exposure, the basic principles for minimizing the dose, and the equipment's dose-control features. This knowledge has to be integrated with decisions regarding patient selection, procedure conduction, and equipment operation. Thus, the physician must possess both fundamental knowledge and machine-specific training in order to optimize the use of radiation use. In addition, the physician is responsible for conducting appropriate communication with the patient concerning the risk of radiation injury.

The ICRP, the FDA, the European Commission, and several scientific and professional societies have provided recommendations for the training in radiation protection of interventionalists. An example of the content of the training is provided in Table 26.5. The recommended time for this training is about 1 h per topic. In the European Union, a training of 20 h is recommended with similar content.

The International Atomic Energy Agency (IAEA) has produced a free training program for the training in radiation protection of operators [69].

If appropriate simulation facilities are available, operators should consider using them to learn and practice new skills before applying them to patients.

The American College of Cardiology (ACC) recommends that institutions that have x-ray fluoroscopic equipment should employ a credentialing process to authorize physicians to operate it. The process should establish required knowledge thresholds that physicians need to be authorized to perform fluoroscopically guided procedures. This will assure optimal patient and staff safety and optimal quality of diagnostic/interventional imaging.

26.2.9 Management Responsibilities

Management should provide an appropriate level of resources, such as staff, facilities, and equipment, to ensure that patient radiation dose is adequately controlled. Quality assurance is an essential component

Chapter 26

Table 26.5   Basic Radiation Physics and Safety Curriculum for Interventional Specialists Who Perform Fluoroscopically Guided Procedures

Imaging physics and technology
X-ray dosimetry concepts
X-ray production and feedback control
Image formation
Fluoroscopic systems
Image handling
Patient and staff radiation management
Radiation risks including pregnancy and heritable concerns
Patient selection, consents, history, physical examinations, follow-up procedures
Review of radiation injury cases
Distance–time–shielding
Situational awareness
Pregnant staff
Operational certification for each fluoroscope used in the laboratory
Location and function of key controls
Available clinical modes and their associated dose rates
Available radiation-shielding devices
Certification examination
Written certification examination with constructive review of responses

and should include the proper selection, installation and maintenance of the radiological equipment, a regular quality control program, patient dose monitoring, and the adoption of optimized clinical protocols. These actions should be in charge of the head of the department with the support of a medical physicist, in accordance with national regulations on radiation protection in medical exposure. In particular, the head of the department is responsible for the procedure protocols and maintenance and proper setup of the equipment. The medical physicist will be also responsible not only for procedure protocols and equipment setup but also for the quality control of the equipment and periodic patient dose assessment.

Patient input doses for fluoroscopy and acquisition modes should be set at the lowest values that are consistent with satisfactory image quality. Operators should recognize that a good image contains a degree of noise and should not request calibrations or imaging parameters that produce completely smooth images.

In summary, recommendations should be followed by interventional centers as an element of a quality assurance program for the optimization of radiation protection of patients:

- DRLs for interventional procedures should be established at the national or regional level using the 75th percentile of the distribution of selected parameters (KAP, kerma at IRP, FT, and number of acquired images) for a sample of procedures performed in a representative sample of centers. DRLs should be reviewed periodically.
- Departments should undertake dose surveys as part of a quality assurance program. The obtained results should be used to compare local practice with that in other centers using the assessed DRLs. An investigation into local practice should occur if median values of local dose distributions exceed the guidance levels or are below the action levels.
- The use of a complexity index is advised. This approach enables centers to be compared on an equitable basis. Further research into the use of complexity indices is needed.
- Dose information recorded in the DICOM RDSR should be recorded in the patient's electronic medical record alongside demographic information. The latter information should be recorded as part of a patient dose survey.

- PSD should be periodically monitored using appropriate means.
- "Trigger levels" should be assessed for each high-dose procedure and used to identify high dose

levels during the procedure, to enable optimization of the technique and to select patients to submit to a follow-up program aimed at identifying skin burns.

References

1. Tsapaki V et al. Radiation exposure to patients during interventional procedures in 20 countries: Initial IAEA project results. *Am J Roentgenol.* August 2009; 193(2):559–569. DOI:10.2214/AJR.08.2115.
2. Miller DL. Overview of contemporary interventional fluoroscopy procedures. *Health Phys.* 2008; 95:638–644.
3. Kim KP et al. Occupational radiation doses to operators performing cardiac catheterization procedures. *Health Phys.* 2008; 94:211–227.
4. Vano E, Gonzalez L, Fernández JM, Haskal ZJ. Eye lens exposure to radiation in interventional suites: Caution is warranted. *Radiology.* 2008; 248:945–953.
5. Kleiman NJ. Radiation cataract. In: *Working Party on Research Implications on Health and Safety Standards of the Article 31 Group of Experts*, ed. Radiation Protection 145. EU Scientific Seminar 2006. New insights in radiation risk and basic safety standards. Brussels, Belgium: European Commission, 2007; pp. 81–95. Available at http://ec.europa.eu/energy/nuclear/radioprotection/publication/doc/145_en.pdf Accessed August 16, 2009.
6. Worgul BV et al. Cataracts among Chernobyl clean-up workers: Implications regarding permissible eye exposures. *Radiat Res.* 2007; 167:233–243.
7. Nakashima E, Neriishi K, Minamoto A. A reanalysis of atomic-bomb cataract data, 2000–2002: A threshold analysis. *Health Phys.* 2006; 90:154–160.
8. Neriishi K et al. Postoperative cataract cases among atomic bomb survivors: Radiation dose response and threshold. *Radiat Res.* 2007; 168:404–408.
9. International Commission on Radiological Protection. ICRP Statement on Tissue Reactions/Early and Late Effects of Radiation in Normal Tissues and Organs – Threshold Doses for Tissue Reactions in a Radiation Protection Context. ICRP Publication 118. *Ann ICRP.* 41(1/2), 2012.
10. Vañó E, González L, Guibelalde E, Fernández JM, Ten JI. Radiation exposure to medical staff in interventional and cardiac radiology. *Br J Radiol.* 1998; 71:954–960.
11. Stoeckelhuber BM et al. Radiation dose to the radiologist's hand during continuous CT fluoroscopy-guided interventions. *Cardiovasc Intervent Radiol.* 2005; 28:589–594.
12. Hohl C et al. Dose reduction during CT fluoroscopy: Phantom study of angular beam modulation. *Radiology.* 2008; 246:519–525.
13. European Commission. Radiation protection 116. Guidelines on education and training in radiation protection for medical exposures. Luxembourg: European Commission. Directorate-General for the Environment, 2000. Available at http://ec.europa.eu/energy/nuclear/radiation_protection/doc/publication/116.pdf.
14. International Commission on Radiological Protection. Conversion coefficients for use in radiological protection against external radiation. Adopted by the ICRP and ICRU in September 1995. *Ann ICRP.* 1996; 26:1–205.
15. National Council on Radiation Protection and Measurements. Use of personal monitors to estimate effective dose equivalent and effective dose to workers for external exposure to low-LET radiation. NCRP Report No. 122. Bethesda, MD: National Council on Radiation Protection and Measurements, 1995.
16. International Commission on Radiological Protection. The 2007 Recommendations of the International Commission on Radiological Protection. ICRP publication 103. *Ann ICRP.* 2007; 37:1–332.
17. International Commission on Radiation Units and Quantities. Measurement of dose equivalents from external photon and electron radiations. ICRU Report 47, Bethesda, MD, 1989.
18. Järvinen H, Buls N, Clerinx P, Jansen J, Miljanić S, Nikodemová D, Ranogajec-Komor M, d'Errico F. Overview of double dosimetry procedures for the determination of the effective dose to the interventional radiology staff. *Radiat Prot Dosimetry.* 2008; 129(1–3):333–339.
19. National Council on Radiation Protection and Measurements. Uncertainties in the measurement and dosimetry of external radiation: Recommendations of the National Council on Radiation Protection and Measurements. NCRP Report No. 158. Bethesda, MD: National Council on Radiation Protection and Measurements, 2008.
20. World Health Organization. *Efficacy and Radiation Safety in Interventional Radiology.* Geneva, Switzerland: World Health Organization, 2000.
21. International Commission on Radiological Protection. Avoidance of radiation injuries from medical interventional procedures. ICRP Publication 85. *Ann ICRP.* 2000; 30:7–67.
22. Miller DL, Vañó E, Bartal G, Balter S, Dixon R, Padovani R, Schueler B, Cardella J, de Baère T. Occupational radiation protection in interventional radiology: A joint guideline of the cardiovascular and interventional radiology Society of Europe and the Society of Interventional Radiology. *J Vasc Interv Radiol.* 2010; 21:607–615.
23. Whitby M, Martin CJ. A study of the distribution of dose across the hands of interventional radiologists and cardiologists. *Br J Radiol.* 2005; 78:219–229.
24. National Council on Radiation Protection and Measurements. Limitation of Exposure to Ionizing Radiation. NCRP Report No. 116. Bethesda, MD: National Council on Radiation Protection and Measurements, 1993.
25. Balter S, Lamont J. Radiation and the pregnant nurse. *Cath Lab Digest.* 2002; 10:e1. Available at http://www.cathlabdigest.com/article/357 Accessed August 16, 2009.

Chapter 26

26. Tsapaki V et al. Occupational dose constraints in interventional cardiology procedures: The DIMOND approach. *Phys Med Biol*.2004; 49:997–1005.

27. Dendy PP. Radiation risks in interventional radiology. *Br J Radiol*. 2008; 81:1–7.

28. Stratakis J, Damilakis J, Hatzidakis A, Theocharopoulos N, Gourtsoyiannis N. Occupational radiation exposure from fluoroscopically guided percutaneous transhepatic biliary procedures. *J Vasc Interv Radiol*. 2006; 17:863–871.

29. Layton KF, Kallmes DF, Cloft HJ, Schueler BA, Sturchio GM. Radiation exposure to the primary operator during endovascular surgical neuroradiology procedures. *AJNR Am J Neuroradiol*. 2006; 27:742–743.

30. Stavas JM, Smith TP, DeLong DM, Miller MJ, Suhocki PV, Newman GE. Radiation hand exposure during restoration of flow to the thrombosed dialysis access graft. *J Vasc Interv Radiol*. 2006; 17:1611–1617.

31. Lipsitz EC, Veith FJ, Ohki T, et al. Does the endovascular repair of aortoiliac aneurysms pose a radiation safety hazard to vascular surgeons? *J Vasc Surg*. 2000; 32:704–710.

32. Buls N, Pages J, Mana F, Osteaux M. Patient and staff exposure during endoscopic retrograde cholangiopancreatography. *Br J Radiol*. 2002; 75:435–443.

33. Hellawell GO, Mutch SJ, Thevendran G, Wells E, Morgan RJ. Radiation exposure and the urologist: What are the risks? *J Urol*. 2005; 174:948–952.

34. Harstall R, Heini PF, Mini RL, Orler R. Radiation exposure to the surgeon during fluoroscopically assisted percutaneous vertebroplasty: A prospective study. *Spine*. 2005; 30:1893–1898.

35. Synowitz M, Kiwit J. Surgeon's radiation exposure during percutaneous vertebroplasty. *J Neurosurg Spine*. 2006; 4:106–109.

36. Shortt CP, Al-Hashimi H, Malone L, Lee MJ. Staff radiation doses to the lower extremities in interventional radiology. *Cardiovasc Intervent Radiol*. 2007; 30:1206–1209.

37. Maeder M et al. Impact of a lead glass screen on scatter radiation to eyes and hands in interventional cardiologists. *Catheter Cardiovasc Interv*. 2006; 67:18–23.

38. Thornton RH, Altamirano J, Dauer L. Comparing strategies for IR eye protection [abstract]. *J Vasc Interv Radiol*. 2009; 20:S52–S53.

39. Vañó E, González L, Beneytez F, Moreno F. Lens injuries induced by occupational exposure in non-optimized interventional radiology laboratories. *Br J Radiol*. 1998; 71:728–733.

40. Dromi S, Wood BJ, Oberoi J, Neeman Z. Heavy metal pad shielding during fluoroscopic interventions. *J Vasc Interv Radiol*. 2006; 17:1201–1206.

41. King JN, Champlin AM, Kelsey CA, Tripp DA. Using a sterile disposable protective surgical drape for reduction of radiation exposure to interventionalists. *Am J Roentgenol*. 2002; 178:153–157.

42. Klein LW et al. Occupational health hazards in the interventional laboratory: Time for a safer environment. *J Vasc Interv Radiol*. 2009; 20:147–152; quiz 153.

43. Detorie N, Mahesh M, Schueler BA. Reducing occupational exposure from fluoroscopy. *J Am Coll Radiol*. 2007; 4:335–337.

44. Christodoulou EG, Goodsitt MM, Larson SC, Darner KL, Satti J, Chan HP. Evaluation of the transmitted exposure through lead equivalent aprons used in a radiology department, including the contribution from backscatter. *Med Phys*. 2003; 30:1033–1038.

45. Wagner LK, Mulhern OR. Radiation-attenuating surgical gloves: Effects of scatter and secondary electron production. *Radiology*. 1996; 200:45–48.

46. Finnerty M, Brennan PC. Protective aprons in imaging departments: Manufacturer stated lead equivalence values require validation. *Eur Radiol*. 2005; 15:1477–1484.

47. Marshall NW, Faulkner K, Clarke P. An investigation into the effect of protective devices on the dose to radiosensitive organs in the head and neck. *Br J Radiol*. 1992; 65:799–802.

48. Moore WE, Ferguson G, Rohrmann C. Physical factors determining the utility of radiation safety glasses. *Med Phys*. 1980; 7:8–12.

49. IEC. Report 60601 Medical electrical equipment. Part 2-43. Particular requirements for the safety of x-ray equipment for interventional procedures, 2nd edn. Geneva, Switzerland: International Electrotechnical Commission, 2010.

50. Neofotistou V et al. Preliminary reference levels in interventional cardiology. *Eur Radiol*. 2003; 13(10):2259–2263.

51. Brambilla M et al. Patient radiation doses and references levels in interventional radiology. *Radiol Med* (Torino). 2004; 107(4):408–418.

52. Padovani R et al. Reference levels at European level for cardiac interventional procedures. *Radiat Prot Dosimetry*. 2008; 129:104–107.

53. Vano E et al. Patient dose reference levels for interventional radiology: A national approach. *Cardiovasc Intervent Radiol*. 2009; 32:19–24.

54. Miller DL, Kwon D, Bonavia GH. Reference levels for patient radiation doses in interventional radiology: Proposed initial values for U.S. practice. *Radiology*. 2009; 253:753–764.

55. IAEA. Establishing guidance levels in x ray guided medical interventional procedures: A pilot study. Safety Report Series No. 59. Vienna, Austria: International Atomic Energy Agency, 2009; p. 56.

56. Bogaert, E. et al. A large-scale multicentre study of patient skin doses in interventional cardiology: Dose-area product action levels and dose reference levels. *Br J Radiol*. 2009; 82:303–312.

57. International Commission on Radiological Protection. Avoidance of radiation injuries from medical interventional procedures. ICRP Publication 85. *Ann ICRP*. 30 (2), page 11 (6), 2000.

58. International Commission on Radiological Protection. Diagnostic reference levels in medical imaging: Review and additional advice. *Ann ICRP*. 2001; 31(4), page 48 (14).

59. ICRP. The 2007 Recommendations of the International Commission on Radiological Protection. ICRP Publication 103. *Ann ICRP*. 2007; 37:1–332.

60. International Commission on Radiation Units and Quantities. ICRU. Patient dosimetry for X-rays used in medical imaging. ICRU Report 74, Bethesda, MD, 2005.

61. Kuon E et al. Radiation exposure to patients undergoing percutaneous coronary interventions: Are current reference values too high? *Herz.* 2004; 29(2):208–217.

62. Vano E et al. Patient dose in interventional radiology: A European survey. *Radiat Prot Dosimetry.* 2008; 129(1–3):39–45.

63. Prieto C, Vano E, Fernandez JM, Galvan, C, Sabate M, Gonzalez L, Martinez D. Six years experience in intracoronary brachytherapy procedures: Patient doses from fluoroscopy. *Br J Radiol.* 2006; 79(945):730–733.

64. Aroua A, Rickli H, Stauffer JC, Schnyder P, Trueb PR, Valley JF, Vock P, Verdun, FR. How to set up and apply reference levels in fluoroscopy at a national level. *Eur Radiol.* 2007; 17(6):1621–1633.

65. Balter S, Miller DL, Vano E, Ortiz Lopez P, Bernardi, G, Cotelo E, Faulkner K, Nowotny R, Padovani R, Ramirez, A. A pilot study exploring the possibility of establishing guidance levels in x-ray directed interventional procedures. *Med Phys.* 2008; 35:673–680.

66. Peterzol A et al. Reference levels in PTCA as a function of procedure complexity. *Radiat Prot Dosimetry.* 2005; 117(1–3):54–58.

67. Padovani R, Bernardi G, Padovani R, Morocutti G, Vano E, Malisan MR, Rinuncini M, Spedicato L, Fioretti PM. Patient dose related to the complexity of interventional cardiology procedures. *Radiat Prot Dosimetry.* 2001; 94(1–2):189–192.

68. Bernardi G et al. Clinical and technical determinants of the complexity of percutaneous transluminal coronary angioplasty procedures: Analysis in relation to radiation exposure parameters. *Catheter Cardiovasc Interv.* 2000; 51(1):1–9.

69. IAEA training material for radiation protection in interventional radiology. Available at https://rpop.iaea.org/RPOP/RPoP/Content/AdditionalResources/Training/1_TrainingMaterial/Cardiology.htm. April 20, 2015.

Chapter 26

Trends

27. Current and Future Trends

Stephen Rudin

There were a number of topics that could not be covered in detail in this book. In this section, we have provided an overview of current and future trends including some of the imaging modalities and methods that might not have been covered in this book as well as some views and perhaps speculation on the current and future trends in neurovascular and cardiovascular imaging. These discussions include such topics as higher spatial resolution, advancing imaging methods and new modalities, advanced x-ray sources, improved and more informative image processing, improved image-guided therapeutic treatments, and safety considerations.

27.1 Higher Image Resolution

When CT first became available, what was revolutionary was not so much the spatial or temporal resolution since neither of these was particularly exciting; it was the contrast resolution and the ability to distinguish soft tissue differences such as white and gray matter in the brain. It is a well-established fact that the spatial resolution in the z-direction contributed to the revolutionary nature of this new modality, but tomography in other forms did exist prior to CT. Since the original EMI scanner, however, spatial and temporal resolution has improved while single-slice scanners have evolved into many multislice machines that acquire images at increasing speeds.

For neurovascular and cardiovascular applications, we expect this trend toward improved spatial and temporal resolution to continue not only in CT but in other important modalities as well. Part of the reason for this trend appears to be the continuous progression from invasive surgical procedures to minimally invasive surgical and endovascular procedures as well as the consequent continuing progress in finer and more complex endovascular devices that are deployed under image guidance. Some examples of this were discussed in Chapters 4 and 13. In addition to the trend of finer treatments, there is also a trend toward moving pathology diagnosis into the patient with the increasing movement toward digital pathology. Evidence for this trend is the newly created *Symposium on Digital Pathology at the Annual SPIE Medical Imaging Conference*, one of the most important international meetings for medical imaging of all kinds. It seems obvious that there will be a trend toward obtaining and even interpreting the very-high-resolution images required by pathologists but now in real time from new digital imaging acquisition tools entering the patient. Some of the ability to move toward higher resolution will depend upon advances in the acquisition methods and modalities.

Chapter 27

27.2 Advancing Acquisition Imaging Methods and New Modalities

Trends in acquisition can be divided into three parts: those for nonionizing radiation modalities, those for ionizing modalities, and those for hybrid modalities involved more than one modality. For nonionizing modalities, the most important are MRI, ultrasound, and the newest one, optical coherent tomography (OCT). When MRI made its first appearance, the expectation was that it would replace many of the CT procedures because of the improved safety associated with nonionizing rather than ionizing radiation. However, almost the opposite occurred while CT procedures expanded rapidly becoming in many cases the imaging modality of choice because of its speed, availability, and ease of use compared to MRI. Most recently, MRI has been taking an increasing role in vascular imaging, and in many instances, magnetic resonance angiography (MRA) appears to be replacing CT angiography (CTA). Additionally, MRI can provide information on blood flow characteristics that is not available with CT. Nevertheless, it is clear that both modalities will retain their importance in neurovascular and cardiovascular imaging for the foreseeable future.

Although not discussed in detail, the nonionizing modalities of intravascular ultrasound (IVUS) and OCT will continue to be prominent as far as vascular imaging is concerned. Both have the advantage of being able to visualize the intricacies of the lumen of vessels, thereby enabling evaluation of the vessel wall. IVUS has earned an important place in diagnosing coronary and carotid artery disease; however, because of the relatively large size of the devices at the end of the catheter, they could not be used for smaller vessels and especially for the important vessels in and beyond the circle of Willis in the brain. Although OCT has gained a place in ophthalmology, it is only recently beginning to replace some IVUS uses because of the exceptional spatial resolution capabilities. Nevertheless, because of the opacity of blood to light but not to ultrasound, there remain some operational advantages to IVUS compared with OCT.

As discussed in Chapters 12 and 13, modalities where ionizing radiation is used have also progressed. The single-photon counting (SPC) methods that are so familiar in the quantum starved modalities of nuclear medicine imaging (Chapter 8), are now entering the x-ray imaging domain where quanta are detected so rapidly that overlap can be a problem as described recently by K. Taguchi at Johns Hopkins University for new SPC CT scanners. However, as mentioned in Chapter 13, there are potentially great advantages in improved resolution and signal-to-noise ratio (SNR). Also the advantages of spectral imaging as discussed in Chapter 12 may be enhanced when SPC becomes increasingly available.

The new modality involving ionizing radiation of phase contrast imaging is being widely and actively researched and may eventually have a place in vascular imaging. In phase contrast imaging, x-rays are made to form a coherent image using fine line gratings and specially tailored sources. There may be future clinical application to 2D and CT vascular imaging because of the possibilities for edge enhancement; however, at present, most of the applications seem to be for preclinical animal tumor models and mammography.

Finally, there are hybrid modalities combining two different modalities. In nuclear medicine imaging, PET–CT and PET–MRI were briefly discussed in Chapter 8. At first, such hybrid scanning was done by placing a PET scanner and either a CT or MRI scanner next to one another with a common axis. The patient could receive the first scan, and then have the support bed translated along the common axis to receive the other scan. While integral PET–CT scanners were built so that minimal patient movement was necessary and the registration of images from the two modalities would be well maintained, this was more difficult to accomplish with PET–MRI scanner designs. For PET–MRI machines, the design of the gamma-ray detector modules has to be made so that they are not sensitive to the high magnetic fields of the MRI unit. Sensors involving volumes where the paths of electrons are long such as in photomultiplier tubes would be prohibitive. Additionally, x-ray tubes used near MRI units must be redesigned because of the extended electron path within the x-ray tube that could be greatly affected by the magnetic field of the MRI. Although these hybrid scanners can have a major impact in diagnosing cancers, the impact on vascular imaging remains unclear; CT adds much value to the hybrid scanner by improving the attenuation needed for improving the PET image acquisition.

Perhaps the newest hybrid modality is photoacoustic imaging. Here a laser is used to activate vibrations that are sensed by an ultrasound transducer to produce a rather high-resolution image. Because of the short range of the light photons from the laser, however, this hybrid modality only appears to have potential impact on preclinical tumor evaluation in small animals rather than clinical patient applications presently.

27.3 Advanced X-Ray Sources

The basic method for generating x-ray spectra using the Bremsstrahlung mechanism has been used since the discovery of x-rays, so recent advances in x-ray source design have been centered around increasing output such as for use in modern multislice CT machines. Innovative designs such as the distributed anode described in Chapter 11 are on the verge of commercialization, and previously distributed anode designs for CT such as the Imatron Electron Beam CT (see Chapter 15) have been commercialized in the past but have not persisted. The possibility of having miniature low output sources that might be inserted within the patient have also been explored but without a clear endovascular advantage. Other changes in the standard rotating anode structure of x-ray tube inserts have appeared in recent times such as Siemen's Straton, and the use of liquid metal bearings has contributed to greater tube output and longer tube lifetimes. Further advances using relatively simple modifications of current tube design appear possible to satisfy new requirements for higher output while maintaining very small focal spots by decreasing the anode angle and extending the filament length as indicated in Chapter 4. However, radically different methods for generating x-ray spectra such as the use of synchrotron radiation or the use of free electron lasers for monochromatic x-ray beams have not yet been widely available. Although the potential advantage of monochromatic x-ray beams in angiography where the photon energy can be tailored to be just above the absorption edge of iodine-containing contrast media has been considered, the cost and size of the facilities needed to produce such an appropriate monochromatic source has been the major limiting factor. The demand for these sources may continue to evolve with the further development of phase contrast imaging methods as well.

27.4 Improved and More Informative Image Processing

With the continual advancement of computer power has come increased imaging processing capability. Even complex calculations such as solutions of the Navier–Stokes equation for blood flow as discussed in Chapter 21, the calculations required for parametric and perfusion imaging described in Chapters 17 and 18, and new CT reconstruction methods described in Chapter 14 should all be speed up in the not too distant future so as to provide even real-time information to clinicians conducting endovascular image-guided interventions. Alternatively, the increasing computer capability should provide improved accuracy in both these and other areas where intensive computation may be required.

27.5 Improved Image-Guided Therapeutic Treatments

As image guidance improves, there should be improvements in the accuracy of both radiotherapeutic (Chapter 23) and endovascular treatments. With the attractiveness of minimally invasive endovascular interventions compared to invasive surgical procedures, there is a huge effort to improve and develop new endovascular devices. As these devices and delivery systems are made smarter and finer, image guidance will have to improve as demonstrated in Chapters 4 and 13. Although currently new designs for devices and their delivery systems seem to involve mostly contributions from the field of mechanical engineering, it should be possible to add electronic remote controls to accomplish a variety of missions such as steering, tissue analysis, and even treatment. One method, for example, where there has been a long-term effort to develop magnetic guidance of endovascular devices by Stereotaxis Corporation has now had limited commercial success in cardiology, especially in the guidance of electrophysiology treatments in cardiology.

27.6 Safer Procedure Considerations

As endovascular interventions become more complex with increased requirements on image guidance for improved spatial and temporal resolution, safety will also become increasingly a concern. If modalities with nonionizing radiation are used, then the safety concerns tend to be more mechanical or thermal such as the effect of the magnetic field or the RF field of an MRI unit on metallic devices. If magnets with fields higher than the current commonly available 3 T units and consequently higher-frequency RF radiation become more widespread, then thermal injury will be of greater concern during even diagnostic vascular procedures. While MRI has been used for biopsy guidance and the use of a split magnet was explored for a while by GE at Stanford University for image guidance of surgical procedures, there has not been wide use of MRI for real-time guidance of vascular interventions. Also safety concerns of ultrasound for vascular applications would appear to be limited except perhaps for IVUS and intravascular OCT.

For modalities involving ionizing radiation such as CT and 2D angiography, reduction or at least minimization of radiation dose becomes important limiting factors in the use of these modalities. New methods of dose saving are currently being, and will be in the future, actively developed such as the use of statistical iterative reconstruction methods for CT as described in Chapter 14. Similarly, in an attempt to minimize the radiation dose to the patient, beam modulation such as the use of x-ray attenuators in the periphery of a region of interest at the site of an intervention, yet without appreciably degrading the image guidance, is an active area of research as described in Chapter 4. Also detailed monitoring of radiation dose with a real-time dose-tracking system (DTS) that can provide the interventionalist with dose distribution information, enabling the possibility of dose spreading to reduce maximum skin doses, has recently become a reality as described in Chapter 24. Ultimately if SPC and precise control of the x-ray exposure location such as for the distributed anode source described in Chapter 11 become widely available, details of dose distribution can be controlled as a function of the real-time requirements of the interventional procedure, hence optimizing the radiation dose to the patient.

27.7 Conclusion

The aim of this book was to demonstrate the importance of understanding the prominent roles of physics and technology in cardiovascular and neurovascular imaging. It appears that rapid progress will continue to be made based on the foundation built by the methods mentioned in this volume.

Index

Printed and bound by CPI Group (UK) Ltd, Croydon, CR0 4YY

23/10/2024

01778257-0009